Portable Biosensing of Food Toxicants and Environmental Pollutants

Series in Sensors

Series Editors: Barry Jones and Haiying Huang

Portable Biosensing of Food Toxicants and Environmental Pollutants

Edited by
Dimitrios P. Nikolelis
Theodoros Varzakas
Arzum Erdem
Georgia-Paraskevi Nikoleli

CRC Press
Taylor & Francis Group
Boca Raton London New York

CRC Press is an imprint of the
Taylor & Francis Group, an **informa** business

First published in paperback 2024

First published 2013
by CRC Press
2385 NW Executive Center Drive, Suite 320, Boca Raton FL 33431

and by CRC Press
4 Park Square, Milton Park, Abingdon, Oxon, OX14 4RN

CRC Press is an imprint of Taylor & Francis Group, LLC

Library of Congress Cataloging-in-Publication Data

Portable biosensing of food toxicants and environmental pollutants / editors, Dimitrios P. Nikolelis, Theodoros Varzakas, Arzum Erdem, Georgia-Paraskevi Nikoleli.
    pages cm. -- (Series in sensors)
"A CRC title."
Includes bibliographical references and index.
ISBN 978-1-4665-7632-2 (hardcover : alk. paper)
1. Biosensors. 2. Food poisoning--Prevention. 3. Environmental health. 4. Environmental toxicology. I. Nikolelis, Dimitrios P., editor of compilation. II. Varzakas, Theodoros, editor of compilation. III. Erdem, Arzum, editor of compilation. IV. Nikoleli, Georgia-Paraskevi, editor of compilation. V. Title: Biosensing of food toxicants and environmental pollutants.

R857.B54P67 2014
610.28--dc23   2013032500

ISBN: 978-1-4665-7632-2 (hbk)
ISBN: 978-1-03-291747-4 (pbk)
ISBN: 978-0-429-10151-9 (ebk)

DOI: 10.1201/b15589

Visit the Taylor & Francis Web site at
http://www.taylorandfrancis.com

and the CRC Press Web site at
http://www.crcpress.com

Contents

Preface

A chemical sensor is a device that transforms chemical information, ranging from the concentration of a specific sample component to total composition analysis, into an analytically useful signal. Chemical sensors usually contain two basic components connected in series: a chemical recognition element ("receptor") and a physicochemical transducer. The recognition system translates the chemical information (i.e., concentration of the analyte) into a chemical or physical output signal. The transducer (i.e., a physical detection system) serves to transfer the signal from the output domain of the recognition element to the electrical, optical, or piezoelectric domain. A biosensor is a self-contained integrated device that is capable of providing specific quantitative analytical information using a biological recognition element (e.g., enzymes, antibodies, natural receptors, and cells), which is retained in direct spatial contact with a transduction element. Recent advances in the technology of artificial receptors have prompted a clear distinction between chemical sensors and biosensors. The latter utilize a transduction element of biological origin; however, since there has not been much development in engineered molecules, both terms are and can be used in the literature for this class of devices. The chemical sensors should be clearly distinguished from an analytical system that incorporates additional separation steps, such as liquid chromatography (LC), or additional hardware and/or sample processing, such as specific reagent introduction, for example, flow injection analysis (FIA). Biosensors have not yet made a large impact in the area of environmental, food, and biomedical applications but clearly offer advantages in comparison to standard analytical methods, such as minimal sample preparation and handling, real-time detection, rapid detection of the analytes of concern, and use of nonskilled personnel. Biosensors have the ability to be repeatedly calibrated; the term "multiple-use biosensor" is limited to devices suitable for monitoring both the increase and the decrease of analyte concentrations. Devices that cannot rapidly be regenerated should be named "single-use biosensors."

This book aims to provide both the basic knowledge regarding biosensor technology at a postgraduate level (MSci or PhD) as well as recent advances in chemical sensor technology and thus can be utilized by researchers. It covers the major areas of biosensors as follows:

1. Transducer schemes, that is,
    a. Optical photonic crystal waveguide sensors
    b. Electrochemical sensors
    c. Piezoelectric sensors
    d. Surface-enhanced Raman spectroscopy (SERS)–based biosensors

2. Recognition element, that is,
   a. Enzyme sensors
   b. Antibody-based biosensors
   c. Ion-channel switch and lipid film–based biosensors
   d. Nucleic acid biosensors
   e. *De novo* DNA synthesis and its biosensor detection
   f. Aptasensors
   g. Tissue, microorganisms, organelles, and cell-based biosensors
3. Recent trends in biosensing, that is,
   a. Bioelectronic tongue
   b. Nanotechnology and nanofabrication
   c. Molecular imprinting
   d. Lab-on-a-chip
4. Biosensor applications, that is,
   a. Biosensors in quality assurance of dairy products
   b. Water-soluble vitamin and drug residue determination
   b. Optical biosensors in food safety and control
   c. Biosensors in express control of quality assurance of products
   d. Efficiencies of biosensors in environmental monitoring
   e. Oligonucleotide and DNA microarrays as versatile tools for rapid diagnostics
   f. Biosensors for pesticides and foodborne pathogens
   g. Micro- and nanopatterning for bacteria- and virus-based biosensing applications
   h. Electrochemical DNA biosensors in food safety determination of phenolic compounds and antioxidant capacity in foods and beverages
   i. Biosensors in quality of meat products
   j. Microbial cells and enzymes for assaying the fermentation processes of alcohol production—starch, glucose, ethanol, BOD
   k. Biosensors in control of biochemical quantities to diagnose diseases

Chapter 1 introduces the basic principles of biosensing that will be useful to newcomers to this technology. Chapter 2 explains how the incorporation of a "receptor" can provide analytically useful information. Chapter 3 describes recent trends in biosensing and how a small-sized device can have portability for *in situ* determination of toxicants. Finally, Chapter 4 provides several examples for the determination of toxicants in food and environmental samples. The book is organized in such a way as to introduce to both newcomers and experts a roadmap to this technology by providing useful information on recent trends in biosensing devices.

Environmental and food biosensors come in thousands of forms and types based on a wide range of physical and chemical principles with varying types of usable outputs. Typical environmental contaminants monitored include metals, volatile organic compounds, biological contaminants, and radioisotopes. The field applications of sensors are also extremely varied. Among the key trends in the environmental and food sensor business is miniaturization down to the nanoscale, continuous and/or real-time sensing capabilities, wireless networked operation, rapid processing, and increased sensitivity or flexibility. Areas of environmental focus include vehicular emissions, combustion of fossil fuels, agricultural runoff, industrial and mine waste disposal, ocean spills and dumping, and climate change and weather monitoring. Very few examples of portable biosensors which can rapidly detect environmental pollutants directly in the gas phase have been reported in the literature.

Progress in nanosciences has led to a range of new technologies that allows us to drastically improve and even rethink and create industrial processes and products offering new functionalities. Sensors are core elements in any intelligent system for monitoring and controlling natural and industrial environments, and nanotechnology offers new possibilities for sensors, sensing-based systems, and applications. Recent trends in integrated electronics have started a revolution in this field, allowing the shrinking of very complex electronic systems into millimeter-sized squares. This would allow implementing complex and sophisticated instrumentation in cheap and portable devices for rapid detection of harmful and toxic agents.

The specific objective of this book is to tap into the progress made in nanosciences to deploy nanotechnology in affordable, mass-produced sensors and to integrate these into components and systems (including portable ones) for mass market applications in environmental and food monitoring. Sensing includes chemical and microbiological environmental toxicants, such as chlorinated hydrocarbons, heavy metals, polychlorinated dibenzo-dioxins (commonly called dioxins), nitrogen oxides (NO and NO_2), components of the photochemical smog that is a noxious mixture of air pollutants (SO_2, aldehydes, nitrogen oxides, peroxyacyl nitrates, etc.), chlorofluorocarbons (CFCs), hydrogen sulfide, water pollution (by the discharge of wastewater from commercial and industrial waste), discharges of untreated domestic sewage, as well as insecticides, pesticides, and herbicides.

The book also aims to make the rapid detection of bioterrorism weapons possible. The objective of this book is also to present advances in the development of portable chemical sensors for the rapid detection of biotoxicants in the environment; the scope encompasses the advances of devices that can be used for the rapid real-time detection of toxicants such as microbes, pathogens, toxins, and nerve gases, for example, *Clostridium botilinum* toxin, *Escherichia coli*, *Klebsiella pneumoniae*, sarin, VX, *Listeria monocytogenes*, *Salmonella*, marine biotoxins (such as palitoxins and spirolides), staphylococcal enterotoxin B, saxitoxin, gonyautoxin (GTX5), Francisella spore virus, *Bacillus subtilis*, and ochratoxin.

Topics include sensor design and fabrication (including the use and development of modeling tools), a technology demonstrator, and a positive production capacity feasibility study (including economic assessment) and plans for their commercial implementation.

Systems integration aspects to consider include easy and fast (multi)sensor interrogation and interfacing with monitoring and control functions. Reliability is required within the foreseen operating environment, considering temperature, humidity, and other parameters that affect stability. Initiation (resetting) and calibration require special attention.

The functionality should be demonstrated by integrating the developed sensor element into an existing or prototype system for validating its industrial relevance. The target is to realize the market potential of the results of the existing research, which will lead to improved performance of applications in the fields of environmental monitoring and provide significant benefits to citizens and for the environment.

Topics are focused toward the rapid detection of indoor and outdoor pollutants such as heavy metals, nitrogen oxide dioxins, CO_2, sulfur dioxide, volatile organic compounds (VOCs), low levels of methane, insecticides, pesticides, plant and vegetable hormones, etc.

Although biosensors exhibit double-digit growth rates, they still have to overcome a number of challenges, including the following:

- New research focuses less on fundamental research due to increasing challenges in applications.
- Development of a single biosensor platform with multipurpose diagnostic capability has restricted biosensor application.
- A number of problems prohibit the successful commercialization of biosensors. These problems have to be encountered in their development strategies.
- Competition from nonbiosensor technologies has hindered revenue growth.
- A low rate of technology transfer and lower level of development have deterred the development of new biosensors.
- The fabrication of portable chemical sensors in field uses has seen very little development.

In recent times, the market thrust has shifted to nanosensors' capabilities to detect environmental toxicants and food toxicants rapidly. A recent report from the market research firm In-Stat revealed that the media spotlight on this application may be premature: Despite the public's anticipation that biosensors with real-time detection will be able to monitor biochemical environmental toxicants, the technology has not matched expectations. Presently, biosensors in environmental monitoring stations nationwide can detect compounds like anthrax—but detection can take 12 to 24 h. The best ones in the market take 20 min.

Gas phase chemical detection is of critical importance for the sensing of highly toxic molecules, such as chemical warfare agents (CWAs) and toxic industrial chemicals (TICs). Beyond the ability to detect CWAs in a laboratory, that is, in the context of large, technical, and relatively expensive apparatus, of significant interest is the detection of CWAs with SERS in the field. The advent of small, portable Raman spectrometers with dimensions close to that of a smartphone, such as the ReporteR spectrometer (Intavec, Inc.) or the First Defender (Thermo Scientific, Inc.), and the development of stand-off SERS detection have begun the transition from the lab to the field.

In this book, portable and handheld nanosensors, for example, dynamic DNA and protein arrays for rapid and accurate detection of environmental pollutants, pathogens, etc., are described. The following challenges are focused on

1. High sensitivity—*detecting* very small amounts of environmental pollutants such as heavy metals, pathogens, toxins, and chemical toxicants in the environment
2. High selectivity—*discriminating* targets from other materials
3. High parallelism—*detecting* multiple pathogens, *minimizing* false positives, and *having* rapid response without sample preparation
4. High transportability—*being* nanosized and transportable yet robust and easy to operate
5. High affordability—*being* made up of inexpensive materials
6. High adaptability—*being* adaptable to new biotoxicants and integrated with chemo/biosensors
7. High precision—*allowing* for the detection of single molecules

Therefore, the state of the art of our devices is summarized as follows: Nanosensors will be highly sensitive, selective, rapidly responding, real-time, massively parallel, with no or minimum sample preparation, on a platform suited to portable and handheld nanosensors for the rapid detection of environmental pollutants for in field uses even by non-skilled personnel.

Dimitrios P. Nikolelis
Theodoros Varzakas
Arzum Erdem
Georgia-Paraskevi Nikoleli

Acknowledgments

This book has benefited from contributions by well-known and distinguished researchers worldwide. We would like to acknowledge their valuable contribution and cooperation. It has been an honor for us to edit this book. Two of the editors (D. P. Nikolelis and G.-P. Nikoleli) express their gratitude to the Greek Ministry of Development, General Secretariat of Research and Technology, and the European Commission (in particular the European Regional Development Fund and National Resources) (Contract 12SLO_ET30_1036) for their financial assistance.

Editors

Dimitrios P. Nikolelis earned his PhD from the University of Athens in 1976. He currently serves as a professor of environmental chemistry in the Department of Chemistry, University of Athens. He has coordinated three European projects on environmental biosensors (CIPA96-0231, IC15-CT96-0804, and QLK3-2000-01311) and has served twice as a NATO director in the following advanced research workshops: (1) Biosensors for Direct Monitoring of Environmental Pollutants in Field, Smolenice, Slovakia, in May 1996 and (2) Portable Biosensors for the Rapid Detection of Biochemical Weapons of Terrorism, Lunds, Sweden, in July 2012. Prof. Nikolelis has published over 200 scientific papers in scientific journals and conferences. In addition to this book, he is the editor of two other books on biosensors: *Biosensors for Direct Monitoring of Environmental Pollutants in Field* (Springer, 1996) and *Portable Chemical Sensors—Weapons against Bioterrorism* (Springer, 2012). He is also the editor of the scientific journal *Chemical Sensors*. His research is targeted on the fabrication of portable biosensors for uses in the field, and he has authored a large number of scientific papers on the detection of environmental pollutants such as hydrazines, dioxins, insecticides, and toxins. His current interests include the construction of novel chemical nanosensors that can be used for the rapid detection of environmental pollutants directly in the gas phase. Professor Nikolelis' group has recently utilized graphene and ZnO electrodes to develop chemical nanosensors for species of clinical or food significance such as urea, uric acid, and cholesterol. Work in progress includes evaluation and validation of nanosensors that are based on stabilized lipid films supported on glass fiber filters. This is done by using a dry spot test and optical methods of analysis (i.e., fluorescence) to detect environmental pollutants such as insecticides, plant hormones, toxins, and hydrazines.

Theodoros Varzakas earned his bachelor's degree (honors) in microbiology and biochemistry (1992) and his PhD in food biotechnology and MBA in food management from Reading University, United Kingdom (1998). He worked as a postdoctoral research staff at the same university. He has also worked in large pharmaceutical and multinational food companies in Greece for 5 years and has 14 years of experience in the public sector. Since 2005, he has been serving as assistant professor in the Department of Food Technology, Technological Educational Institute of Kalamata, Greece, specializing in issues of food technology, food processing, food quality, and safety. In 2012, he was elected as associate professor. He has served as a reviewer in many international journals such as the *International Journal of Food Science & Technology*, the *Journal of Food Engineering*, *Waste Management*, *Critical Reviews in Food Science and Nutrition*, the *Italian Journal of Food Science*, the *Journal of Food Processing and Preservation*, the *Journal of Culinary Science* *and Technology*, the *Journal of Agricultural and Food Chemistry*, the *Journal of Food Quality*, and *Food Chemistry*. He has also written more than 80 research papers and reviews and has presented more than 80 papers and posters at national and international conferences. He has published two books in Greek, one on genetically modified food and the other on quality control in food. He has also edited a book on sweeteners, which was published in 2012 by CRC Press.

Dr. Varzakas has participated in many European and national research programs as a coordinator or scientific member. He has been a fellow of the Institute of Food Science & Technology since 2007.

Arzum Erdem is a professor at the Analytical Chemistry Department in the Faculty of Pharmacy of Ege University in Turkey. She earned her PhD in analytical chemistry from Ege University in 2000. She was recognized as a highly skilled young scientist in 2001 by the Turkish Academy of Sciences (TUBA) and also received the Junior Science Award in 2006 from The Scientific and Technological Research Council of Turkey (TUBITAK). Dr. Erdem has initiated several international collaborative research projects on the development and applications of electrochemical (bio)sensors based on drug, enzyme, and nucleic acids. Her recent research is centered on the development of novel transducers and chemical and biological recognition systems using different nanomaterials (e.g., magnetic nanoparticles, carbon nanotubes, gold and silver nanoparticles, and nanowires) designed for the electrochemical sensing of nucleic acid (DNA, RNA) hybridization, and also the specific interactions between drug and DNA, or protein and DNA and aptamer–protein, as well as the development of integrated analytical systems for environmental, industry, and biomedical monitoring.

Georgia-Paraskevi Nikoleli earned her BSc from the University of Athens in 2007. She is currently pursuing her PhD in the Laboratory of Inorganic and Analytical Chemistry, School of Chemical Engineering, Chemical Sciences, National Technical University of Athens, Greece. Her research interests include environmental and food analysis and nutrition. Her research is targeted on the fabrication of portable biosensors for in-the-field uses. She has authored a large number of scientific papers on the detection of environmental pollutants such as insecticides, toxins, and dioxins. Her current interests include the construction of novel chemical nanosensors that can be used for the rapid detection of environmental pollutants directly in the gas phase and for real-time monitoring of environmental pollutants in waters. Her other interests include waste recycling.

Contributors

Salvador Alegret
Grup de Sensors i Biosensors
Departament de Química
Universitat Autónoma de Barcelona
Barcelona, Spain

V.A. Alferov
Federal State Budgetary Educational
    Institution of Higher Professional
    Education
Tula State University
Tula, Russia

V.A. Arlyapov
Federal State Budgetary Educational
    Institution of Higher Professional
    Education
Tula State University
Tula, Russia

Evangelos Bakeas
Laboratory of Analytical Chemistry
Department of Chemistry
University of Athens
Athens, Greece

Camelia Bala
Department of Analytical
    Chemistry
University of Bucharest
Bucharest, Romania

Madalina M. Barsan
Faculty of Sciences and Technology
Department of Chemistry
University of Coimbra
Coimbra, Portugal

Sergey O. Boyarintsev
Institute for Theoretical and
    Applied Electromagnetics
Russian Academy of Sciences
Moscow, Russia

Christopher M.A. Brett
Faculty of Sciences and Technology
Department of Chemistry
University of Coimbra
Coimbra, Portugal

Igor A. Budashov
Department of Chemistry
Lomonosov Moscow State
    University
Moscow, Russia

Susana Campuzano
Faculty of Chemistry
Department of Analytical
    Chemistry
Complutense University of Madrid
Madrid, Spain

Xavier Cetó
Sensors and Biosensors Group
Department of Chemistry
Autonomous University of Barcelona
Barcelona, Spain

Gulsah Congur
Faculty of Pharmacy
Department of Analytical
    Chemistry
Ege University
Izmir, Turkey

Boris B. Dzantiev
A.N. Bach Institute of Biochemistry
Russian Academy of Sciences
Moscow, Russia

Anastasios Economou
Department of Chemistry
University of Athens
Athens, Greece

Arzum Erdem
Faculty of Pharmacy
Department of Analytical
    Chemistry
Ege University
Izmir, Turkey

Gennady A. Evtugyn
Department of Analytical
    Chemistry
A.M. Butlerov Chemistry Institute
Kazan Federal University
Kazan, Russian Federation

M. Emilia Ghica
Faculty of Sciences and Technology
Department of Chemistry
University of Coimbra
Coimbra, Portugal

Manuel Gutiérrez-Capitán
Institute of Microelectronics of
    Barcelona
Higher Council of Scientific Research
Barcelona, Spain

Tibor Hianik
Faculty of Mathematics, Physics
    and Informatics
Comenius University
Bratislava, Slovakia

Zafar H. Ibupoto
Department of Science and
    Technology
Linköping University
Linköping, Sweden

Ilya N. Kurochkin
Department of Chemistry
Lomonosov Moscow State
    University
Moscow, Russia

Andrey N. Lagarkov
Institute for Theoretical and
    Applied Electromagnetics
Russian Academy of Sciences
Moscow, Russia

Stelios Liodakis
Laboratory of Inorganic and
    Analytical Chemistry
Department of Chemical Sciences
School of Chemical Engineering
National Technical University
    of Athens
Athens, Greece

Sergey S. Maklakov
Institute for Theoretical and
    Applied Electromagnetics
Russian Academy of Sciences
Moscow, Russia

A.V. Malinin
Research-Educational Institute of
    Optics and Biophotonics
N.G. Chernyshevsky Saratov State
    University
and
SPE LLC Nanostructed Glass
    Technology
Saratov, Russian Federation

Jean-Louis Marty
Laboratory IMAGES
University of Perpignan
Perpignan, France

Mykola Mel'nichenko
Department of Physics
Taras Shevchenko Kiev National
    University
Kiev, Ukraine

Mihrican Muti
Faculty of Science
Department of Chemistry
Adnan Menderes University
Aydın, Turkey

Georgia-Paraskevi Nikoleli
Laboratory of Inorganic and
    Analytical Chemistry
Department of Chemical Sciences
School of Chemical Engineering
National Technical University
    of Athens
Athens, Greece

Dimitrios P. Nikolelis
Laboratory of Environmental
    Chemistry
Department of Chemistry
University of Athens
Athens, Greece

Julia O. Ogorodnijchuk
Department of Molecular Biology,
    Microbiology and Biosafety
National University of Life and
    Environmental Sciences
Kiev, Ukraine

Andreea Olaru
R&D Department
RACC
Bucharest, Romania

Ilaria Palchetti
Department of Chemistry
University of Florence
Florence, Italy

María Pedrero
Faculty of Chemistry
Department of Analytical
    Chemistry
Complutense University of Madrid
Madrid, Spain

José M. Pingarrón
Faculty of Chemistry
Department of Analytical
    Chemistry
Complutense University of Madrid
Madrid, Spain

María Isabel Pividori
Grup de Sensors i Biosensors
Departament de Química
Universitat Autónoma de Barcelona
Barcelona, Spain

Anna V. Porfireva
Department of Analytical Chemistry
A.M. Butlerov Chemistry Institute
Kazan Federal University
Kazan, Russian Federation

Nikolas Psaroudakis
Laboratory of Inorganic Chemistry
Department of Chemistry
University of Athens
Athens, Greece

Vassilios N. Psychoyios
Laboratory of Inorganic and
  Analytical Chemistry
Department of Chemical Sciences
School of Chemical Engineering
National Technical University
  of Athens
Athens, Greece

Anatoly N. Reshetilov
Federal State Budgetary Institution
  of Science
G.K. Skryabin Institute of
  Biochemistry and Physiology
  of Microorganisms
Russian Academy of Sciences
and
Pushchino State Institute of Natural
  Sciences
Pushchino State University
Moscow, Russia

Ilya A. Ruzhikov
Institute for Theoretical and
  Applied Electromagnetics
Russian Academy of Sciences
Moscow, Russia

Andrey K. Sarychev
Institute for Theoretical and
  Applied Electromagnetics
Russian Academy of Sciences
Moscow Region, Russia

Audrey Sassolas
Laboratory IMAGES
University of Perpignan
Perpignan, France

J.S. Skibina
Research-Educational Institute of
  Optics and Biophotonics
N.G. Chernyshevsky Saratov State
  University
and
SPE LLC Nanostructed Glass
  Technology
Saratov, Russian Federation

Petr Skládal
Department of Biochemistry
Masaryk University
Brno, Czech Republic

Nelja F. Slishek
Department of Molecular Biology,
  Microbiology and Biosafety
National University of Life and
  Environmental Sciences
Kiev, Ukraine

Nickolaj F. Starodub
Department of Molecular Biology,
  Microbiology and Biosafety
National University of Life and
  Environmental Sciences
Kiev, Ukraine

V.V. Tuchin
Research-Educational Institute of
  Optics and Biophotonics
N.G. Chernyshevsky Saratov State
  University
and
Institute of Precise Mechanics and
  Control
Russian Academy of Sciences
Saratov, Russian Federation

and

Optoelectronics and Measurement
  Techniques Laboratory
University of Oulu
Oulu, Finland

Nikolaos Tzamtzis
Laboratory of Inorganic and
  Analytical Chemistry
Department of Chemical Sciences
School of Chemical Engineering
National Technical University
  of Athens
Athens, Greece

Manel del Valle
Sensors and Biosensors Group
Department of Chemistry
Autonomous University of Barcelona
Barcelona, Spain

Theodoros Varzakas
Department of Food Technology
Higher Institute of Kalamata
Kalamata, Greece

Magnus Willander
Department of Science and
  Technology
Linköping University
Linköping, Sweden

M.G. Zaitsev
Federal State Budgetary Educational
  Institution of Higher Professional
  Education
Tula State University
Tula, Russia

A.A. Zanishevskaya
Research-Educational Institute of
  Optics and Biophotonics
N.G. Chernyshevsky Saratov State
  University
and
SPE LLC Nanostructed Glass
  Technology
Saratov, Russian Federation

Lyudmila A. Zheleznaya
Federal State Budgetary Institution
  of Science
Institute of Theoretical and
  Experimental Biophysics
Russian Academy of Sciences
Pushchino, Russia

Anatoly V. Zherdev
A.N. Bach Institute of Biochemistry
Russian Academy of Sciences
Moscow, Russia

Nadezhda V. Zyrina
Federal State Budgetary Institution
  of Science
Institute of Theoretical and
  Experimental Biophysics
Russian Academy of Sciences
Pushchino, Russia

1

Photonic Crystal Waveguide Sensing

J.S. Skibina, A.V. Malinin, A.A. Zanishevskaya, and V.V. Tuchin

CONTENTS

1.1 Introduction

The previous decade brought photonic crystal waveguides to the attention of experts in the field of nanobiophotonics—science and technology related to the design and manufacturing of nanostructure devices for generation, amplification, modulation, transmission, and detection of electromagnetic radiation for applications in biology and medicine [1–3]. In order to gain a deeper understanding of the utility of structures, called photonic crystal waveguides (PCWs) or photonic crystal fibers (PCFs), we consider their optical and, more specifically, sensing properties, which inspire researchers to carry out further investigation of PCWs and to search for applications of their newly-developed types. The unique properties of PCWs, such as tight localization of electromagnetic radiation and high optical nonlinearity, allow one to classify PCWs not only as optical waveguides but also as effective optical sensors with extraordinary sensing abilities [3].

A specific type of sensors, developed for biomedical applications and called biosensors, could be significantly advanced in its efficiency by use of PCWs as sensitive elements. Optical biosensors possess high sensitivity and resolution for detection and quantitative assessment of chemical substances and biological processes [1,3]. Depending on the type of the optical sensors, their actions are based on different optical phenomena, including absorption and reflection of the incident light, light interaction with luminescence or surface plasmon resonance (SPR) excitation, and photonic bandgap formation. The signal of an optical biosensor can be described in terms of the variation of the wavelength, intensity, phase, and polarization of light, depending on the concentration of the detected substances and the influence of the environment. Typically, the biosensor includes a light source, a unit introducing radiation into the sensor, and a device for analyzing the radiation spectrum, mode profile, etc. [1,3].

A brief analysis of the state of the art in optical sensing reveals two rapidly growing branches: SPR-based sensors and optical waveguide–based sensors. SPR-based sensors allow real-time analyte detection without using fluorescent probes [4,5]. The plasmon resonance, that is, the surface excitation of plasmons by light, occurs at the surface of metals under the conditions of total internal reflection and is defined by the characteristic refractive index and reflection angle. The measurements are based on the migration of the plasmon resonance energy from the surface of the metallic film into the solution, where, due to the intermolecular interaction, the change in the resonance angle and, therefore, in the refractive index in the near-surface layer occurs. Based on the change in the latter, one can judge the interaction between the biological molecules. The variations in the refractive index are largely affected by the medium surrounding the metallic nanostructure. The change in the refractive index is detected in the immediate vicinity of the nanostructure, owing to which the phenomenon of local plasmon resonance may be used for the label-free assay.

In most cases, fiber-based sensors involve the use of exponentially decaying (evanescent) waves, localized in a cladding of solid-core microstructured fiber [3,6]. These waveguides consist of a cylindrical core (silica with high refractive index) and a low-index cladding (air-filled channels). The radiation propagates in the core of the fiber due to total internal reflection. In the cladding, the penetration depth of the evanescent field is on the order of the wavelength and depends on the refractive indices and the cladding material. By partially removing the cladding material, one can provide direct interaction of the evanescent field with the biological analyte, which allows for plotting the evanescent field intensity versus the analyte refractive index [6]. A new type of optical fiber, which enables interaction of the liquid analyte with fundamental modes, propagating through the fiber's core, has a great potential to become the basis for a highly sensitive optical analyzer.

This type of a fiber is called hollow-core PCW and is composed of a two-dimensional photonic crystal and a central defect, which inherently serves as a waveguide [1,3].

1.2 Hollow-Core Photonic Crystal Waveguides

What does the PCW represent? The internal geometry of hollow-core PCW is a two-dimensional dielectric lattice, which is completely analogous to structures theoretically predicted and later realized by Yablonovitch in 1987 [7].

The authors of the book [8] give us a good general definition of photonic crystal: photonic crystal is the sort of material that affords us complete control over light propagation. For better understanding of a photonic crystal's nature, we can consider it as an analogue of a conventional crystal. The main difference between the two is in different physical parameters characterizing their structures. The conventional crystal is a periodic arrangement of atoms, representing, in other words, periodic potential for a propagating particle (electron). On the other hand, the structure of photonic crystal implies a periodic change of dielectric constant for a particle (photon), propagating through the photonic crystal. Developing this analogy, we can state that only the photons with certain energies are allowed to pass through the structure of photonic crystal. Thus, for a certain crystal structure, there are photons in the incident beam that cannot overcome the air–crystal interface and are allowed only to reflect back. The energy range corresponding to the energies of these photons is called photonic bandgap.

With respect to the character of the refractive index variation, photonic crystals (PCs) may be divided into three basic classes, namely, one-, two-, and three-dimensional PCs. In one-dimensional crystals, the refractive index periodically varies in one spatial direction. Such photonic crystals consist of parallel layers of different materials with different refractive indices and can manifest their properties in the only direction perpendicular to the layers. Bragg structures, that is, periodical structures of dielectric layers with thickness $\lambda/4$ and two alternating values of the refractive index are an example of one-dimensional photonic crystals. Two-dimensional photonic crystals have the refractive index that periodically changes in two spatial directions, and they can also manifest their properties in two spatial directions. In three-dimensional photonic crystals, the refractive index periodically changes in three spatial directions, and hence they are able to exhibit their properties in three dimensions; they can be represented as arrays of spatial bodies (spheres, cubes, etc.) arranged in a three-dimensional crystal lattice.

Similar to electric media, photonic crystals can be divided into conductors, that is, the substances that can transmit light along paths of large length with minor losses; dielectrics, that is, next-to-perfect mirrors; semiconductors, that is, the substances able, for example, to reflect selectively the photons with a definite wavelength; and superconductors, in which due to collective phenomena the photons can propagate along the paths of practically unlimited length [9].

Periodic structures with pronounced coloring often occur in nature [10–12]. Examples of one-dimensional periodic structures are the layers coating the wings of some butterflies, the peacock tail feathers, and the shells of some bugs. The role of interference in the coloring of the peacock feathers was noticed as far back as in 1730 by Isaac Newton. The structures with two-dimensional periodicity are present in the eyes of some insects, for example, the moth, as well as humans and other mammals, and also in some kinds of algae [13]. Two-dimensional periodicity is inherent in natural pearl, consisting of packed cylindrical elements. Three-dimensional periodic structures occur in nature in the form of colloid crystals [14]. They were first observed in the studies of viruses [15]. The semiprecious stone opal is a colloid crystal, consisting of monodisperse spherical globules of silicon oxide [16]. It is just the light interference in the three-dimensional periodic structure that determines its sparkling color, depending on the angles of incidence and observation.

The appearance of photonic bandgaps makes photonic crystals a unique material. Artificial photonic crystals are the basis for light controlling devices, as this allows us to filter or localize the electromagnetic waves. The localization of electromagnetic field in a photonic crystal is possible in the case of

FIGURE 1.1
Example of a hollow-core PCW.

defect appearance in the crystal's structure. An electromagnetic wave with the wavelength corresponding to the bandgap of a photonic crystal can be tightly confined in the defect, which is inherently a local change in crystal periodicity. Considering a hollow-core PCW, we can see that the hollow core (Figure 1.1) is nothing but a light guiding defect in a photonic crystal. Before we discuss optical and sensing properties of PCWs, let us briefly consider the concept of their manufacturing.

1.3 Manufacturing of PCWs

PCWs are produced with a fiber drawing technology and represent micro- or nanostructured fibers of glass or silica. The term "photonic crystal fiber" was originally proposed by Russel in 1995 [2,17,18]. The difference between the technology of production of conventional optical fibers and PCFs (or waveguides) is in the procedure of a preform preparation. In the case of PCWs, it is a much more complicated process.

PCWs consist of regularly stacked capillaries or rods. The geometric parameters both of individual elements and of the whole structure may be arbitrary. In PCWs, various types of glasses, differing in optical, electrical, and chemical properties, may be combined as well.

The first stage of manufacturing of the PCW is producing of the thick wall glass capillaries. These capillaries are assembled in a multicapillary stack. The stack is heated up to the glass softening temperature and can be drawn to a multicapillary structure of a much smaller size. The stability and precision of keeping the furnace temperature constant and the rotational velocity of the drive of the drawing unit are largely responsible for the stability and precision of the geometric dimensions of the glassware cross section. Mathematically, the procedure is described as

$$D_0^2 V = d_0^2 \upsilon, \qquad (1.1)$$

where
  D_0 is the initial size of the preform (diameter, diagonal length of a hexagonal package, or its size along double apothem)
  V is the speed of delivery into the furnace
  υ is the speed of drawing
  d_0 is the resulting size of the element

The next stage of fabrication is packing a preform with the required multicapillary arrangement. The shape of the glassware is determined by the shape of the preform, while its geometric size depends on the delivery-to-drawing

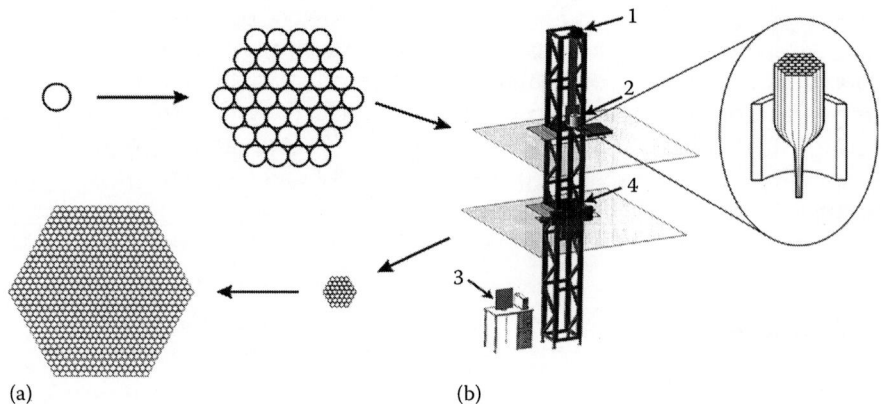

(a)   (b)

FIGURE 1.2
(a) Schematic diagram of PCW manufacturing and (b) the setup for PCW drawing: (1) mechanism for the delivery of glass elements into the furnace, (2) furnace, (3) computer-controlled system, and (4) chain drawing mechanism.

speed ratio. The bundle is then heated up to the glass softening point, pulled off with pincers, and gripped by the feed motion that delivers it into the furnace. Schematic diagrams of the waveguide manufacturing and the drawing setup are shown in Figure 1.2. The multistage technology of manufacturing the multichannel preforms consisting of many individual capillaries improves the regularity of stacking and allows for the production of glassware of arbitrary size and with arbitrary channel diameters. The PCW cross-section examples are shown in Figure 1.3. Images, shown in Figure 1.3, were acquired using a Carl Zeiss scanning electron microscope.

Waveguides, presented in Figure 1.3, were produced from C89 glass, which is a soft optical glass composed of 72% silica, 16% Na_2O, and smaller amounts of CaO, MgO, and BaO. This optical glass exhibits a refractive index $n = 1.519$, which makes its optical properties nearly identical to Schott crown glass K7.

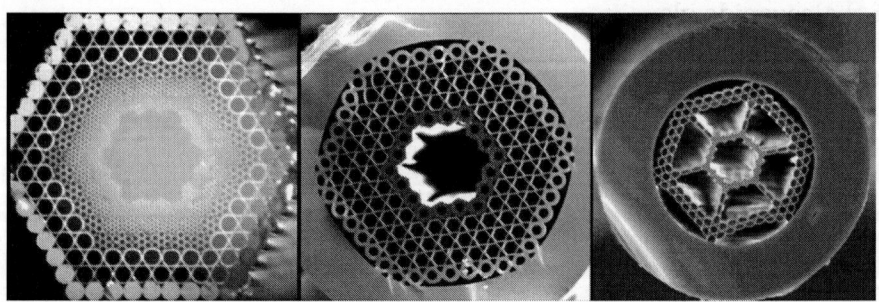

FIGURE 1.3
SEM images of the cross-sections of the hollow-core PCW with different internal geometry.

Currently, many different materials are used for PCW production. Fused silica (SiO$_2$), soft glasses, doped glasses, and even polymers are among them.

Fused silica is the dominating material due to its remarkable optical and mechanical properties, such as good transparency in the visible and NIR wavelength range, high mechanical strength against pulling and bending, and high damage threshold. On the other hand, different types of soft glasses may be suitable for making a preform of PCWs, since they have a much lower melting point, which is near 600°C, and such glasses are, commonly, cheaper.

Another type of glass material, fluoride glass, provides big opportunities for the creation of waveguides for mid-infrared transmission (2–9 μm). Such glasses contain metals, for example, aluminum or zirconium, in the form of fluorides (AlF$_3$ and ZrF$_4$, respectively). This relatively expensive type of glass exhibits a few advantages in comparison with conventional silica material: low transmission losses beyond 2 μm wavelengths and ability for doping with rare earth elements, thereby strong reduction of ability of quenching.

Phosphate glasses, based on phosphorous pentoxide (P$_2$O$_5$), provide strong ability for doping with rare earth ions as well. Moreover, concentrations of incorporated ions are allowed to be much higher in comparison with fluoride glasses. Glass types, doped with erbium, neodymium, and ytterbium, are the basis for short fiber lasers, implying high absorption of the pump radiation and large gain bandwidth. As disadvantages of this kind of glass material, we can mention slightly narrower transmission band in comparison with fused silica (0.4–2 μm) and low melting temperature (370°C), which prevents application in high-power lasers.

Chalcogenide glasses, containing a minimum of one chalcogenide element (except oxygen) such as selenium, sulfur, and tellurium, provide a wide transmission band in the mid-infrared region, covering two atmospheric windows from 3 to 5 and from 8 to 12 μm [19]. Due to its strong optical nonlinearity, chalcogenide glass can be used to produce microstructured fibers for the generation of supercontinuum, that is, generation of coherent optical radiation with ultra-broad spectrum (few optical octaves). The strong optical nonlinearity of chalcogenide glasses is accompanied by a high refractive index (~2.7) of these materials. The major disadvantages of chalcogenide glasses are high brittleness and toxicity. However, in the last few years, nontoxic chalcogenide glasses have appeared on the market, making this kind of material even more suitable for processing and fiber drawing.

Polymer materials, which are, at the first sight, poorly suitable for fiber production due to high attenuation of optical signal, are still exploited for the creation of mechanically robust and flexible optical waveguides. Although the development of microstructured polymer fibers is a completely new direction, the fabrication of conventional plastic optical fibers is already in practice. Polymer fibers provide high numerical aperture, which makes them well suited for combining with light-emitting diodes for data transmission. Furthermore, polymer fibers are cheap, light, and easy to use.

1.4 Photonic Crystal Sensors

At present, PCWs are considered as the most promising sensitive elements of fiber-optical sensors of physicochemical quantities, the use of which would substantially improve the performance (Table 1.1) capabilities of biosensors. Their basic advantages include the electromagnetic noise immunity (as in many fiber-optical sensors), high sensitivity, reliability, reproducibility, wide dynamic range of measurements, the possibility of spectral and spatial multiplexing of sensitive elements located in one or a few waveguides, short time of response to the variation in the measured quantity, small outer dimensions, and the possibility to combine waveguide principles and microfluidistics.

Since the first publication in 1996 on PCFs [20], the optical fiber community has studied the optical properties and fabrication of these new classes of fibers. It was not possible to mention every publication or patent or every detail of each sensor based on PCW. The application of PCW-based sensors has become an area of significant interest in many countries: Cox et al. from Optical Fibre Technology Centre (University of Sydney, Australia) [21,22], Wolfbeis from Institute of Analytical Chemistry (University of Regensburg, Germany) [23,24], Mignani et al. from Istituto di Fisica Applicata (Italy) [25,26], Pinto and Lopez-Amo from Departamento de Ingeniería Eléctrica y Electrónica (Universidad Pública de Navarra, Spain) [20], Zheltikov from International Laser Center (Moscow State University, Russia) [3], Skibina et al. from Saratov State University and SPE LLC Nanostructured Glass Technology (Saratov, Russia), and many others are involved in this prospective field of research and design.

The progress made during the past decade in the field of optical fiber biosensors is analyzed in a few review papers [20–29]. Wolfbeis [23,24] in his reviews covers the time period from 2000 to 2008 and tells us about numerous applications of optical fibers for sensing gases, vapors, and humidity; pH and ions; organic chemicals; DNA; bacteria; and index of refraction.

Pinto and Lopez-Amo presented a review [20] that includes a range of applications in which PCWs offer novel alternatives for quantitative sensing. It follows from their review that PCW is a technology with the outstanding potential for sensing confirmed by many approved patents, which unlock the path for a commercial scenario.

Another review by Bosch et al. [27] discussed advantages of optical fiber biosensors, such as excellent light delivery, long interaction length, low cost, the ability not only to excite the target molecules but also to capture the emitted light from the targets, and easy combination with spectroscopic techniques (absorption, fluorescence, phosphorescence, SPR, etc.).

Passaro et al. [28] discussed the possibility of using hollow and anti-resonant waveguides and Bragg gratings to realize very sensitive and selective, ultracompact, and fast biosensors. It was stated that photonic and electronic micro- and nanosensors are emerging as very attractive devices to be employed in

TABLE 1.1

Optical Sensing Methods

Design Concept	Method	Analyte
Surface Plasmon resonance (SPR)	SPR	Solutions
	Broadband SPR	Solutions
	SPR tomography	Solutions
	SPR tomography	Proteins
	Optical heterodyne SPR	Proteins
	Phase-sensitive SPR	Proteins
	Wavelength-modulated SPR	DNA
	SPR tomography	DNA and RNA
	Flow-injection SPR	DNA
	Angle-modulated SPR	Protein (PSA)
	SPR	Protein (CA19-9)
	SPR	Protein (*a*-fetoprotein)
	Prism-based SPR	Bacteria (*E. coli*)
	Prism-based SPR	Bacteria (*Salmonella typhimurium*)
	SPR BIAcore 2000	Bacteria
Interferometry	Mach–Zehnder interferometer	Solutions
		Proteins
	Young interferometer	Solutions
		Viruses
	Hartman interferometer	DNA
		Proteins
		Viruses
		Bacteria
	Porous silicon	DNA
	Porous silicon	Proteins
	Microchannel back scattering	Solutions
		Proteins
Waveguides	Resonance mirror	Proteins
		Cells
	Waveguide with metallic cladding	Bacterial spores
		Cells
	Reverse-symmetry waveguide	Cells
	Waveguide with symmetric metal cladding	Solutions
Ring resonator	Resonance mirror	Solutions
		DNA
		Proteins
		Bacteria
	Dielectric microsphere	Solutions
		DNA

(continued)

TABLE 1.1 (continued)

Optical Sensing Methods

Design Concept	Method	Analyte
		Proteins
		Viruses
		Bacteria
	Optical capillary liquid ring	Solutions
		DNA
		Proteins
		Viruses
Optical fibers	Fiber Bragg grating	Solutions
		DNA
	Large-period grating	Solutions
		Proteins
	Nanofibers	Solutions
	Fabry–Perot fiber resonator	Solutions
		DNA
		Proteins
Photonic crystals	2D PC	Solutions
		Proteins
	2D PC resonator	Solutions
		Proteins
	2D PCW	Proteins
	One-dimensional micro-cavity array based on PCW	Solutions
	PCW	Solutions
		Proteins

Source: Skibina, Yu.S. et al., *Quant. Electron.,* 41(4), 284, 2011. With permission.

a great number of application fields such as medicine, microbiology, particle physics, automotive technology, environmental safety, and defense.

The review by Righini et al. [26] summarizes the fundamentals of light propagation in fiber and integrated optics and explains the basic working principles of optical sensors making use of these waveguides. They discussed the main reasons that make guiding wave optics attractive for sensing and said that further advances in the fabrication and understanding of microstructured fibers and photonic crystal structures will provide a platform for new sensors, aiming at being alternatives for standard sensing technologies.

There are many papers describing PCW-based sensors. Marques et al. [29] presented an interesting work on optical sensing of gases using a hollow-core PCW, reporting the preferences for it as compared with conventional fibers. The development of PCW gas sensors is also described by Cox et al. [21] and Hoo et al. [30]. They reported results on modeling of all-fiber PCW gas detectors.

The sensor based on hollow-core photonic bandgap fibers for methane detection was described by Cubillas et al. [31].

Some articles present evanescent-wave fiber-optic biosensors [32–34]. Cordeiro et al. [32] have shown a new approach to enhance the evanescent field when using microstructured optical fibers for gas/liquid sensing. Jensen et al. have demonstrated highly efficient evanescent-wave detection of fluorophore-labeled biomolecules in aqueous solutions positioned in the air holes of the microstructured part of a PCW [33]. Hoo et al. [34] have reported the experimental demonstration of evanescent-wave gas detection with a silica–air microstructured fiber.

Fiber-optic sensing using absorption spectroscopy principles has been reviewed in some papers [25,35]. It should be mentioned that Yua et al. have reported a highly sensitive evanescent field–based absorption spectroscopy using an optimized structure of PCW with a defected core. They obtained an excellent linear relationship between absorption coefficient and analyte concentration in water solution, thus demonstrating that PCWs are suitable for off-chip or on-chip analyte quantification.

Highly efficient fluorescence sensing with PCW was demonstrated in some papers [36,37]. Konorov et al. [36] have demonstrated a high potential of PCWs for optical sensing and sample collection. Their PCW design integrates an optical sensor with a micropipette, a sample collector, and a microfluidic polycapillary array in a single fiber-optic component, which is ideally suited for lab-on-chip applications, biosensing, and *in vivo* biomedical studies. The potential application of microstructured optical fibers for a highly sensitive absorption and fluorescence measurements was investigated by Smolka et al. [37] with infiltration of dye solution into the holey structure.

The refractive index sensors are reviewed in Refs. [36,38,39]. A finite element computing package to analyze cladding mode field extensions into the air holes of PCW aimed for refractive index sensing was developed by Nguyen et al. [38]. Skivesen et al. have presented the PCW sensor for refractive index measurements and detection of protein concentration [39].

The PCF made of different materials has been under study during the last years. Cox et al. have demonstrated that hollow-core microstructured polymer optical fibers can be used for sensing of index of refraction of liquids [21]. Brilland et al. have presented the fabrication process of chalcogenide microstructured optical fibers and their use for detection of dioxide by detection of the absorption band near 4.2 μm [40].

Russel et al. have introduced PCFs for laser propelling of individual cells over distances of tens of centimeters through a stationary liquid in a microfluidic channel [41]. The liquid-filled hollow core of the PCF was used as a channel for red blood cell transport. The described system was proposed as a new tool for the study of single cell biomechanics, since it allows one to register the velocity of the cell in the microfluidic channel as well as the change in the cell's shape.

Recent experiments have revealed that transmission of the optical radiation and the photonic bandgap of a hollow PCW are largely affected by the refractive index of the medium, filling up the space in the hollow core and the channels of the cladding [42,43]. The spectral sensitivity of the PCW to the physical properties of the medium (gas or liquid) filling up the PCW's internal structure determines the promising applications of waveguides in analytic devices (chips and sensors) for biology and medicine. The possibility of using PCWs as sensors was first demonstrated in Ref. [34], where the absorption spectra of acetylene, filling up the waveguide under pressure, were measured. A photonic crystal sensor that allows for the detection of individual molecules via their two-photon fluorescence was described [44]. The possibility of selective detection of antibodies in a hollow PCW was also shown [45]. Two protocols of implementing biological PCW-based sensors were demonstrated [36]. The first protocol is based on the waveguide with a double periodic structured cladding and core. The radiation from a diode-pumped solid state laser (532 nm) is introduced into the core of the waveguide and is captured there due to the guiding properties of the first structured cladding. The radiation is transported to the probed sample, and the excited fluorescence is collected by the second structured cladding. In the second protocol the biological sample, marked with a fluorescent dye, is introduced into the structured cladding of the waveguide. The laser radiation is focused into the core, and the radiation, passing through the waveguide, is detected. As shown in Ref. [36], these two protocols are completely identical with respect to the character of the acquired data about the biological analyte. Experiments in the last few years proved the high potential of PCWs/PCFs with a big grating period as the optical sensors [42,43,46–49].

1.5 Hollow-Core PCWs with a Big Grating Period

In practice, the hollow defect of PCW has a diameter that is a few times bigger than the surrounding crystal lattice period. The modulation period of dielectric constant in photonic crystal is comparable to one or a few wavelengths of the incident light, thus the light guiding diameter usually does not exceed a few tens of microns. In this section, we will consider a special type of PCWs with a big grating period. The internal geometry of such waveguides is characterized by a large hollow-core diameter, more than 100 μm, and proportionally large diameter of capillaries in cladding. Sensing properties of such fibers were theoretically predicted [50] and experimentally tested [42,43,46–49].

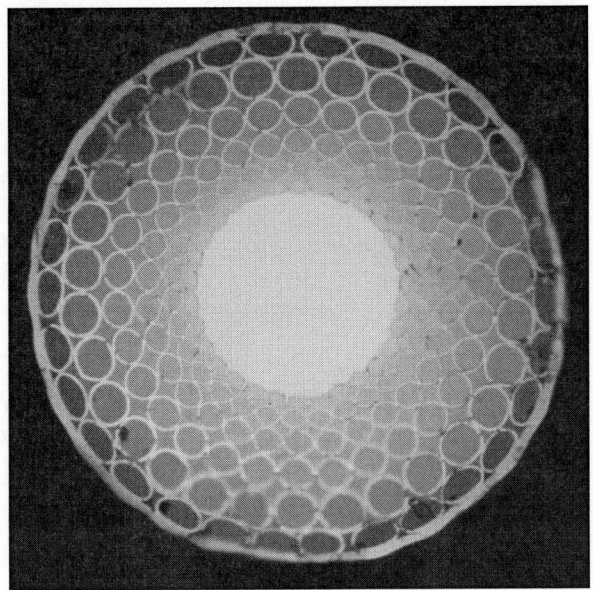

FIGURE 1.4
Microphotograph of the internal structure of BGP PCW. Hollow-core diameter is 200 μm and width of the wall of internal layer of capillaries is 2 μm.

Further in this section we discuss optical properties of PCW preforms with the internal geometry proposed in Ref. [51] (a chirped fiber). These preforms (Figure 1.4) were successfully adapted for sensing applications. The waveguides provide similar transmission properties as the classical PCW with one significant difference, PCW with a big grating period (BGP PCW) possesses a few narrow and smooth transmission bands in the visible/NIR wavelength range.

Guidance properties of BGP PCWs were discussed in detail in Refs. [45,52]. The fundamental modes propagating along the hollow-core axis of the waveguide are antiresonant to its cladding modes [17]. The BGP PCWs designed and produced by our group were characterized using a setup (Figure 1.5a and b) containing a broadband light source (halogen lamp), optical elements for launching and collecting light, adjusting optomechanics, a spectrometer Ocean Optics HR4000, operating in the visible/NIR range, and a PC. The samples were placed into specially designed cuvettes made on the basis of BGP PCW (Figure 1.5c).

The spectral characteristics acquired for the BGP PCW with wide capillary walls (2 μm) and large (210 μm) hollow core (shown in Figure 1.6) prove the assumption that internal glass walls represent a built-in Fabry–Perot interferometer. Its transmitting bands' appearance mechanism is similar to the formation of the reflected spectrum of the Fabry–Perot resonator.

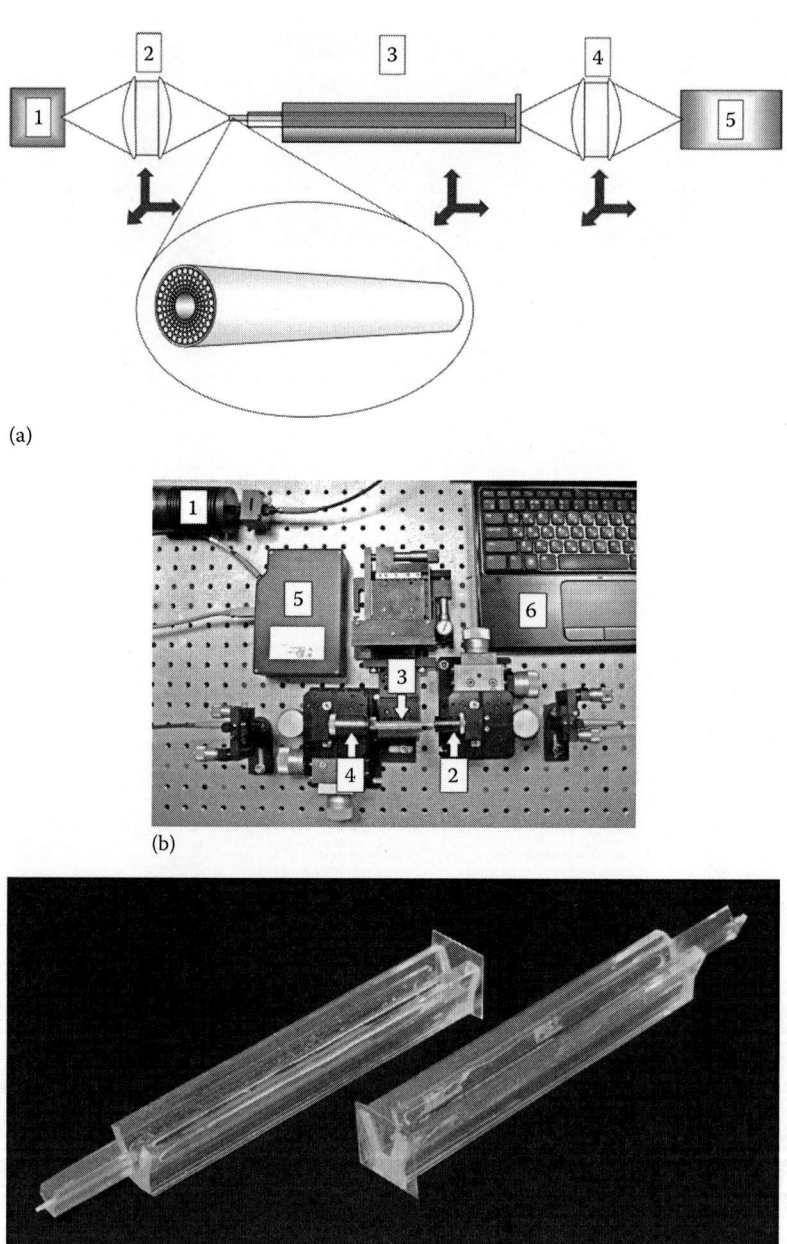

FIGURE 1.5
(a) Schematics and (b) general view of experimental setup for spectral characterization of PCWs: (1) broadband light source (halogen lamp), (2, 4) optical elements for launching and collecting of light and adjusting optomechanics, (3) PCW integrated with a special cuvette, (5) spectrometer Ocean Optics HR4000, (6) personal computer. (c) Special glass cuvettes integrated with the PCWs.

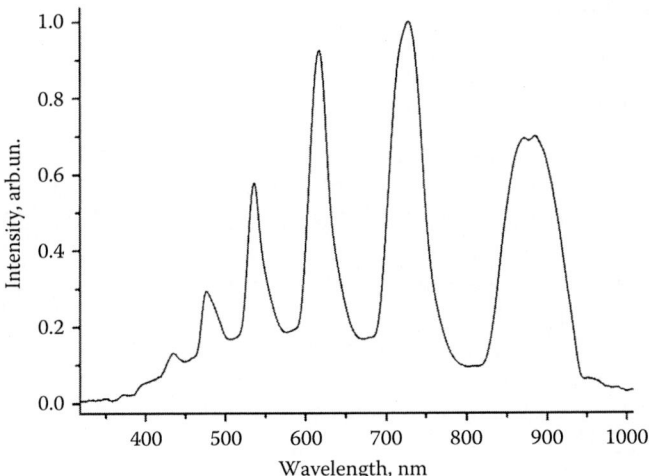

FIGURE 1.6
Transmission spectrum of BGP PCW with hollow-core diameter 210 µm.

An equation for the location of intensity maxima in the transmission spectrum of the waveguide with the built-in Fabry–Perot interferometer was theoretically obtained in [52]:

$$\lambda_j = \frac{4d}{2j+1}\left(n_2^2 - n_1^2\right)^{1/2}, \tag{1.2}$$

where
  j is the integer
  d is the capillary wall thickness in cladding
  n_2 is the glass refractive index
  n_1 is the refractive index of the medium filling up the internal channels of the waveguide

The fulfillment of the condition (1.2) corresponds to the case of the antiresonance between Fabry–Perot modes in cladding and the core modes of the waveguide. Thus, the reflectivity of the structured cladding rises to its maximum providing maximum of the waveguide's transmittance ability.

The found sensitivity of PCW transmission spectra to changes in the refractive index of the medium inside the hollow core and cladding channels induced a great interest for studies of waveguide sensing properties in the last decade [1,42,43]. In the next section, we discuss results of experimental studies of BGP PCW guidance abilities and spectral characteristic alterations due to variations in refractive index, absorption, and scattering. The possible applications of PCW sensors are also considered.

1.6 Experimental Study of PCW Sensing Properties

1.6.1 Refractive Index Sensing

The spectral properties of a PCF/PCW depend on both the geometry of its internal structure and on the variation in the refractive index of the fluid inside the core and/or channels of the waveguide [42,43,46]. In Figure 1.7, experimental transmission spectra of the PCW filled up by model liquids (water solutions of glucose) with different refractive indices for testing of the BGP PCW sensor are presented.

The analysis of the spectra (Figure 1.7) shows that with the increase in the medium refractive index all transmission bands are shifted toward the shorter wavelength region, and the magnitude of this displacement linearly depends on refraction ability. This linear dependence (the calibration curve) is presented in Figure 1.8.

The BGP PCW sensor described allows one to get an instant optical response to the change in refractive index of the analyte under study with the refractive index accuracy up to the third decimal digit. However, there are good prospects for enhancement of the measuring accuracy. Depending on the core diameter of the PCW used, the sensitivity can be optimized. For a PCW with a core diameter of 280 µm, the slope angle of calibration curve is equal to 63°, and with a core diameter of 220 µm, it is equal to 70°. Another essential advantage of using PCWs is the possibility of analyzing a small volume of a liquid. The use of a standard photometric cuvette with 10 mm thickness requires 2–3 mL of the liquid for a single measurement, while only 10–15 µL

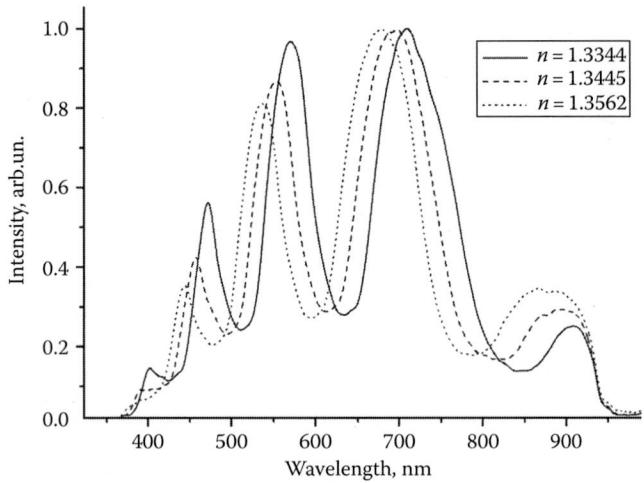

FIGURE 1.7
Transmission spectra of the chirped PCW sensing element filled up with aqueous solutions of glucose at different concentrations, that is, refractive indices. Core diameter is 210 µm.

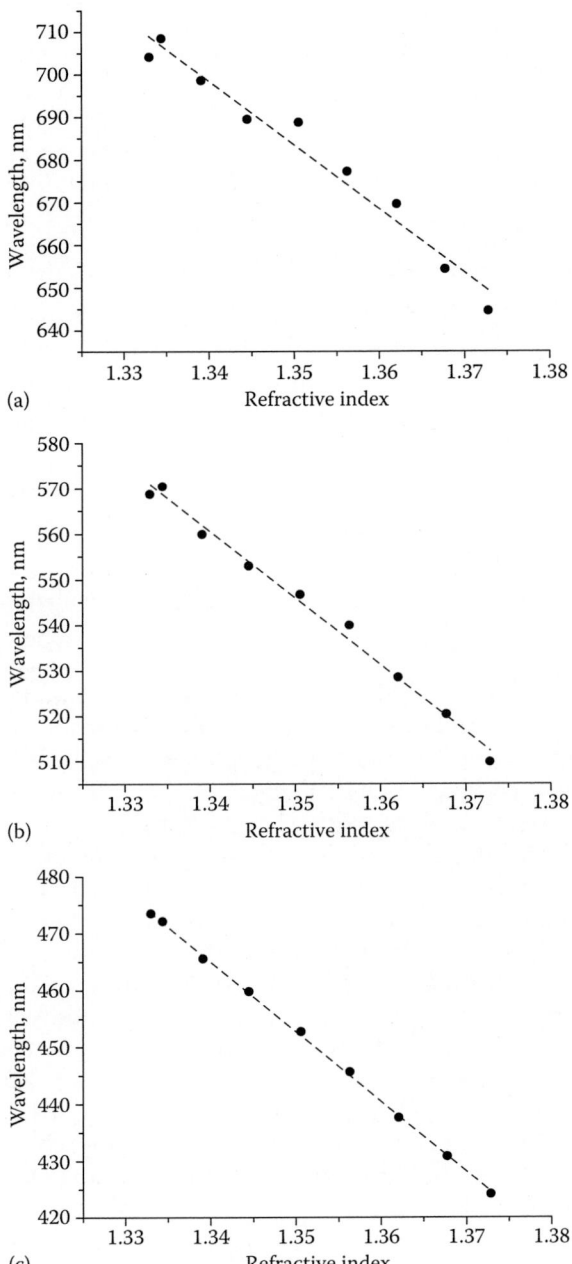

FIGURE 1.8

Shifting of the maxima of the transmission spectrum of PCF/PCW with the refractive index increase of the glucose solution in water presented in Figure 1.7. The PCW peak responses on index of refraction increase at three wavelength ranges are shown: (a) 704–645 nm, (b) 568–510 nm, and (c) 474–425 nm.

is required for filling up a PCW sensor of 60 mm length. And the third argument in favor of using PCWs is a possibility to provide a multiparameter sensing. For the present, we have considered only one parameter—index of refraction by measuring the wavelength dependence of transmission band maximum. In the next section, we consider PCW transmittance change, which is defined by the absorption coefficient of the analyte.

1.6.2 Absorption of the Guided Light

The attenuation of the intensity of a collimated beam passing through a layer of absorbing medium is described by the Beer law:

$$I(l) = I_0 e^{-\varepsilon c l} \tag{1.3}$$

where
 I_0 is the intensity of the incident light
 l is the thickness of the analyte layer
 ε is the molar absorption coefficient of the absorbing substance
 c is its concentration

According to Equation 3.3, the increase in probing length l allows for the detection of the lower absorption coefficient, that is, the lower concentration of absorbing molecules. Thus, a PCW-based sensor with the extent length is suitable for low concentration sensing, which in addition provides small volume probing of the analyte.

The chirped-cladding waveguide consisting of five concentric layers of small diameter capillaries and the hollow core with a diameter of 284 μm was used to study dependence of PCW spectral properties on an absorber concentration in a solution [1]. As the model liquid, the aqueous solutions of riboflavin were examined. Riboflavin at UV/visible excitation is used as a photodynamic dye for generating singlet oxygen and other superoxide free radicals, for example, to provide the extracorporal photodynamic therapy of the whole blood and its individual components [53]. Aqueous solutions of riboflavin have two broad absorption bands at 370 and 450 nm; the molar extinction coefficient for 450 nm is equal to 9800 L/mole·cm.

As the transmission spectrum of PCW with a hollow-core diameter of 284 μm has a transmission band at 450 nm that is a good match to the intrinsic absorption band of riboflavin, it is reasonable to expect the best sensitivity of PCW to variations in concentration of riboflavin for this particular band.

The PCW transmission spectra for four aqueous solutions of riboflavin at concentrations 0.001, 0.003, 0.005, and 0.01 mg/mL, and water as a reference, are presented in Figure 1.9. It is easily seen that with concentration increase the transmission spectrum is apparently transformed. The height of the peak at 450 nm decreases exponentially with riboflavin concentration (Figure 1.10).

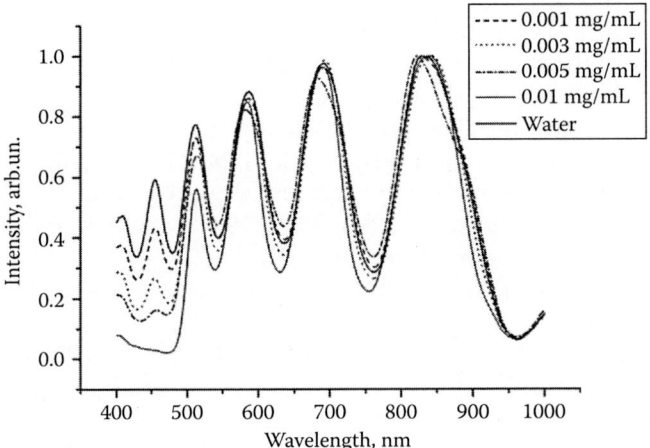

FIGURE 1.9
PCW transmission spectra for four aqueous solutions of riboflavin at different concentrations and water as a reference. (From Skibina, Yu.S. et al., *Quant. Electron.*, 41(4), 284, 2011. With permission.)

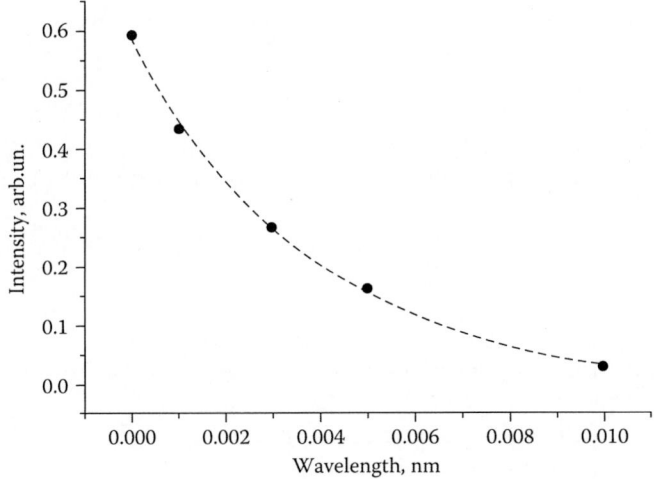

FIGURE 1.10
PCW transmittance at 450 nm versus concentration of riboflavin solution. (From Skibina, Yu.S. et al., *Quant. Electron.*, 41(4), 284, 2011. With permission.)

On the contrary, the intensity maxima at the longer wavelengths are much less sensitive to variations in riboflavin concentration. The longer wavelength response of PCW, far away from the absorption band, may be caused by the index of refraction change.

Thus, a PCW-based sensor allows for detection of small amounts of absorbing agents at the level of $\sim 10^{-8}$ mol/L.

1.6.3 Scattering of the Guided Light

The presence of the scattering particles in the medium filling the PCW's core and channels in cladding also leads to the transformation of its transmittance in the visible/NIR. The photons that have changed the direction of propagation after scattering collisions either propagate within the core in the backward direction or leave the core and enter the cladding. The effect of light scattering on PCW's light guidance ability was studied in designing the technique for blood type determination [47].

There are many blood compounds that affect scattering of photons passing through a blood sample: red blood cells (RBCs), leukocytes, platelets, lymphocyte, etc. About 99% of blood cells are RBCs (or erythrocytes), so they mainly determine optical properties of blood. The RBCs have the shape of a biconcave disc with a diameter of 5.7–9.3 μm [54] and thickness of 1.7–2.4 μm [55]. The RBC surface contains different antigens that form the group systems, among which the ABO system is of primary importance for medical practice. This group system comprises two erythrocyte antigens, denoted as A and B, and two types of plasma antibodies, (a) anti-A and (b) anti-B.

The blood of a human cannot contain antigen A and antibodies a (or antigen B and antibodies b) simultaneously, because the interaction of these nonspecific proteins would cause agglutination reaction when cell–cell adhesion of erythrocytes occurs and complexes of a few units to a few tens of RBCs are formed. In the classical method of blood typing, the formation or the absence of such complexes after adding specific agglutinating serums to the whole blood is a signature of the blood belonging to a definite group.

One of the approaches to the description of scattering in the blood is the radiation transport theory [2,56]. In this theory, the parameters used are the absorption coefficient μ_a, the scattering coefficient μ_s, and the anisotropy factor g (the mean cosine of the scattering angle). These parameters are determined by the RBC size and real and imaginary parts of the complex refractive indices $(n + i\chi)$ of RBCs and blood plasma [2,54,55].

According to [57], the scattering coefficient is expressed as

$$\mu_s = (1-H)\sum_{i=0}^{M} N_i \sigma_{s_i},$$

(1.4)

where
  H is the blood hematocrit
  M is the number of volume fractions of RBCs
  N_i is the number of particles with the ith diameter in the unit volume of the medium
  σ_{s_i} is the scattering cross section of the particles of the ith diameter

When the agglutination reaction of RBCs is positive, a large number of produced RBC complexes subside at the bottom of the agglutination tube. As a consequence, the number of RBCs suspended in the solution is reduced, which in accordance with Equation 1.4 results in the reduction of the scattering coefficient [42].

Thus, human blood is a perfect model of a scattering medium. In the experiments described in [42,47], the mixture consisting of blood diluted with saline and agglutinating serum was introduced into the PCW testing tube. The waveguide sample was subjected to irradiation from a broadband light source, and the output radiation intensity was measured in the wavelength range 400–700 nm. The dilution of whole blood with saline up to hematocrit less than 1% was undertaken mainly for getting the output intensity sufficient for detection. The first solution was prepared by adding the agglutinating serum taken from the A group to the test tube with whole blood taken from the B group (positive agglutination reaction) followed by dilution with saline up to the hematocrit of 0.8% after 10 min of incubation. The second solution was prepared following a similar protocol but using the agglutinating serum B (negative agglutination reaction in the test tube).

Figure 1.11 represents the averaged transmission spectra of the identical PCW sensors (testing tubes) with the hollow-core diameter of 270 μm filled up with the prepared solutions. For each solution, the averaging was done over five measured transmission spectra for the identical PCW testing tubes made from 50 mm long PCW. From data presented in Figure 1.11, it is seen that in the case of positive agglutination reactions the maximal transmittance at the wavelength 662 nm reaches 55 units, while in the case of negative reaction it is only 29 units.

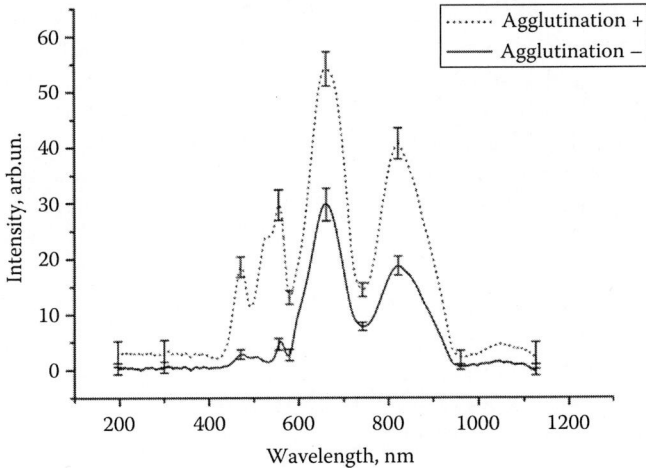

FIGURE 1.11
Averaged transmission spectra of identical PCW testing tubes with the diameter of the hollow core being 270 mm, filled up by the products of positive (+) and negative (–) reactions of blood agglutination. (From Malinin, A.V. et al., *Quant. Electron.*, 41(4), 302, 2011. With permission.)

For the differentiation of these two reactions, one can use the rule of three standard deviations and introduce the interval δ, defined by inequality [42]:

$$< Y > -3 \times S_n \leq \delta \leq < Y > + 3 \times S_n. \tag{1.5}$$

According to Equation 1.5, at wavelength 662 nm, the agglutination reaction should be considered positive if the measured transmittance falls within the interval from 51.6 to 58.6 and negative within the interval from 26.6 to 33.3.

For PCW filled up with the scattering medium, that is, corresponding to negative agglutination reaction, the transmittance is on average 1.9 times lower in comparison with the positive agglutination case. This opens up the possibility of applying PCW testing tubes to blood typing and suggests a novel high-precision protocol allowing for automatic blood control, fast response to even weak agglutination reactions, and significant reduction of the trial blood volume and the agglutinating serum consumption.

1.7 Applications of PCWs

We have already described some applications of PCW sensors based on refractometry and detection of light absorption and scattering. PCW sensors potentially have much broader applications, including detection of DNA, proteins, bacteria and viruses (see Table 1.1), blood glucose sensing, food quality analysis, etc. In this section, we analyze a few more examples of PCW application for sensing.

1.7.1 Example 1: Quality Analysis of Drinks

One of the most important parameters for juice or wine quality evaluation is sugar fraction quantification and detection of the presence of additives, such as artificial sweeteners. For example, the average sugar content in apple juice is 7–12 mg/100 mL [58]. In the food industry, refractometry is typically used for sugar determination. The refractive index (RI) of aqueous solution of any carbohydrate can be calculated using the following expression:

$$n = n_w + \alpha C, \tag{1.6}$$

where
n_w is the RI of water (25°C)
α is the specific increment of a refraction being equal to 0.00143 at 25°C
C is the concentration of sugar in g/100 mL

Sucrose, glucose, and fructose have almost the same specific refractive index increment α.

During the process of fruit juice production, different synthetic sweeteners can be added to drinks in order to improve their taste. Such sweeteners are

10–100 times sweeter than glucose, fructose, and sucrose. One of the most popular and cost-saving sweeteners is a mixture of cyclamate (E952) and saccharin (E954). The coefficient of sweetness for this mixture is 120–210. The value means that this mixture is roughly 120–210 times sweeter than sucrose. The required amount of the sweetener M can be calculated using the following equation:

$$M = \frac{C_{sw}}{K_{sw}},$$ (1.7)

where
   C_{sw} is the amount of substitutable sugar and
   K_{sw} is the coefficient of sweetness [49]

Obviously, solutions containing artificial sweeteners have a lower refractive index than those for natural sugar, because of the considerably lower concentration that is needed to provide a similar sweetness. In fact, the RI of glucose solution of concentration 7 mg/100 mL is 1.3534 ($T = 25°C$), but the concentration of a sweetener solution with similar sweetness is only 0.058 mg/100 mL, and correspondingly its RI is 1.3329 ($T = 25°C$) [49]. In our experiments, we have detected the difference in spectral characteristics of PCW testing tubes filled up with liquids containing natural sugar and mixture of sweeteners. We have obtained transmission spectra of the PCW filled up with three different samples of drinks: natural apple juice, carbonated soft drinks with artificial sweetener, and natural sugar. The same experiment was demonstrated in Ref. [49].

The transmission spectra plotted in Figures 1.12 and 1.13 demonstrate that the PCW transmission bands for natural apple juice for babies and

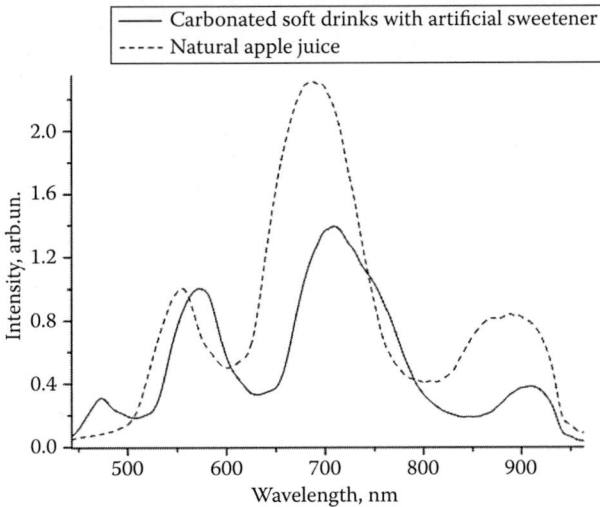

FIGURE 1.12
Normalized transmission spectra of the PCW testing tubes filled up with natural apple juice for babies and carbonated soft drink with artificial sweetener.

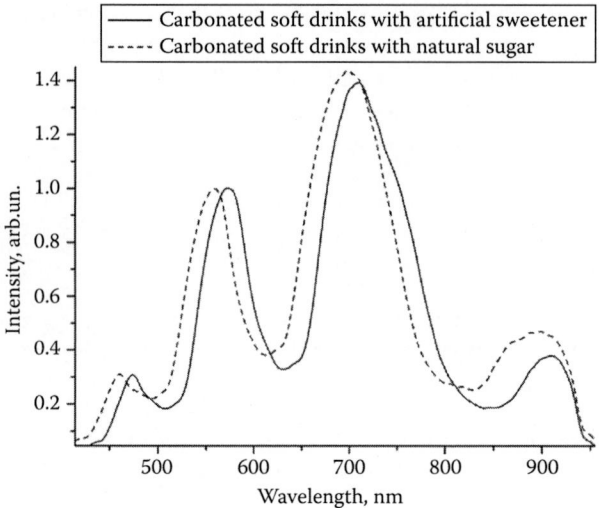

FIGURE 1.13
Normalized PCW transmission spectra for carbonated soft drinks containing artificial sweetener and natural sugar.

carbonated soft drinks containing natural sugar are shifted to the shorter wavelength range relative to the transmission band of the carbonated soft drink with the artificial sweetener.

As was shown earlier (see Figure 1.8), the dependence of the position of the central wavelength of the transmission band on glucose concentration can be used as a calibration curve to evaluate the total sugar concentration. Using this calibration curve, the total natural sugar concentration was calculated for three types of drinks (Figure 1.14 and Table 1.2).

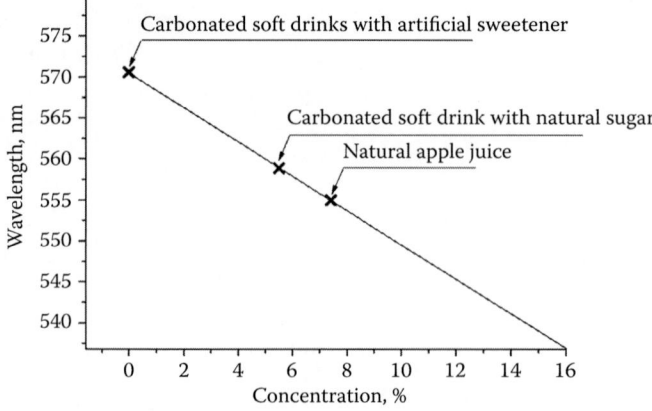

FIGURE 1.14
Dependence of the central wavelength of optical transmittance band for PCW on glucose concentration; experimental data for the apple juice and soft drink.

TABLE 1.2

PCW-Based Estimation of Carbohydrate Concentration in Drinks

Drink	Central Wavelength of Transmittance Band (nm)	Concentration (%)
Natural apple juice	554.98	7.4
Carbonated with natural sugar	558.88	5.5
Carbonated with artificial sweetener	571.59	0

1.7.2 Example 2: Selective Sugar Determination

The sweetness of juices or other drinks depends on the content of different natural sugars (sucrose, glucose, and fructose), artificial sweeteners, sugar substitutes, or their different combinations. Sweetness should be objectively controlled at drink production, and the selective determination of natural sugars and artificial sweeteners should be provided at the output of drink production because of variability of sugar content in fruits and other materials used in drink production technology. The use of hollow-core PCWs and polarization optical measurements might be one of the prospective approaches to solve this problem.

PCW-based selective sugar determination technique combines polarimetric and refractometric approaches. The refractometric technique has been discussed earlier. As it follows from the calibration curve (Figure 1.14), we can find only total sugar in a liquid (Figure 1.15), because sucrose, glucose, and fructose have almost the same specific refractive index increment.

The polarimetric technique is based on the molecular optical activity phenomenon when glucose and fructose rotate the azimuthal angle of the plane

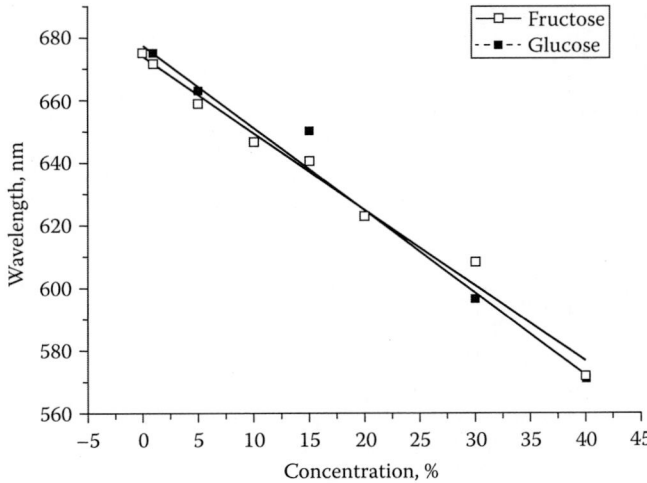

FIGURE 1.15

Dependence of the shift of the central wavelength of transmission band PCW sensor filled up by aqueous solutions of glucose or fructose on their concentration.

of polarization of a propagating linear polarized light beam by an amount α that is proportional to their concentration C and path length l. The particular molecule's unique specific rotation is dependent on the light wavelength λ, the *pH*, and the temperature of the sugar aqueous solution [59]:

$$[\alpha]_{\lambda, pH}^{T}\left[\frac{\text{Angular degrees}}{(\text{dm} \cdot \text{g/mL})}\right] = \frac{\alpha}{lC}. \tag{1.8}$$

For aqueous solutions of glucose at 589.3 nm and $T = 25°C$, $[\alpha] = 52.7°/(\text{dm} \cdot \text{g/mL})$. At 656 nm, it is 41.9, at 535 nm 65.4, at 508 nm 73.6, at 479 nm 83.9, and at 447 nm 96.6°/(dm·g/mL). This optical rotator dispersion (ORD) depends on molecular structure and bonds, and its wavelength dependence is the unique characteristic of optically active molecules and gives the basis for multispectral polarimetry of sugars.

Fructose, in particular, is strongly levorotatory—$[\alpha]_{589.3} = -92°/(\text{dm} \cdot \text{g/mL})$ [60]. The ratio for rotation of azimuthal angles of glucose and fructose is –1.75 at 589.3 nm and $T = 25°C$.

In our experiments, we measured the azimuthal angle of the plane of polarization of a propagating linear polarized light beam, which was transmitted through the PCWs filled up with aqueous solutions of glucose and fructose with different concentrations ($T = 20°C$, $pH = 7.5$, λ = 570–670 nm). The results are demonstrated in Figure 1.16.

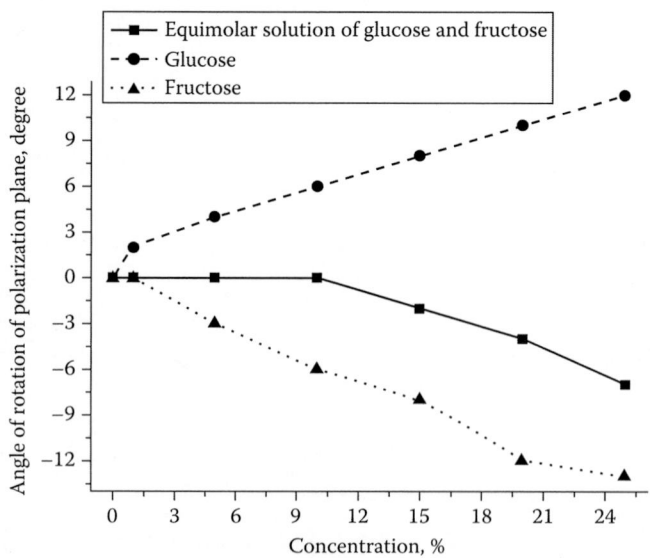

FIGURE 1.16
Dependence of rotation angle of polarization plane for glucose, fructose, and equimolar mixture of glucose and fructose on their concentration in aqueous solution.

Thus, by simultaneously measuring total sugar and angles of rotation of polarization plane, one can precisely estimate the percentage composition of glucose and fructose and define their concentration.

It should be noted that PCWs have some depolarization ability due to optical reflections from their structure at light propagation, but depolarization is automatically accounted for at calibration stage using an empty or water filled PCW, thus does not affect the accuracy of measurement.

1.7.3 Example 3: Oxidase Glucose Sensing in Blood

The possibility of developing a PCW-based glucose sensor was discussed previously. We considered the concept of the use of PCW as a refractometer and "smart" photometric cuvette. To enhance glucose sensing ability to tenths of millimoles per liter, an oxidase method for glucose determination is often used [46]. This technique implies the use of commercially available reagents, such as glucose oxidase, peroxidase, and chromogen. After the mixing of aqueous solution of glucose and the reagents, the dissolved glucose is oxidized by the enzyme glucose oxidase to gluconic acid and equimolar quantity of peroxide. Peroxide is converted to water and oxygen by the enzyme peroxidase. Free oxygen oxidizes chromogen, which can be measured by the evaluation of solution optical density change at wavelength 515 nm. Initially transparent, the glucose indicator becomes deep red colored. This allows one to detect glucose at a concentration level of millimoles per liter.

In our experiments, we used nine aqueous solutions of glucose with concentrations from 1.0 to 9.0 mmol/L and liquid reagent indicator. The mixture of a buffer (water) and the indicator was used as well. Each aqueous solution of glucose was mixed with liquid reagent at 10:1 ratio. Mixtures were incubated for 10 min before their injection into the chirped PCW. Figure 1.17 demonstrates

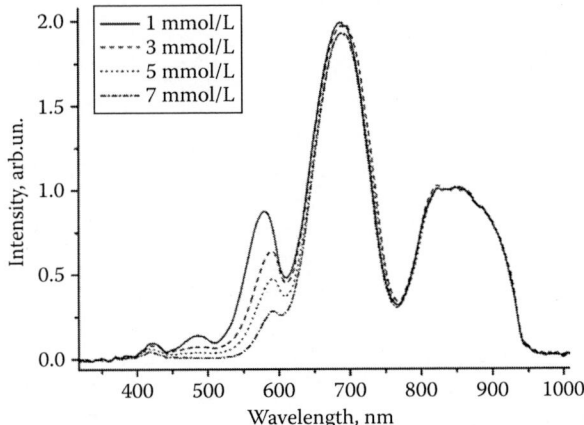

FIGURE 1.17
Normalized transmission spectra of PCW smart cuvettes filled up with aqueous solutions of glucose and indicator.

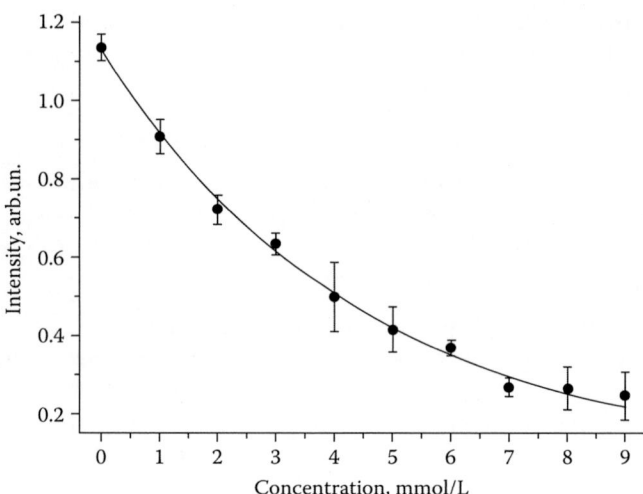

FIGURE 1.18
Transmitted intensity at 575 nm versus concentration of glucose in the solution.

four transmission spectra of the PCWs filled up with glucose/glucose indicator mixture. Spectra were normalized to 835 nm peak value since there is no expected absorption at this wavelength.

In Figure 1.18, the dependence of transmitted intensity at 575 nm on glucose concentration is presented. Measured values were approximated by the exponential function. An adequate optical response to increase in glucose concentration approximated by a smooth curve with low deviation was obtained. The measurement error was calculated for three measurements.

1.8 Summary

In this chapter, we reviewed the recent progress made in the understanding of the guidance properties of PCWs, in fabrication of such waveguides, studies of their sensing properties, and, finally, in their applications in refractometry of biological liquids, blood typing, glucose sensing in blood, and quality analysis of food drinks. We presented some experiments that demonstrated the efficiency and high sensitivity of PCWs.

The chapter summarizes the recent experimental studies on the sensing properties of the hollow-core PCWs with a large grating period. The impact of changes in refractive index, absorption, and scattering properties on guidance abilities and spectral characteristics of PCWs was discussed. It was found that the increase in the refractive index of the tested solution leads to the shift of the PCW transmission band. The presence of even small amounts of the light absorbing solution in the hollow core of the PCW causes strong attenuation of the radiation in the spectral range, corresponding to the absorption band. The scattering increase, instead, results in PCW transmittance decrease at all wavelengths. Based on the obtained experimental data, some possible applications of PCW sensors were discussed.

The unique features of PCWs make fiber-based devices ultracompact and easy to use with high performance ability. These waveguide structures allow one to make different complex sensing systems easy to design and construct.

Acknowledgments

This study was supported in part by SPE LLC Nanostructured Glass Technology, Saratov, RF; grants 1177.2012.2 of the President of RF "Supporting of Leading Scientific Schools"; 1.4.09 of RF Ministry of Education and Science; RF Governmental contracts 14.B37.21.0728 and 14.B37.11.0563; 224014, PHOTONICS4LIFE of FP7-ICT-2007-2; SCOPES EC, Uzb/Switz/RF, Swiss NSF, IZ74ZO_137423/1; FiDiPro (TEKES, 40111/11), Finland. Authors of the chapter thank the director of LLC SPE Nanostructured Glass Technology, Valentin I. Beloglazov, and the executive director of LLC SPE Nanostructured Glass Technology, Nina B. Skibina, for their invaluable help in the production of photonic crystal waveguides and productive discussion of the experimental results. Authors also thank Dr. Günter Steinmeyer (Max-Born Institute, Berlin, Germany; Optoelectronics Research Centre, Tampere, Finland) for his ideas in PCW design, helpful discussion of the results, and preparation of publications.

References

1. Skibina Yu.S., Tuchin V.V., Beloglazov V.I., Steinmeyer G., Bethge J., Wedell R., and Langhoff N. 2011. Photonic crystal fibres in biomedical investigations. *Quant. Electron.*, 41(4): 284–301.
2. Tuchin V.V. 2007. *Tissue Optics: Light Scattering Methods and Instruments for Medical Diagnosis*, 2nd edn. Bellingham, WA: SPIE Press.
3. Zheltikov A.M. 2011. *Microstructure Fibers in Biophotonics in Handbook of Biophotonics*, eds. Popp J., Tuchin V.V., Chiou A., and Heinemann S.H. Weinheim, Germany: Wiley-VCH Verlag GmbH & Co. KGaA.
4. Hoa X.D., Kirk A.G., and Tabrizian M. 2007. Towards integrated and sensitive surface plasmon resonance biosensors: A review of recent progress. *Biosens. Bioelectron.*, 23(2): 151–160.

5. Homola J., Yee S.S., and Gauglitz G. 1999. Surface plasmon resonance sensors: Review. *Sens. Actuators B*, 54: 3–15.
6. Leung A., Shankar P.M., and Mutharasan R. 2007. A review of fiber-optic biosensors. *Sens. Actuators B*, 125: 688–703.
7. Yablonovitch E. 1987. Inhibited spontaneous emission in solid-state physics and electronics. *Phys. Rev. Lett.*, 58: 2059–2062.
8. Joannopoulos J., Meade R., and Winn J. 2008. *Photonic Crystals: Molding the Flow of Light*. Princeton, NJ: Princeton University Press.
9. Badding J.V., Gopalan V., and Sazio P.A. 2006. Building semiconductor structures in optical fiber. *Photon. Spectra*, 40(8): 80–86.
10. Gaponenko S.V., Rosanov I.N., Ivchenko E.L. et al. 2005. *Optika nanostruktur (Nanostructure Optics)*. Saint-Petersburg, Russia: Nedra.
11. Srinivasarao M. 1999. Nano-optics in the biological world: Beetles, butterflies, birds, and moths. *Chem. Rev.*, 99(7): 1935–1962.
12. Parker A.R. 2000. 515 million years of structural colour. *J. Opt. A Pure Appl. Opt.*, 6(2): R15.
13. Ameen D.B., Bishop M.F., and McMullen T. 1998. A lattice model for computing the transmissivity of the cornea and sclera. *Biophys. J.*, 75(5): 2520–2531.
14. Pieranski P. 1983. Colloidal crystals. *Contemp. Phys.*, 24(1): 25–73.
15. Williams R.C. and Smith K. 1957. A crystallizable insect virus. *Nature*, 45: 119–120.
16. Deniskina N.D., Kalinin D.V., and Kazantseva L.K. 1987. *Blagorodnye opaly (prirodnye i sinteticheskiye) Noble Opals (Natural and Synthetic)*. Novosibirsk, Russia: Nauka.
17. Russell P.St.J. 2006. Photonic-crystal fibers. *J. Lightwave Technol.*, 24(12): 4729–4749.
18. Yeh P., Yariv A., and Marom E. 1978. Theory of Bragg fiber. *JOSA*, 68(9): 1196–1201.
19. Bureau B., Zhang X.H., Smektala F. et al. 2004. Recent advances in chalcogenide glasses. *J. Non-Cryst. Solids* 345 and 346: 276–283.
20. Pinto A.M.R. and Lopez-Amo M. 2012. Photonic crystal fibers for sensing applications. *J. Sens.*, 1: 1–21.
21. Cox F.M., Argyros A., and Large M.C.J. 2006. Liquid-filled hollow core microstructured polymer optical fiber. *Opt. Express*, 14(9): 4135–4140.
22. Cox F.M., Lwin R., Large M.C.J., and Cordeiro C.M.B. 2007. Opening up optical fibres. *Opt. Express*, 15(19): 11843–11848.
23. Wolfbeis O.S. 2004. Fiber-optic chemical sensors and biosensors. *Anal. Chem.*, 76: 3269–3284.
24. Wolfbeis O.S. 2008. Fiber-optic chemical sensors and biosensors. *Anal. Chem.*, 80: 4269–4283.
25. Mignani A.G., Ciaccheri L., Ottevaere H., Thienpont H., Conte L., Marega M., Cichelli A., Attilio C., and Cimato A. 2010. Diffuse-light absorption spectroscopy by fiber optics for detecting and quantifying the adulteration of extra virgin olive oil. *Proc. SPIE*, 7653: 76531C-1–76534C-1.
26. Righini G.C., Mignani A.G., Cacciari I., and Brenci M. 2009. Fiber and integrated optics sensors: Fundamentals and applications. In *An Introduction to Optoelectronic Sensors*, eds. G.C. Righini, A. Tajani, and A. Cutolo. Singapore: World Scientific Publisher, pp. 1–33.
27. Bosch M.E., Sánchez A.J.R., Rojas F.S., and Ojeda C.B. 2007. Recent development in optical fiber biosensors. *Sensors*, 7: 797–859.

28. Passaro V.M.N., Dell'Olio F., Casamassima B., and Leonardis F. 2007. Guided-wave optical biosensors. *Sensors*, 7: 508–536.

29. Marques M.B., Magalhaes F., Carvalho J.P., Frazao O., Araujo F.M., Santos J.L., and Ferreira L.A. 2008. Recent advances on optical sensing using photonic crystal fibers. *AIP Conf. Proc.* 1055: 39–42.

30. Hoo Y.L., Jin W., Shi C., Ho H.L., Wang D.N., and Ruan S.C. 2003. Design and modeling of a photonic crystal fiber gas sensor. *Appl. Opt.*, 42(18): 3509–3515.

31. Cubillas A.M., Lazaro J.M., Conde O.M., Petrovich M.N., and Lopez-Higuera J.M. 2009. Gas sensor based on photonic crystal fibres in the $2\upsilon3$ and $\upsilon2 + 2\upsilon3$ vibrational bands of methane. *Sensors*, 9: 6261–6272.

32. Cordeiro C.M.B., Franco M.A.R., Chesini G., Barretto E.C.S., Lwin R., Brito Cruz C.H., and Large M.C.J. 2006. Microstructured-core optical fibre for evanescent sensing applications. *Opt. Express*, 14(26): 13056–13066.

33. Jensen J.B., Pedersen L.H., Hoiby P.E. et al. 2004. Photonic crystal fiber based evanescent-wave sensor for detection of biomolecules in aqueous solutions. *Opt. Lett.*, 29(17): 1974–1976.

34. Hoo Y.L., Jin W., Ho H.L., Wang D.N., and Windeler R.S. 2002. Evanescent-wave gas sensing using microstructure fiber. *Opt. Eng.*, 41(1): 8–9.

35. Yua X., Zhanga Y., Kwokb Y.C., and Shumc P. 2010. Highly sensitive photonic crystal fiber based absorption spectroscopy. *Sens. Actuat. B Chem.*, 145: 110–113.

36. Konorov S.O., Zheltikov A.M., and Scalora M. 2005. Photonic-crystal fiber as a multifunctional optical sensor and sample collector. *Opt. Express*, 13(9): 3454–3459.

37. Smolka S., Barth M., Benson O. 2007. Highly efficient fluorescence sensing with hollow core photonic crystal fibers. *Opt. Express*, 15(20): 12783–12791.

38. Nguyen K.N., Alameh L.V., and Chung Y. 2010. Cladding modes analysis of photonics crystal fiber for refractive index sensors using finite element method. *Proceedings of CLEO*, San Jose, CA, pp. JWA59–JWA60.

39. Skivesen N., Têtu A., Kristensen M., Kjems J., Frandsen L.H., and Borel P.I. 2007. Photonic-crystal waveguide biosensor. *Opt. Express*, 15(6): 3169–3176.

40. Brilland L., Charpentier F., Troles J., Bureau B., Boussard-Plédel C., Adam J.L., Méchin D., and Trégoat D. 2009. Microstructured chalcogenide fibers for biological and chemical detection. Case study: A CO_2 sensor. *Proc. SPIE*, 7503: 750358.

41. Unterkofler S., Garbos M., Euser T. et al. 2013. Long-distance laser propulsion and deformation monitoring of cells in optofluidic photonic crystal fiber. *J. Biophoton.*, 6: 1–10. doi: 10.1002/jbio.201200180.

42. Malinin A.V., Skibina Yu.S., Tuchin V.V., Chainikov M.V., Beloglazov V.I., Silokhin I.Yu., Zanishevskaya A.A., Dubrovskii V.A., and Dolmashkin A.A. 2011. The use of hollow-core photonic crystal fibres as biological sensors. *Quant. Electron.*, 41(4): 302–307.

43. Malinin A.V., Skibina Yu.S., Mikhailova N.A., Silokhin I.Yu., and Chainikov M.V. 2010. Biological sensor based on a hollow-core photonic crystal fiber. *Tech. Phys. Lett.*, 36(4): 362–364.

44. Myaing M.T., Ye J.Y., Norris T.B., Thomas T., Baker J., Wadsworth W.J., Bouwmans G., Knight J.C., and Russell P.S.J. 2003. Enhanced two-photon biosensing with double-clad photonic crystal fibers. *Opt. Lett.*, 28(14): 1224–1226.

45. Jensen J., Hoiby P., Emiliyanov G., Bang O., Pedersen L., and Bjarklev A. 2005. Selective detection of antibodies in microstructured polymer optical fibers. *Opt. Express*, 13(15): 5883–5889.

46. Malinin A.V., Zanishevskaja A.A., Tuchin V.V., Skibina Yu.S., and Silokhin I.Yu. 2012. Oxidase method for glucose determination using long-period grating waveguide. *Proc. SPIE*, 8222: 82221B.
47. Malinin A.V., Zanishevskaja A.A., Skibina Yu.S., Silokhin I.Yu., Tuchin V.V., Dubrovskiy V.A., and Dolmashkin A.A. 2011. Determination of blood types using a chirped photonic crystal fiber. *Proc. SPIE*, 7898: 78981A-1–78987A-1.
48. Malinin A.V., Zanishevskaja A.A., Tuchin V.V., Skibina Yu.S., and Silokhin I.Y. 2012. Photonic crystal fibers for food quality analysis. *Proc. SPIE*, 8427: 842746-1–842748-1.
49. Zanishevskaya A.A., Malinin A.V., Tuchin V.V., Skibina Yu.S., and Silokhin I.Yu. 2013. Photonic crystal waveguide biosensor. *J. Innov. Opt. Health Sci.*, 6(2): 1350008-1-6.
50. Fedotov A.B., Beloglazov V.I., and Zheltikov A.M. 2012. Structure-integrated arrays of hollow waveguides for sensor devices. *Nanotechnol. Russ.*, 3(1): 58–63.
51. Skibina J.S., Iliew R., Bethge J., Bock M., Ficher D., Beloglazov V. I., Wedell R., and Scheinmeyer G. 2008. A chirped photonic crystal fiber. *Nat. Photon.*, 2: 679–683.
52. Zheltikov A.M. 2008. Colors of thin films, antiresonance phenomena in optical systems, and the limiting loss of modes in hollow optical waveguides. *Phys. Usp.*, 51: 591–600.
53. Zheng H.J. 2009. Photodynamic therapy–An update on clinical applications. *J. Innov. Opt. Health Sci.*, 2(1): 73.
54. Borovoi A.G., Naats E.I., and Oppel U.G. 1998. Scattering of light by a red blood cell. *J. Biomed. Opt.*, 3(3): 364–372.
55. Kirillin M.Yu. and Priezzhev A.V. 2002. Monte Carlo simulation of laser beam propagation in a plane layer of the erythrocyte suspension: Comparison of contributions from different scattering orders to the angular distribution of light intensity. *Quant. Electron.*, 32(10): 883–887.
56. Tuchin V.V. 1997. Light scattering study of tissues. *Phys. Usp.*, 40: 495–515.
57. Schmitt J.M. and Kumar G. 1998. Optical scattering properties of soft tissue: A discrete particle model. *Appl. Opt.*, 37(13): 2788–2797.
58. Markowski J., Baron A., Mieszczakowska M., and Płocharski W. 2009. Chemical composition of French and Polish cloudy apple juices. *J. Horticult. Sci. Biotechnol.*, 84(6): 68–74.
59. Cote G.L. and Cameron B.D. 2009. A noninvasive glucose sensor based on polarimetric measurements through the aqueous humor of the eye. In *Handbook of Optical Sensing of Glucose in Biological Fluids and Tissues*, ed. V.V. Tuchin. London, U.K.: CRC Press, pp. 183–211.
60. Yamamoto A., Matsunaga A., and Miyazaki M. 1991. Polarized photometric chromatography. *Sciences*, 7: 719–721.

2

Electrochemical Biosensors

Madalina M. Barsan, M. Emilia Ghica, and Christopher M.A. Brett

CONTENTS

2.1 Introduction

The purpose of this chapter is to introduce the principles of electrochemical transduction in biosensors. After a brief introduction, the different types of biosensor will be described in turn, with emphasis on recent advances in their conceptualization and application.

A chemical sensor is generally defined as a device that transforms chemical information into an analytically useful signal [1]. A biosensor is defined as a device that uses specific biochemical reactions mediated by enzymes, immunosystems, tissues, organelles, or whole cells to detect chemical compounds, usually by electrical, thermal, or optical signals [2]. Considering the transduction principle of the biosensor, a more complete classification is

- Electrochemical: potentiometric, amperometric/voltammetric
- Electrical: ion sensitive, conductometric/impedimetric
- Optical
- Piezoelectric: acoustic, microcantilever
- Thermometric

This chapter will focus on the first two of these. The reason for this is that the basic principle for electrochemical biosensors is due to the fact that due to chemical reactions between immobilized biomolecules, the biorecognition element, and a target analyte, ions or electrons are produced or consumed, which affects the measurable electrical properties of the solution, such as electric current or potential [3]. Electrochemical biosensors can be categorized according to the measured electrical parameter:

- Potentiometric biosensors, which measure the difference between two potentials at equilibrium (in volts)
- Amperometric (or voltammetric) biosensors, which measure the current (or charge) due to electron transfer (in amperes or coulombs)
- Conductometric biosensors (chemiresistors), which measure the resistance (in ohms) or conductance (the reciprocal of resistance)
- Impedimetric biosensors, which measure the impedance (in ohms)

The biosensing elements can be tissue parts, microorganisms, organelles, cell receptors, enzymes, antibodies (Abs), nucleic acids, aptamers, recombinant Abs, synthetic receptors, biomimetic catalysts, combinatorial ligands, and also imprinted polymers as biomimetic elements. The specific molecular recognition that leads to high selectivity and specificity is based on the

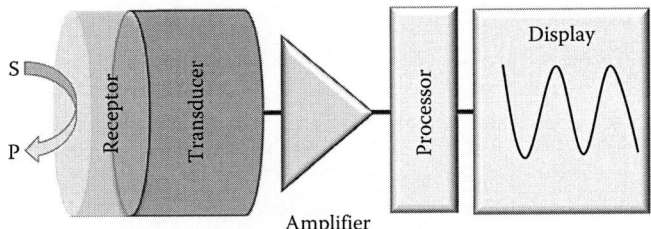

FIGURE 2.1
Schematic diagram showing the main components of an electrochemical biosensor.

affinity between complementary structures such as enzyme with substrate or Ab with antigen (Ag). The result of this interaction is the conversion of a biochemical signal into a quantifiable electrical response, which is a function of concentration [4–6].

Figure 2.1 shows the key elements of a biosensor: the receptor that contains the biological sensing element converting the substrate to products, the transducer, the amplifier, and the signal processor.

This chapter will describe the three groups of electrochemical sensors, potentiometric, amperometric/voltammetric, and conductometric/impedimetric separately. All have two things in common: the measurement must be done with a closed electrical circuit and electrical neutrality must be preserved.

A short history regarding the discovery and developments of electrochemical biosensors is given in [7] and important landmarks are shown in Table 2.1 [8–18].

2.2 Potentiometric Biosensors

2.2.1 Basic Principles

Potentiometric biosensors, which measure a potential difference at zero current, that is, at equilibrium, are of two types, in which either ion-selective electrodes (ISEs) or electrolyte–insulator–semiconductors (EISs) are used in order to transduce the biological reaction into an electric signal [19]. For both of these transducer types, the key part is an ion-selective membrane in contact with the analyte-containing solution, the membrane being responsible for the selectivity to target ions in the presence of interfering ions in the sample being analyzed. The main phenomenon responsible for generation of the response, that is, the potential difference across the membrane, is ion exchange between the two membrane/solution phases, which depends

TABLE 2.1

Historical Landmarks in Electrochemical Biosensor Evolution

Year	Event	References
1922	First glass pH electrode—Hughes	[8]
1925	First blood pH electrode—Kerridge	[9]
1954	Invention of the pCO_2 electrode—Stow and Randall	[10]
1956	Invention of the oxygen electrode—Clark	[11]
1962	First description of a biosensor: an amperometric enzyme electrode for glucose—Clark and Lyons	[12]
1969	First potentiometric biosensor: urease immobilized on an ammonia electrode to detect urea—Guilbault and Montalvo	[13]
1970	Invention of the ISFET—Bergveld	[14]
1972/5	First commercial biosensor Yellow Springs Instruments glucose biosensor	
1975	First bacteria-based biosensor for the measurement of alcohol—Diviès	[15]
	First immunosensor: ovalbumin on a platinum wire—Janata	[16]
1976	LaRoche Lactate Analyser LA640 using hexacyanoferrate soluble mediator	
1982	The first needle-type enzyme implantable electrode—Shichiri	[17]
1984	First mediated amperometric biosensor: ferrocene and glucose oxidase—Cass	[18]
1987	Launch of the MediSense ExacTech™ pen-sized glucose sensor based on SPE	
1992	Launch of the *i*-STAT handheld blood analyzer	
1996	Launch of the Glucocard®	
1998	Launch of the LifeScan FastTake® blood glucose biosensor	
Present	Quantum dot-, nanoparticle-, magnetic nanoparticle-, nanotube-, and nanowire-based biosensors	

on the activity of the target ion (analyte) in these phases [20]. This potential is described by the relevant Nernst equation:

$$E = E_0 + \frac{RT}{zF} \ln a$$

where
  E is the measured potential (in V)
  E_0 is a characteristic constant for the ion-selective/external electrode system
  R is the gas constant
  T is the absolute temperature (K)
  z is the signed ionic charge
  F is the Faraday constant
  a is the activity of the ionic species of interest

The logarithmic dependence of the potential on the ionic activity, often replaced by concentration, is responsible for the wide analytical range and for the high accuracy and precision of these sensors.

Interference effects are commonly described by the semiempirical Nicolsky–Eisenman equation, an extension of the Nernst equation, which is given by

$$E = E_0 + \frac{RT}{z_i F} \ln\left[a_i + \sum_j \left(k_{ij} a_j^{z_i/z_j} \right) \right]$$

where
   i is the ion of interest
   j is the interfering ion
   k_{ij} is the potentiometric selectivity coefficient

The smaller the value of the selectivity coefficient, the less is the interference by j, although the extent of the interference also depends on the activity of the interfering ion.

2.2.2 Ion-Selective Electrodes

For an ISE and in the simplest case, the potential difference across the membrane is measured by an internal reference electrode immersed in the inner solution whose composition is constant and an external reference electrode that is immersed in the analyte-containing solution. An ISE biosensor can consist, for example, of a membrane with immobilized enzyme surrounding a pH electrode [21], the enzymatic reaction occurring next to the thin sensing glass membrane causing a change in pH. A simple potentiometric biosensor scheme is presented in Figure 2.2. As seen, a semipermeable membrane (1) surrounds the biocatalyst (2) entrapped next to the active glass membrane (3) of a pH probe (4). The electrical potential is measured between the internal reference electrode (RE_{int}) bathed in dilute HCl and an external reference electrode (RE_{ext}).

There are three types of ISEs that are of use for biosensors:

1. *Glass electrodes for cations*, in which the sensing element is a very thin hydrated glass membrane that generates an electrical potential due to the concentration-dependent competition between the cations for specific binding sites [22]. Originally the glass consisted of a mixture of Na_2O, Al_2O_3, and SiO_2. Apart from the most common H^+, these electrodes can be made responsive to univalent cations such as Na^+, K^+, Li^+, Rb^+, Cs^+, Ag^+, and NH_4^+, depending on the doping of the glass membrane.

2. *Glass pH electrodes coated with a gas-permeable membrane* selective for CO_2, NH_3, NO, NO_2, CO, CO_2, H_2S, etc. The diffusion of the gas through this membrane causes a change in pH of a sensing solution between the membrane and the electrode, which is then determined [23,24].

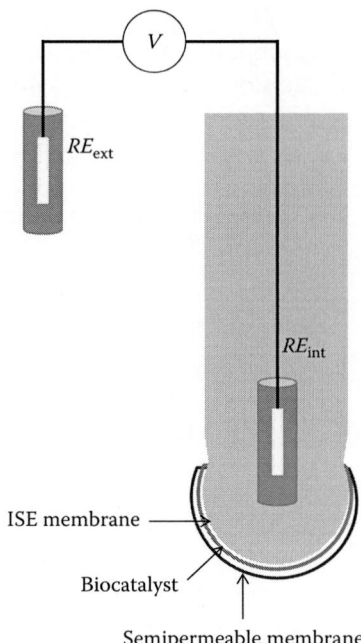

FIGURE 2.2
Schematic representation of a potentiometric biosensor.

3. *Solid-state electrodes* where the glass membrane is replaced by a thin membrane of a specific ion, usually for halides. The F^- electrode consists in a membrane of crystalline LaF_3, which has very low solubility in water. The chemistry at the membrane surface is

$$LaF_3(s) \rightarrow LaF_2(bound) + F^-(aq)$$

Silver sulfide is also used for ion sensing, its design being similar to the LaF_3 electrode. The membrane alone can sense Ag^+ and S^{2-} and, if it is mixed with a silver halide (AgX), can sense halide ions (e.g., F^-, Cl^-, I^-, Br^-).

Membranes may contain fixed binding sites or have mobile ion exchangers or ionophores (carriers). The binding sites are incorporated in the membrane matrix, which determines the internal polarity, lipophilicity, transport, and other mechanical properties of the membrane.

Depending on the membrane, ISE-based biosensors may be sensitive to H^+, F^-, I^-, and Cl^- ions in addition to gases such as CO_2 and NH_3 and, in the context of this chapter, most importantly to enzyme systems that change the concentration of any of these ions or gases.

The majority of potentiometric biosensors for detection of environmental pollutants have used enzymes that catalyze the consumption or production of

TABLE 2.2

Examples of Microorganism Biosensors for Toxicant
Detection Based on Potentiometric Transducers

Target	Microorganism	Transducer Type
Organophosphates	*Flavobacterium* sp.	pH electrode
	Recombinant *E. coli*	
Penicillin	Recombinant *E. coli*	pH electrode
Urea	*Bacillus* sp.	NH_4 electrode
Trichloroethylene	*P. aeruginosa*	Chloride electrode
Ethanol	*S. ellipsoideus*	Oxygen sensor

Source: Adapted from Lei, Y. et al., *Anal. Chim. Acta*, 568, 200,
2006. With permission.

protons. Biosensors based on ISEs were reported for detection of insecticides
such as aldicarb, carbaryl, carbofuran, and dichlorvos, the principle consisting
in cholinesterase enzyme inhibition [25]. Heavy metal ions can also be measured using the enzyme urease coupled to an ammonium ion sensor. Because
the activity of urease is sensitive to heavy metal ions, inhibition of enzyme
activity can be used to estimate the total concentration of these ions [26].

A few examples of microorganism biosensors for toxicant detection based
on potentiometric transducers are summarized in Table 2.2 [27].

2.2.3 Electrolyte–Insulator–Semiconductors

EISs are the basis for silicon field-effect chemical sensors and include ion-
selective field-effect transistors (ISFETs) [28], light-addressable potentiometric sensors (LAPSs) [29], and capacitive EIS sensors [30]. Their advantages
are small size, potential for mass production at low cost, and possibility of
on-chip circuitry. There are two fundamental models explaining chemical
sensitivity of the potential drop at the electrolyte–insulator interface:

1. Ion exchange, the same basis as for the theory of the glass electrode [20]
2. Adsorption of potential-determining ions, as proposed in colloid chemistry to describe surface charging of oxides in electrolyte solutions [31]

EIS biosensors are based on the measurement of pH and/or its change due to
enzyme-catalyzed reactions or cellular metabolism. Thin insulating oxide and
nitride films produced by conventional methods of microelectronics are used
as pH-sensitive gate dielectrics. Chalcogenide films have promising potential for the detection of heavy metal ions in solutions. The recently developed
pulsed laser deposition (PLD) method opens prospects of multicomponent
analysis by thin chalcogenide film sensor arrays, using the principles of the
"electronic tongue" [32].

Each of ISFETs, LAPS, and capacitive EIS will now be described in turn.

2.2.3.1 Ion-Selective Field-Effect Transistors

The ISFET principle is based on a local potential generated by surface ions from a solution [33], being able to measure the conductance of a semiconductor as a function of an electrical field perpendicular to the gate oxide surface. In the simplest version (i.e., an *n*-channel metal oxide semiconductor field-effect transistor [MOSFET]), a *p*-type silicon substrate (bulk) contains two *n*-type diffusion regions (source and drain). The structure is covered with a silicon dioxide insulating layer on top of which a metal gate electrode is deposited (see Figure 2.3). When a positive voltage with respect

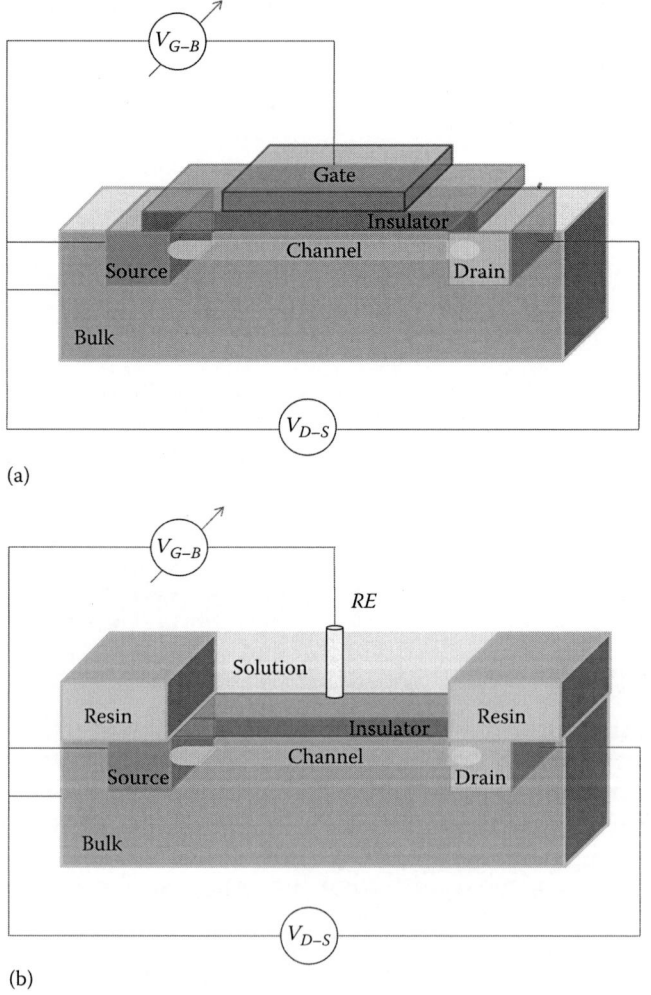

FIGURE 2.3
Schematic representation of (a) metal oxide semiconductor FET (MOSFET) and (b) ion sensitive FET (ISFET).

to the *p*-type silicon is applied to the gate electrode, a conducting channel is created between the source and the drain, near the silicon dioxide interface. The conductivity of this channel can be modulated by adjusting the strength of electrical field between the gate electrode and the silicon (V_{S-G}), perpendicular to the substrate surface. At the same time, a voltage can be applied between the drain and the source (V_{S-D}), which results in a drain current between the *n*-regions.

In the case of the ISFET, the gate metal electrode of the MOSFET is replaced by an electrolyte solution in which a reference electrode is immersed, so that the SiO_2 gate oxide is placed directly in electrolyte solution (Figure 2.3b), the reference electrode being considered the gate of the MOSFET [14]. Thus, in ISFET-based biosensors, the channel is in direct contact with the environment, the gate-source potential V_{S-G}, being determined by the surface potential at the insulator/electrolyte interface.

When SiO_2 is used as the insulator, the chemical nature of the interface oxide is reflected in the measured source–drain current V_{S-D}. The surface of the SiO_2 contains OH functionalities, which are in electrochemical equilibrium with hydrogen and hydroxyl ions in the sample solution, and so the surface potential is pH dependent. A site-dissociation model describes the signal transduction as a function of the ionization state of the amphoteric Si–OH-containing surface [34,35]. Typical pH sensitivities measured with SiO_2 ISFETs are 37–40 mV/pH unit.

In order to improve performance, other inorganic insulators for pH sensors such as Al_2O_3, Si_3N_4, and Ta_2O_5 have been deposited on top of the first layer of SiO_2 by chemical vapor deposition (CVD) in order to increase sensitivity and decrease the hysteresis and drift of SiO_2-based ISFETs. Nevertheless, SiO_2 is still being used for the fabrication of biosensors, since the Si–OH groups can be used for covalent attachment of organic molecules and polymers.

ISFET biosensors can be constructed by covering the sensor electrode with a selectively permeable polymer layer, through which ions may diffuse and cause a change in the FET surface potential. The first suggestion for using ISFETs for enzyme biosensor construction was reported in [36], and the first results were for penicillin determination [37]. The device consisted of two pH-sensitive ISFETs, one of which contain a membrane with covalently bound penicillinase and albumin, while the other contained only albumin, the device being called "enzyme field-effect transistor (ENFET)." When penicillin was present in the solution, penicillinase in the membrane catalyzed penicillin hydrolysis that resulted in the production of protons and therefore in a local pH decrease in the gate area, registered by the ISFET. ISFET-based biosensors for the determination of many pollutants and food toxicants were reviewed by Dzyadevych et al. in [38], a list of which is presented in Table 2.3.

Nowadays, nanomaterials are being used to obtain ultrasensitive biosensors, especially using carbon nanomaterials: single-walled carbon

TABLE 2.3

ISFET-Based Biosensors for Determination of Some Pollutants
and Food Toxicants

Target	Enzyme
4-Chlorophenol	Tyrosinase
Formaldehyde	Alcohol oxidase, aldehyde dehydrogenase
Hypochlorite	Acetylcholinesterase
Organophosphorus pesticides	Acetylcholinesterase, OPH, organophosphorus acid anhydrolase
Heavy metal ions	Urease
Cyanide	Peroxidase

Source: Adapted from Dzyadevych, S.V. et al., *Anal. Chim. Acta*, 568, 248, 2006. With permission.

nanotube (SWCNT) and graphene [39]. There are strategies to increase sensitivity, dynamic detection in cells and liquid environment, DNA hybridization, and single-molecule detection. For example, the integration of supported lipid bilayers with SWCNT FETs showed the possibility of electrical detection of specific binding of streptavidin to biotinylated lipids [40]. The formation of fluidic supported lipid bilayers on SWCNTs allowed the study of lipid–SWCNT interactions and sensing of analytes binding to specific receptors embedded in the supported lipid bilayers [41]. Functionalization of SWCNT FETs with RNA-based *Escherichia coli* aptamers can selectively recognize *E. coli*, demonstrating a reliable screening tool for microorganisms such as *E. coli* [42] or *Salmonella infantis* [43]. These studies suggest that functionalized SWCNT FETs promise to be extremely useful platforms in the future.

2.2.3.2 Light-Addressable Potentiometric Sensors

A LAPS is a semiconductor device first developed by Hafeman in 1988 [29] and is now as commonly used as ISFETs. The LAPS principle is based on semiconductor activation by a light-emitting diode (LED) or a laser [44], as shown in Figure 2.4.

A LAPS usually consists of a metal insulator–semiconductor (MIS) or EIS structure. Its heterostructure of silicon/silicon oxide/silicon nitride $(Si/SiO_2/Si_3N_4)$ can be excited by a modulated light source, producing a photocurrent, the amplitude of which is sensitive to the surface potential, LAPS thus being able to detect the potential variation caused by an electrochemical event. The basic function of LAPS is for pH detection, via the H^+-sensitive Si_3N_4 layer fabricated on the LAPS surface. LAPSs show some advantages compared to ISFET, the fabrication process being fully compatible with standard microelectronics facilities. The encapsulation of LAPS is much less critical since no metal contact is formed on the surface and the extremely flat surface

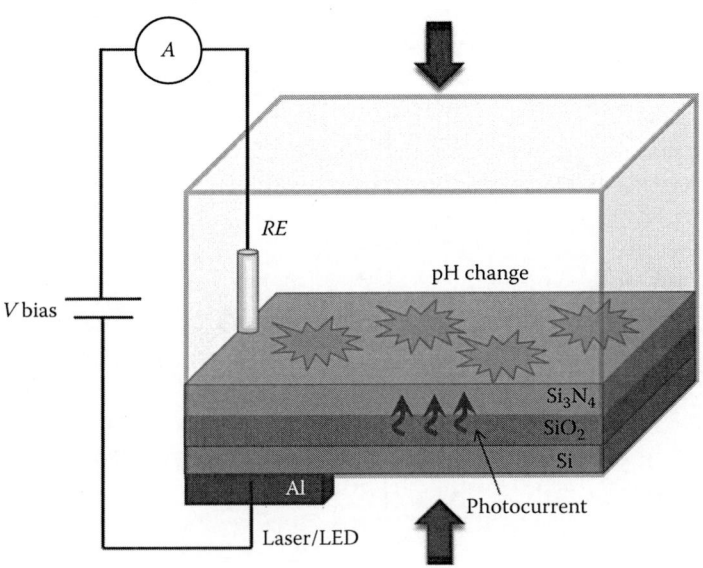

FIGURE 2.4
Working principle of LAPS.

makes it possible to incorporate into a very small volume chamber, which is important for small dose measurement [45]. Besides pH detection, it has been used for the detection of extracellular events. After cells are cultured on the LAPS, a focused laser is used to illuminate the front side of the chip to address the cells to be monitored. Excitable cells can generate an extracellular action potential, which subsequently changes the amplitude of the photocurrent [46]. Applications of LAPS in cell-based biosensors have been introduced in cell biology, pharmacology, toxicology, and environmental measurement. Using living cells, especially mammalian cells, LAPSs are able to detect toxicities of heavy metals as a result of cellular physiological changes [47–49].

2.2.3.3 Capacitive EIS

Capacitive EISs are based on the charge carrier distribution at an insulator–semiconductor interface, which is controlled by an external voltage (V_{S-D}), with a superimposed voltage (V_{S-G}) then being able to measure the capacity of the space-charge layer (see Figure 2.5). Depending on the applied and superimposed voltage, a characteristic integral capacitance–voltage ($C–V$) curve can be obtained.

Electrochemical interaction at the phase boundary between sensing layer and electrolyte leads to a horizontal shift of the $C–V$ characteristic, which corresponds to the change of ion activity in the solution; this shift can be used as a quantitative sensor signal in the linear range of the $C–V$ curve. Dielectric oxide materials such as Al_2O_3 and Ta_2O_5 can be utilized as pH-sensitive gate

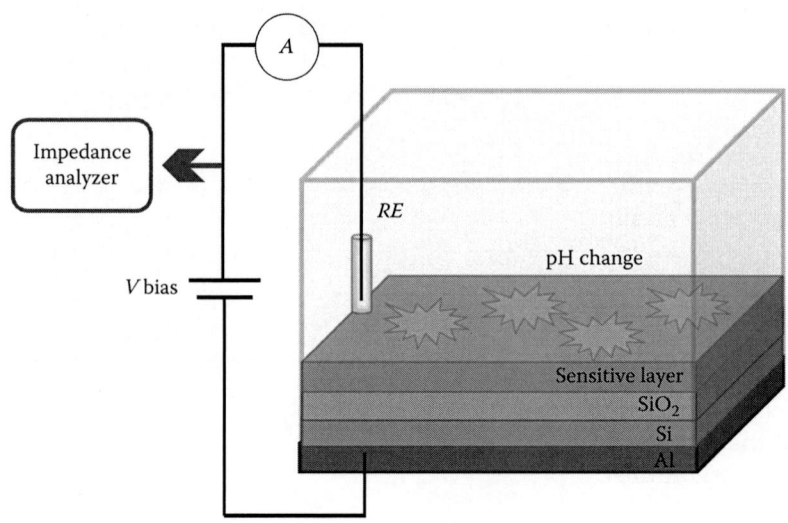

FIGURE 2.5
Working principle of capacitive EIS.

insulators for such capacitive EIS structures [30]. These materials can be produced by the PLD technique [50–52].

Some examples from the literature using this biosensor strategy are as follows. Penicillin sensors were fabricated by immobilizing the enzyme penicillinase on top of the pH-sensitive EIS structure [53,54]. Layer-by-layer (LbL) films made with alternating layers of SWCNT and polyamidoamine (PAMAM) dendrimers have been used for the fabrication of devices with enhanced performance [55]. A cyanide biosensor based on a pH-sensitive structure with an immobilized enzyme (cyanidase) was prepared [56]. Thirdly, a dual amperometric/potentiometric biosensor chip, adapted to a flow injection analysis (FIA) system, for the determination of organophosphorus nerve agents paraoxon, dichlorvos, parathion, and diazinon based on the enzyme organophosphorus hydrolase was developed, the potentiometric sensor being a conductive field-effect-based EIS structure [57].

2.3 Voltammetric/Amperometric Biosensors

2.3.1 General Aspects

In voltammetric and amperometric sensors, the measurement is based on the transfer of electrons across the sensor–electrolyte interface, the current measured being directly proportional to the number of electroactive species, which are being oxidized or reduced.

For *voltammetric sensors*, a current–voltage profile is recorded, the current being measured as a function of applied potential using a potentiostat. Occasionally, the potential response to an applied current is recorded using a galvanostat. Unlike potentiometric sensors, where a potential difference is measured at equilibrium, usually three electrodes are necessary: working (sensor), auxiliary, and reference electrodes; the reference electrode can also serve as auxiliary electrode if the currents are very small [58]. The applied potential is the driving force for the electron-transfer reaction and the current results from the electrochemical oxidation or reduction of the electroactive species. The current should be directly proportional to the bulk concentration of the electroactive species. Voltammetric techniques are commonly employed including linear sweep and cyclic voltammetry, differential pulse voltammetry (DPV), and square wave voltammetry (SWV); a description of the fundamentals of these voltammetric techniques can be found in [59]. With voltammetric sensors, the information obtained is diverse, detection limits can be low (down to nM with preconcentration), and more than one electroactive species, if each reacts at different applied potentials, can be determined in the same experiment. Thus, in the most easily applicable situations for these types of sensor, there is no need for prior separation of components of complex mixtures.

Amperometric sensors operate at fixed applied potential, thus constituting a special class of voltammetric sensors, which are more common for biosensor applications. The current is monitored as a function of time, often reaching a steady state [3,60], and is due to the reduction or oxidation of a component in the biosensor assembly that has interacted with the analyte in solution. Since biocatalytic reaction rates are often chosen to have a first-order dependence on bulk analyte concentration, such steady-state currents are usually proportional to the bulk analyte concentration [3].

2.3.2 Clark Electrode

The most common example of an amperometric biosensor is based on the oxygen Clark electrode (Figure 2.6) that has been routinely used for monitoring the concentration of dissolved oxygen in fluids [61]. It consists of a platinum cathode and a silver anode both in contact with an electrolyte solution, the cathode covered by a thin oxygen-permeable membrane, which is impermeable to other species in solution, such as ions.

Oxygen reduction at the cathode, caused by an appropriate potential being applied between working and reference electrode, generates an electric current. The O_2 diffusion rate to the cathode is proportional to partial pressure of O_2 in the sample. The amperometric signal is proportional to the partial pressure of oxygen and thence to concentration [62], the change in oxygen concentration being due, for example, to its consumption in an oxidase enzyme reaction occurring in a second enzyme-containing membrane placed over the oxygen-permeable membrane.

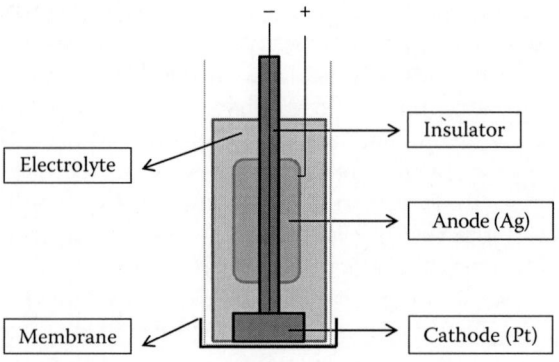

FIGURE 2.6
Schematic representation of the oxygen Clark electrode.

2.3.3 Enzyme Biosensors

Amperometric biosensors are mostly based on enzymes that either consume oxygen (e.g., all the oxidases), produce hydrogen peroxide (excluding oxidases that produce water), or produce (indirectly) the reduced form of β-nicotinamide adenine dinucleotide (phosphate) (NAD(P)H) (e.g., dehydrogenases), during the course of the catalytic reaction on the substrate of interest [63]. The general equations of these types of amperometric biosensor are

$$\text{Substrate} + O_2 \xrightarrow{\;Enzyme\;} \text{Product} + H_2O \tag{2.1}$$

$$\text{Substrate} + O_2 \xrightarrow{\;Enzyme\;} \text{Product} + H_2O_2 \tag{2.2}$$

$$\text{Substrate} + NAD(P)^+ \xrightarrow{\;Enzyme\;} \text{Product} + NAD(P)H \tag{2.3}$$

2.3.3.1 First-Generation Biosensors

The consumption of oxygen (Equations 2.1 and 2.2) can be measured by a Clark-type electrode at a platinum cathode at −0.6 V versus Ag/AgCl reference electrode, according to

$$O_2 + 2H_2O + 4e^- \xrightarrow{\;-0.6\,V\,vs.\,Ag/AgCl\;} 4OH^- \tag{2.4}$$

Hydrogen peroxide generated in Equation 2.2 can be measured amperometrically by electrooxidation at a solid electrode (e.g., platinum, carbon) polarized at +0.65 V, according to

$$H_2O_2 \xrightarrow{\;+0.65\,V\,vs.\,Ag/AgCl\;} O_2 + 2H^+ + 2e^- \tag{2.5}$$

Biosensors based on detection of either consumed oxygen or produced hydrogen peroxide are included in the class of first-generation biosensors. At the potential where anodic oxidation of peroxide takes place, other organic compounds, such as ascorbic and uric acid, are oxidized, which can lead to poor selectivity. Moreover, prolonged use in biological matrices results in fouling of the electrode surface.

2.3.3.2 Second-Generation Biosensors

In order to reduce the problems associated with the use of high potentials, redox mediators were introduced, which constitute second-generation biosensors. The function of the redox mediator is to shuttle electrons between enzyme redox centers and the electrode surface. A good redox mediator should [64]

- React rapidly with the enzyme and at the electrode
- Exhibit reversible electron-transfer kinetics
- Undergo regeneration at low potential
- Possess electrochemical properties that are not pH dependent
- Exhibit chemical stability in both oxidized and reduced forms
- Not undergo secondary reactions
- Be nontoxic and amenable to immobilization

The most common redox mediators include ferrocene and its derivatives, metal hexacyanoferrates, phenazine dyes (Meldola blue, Nile blue, methylene blue, neutral red, etc.), and quinones [63].

Mediator-based biosensors function by monitoring H_2O_2 or NAD(P)H according to

$$H_2O_2 + Med_{red} \xrightarrow{\text{Peroxidase}} Med_{ox} + H_2O \qquad (2.6)$$

$$NAD(P)H + Med_{ox} + H^+ \xrightarrow{\text{Diaphorase}} Med_{red} + NAD^+ \qquad (2.7)$$

$$Substrate + Med_{ox} \xrightarrow{\text{Enzyme}} Product + Med_{red} \qquad (2.8)$$

The functioning of an amperometric biosensor using a redox mediator is shown in Figure 2.7.

2.3.3.3 Third-Generation Biosensors

A special class of amperometric biosensors is when there is direct electron transfer between the enzyme and the electrode surface. This is usually achieved by "wiring" electrodes such that the shell of the enzymes is modified with redox polymers that contact the electrode substrate in order to make it electrically conductive. In this way, the redox center of the enzyme can directly exchange electrons with the electrode on which it is adsorbed [65].

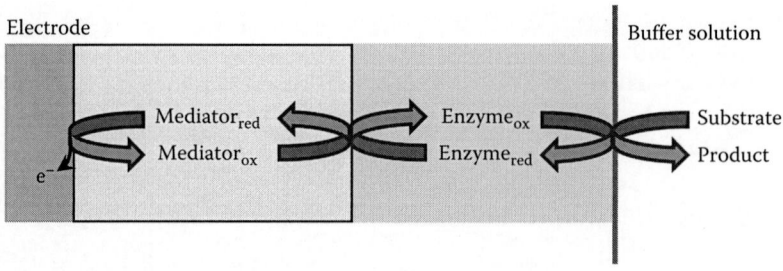

Electrode | Buffer solution

FIGURE 2.7
Schematic representation of the sequence of steps that occurs in second-generation ampero-metric biosensors (oxidation of the mediator).

2.3.3.4 Enzyme Immobilization

Immobilization of enzymes can be irreversible or reversible. If irreversible, once the biocatalyst is attached to the support, it cannot be detached without destroying either the biological activity of the enzyme or the support [66]. The most common procedures for irreversible immobilization are covalent coupling, entrapment, microencapsulation, and cross-linking.

The use of reversible enzyme immobilization methods is highly attractive, mostly for economic reasons because when the enzymatic activity decays, the support can be regenerated under gentle conditions and reloaded with fresh enzyme. Reversible methods of enzyme immobilization include adsorption, ionic binding, affinity binding, chelating or metal binding, and disulfide bonds.

Among other immobilization methods for enzymes, one that has been much employed in recent years is that of self-assembled monolayers (SAMs), normally carried out by immersion in a solution carrying the charged species of interest. This technique presents several advantages [67]:

1. The technology is easy and straightforward to use because the formation of SAMs is relatively fast.
2. Enzymes can be adsorbed on SAMs provided the enzyme has an overall surface charge opposite to that of the outer layer of the SAM.
3. Covalent attachment of redox enzymes is possible.
4. The design of anchoring alkanethiols, attached to the electrode substrate, can be tailored to particular needs (e.g., length of spacer, type of functional groups, mixture of alkanethiols with different properties for integrating different chemical functionalities, generation of multilayers).
5. SAMs are sufficiently stable with respect to temperature and pH and can be operated in a rather broad potential range (about −1.4 to +0.8 V vs. SCE).

In order to increase biosensor selectivity, permselective membranes might be applied on top of the enzyme layer; these are based on different transport mechanisms: size exclusion (e.g., cellulose acetate), charge exclusion (e.g., Nafion), polarity (e.g., phosphatidylcholine), mixed control (cellulose acetate/Nafion), etc. Membrane-based electrochemical sensors exhibit various advantages, such as retention of mediators, cofactors, permeability of the membrane to the analyte, and extension of the linear range; however, under diffusion limitations, the sensitivity of these devices decreases [63].

2.3.3.5 Electrode Materials

Electrode materials used for the development of enzyme biosensors are solid [68], relatively inert, and highly conductive, resulting in low background currents, and their surface can be easily modified. For the oxidation of hydrogen peroxide, platinum, carbon, or gold electrodes are normally used for the oxidation of reduced form of nicotinamide adenine dinucleotide (NADH) different types of carbon: graphite, glassy carbon, carbon paste, etc. Other solid electrode materials used are semiconductors, for example, metal oxides and conductive organic salts. Graphite powder with different metal micro-/nanoparticles (palladium, platinum, ruthenium, cobalt) has been made into electrodes together with an inert matrix binder for NADH and H_2O_2 determination.

2.3.3.6 Advantages and Disadvantages

The advantages and disadvantages of employing enzymes in biosensor architectures are summarized in [67]:

Advantages

1. A very high catalytic activity with a turnover that makes them exceptional bioelectrocatalysts for signal amplification. Good turnover frequencies k_{cat} are up to more than 100/s.
2. Enzymes typically have a very high selectivity for their substrates.
3. The driving force, that is, the redox potential that is needed to achieve enzymatic biocatalysis, is often very close to that of the enzyme substrate. This means that biosensors can operate at moderate potentials.
4. In a number of cases, an improvement in enzyme stability was found when enzymes were immobilized on transducer surfaces.

Disadvantages

1. The fact that enzymes are rather large molecules means that, despite the high catalytic turnover at the enzyme active site, the overall catalytic density is low.
2. Often the active site of the enzyme is deeply buried within the surrounding protein shell. Thus, direct electron transfer is often not possible and artificial redox mediators are needed.
3. Enzymes have a limited lifetime, and therefore, biosensors exhibit only a limited long-term stability. So far, operational lifetimes of biosensors are up to 30–60 days.

2.3.3.7 Applications

Direct monitoring of analytes has been the major application of biosensors, the most important application being glucose. Biosensors have also been developed for the indirect monitoring of enzyme-toxic species, known as enzyme inhibition biosensors. Unfortunately, the inhibition is often irreversible so that the original biological activity can be restored only after chemical treatment. Their potential use, in such cases, is thus often more as a warning system, not requiring exact measurement of the analyte concentration [3].

Enzyme biosensors have been used in food toxicity and environmental monitoring. The main analytes monitored by enzymatic biosensors for environmental and food applications are summarized in Table 2.4. Some applications in these areas are as follows:

Heavy metals: Inhibition-based biosensors for heavy metal determination have been the subject of reviews [69,70]. An amperometric biosensor based on urease has been developed for the detection of mercury, cadmium, and arsenic at a screen-printed electrode (SPE) with rhodinized carbon [71]. Cadmium, copper, lead, and zinc have been determined in the micromolar range by inhibition of glucose oxidase at a poly(neutral red) carbon film modified electrode [72] for detection in milk.

Pesticides and herbicides: Enzymatic biosensors based on inhibition are also used to measure pesticides, mostly acetylcholinesterase in combination with choline oxidase [73–75]. A remote biosensor for field monitoring of

TABLE 2.4

Application of Amperometric Enzyme Biosensors in Environmental and Food Toxicity

Target	Enzyme	References
Cd^{2+}, As^{2+}, Hg^{2+}	Urease	[71]
Cd^{2+}, Co^{2+}, Pb^{2+}, Zn^{2+}	Glucose oxidase	[72]
Methyl paraoxon	Acetylcholinesterase	[75]
Organophosphate	OPH	[76]
Cyanide	Tyrosinase	[77]
Hydrazine	Tyrosinase	[78]
AF B1	Acetylcholinesterase	[79]
Phenol, 4-chlorphenol, catechol	Tyrosinase	[80]
E. coli	Tyrosinase	[82]
Biogenic amines	Polyamine oxidase	[84]
Ethanol	Alcohol oxidase	[85]
Acetaldehyde	Aldehyde dehydrogenase	[86]
Hypoxanthine	Xanthine oxidase	[87]

organophosphate pesticides has been developed using organophosphate hydrolase (OPH) [76], and the capability of various pesticides to inhibit tyrosinase has been reported [77,78].

Toxins: Aflatoxins (AFs) are a group of highly toxic and carcinogenic secondary metabolites produced by *Aspergillus flavus* and *Aspergillus parasiticus* fungus; an AF B_1 sensor has been developed for measurement in olive oil based on inhibition of acetylcholinesterase [79].

Phenolic compounds: A considerable number of organic pollutants, which are found widely distributed in the environment, have phenolic structures. Phenol, 4-chlorphenol, and catechol were determined by a tyrosinase-containing carbon ink printed on a wearable neoprene [80]. Phenolic compounds in olive oil have been measured by DPV using an SPE with immobilized tyrosinase [81].

Microorganisms: Bacteria, viruses, and other microorganisms are found widely in polluted untreated and treated waters and constitute hence a public health problem. An amperometric biosensor for the detection of *E. coli* in wastewater was developed based on tyrosinase-catalyzed oxidation of polyphenolic compounds, which are produced microbiologically from salicylic acid [82].

Biogenic amines: Biogenic amines, mainly putrescine, cadaverine, spermidine, histamine, and tyramine, are not only biosynthesized in animal and plant cells but also produced by microbial decarboxylation of amino acids. Amperometric biosensors for amines were reported for determination in fish using SPEs with monoamine oxidase and putrescine oxidase [83] and in fruits [84] using diamine oxidase and polyamine oxidase immobilized on polymeric membranes.

2.3.4 Immunosensors

Immunoassay (IMA) technology originated in the late 1950s when Yalow and Berson [88] reported an immunological assay that detects human insulin in blood samples at the picogram level. IMA employs Abs to detect and quantify antigens (Ags) [89]. An immunosensor is therefore a sensor that uses an Ab as a biorecognition element. Abs are a class of proteins, known as immunoglobulins (Ig), generated by the immune system as a response to foreign species, such as bacteria, viruses, and parasites [90]. An Ag or immunogen is a molecule that can be recognized by the immune system and that can be bound specifically to an Ab. Haptens are molecules that are recognized by Abs but do not produce an immune response.

The Abs can be derived from antiserum–polyclonal antibodies (PAbs) or from hybridomas–monoclonal antibodies (MAbs). There are several classes of natural Abs (IgG, IgM, IgA, IgE) that provide animals with key defenses against pathogenic organisms and toxins. Most IMA systems rely on IgG, which is bivalent and has the ability to bind to two antigenic sites [90].

Besides Abs, fragments of Abs or aptamers can be used as recognition elements. These fragments provide the same specificity as the whole Ab but are smaller, which can be an advantage in biosensor applications [91]. Aptamers are folded, single-stranded DNA or RNA oligonucleotides, which specifically bind to molecular targets, such as proteins and haptens, and can be regarded as synthetic Abs. Using aptamers as the biosensor recognition element in place of proteic Abs makes label-free biosensors possible [90]. Advantages over Abs are easier deposition on the sensing surface, higher reproducibility, longer shelf life, easier regeneration, and a higher resistance to denaturation [91].

In the electrochemical context, normally, enzymes are used as markers in an enzyme immunoassay (EIA), that is, an electrochemical enzyme biosensing strategy. The most widespread amperometric enzyme IMA is enzyme-linked immunosorbent assay (ELISA) for the determination of the human chorionic gonadotropin (hCG) [92]. In this immunosensor, MAb specific to hCG is immobilized on the membrane of an oxygen electrode, to which hCG in the sample and catalase-labeled hCG can compete for binding.

Immunosensors can be

- Homogeneous: Abs, Ags, and labeled Ags are mixed.
- Heterogeneous (the most used): Either Ab or Ag is immobilized on a solid carrier.
- Sandwich: Ab is immobilized, and after addition of the sample containing Ag, a labeled secondary Ab is added.

IMA can be competitive or noncompetitive. In the former, competition takes place between free and bound Ag for a limited amount of labeled Ab or between Ag (the sample) and labeled Ag for a limited amount of Ab [93]. The detection can be achieved directly or indirectly. A schematic representation of IMAs is shown in Figure 2.8.

IMAs generally have high sensitivity and selectivity, wide applicability, and rapidity and are easy to use, ideal for large sample load, easily automated, and suitable for both lab and field use and for cost-effective analysis of small-volume samples [89,94]. However, disadvantages are linked to difficulty in the synthesis of hapten and vulnerability to cross-reacting compounds and nonspecific interactions and are not suitable for small load or multiresidue determination.

IMA techniques for ecosystem contaminants were introduced in the early 1970s [95]. There are several reviews on the application of immunosensors for environmental contaminants and food analysis [89,96,97]. Applications of amperometric immunosensors using different labeling systems include determination of herbicides (2,4-dichlorophenoxyacetic acid [2,4-D]) [98,99], pesticides (atrazine) [100,101], and a multianalyte immunosensor to measure human gonadotropin hormones (follicle-stimulating hormone and luteinizing hormone) using two horseradish peroxidase-labeled Abs in a ferrocene-mediated system [102].

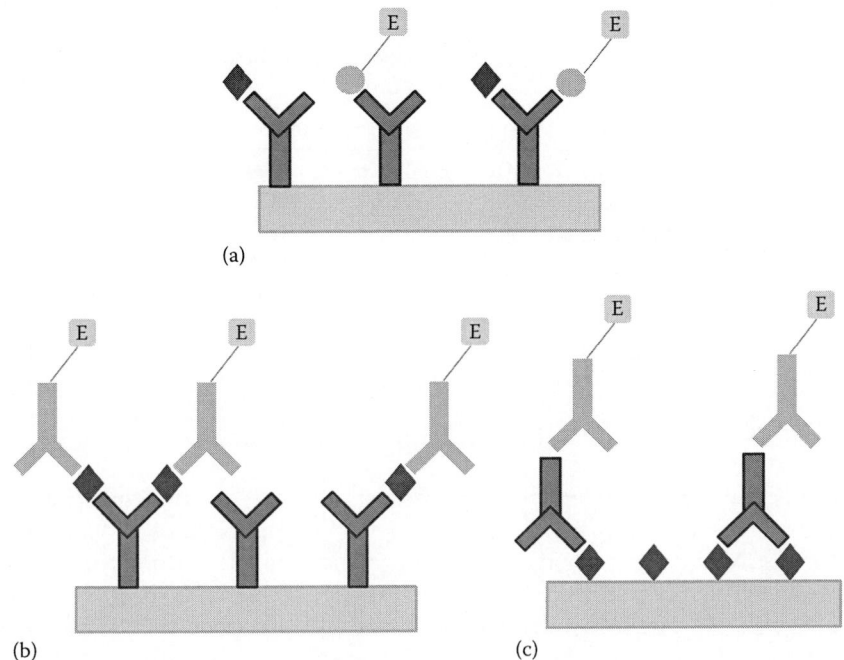

FIGURE 2.8

Representation of IMA: (a) direct competitive; (b) direct sandwich; (c) indirect. ⅄, Ab; ⅄, labeled Ab; ◆, Ag; ●, labeled Ag.

Examples of other hormones detected by amperometric immunosensors are estrogen [103] and progesterone in cow's milk [104].

Environmental IMAs have been developed and evaluated for analytes including major classes of organic compounds such as polychlorinated biphenyls (PCBs) [105], polyaromatic hydrocarbons (PAHs) [106,107], toxins (AF) [108,109], as well as heavy metals [110,111], and microbial toxins (*Salmonella*) [112].

Table 2.5 summarizes selected applications of immunosensors in food and environmental monitoring.

It is expected that nanomaterials will provide new tools to improve the performance of immunosensors in the future [113]. There is much interest in including carbon nanotubes [114], as well as nanoparticles [115], reducing the incubation time to a few minutes.

2.3.5 Cell-Based Biosensors

Cell-based biosensors employ immobilized living cells as sensing elements, combined with electrochemical transducers to detect the intracellular and extracellular microenvironment conditions and physiological parameters, and produce

TABLE 2.5

Applications of Immunosensors in Food
and Environmental Monitoring

Target	References
2,4-Dichlorophenoxyacetic acid	[98]
Atrazine	[100]
Estrogen	[103]
Progesterone	[104]
PCBS	[105]
PAHs	[107]
AF	[109]
Cd^{2+}, Co^{2+}, Pb^{2+}	[111]
Salmonella	[112]

responses through the interaction between stimulus and cells [116]. They can incorporate plant or animal isolated cells as well as microbial cells. Most of the cell-based biosensors developed up to now are microbial biosensors, owing to the simplicity of microorganism cultivation and their viability when immobilized as biocatalysts [117]. The first microbial biosensor was described in 1975 by Diviès and was based on the use of *Acetobacter xylinum* and an oxygen electrode [15].

Cell biosensors consist of two main parts. One is from living cells, which is the primary transducer used in the first sensing element receiving and producing signals, and the other belongs to the secondary transducers used in converting the physiological signals to electrical signals.

Amperometric microbial biosensors comprise three generations, similar to enzyme biosensors. The first-generation biosensors are based on the recording of the consumption of a co-substrate (normally oxygen) during analyte oxidation or generation of reaction product [117]. The most typical design is based on the Clark-type electrode and is called a "respiratory electrode." The approach has been used for the development of microbial biosensors based on oxygen consumption [118]. In second-generation microbial biosensors, oxygen is substituted by a redox mediator. This can be realized only if the analyte conversion involves easily accessible cell surface-localized enzymes or if the cell wall and membrane are permeable to mediator in both directions. This type of biosensor has been developed with different mediators [119]. Direct electron transfer, specific to third-generation microbial biosensors, was also reported for biosensors based on fuel cells [120].

Microbial fuel cells (MFCs) have been studied as a biochemical oxygen demand (BOD) sensor for a long time. Since Karube et al. reported a BOD sensor based on MFC using the hydrogen produced by *Clostridium butyricum* immobilized on the electrode in 1977 [121], a variety of MFC BOD sensors with the use of electron mediators have been developed [122–124]. Even though the inclusion of mediators can enhance the electron-transfer rate, these biosensors have poor stability because of mediator toxicity. Recently, mediator-less

MFCs have been exploited to fabricate novel BOD sensors for continuous and real-time monitoring [125,126], and the performance of an MFC as BOD sensor was improved using respiratory inhibitors [127].

Immobilization of microbes can be achieved by both physical and chemical methods. Physical methods include adsorption and entrapment (e.g., on dialysis membranes), while chemical methods include covalent binding and cross-linking. Whereas covalent binding has not been useful with very low cell loading being achieved [128,129], cross-linking using bifunctional agents, such as glutaraldehyde, has been successfully used for the immobilization of cells on supports.

Applications: The main application of microbial biosensors is in the environmental field, especially to assay BOD in wastewater. BOD microbial biosensors take advantage of the high reaction rates of microorganisms interfaced to electrodes to measure oxygen depletion rates [129]. Microbial biosensors have been investigated for the detection of environmental pollutants, for example, PAHs. An amperometric sensor for naphthalene has been developed using either *Sphingomonas sp.* B1 or *Pseudomonas fluorescens* WW4 [130]. Other biosensors were developed for phenolic compounds [131], organophosphates [132], and cyanide [133]. Alcohol determination with microorganism biosensors is also employed, for instance, based on *Saccharomyces ellipsoideus* [134] and *Gluconobacter oxydans* [135]. Heavy metal determination using microbial biosensors is more recent and includes, for example, detection of cadmium [136], chromium [137], copper [138], and zinc [139].

The main applications of microbial biosensors in food and environmental monitoring are shown in Table 2.6 [131–144].

TABLE 2.6

Applications of Microbial Biosensors in Food and Environmental Monitoring

Target	Microorganism	References
Phenolic compounds	*Pseudomonas putida*	[131]
Organophosphate	Recombinant *P. putida* JS 444	[132]
Cyanide	*Thiobacillus ferrooxidans*	[133]
Ethanol	*S. ellipsoideus*	[134]
Ethanol	*G. oxydans*	[135]
Cd^{2+}	Recombinant *E. coli*	[136]
Cr^{VI}	*Cytochrome c3* from *Desulfomicrobium norvegicum*	[137]
Cu^{2+}	*Saccharomyces cerevisiae*	[138]
Zn^{2+}	*Chlorella* sp.	[139]
BOD	*Arxula adeninivorans* LS3	[140]
BOD	*Microbial consortium*	[141]
p-Nitrophenol	*Arthrobacter* JS 443	[142]
p-Nitrophenol	*Moraxella* sp.	[143]
2,4-Dinitrophenol	*Rhodococcus erythropolis*	[144]

Future trends: With a better understanding of the genetic information of microbes and the development of improved recombinant DNA technologies, different enzymes and proteins have been expressed on the cell surface through surface expression anchors [27]. Another trend is to develop biosensors for application in extreme conditions, such as highly acidic, alkaline, saline, extreme temperature, and organic solvent environment because more and more detections will involve such unfriendly conditions. Additionally, the microbes are immobilized on the surface of chips in order to develop microelectronic mechanical systems (MEMSs).

2.4 Conductometric and Impedimetric Biosensors

2.4.1 Principles

The measurement of the electric properties of a biosensor constitutes either a conductometric or an impedimetric biosensor.

2.4.1.1 Conductometric Sensing

The conductivity of electrolyte solutions, due to ion migration in an electrical field, is measured by applying a potential difference between two electrodes: ions with negative charge move toward the anode, while positively charged ones move toward the cathode. The conductivity of the electrolyte solution depends on the ion concentration and mobility. The resistance, R, the reciprocal of the conductivity, is directly proportional to the distance l between the immersed electrodes and reciprocal to their area A; therefore,

$$R = \rho \frac{l}{A}$$

where ρ is the resistivity. The conductometric measurement usually consists in determining the conductivity of a solution between two parallel electrodes, its value being due to the contributions of all the ions within the solution tested.

The conductometric transducer, which is a specific case of this general measurement procedure for application in sensors, is a miniature two-electrode device designed to measure the conductivity of the thin electrolyte layer adjacent to electrode surfaces. The best design for the development of conductometric electrodes is an interdigitated structure [145–147].

Conductometric transducers are mostly manufactured by microelectronic techniques—photolithography and vacuum spraying [148] or thick-film printing technology [149,150]—which allow miniaturization of the transducer. Various materials have been tested for the construction of conductometric interdigital biosensors: Pt, Au, Al, Ni, Cu, Ti, Cr, Ta_2O_5, Ag, and carbon.

In general, all these materials are suitable, especially when high-frequency alternating current (AC) is used; however, electrodes made of noble metals have better characteristics. Titanium, chromium, and aluminum electrodes have been shown to be undesirable for operation with biological liquids since they have low sensitivity to changes in the ion strength of solutions and reach conductivity saturation in a short time.

Biosensors based on the conductometric principle present a number of advantages: (1) Thin-film electrodes are suitable for miniaturization and large-scale production using inexpensive technology, (2) they do not require any reference electrode, (3) transducers are not light sensitive, (4) the driving voltage can be sufficiently low to decrease significantly the power consumption, and (5) a large spectrum of compounds of different nature can be determined on the basis of various reactions and mechanisms.

2.4.1.2 Capacitive and Impedimetric Sensing

Measurements of sensor interfacial capacitance and impedance can be done using voltammetric or impedance techniques. There are several different ways to measure capacitance using controlled potential techniques. It can be measured from cyclic voltammetry, by electrochemical impedance measurements, or by perturbation with a potential pulse or other wave form.

Using cyclic voltammetry, the well-known expression relating the current to the scan rate is used in a region where there is no faradic contribution according to $I = C\,(dE/dt)$, where dE/dt is the scan rate and C is the capacitance. Thus, low scan rates give low currents making it difficult to accurately evaluate the capacitance, while higher scan rates may be limited by the timescale of the response of the instrument's electronic components. Electrochemical impedance spectroscopy (EIS) is a more advanced and accurate technique for measuring capacitance and, additionally, can be used for more complete characterization and quantitative determination of parameters linked to the interfacial region of sensors and changes in their values in the presence of analyte.

EIS is an AC method that records the response of an electrochemical cell to a small amplitude sinusoidal voltage signal as a function of frequency. The resulting current sine wave differs in time (phase shift) with respect to the perturbing (voltage) signal, and the ratio $E(t)/I(t)$ is defined as the impedance (Z).

The excitation signal, expressed as a function of time, has the form

$$E_t = E_0\sin(\omega t)$$

where
E_t is the potential at time t
E_0 is the amplitude of the signal
ω is the radial frequency

The relationship between radial frequency ω (expressed in radians/second) and frequency f (expressed in Hertz) is ω = 2πf.

In a linear system, the response signal, I_t, is shifted in phase (φ):

$$I_t = I_0 \sin(\omega t + \varphi)$$

An expression analogous to Ohm's law allows the calculation of the impedance as

$$Z_t = \frac{E_t}{I_t} = \frac{E_0 \sin(\omega t)}{I_0 \sin(\omega t + \varphi)} = Z_0 \frac{\sin(\omega t)}{\sin(\omega t + \varphi)}$$

The impedance is therefore expressed in terms of a magnitude, Z_0, and a phase shift, φ. Furthermore, considering Euler's equation ($\exp(ix) = \cos(x) + i\sin(x)$), the impedance can be expressed as

$$Z_t = Z_0(\cos\varphi + i\sin\varphi)$$

so that the impedance is composed of a real and an imaginary part. For sensor applications, the data presentation is usually as a complex plane plot, with the real part plotted on the x-axis and the imaginary part plotted on the y-axis, also referred to as a "Nyquist plot" (see Figure 2.9).

EIS data are commonly analyzed by fitting the EIS spectra using an equivalent electrical circuit model, which contains circuit elements that are common electrical elements such as resistors, capacitors, and inductors. EIS models usually consist of a number of elements in a network, both connected in series and/or in parallel. Charge transfer is represented by resistances and charge separation (e.g., the electrical double layer) by capacitors.

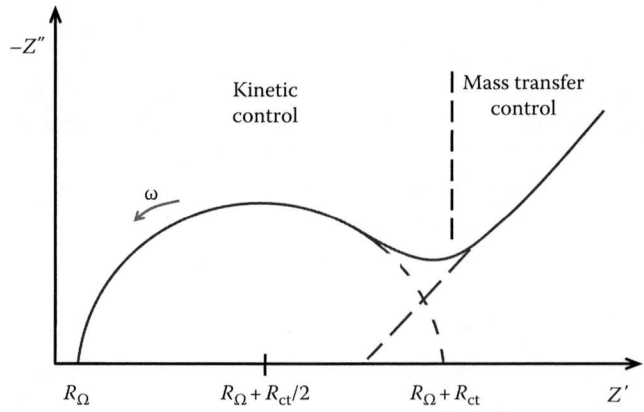

FIGURE 2.9
Schematic complex plane impedance plot for a simple electrochemical system: R → O + ne^-.

Further details on representation and analysis of impedance spectra, especially in more complicated situations, may be found in [151,152].

2.4.2 Applications in Biosensors

Conductometric transducers for biosensing devices were introduced by Watson et al. [153], the device consisting of oxidized silicon wafer with serpentined, interdigitated gold electrode pairs on one surface in a planar configuration. The close proximity of the electrodes in each pair allows a small amplitude sine wave (1 kHz, 10 mV peak to peak) to induce an AC response. After DC conversion, the resulting signal is linearly proportional to solution conductance. The transduction was tested with urease immobilized by cross-linking with glutaraldehyde and albumin. Almost simultaneously, Newman [154] and Bataillard [155] reported the concept of capacitive or impedimetric-based immunosensors, which exploit the change in dielectric properties and/or thickness of the dielectric layer at the electrolyte–electrode interfaces, due to the Ab–Ag interaction.

Five factors regarding enzymatic reactions allow the application of conductometric methods to monitor enzymatic reactions: (1) the generation of ionic groups (e.g., amidases), (2) the separation of different charges (e.g., dehydrogenases, decarboxylases), (3) ion migration (e.g., esterases), (4) changes in the degree of association of ionic groups resulting from chelation (e.g., kinases), and (5) changes in the sizes of charge-carrying groups (e.g., phosphatases) [156]. Because all charge-carrying species are detected simultaneously, conductometric methods are relatively nonselective. Buffers of low ionic strength must be used for the detection of low levels of substrate, since detection limits are ultimately controlled by the ratio $\Delta G/G$, where G is the conductance of the medium and ΔG is the conductance change due to enzymatic reaction [157]. However, in the case of an integrated microbiosensor, most difficulties can be overcome using a differential measuring scheme, which compensates for changes in background conductivity and the influence of temperature variations and of other factors.

Developments in the field of conductometric enzyme biosensors for environmental monitoring during recent decades were reviewed by Renault and Dzyadevych [158] and are summarized in Table 2.7.

Recently impedimetric biosensors for the detection of foodborne pathogenic bacteria were reviewed [159].

2.4.2.1 Immunosensors

Conductometric immunosensors monitor the change in dielectric properties and/or thickness of the dielectric layer at the electrolyte/electrode interface, due to the Ab–Ag interaction. This allows the detection of a receptor-specific analyte that has been immobilized on the insulating dielectric layer, previously deposited on the surface of the working electrode (Figure 2.10) [160].

TABLE 2.7

Applications of Conductometric Enzyme Biosensors for Environmental Monitoring

Target	Enzyme
Organophosphorus pesticides	Acetylcholinesterase, butyrylcholinesterase
Heavy metal ions	Urease, alkaline phosphatase
Formaldehyde	Alcohol oxidase
Chlorophenol, triazine herbicides, diuron, atrazine, Cu^{2+}	Tyrosinase
Carbamate pesticides	Acetylcholinesterase
Nitrate	Nitrate reductase
Proteins as marker of DOC	Proteinase K

Source: Jaffrezic-Renault, N. and Dzyadevych, S.V., *Sensors*, 8, 2569, 2008.

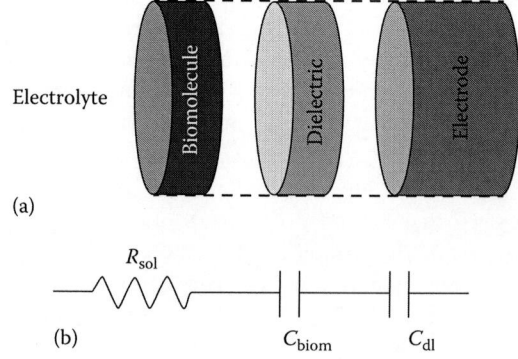

FIGURE 2.10

(a) Scheme of a capacitive biosensor and (b) corresponding electrical circuit model.

This assembly corresponds to two capacitors in series; the inner one corresponds to the dielectric (C_{dl}) layer and the outer one corresponds to the biomolecule layer (C_{biom}). The total capacitance C_t can be described by

$$\frac{1}{C_t} = \frac{1}{C_{dl}} + \frac{1}{C_{biom}}$$

In a useful sensitive impedimetric immunosensor, capacitance changes due to the binding of the analyte to the receptor have to dominate the total capacitance, so that the first contribution to the total capacitance has to be small, that is, the insulating layer should be thin enough and/or have a high dielectric constant. In addition, the insulating layer should be complete and time stable and provide functional groups for the immobilization of the receptor [161]. Signal changes in capacitive immunosensors can also be induced by changes in dielectric properties, charge distribution, or even conformational changes of the immobilized biomolecule layer upon its interaction with the analyte.

These capacitors are usually not ideal, in the working electrode/dielectric/electrolyte configuration, due to nonhomogeneity and/or porosity of the sensor assembly interfacial region. For modeling the impedance spectra, a constant phase element (CPE), where CPE = $\{(Ci\omega)^\alpha\}^{-1}$, is therefore needed as a nonideal capacitor, for adjusting deviations from an ideal capacitor. Additionally, the presence of defects and of water molecules within the protein structure of the insulating and biomolecule layers can be represented by a resistance as a parallel element to each of the corresponding capacitors.

Thus, impedimetric immunosensors can be classified into two main categories:

1. Capacitive, where the surface of the electrode is completely covered by a dielectric layer and the whole electrode assembly behaves as an insulator. No redox probe is present in the measuring solution, and the capacitive current is measured under a small amplitude sinusoidal voltage signal, typical excitation frequencies being 10–1000 Hz.

2. Faradic impedimetric, where the surface of the electrode, which is partially or wholly covered by a noninsulating layer or partially covered by an insulating layer, is able to oxidize or reduce a redox probe, which is added to the measuring solution. In this case, the measured parameter is the charge transfer resistance (the real component of impedance at low frequency values, typically 0.1–1.0 Hz). Ab–Ag interactions are expected to cause an increase in its value as the faradic reaction becomes increasingly hindered. In general, these immunosensors exhibit a higher sensitivity to Ab–Ag interactions. However, the redox species may have an effect on both the stability and the activity of the electrode assembly.

Potential problems are that the dissociation constant for the Ab–Ag complex is low; immunosensor regeneration is required, such as by the use of very acidic or alkaline buffer solutions; and high ionic strength solutions or nonpolar water-soluble solvents normally have an influence on both the activity of the biorecognition layer and the stability of the inner dielectric layer. This makes the immunosensor difficult to reuse in a reproducible manner [162].

The means for improving impedimetric immunosensor efficiency are (1) improving Ab immobilization methods on the electrode surface, (2) improving electrode performance to enhance sensitivity, (3) using enzyme labeled and nanomaterials to amplify the detection signal, (4) using optimized equivalent circuit fitting for analyzing impedance changes, and (5) concentrating samples by dielectrophoresis.

A wide variety of impedimetric immunosensors reported for different bacteria detection, mainly *E. coli*, based on the immobilization of the corresponding Abs, were reported during the past decade [163–168].

Nucleic acid-based biosensors have been in continuous development during recent years, and their use in analysis has become an important tool for identification of disease-causing microorganisms in food and environment.

The principle of a nucleic acid bioreceptor for pathogen detection is based on the identification of a target analyte's nucleic acid, achieved by matching the complementary base pairs that are often the genetic components of an organism. This causes a shift in impedance or change in capacitance or admittance at the bulk of the electrode interface due to the insulating properties. Since each organism has unique nucleic acid sequences, any self-replicating microorganism can be easily identified. They have been used to detect different pathogen bacteria, for example, *Salmonella* and *Listeria* [169] fecal bacteria [170], uropathogens [171], and *Bacillus anthracis* [172].

2.5 Future Perspectives

In spite of the past and current large amount of research in biosensor development, there is still a challenge to create improved and more reliable devices.

2.5.1 Nanotechnology

Nanomaterials are natural or engineered materials that have at least one dimension at the nanometer scale ($\leq$100 nm). Nanomaterials possess completely new and enhanced properties compared with the parent bulk materials in terms of increased sensitivity, which, in electrochemical sensors, improves the signal-to-noise ratio and lowers the detection limit and possibly permits electrocatalytic effects. Examples of advanced nanomaterials include metallic, metal oxide, polymeric, semiconductor and ceramic nanoparticles, nanowires, nanotubes, quantum dots, nanorods, and composites of these materials. The unique properties of these materials are attributed to the extremely high surface area per unit weight and their mechanical, electrical, optical, and catalytic properties. These properties offer a wide range of opportunities for the detection of environmental contaminants and toxins in addition to their remediation. Applications in environmental monitoring using nanotubes and nanoparticle-based electrochemical sensors have been the subject of reviews [173–175].

2.5.2 Self-Assembled Monolayers

Inspired by nature, molecular self-assembly has been proposed for the synthesis of nanostructures capable of performing unique functions. Self-assembly is the formation of organized, patterned structures without external direction. Biomaterials, such as proteins, lipids, and nucleic acids, can self-assemble [176,177]. Such nanostructures are applied for the development of amperometric biosensors [178,179].

2.5.3 Miniaturization and Chips

Many current trends in biosensor development are conducted through small, easy-to-use, and fast sensors, so-called "smart" systems. A key issue to be addressed in the future is the increasing demand for higher sensitivity and selectivity that will allow molecules to be monitored in real time at a minimal cost. The future biosensor is expected to function on the principle of "laboratory on a chip" [180], having all the essential components microfabricated on a chip with the aim to simplify and extend reliable monitoring of the analytes outside the laboratory [181]. Compact and portable devices will constitute another future area of intensive multidisciplinary sensor research.

There is a growing tendency toward miniaturization of analytical systems, since it allows the handling of low-volume samples, a reduction in reagent consumption and waste generation, and increases sample throughput [182]. Taking advantage of the benefits of miniaturization, sensors and biosensors can become inexpensive and easy-to-handle analytical devices for fast, reliable measurements of chemical species that can be commercialized and be readily used by nonskilled personnel for field measurements. This is important for environmental monitoring. Development of sensors capable of determining several analytes simultaneously can represent an interesting tool in environmental monitoring and screening [182], allowing a reduction in analysis time and total sample volume and of other reagents required. The development of large-scale biosensor arrays composed of highly miniaturized signal transducer elements, for example, enables the real-time parallel monitoring of multiple species. Multichannel biosensors are required for direct detection in high-throughput screening systems.

2.5.4 Screen-Printed Sensors and Biosensors

A range of environmentally friendly electrochemical sensors for water monitoring is available and a number of them are now in advanced prototype stage. Conventional electrochemical cells are now being replaced with SPEs connected to miniaturized potentiostats. SPEs are finding uses in major analytical laboratory equipment and as handheld field devices. A number of SPEs are commercially available, and it is possible to manufacture them in the laboratory for research applications since screen-printing technology is becoming cheaper and more easily available, as explained in recent reviews [183–185]. Advances in all these areas can be expected in the near future.

Acknowledgments

Financial support from Fundação para a Ciência e a Tecnologia (FCT), Portugal, PTDC/QUI-QUI/116091/2009, POCH, POFC-QREN (co-financed by FSE and European Community Fund FEDER/COMPETE), and FCT project *PEst-C/EME/UI0285/2011* (CEMUC®—Research Unit 285), Portugal, is gratefully acknowledged. M.M.B. and M.E.G. thank FCT for postdoctoral fellowships SFRH/BPD/72656/2010 and SFRH/BPD/36930/2007, respectively.

References

1. A. Hulanicki, S. Geab, F. Ingman, *Pure Appl. Chem.* 63 (1991) 1247–1250.
2. B. Nagel, H. Dellweg, L.M. Gierasch, *Pure Appl. Chem.* 64 (1992) 143–168.
3. D.R. Thévenot, K. Toth, R.A. Durst, G.S. Wilson, *Pure Appl. Chem.* 71 (1999) 2333–2348.
4. A.P.F. Turner, I. Karube, G.S. Wilson, *Biosensors-Fundamentals and Applications*, Oxford University Press, New York, (1987).
5. R.M. Buch, G.A. Rechnitz, *Biosensors*, 4 (1989) 215–230.
6. L.J. Blum, P.R. Coulet, *Biosensor Principles and Applications*, Marcel Dekker, New York, (1991).
7. R.F. Taylor, J.S. Schultz, *Handbook of Chemical and Biological Sensors*, IOP Publishing Ltd, Bristol, PA, (1996).
8. W.S. Hughes, *J. Am. Chem. Soc.* 44 (1922) 2860–2867.
9. P.T. Kerridge, *Biochem. J.* 19 (1925) 611–617.
10. R.W. Stow, B.F. Randall, *Am. J. Physiol.* 179 (1954) 678.
11. L.C. Clark Jr., *Trans. Am. Soc. Artif. Int. Organs* 2 (1956) 41–48.
12. L.C. Clark Jr., C. Lyons, *Ann. NY Acad. Sci.* 102 (1962) 29–45.
13. G.G. Guilbault, J. Montalvo, *JACS* 91 (1969) 2164–2569.
14. P. Bergveld, *IEEE Trans. Biomed. Eng.* BME 17 (1970) 70–71.
15. C. Diviès, *Ann. Microbiol. A.* 126 (1975) 175–186.
16. J. Janata, *J. Am. Chem. Soc.* 97 (1975) 2914–2916.
17. M. Shichiri, R. Kawamori, R. Yamaski, Y. Hakai, H. Abe, *Lancet* 2 (1982) 1129–1131.
18. A.E.G. Cass, D.G. Francis, H.A.O. Hill, W.J. Aston, I.J. Higgins, E.V. Plotkin, L.D.L. Scott, A.P.F. Turner, *Anal. Chem.* 56 (1984) 667–671.
19. J.M. Kauffmann, G.G. Guilbault, *Bioprocess Technol.* 15 (1991) 63–82.
20. G. Eisenman, *Glass Electrodes for Hydrogen and Other Cations*, Marcel Dekker, New York, (1967).
21. J.D. Newman, S.J. Setford, *Mol. Biotechnol.* 32 (2006) 249–268.
22. G. Eisenman, *Biophys. J.* 2 (1962) 259–323.
23. J. Maier, *Solid State Ionics* 62 (1993) 105–111.
24. N. Yamazoe, N. Miura, *J. Electroceram.* 2 (1998) 243–255.
25. J.-L. Marty, D. Garcia, R. Rouillon, *Trends Anal. Chem.* 14 (1995) 329–333.
26. R.B. Shi, K. Stein, G. Schwedt, *Fresenius J. Anal. Chem.* 357 (1997) 752–755.
27. Y. Lei, W. Chen, A. Mulchandani, *Anal. Chim. Acta* 568 (2006) 200–210.
28. P. Bergveld, A. Sibbald, *Analytical and Biomedical Applications of Ion-Selective Field-Effect Transistors*, Elsevier, Amsterdam, the Netherlands, (1988).
29. D.G. Hafeman, J.W. Parce, H.M. McConnell, *Science* 240 (1988) 1182–1185.
30. M.J. Schöning, H. Luth, *Phys. Stat. Sol. A*, 185 (2001) 65–77.
31. R.J. Hunter, *Foundations of Colloid Science*, Vol. 2, Clarendon Press, Oxford, U.K., (1989).

32. Y.G. Vlasov, Y.A. Tarantov, P.V. Bobrov, *Anal. Bioanal. Chem.* 376 (2003) 788–796.
33. M. Yuqing, G. Jianquo, C. Jianrong, *Biotechnol. Adv.* 21 (2003) 527–534.
34. L. Bousse, P. Bergveld, *Sens. Actuators* 6 (1984) 65–78.
35. A. van den Berg, P. Bergveld, D.N. Reinhoudt, E.J.R. Sudholter, *Sens. Actuators* 8 (1985) 129–148.
36. J. Janata, S.D. Moss, *Biomed. Eng.* 6 (1976) 241–245.
37. S. Caras, J. Janata, *Anal. Chem.* 52 (1980) 1935–1937.
38. S.V. Dzyadevych, A.P. Soldatkin, A.V. Elskaya, C. Martelet, N. Jaffrezic-Renault, *Anal. Chim. Acta* 568 (2006) 248–258.
39. S. Liu, X. Guo, *NPG Asia Mater.* 4 (2012) e23, doi:10.1038/am.2012.42.
40. X.J. Zhou, J.M. Moran-Mirabal, H.G. Craighead, P.L. McEuen, *Nat. Nanotechnol.* 2 (2007) 185–190.
41. T.H. Kim, S.H. Lee, J. Lee, H.S. Song, E.H. Oh, T.H. Park, S. Hong, *Adv. Mater.* 21 (2009) 91–94.
42. H.-M. So, D.-W. Park, E.-K. Jeon, Y.-H. Kim, B.S. Kim, C.-K. Lee, S.Y. Choi, S.C. Kim, H. Chang, J.-O. Lee, *Small* 4 (2008) 197–201.
43. R.A. Villamizar, A. Maroto, F.X. Rius, I. Inza, M.J. Figueras, *Biosens. Bioelectron.* 24 (2008) 279–283.
44. T. Yoshinobu, H. Iwasaki, Y. Ui, K. Furuichi, Y. Ermolenko, Y. Mourzina, T. Wagner, N. Nather, M.J. Schöning, *Methods* 37 (2005) 94–102.
45. P.A. Serra, *Biosensors for Health, Environment and Biosecurity*, InTech, New York, (2011).
46. G.X. Xu, X.S. Ye, L.F. Qin, Y. Xu, Y. Li, R. Li, P. Wang, *Biosens. Bioelectron.* 20 (2005) 1757–1763.
47. J.W. Parce, J.C. Owicki, K.M. Kercso, G.B. Sigal, H.G. Wada, V.C. Muir, L.J. Bousse, K.L. Ross, B.I. Sikic, H.M. McConnell, *Science* 246 (1989) 243–247.
48. H.M. McConnell, J.C. Owicki, J.W. Parce, D.L. Miller, G.T. Baxter, H.G. Wada, S. Pitchford, *Science* 257 (1992) 1906–1912.
49. H.G. Wada, J.C. Owicki, L.H. Bruner, K.R. Miller, K.M. Raley-Susman, P.R. Panfili, G. Humphries, J.W. Parce, *AATEX* 1 (1992) 154–164.
50. D.B. Chrisey, G.H. Hubler, *Pulsed Laser Deposition of Thin Films*, John Wiley & Sons, New York, (1994).
51. M.J. Schöning, D. Tsarouchas, L. Beckers, J. Schubert, W. Zander, P. Kordos, H. Lüth, *Sens. Actuators B* 35 (1996) 228–233.
52. M.J. Schöning, C. Schmidt, J. Schubert, W. Zander, S. Mesters, P. Kordos, H. Lüth, A. Legin, B. Seleznev, Yu.G. Vlasov, *Sens. Actuators B* 68 (2000) 254–259.
53. M. Thust, M.J. Schöning, J. Vetter, P. Kordos, H. Lüth, *Anal. Chim. Acta* 323 (1996) 115–121.
54. M.H. Abouzar, A. Poghossian, J.R. Siqueira Jr., O.N. Oliveira Jr., W. Moritz, M.J. Schöning, *Phys. Status Solidi A* 207 (2010) 884–890.
55. J.R. Siqueira Jr., M. Bäcker, A. Poghossian, V. Zucolotto, O.N. Oliveira Jr., M.J. Schöning, *Phys. Status Solidi A* 207 (2010) 781–786.
56. M. Turek, L. Ketterer, M. Claßen, H.K. Berndt, G. Elbers, P. Krüger, M. Keusgen, M.J. Schöning, *Sensors* 7 (2007) 1415–1426.
57. M.J. Schöning, R. Krause, K. Block, M. Musahmeh, A. Mulchandani, J. Wang, *Sens. Actuators B* 95 (2003) 291–296.
58. C.M.A. Brett, A.M. Oliveira-Brett, *J. Solid State Electrochem.* 15 (2011) 1487–1494.
59. C.M.A. Brett, A.M. Oliveira Brett, *Electroanalysis*, Oxford University Press, Oxford, U.K., (1998).

60. J. Wang, *Analytical Electrochemistry*, VCH Publishers, New York, (1994).
61. G. Hanrahan, D.G. Patila, J. Wang, *J. Environ. Monit.* 6 (2004) 657–664.
62. G.L. Turdean, S.E. Stanca, I.C. Popescu, *Biosenzori amperometrici. Teorie si aplicatii.* Editura Presa Universitara Clujeana, Cluj-Napoca, Romania, (2005).
63. M.I. Prodromidis, M.I. Karayannis, *Electroanalysis* 14 (2002) 241–261.
64. A.P.F. Turner, in *Methods in Enzymology*, Vol. 137, ed. K. Mosbach, Academic Press, San Diego, CA, (1988).
65. A. Heller, *Accounts Chem. Res.* 23 (1990) 128–134.
66. B.M. Brena, F. Batista-Viera, in *Methods in Biotechnology: Immobilization of Enzymes and Cells*, ed. G. M. Guisan, Humana Press Inc., Totowa, NJ, (2006).
67. S. Borgmann, A. Schulte, S. Neugebauer, W. Schuhmann, in *Advances in Electrochemical Science and Engineering*, eds. R. C. Alkire, D. M. Kolb, and J. Lipkowski, Wiley-Verlag, Weinheim, Germany, (2011).
68. R.N. Adams, in *Electrochemistry at Solid Electrodes*, Marcel Dekker, New York, (1969).
69. A. Amine, H. Mohammadi, I. Bourais, G. Palleschi, *Biosens. Bioelectron.* 21 (2006) 1405–1423.
70. G.L. Turdean, *Int. J. Electrochem.* ID 343125 (2011) 1–15.
71. P. Pal, D. Bhattacharyay, A. Mukhopadhyay, P. Sarkar, *Environ. Eng. Sci.* 26 (2009) 25–32.
72. M.E. Ghica, C.M.A. Brett, *Microchim. Acta* 163 (2008) 185–193.
73. J.-W. Choi, Y.-K. Kim, I.-H. Lee, J. Min, W.H. Lee, *Biosens. Bioelectron.* 16 (2001) 937–943.
74. V.G. Andreou, Y.D. Clonis, *Biosens. Bioelectron.* 17 (2002) 61–69.
75. A. Arvinte, L. Rotariu, C. Bala, in *Chemicals as Intentional and Accidental Global Environmental Threats*, eds. L. Simeonov and E. Chirila, Springer, Dordrecht, The Netherlands, (2006).
76. J. Wang, L. Chen, A. Mulchandani, W. Chen, *Electroanalysis* 11 (1999) 866–890.
77. M.H. Smit, G.A. Rechnitz, *Anal. Chem.* 65 (1993) 380–385.
78. J. Wang, L. Chen, *Anal. Chem.* 67 (1995) 3824–3827.
79. I. Ben Rejeb, F. Arduini, A. Arvinte, A. Amine, M. Gargouri, L. Micheli, C. Bala, D. Moscone, G. Palleschi, *Biosens. Bioelectron.* 24 (2009) 1962–1968.
80. K. Malzahn, J.R. Windmiller, G. Valdes-Ramirez, M.J. Schöning, J. Wang, *Analyst* 136 (2011) 2912–2917.
81. C. Capannesi, I. Palchetti, M. Mascini, A. Parenti, *Food Chem.* 71 (2000) 553–562.
82. Y. Hasebe, K. Yokobori, K. Fukasawa, T. Kogure, S. Uchiyama, *Anal. Chim. Acta* 357 (1997) 51–54.
83. G.C. Chemnitius, U. Bilitewski, *Sens. Actuators B* 32 (1996) 107–113.
84. M. Esti, G. Volpe, L. Massignan, D. Compagnone, E. LaNotte, G. Palleschi, *J. Agric. Food Chem.* 46 (1998) 4233–4237.
85. M.M. Barsan, C.M.A. Brett, *Talanta* 74 (2008) 1505–1510.
86. M.E. Ghica, R. Pauliukaite, N. Marchand, E. Devic, C.M.A. Brett, *Anal. Chim. Acta* 521 (2007) 80–86.
87. A.C. Torres, M.E. Ghica, C.M.A. Brett, *Anal. Bioanal. Chem.* 485 (2013) 3813–3822.
88. R.S. Yalow, S.A. Berson, *Nature* 184 (1959) 1648–1649.
89. G. Płaza, K. Ulfig, A.J. Tien, *Pol. J. Environ. Stud.* 9 (2000) 231–236.
90. C. Moina, G. Ybarra, *Advances in Immunoassay Technology*, eds. N. H. L. Chiu and T. K. Christopoulos, InTech, Rijeka, Croatia, (2012).

91. D. Grieshaber, R. MacKenzie, J. Vörös, E. Reimhult, *Sensors* 8 (2008) 1400–1458.
92. M. Aizawa, A. Morioka, S. Suzuki, Y. Nagamura, *Anal. Biochem.* 94 (1979) 22–28.
93. P. Tijsen, *Principles of Immunoassay: Enzymes*, eds. W. H. W. Alber, R. F. Masseyeff, and N.A. Staines, VCH Publishers, Weinheim, Germany, (1993).
94. J. Sherry, *Crit. Rev. Anal. Chem.* 23 (1992) 217–300.
95. C.D. Ercegovich, in *Pesticides Identification at the Residue Level*, ed. R. F. Gould, American Chemical Society, Washington, DC, (1971).
96. B. Hock, A. Dankwardt, K. Kramer, A. Marx, *Anal. Chim. Acta* 311 (1995) 393–405.
97. F. Ricci, G. Volpe, L. Micheli, G. Palleschi, *Anal. Chim. Acta* 605 (2007) 111–129.
98. M. Wilmer, D. Trau, R. Renneberg, F. Spener, *Anal. Lett.* 30 (1997) 515–525.
99. T. Kalab, P. Skladal, *Anal. Chim. Acta* 304 (1995) 361–368.
100. L. Campanella, S. Eremin, D. Lelo, E. Martini, M. Tomassetti, *Sens. Actuators B: Chem.* 156 (2011) 50–62.
101. K. Grennan, G. Strachan, A.J. Porter, A.J. Killard, M.R. Smyth, *Anal. Chim. Acta* 500 (2003) 287–298.
102. D.J. Pritchard, H. Morgan, J.M. Cooper, *Anal. Chim. Acta* 310 (1995) 251–256.
103. M. Murata, M. Nakayama, H. Irie, K. Yakabe, K. Fukuma, Y. Katayama, M. Maeda, *Anal. Sci.* 17 (2001) 387–390.
104. Y.F. Xu, M. Velasco-Garcia, T.T. Mottram, *Biosens. Bioelectron.* 20 (2005) 2061–2070.
105. F.E. Ahmed, *Trends Anal. Chem.* 22 (2003) 170–185.
106. K.A. Fahnrich, M. Pravda, C.G. Guilbault, *Biosens. Bioelectron.* 18 (2003) 73–82.
107. M. Lin, Z. Yang, Y. Huang, Z. Sun, Y. He, C. Ni, *Int. J. Electrochem. Sci.* 7 (2012) 965–978.
108. L. Micheli, R. Grecco, M. Badea, D. Moscone, G. Palleschi, *Biosens. Bioelectron.* 21 (2005) 588–596.
109. J.H.O. Owino, O.A. Arotiba, N. Hendricks, E.A. Songa, N. Jahed, T.T. Waryo, R.F. Ngece, P.G.L. Baker, E.I. Iwuoha, *Sensors* 8 (2008) 8262–8274.
110. S.S. Babkina, N.A. Ulakhovich, *Bioelectrochemistry* 63 (2004) 261–265.
111. D.A. Blake, R.M. Jones, R.C. Blake II, A.R. Pavlov, I.A. Darwish, H. Yu, *Biosens. Bioelectron.* 16 (2001) 799–809.
112. C. Singh, G.S. Agarwal, G.P. Rai, L. Singh, V.K. Rao, *Electroanalysis* 17 (2005) 2062–2067.
113. H. Chen, C. Jiang, C. Yu, S. Zhang, B. Liu, J. Kon, *Biosens. Bioelectron.* 24 (2009) 3399–3411.
114. C.B. Jacobs, M.J. Peairs, B.J. Venton, *Anal. Chim. Acta.* 662 (2010) 105–127.
115. A. de la Escosura-Müniz, C. Parolo, A. Merkoci, *Mater. Today* 13 (2010) 24–34.
116. P. Wang, Q. Liu, in *Cell-Based Biosensors: Principles and Applications*, Artech House, Norwood, MA, (2010).
117. A.N. Reshetilov, P.V. Iliasov, T.A. Reshetilova, in *Intelligent and Biosensors*, ed. V. S. Somerset, Intech, New York, (2010).
118. M. Held, W. Schuhmann, K. Jahreis, H.-L. Schmidt, *Biosens. Bioelectron.* 17 (2002) 1089–1094.
119. J. Katrlik, I. Vostiar, J. Sefcovicova, J. Tkac, V. Mastihuba, M. Valach, V. Stefuca, P. Gemeiner, *Anal. Bioanal. Chem.* 388 (2007) 287–295.
120. S.K. Chaudhuri, D.R. Lovley, *Nat. Biotechnol.* 21 (2003) 1229–1232.
121. I. Karube, T. Matsunaga, S. Mitsuda, S. Suzuki, *Biotechnol. Bioeng.* 19 (1977) 1535–1547.
122. T. Matsunaga, I. Karube, S. Suzuki, *Eur. J. Appl. Microbiol. Biotechnol.* 10 (1980) 235–243.

123. J.L. Stirling, H.P. Bennetto, G.M. Delaney, J.R. Mason, S.D. Roller, K. Tanaka, C.F. Thurston, *Biochem. Soc. Trans.* 11 (1983) 451–453.
124. C.F. Thurston, H.P. Bennetto, G.M. Delaney, J.R. Mason, S.D. Roller, J.L. Stirling, *J. Gen. Microbiol.* 131 (1985) 1393–1401.
125. I.S. Chang, J.K. Jang, G.C. Gil, M. Kim, H.J. Kim, B.W. Cho, B.H. Kim, *Biosens. Bioelectron.* 19 (2004) 607–613.
126. B.H. Kim, I.S. Chang, G.C. Gil, H.S. Park, H.J. Kim, *Biotechnol. Lett.* 25 (2003) 541–545.
127. I.S. Chang, H. Moon, J.K. Jang, B.H. Kim, *Biosens. Bioelectron.* 20 (2005) 1856–1859.
128. S.F. D'Souza, *Biosens. Bioelectron.* 16 (2001) 337–353.
129. S. Dolatabadi, D. Manjulakumari, *Res. J. Biotechnol.* 7 (2012) 102–108.
130. S.F. D'Souza, *Appl. Biochem. Biotechnol.* 96 (2001) 225–238.
131. S. Timur, N. Pazarlioglu, R. Pilloton, A. Telefoncu, *Talanta* 61 (2003) 87–93.
132. Y. Lei, P. Mulchandani, W. Chen, A. Mulchandani, *J. Agric. Food Chem.* 53 (2005) 524–527.
133. M. Okochi, K. Mima, M. Miyata, Y. Shinozaki, S. Haraguchi, M. Fujisawa, M. Kaneko, T. Masukata, T. Matsunaga, *Biotechnol. Bioeng.* 87 (2004) 905–911.
134. L. Rotariu, C. Bala, *Anal. Lett.* 36 (2003) 2459–2471.
135. J. Tkac, I. Vostiar, P. Gemeiner, E. Sturdik, *Bioelectrochemistry* 56 (2002) 127–129.
136. I. Biran, R. Babai, K. Levcov, J. Rishpon, E.Z. Ron, *Environ. Microbiol.* 2 (2000) 285–290.
137. C. Michel, A. Ouerd, F. Battaglia-Brunet, N. Guiques, J.P. Grasa, M. Bruschi, I. Ignatiadis, *Biosens. Bioelectron.* 22 (2006) 285–290.
138. K. Tag, K. Riedel, H.J. Bauer, G. Hanke, K.H.R. Baronian, G. Kunze, *Sens. Actuat. B,* 122 (2007) 403–409.
139. J. Singh, S.K. Mittal, *Anal. Methods* 4 (2012) 1326–1331.
140. C. Chan, M. Lehmann, K. Chan, P. Chan, C. Chan, B. Gruendig, G. Kunze, R. Renneberg, *Biosens. Bioelectron.* 15 (2000) 343–353.
141. J. Liu, G. Olsson, B. Mattiasson, *Biosens. Bioelectron.* 20 (2004) 571–578.
142. Y. Lei, P. Mulchandani, W. Chen, J. Wang, A. Mulchandani, *Electroanalysis* 16 (2004) 2030–2034.
143. P. Mulchandani, M.H. Carlos, Y. Lei, W. Chen, A. Mulchandani, *Biosens. Bioelectron.* 21 (2005) 523–527.
144. E.V. Emelyanova, A.N. Reshetilov, *Process Biochem.* 37 (2002) 683–692.
145. S.V. Dzyadevych, A.A Shulga, S.V. Patskovsky, V.N. Arkhipova, A.P. Soldatkin, V.I. Strikha, *Rus. J. Electrochem.* 30 (1994) 887–891.
146. W. Olthuis, A.Volanschi, J.G. Bomer, P. Bergveld, *Sens. Actuators B* 13–14 (1993) 230–233.
147. N.F. Sheppard, R.C. Tucker, C. Wu, *Anal. Chem.,* 65 (1993) 1199–1202.
148. S.V. Dzyadevych, A.P Soldatkin, V.N. Arkhypova, C. Martelet, N. Jaffrezic-Renault, A.V. Elskaya, *Electrochemical Enzyme Biosensors,* IMBG Press, Kyiv, Ukraine, (2006).
149. U. Bilitewski, W. Drewes, R.D. Schmid, *Sens. Actuators B* 7 (1992) 321–326.
150. W.O. Ho, S. Krause, C.J. McNeil, J.A. Pritchard, R.D. Armstrong, D. Athey, K. Rawson, *Anal. Chem.* 71 (1999) 1940–1946.
151. C.M.A. Brett, A.M. Oliveira Brett, *Electrochemistry. Principles, Methods and Applications,* Oxford University Press, Oxford, U.K., (1993).
152. M.E. Orazem, B. Tribollet, *Electrochemical Impedance Spectroscopy,* Wiley, Hoboken, NJ, (2008).

153. L.D. Watson, P. Maynard, D.C. Cullen, R.S. Sethi, J. Brettle, C.R. Lowe, *Biosensors* 3 (1987/88) 101–115.
154. A.L. Newman, K.W. Hunter, W.D. Stanbro, *Proc. 2nd Int. Meet. Chem. Sens.* (1986) 596–598.
155. P. Bataillard, F. Gardies, N.J. Renault, C. Martelet, B. Colin, B. Mandrand, *Anal. Chem.* 60 (1988) 2374–2379.
156. A.J. Lawrence, G.R. Moores, *Eur. J. Biochem.* 24 (1972) 538–546.
157. S.R. Mikkelsen, G.A. Rechnitz, *Anal. Chem.* 61 (1989) 1737–1742.
158. N. Jaffrezic-Renault, S.V. Dzyadevych, *Sensors* 8 (2008) 2569–2588.
159. Y. Wang, Z. Ye, Y. Ying, *Sensors* 12 (2012) 3449–3471.
160. E. Katz, I. Willner, *Electroanalysis* 15 (2003) 913–947.
161. A. Gebbert, M. Alvarez-Icaza, W. Stocklein, R.D. Schmid, *Anal. Chem.* 64 (1992) 997–1003.
162. M.I. Prodromidis, *Electrochim. Acta*, 55 (2010) 4227–4233.
163. D.J. Li et al., *Anal. Chim. Acta* 687 (2011) 89–96.
164. D.J. Li et al., *Trans. ASABE* 51 (2008) 1847–1852.
165. L.J. Yang, Y.B. Li, G.F. Erf, *Anal. Chem.* 76 (2004) 1107–1113.
166. M.B. dos Santos et al., *J. Procedia Chem.* 1 (2009) 1291–1294.
167. V. Escamilla-Gomez, S. Campuzano, M. Pedrero, J.M. Pingarron, *Biosens. Bioelectron.* 24 (2009) 3365–3371.
168. P. Geng, X. Zhang, W. Meng, Q. Wang, W. Zhang, L. Jin, Z. Feng, Z. Wu, *Electrochim. Acta* 53 (2008) 4663–4668.
169. F. Farabullini, F. Lucarelli, I. Palchetti, G. Marrazza, M. Mascini, *Biosens. Bioelectron.* 22 (2007) 1544–1549.
170. M.J. LaGier, C.A. Scholin, J.W. Fell, J. Wang, K.D. Goodwin, *Mar. Pollut. Bull.* 50 (2005) 1251–1261.
171. J.C. Liao et al., *J. Clin. Microbiol.* 44 (2006) 561–570.
172. P. Kara, B. Meric, M. Ozsoz, *Electroanalysis* 20 (2008) 2629–2634.
173. C.W. Tan, K.H. Tan, Y.T. Ong, A.R. Mohamed, S.H.S. Zein, S.H. Tan, *Environ. Chem. Lett.* 10 (2012) 265–273.
174. G.A. Rivas, M.D. Rubianes, M.C. Rodriguez, N.F. Ferreyra, G.L. Luque, M.L. Pedano, S.A. Miscoria, C. Parrado, *Talanta* 74 (2007) 291–307.
175. L. Rassaei, M. Amiri, C.M. Cirtiu, M. Sillanpää, F. Marken, M. Sillanpää, *Trends Anal. Chem.* 30 (2011) 1704–1715.
176. M.M. Barsan, E.M. Pinto, C.M.A. Brett, *Electrochim. Acta* 55 (2010) 6358–6366.
177. E.C. Wurster, A. Elbakry, A. Göpferich, M. Breunig, *Methods Mol. Biol.* 948 (2013) 171–182.
178. F.N. Crespilho, M.E. Ghica, C. Gouveia-Caridade, O.N. Oliveira Jr., C.M.A. Brett, *Talanta* 76 (2008) 922–928.
179. C. Yu, Y. Wang, Z. Zhu, N. Bao, H. Gu, *Colloid Surf. B Biointerfaces* 103C (2012) 231–237.
180. J. Wang, *Nucleic Acid Res.*, 28 (2000) 3011–3016.
181. J.E. Pearson, A. Gill, P. Vadgama, *Ann. Clin. Biochem.* 37 (2000) 119–145.
182. S. Rodriguez-Mozaz, M.-P. Marco, M.J. Lopez de Alda, D. Barceló, *Pure Appl. Chem.* 76 (2004) 723–752.
183. M. Li, Y.-T. Li, D.-W. Li, Y.-T. Long, *Anal. Chim. Acta*, 734 (2012) 31–44.
184. K.C. Honeychurch, *Insciences J.* 2 (2012) 1–51.
185. J.C. Quintana, F. Arduini, A. Amine, K. van Velzen, G. Palleschi, D. Moscone, *Anal. Chim. Acta* 29 (2012) 92–99.

3

Piezoelectric Sensors

Petr Skládal and Tibor Hianik

CONTENTS

3.1 Introduction

The direct label-free detection methods for bioanalytical applications employ much simplified assay formats compared to traditional approaches based on labels as enzymes, fluorophores, and radioactivity. The continuous evaluation of the progress of the affinity binding interaction provides more detailed information compared to traditional techniques that measure the amount of the surface-bound label only at the end of the interaction. Furthermore, the direct sensors can be used repeatedly for many assays, thus reducing the running costs. Several physical transducers are capable of measuring surface mass changes resulting from the formation of biomolecular complexes at the sensitive area. Although mostly advanced optical systems are utilized, the piezoelectric (PZ) and acoustic devices [1] represent a similar but significantly less expensive alternative.

Several anisotropic crystals exhibit the PZ effect—mechanical deformation results in oriented dipoles and electric voltage. In the opposite, applying alternating voltage excites vibrations, and at resonant frequency equal to the natural vibrations, the transfer of energy from the electric field to the crystal is the most efficient and the energy remains conserved in the oscillating system.

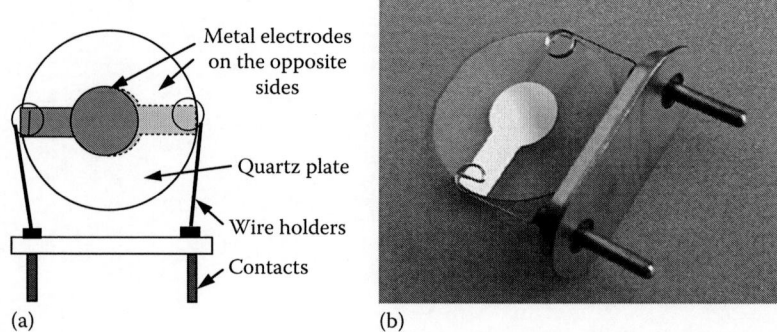

(a)   (b)

FIGURE 3.1
PZ quartz crystal resonator: (a) schematic description and (b) photo of the polished smooth plate with metal electrode on top.

The typical PZ quartz crystal is shown in Figure 3.1a. A thin plate of quartz (temperature compensated quartz cut [At-cut]) is covered with metal (gold) electrodes on both sides and inserted in a holder for simplified manipulation and contacting.

The PZ transducers are used as chemical sensors since the discovery of the relationship between the mass of adsorbed films and the resonant frequency by Sauerbrey [2]:

$$\Delta f = -\frac{2f_0^2 \Delta m}{A\sqrt{\rho_q \mu_q}} = -2.26 \cdot 10^6 f_0^2 \frac{\Delta m}{A} \tag{3.1}$$

The change Δf of the resonant frequency f_0 is directly proportional to the mass change Δm; the numeric constant applies to calculations using Δf in Hz, f_0 in MHz, and Δm in $g \cdot cm^{-2}$. The typical working frequencies are from 5 to 20 MHz. The amount of molecules bound on the sensitive area of electrodes can be easily quantitatively measured as a decrease of frequency; according to recent studies, the microbalance response is more complex [3].

The immobilization of biomolecules in the sensing area provides affinity biosensors; when the immobilized molecule A is either antigen or antibody, the immunosensor specific to the corresponding binding partner B is obtained. The immobilization of bioligands employs generally applicable protocols [4]. The PZ affinity sensors were successfully used for measuring concentration of many different analytes [5,6]. The working formats include direct, competitive, and displacement assays; viruses and bacteria can be conveniently measured directly. The direct assay is generally suitable only for analytes with molecular weight above 20 kDa; otherwise the measured changes of frequency will become too small. The differential measurement ensures that different compositions of samples will not affect the measured signal. For small analytes, the competitive assay provides higher changes of frequency; the analyte (e.g., hapten) is mixed with a tracer molecule (e.g., antibody) to form complexes, and the remaining free binding sites of the tracer subsequently interact with the surface modified with a derivative of the analyte.

3.2 Experimental Procedures

3.2.1 Measuring Setup

The PZ crystals with gold electrodes with the basic resonant frequency of 10 MHz are most typical. The crystals with higher frequencies (up to 20 MHz) will provide a higher sensitivity (Equation 3.1), but manipulation with a very thin plate is not easy. PZ crystals, oscillators, and other instrumentations for microbalance might be obtained commercially: International Crystal Manufacturing Company (ICM, www.icmfg.com), Maxtek (www.maxtekinc.com/), CH Instruments (chinstruments.com), ELCHEMA (www.elchema.com), SEIKO EG & G, Attana (www.attana.com/), Q-Sense (www.q-sense.com), and QCM lab (www.qcmlab.com). The comparison of other PZ devices was published in the literature [7].

The crystal is placed in a suitable cell and operated typically in flow conditions (Figure 3.2). Flow-through measuring cell should allow contact of one side of the crystal with working solutions. The crystal is usually sandwiched between two soft rubber O-rings; reasonable force should be applied to prevent damage.

The simplest drive for the PZ crystal is the gate oscillator based on 74LS01 [9]; the integrated version 74LS320 provides higher energy to the crystal working under variable conditions [8,10,11] (Figure 3.1b). Excellent time stability

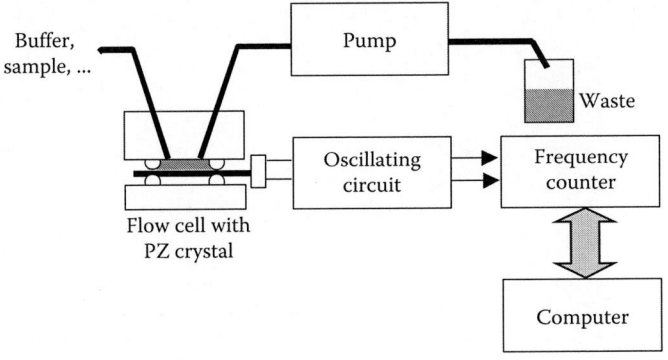

FIGURE 3.2
Measuring setup for QCM sensors.

of frequency and additional resistance (corresponds to dissipation) can be obtained using advanced lever oscillators balancing the energy pumped to the resonator according to external dumping (e.g., ICM). A counter for measuring changes of the resonant frequency is a common device used in electronics. The resolution of frequency should be at least 1 Hz within 1 s interval; higher resolution is commonly achieved using the indirect counting (stable high frequency is "clocked" by the measured signal), using the beat method or after conversion of frequency to voltage. The connection of the counter to computer allows real-time display of signal and convenient evaluation.

3.2.2 Bioanalytical Applications

The direct assay of analytes is most typical for immunosensors; an example of repeated analysis of antigen using antibody-modified PZ sensor is shown in Figure 3.3. The reuse of sensors is of significant advantage; however, the regeneration of the sensitive layer must be optimized; otherwise the binding capacity becomes gradually reduced.

The competitive assay formats need preincubation of the sample with a fixed concentration of a tracing molecule (e.g., antibody), and this mixture is then measured using a crystal modified with the immobilized version of analyte. The measured change of frequency will be decreasing with increasing concentration of the analyte.

Displacement assays employ the crystal modified with the analyte similarly as for the competitive version, but the surface is also presaturated with the tracer. In the presence of the measured free analyte, the bound tracer becomes displaced resulting in a decrease of signal.

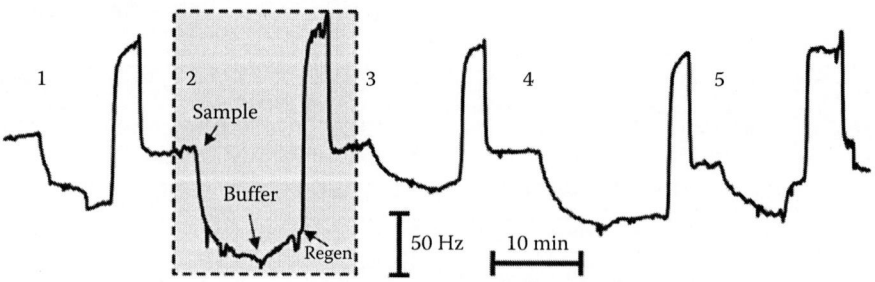

FIGURE 3.3
Experimental traces (frequency vs. time) for five sequential measurements of serum samples. The marked part indicates steps of the assay cycle (2 min baseline, 10 min interaction with sample, 5 min buffer zone for signal readout, 4 min regeneration, and 2 min baseline). The PZ sensor with immobilized antiosteoprotegerin Ab and osteoprotegerin in serum was analyzed (for detailed procedure, see [11]). (Reproduced from *Biosens. Bioelectron.*, 20, Skládal, P., Jílková, Z., Svoboda, I., and Kolář, V., 2027, Copyright 2006, with permission from Elsevier.)

3.2.3 Affinity Kinetic Studies

The direct evaluation of the interaction between two complementary molecules (one immobilized at the sensing surface, the other one free in solution) is a very attractive option provided by PZ biosensors. The formation of the complex antibody (AB) at the surface of the PZ crystal is characterized by the kinetic association k_a and dissociation k_d rate constants:

$$\left|-A+B\underset{k_d}{\overset{k_a}{\rightleftharpoons}}\right|-AB$$

The rate of formation of the complex $d[AB]/dt$ can be expressed using the measured frequency f and concentration c for the free partner B:

$$\frac{d[AB]}{dt} = -\frac{df}{dt} = k_a(f_{max}-f)c-k_d f \tag{3.2}$$

The binding curves (f vs. t dependencies) were usually transformed to obtain df/dt versus f plots that subsequently provide kinetic constants from Equation 3.2 using linear regression [12]. A more elegant and precise method is based on the integration of Equation 3.2 and then substituting f_{eq} and k_{obs} [13]. The dependence of frequency f on time t can be fitted to the kinetic equation similarly as described for the optical biosensors [14]:

$$f = \frac{k_a c f_{max}}{k_a c + k_d}\left\{1-\exp\left[-(k_a c + k_d)t\right]\right\} = f_{eq}\left[1-\exp(-k_{obs}t)\right] \tag{3.3}$$

where f_{max} represents the binding capacity (maximum change of frequency in the saturation of all binding sites). In this way, the binding curves can be used directly for the calculation of parameters using nonlinear regression. The plot of k_{obs} against concentration c provides values of the rate constants. A typical affinity experiment—antibody–antigen interaction—is schematically shown in Figure 3.4.

The kinetic experiment consists of association and dissociation phases. The immobilized antibodies with free binding sites interact with the free antigen, and a decrease of frequency is observed as surface mass on the crystal increases (the equilibrium change f_{eq} can be achieved eventually). Then, the buffer is injected again and the dissociation of immunocomplexes is observed. From this phase, the dissociation constant k_d can be obtained independently (Equation 3.4); f_a represents the amount of surface-bound immunocomplex at the beginning of the dissociation:

$$f = f_a\exp(-k_d t) \tag{3.4}$$

The example of experimental kinetic study is shown in Figure 3.5a where immobilized antigen was interacting with five different concentrations of

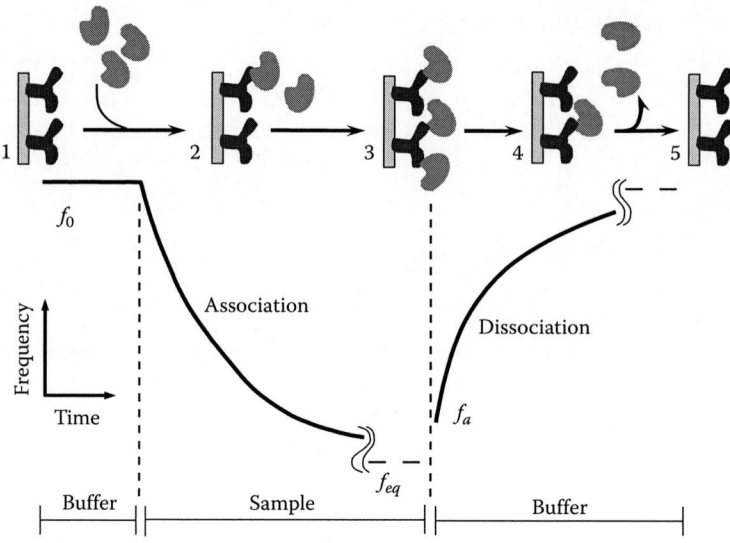

FIGURE 3.4
Characterization of kinetic properties of antibodies. First, the background signal is recorded using the PZ sensor with immobilized antibody (1) in the carrier buffer only. The association reaction follows in the presence of the sample containing antigen—(2) the formation of immunocomplexes at the sensing surface, (3) resulting eventually in equilibrium. (4) Afterward, a spontaneous dissociation of the immunocomplexes is observed in the absence of the sample (5) and eventually all immunocomplexes become dissociated.

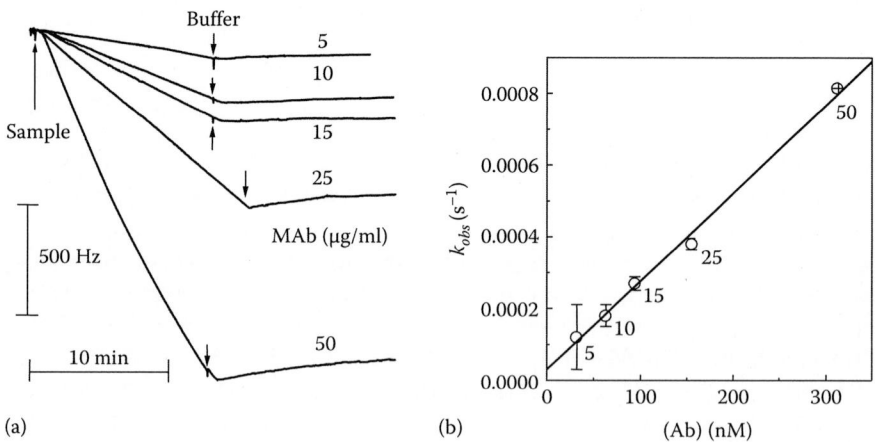

FIGURE 3.5
(a) Experimental curves from the piezosensor analysis of antibody binding on the sensing surface with immobilized antigen—secalins. The up arrow indicates the beginning of flow of antibodies (in the association phase, the concentrations of monoclonal antibody (MAb) are shown close to the curves); down arrows mark the beginning of buffer flow only (dissociation phase). (b) Plot of the k_{obs} constant obtained by nonlinear fitting of the association parts of the binding curves against the corresponding concentrations of the antibody (M_r of antibody was assumed to be 160 kDa).

the studied antibody. For each concentration, the association curve was fitted to Equation 3.3, and thus, the obtained k_{obs} was subsequently plotted (Figure 3.5b) against the molar concentration of the antibody. The linear fit provided k_a (slope) and k_d (intercept) (Equation 3.3).

The value of k_d can be obtained independently from the dissociation trace by fitting to Equation 3.4. The technique described earlier was used to characterize several monoclonal [15] as well as recombinant [16] antibodies. In the case of polyclonals, exact kinetic parameters are not obtained; nevertheless, the quantitative comparison of different products is possible. The fast determination of affinities is useful during screening and development of ligands and receptors, and the kinetic parameters enable to select the most suitable binders for the planned purpose.

3.3 Recent Applications in Environmental and Food Analysis

3.3.1 Immunosensors

Table 3.1 summarizes the recent most interesting applications of PZ quartz sensors in the areas of interest. In addition to immunosensors, the examples of deoxyribonucleic acid (DNA) sensors are presented. The principles of DNA sensors are discussed in part 3.2. Detailed reviews of literature focused on piezosensors from previous years were compiled elsewhere [17,18].

The presented examples demonstrate that PZ biosensor can be combined with different recognition strategies not limited to classic immunosensors and hybridization sensors. The list of measured compounds and biological objects is quite illustrative.

3.3.2 DNA Sensors

The QCM method is very effective for the development of genosensors for biomedical and food industry applications. The interest to QCM was raised also in connection with the necessity to screen genetically modified organisms (GMOs) or distinguish not allowed additives into the pork or beef and the meat from other animals. Typically, such a sensor is fabricated by the immobilization of an oligonucleotide, that is, a short-chain fragment of single-stranded DNA (ssDNA), probe onto the surface of a solid support, for example, gold, carbon paste, or screen-printed carbon electrode (see [35] for a review of different methods of ssDNA immobilization). Usually, this oligonucleotide probe is composed of 15–20 nucleotides with the sequence characteristic for a gene fragment to be determined (target). If the target, containing a gene fragment of the nucleotide sequence complementary to that of the probe, is present in the test solution, then it will hybridize with the probe at the surface. This hybridization can be detected by different techniques, for example, chronoamperometry, UV–VIS or fluorescence spectroscopy, and QCM. In PZ microgravimetry, the

TABLE 3.1

Selected PZ Biosensors for Assay of Environmental and Food Samples

Analyte	LOD	Time	Sensing Principle/ Sample Matrix	References
Alcohol volatiles (*Salmonella* produced)	<5 ppm	10 min	Biomimetic peptide (odorant binding protein), gas analysis/beef	[19]
Chloramphenicol	0.2 ng·mL^{-1}	30 min	PPy–chloramphenicol, competitive immunosensor/milk, meat, honey	[20]
Cryptosporidium parvum	3·10^5 cells·mL^{-1}	1.5 h	Direct immunosensor	[21]
Cymbidium mosaic potexvirus	25 ng·mL^{-1}	5 min	SWCN–Tween–Ab, direct immunosensor/orchid leaves	[22]
Desulfotomaculum sp.	1.8·10^4 CFU·mL^{-1}	70 min	Magn NP–vancomycin label cells, Magn attached to PZ/sea mud	[23]
Enterotoxin A	7 ng·mL^{-1}	25 min	Sandw immunosensor/milk	[24]
Escherichia coli	10^7 CFU·mL^{-1}	5 h	Growth curve/milk, chicken stock	[25]
E. coli O157:H7	53 CFU·mL^{-1}	2 h	Magn NP–Ab capture, sandw immunosensor, AuNP–Ab2, particle enlargement/milk	[26]
Formaldehyde	1 ppm	<10 min	Polyethylene/nanofiber cellulose/air	[27]
GMO (*pflp* gene)	0.25 ng·µL^{-1}	PCR + 1 h	Thiol–oligonucleotide probe, PCR, hybridization/tobacco	[28]
Hemorrhagic septicemia virus	1.6 nM	PCR + 20 min	SA/biotin–probe, RT–PCR, hybridization/ fish	[29]
Lactobacillus sp.	100 CFU·mL^{-1}	4 h	Microbe-induced coagulation/milk	[30]
Microcystin-LR	1 ng·mL^{-1}	3 h	Vesicles–dendrimer–Ab, sandw immunosensor, Ab2–AuNP/surface water	[31]
Shrimp allergen	330 ng·mL^{-1}	10 min	AuNP–thiol–Ab, direct immunosensor/food	[32]
Toxicity of nanomaterials	3 µg·mL^{-1}	1–6 h	DH82 macrophage cells, phagocytosis, apoptosis/cytotoxicity	[33]
Vibrio harveyi	10^3 CFU·mL^{-1}	<30 min	Direct immunosensor/ shrimp culture	[34]

Abbreviations: Ab, antibody; Ab2, secondary; Ab, GMO, genetically modified organisms; magn, magnetic; NP, nanoparticle; sandw, sandwich immunocomplex; SA, streptavidin; SWCNT, single-walled carbon nanotubes.

oligonucleotide probe is immobilized onto the surface of a gold film electrode coating an AT-cut quartz crystal transducer. The hybridization results in an increase of the total mass of the crystal, which causes the decrease of crystal resonant frequency. The hybridization experiments involving QCM can be performed under either batch [36,37] or flow-injection analysis (FIA) [38] conditions by using quartz crystals of the 5–10 MHz fundamental frequency.

So far, the reported genosensors were tested mostly for the detection of short-chain nucleotide targets of the chain length comparable to those of the oligonucleotide probe (15–20 nucleotides). The changes of resonant frequency of quartz crystal transducer are measured following an addition of the probe. The example of genosensor is presented in Figure 3.6a. It is composed of a self-assembled monolayer (SAM) formed on a clean gold surface of a QCM transducer by a mixture of 11-mercaptoundecanoic acid (MUA) and 1-dodecanethiol (DT) dissolved in ethanol (molar ratio MUA/DT = 10:1). The carboxylic groups of MUA are activated by 2 mmol·L^{-1} N-ethyl-N'-(3-dimethylaminopropyl)carbodiimide hydrochloride (EDC) and 5 mmol·L^{-1} in N-hydroxysuccinimide (NHS). Such a surface modification was demonstrated earlier to be optimal for the immobilization of proteins [39]. After this, 1 mg·mL^{-1} avidin dissolved in deionized water was added, which resulted in its covalent immobilization onto SAM. Subsequently, the crystal was rinsed with the Millipore water and exposed to a 1 µmol·L^{-1} solution of a 19-base biotinylated oligonucleotide probe (5'-CCC TTT TAA AGT CGT TCC A-3'-biotin). The sequence of the probe corresponds to the characteristic part of the *Salmonella typhimurium* gene. Due to the high affinity of avidin to biotin, a stable DNA array has been obtained. The probe can be, however, immobilized also in a more simple way, either by immobilization of biotinylated probe onto the neutravidin layer chemisorbed at gold [40] (Figure 3.6b) or by chemisorption of thiolated ssDNA onto a clean gold surface [35] (Figure 3.6c).

An example of sensor response following the addition of a 19 mer complementary target is shown in Figure 3.6d. It can be seen that the addition of the target ssDNA in increased concentrations resulted in decrease of resonant frequency. After each target addition, the sensor surface was washed by the buffer (0.5 M NaCl, 10 mM *Tris*-HCl, 1 mM EDTA, pH = 7.6). At higher concentrations of the target (>2 µM), the saturation of the resonant frequency changes took place as it is evident from inset in Figure 3.6d. This method provides reliable results for short nucleotides and can distinguish ssDNA even with single mismatches [41]. A similar approach has been used for the gravimetric detection of β-thalassemia, which is a hereditary disorder caused by mutations in the gene of the β-globin chain of hemoglobin [42]. The gravimetric detection of the α-thalassemia gene has been reported by Chomean et al. [43]. They immobilized biotinylated probe to a gold layer of a QCM transducer covered by avidin. The sensor selectively detects the α-thalassemia ssDNA probe, but not the β-thalassemia. The identification of α-thalassemia was comparable with standard electrophoresis. Usually the sensitivity of direct detection of DNA hybridization at surface is in nmol·L^{-1} level. However, Willner et al. [44] proposed a

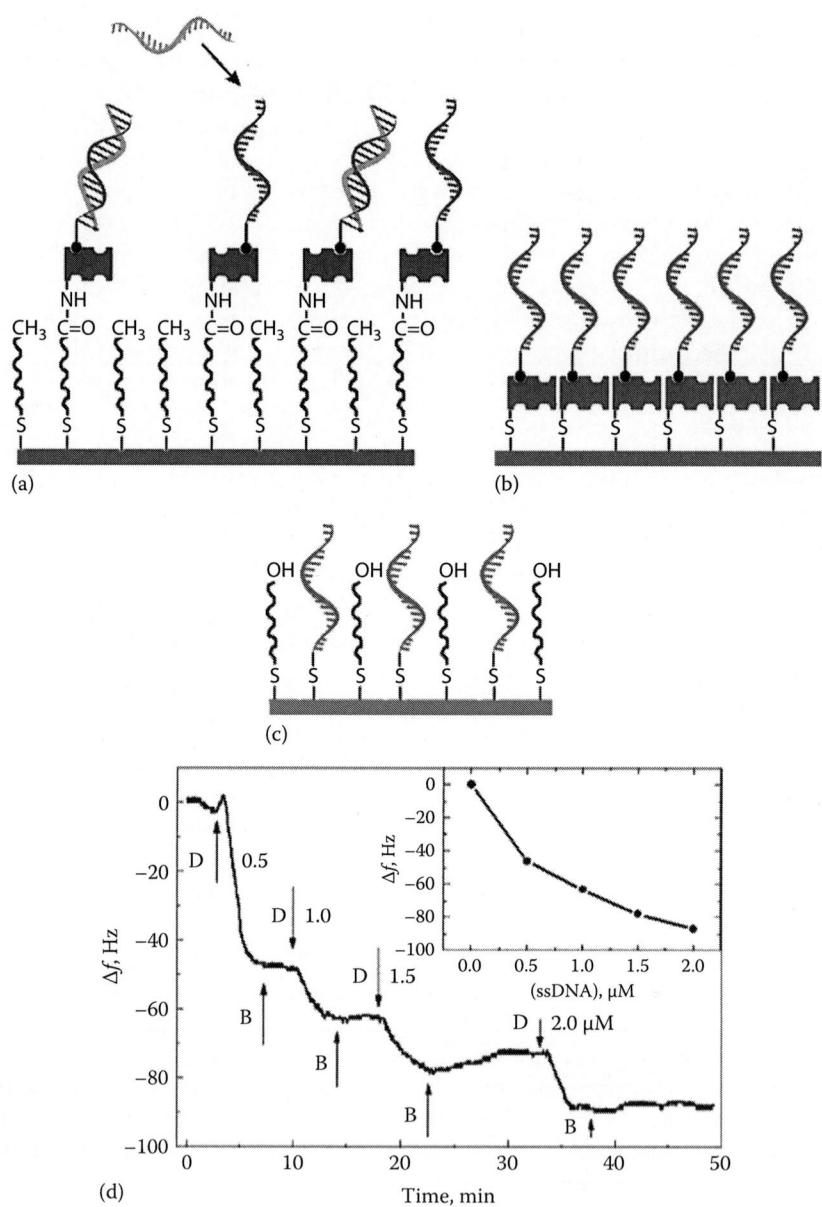

FIGURE 3.6

Scheme of most common methods of immobilization of ssDNA in genosensors: (a) Biotinylated ssDNA probe is immobilized on the avidin layer covalently attached to the SAM composed of 11-MUA and 1-DT, (b) biotinylated ssDNA is immobilized on chemisorbed neutravidin layer, and (c) thiolated ssDNA is chemisorbed on a gold support. The mercaptoethanol is used for blocking the naked gold surface. (d) The kinetics of frequency changes following the addition of target ssDNA (D) in increased concentrations and washing the sensor surface by buffer flow (B). Inset: the plot of frequency changes versus the concentration of target ssDNA.

method with substantially improved sensitivity of detection down to 10^{-15} M. In this work, ssDNA probe has been chemisorbed onto the gold surface of a QCM transducer. After an addition of target—circular DNA—additional short biotinylated ssDNA complementary to certain part of target was added. The amplification of a gravimetric detection has been achieved by the addition of avidin-coated gold nanoparticles. A slightly modified scheme was used also for the detection of linear ssDNA [45]. In this case instead of circular DNA, a linear ssDNA target was added following an addition of biotinylated ssDNA complementary to the terminal part of the target. Similarly to previous work, the gravimetric amplification has been achieved by an addition of avidin-coated gold nanoparticles. The amplification of ssDNA detection was reported also for hairpin-like probe immobilized on a gold surface of a QCM transducer [46] with a similar sensitivity like in Ref. [44,45]. The immobilization of the DNA probe on the gold nanoparticles substantially improved the detection of DNA hybridization as it is evident from Ref. [47]. In this work, the detection of P53 gene by hybridization using multichannel QCM revealed approx. 500 times higher hybridization constant for probe immobilized on gold nanoparticles in comparison with that without nanoparticles. Rather a high sensitivity of detection has been reported by Lien et al. [48], who immobilized the ssDNA probe to a surface of multiwalled carbon nanotubes (MWCNTs) doped by polypyrrole (PPy). The sensor was able to detect by QCM and electrochemical microscopy the DNA from GMO in a target concentration as low as 4 pM. Similar sensitivity was achieved by the immobilization of the ssDNA probe onto the QCM transducer covered by polystyrene-coated nanoparticles with a diameter of 12 nm [49]. The substantially improved sensitivity of detection hybridization has been achieved by the immobilization of ssDNA on the gold nanoparticle–liposome composites. The lowest target DNA concentration detected was 0.1 pmol·L^{-1} [50].

In addition to the immobilization of ssDNA to a solid support (Figure 3.6a through c), also supported lipid films can be used for probe anchoring. Gambinossi et al. [51] synthesized ssDNA modified by cholesterol that was incorporated into the supported phospholipid films. Using QCM method, they demonstrated hybridization with target ssDNA. We should, however, mention that earlier the hybridization at the surface of lipid monolayers [52] and bilayers [53] was clearly shown for probe anchored into the lipid film by either hydrophobic chain or cholesterol. In addition, we studied also the effect of orientation of the ssDNA relatively to the lipid layer [52]. For this purpose, 19 mer oligonucleotides were modified by oleylamine at both (3′ and 5′) terminals or only at one (3′) terminal. The interaction of single-stranded (19 mer) oligonucleotides without oleylamine with dioleoylphosphatidylcholine (DOPC) monolayers resulted only in a slight increase of surface pressure and the area per phospholipid molecule, while more substantial and significant increase of these values was observed following the incorporation of oligonucleotides modified by oleylamine. This influence is similar for both types of oligonucleotide modifications; however, considerable differences in changes of monolayer properties took place after the hybridization with complementary oligonucleotides. The hybridization of oligonucleotides with the DNA modified by oleic acid at both 3′ and 5′ terminals at the surface of lipid monolayer resulted in further increase of surface pressure and in the increase of the area per phospholipid molecule, while a decrease of both the surface pressure and the area per phospholipid molecules was observed for hybridization with DNA modified by oleic acid at 3′ terminal. It is possible that in the latter case, the hybridization caused the loss of hybridized molecules from monolayers. The interaction of noncomplementary chains with DOPC mono-layers with incorporated oleyl acid-modified DNA surface also influenced the properties of monolayers, but the effect was weaker in comparison with that observed for complementary chains. Another approach for immobilization of the oligonucleotide parallel to the lipid film surface was proposed recently by Woller et al. [54]. In this work, 39 mer ssDNA was anchored to a lipid layer by zinc-porphyrin anchor linked to the DNA via two or three phenylethynylene moieties. The binding and subsequent hybridization of ssDNA on the surface were demonstrated and were coverage dependent. At low coverage, the hybrid-ization results in an increase in mass and has been accompanied by a slight increase in the rigidity of the DNA layer. At high coverage, hybridization expels molecules from the membrane. Su et al. [55] compared the sensitivity of detec-tion in DNA hybridization by simultaneous application of QCM and surface plasmon resonance (SPR) methods. Although the SPR revealed a more precise determination of the surface density of the molecules, QCM method provided comparable results in hybridization assay.

However, the ssDNA targets encountered in real samples are considerably lon-ger and may even contain several hundred nucleotides. PZ genosensors have been used for detecting the presence of characteristic sequences of oligonucleotides (18–23 mers) within a relatively long chain of DNA. Tombelli et al. [36] reported a genosensor composed of a 23 mer probe immobilized on a streptavidin-coated gold surface to monitor interactions with a specific gene in *Aeromonas hydrophila*. The characteristic 23 mer target was located close to one terminal of a longer 205 mer chain. The analysis of different immobilization strategies of the probe for detection of *A. hydrophila* showed that the immobilization of biotinylated probe onto the streptavidin layer is preferable. The important step consisted also in preparation of polymerase chain reaction (PCR)-amplified samples that have to be treated by a denaturation and addition of blocking oligonucle-otides [56]. Zhou et al. [37] reported a genosensor based on 18 mer DNA immo-bilized on varied types of mono- and multilayered films for the detection of β-thalassemia mutations. The characteristic 18 mer sequence was within the tar-get 443 mer DNA. In both reports, the presence of a target long-chain oligonucle-otide resulted in a decrease of the oscillation frequency of the quartz by less than 250 Hz, which is lower than expected for tight attachment of the target to the sensor surface. Karamollaoglu et al. [28] reported a genosensor based on chemi-sorbed 25 mer thiolated ssDNA probe onto the gold electrode of the QCM trans-ducer. The probe was complementary to the part of ferredoxin-like protein gene inserted into tobacco plants. Thus, the aim of this work was to demonstrate the possibility of application of QCM for the detection of GMO. They fragmented the genomic DNAs by digestion with restriction endonucleases and by ultra-sonication. They showed that the fragmentation allows the detection of target sequence directly in nonamplified genomic DNA. The approach developed by Willner et al. [44] of the amplification of QCM detection of DNA hybridization by means of gold nanoparticles has been applied also for longer target corresponding to the *Bacillus anthracis* gene [57]. The 168 base pair (bp) fragment of the Ba813 gene in chromosomes and the 340 bp fragment of the pag gene in plasmid pXO1 were used as the target. They showed that the biosensor specifically recognized the target DNA fragment of *B. anthracis* from that of its closest species, such as *Bacillus thuringiensis*, with the limit of detection (LOD) reaching 3.5×10^2 colony forming units $(CFU) \cdot mL^{-1}$ of *B. anthracis* vegetative cells. In this work, only asymmetric PCR amplification was used without culture enrichment. A similar approach was used for the development of multichannel bacterial gene detection. For example, the *Pseudomonas aeruginosa* gene was detected with LOD corresponding to 1.5×10^2 $CFU \cdot mL^{-1}$ of bacteria cells [58]. The QCM method has been shown as efficient for the detection of malaria infection with the *Plasmodium falciparum* gene [59] and vaccinia virus DNA [60]. In our work [41], we analyzed also the hybridization of a long-chain ssDNA target containing 454 bases with 19 mer probe at the QCM transducer surface (see preceding text). However, in addition to the frequency response, we measured also the motional resistance that measures the surface viscosity contribution. The viscoelastic contribution caused by the molecular slip also contributed to the frequency changes and can result in both increase and decrease of the frequency [61,62]. The sensor response following an addition of a long-chain target onto the sensor surface is presented in Figure 3.7.

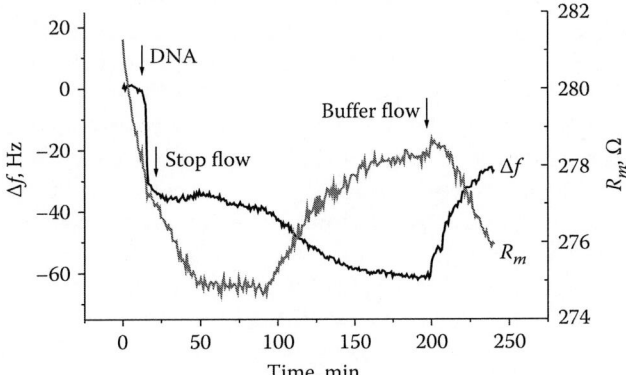

FIGURE 3.7
The kinetics of the changes of resonant frequency, Δf, and motional resistance, R_m, for the biosensor composed of a neutravidin layer chemisorbed at gold surface of QCM transducer and 19 mer biotinylated probe corresponding to the characteristic sequence of the *S. typhimurium* gene following the addition of ssDNA target containing 454 bases with complementary part of the probe. The stop flow and washing of the surface by buffer are shown by arrows. (From Hianik, T. et al., unpublished results.)

The addition of the target in a continuous flow resulted in a sharp decrease of resonant frequency and decrease of motional resistance, R_m. After the stop of the flow, a further decrease of the resonant frequency took place. However, the changes of motional resistance were more complicated. First, almost a steady-state R_m value was achieved, but then an increase of motional resistance was observed. Washing of the surface by a buffer resulted in an increase of frequency and a decrease of motional resistance, which corresponds probably to the removal of not hybridized target molecules from the sensor surface. The resulting changes of frequency (26 Hz) were much lower in comparison with that expected considering the high molecular mass of the target. It can be seen also comparing the results for a short-chain (19 mer) target presented in Figure 3.6d. The addition of comparable concentration ($2 \mu mol \cdot L^{-1}$) of a target resulted in much higher frequency changes. At the same time, the motional resistance is lower in comparison with that for the biosensor prior to the addition of the target. It is likely that the complexity between the probe and target is rather soft, which caused an increase of molecular slip and, as a result, a decrease of motional resistance.

The variation in the changes of frequency and dissipation, which is also a measure of the viscosity contribution, has been reported also in other works. Tsortos et al. [63,64] analyzed the effect of DNA conformation on the dissipation of acoustic waves. They showed that by measurement of changes in resonant frequency and dissipation of acoustic waves, it is possible to distinguish between double-stranded DNAs (dsDNAs) of the same shape (straight rod) but various sizes (from 20 to 198 bp) and the same mass and size (90 bp) but various shapes ("straight," "bent," "triangle"). These principles resulted also in the development of mass-independent sensing principle of hybridization using QCM-D technique (simultaneous measurement of resonant frequency and dissipation) [65]. The measurement of the acoustic ratio $\delta D/\delta F$ (dissipation to frequency changes) together with discrete molecule binding allowed the detection of hybridization of ssDNA with specific lengths. Aung et al. [66] demonstrated that dsDNA is more dissipative in comparison with ssDNA. They also found a direct correlation between dissipation and motional resistance. Thus, the surface viscosity is rather a sensitive parameter to the conformation and DNA hybridization at surface. For recent review on application of QCM in genosensors, see also Ref. [18].

Except the detection of DNA hybridization, the QCM method has been used also for the detection of the point mutations in DNA [67]. In this work, the biotinylated ssDNA probe has been immobilized onto magnetic streptavidin-coated beads. The beads were adsorbed to a QCM surface by magnet, which has been followed by DNA hybridization with complementary strand. Single-base-coded cadmium telluride (CdTe) nanoprobes (A-CdTe, T-CdTe, C-CdTe, and G-CdTe, respectively) were used as the detection probes. The point mutation has been detected by a decrease of resonant frequency of the quartz crystal. However, even common immobilization of biotinylated ssDNA probe onto the neutravidin layer and application of

QCM together with SPR detection allowed an efficient determination of single point mutation in TP53 gene, responsible for lung cancer [68]. The original method of detection of point mutation in p53 tumor suppressor gene based on the conformational changes of hairpin-like DNA capture probe was proposed in Ref. [69]. The thickness shear mode (TSM) method with simultaneous detection of series resonant frequency and motional resistance is a sensitive tool for the detection of the presence of abasic sites in dsDNA [40]. Aldehydic apurinic or apyrimidinic sites (AP) that lack a nucleobase moiety are one of the most common forms of toxic lesions in DNA. We synthesized a close structural analog of a native AP site, the 2-(hydroxymethyl)tetrahydrofuranyl residue. The prepared oligodeoxyribonucleotides (ODNs) containing one, two, or three abasic sites were hybridized to complementary sequences immobilized on a gold surface by means of biotin–neutravidin technology. The changes of mass and motional resistance measured by the acoustic wave method correlated well with the thermostability of DNA duplexes in solution. With respect to the latter, UV-monitored melting curves indicate that both the number of AP sites and their localization in the double-stranded structure influence the amount by which a 19 bp duplex is destabilized. The presence of three abasic sites completely destabilized the DNA duplex.

The TSM method has been efficient for detecting DNA damage connected with the appearance of the thymidine glycols (TGs). TG is the product of oxidation of thymidine caused by ionizing radiation among a number of factors. We synthesized short 19–31 mer oligonucleotides containing one or two TG residues. Melting studies reveal that the presence of TG in dsDNA reduces the melting temperature by 8°C–23°C depending on the number of TG residues (one or two), the length of the nucleic acid strand, and the localization of the residue in such a strand. The cooperativity of the melting transition of DNA containing two TG residues was considerably lower in comparison with those with one TG residue present. The effect of TG was also observed from the study of DNA hybridization at the surface of TSM device. A decrease of the rate of hybridization took place for the system in which one strand contained one or two TG residues. This is likely caused by the different structures of the single strand, which contains a hydrophilic TG residue in comparison with undamaged oligonucleotide [70].

3.3.3 Aptamer-Based Biosensors

DNA and RNA aptamers are single-stranded oligonucleotides, which, under certain conditions, fold into 3D structures containing specific binding sites for low or macromolecular compounds of various types, including cells, cell surface proteins, bacteria, and viruses. Their specificity is comparable to and, in certain cases, even higher than those of antibodies [71]. In contrast to antibodies, aptamers are prepared by an *in vitro* selection procedure developed independently in the early 1990s by three separate groups of investigators. Robertson and Joyce [72] have described the method of RNA selection with improved enzymatic activity to cleave DNA. Tuerk and Gold [73] patented the process of the selection of DNA ligands as a target for T4 RNA polymerase. This method is known as systematic evolution of ligands by exponential enrichment (SELEX). Ellington and Szostak [74] reported a method of *in vitro* selection of RNA, which binds specifically to organic dyes. They also introduced the term *"aptamer"* (from the Latin *aptus*, meaning "to fit," and Greek *meros*, meaning "the part"). The identification of aptamers is based on a combinatorial approach. Specific oligonucleotides are isolated from complex libraries of synthetic nucleic acids. For this purpose, random-sequence DNA libraries are obtained by automated DNA synthesis. The size of a randomized region can vary from 30 to 60 nucleotides, flanked on both sides with a specific, unique DNA sequence for polymerase chain reaction (PCR) amplification. The theoretical diversity of individual oligonucleotides in these random DNA libraries is relatively wide; for example, $4^{40} = 1.2 \times 10^{24}$ in the case of oligonucleotides composed of 40 bases. In practice, however, a considerably smaller library of approximately 10^{13}–10^{15} molecules is used [75]. The selection consists of DNA binding with immobilized ligands, such as proteins or other compounds. However, because a smaller library in comparison with all possibility of sequence is used, the selected sequence should not have maximal affinity to the target. Therefore, an additional post-SELEX modification could improve the affinity properties of the aptamers [76]. The stability of complexes is characterized by the apparent dissociation constant, K_D. For aptamer–protein complexes, K_D varies within the 1–100 nM range, which is similar to that of antibody–antigen complexes. Unbound DNA/RNA molecules are eluted from the column, while bound aptamers are isolated from the complex and then amplified by PCR. This cycle is repeated several times (around 6–10), and as a result, the DNA or RNA sequence with a high affinity to the target ligand is obtained. The SELEX technique has been discussed in detail in a number of papers and reviews [77,78].

Initially the aptamers were based on RNA and were selected against bacteriophage T4 DNA polymerase [73] or fluorescent dyes [74]. However, Bock et al. [79] selected a DNA aptamer specific to human α-thrombin-binding aptamer (TBA). TBA is a 15 mer guanine-rich aptamer of the sequence (GGTTGGTGTGGTTGG), which selectively binds to the fibrinogen binding site of thrombin. This binding site is responsible for the cleavage of the fibrinogen, which results in the formation of fibrin clots in blood. Nuclear magnetic resonance (NMR) studies have revealed that TBA in the presence of K⁺ ions folds into guanine quadruplexes (G-quadruplexes) composed of two guanine tetrads connected by one TGT and two TT loops. Each tetrade is stabilized by Hoogsteen bonds (Figure 3.8) [80,81].

G-quadruplexes have been found also in other DNA aptamers, although some aptamers do not contain these structures, for example, aptamers sensitive to cellular prions [82]. The importance of G-quadruplexes lies in the stability of the 3D aptamer structure and in the improvement of electrostatic interactions to the positively charged binding site at the ligands. This is largely due to the fact that the negative charge density of G-quadruplexes is twice as high as that of the linear DNA [83], although other forces, such as van der Waals, π–π stacking, and

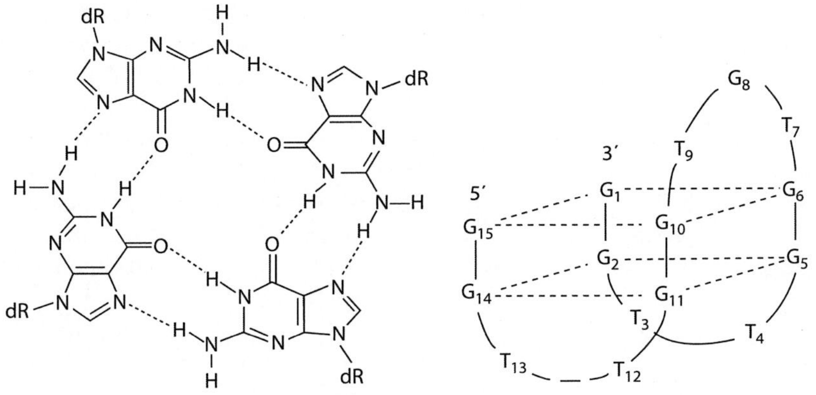

(a)　　　　　　　　　　　　　　(b)

FIGURE 3.8
(a) Scheme of the G-quartet structure. (b) G-quadruplex of TBAs.

hydrophobic interactions, are also important. The stability of the G-quadruplex is crucial in providing a high affinity of aptamers to the ligands.

Due to the high affinity of the aptamers to the ligands, they found a wide range of applications in medical therapy, targeted drug delivery, molecular imaging, as well as receptors in biosensors, in particular also for food safety applications [84,85]. The advantages of aptamers over antibodies are high stability, easy regeneration without aptamer damage, high reproducibility in synthesis, and lower cost. Aptamers are rather flexible. Using simple molecular engineering methods, it is possible to substantially improve their binding properties and to design aptamers that transduce the recognition event into a readily detectable electronic, optical, or acoustical signal through binding-induced conformational changes. DNA or RNA aptamers can be immobilized on various surfaces like other oligonucleotides (see previous section). Once the aptasensor is prepared, it can be stored in a refrigerator in a dry condition for several months without loss of the affinity to the target [85].

Despite high advantages of aptamers, they have been used only relatively recently for the development of biosensors [86]. However, currently the application of aptamers in biosensing grows almost exponentially (see [85,87]). In this chapter, we will focus on the application of aptamers for the development of biosensors based on QCM.

According to the Web of Science database, the first report on the application of the QCM method for the detection of protein–aptamer interactions appeared in 2000 [88]. In this work, the R5 helix peptide was immobilized on a QCM transducer and the interaction with RNA aptamer was monitored by quartz crystal. The approach was used for the purpose of the control of affinity properties of the aptamer obtained by SELEX. The biosensor format when aptamer as receptor was immobilized on the QCM transducer was reported a bit later. Tassew and Thompson [89] reported the study of the binding of the human immunodeficiency virus type 1 *trans*-activator of transcription (Tat) protein with RNA immobilized via biotin–neutravidin technology to the TSM quartz crystal surface. They investigated also the effect of RNA mutation and inhibition of the peptide–RNA interaction induced by neomycin. They clearly demonstrated the efficiency of the TSM method for binding the small peptide to the RNA aptamers. At the same time also, QCM biosensor based on DNA aptamers specific to human immunoglobulin E (IgE) was reported [90]. In this work, the possibility of optimization of the DNA aptamers by involving an extension in the oligonucleotide chain that resulted in improved affinity to IgE has also been demonstrated. Minunni et al. [91] reported the development of RNA aptamer-based QCM biosensor for the detection of Tat protein. They used a similar method of aptamer immobilization as in [89], but instead of neutravidin, the streptavidin was adsorbed at the gold layer of the QCM transducer. They compared the detection of Tat by aptamers with that based on specific antibodies and confirmed the efficiency of aptamers as recognition elements. Moreover, the possibility of regeneration of the aptamer-sensing surface by high ionic strength water solution has been shown. In a further study, the same aptamers were applied for comparison of the detection of the Tat protein by QCM and SPR techniques, and similar results were obtained [92]. In our work [93], we developed QCM biosensor for the detection of thrombin. The 32 mer biotinylated DNA aptamers sensitive to heparin/binding site to a thrombin were immobilized to the avidin layer, covalently attached to the gold surface of a QCM transducer (Figure 3.9a). An AT-cut crystal of fundamental frequency of 9 MHz was used for the detection of the interaction of thrombin with the

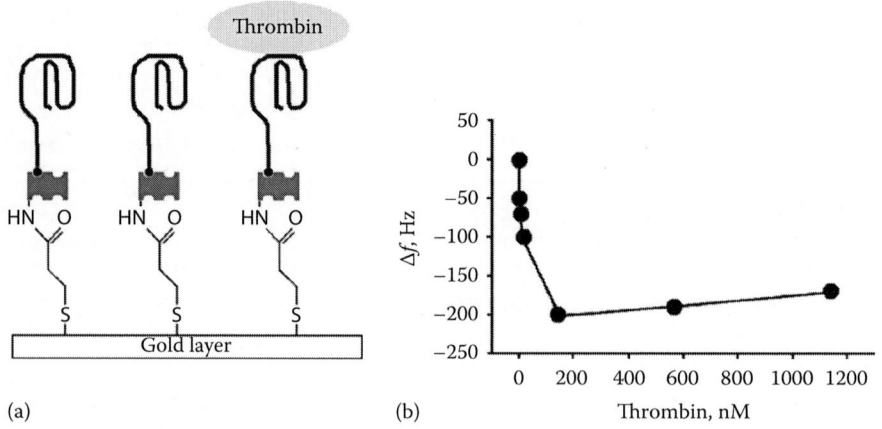

(a)   (b)

FIGURE 3.9
(a) Scheme of biosensor for the detection of thrombin. Biotinylated aptamers are immobilized on the avidin layer covalently attached to a gold surface through 3,3′-dithiopropionic acid-di(N-succinimidyl ester) (DSP). (b) The plot of the changes of resonant frequency of quartz crystal versus concentration of the thrombin. (Reproduced from *Bioorg. Med. Chem. Lett.*, 15, Hianik, T., Ostatna, V., Zajacova, Z., Stoikova, E., and Evtugyn, G., 291, Copyright 2005, with permission from Elsevier.)

aptamers by measurement changes of resonant frequency. Rather a sharp decrease of the frequency was observed at relatively low thrombin concentrations (Figure 3.9b).

The LOD of the thrombin with this acoustic sensor was 10 nM, which was comparable with electrochemical method of detection based on the electrochemical indicator methylene blue (MB) reported also in this chapter. Further, we determined also kinetic and binding constants of thrombin to the two types of DNA aptamers—conventional linear aptamers and aptamer beacon [94]. Both aptamers revealed similar binding properties. This means that an addition of the thrombin resulted in conformational changes of molecular beacon into the configuration typical for conventional aptamers, that is, containing a G-quadruplex motif. We also showed that the increased concentration of NaCl resulted in the weakening of the binding of thrombin to the aptamers, probably due to shielding effect of Na^+ ions. The binding of the thrombin to the aptamer depended on electrolyte pH, which is presumably connected with maintaining the 3D aptamer configuration, optimal for binding the protein. Similar sensitivity of thrombin detection (LOD 10 nM) by means of the QCM sensor was obtained using another type of aptamer immobilization—on the polymeric layer formed by electropolymerization of phenothiazine dyes, MB, and methylene green [95].

The advantage of simultaneous measurements of series resonant frequency and motional resistance for the detection of protein aptamer interaction has been shown in our work [96]. We applied TSM method for the study of the surface properties of a DNA aptasensor that specifically binds IgE. The biotinylated 45 mer DNA aptamers were immobilized on the surface of a self-assembled layer composed of a mixture of polyamidoamine dendrimers of the fourth generation with 1-hexadecanetiol covered by neutravidin. Using the TSM method, we studied the kinetics of changes of the series resonant frequency, f, and the motional resistance, R_m, of a quartz crystal transducer, used as a support for the formation of the sensing layer. We have shown that the attachment of the biotinylated DNA aptamers onto the surface covered by neutravidin resulted in a decrease of f but in an increase of R_m. Similar changes of f and R_m were observed following an addition of IgE. This suggests the contribution of friction forces to the crystal oscillation, which was taken into account in the calculation of the mass changes at the sensor surface following binding processes. Later, we used a similar approach for analyzing the binding of thrombin to the DNA aptamers of various configurations such as conventional aptamer, aptamer with rigid supporting part provided by short dsDNA spacer, and aptamer homo- and heterodimers (Figure 3.10a) [97].

We have shown that the rigid supporting part that provides better aptamer orientation and access to the target protein resulted in a higher binding of thrombin in comparison with conventional DNA aptamers. A further increase of binding sensitivity was obtained for aptamer heterodimers contained in

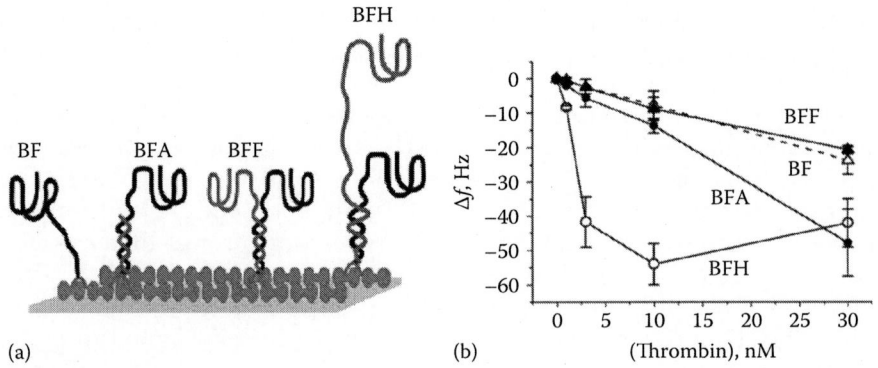

FIGURE 3.10
(a) The scheme of immobilization thrombin-sensitive aptamers of various configurations on a gold surface covered by neutravidin. BF, conventional aptamer; BFA, conventional aptamer with rigid supporting part formed by dsDNA spacer; BFF, aptamer homodimer with two identical binding sites to fibrinogen epitope at thrombin; BFH, aptamer heterodimer with two binding sites sensitive to fibrinogen (black) and heparin (gray) epitopes at thrombin. (b) The plot of the changes of resonant frequency versus thrombin concentration for aptasensor composed of aptamers of various configurations. BF, BFA, BFF, and BFH, respectively. The results are mean ± S.D. obtained from five independent experiments in each series. From Hianik, T., Grman, I., and Karpisova, I., *Chem. Commun.*, 41, 6303, 2009. Reproduced by permission of The Royal Society of Chemistry.

two binding sites for fibrinogen and heparin epitopes at thrombin and connected by their complementary supporting part (Figure 3.10b). The obtained data correlated well with atomic force spectroscopy studies [98].

QCM aptasensor based on either RNA or DNA aptamers has been shown as a very sensitive tool for the detection of interferon (IFN)-γ that is considered as a selective marker for tuberculosis [99]. RNA-based aptasensor revealed the sensitivity of detection at 100 fmol·L^{-1} of IFN, which was 10 times better than that for DNA-based aptasensor. Cantilever- and QCM-based aptasensor revealed a comparable sensitivity in the detection of CDK2 protein from yeast cell lysate [100]. The efficiency of the detection of IgE in a human serum by QCM aptasensor was approved by Yao et al. [101]. The QCM magnetic sensor utilizing the magnetic bead–coated aptamers has been used for selective detection of human acute leukemia cells with a detection limit of 8 × 10^3 cells·mL^{-1} [102]. The satisfactory results were obtained in detecting these cells from a cell mixture. Thus, these works demonstrate the possibility of using aptamers for the detection of proteins in complex biological liquids. The detection of viruses is also possible. Using QCM aptasensor, the detection of avian influenza virus (AIV) H5N1 has been demonstrated [103].

The advantage of aptamers consists in their selection to different binding sites at the target. Typical examples are aptamers sensitive to fibrinogen and heparin binding sites at thrombin. Having these two aptamers, it is possible to amplify the detection of the target. This method has been reported in paper by Pavlov et al. [104] who showed that after binding the thrombin to the DNA aptamer array at gold layer of quartz crystal transducer, the addition of the gold nanoparticles modified by aptamers resulted in additional remarkable decrease of the resonant frequency. They observed a substantial amplification even when using two identical fibrinogen-specific aptamers. Similar approach was later published in Ref. [105] using QCM-D technique. The QCM aptasensor was rather effective also in the detection of small molecules like daunomycin with LOD of 52.3 ± 2.1 pM [106], ATP [107], or even mercury (II) with LOD of 0.24 ± 0.06 nM [108].

Recently also, the detection of mycotoxin ochratoxin A (OTA) has been proposed by QCM aptasensor [109]. We showed that an addition of OTA to a sensor surface with immobilized biotinylated DNA aptamers resulted in a decrease of resonant frequency, f, and an increase of motional resistance, R_m. We were able to detect OTA with LOD of 30 nM and determined the equilibrium dissociation constant $K_D = 43.9 \pm 30$ nM. The OTA interacted with aptamer only in the presence of calcium ions. Therefore, binding studies were performed in the presence of 20 mM Ca^{2+}. No significant changes of f and R_m were observed without calcium. We analyzed also the changes of acoustic parameters in the presence of possible interference N-acetyl-L-phenylalanine (NAP). The addition of NAP in a concentration range of 25–740 nM resulted in similar frequency changes like that induced by OTA; however, significant but much lower changes of motional resistance were observed only at the highest NAP concentration analyzed (740 nM).

In addition to DNA/RNA aptamers, also peptide aptamers are perspective as receptors in QCM biosensors. For example, Zhou et al. [110] developed peptide aptamer-based biosensor for the detection of protein kinase A.

The combination of QCM technique with electrochemical measurements, the so-called electrochemical quartz crystal microbalance (EQCM), has been shown as very effective in protein detection. Recently we applied this method in the thrombin-specific aptasensors based on MWCNT–MB composite [111]. Simultaneously with measurements in the changes of the oscillation frequency of quartz crystal, which reflect both mass and viscosity changes at the crystal surface, we performed the cycling of the voltage in the range of -0.7 to $+0.6$ V (vs. Ag/AgCl reference electrode). This voltage cycling provided a better condition for the interaction of the thrombin with DNA aptamers. This method has improved detection limit substantially. While conventional QCM method with aptamer immobilization by avidin–biotin technology resulted in the detection of thrombin with LOD of 10 nM [93], the immobilization of aptamers to the MWCNT–MB composite layer and simultaneous cycling of the voltage allowed to detect thrombin with LOD of 0.3 nM [112] and cellular prions with LOD of $50 \text{ pmol} \cdot \text{L}^{-1}$ [113].

3.4 Conclusion

PZ sensors were previously considered as not being sensitive enough compared to other types of transducers. As theoretical backgrounds for operation of piezosensors in liquids became completed and many successful applications were reported, this situation changed. The PZ biosensors represent a convenient tool for fast and simple determination of viruses and bacteria, proteins, nucleic acids, and small molecules such as drugs, hormones, and pesticides. Usually, no additional unstable reagents are required for the analysis. The traditionally interpreted mass changes (microbalance operation) are currently supplemented with dissipation and viscoelastic changes—the presence of analyzed molecules initiated agglutination reactions and conformation changes of attached biolayers, and enhanced responses are due to nanostructured surfaces and application of nanoparticles. This provides novel ways to improve the sensitivity of measurements and recently the studies of eukaryotic cells (adhesion, morphology, apoptosis, etc.) became frequent.

Piezosensors are also useful as a research tool for biochemistry and biology. The direct label-free and real-time monitoring of affinity interactions with piezosensors represents an economic alternative to the overpriced optical systems based on surface plasmon resonance. The valuable characteristics of affinity binding reaction can be obtained easily and quickly.

PZ system remains an open platform and moderately skillful researchers can build the experimental devices themselves from components of the shell. This results in quite a wide application field, fruitful combinations together with other sensing technologies, and very interesting scientific achievements.

References

1. J.W. Grate, S.J. Martin, R.M. White, *Anal. Chem.* 65 (1993) A940–A987.
2. G. Sauerbrey, *Z. Phys.* 155 (1959) 206.
3. R. Lucklum, P. Hauptmann, *Anal. Bioanal. Chem.* 384 (2006) 667–682.
4. G.T. Hermanson, A.K. Mallia, P.K. Smith, *Immobilized Affinity Ligand Techniques*, Academic Press, San Diego, CA (1992).
5. A.A. Suleiman, G.G. Guilbault, *Analyst* 119 (1994) 2279.
6. R.L. Bunde, E.J. Jarvi, J.J. Rosentreter, *Talanta* 46 (1998) 1223.
7. C. Henry, *Anal. Chem.* 68 (1996) 625A.
8. Texas Instruments, SN74LS320 crystal-controlled oscillators D2418 (1981) 3–801.
9. T. Nomura, M. Watanabe, T.M. West, *Anal. Chim. Acta* 175 (1985) 107.
10. P. Skládal, J. Horáček, *Anal. Lett.* 32 (1999) 1519.
11. P. Skládal, Z. Jílková, I. Svoboda, V. Kolář, *Biosens. Bioelectron.* 20 (2006) 2027.

12. P. Skládal, M. Minunni, M. Mascini, V. Kolář, M. Fránek, *J. Immunol. Meth.* 176 (1994) 117.
13. D.J. O'Shannessy, *Curr. Opin. Biotechnol.* 5 (1994) 65–71.
14. T.A. Morton, D.G. Myszka, I.M. Chaiken, *Anal. Biochem.* 227 (1995) 176.
15. J. Horáček, P. Skládal, *Anal. Chim. Acta* 347 (1997) 43.
16. J. Horáček, P. Skládal, *Food Agric. Immunol.* 10 (1998) 363.
17. B. Becker, M.A. Cooper, *J. Mol. Recognit.* 24 (2011) 754.
18. R.E. Speight, M.A. Cooper, *J. Mol. Recognit.* 25 (2012) 451.
19. S. Sankarana, S. Panigrahi, S. Mallik, *Biosens. Bioelectron.* 26 (2011) 3103.
20. N.A. Karaseva, T.N. Ermolaeva, *Talanta* 93 (2012) 44.
21. C. Poitras, J. Fatisson, N. Tufenkji, *Water Res.* 43 (2009) 2631.
22. Y.S. Chen, Y.C. Hung, J.C. Chiou, H.L. Wang, H.S. Huang, L.C. Huang, G.S. Huang, *Jpn. J. Appl. Phys.* 49 (2010) 105103.
23. Y. Wan, D. Zhang, B. Hou, *Biosens. Bioelectron.* 25 (2010) 1847.
24. M. Salmain, M. Ghasemi, S. Boujday, C.M. Pradier, *Sens. Actuators B* 173 (2012) 148.
25. M.R. Plata, A.M. Contento, A. Ríos, *Microchim. Acta* 172 (2011) 447.
26. Z.Q. Shen, J.F. Wang, Z.G. Qiu, M.J.X.W. Wang, Z.L. Chen, J.W. Li, F.H. Cao, *Biosens. Bioelectron.* 26 (2011) 3376.
27. W. Hu, S. Chen, L. Liu, B. Ding, H. Wang, *Sens. Actuators B* 157 (2011) 554.
28. I. Karamollaoglua, H.A. Öktema, M. Mutlu, *Biochem. Eng. J.* 44 (2009) 142.
29. S.R. Hong, H.D. Jeong, S. Hong, *Talanta* 82 (2010) 899.
30. H.D. Jang, K.S. Chang, Y.G. Lee, S.J. Lee, C.L. Hsu, *Eur. Food Res. Technol.* 229 (2009) 349.
31. Y. Xia, J. Zhang, L. Jiang, *Coll. Surf. B* 86 (2011) 81.
32. S. Xiulan, Z. Yinzhi, S. Jingdong, S. Liyan, Q. He, Z. Weijuan, *Eur. Food Res. Technol.* 231 (2010) 563.
33. G. Wang, A.H. Dewilde, J. Zhang, A. Pal, M. Vashist, D. Bello, K.A. Marx, S.J. Braunhut, J.M. Therrien, *Particle Fibre Toxicol.* 8 (2011) 4.
34. S. Buchatip, C. Ananthanawat, P. Sithigorngul, P. Sangvaniche, S. Rengpipat, V.P. Hoven, *Sens. Actuators B* 145 (2010) 259.
35. M.I. Pividori, A. Merkoci, S. Alegret, *Biosens. Bioelectr.* 15 (2000) 291.
36. S. Tombelli, M. Mascini, C. Sacco, A.P.F. Turner, *Anal. Chim. Acta* 418 (2000) 1.
37. X.C. Zhou, L.Q. Huang, S.F.Y. Li, *Biosens. Bioelectr.* 16 (2001) 85.
38. T. Ketterer, H. Stadler, J. Rickert, E. Bayer, W. Göpel, *Sens. Actuat. B* 65 (2000) 73.
39. M. Šnejdárková, L. Csaderová, M. Rehák, T. Hianik, *Electroanalysis* 12 (2000) 940.
40. T. Hianik, X. Wang, S. Andreev, N. Dolinnaya, T. Oretskaya, M. Thompson, *Analyst* 131 (2006) 1161.
41. T. Hianik, V. Gajdos, A. Kochman, W. Kutner, H. Drahovska, J. Turna, P. Vadgama, unpublished results.
42. M. Minunni, S. Tombelli, R. Scielzi, I. Mannelli, M. Mascini, C. Gaudiano, *Anal. Chim. Acta* 481 (2003) 55.
43. S. Chomean, T. Potipitak, C. Promptmas, W. Ittarat, *Clin. Chem. Lab. Med.* 48 (2010) 1247.
44. I. Willner, F. Patolsky, A. Lichtenstein, *Anal. Sci.* 17 (2001) i351.
45. Y. Weizmann, F. Patolsky, I. Willner, *Analyst* 126 (2001) 1502.
46. Y. Fei, X.-Y. Jin, Z.-S. Wu, S.-B. Zhang, G. Shen, R.-Q. Yu, *Anal. Chim. Acta* 691 (2011) 95.
47. S. Li, X. Li, J. Zhang, Y. Zhang, J. Han, L. Jiang, *Sens. Actuators A.* 364 (2010) 158.

48. T.T.N. Lien, T.D. Lam, V.T.H. An, T.V. Hoang, D.T. Quang, D.Q. Khieu, T. Tsukahara, Y.H. Lee, J.S. Kim. *Talanta* 80 (2010) 1164.
49. S. Li, Y. Xia, J. Zhang, J. Han, L. Jiang, *Electrophoresis* 31 (2010) 3090.
50. M. Bhuvana, J.S. Narayanan, V. Dharuman, W. Teng, J.H. Hahn, K. Jayakumar, *Biosens. Bioeelctr.* 41 (2013) 802.
51. F. Gambinossi, M. Banchelli, A. Durand, D. Berti, T. Brown, G. Caminati, P. Baglioni, *J. Phys. Chem. B.* 114 (2010) 7338.
52. T. Hianik, P. Vitovič, D. Humeník, S.J. Andreev, T.S. Oretskaya, E.A.H. Hall, P. Vadgama, *Bioelectrochemistry* 59 (2003) 35.
53. T. Hianik, M. Fajkus, B. Sivák, I. Rosenberg, P. Koiš, J. Wang, *Electroanalysis* 12 (2000) 495.
54. J.G. Woller, K. Börjesson, S. Svedhem, B. Albinsson, *Langmuir* 28 (2012) 1944.
55. X. Su, Y.-J. Wu, W. Knoll, *Biosens. Bioelectr.* 21 (2005) 719.
56. S. Tombelli, M. Minunni, A. Santucci, M.M. Spiriti, M. Mascini, *Talanta* 68 (2006) 806.
57. R.-Z. Hao, G.-M. Song, G.-M. Zuo, R.-F. Yang, H.-P. Wei, D.-B. Wang, Z.-Q. Cui, Z. Zhang, Z.-X. Cheng, X.-E. Zhang, *Biosens. Bioelectr.* 26 (2011) 3398.
58. J. Cai, C. Yao, J. Xia, J. Wang, M. Chen, J. Huang, K. Chang, C. Liu, H. Pan, W. Fu, *Sens. Actuators B* 155 (2011) 500.
59. T. Potipitak, W. Ngrenngarmlert, C. Promptmas, S. Chomean, W. Ittarat, *Clin. Chem. Lab. Med.* 49 (2011) 1367.
60. K. Kleo, A. Kapp, L. Ascher, F. Lisdat, *Anal. Biochem.* 418 (2011) 260.
61. B.A. Cavic, G.L. Hayward, M. Thompson, *Analyst* 124 (1999) 1405.
62. J.S. Ellis, M. Thompson, *Phys. Chem. Chem. Phys.* 6 (2004) 4928.
63. A. Tsortos, G. Papadakis, E. Gizeli, *Biosens. Bioelectr.* 24 (2008) 836.
64. A. Tsortos, G. Papadakis, K. Mitsakakis, K.A. Melzak, E. Gizeli, *Biophys. J.* 94 (2008) 2706.
65. G. Papadakis, A. Tsortos, F. Bender, E.E. Ferapontova, E. Gizeli, *Anal. Chem.* 84 (2012) 1854.
66. K.M.M. Aung, X. Ho, X. Su, *Sens. Actuators B* 131 (2008) 371.
67. Y. Zhang, F. Lin, Y. Zhang, H. Li, Y. Zeng, H. Tang, S. Yao, *Anal. Sci.* 27 (2011) 1229.
68. Z. Altintas, I.E. Tothill, *Sens. Actuators B* 169 (2012) 188.
69. D. Wang, W. Tang, X. Wu, X. Wang, G. Chen, Q. Chen, N. Li, F. Liu, *Anal. Chem.* 84 (2012) 7008.
70. F. Yang, E. Romanova, E. Kubareva, N. Dolinnaya, V. Gajdos, O. Burenina, E. Fedotova et al., *Analyst* 134 (2009) 41.
71. V. Viglasky, T. Hianik, *Gen. Physiol. Biophys.* 32 (2013) 149.
72. D.L. Robertson, G.F. Joyce, *Nature* 344 (1990) 467.
73. C. Tuerk, L. Gold, *Science* 249 (1990) 505.
74. A.D. Ellington, J.W. Szostak, *Nature* 346 (1990) 818.
75. S.D. Jayasena, *Clin. Chem.* 45 (1999) 1628.
76. Y. Nonaka, W. Yoshida, K. Abe, S. Ferri, H. Schulze, T.T. Bachmann, K. Ikebukuro, *Anal. Chem.* 84 (2012) 8259.
77. A.D. Keefe, S.T. Cload, *Curr. Opin. Chem. Biol.* 12 (2008) 448.
78. B. Strehlitz, R. Stoltenburg. In: *Aptamers in Bioanalysis* (Ed. M. Mascini), pp. 31–59, J. Wiley & Sons, Inc., Hoboken, NJ, (2009).
79. L.C. Bock, L.C. Griffin, J.A. Latham, E.H. Vermaas, J.J. Toole, *Nature* 355 (1992) 564.
80. Y. Wang, D.J. Patel, *Structure* 1 (1993) 263.
81. M. Adrian, B. Heddi, A.T. Phan, *Methods Enzymol.* 338 (2012) 341.

82. D. Ogasawara, N.S. Hachiya, K. Kaneko, K. Sode, K. Ikebukuro, *Biosens. Bioelectr.* 24 (2009) 1372.
83. B. Gatto, M. Palumbo, C. Sissi, *Curr. Med. Chem.* 16 (2009) 1248.
84. D.-M. Xu, M. Wu, Y. Zou, Q. Zhang, C.-C. Wu, Y. Zhou, X.-J. Liu, *Chin. J. Anal. Chem.* 39 (2011) 925.
85. T. Hianik, J. Wang, *Electroanalysis* 21 (2009) 1223.
86. F.W. Scheller, U. Wollenberg, A. Warsinke, F. Lisdat, *Curr. Opin. Biotechnol.* 12 (2001) 15.
87. W.O. Tucker, K.T. Shum, J.A. Tanner, *Curr. Pharm. Des.* 18 (2012) 2014.
88. S. Fukusho, H. Furusawa, Y. Okahata, *Nucleic Acids Symp. Ser.* 44 (2000) 187.
89. N. Tassew, M. Thompson, *Anal. Chem.* 74 (2002) 5313.
90. M. Liss, B. Petersen, H. Wolf, E. Prohaska, *Anal. Chem.* 74 (2002) 4488.
91. M. Minunni, S. Tombelli, A. Gullotto, E. Luzi, M. Mascini, *Biosens. Bioelectr.* 20 (2004) 1149.
92. S. Tombelli, M. Minunni, E. Luzi, M. Mascini, *Bioelectrochemistry* 67 (2005) 135.
93. T. Hianik, V. Ostatna, Z. Zajacova, E. Stoikova, G. Evtugyn, *Bioorg. Med. Chem. Lett.* 15 (2005) 291.
94. T. Hianik, V. Ostatná, M. Sonlajtnerova, I. Grman, *Bioelectrochemistry* 70 (2007) 127.
95. A. Porfirieva, G. Evtugyn, T. Hianik, *Electroanalysis* 19 (2007) 1915.
96. M. Šnejdárková, L. Svobodová, V. Polohová, T. Hianik, *Anal. Bioanal. Chem.* 390 (2008) 1087.
97. T. Hianik, I. Grman, I. Karpisova, *Chem. Commun.* 41 (2009) 6303.
98. I. Neundlinger, A. Poturnayova, I. Karpisova, C. Rankl, P. Hinterdorfer, M. Snejdarkova, T. Hianik, A. Ebner, *Biophys. J.* 101 (2011) 1781.
99. K. Min, M. Cho, S.-Y. Han, Y.-B. Shim, J. Ku, C. Ban, *Biosens. Bioelectr.* 23 (2008) 1819.
100. W. Shu, S. Laurenson, T.P.J. Knowles, P. Ferrigno, A.A. Seshia, *Biosens. Bioelectr.* 24 (2008) 233.
101. C. Yao, Y. Qi, Y. Zhao, Y. Xiang, Q. Chen, W. Fu, *Biosens. Bioelectr.* 24 (2009) 2499.
102. Y. Pan, M. Guo, Z. Nie, Y. Huang, C. Pan, K. Zeng, Y. Zhang, S. Yao, *Biosens. Bioelectr.* 25 (2010) 1609.
103. R. Wang, Y. Li, *Biosens. Bioelectr.* 42 (2013) 148.
104. V. Pavlov, Y. Xiao, B. Shlyahovsky, I. Willner, *J. Am. Chem. Soc.* 126 (2004) 11768.
105. Q. Chen, W. Tang, D. Wang, X. Wu, N. Li, F. Liu, *Biosens. Bioelectr.* 26 (2010) 575.
106. P. Chandra, H.-B. Noh, M.-S. Won, Y.-B. Shim, *Biosens. Bioelectr.* 26 (2011) 4442.
107. V.C. Özalp, *Analyst* 136 (2011) 5046.
108. Z.-M. Dong, G.-C. Zhao, *Sensors (Switzerland)* 12 (2012) 7080.
109. I. Lambert, L. Mosiello, T. Hianik, *Chem. Sens.* 1 (2011) 1.
110. X. Xu, J. Zhou, X. Liu, Z. Nie, M. Qing, M. Guo, S. Yao, *Anal. Chem.* 84 (2012) 4746.
111. G. Evtugyn, A. Porfireva, M. Ryabova, T. Hianik, *Electroanalysis* 20 (2008) 2310.
112. G.A. Evtugyn, A.V. Porfireva, T. Hianik, M.S. Cheburova, H.C. Budnikov, *Electroanalysis* 20 (2008) 1300.
113. T. Hianik, A. Porfireva, I. Grman, G. Evtugyn, *Prot. Peptide Lett.* 16 (2009) 363.

4

Surface-Enhanced Raman Scattering–Based Biosensors

Ilya N. Kurochkin, Andrey K. Sarychev, Ilya A. Ruzhikov, Igor A. Budashov, Sergey S. Maklakov, Sergey O. Boyarintsev, and Andrey N. Lagarkov

CONTENTS

Giant Raman scattering in the metal nanostructures that is known as surface-enhanced Raman scattering (SERS) was discovered in the last quarter of the twentieth century. It might be the most important discovery in optics at the end of the century. SERS effect is observed when the investigated substance is deposited onto metal nanoparticles (NPs), clusters of the particles, rough metal surfaces (e.g., chemical etching electrodes), and semicontinuous metal films. SERS has witnessed many successes over the past three decades, owing particularly to its simplicity of use as well as its highly multiplexing capability in biomedical investigations [1]. SERS has emerged as a promising nanosensing technique for *in vivo* diagnostics. This chapter provides an overview of SERS and its applicability in the field of biosensing.

This chapter is divided up as follows: First, the fundamental mechanisms of Raman scattering and SERS are briefly discussed. Second, an assortment of preclinical and clinical SERS applications is discussed. Third, different technologies for SERS-active plasmonic substrate preparation are described. Finally, we consider a new phenomenon associated with the giant enhancement of Raman scattering in the profiled dielectric film.

4.1 Raman Scattering and SERS

Raman scattering was independently discovered in 1928 by Mandelstam and Landsberg in Moscow [2] and Raman in Calcutta [3].

They have noticed that in scattered light, along with spectrum lines of incident light, weak shifted lines appear. The number and frequency of these lines are characteristics of a scattering substance. It turned out that frequency difference of shifted lines and incident light corresponds to frequency of vibration of atoms in molecules. These frequencies appear in the infrared spectra of a substance.

Consider the plane monochromatic wave of frequency ω_0 lights the molecule. Most part of the light will pass without distortion, but a small part will scatter. If one will explore spectra of scattered light, then waves of shifted frequency will be discovered. Usually they appear in pairs with frequencies $\omega_0 \pm \omega_M$. Scattering without change of frequency is called Rayleigh, and with shift of frequency, Raman scattering. In molecular systems, the frequency ω_M primarily falls in range of energy, which corresponds to the transfer between rotational, vibration, and electronic levels. Negative frequency shift is usually referred as Stokes and positive shift is called anti-Stokes shift. Often the polarization of scattered light differs from the polarization of incident light; also intensity and polarization of scattered light can depend on the direction of observation.

It is worth noting that during scattering, the energy of the incident photon doesn't have to correspond to energy transfer between two energy levels.

Nevertheless, as the energy of the photon gets closer to the electronic transfer energy, the cross section of scattering increases. This phenomenon is called resonance scattering.

Let's consider the simplest model of Raman scattering. We suppose that before the interaction of light and molecule, only photons with energy $\hbar\omega_0$ existed and molecule energy was E_i. During light–molecule interaction, the photon with energy $\hbar\omega_0$ disappears, the photon with energy $\hbar\omega_s$ is created, and the molecule moves to the state with energy E_f. The energy conservation law for this process is written down as follows: $E_f - E_i = \hbar(\omega_0 - \omega_s)$ or $\omega_s = \omega_0 - \omega_{fi}$. But energy $\hbar\omega_0$ doesn't correspond to any transition of molecule. It means that the photon is not actually absorbed. It activates the molecule and makes possible the transition to the virtual state.

In Figure 4.1, energy level diagrams for Stokes and anti-Stokes shift are presented. Arrows directed up stand for annihilation of photon, while arrows directed down present a birth of photon. Solid horizontal lines stand for discrete quantum state of molecule and dashed lines for virtual state. The energy level, in which the absorption of photons conservation law is not satisfied, is called virtual and a final state is also called a virtual. This energy level diagram is presented here only for general understanding of photon and molecule energy. It doesn't give any information about the mechanism of photon–molecule interaction.

Briefly and qualitatively, photon–molecule interaction can be described as follows. Incident light $\mathbf{E} = \mathbf{E}_0\cos(\omega t)$ will move the molecule's electron density with frequency ω and induce moment $\mathbf{P} = \alpha_0\cos(\omega t)$ (α_0—constant), which defines scattered light of frequency ω. But in reality the mechanism will be different. Molecule consists of atoms and atoms always move; consequently, induced moment is described by the following:

$$\mathbf{P}(q, \omega) = \alpha(q)\mathbf{E}_0\cos(\omega t),$$

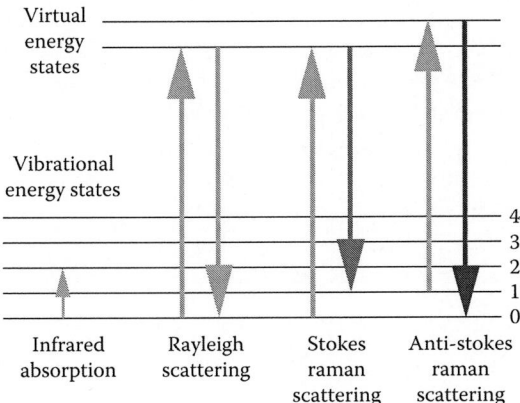

FIGURE 4.1
Energy level diagram for Stokes and anti-Stokes Raman scattering.

where $\alpha(q)$ is a polarizability function of a coordinate. While atom deflection from the equilibrium state is small, one can expand polarizability:

$$\alpha(q) = \alpha(0) + \frac{\partial \alpha}{\partial q} q \cos(\Omega t),$$

where Ω is a frequency of molecule movement. Consequently, induced moment is the following:

$$\mathbf{P}(q\omega) = \alpha(0)\mathbf{E}_0 \cos(\omega t) + \frac{1}{2}\frac{\partial \alpha}{\partial q} q\mathbf{E}_0 \{\cos[(\omega+\Omega)t+\varphi] + \cos[(\omega-\Omega)t-\varphi]\}$$

The intensity and spectra of scattered light will be defined by induced moment. Thus, in the spectrum of scattered light along with frequency ω, Stokes $\omega - \Omega$ and anti-Stokes $\omega + \Omega$ frequencies will be present. One can look through the quantitative description in [4].

Raman scattering is incoherent, and the intensity of N non-interacting molecules will be equal to the sum of intensities of each molecule and doesn't depend on their distribution.

As light shift corresponds to the characteristic frequency of a molecule, exploring Raman spectra is a powerful molecule identification technique. The biggest disadvantage and limitation of this method is a very small cross section, which makes it impossible to study the small concentration of substance. Thus, mechanisms that increase intensity of the Raman spectra are of great interest. The most widely-used and well-known method is SERS.

4.1.1 SERS

SERS provides up to 10^{12}. SERS increases the cross section of molecules adsorbed in a nanostructured system. The enhancement is induced by the fluctuation of local electromagnetic (EM) field. This allows one to explore the structure [5]. SERS was discovered by Fleishman et al. in 1974 [6]. A lot of approximation and theories have been developed [7,8]. The main contribution of SERS is electromagnetic enhancement. Another mechanism of enhancement is chemical, which is caused by (1) enhancement related to the chemical interaction of a molecule and NP in ground state and (2) resonance enhancement. Theoretically, electromagnetic enhancement reaches and overcomes 8–10 orders of magnitude, while chemical enhancement is limited by 1–2 orders.

Electromagnetic enhancement comes from surface plasmon excitation, which leads to local electromagnetic field enhancement. It is easy to show this with an isolated metal sphere—electric field on the surface in quasistatic approximation can be estimated as follows: $E \sim 3E_0/(\varepsilon_m(\omega)+2)$, where $\varepsilon_m(\omega)$ is the dielectric permeability and E_0 is the outside temperature. Electric field will be resonantly enhanced if $\varepsilon_m(\omega) \approx -2$. If a system consisting of several metal particles is induced by light, then local field is enhanced. Molecules under investigation experience this enhanced local field and light is scattered inelastically. Scattered light has different frequency, so it does not interact with incident light. The field created by scattered light again induces currents in the system. The light scattered by induced currents can be registered. Spectra of scattered light contain information about the structure of the molecule under investigation. One can compare obtained spectra with etalons and identify the molecule. In rough approximation, overall enhancement is proportional to the fourth power of local field E_{loc} [6,7]:

$$G_R = \frac{\left\langle |\varepsilon(r)|^2 |E_{loc}(r)|^4 \right\rangle}{\left(\varepsilon_d^2 |E_0|^4 \right)}, \tag{4.1}$$

where triangular brackets stand for averaging over the system volume. Despite the fact that this expression is accurate only if Stokes shift is negligible, this expression is used for quantitative estimations often. Thus, for SERS one should create the system of interacting metal nanostructures, where the local field will be enhanced. An ideal SERS substrate should have the following properties:

1. It has to show a large enhancement factor (EF) up to a single-molecule sensitivity.
2. The signal enhancement has to be homogenous across the surface and reproducible from substrate to substrate.
3. It should be long, stable, and insensitive to external influences.
4. The biocompatibility of the substrate plays an important role in biomedical applications.
5. A SERS substrate should be easy and cheap in production.

By the moment of publication, no ideal substrate suitable for all applications has been created. Thus, for each particular application, one has to sacrifice something. Historically the first system to report SERS was roughened silver electrode. Such systems usually have fractal morphology [9–11]. Using quasistatic approach, Stockman has developed the theory of SERS in fractals [12–14]. Enhancement in this type of structures reaches 10^{12}, which is enough for single-molecule detection [5,15,16]. The biggest disadvantage is the lack of reproducibility.

Randomly distributed NPs [17,18] are one of the most popular SERS substrates. Despite a simple and cost-efficient production, these substrates have strong enhancement in a wide range of frequencies. Enhancement over 10^{10} has been reported for golden and silver NPs of size 50–100 nm, which is enough for single-molecule sensitivity [19,20]. It is hard to improve reproducibility, because the field is enhanced only in small randomly distributed regions called hot spots. Semicontinuous metal films have very similar properties. One of the ways to improve this kind of substrates is self-assembling. For example, DNA-based assembly leads to clusters of NPs with nanosized gaps [21].

Periodic structures [22,23] have completely different properties. They have less enhancement, compared to random structures, and plasmon resonance is observed only in a narrow frequency region. This allows designing sensor for a particular type of substance. Also it is worth noting that the local field is enhanced in known positions, which leads to high reproducibility of results. The moderate enhancement of 10^5–10^6 for periodic structures has been reported. But even the latest techniques fail to make controllable small gaps in several nanometers between metal structures.

Interest in the roughened NPs has increased in the last couple of years. SERS enhancement of 10^6 and relatively high reproducibility have been reported for golden mezoparticles [24]. Moreover, such systems are quite cheap and simple in terms of preparation.

4.2 SERS Biosensor Systems for Biomedical Application (Biosensing)

SERS detection does not require the adsorbate to be in direct contact with the metal surface. At the same time the electromagnetic part of enhancement sharply decreases as the distance between the analyte molecule and the surface increases. As shown in the article by Camden and coauthors a 2.8 nm gap between the analyte molecule and the silver surface decreased the SERS signal intensity tenfold [25]. Practically important to understand that to measure intense SERS signals, analytes must be within a few nanometers from the surface of metal substrate. That is why in SERS studies seek to use molecules that are capable of forming covalent bonds to the metal surface [26]. To take advantage of the sensitivity and selectivity of SERS for the detection of molecules unlikely to dwell within 2–4 nm of the nanoscale roughness features, an alkanethiol self-assembled monolayer (SAM) has been employed on the surface of the substrate to facilitate the approach and concentration of the analyte within the zone of enhancement [27,28]. Though this approach has not been broadly applied to biomolecule analytes, there are a few examples where a partition layer has been employed to model lipid bilayers and as a tether in aptamer-based sensors [29,30].

Analytes can also be captured within a few nanometers of the substrate surface using tip-enhanced Raman spectroscopy (TERS) approach [26]. TERS is an emerging new branch of SERS, wherein the substrate is fabricated on or mounted to the probe of a scanning probe microscope (SPM). In this configuration, the apex of a laser-irradiated metal tip captures incident light and generates a strong localized surface plasmon resonance (LSPR) for SERS [31]. Typically, the tip is a metal-coated SPM tip or a thin silver or gold wire. Single-crystalline silver wires have also been used [26]. The radius of the tip is much smaller than the diffraction-limited spot size, allowing ultrahigh-resolution SERS imaging. This combination of SERS chemical fingerprinting and SPM imaging is advantageous because it can provide high sensitivity as well as contrast results with a lateral resolution down to a few tens of nanometers with a few seconds of acquisition time. Furthermore, TERS is a versatile technique that can extend the utility of SERS because the analyte does not need to be in direct contact with the SERS-active substrate. The presence of the tip is directly measurable, since when inserted, Bulgarevich and Futamata showed that the Raman signal from an isolated diamond particle can be enhanced 1000-fold [32]. Theoretical modeling predicts that bringing a tip close to a metallic substrate can increase the local electric field intensity 5000-fold [33]. When a gold tip is brought to within one nanometer of a gold surface coated with nonresonant benzenethiol molecules, EFs of 106–108 have been reported [26]. In addition, a series of recent studies have established TERS as a promising detection technique with single-molecule sensitivity [26,34].

SERS detection of biomolecules has been accomplished in both intrinsic and extrinsic formats. In intrinsic SERS biosensing, the molecular signature for the analyte of interest, such as small molecule, DNA strand, or protein, is acquired directly. In extrinsic SERS, the analyte or interaction of interest is associated with a molecule with an intense and distinguishable Raman signature, traditionally a commercially available fluorescent dye, and it is the SERS spectrum of the tag that is used for sensing or quantification. In either format, SERS has unique advantages for biosensing.

In the past decade [26,35,36], a variety of SERS and TERS substrates have been employed for the detection of small molecules, DNA/aptamers, proteins/peptides, enzymes, and microorganisms. While small-molecule detection with SERS has been predominantly accomplished with intrinsic SERS, DNA/aptamers, proteins/peptides, enzymes, and microorganisms have been detected directly with intrinsic and extrinsic formats.

4.2.1 Small-Molecule SERS Biosensing

There are a number of examples in the recent literature where SERS has been applied for the detection of biologically relevant small molecules. These small molecules range from antioxidants, like glutathione, to glucose, to small-molecule markers for biowarfare agents such as anthrax. In another example of endogenous small-molecule detection, Gogotsi et al. used a glass substrate coated with gold NPs for SERS detection of nicotinic acid adenine dinucleotide phosphate (NAADP), a calcium secondary messenger that plays a crucial role for intracellular Ca^{2+} release.

To demonstrate multianalyte sensing capabilities with partition layer-modified substrates, some researchers have used AgFONs with DT/MH mixed

SAMs to detect both glucose and lactate, which is an important indicator of potential mortality in intensive care patients [26]. Like earlier studies on glucose alone, they showed both partitioning and departitioning of lactate from the SAM layer. Upon partitioning into the SAM, lactate bands at 1463, 1422, 1272, 1134, 1094, 1051, 936, and 868 cm^{-1} were readily apparent. Using PLS-LOO methods, they demonstrated quantitative analysis of lactate in the concentration range of 10–240 mg dL^{-1}. Sequential injection of lactate and glucose into a flow cell was used to demonstrate the capability of the sensor to discriminate between the two analytes.

Another category of biomolecules directly detectable with SERS is lipids. Lipids display a large structural diversity, from amphiphilic structures with glycerol backbones like phospholipids to multiple ring structures like steroids. Groups have employed Raman microscopy for investigations of lipid bilayers, but SERS for studying lipids is a newly emerging area. Most applications of SERS to lipid sensing have focused on investigations of phospholipid bilayers and their properties, as well as their interactions with various molecules of interest. The limited use of extrinsic SERS for small molecules is due in large part to the ease of obtaining the structural vibrations and rotations directly from small molecules as opposed to more complex systems like DNA and proteins, which are discussed later in this chapter.

However, some extrinsic small-molecule SERS detection schemes do exist. Also focused on anthrax biosensing, Chung et al. used Au NPs modified with a 16 amino acid peptide or antibody coupled to the Raman reporter 5, 5'-dithiobis(succinimidyl-2-nitrobenzoate) (DSNB) to detect a different anthrax biomarker, protective antigen (PA). In this example, either the peptide or antibody that binds to PA and DSMB is detected by its characteristic band at 1336 cm^{-1}. The peptide-binding partner was shown to be as efficient as an antibody-binding partner, and LODs for their biomarker were in the low fM range. As discussed in the intrinsic SERS section, SERS detection can be employed to detect endogenous species like glutathione as well as previously mentioned examples of exogenous species. Ozaki et al. developed an extrinsic SERS method for detection of glutathione. In this "reversed reporting agent" scheme, SERS signal of reporting agent-capped Ag colloids is reduced upon the addition of glutathione, which induces aggregation of the Ag colloids. Various reporting agents were tested, but 5 µM R6G was the only viable candidate for glutathione sensing. The LOD using this scheme was 1 µM, which is much higher than for their heat-induced SERS sensing of glutathione.

4.2.2 DNA/Aptamer SERS Biosensing

4.2.2.1 Direct Detection Format

Direct SERS and TERS measurements can also be employed beyond small-molecule detection for DNA and aptamer biosensing. Recent advances in achieving low limits of detection and good reproducibility make SERS a useful tool for either single- or double-stranded thiolated DNA oligomer detection. For example, Halas et al. used Au nanoshells bound to glass substrates to obtain SERS spectra of DNA, which were dominated by adenine vibrational bands at 729 cm^{-1}. Moreover, they observed changes in the dsDNA spectrum upon interaction with cisplatin and transplatin, *cis* and *trans* forms of common chemotherapy agent, revealing an opportunity for SERS to contribute to pharmaceutical research. Critical to the achievement of highly reproducible DNA spectra in this work was a thermal pretreatment that promoted extended linear conformation of ssDNA and dsDNA on the Au nanoshell substrate.

TERS has also been proposed for direct DNA and RNA sequencing, identification of biomacromolecules, and characterization of single viruses at the molecular level. Bailo and Deckert used TERS for direct, label-free detection of a single-stranded RNA cytosine homopolymer. All spectra obtained along the length of a 20 nm long single strand of RNA showed spectral features of cytosine with little variation in band intensities or positions. Moreover, as deduced from this experiment, it is possible to find the sequence of RNA with controlled movement of the TERS probe from base to base.

4.2.2.2 SERS Tag Format

While there are relatively few examples of intrinsic SERS for DNA and aptamer detection, DNA and aptamer monitoring via SERS tag format has been employed extensively. A common detection scheme for DNA binding events is to functionalize a Au or Ag NP with a reporter molecule (usually a fluorescent dye) and a single-stranded piece of DNA. Upon hybridization with a complementary strand of DNA, typically bound to another Au or Ag surface, the SERS or surface-enhanced resonant Raman scattering (SERRS) signal of the reporter molecule is observed. The area where SERS tag format is the most promising is in the gene detection schemes described by Haynes and Graham [26,36].

4.2.3 Protein/Enzyme/Peptide/Antibody SERS Biosensing

4.2.3.1 Direct Detection Format

While direct SERS measurements are relatively straightforward for small molecules, lipids, and DNA where each target has a small number of vibrational modes, it is also feasible to perform SERS biosensing on more complicated, larger molecules such as peptides, proteins, enzymes, and antibodies. Unlike the more tractable small molecules, band assignments for spectra from these larger species are often based on general motifs seen through the molecules rather than localized vibrational modes. One commonly used example is the characteristic aromatic amino acid bands present at ~950 cm^{-1}, for the CCOO– stretch, and ~1400 cm^{-1}, for the

COO– symmetric stretch. Another commonly used spectral feature is the broad amide (CO–NH) I and III bands, at 1600–1700 and 1200–1350 cm^{-1}, respectively, used in intrinsic peptide and protein analysis to describe both primary and secondary structure characteristics. SERS sensing of the simplest amino acid structures, peptides, has been pursued both from a fundamental perspective, with the goal of assigning Raman bands to particular amino acids, and from an applied perspective, to sense particular biomarkers.

4.2.3.2 SERS Tag Format

To address the challenges of direct SERS detection of proteins, Raman reporter molecules have been used when specific protein spectra are not necessary, and the goal is to quantitatively detect either individual proteins or total protein content. For SERS tag format detection of proteins, reporter molecules must (1) maintain stability during tagging and measurement and (2) generate a robust and consistent Raman spectrum under the conditions of the SERS measurement [34]. For protein detection, commonly used SERS reporter molecules include DSNB with intensity monitored at 1336 cm^{-1}, 4-mercaptobenzoic acid (MBA) with intensity monitored at 1585 cm^{-1},117 4-nitrobenzenethiol (4-NBT) with intensity monitored at 1336 cm^{-1}, 2-methoxybenzenethiol (2-MeOBT) with intensity monitored at 1037 cm^{-1}, 3-methoxybenzenethiol (3-MeOBT) with intensity monitored at 992 cm^{-1}, and 2-napthalenethiol (NT) with intensity monitored at 1384 cm^{-1}. Also used as Raman labels for SERS are 4,4′-bipyridine (BiPy), with strong bands present at 1609, 1227, and 1291 cm^{-1}; thiophenol (TP), with strong bands at 994 and 1570 cm^{-1}; and p-isothiocyanate (FITC) and malachite green isothiocyanate (MGITC), which are widely used fluorescence tags and have also been used as common Raman reporter molecules for SERRS and SERS detection of proteins. FITC intensity is monitored at 1630 cm^{-1} following excitation at 514.5 nm, and MGITC intensity is monitored at 1615 cm^{-1} when excitation at 647.4 nm is used. Many of the aforementioned extrinsic Raman labels have been used for immunoassays, which rely on the specificity of the interaction between antibodies and their corresponding antigens [44]. Multiple formats have been employed for SERS-based immunoassays, including the use of silver island films, colloids in solution, immobilized colloids, and modified colloids as probe molecules [44]. Multiplexing is also possible with SERS immunoassays.

The use of Raman labels for protein detection is not limited to immunoassays. A protein concentration assay developed by Han et al. [26] uses the SERS signal of Coomassie Brilliant Blue dye adsorbed nonspecifically to silver colloids to monitor the total protein concentration. The SERS signal of Coomassie Brilliant Blue displayed a linear and inverse relationship to protein concentration over a bovine serum albumin concentration range of 10^{-5}–10^{-9} g mL^{-1}. Clearly, SERS tag format has enabled protein sensing *ex vitro*; in fact, a similar approach can be adopted to perform SERS detection inside cells as well.

Direct SERS detection of proteins is limited by the spectral similarities of many proteins. Indirect SERS detection of protein concentration through immunoassays is promising, but the use of antibodies and detection limits on the order of ELISA assays with fluorescence detection prevent SERS-based immunoassays from replacing the more widespread enzyme-based immunoassays for clinical and diagnostic use.

4.2.4 Cellular and *in vivo* Sensing

4.2.4.1 Direct Detection Format

In a relatively small number of cases, SERS biosensing of proteins, DNA, and small molecules has been extended from the aforementioned examples toward cellular and *in vivo* systems. However, there are only a few where direct, intrinsic SERS detection is performed in cells: this is due to the complex biological environment that can mask signals from analytes of interest or cause fluorescence. The Van Duyne group first demonstrated *in vivo* application of direct detection SERS by measuring the glucose concentration from the interstitial fluid of a rat. A SAM-functionalized AgFON substrate was surgically implanted under the skin of a rat such that it was in contact with the interstitial fluid and optically addressable through a glass window placed along the midline of the rat's back. The SERS spectra from glucose were acquired through the window using a 785 nm Ti–sapphire laser with a power of 50 mW for 2 min. The glucose concentrations obtained from the implanted AgFON sensor matched the data from a commercial glucometer. With further refinements and miniaturization in the system, SERS biosensors could help the treatment and care of diabetics or other conditions that would benefit from time-lapse monitoring. SERS spectra have also been successfully measured from endosomes in live cells (rat renal proximal tubule cells and mouse macrophages) using gold NPs. Both cell lines were exposed for 30 min to gold NPs, which were then localized in the endosomes of the cells. In this case, cells were raster-scanned with a low-power (2 mW) near infrared (NIR) excitation laser for 1 s to prevent interference from normal Raman scattering and possible change in the live cells due to laser illumination. The strongest SERS signal was acquired 120 min after exposure to NPs, when gold aggregates formed in the lysosomes. Gold aggregates in both cell lines produced tentatively assigned bands in the spectra that indicate the presence of proteins, lipids, carbohydrates, and nucleotides in chemical nano-environments of lysosomes. In addition, the spectral signature of adenosine phosphate was detected in macrophage endosomes, which indicates differences in the endosomes of the two cell lines. To date, direct SERS of the cellular milieu provides general information about the types of biomolecules present but provides limited information about specific biomarkers of interest.

4.2.4.2 SERS Tag Format

The use of SERS tag format in cellular and *in vivo* sensing is extensive due to the ability of Raman labels to overcome background signals from a complex biological matrix and associate with specific biomolecules. Many groups [26] have used SERS tag format for mapping the local pH in cells based on the important role pH plays in regulating cellular function. Kneipp et al. measured pH-dependent SERS spectra of *p*mercaptobenzoic acid (pMBA), a pH-sensitive Raman tag, on aggregated gold NPs. The accessible pH ranges of SERS and surface-enhanced hyper-Raman scattering (SEHRS) were 5.5–8 and 2–8, respectively, but the high laser power (~10 mW) and long collection time (10 s) required for SERS were not suitable for scanning live cells. Moskovits et al. also showed pH dependence of SERS spectra using pMBA-functionalized NP clusters and mapped local pH in live HeLa cells. In this work, Ag clusters were linked using bifunctional hexamethylenediamine (HMD) molecules and encapsulated by polyvinylpyrrolidone (PVP) to prevent aggregation during MBA tag infusion. The SERS-active clusters were also coated with dye-labeled streptavidin with BSA to track the distribution of Ag NP clusters in cells and correlate the fluorescence and SERS pH maps. The MBA molecules infused through the polymer coat into junctions between NPs and facilitated measurement of pH values inside live HeLa cells using a relatively low laser power (1.1 mW) and short integration time (250 ms). These pH mapping probes could be used to understand intracellular interactions, including endocytotic pathways. Vo-Dinh et al. used a Ag-coated submicron-sized fiber-optic probe functionalized by pMBA for similar purposes. The probe was physically inserted into a live cell using a micromanipulator. The intracellular pH value was determined by comparing SERS intensity of pMBA bands to a calibration curve obtained from standard pH solutions ranging from pH 6.0 to 7.5.

Clearly, it is feasible to perform SERS tag format detection in cells using single extrinsic Raman labels, but there is also significant interest in multiplex biomarker detection. In fact, Gambhir et al. have demonstrated multiplexed extrinsic SERS. They used 10 SERS–NP complexes for multiplex imaging, and each particle consisted of a unique Raman reporter molecule layer adsorbed onto a 60 nm diameter Au core coated with silica, making the total diameter about 120 nm. Each molecular layer shows distinguishable SERS spectra, and five of them were used for *in vivo* SERS imaging in a nude mouse. A mixture of four kinds of unique NPs of varying concentrations was injected either intravenously or subcutaneously into the mouse where a linear correlation of SERS signal with the concentrations was measured.

SERS has also been utilized for non-direct detection of cancer markers in a live cell. Oh et al. used Au/Ag core-shell NPs where R6G Raman tags were adsorbed on the gold surface with a BSA layer. Subsequently, the NPs were conjugated with IgG antibodies that selectively bind to phospholipase Cγ1 biomarker proteins (PCγ1) on HEK293 (human embryonic kidney) cells. While no normal Raman signal from R6G was measured from a control cell, the cancer cell showed significant Raman signal, based on the 1650-cm^{-1} R6G peak that correlated well with quantum dot-labeled fluorescence images. Moving from the imaging of cancer markers in live cells to tumor targeting, Nie et al. used PEGylated gold NPs as extrinsic SERS labels with tumor-targeting ligand to identify tumors both *in vitro* and *in vivo*. A core size of 60–80 nm diameter was chosen to position the LSPR peaks within the "water window" (630–785 nm) where the optical absorption of water is minimal. SERS tags were over 200 times brighter than NIR-emitting quantum dots. They could measure SERS spectra at targeted tumor sites up to 2 cm below the skin. Moreover, Stone et al. proposed deep Raman spectroscopy with citrate-reduced Ag-conjugated NPs for detecting low concentration of molecules through tissues up to 25 mm thick. It demonstrated a potential use of SERS NPs for detecting target molecules deeply buried within tissues. Because of the high sensitivity, label-free detection, and multiplex capability, applications of SERS targeting in live cells, tissues, and *in vivo* detection, will continue to expand. At the same time, many hurdles need to be overcome, which include the stability as well as toxicity of SERS tags and the background noise resulting from the structural similarity of many proteins and molecules.

4.3 SERS-Active Substrates: Types and Preparation Techniques

For the most part, Au or Ag has been used for making SERS substrates, although other metals such as Al, In, Cu, and Ga can also support plasmon resonances in the UV–vis–NIR range. Among these, Al and Cu are of potential interest for SERS and plasmonic biosensing because of their abundance and low cost. Furthermore, aluminum has plasmon resonances in the UV regime and thus can extend the spectral range of Au- or Ag-based devices [2].

The SERS-active structures can be divided as follows.

4.3.1 Single-Layer Structures

- Island-type films with random distribution of metallic nanoislands deposited onto a dielectric substrate
- Island-type films with regular arrangement of nanoislands deposited onto a dielectric substrate
- Thin metal films with nonregular subwavelength openings

- Thin metal films with regular subwavelength openings
- Metal nanostructured surfaces with arranged projections of cone or pyramid shape, which provide local concentration of electromagnetic field
- NPs of noble metals, deposited onto dielectric substrates

4.3.2 Multilayer Structures

- Nonresonant multilayer structures: dielectric substrate–metal mirror–dielectric layer-arranged metal NPs.
- Interference multilayer structures: dielectric substrate–glassy metal–dielectric layer-arranged metal NPs.

The structure of an outer metal layer can appear regular as well as random, like in a case of a single layer.

4.3.3 Techniques of Preparation of SERS-Active Structures

Common techniques to produce SERS-active substrates are deposition of thin films in vacuum and electrochemical treatment of noble electrodes. Nowadays, a large amount of methods of surface patterning is used [37]. In most cases, sample preparation includes combination of photo- and nanolithography techniques proceeded with chemical etching [38]. Nanolithography includes electron- and ion-beam techniques, which result in porous matrices with given pore shape and size. These matrices can be applied as substrates for immobilization of noble metal NPs and as a basis for a template synthesis [39]. A sequence of technological steps may contain a large number of techniques [40,41]. Also, thin uniform metal film can be subjected to a microwave heating that affects the topography of the surface [42].

Further research showed that a substrate also plays an important role in the enhancement of Raman scattering. Particularly, the use of a glassy metal surface increases probability of interaction of a target molecule with excitation radiation. In such a case, a transparent dielectric coating is needed to prevent appearance of metal conductivity. Optical interference provides an auxiliary contribution to SERS magnitude (IE-SERS) [43]. Further optimization of the structures revealed periodical dependence of SERS magnitude on thin film thickness [44]. It has been shown that a dielectric type and method of dielectric layer obtainment influence the magnitude of the signal.

It is well known that under vacuum deposition of metal films, early stage (below 10 nm) produces coatings of island type. The increase in film thickness results in hopping conductivity proceeded with tunnel conductivity; uniform film possesses metal conductivity. In several cases, polycrystalline structure with subwavelength grain boundaries is formed. Uniform metal films usually do not demonstrate enhancement of Raman scattering.

In brief, a technological route to SERS-active substrate consists of the following:

- Pretreatment of surface of a substrate by chemical processing or ion etching
- Deposition of glassy metal (Al) layer (100–200 nm) by means of magnetron sputtering or electron-beam evaporation
- Dielectric (MO_x) layer deposition with required thickness, roughness, and surface morphology
- Development of a SERS-active metallic layer according to a previous classification

4.3.4 Features of Technological Processes for SERS-Active Substrates Manufacturing

Substrate requirements: The substrate requirements are resistance to chemically active media, specified roughness, and high adhesion to upper dielectric or metal layer. The use of monocrystals allows developing of anisotropy within upper layers. The most commonly used materials are Si, polycrystalline SiO_2 and Al_2O_3, SiO_2 monocrystals, amorphous SiO_2, and polished Al. The substrate applied should not show Raman scattering within all the wavelength range under study.

Deposition of thin films in vacuum: We may divide techniques of vacuum deposition as follows: electron-beam or resistive evaporation and ion-beam or magnetron sputtering for metals and electron-beam and ion-beam evaporation and high-frequency magnetron sputtering for nonmetals. The ability to control film thickness precisely is important. For that purpose, an optical interferometer is widely used in vacuum systems. In such case, a reference sample "witness" is applied to estimate the absolute value of thickness. Several vacuum devices allow to apply more than one sample "witness," which can be replaced in optical scheme by means of a selector. Another thickness control technique is a quartz crystal microbalance that is based on a measure of frequency of a quartz crystal resonator. The frequency varies with the thickness of a deposited layer. Also, an electric conductance can be applied to investigate metal film thickness.

A large amount of reviews and research articles are published, which are devoted to nano-patterned SERS-active substrates (see, for instance, [45,46]). Questions are well defined of an influence of size, shape, and interposition of metal NPs on a magnitude of Raman scattering bands of vast selection of molecules. Criteria for selection of SERS-active spectral bands are revealed. Mechanisms of interaction of a substance analyzed with NPs are described [47]. Although there is no clear understanding of how does nature of chemical bond between Raman molecule and NPs influences on the intensity of Raman maximum.

As a result of an interaction between plasmonic NPs and adsorbed molecule, a charge transfer complex is formed, which is characterized by a dipole moment. Some part of the energy that is received by the molecule was realized in the form of radiation with a Raman frequency. To perform a comparative SERS enhancement study factor, one needs to apply molecules, in which oscillations do not change under absorption conditions to keep the form of the Raman spectra. This assumption is based on the results of some researches where different molecules with the same oscillations under absorption show totally diverse Raman intensities for identical oscillations.

4.4 Giant Enhancement of Raman Scattering in the Profiled Dielectric Film

As was mentioned earlier, SERS effect is observed when the investigated substance is deposited onto metal NPs, clusters of particles, rough metal surfaces (e.g., chemical etching electrodes), and semicontinuous metal films. The semicontinuous metal films are traditional objects for the investigation in the modern plasmonics. The films are being investigated for many years since they have appeal for numerous important applications such as optical filters and solar energy concentrators. However, the most intriguing and important discovery has come up recently first in microwave and then in optical experiments. It was discovered that the local electric field is a subject of the giant fluctuation.

In this section, we deal with completely dielectric films. The optical properties of thin random dielectric films have been studied both experimentally and theoretically. Thin dielectric films are usually produced by thermal evaporation or spattering onto a substrate. We considered here thin dielectric films deposited onto the metal substrate. As the film grows on the substrate, the filling factor increases and irregularly shaped clusters are formed on the substrate. The clusters grow together resulting in occurrence of 2D fractal structures. At higher surface coverage, the film is mostly dielectric with an irregular-shaped surface as it is schematically shown in Figure 4.2.

We propose that the local electric fields fluctuate in the profiled dielectric films on the giant scales as it is demonstrated in Figure 4.3. The local intensities $I(x,y) = |E(x,y)|^2$ may exceed by orders of magnitude the intensity $I_0 = |E_0|^2$ of the incident field, where x and y are the coordinates in the film interface.

Giant electromagnetic field fluctuations result in the enhancement of the Raman scattering in the molecules of the analyte deposited on the film. The nonlinear optical phenomena can be also much enhanced. Therefore, the rough dielectric films could become an area of active studies because of many fundamental problems involved and high potential for various applications.

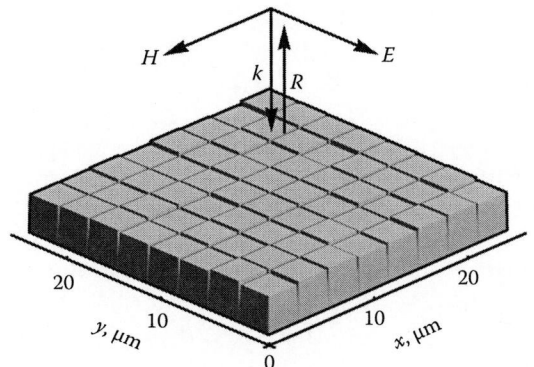

FIGURE 4.2
Dielectric film deposited onto the metal substrate; electromagnetic wave is incident normal to the film along the "z" axis; metal substrate placed at "z = 0" plane.

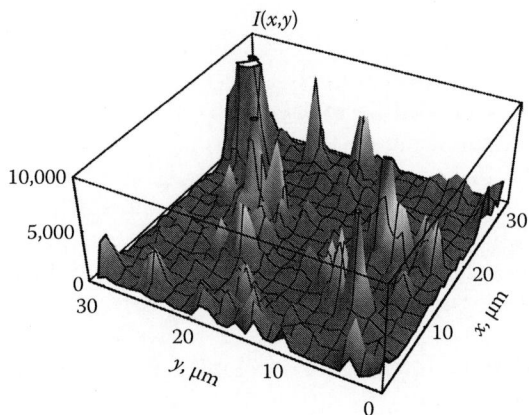

FIGURE 4.3
Local intensity $I(x,y) = |E(x,y)/E_0|^2$ of electric field; wavelengths $\lambda = 800$ nm; computer simulations.

We show that the displacement current and electric fields are concentrated in a few "hot" spots. This allows altering the local permittivity significantly due to the large local fields while the external field can be relatively small. In general, a system with large internal field could be very sensitive to the external electric field when their optical properties are determined by the sparse network of field maxima. This implies that composite materials have much larger nonlinear susceptibilities with respect to the susceptibility of the constituent materials. In optics, even weak nonlinearity can lead to qualitatively new effects. For example, the generation of higher harmonics can be strongly enhanced in the random dielectric, and bistable behavior of the effective permittivity occurs when the permittivity switches between two stable values. We also anticipate that the random dielectric films, prepared from an active (laser) medium, could revive random lasing with much decreased threshold due to the pumping of the enhanced local electric field. Yet, we believe that the most important usage of the films is the Raman sensors.

It is well known that the electric field can be strongly enhanced in a dielectric resonator in the optical, infrared, and microwave spectral regions. For example, in the Fabry–Perot cavity, the internal field could arbitrarily be large in an antinode. However, at the interface of the cavity, where the external electromagnetic wave is incident, the electric field equals to $E = E_0 \exp(i\,k\,z \sin \theta)(1 + r)$, where r is the complex reflection coefficient and $k = \omega/c = 2\pi/\lambda$ is the wave vector and is the angle of incidence. Therefore, the intensity of the surface field cannot be larger than $4I_0$. We can compare this result with the electric field distribution shown in Figure 4.3.

Since the local thickness of the random dielectric film could be about the wavelength, we attempt to expand the theoretical recipes beyond the quasistatic approximation. Our approach is based on the full set of Maxwell's equations. It does not invoke the quasistatic approximation, as the fields are not assumed to be curl-free in the film. Although the theory was developed mainly for the dielectric thin films, it is in fact quite general and can be applied to any kind of inhomogeneous film under appropriate conditions. The suggested theory is similar to "generalized Ohm's law" used to describe semicontinuous metal films.

We will restrict ourselves to the case where all the external fields are parallel to the film plane. This means that the incident and reflected waves are travelling in the direction perpendicular to the film plane. The consideration is focused on the electric and magnetic field magnitudes at the film interface. We relate the interface fields to the displacement currents flowing inside the film. Since we neglect the magnetic field, which is perpendicular to the film, the electric field $\mathbf{E}(x,y)$ at the interface of the film is curl-free rot $\mathbf{E}(x,y) = 0$ and can be expressed as the gradients of a potential field $\varphi(x,y)$. The electric displacement $\mathbf{D}(x,y)$, averaged over the film thickness, obeys the usual 2D continuity equation. Therefore, the equations for the field $\nabla \times \mathbf{E} = 0$ and the equations for the displacement $\nabla \cdot \mathbf{D} = 0$ are the same as in the quasistatic case. The only difference is that the field and the current are now related by a new material equation. To determine these new material equations, we first find the electric and magnetic field distributions inside the dielectric film. The boundary conditions completely determine the solution of Maxwell's equations for the fields in the film when the frequency is fixed. Therefore, the internal fields that change in a "z" direction (perpendicular to the film surface) depend linearly on the electric field $\mathbf{E}(x,y)$ at the interface of the film. The displacement is a linear function of the local internal fields given by the usual local constitutive equation $\mathbf{D} = \varepsilon(x,y)\mathbf{E}$, where $\varepsilon(x,y)$ is the local permittivity. Therefore, the displacement currents flowing inside the film also depend linearly on the electric field $\mathbf{E}(x,y)$ at the external interface of the film. We employ this dependence to estimate the electric field fluctuations at the interface and the enhancement of the Raman scattering in the rough dielectric films.

4.4.1 Basic Equations and Computer Simulations

We consider, for simplicity, the case when all the external fields are parallel to the "*x,y*" plane (see Figure 4.2), meaning that the plane wave propagates normally to the film's surface along the "*z*" axis. The film has thickness *d*. The external electromagnetic wave is incident along the "*z*" axis on the film interface at $z = -d$. A typical roughness δ is assumed to be smaller than the free-space wavelength $\delta < \lambda$.

First, we consider the electric and magnetic fields at the film's interface $z = -d$:

$$\mathbf{E}_1(\mathbf{r}) = \mathbf{E}(\mathbf{r},-d), \quad \mathbf{H}_1(\mathbf{r}) = \mathbf{H}(\mathbf{r},-d). \tag{4.2}$$

The electric field at the surface of the metal substrate $\mathbf{E}(\mathbf{r}, 0) = 0$. All the fields are monochromatic with the usual $\exp(-i\omega t)$ time dependence; the vector $\mathbf{r} = \{x,y\}$ is a 2D vector in the "*x,y*" plane.

We introduce the average electric displacement:

$$\mathbf{D}(\mathbf{r}) = k \int_{-d}^{0} \mathbf{D}(\mathbf{r}, z)dz = k \int_{-d}^{0} \varepsilon(\mathbf{r}, z)\mathbf{E}(\mathbf{r}, z)dz \tag{4.3}$$

where $k = 2\pi/\lambda = \omega/c$ is the free-space wave vector. In the case of laterally inhomogeneous film, the vector $\mathbf{D}$ is dependent on $\mathbf{r}$. The vector $\mathbf{D}(\mathbf{r})$, introduced by Equation 4.3, has the same dimension as the electric displacement. The permittivity $\varepsilon(\mathbf{r}, z)$ takes different values for different regions of the film:

$$\varepsilon(r, z) = \{\varepsilon_d, \text{ if } -d + \delta_l(\mathbf{r}) < z < 0; 1, \text{ if } -d > z > -d + \delta_l(\mathbf{r})\}, \tag{4.4}$$

where the local roughness $\delta_l(\mathbf{r})$ takes values $0 < \delta_l(\mathbf{r}) < \delta$. Therefore, the simplest model of the roughness is adopted: The film thickness changes from *d* to $d - \delta_l(\mathbf{r})$. The average film thickness equals to $d - \delta/2$ and the mean square of the roughness equals to $\delta^2/12$. We assume hereafter for simplicity that the permittivity is a scalar and the permeability is a unit everywhere. We use Gaussian units.

In our approximation, the local electromagnetic field is a superposition of two plane waves propagating in the $+z$ and $-z$ directions. This superposition is different for different regions of the film. We neglect scattered and evanescent waves, which propagate in the "*x, y*" plane since their amplitudes are proportional to $\delta/\lambda < 1$. This allows us to use a two-plane-wave approximation when both electric and magnetic fields in the film have components in the "*x, y*" plane only. After that, Maxwell's equations **rot** $\mathbf{E}(\mathbf{r},z) = ik\mathbf{H}(\mathbf{r},z)$ and **rot** $\mathbf{H}(\mathbf{r},z) = -ik\mathbf{D}(\mathbf{r},z)$ take the following forms:

$$\frac{d[\mathbf{n} \times \mathbf{E}(\mathbf{r}, z)]}{dz} = ik\mathbf{H}(\mathbf{r}, z) \tag{4.5a}$$

$$\frac{d[\mathbf{n} \times \mathbf{H}(\mathbf{r}, z)]}{dz} = -ik\varepsilon(\mathbf{r}, z)\mathbf{E}(\mathbf{r}, z) \tag{4.5b}$$

Unit vector $\mathbf{n} = \{0,0,1\}$ is normal to the film's surface. The permittivity $\varepsilon(\mathbf{r}, z)$ is defined previously by Equation 4.4. The solution of Equations 4.5a and b is the superposition of two plane waves, which is completely determined by the field $\mathbf{E}_1(\mathbf{r}) = \mathbf{E}(\mathbf{r}, -d)$, defined at the front interface of the film, since $\mathbf{E}(\mathbf{r}, 0) = 0$. The electric field $\mathbf{E}_1(\mathbf{r})$ has components only in the "x, y" plane and it is curl-free (otherwise, magnetic field $\mathbf{H}$ would have nonzero "z" components according to Maxwell's equations).

The average electric displacement $\mathbf{D}(\mathbf{r})$ is obtained from Equation 4.3, namely, $\mathbf{D}(\mathbf{r}) = w(\mathbf{r})\,\mathbf{E}_1(\mathbf{r})$, where the effective permittivity

$$w(\mathbf{r}) = \varepsilon_d \frac{\tan[k\delta_l(\mathbf{r})]\tan[knd_l(\mathbf{r})] + [\sec[k\delta_l(\mathbf{r})]\sec[knd_l(\mathbf{r})] - 1]/n}{\tan[knd_l(\mathbf{r})] + n\tan[k\delta_l(\mathbf{r})]} \tag{4.6}$$

where $d_l(\mathbf{r}) = d - \delta_l(\mathbf{r})$ and the refraction coefficient $n = \sqrt{\varepsilon_d}$. The continuity equations for the electric displacement $\mathbf{D}(\mathbf{r})$ is obtained by the integration of the local (3D) equations $\mathbf{div}\,\mathbf{D}(\mathbf{r}, z) = 0$ between the planes $z = -d$ и $z = 0$, which results in $\mathbf{div}\,\mathbf{D}(\mathbf{r}) = 0$, where we take into account the boundary conditions $D_z = 0$ when $z = -d$ and $z = 0$.

4.4.1.1 Equations

$$\mathbf{rot}\,\mathbf{E}(\mathbf{r}) = 0, \tag{4.7a}$$

$$\mathbf{div}\,w(\mathbf{r})\mathbf{E}(\mathbf{r}) = 0 \tag{4.7b}$$

represent the system that connects the electric field, determined in a 2D reference plane, to the average electric displacement (electric current) that flows in the film. Thus, the entire physics of a 3D inhomogeneous film, which is described by the full set of Maxwell's equations, has been reduced to a simple set of quasistatic Equations 4.7. These equations can be solved by standard methods of computer simulation of the metal–dielectric composites [19,20,27]. For example, we can determine currents and local fields by solving Kirchhoff's equations defined on the square lattice. The permittivities of the bonds take values given by Equation 4.6, where the roughness δ_l is distributed with equal probability in the interval $(0, \delta)$. The result of the simulation is presented in Figure 4.3. We indeed observe the giant field fluctuations.

4.4.2 Optical Properties of Rough Dielectric Film

The experimental measurement of the reflectance of the dielectric film prepared from cerium dioxide (CeO_2) is shown in Figure 4.4. (The data were obtained by Gorelik, Physical Institute of Russian Academy of Science.)

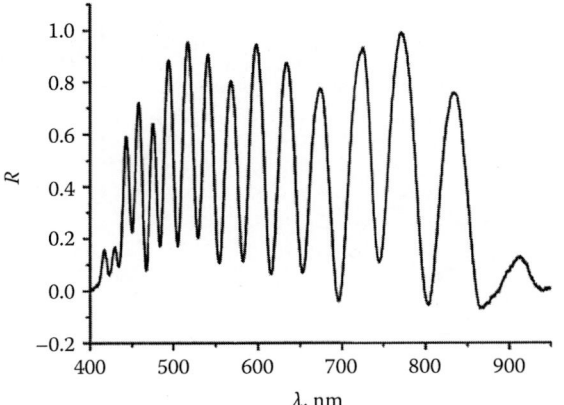

FIGURE 4.4
Reflectance from rough film of cerium deposited on the aluminum substrate (experimental results).

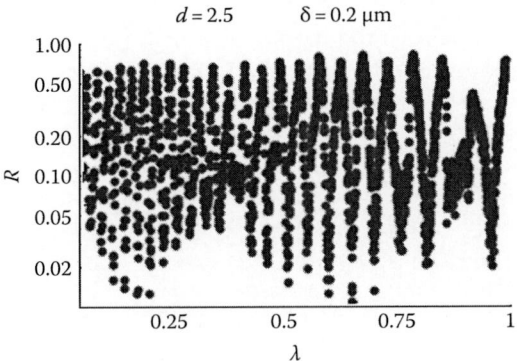

FIGURE 4.5
Reflectance from rough film of cerium oxide (ceria) deposited on metal substrate (computer simulations).

We can observe a very deep minima in the reflectance that cannot be observed in homogenous dielectric films with small loss. Yet, the experimental results are in very good qualitative agreement with our simulations presented in Figure 4.5.

4.4.3 Enhancement of the Raman Scattering

It was shown before that the Raman signal from the analyte, deposited on the film with rough surface, is much enhanced. The enhancement G is approximately proportional to the second moment of the local field intensity $G = \int I^2(\mathbf{r})d\mathbf{r}/(SI_0^2)$, where the integration is over the film surface S and I_0 is the intensity of the incident light. The results of the computer simulations are shown in Figure 4.6.

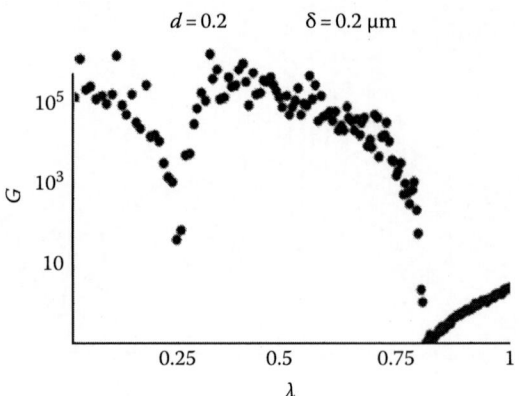

FIGURE 4.6
Enhancement of the Raman scattering in the rough cerium dioxide film (computer simulation) as function of the wavelength λ, μm.

The Raman scattering in the ceria films can be enhanced in hundred and thousand times. We believe that properly prepared profiled dielectric films can be used for design-supersensitive chemical and Raman biosensors.

We have shown that the excitation of new quasiparticles that we name "dirons" in modulated dielectric films could result in the enhanced resonant absorption so that an optically thin film becomes absorptive. The excitation of the localized dirons gives giant local field fluctuations and enhanced Raman scattering. Therefore, we suggest for the further consideration new optical systems—random dielectric films.

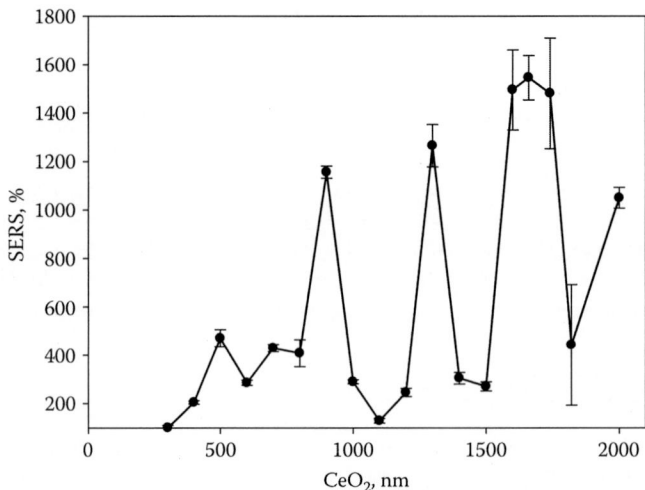

FIGURE 4.7
Enhancement of the Raman scattering signal from gold NPs labeled with DTNB on cerium dioxide films. 100%—Raman signal received at the adsorption of gold NPs labeled with DTNB to the aluminum sublayer.

The confirmation of the previous arguments and computer simulation results were obtained in the following experiment. We measured the intensity of the Raman signal from DTNB that was covalently bound with gold NPs (diameter 56 nm) and adsorbed on cerium dioxide layers with different thickness. Cerium dioxide was deposited on aluminum sublayer. Vacuum deposition of cerium dioxide films was conducted by Afanasiev K.N. (the Institute for Theoretical and Applied Electromagnetics of the Russian Academy of Science). For 100%, we have taken the Raman signal received during the adsorption of gold NPs labeled with DTNB directly to the aluminum sublayer. The results are shown in Figure 4.7. The data show that the increase of the signal films of cerium dioxide is 15 times. The Raman signal as a function of the film thickness of cerium dioxide film oscillates in accordance with the data on the reflection of light (Figures 4.4 and 4.5).

Thus, giant enhancement of Raman scattering in the profiled dielectric film makes it suitable for SERS tag format and direct detection of biomolecules based on Raman spectroscopy.

References

1. K. Kho, C. Fu, J. Dinish, and M. Olivo, Clinical SERS: Are we there yet? *J. Biophotonic*, 4, 667–684 (2011).
2. G. Landsberg and L. Mandelstam, Eine neue Erscheinung bei der Lichtzertreuung, *Naturwissenschaften* Â., 16, S. 557 (1928).
3. C.V. Raman, A new radiation, *Indian J. Phys.*, 2, 387 (1928).
4. D. Stauffer and A. Aharony, *Introduction to Percolation Theory*, London, U.K.: Taylor & Francis Group, 2 edn. (1994).
5. K. Kneipp, Y. Wang, H. Kneipp et al., Single molecule detection using surface-enhanced Raman scattering (SERS), *Phys. Rev. Lett.*, 78, 1667 (1997).
6. M. Fleischmann, P.J. Hendra, and A.J. McQuillan, Raman spectra of pyridine adsorbed at a silver electrode, *Chem. Phys. Lett.*, 26, 163 (1974).
7. M. Moskovits, Surface-enhanced spectroscopy, *Rev. Mod. Phys.* 57, 783 (1985).
8. A.K. Sarychev and V.M. Shalaev, *Electrodynamics of Metamaterials*, Singapore: World Scientific Publishing (2007).
9. K. Kneipp, H. Kneipp, V.B. Kartha, R. Manoharan, G. Deinum, I. Itzkan, R.R. Dasari, and M.S. Feld, Detection and identification of a single DNA base molecule using surface-enhanced Raman scattering (SERS), *Phys. Rev.*, E 57, R6281–R6284 (1998).
10. V. Shalaev, *Optical Properties of Nanostructured Random Media*, Berlin, Germany: Springer-Verlag (2002).
11. Z. Wang, S. Pan, T.D. Krauss, H. Du, and L.J. Rothberg, The structural basis for giant enhancement enabling single-molecule Raman scattering, *PNAS*, 100, 8636–8643 (2003).
12. K. Li, M.I. Stockman, and D.J. Bergman, Self-Similar chain of metal nanospheres as an efficient nanolens, *Phys. Rev. Lett.*, 91, 227402 (2003).

13. J. Dai, F. Cajko, I. Tsukerman, and M.I. Stockman, Electrodynamic effects in plasmonic nanolenses, *Phys. Rev. B*, 77, 115419 (2008).
14. V.M. Shalaev and M.I. Stockman, Optical properties of fractal clusters (susceptibility, surface enhanced Raman scattering by impurities), *Sov. Phys. JETP.*, 65, 287–294 (1987).
15. S.M. Nie and S.R. Emery, Probing single molecules and single nanoparticles by surface-enhanced Raman scattering, *Science*, 275, 1102–1106 (1997).
16. M. Moskovits, L.L. Tay, J. Yang, and T. Haslett, SERS and the single molecule, *Top. Appl. Phys.*, 82, 215–226 (2002).
17. H.X. Xu, J. Aizpurua, M. Kall, and P. Apell, Electromagnetic contributions to single-molecule sensitivity in surface-enhanced Raman scattering, *Phys. Rev. E*, 62, 4318 (2000).
18. F. Le, D.W. Brandl, Y.A. Urzhumov, H. Wang, J. Kundu, N.J. Halas, J. Aizpurua, and P. Nordlander, Metallic nanoparticle arrays: A common substrate for both surface-enhanced Raman scattering and surface-enhanced infrared absorption, *ACS Nano.*, 2, 707 (2008).
19. H.X. Xu, E.J. Bjerneld, M. Kall, and L. Borjesson, Spectroscopy of single hemoglobin molecules by surface enhanced Raman scattering, *Phys. Rev. Lett.*, 83, 4357 (1999).
20. A.M. Michaels, J. Jiang, and L. Brus, Ag nanocrystal junctions as the site for surface-enhanced Raman scattering of single rhodamine 6G molecules, *J. Phys. Chem. B*, 104(11), 965 (2000).
21. N.F. Hulst, J.-A. Veerman, and M.F. Garcia-Parajo, Analysis of individual (macro)molecules and proteins using near field optics, *J. Chem. Phys.*, 112, 7799–7810 (2000).
22. L. Gunnarsson, E.J. Bjerneld, H. Xu, S. Petronis, B. Kasemo, and M. Kall, Interparticle coupling effects in nanofabricated substrates for surface-enhanced Raman scattering, *Appl. Phys. Lett.*, 78, 802 (2001).
23. S.J. Lee, Z.Q. Guan, H.X. Xu, and M. Moskovits, Surface-enhanced Raman spectroscopy and nanogeometry: The plasmonic origin of SERS, *J. Phys. Chem. C*, 111, 17985 (2007).
24. H. Wang and N.J. Halas, Mesoscopic Au "meatball" particles, *Adv. Mater.*, 20, 820 (2008).
25. J.P. Camden, J.A. Dieringer, J. Zhao, and R.P. Van Duyne, Controlled plasmonic nanostructures for surface-enhanced spectroscopy and sensing, *Acc. Chem. Res.*, 41, 1653–1661 (2008).
26. K. Bantz, A. Meyer, N. Wittenberg, H. Im, O. Kurtulus, S. Lee, N. Lindquist, S.-H. Oh, and C. Haynes, Recent progress in SERS biosensing, *Phys. Chem. Chem. Phys.*, 13(24), 11551–11567 (2011).
27. R. Aroca, *Surface-Enhanced Vibrational Spectroscopy*, West Sussex, England: John Wiley & Sons Ltd (2006).
28. J.P. Camden, J.A. Dieringer, Y. Wang, D.J. Masiello, L. Marks, G.C. Schatz, and R.P. Van Duyne, Probing the structure of single-molecule surface-enhanced Raman scattering hot spots, *J. Am. Chem. Soc.*, 130, 12616–12617 (2008).
29. B.J. Kennedy, S. Spaeth, M. Dickey, and K.T. Carron, Determination of the distance dependence and experimental effects for modified SERS substrates based on self-assembled monolayers formed using alkanethiols, *J. Phys. Chem. B*, 103, 3640–3646 (1999).

30. K.C. Bantz and C.L. Haynes, Surface-enhanced Raman scattering detection and discrimination of polychlorinated biphenyls, *Vib. Spectrosc.*, 50, 29–35 (2009).
31. S. Lal, S. Link, and N.J. Halas, Nano- optics from sensing to waveguiding, *Nat. Photon.*, 1, 641–648 (2007).
32. D.S. Bulgarevich and M. Futamata, Apertureless tip-enhanced Raman microscopy with confocal epi-illumination optics, *Appl. Spectrosc.*, 58, 757–761 (2004).
33. R.M. Roth, N.C. Panoiu, M.M. Adams, R.M. Osgood, C.C. Neacsu, and M.B. Raschke, Resonant-plasmon field enhancement from asymmetrically illuminated conical metallic-probe tips, *Opt. Express*, 14, 2921–2931 (2006).
34. A. Hartschun, N. Anderson, and L. Novotny, Near-field Raman spectroscopy using a sharp metal tip, *J. Micros.*, 210, 234–240 (2003).
35. K. Kneipp, M. Moskovits, and H. Kneipp (Eds.), Surface-enhanced Raman scattering—Physics and applications, *Topics Appl. Phys.*, 103, 335–350 (2006).
36. I. Larmour and D. Graham, Surface enhanced optical spectroscopies for bioanalysis, *Analyst*, 136, 3831–3853 (2011).
37. R.J.C. Brown, M.J.T. Milton, Nanostructures and nanostructured substrates for surface—Enhanced Raman scattering (SERS), *J. Raman Spectrosc.*, 39(10), 1313–1326 (2008).
38. M. Fan, G.F.S. Andrade, and A.G. Brolo, A review on the fabrication of substrates for surface enhanced Raman spectroscopy and their applications in analytical chemistry, *Anal. Chim. Acta*, 693(1–2), 7–25 (2011).
39. Y. Wang, T. Gao, K. Wang, X. Wu, X. Shi, Y. Liu, S. Lou, and S. Zhou, Template-assisted synthesis of uniform nanosheet-assembled silver hollow microcubes, *Nanoscale*, 4(22), 7121–7126 (2012).
40. D. Cialla, A. März, R. Böhme, F. Theil, K. Weber, M. Schmitt, and J. Popp, Surface-enhanced Raman spectroscopy (SERS): Progress and trends, *Anal. Bioanal. Chem.*, 403(1), 27–54 (2012).
41. X.-M. Lin, Y. Cui, Y.-H. Xu, B. Ren, and Z.-Q. Tian, Surface-enhanced Raman spectroscopy: Substrate-related issues, *Anal. Bioanal. Chem.*, 394(7), 1729–1745 (2009).
42. C. Yuen, W. Zheng, and Z. Huang, Improving surface-enhanced Raman scattering effect using gold-coated hierarchical polystyrene bead substrates modified with postgrowth microwave treatment, *J. Biomed. Optics*, 13(6), 064040 (2008).
43. G.A.N. Connell, R.J. Nemanich, and C.C. Tsai, Interference enhanced Raman scattering from very thin absorbing films, *Appl. Phys. Lett.*, 36(1), 31–33 (1980).
44. L. Gao, W. Ren, B. Liu, R. Saito, Z.-S. Wu, S. Li, C. Jiang, F. Li, and H.-M. Cheng. Surface and interference coenhanced Raman scattering of graphene, *ACS Nano*, 3(4), 933–939 (2009).
45. X. Ye and L. Qi, Two-dimensionally patterned nanostructures based on monolayer colloidal crystals: Controllable fabrication, assembly, and applications, *Nano Today*, 6(6), 608–631 (2011).
46. M. Culha, B. Cullum, N. Lavrik, and C.K. Klutse, Surface-enhanced Raman scattering as an emerging characterization and detection technique. *J. Nanotechnol.* 2012(97), 15 (2012).
47. J.R. Lombardi and R.L. Birke. A unified approach to surface-enhanced Raman spectroscopy, *J. Phys. Chem. C.* 112(14), 5605–5617 (2008).

5

Enzymatic Biosensors

Anastasios Economou

CONTENTS

5.1 Introduction

5.1.1 Biosensor Concept

The term "biosensor" was introduced by Cammann in 1977 [1]. A stricter definition of the term was set later by the International Union of Pure and Applied Chemistry (IUPAC) [2,3]: a *biosensor* is defined as "a self-contained integrated device that is capable of providing specific quantitative or semi-quantitative analytical information using a biological recognition element (biochemical receptor) which is in direct spatial contact with a transduction element."

Therefore, a biosensor consists of two main components: a *bioreceptor* and a *transducer*. The bioreceptor is composed of a *biomolecule recognition element* (an enzyme, an antibody, a protein receptor, DNA, or whole cells) that recognizes the target analyte, whereas the transducer converts the recognition event into a measurable signal (electrical, optical, thermal, etc). A typical biosensor construct also normally incorporates signal-processing elements (amplification, filtering, data processing, and storage) and a display of the final result. A schematic diagram of a typical biosensor is illustrated in Figure 5.1. Several

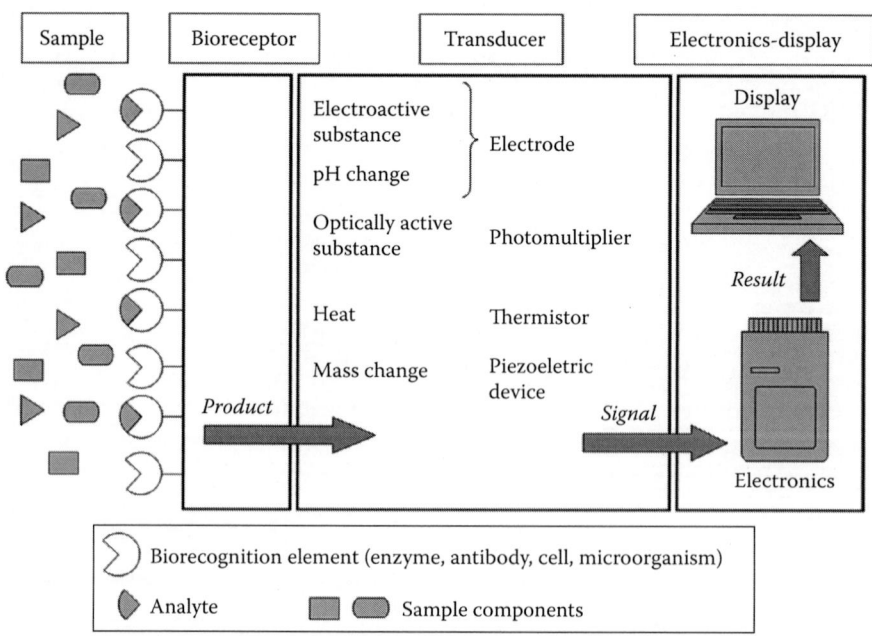

FIGURE 5.1
Schematic diagram describing the parts and operation of a typical biosensing device.

web pages [4,5] and books [6–12] exist dealing with the principles of operation, practical developments, and applications of biosensors.

The uniqueness of a biosensor is that the two components are integrated into one single sensor. One major requirement for a biosensor is that the bioreceptor molecule is immobilized in the vicinity of the transducer, which, in turn, should be capable of converting the biorecognition event into a measurable signal; typically, this is done by monitoring a change that occurs during the interaction between the bioreceptor and the target.

The growth of biosensor market is healthy in every part of the world and involves diverse application areas such as point-of-care testing, home diagnostics, environmental monitoring, research, process industry, and security/biodefense. The global market for biosensors in 2012 is estimated to reach U.S. $ 8.5 billion and is projected to reach U.S. $ 16.8 billion by 2018 [13].

5.1.2 Principles of Enzymatic Biosensors

The specificity of every biosensor is due to the specificity of the bioreceptor molecule used. *Enzymes* have been the most widely used bioreceptor molecules in biosensor applications. Enzymes are globular proteins containing from 62 to over 2500 amino acids and act as catalysts for biochemical reactions. Their activity is determined by their chemical composition and 3D structure. An enzyme is capable of recognizing only a specific target substrate (a substrate is a molecule upon which an enzyme acts) because the enzyme and the substrate have complementary geometric shapes that fit exactly into one another, hence, the specificity of enzymatic biosensors [14]. During interaction, the substrate binds with the enzyme active site, giving rise to an enzyme–substrate complex. The substrate is converted into one or more products that are released from the active site of the enzyme (Figure 5.2), and the enzyme is ready to accept new substrate species. A host of useful information on enzyme technology and applications can be found in a comprehensive tutorial [5].

According to the International Union of Biochemistry and Molecular Biology (IUBMB), enzymes can be classified in one of six classes in terms of the mechanism used in the catalysis [14]: oxidoreductases (dehydrogenases, oxidases, peroxidases, oxygenases) that catalyze oxidation/reduction reactions of the substrate to generate a product via transfer of hydrogen or electrons; transferases that transfer a functional group to the substrate to form a product; hydrolases that catalyze the formation of a product by cleavage of the substrate and the addition of water; lyases that catalyze the formation of a product by cleaving chemical bonds using hydrolysis and oxidation; isomerases that catalyze geometrical or structural changes within a substrate molecule; and ligases that can catalyze the joining of two substrate molecules by forming a new chemical bond by hydrolysis.

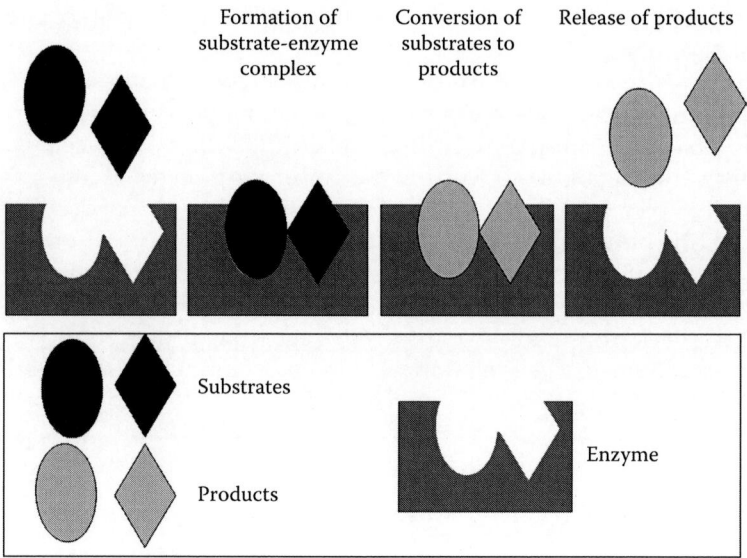

FIGURE 5.2
The mechanism of enzymatic catalysis.

The main reasons accounting for the fact that enzymes are the most commonly used biological components in biosensors are [15,16]:

a. The large number of reactions they catalyze
b. The possibility of detecting a broad range of analytes (substrates, products, inhibitors, and mediators of the catalytic activity)
c. The different transduction principles that can be used to detect the analyte of interest
d. The scope for continuous real-time monitoring of a specific compound since enzymes are not consumed in the reaction
e. The high selectivity as a result of the specific biointeraction
f. The high catalytic activity they exhibit that leads to high sensitivity due to signal amplification
g. The speed of action in comparison with other biological receptors
h. Their commercial availability at high purity

The disadvantages of using enzymes in biosensing devices are the following:

a. Enzymes are rather large molecules, a property that limits the overall catalytic (volume) density.
b. Often the active site of the enzyme is deeply buried within the surrounding protein shell that hinders direct electron transfer in electrochemical sensing, and artificial redox mediators are required.

c. Enzymes have a limited lifetime, and therefore, biosensors exhibit only a limited long-term stability.

d. The enzymatic activity is strongly dependent on pH, ionic strength, chemical inhibitors, and temperature.

e. The cost of some purified enzymes is high.

The underlying detection mechanism that is analytically exploited in enzymatic biosensors involves either the catalytic conversion of the target analyte to a readily detectable product or the detection of an analyte that inhibits or mediates the enzymatic activity. Many enzymatically catalyzed reactions require the presence of other molecules, for example, inorganic ions (e.g., Fe^{2+}, Mg^{2+}, Mn^{2+}, Zn^{2+}), called *cofactors*, or more complex organic molecules (e.g., nicotinamide adenine dinucleotide, NAD^+), called *coenzymes*, that can change chemically during the course of the reaction. Such changes in the optical or redox properties of the co-reactants can also be used to monitor the course of the enzymatic reaction [16].

Different enzymatic reactions commonly exploited for fabrication of biosensors are listed in Table 5.1.

When the reaction, occurring at the immobilized enzyme membrane of a biosensor, is limited by external diffusion of the substrate, the reaction process will be governed by Michaelis–Menten kinetics. In particular,

TABLE 5.1

Common Enzymes and Reactions Used in Enzymatic Biosensors

Enzyme	Enzymatic Reaction
Alcohol oxidase	Ethanol + O_2 → acetaldehyde + H_2O_2
Glucose oxidase	Glucose + O_2 → gluconic acid + H_2O_2
Lactose oxidase	Lactose + O_2 → pyruvate + H_2O_2
Alcohol dehydrogenase	Ethanol + NAD^+ → acetaldehyde + NADH + H^+ Acetaldehyde + NAD^+ → acetic acid + NADH + H^+
Lactate dehydrogenase	Lactate + NAD^+ → pyruvate + NADH
Acetylcholinesterase	Acetylcholine + H_2O → choline + acetic acid
Choline oxidase	Choline + O_2 → betaine aldehyde + H_2O_2
Horseradish peroxidase	H_2O_2 + electron donor → $2H_2O$ + oxidized donor (electron donor = phenols, aromatic amines, luminol, etc.)
Urease	Urea + H_2O → CO_2 + $2NH_3$
Tyrosinase	Phenol + $1/2O_2$ → quinone + H_2O
Lipase	Triglycerides + $3H_2O$ → glycerol + 3 fatty acids
Nitrate reductase	NO_3^- + NADH + H^+ → NO_2^- + NAD^+ + H_2O
Alkaline phosphatase	*p*-nitrophenylphosphate + H_2O → *p*-nitrophenol + $H_2PO_4^-$
Creatinine deaminase	Creatinine + H_2O → NH_4^+ + N-methylhydantoin

the rate of the enzymatic reaction, V, will obey the relationship shown in Equation 5.1 [4,5]:

$$V = \frac{V_{max}[S]}{K_M + [S]} \quad \text{and for} [S] \ll K_M \quad V = \frac{V_{max}[S]}{K_M} \tag{5.1}$$

where
  $[S]$ is the concentration of the substrate
  K_M is a constant (Michaelis constant)
  V_{max} is the maximum rate of reaction (which occurs when the enzyme is completely saturated with substrate)

It follows that the enzyme produces a proportional change in reaction rate in response to the substrate (target species) concentration at lower substrate concentrations. However, the kinetic behavior of an immobilized enzyme can differ significantly from that of the same enzyme in free solution. Control of biosensor response by the external diffusion of the analyte can be controlled by the use of permeable membranes between the enzyme and the bulk solution [5]. Details on the kinetics of surface-bound enzymes and diffusional aspects of the substrate on enzyme kinetics can be found in [4–6]. Regarding the detection analytes that act as inhibitors of enzymatic reactions, appropriate mathematical treatments have been developed to relate the % inhibition of the rate of the enzymatic reaction to the concentration of the inhibitor (target species) [5,6,17].

5.1.3 Origin and Evolution of Enzymatic Biosensors

Professor Leland C. Clark, Jr., (Figure 5.3) is generally acknowledged as the "father" of the biosensor concept. In 1956, Clark published his landmark paper on the oxygen electrode [18]. On the basis of this development, and addressing the desire to expand the range of clinical analytes that could be measured, in a historical address in 1962 at a New York Academy of Sciences symposium, he described "how to make electrochemical sensors (pH, polarographic, potentiometric, or conductometric) more intelligent" by adding "enzyme transducers as membrane-enclosed sandwiches." The first biosensor was first described by Clark and Lyons in 1962, when the term "enzyme electrode" was actually adopted [19].

The original Clark enzyme electrode relied on a thin layer of glucose oxidase entrapped over an oxygen electrode via a semipermeable dialysis membrane. Measurements were made based on the monitoring of the oxygen consumed by the enzyme-catalyzed reaction:

$$\text{glucose} + O_2 \rightarrow \text{gluconic acid} + H_2O_2$$

FIGURE 5.3
Professor Leland C. Clark, Jr., (1918–2005), the "father" of biosensors.

A negative potential was applied to the platinum cathode for reductive detection of the oxygen consumption. The decrease in the oxygen concentration was proportional to glucose concentration in the sample.

The entire field of biosensors can trace its origin to this original glucose enzyme electrode. Clark's original patent covers the use of one or more enzymes for converting various substrates to electroactive products [20]. Updike and Hicks further developed this principle by using two oxygen-working electrodes (one covered with the enzyme) and measuring the differential current in order to correct for the oxygen background variation in samples [21]. In 1973, Guilbault and Lubrano [22] described an enzyme electrode for the measurement of blood glucose based on amperometric (anodic) monitoring of the hydrogen peroxide produced in the reaction shown earlier. This technology was transferred to Yellow Springs Instrument (YSI) Company, which launched the first dedicated glucose analyzer (the Model 23 YSI analyzer) for direct measurement of glucose in whole blood samples in 1975. A key development in the YSI sensor was the employment of membrane technology in order to eliminate interference by other electroactive substances (e.g., ascorbic acid). This was accomplished with an enzyme layer sandwiched between an internal cellulose acetate membrane (allowing the passage of H_2O_2 but preventing the passage of ascorbate or other interferents) and a nucleopore polycarbonate outer membrane (which allowed glucose to

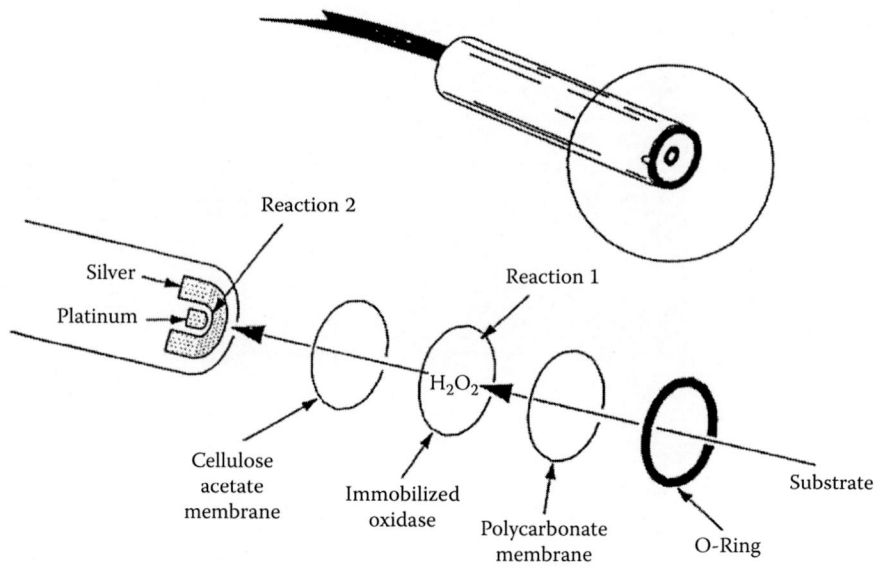

FIGURE 5.4
Sensor probe and immobilized enzyme membrane for the YSI glucose biosensor. (Source: YSI 2300 Stat Instruction Manual.) (Reaction 1: glucose + O_2 → gluconic acid + H_2O_2, Reaction 2: H_2O_2 → $2H^+ + O_2 + 2e^-$).

reach the enzyme layer) (Figure 5.4). Various other approaches developed to minimize the interferences from coexisting electroactive species (e.g., ascorbic acid, uric acid, 4-acetaminophen) in the design of amperometric glucose biosensors have been discussed in detail [23].

The first optical fiber-based enzymatic sensor (optrode) was reported in 1980 by Lubbers and Opitz who fabricated an ethanol sensor by immobilizing alcohol oxidase and an O_2-sensitive indicator at the end of an optical fiber [24].

Some landmarks in the development of biosensors are listed in Table 5.2.

5.1.4 Considerations in Biosensor Development

Once a target analyte has been identified, the major tasks in developing a biosensor involve [4]:

a. Selection of a suitable bioreceptor molecule

b. Selection of a suitable immobilization method

c. Selection of a suitable transducer

d. Designing of biosensor considering measurement range, linearity, and minimization of interference

e. Packaging of the biosensor

TABLE 5.2

Milestones and Achievements Relevant to Biosensor Research and Development

Year	Contribution
1894	Emil Fischer introduced the key-to-lock specific binding between enzyme and substrate
1913	Michaelis and Menten developed the basis for enzyme kinetics
1916	First report on the immobilization of proteins: adsorption of invertase on activated charcoal
1926	Warburg discovered cytochrome c and later he discovered the cofactors (NADH) and the mechanism of dehydrogenases
1956	Leland C. Clark, Jr., presented his first paper about the oxygen electrode (Clark electrode)
1962	First description of an amperometric enzyme electrode for glucose by Clark and Lyons
1960s	H.U. Bergmeyer (Boehringer Mannheim) promoted enzymatic analysis
1963	G.A. Rechnitz and S. Katz achieved the direct potentiometric determination of urea after urease hydrolysis
1967	G.P. Hicks und S.J. Updike introduced the first practical enzyme electrode by immobilizing the enzyme within a gel
1969	G. Guilbault introduced the potentiometric urea electrode
1970s	ELISA was introduced
1973	Ph. Racinee and W. Mindt (Hoffmann La Roche) developed a lactate electrode
1973	G.G. Guilbault and G.J. Lubrano introduced an amperometric glucose enzyme electrode based on the detection of hydrogen peroxide
1975	The first commercial biosensor (YSI analyzer) was introduced
1977	Karl Cammann introduced the term "biosensor"
1980	First optrode for ethanol
1982	First needle-type enzyme electrode for subcutaneous implantation
1982	First fiber optic-based biosensor for glucose
1984	First ferrocene-mediated amperometric glucose biosensor
1987	First electronic blood glucose measuring system (MediSense Inc.)
1988	A. Heller and Y. Degani introduced the electrical connection ("wiring") of redox centers of enzymes to electrodes through electron-conducting redox hydrogels
1988	Direct electron transfer by means of immobilized enzymes was introduced
1980s–1990s	Nanostructured carbon materials such as C60 and nanotubes were discovered
1992	*i*-STAT launched handheld multiparametric blood analyzer
1996	Glucocard launched
1997	IUPAC introduced for the first time a definition for biosensors
1998	Launch of LifeScan FastTake blood glucose biosensor
2007	An implanted glucose biosensor (Freestyle Navigator System) operated for 5 days
2010	An implanted glucose biosensor operated for 1 year in pigs
Current	Flow-through analytical microsystems, nanotechnology, biosensor chips, MEMS, enzyme mimics, paper-based devices

Task (1) requires knowledge in biochemistry and biology; (2) requires knowledge in chemistry and material science; (3) requires knowledge in physical and analytical chemistry and physics; and (4) requires knowledge in analytical chemistry. Once a biosensor has been designed, it has to be put into a package for convenience manufacturing and use. The current trend is miniaturization and mass production. Modern integrated circuit fabrication technology and micromachining technology are used increasingly in fabricating biosensors. Therefore, interdisciplinary cooperation is essential for a successful development of a biosensor.

5.1.5 Requirements for Enzymatic Sensors

To be commercially acceptable, a biosensor has to fulfill the general requirements of commercial sensors. Many aspects related to the properties of a successful biosensor are discussed in [3,4,15,25,26].
These are:

a. Relevance of output signal to measurement environment
b. Accuracy and precision
c. Sensitivity and specificity
d. Dynamic range
e. Speed of response and capability of real-time monitoring
f. Insensitivity to chemical variations (pH, ionic strength) and physical conditions (temperature, stirring) and immunity to electrical and other environmental interferences
g. Long-term stability
h. Amenability to testing and calibration
i. Reliability and self-checking capability
j. Physical robustness
k. Service requirements
l. Capital cost
m. Running costs and lifetime
n. Acceptability by the users
o. Product safety (biocompatibility if the biosensor is to be used for invasive monitoring in clinical situations, and in environmental applications the host system must not be contaminated by the sensor)
p. Small size and portability

5.2 Enzyme Immobilization Approaches

5.2.1 Enzyme Immobilization and Its Advantages

One major requirement for a biosensor is that the bioreceptor molecule is immobilized in the vicinity of the transducer, which, in turn, should be capable of converting the biorecognition event into a measurable signal. The application of immobilized enzymes has several advantages in comparison with free enzymes [27,28]:

a. Reproducibility of response: since enzymes are not normally consumed during the biorecognition reaction, they can be used repeatedly for several measurements, which ensures that the same catalytic activity is present for a series of analyses.

b. Low consumption of the enzyme: although the immobilized enzyme usually has a lower activity than the same mass of enzyme in solution, this can be compensated by using more of the immobilized preparation, knowing that it can be used repeatedly.

c. Stability: many enzymes are intrinsically stabilized by the immobilization process, being active at a wider range of temperatures and pH. Even where this does not occur, there is usually considerable apparent stabilization. It is normal to use an excess of the enzyme within the immobilized sensor system. This gives a catalytic redundancy that is sufficient to ensure an increase in the apparent stabilization of the immobilized enzyme.

d. Simplicity: the analysis procedure is simplified and the analysis time shortened because the use of reagents is not necessary.

An ideal immobilization procedure of a given enzyme should ensure that:

a. The immobilized biomolecules maintain their structure and retain their biological activity.

b. The enzyme remains tightly bound to the surface and is not desorbed during the use of the biosensor.

c. The enzyme retains a high catalytic activity over time.

The type of immobilization method affects the activity and the stability of enzymatic biosensors and drastically influences factors such as accuracy of measurements, sensor-to-sensor reproducibility, and operational

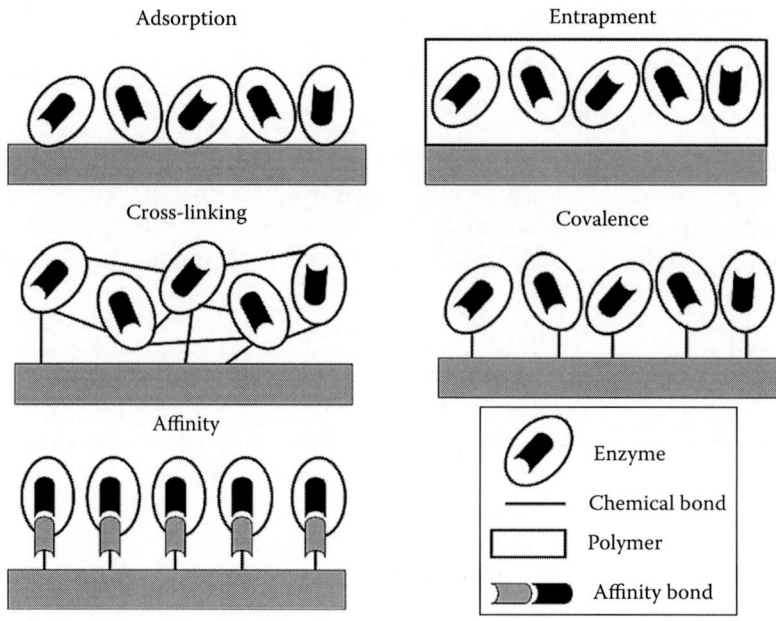

FIGURE 5.5
Schematic representation of the main immobilization methods.

lifetime. Since the analytical performance of a biosensor is strongly affected by the immobilization process, intensive efforts have been made to develop successful immobilization strategies in order to assure greater sensitivity and stability of biosensors. There are five main immobilization methods: adsorption, entrapment, cross-linking, covalent bonding, and affinity attachment. Figure 5.5 schematically illustrates the main enzyme immobilization methods, and Table 5.3 lists the main advantages and drawbacks of each of them; in some cases, enzyme immobilization protocols are also based on the combination of several immobilization approaches. The various enzyme immobilization strategies have been comprehensively covered in previous reviews [5,29–31].

5.2.2 Adsorption

Enzyme adsorption onto solid supports represents the easiest strategy of physical immobilization and has widely been used to develop enzymatic biosensors. Enzyme is dissolved in solution and the solid support is placed in contact with the enzyme solution for a fixed period of time. The adsorption mechanisms are based on weak bonds (such as van der Waals forces), electrostatic or hydrophobic interactions, or a combination of them.

TABLE 5.3

Comparison of the Characteristics of Different Immobilization Approaches

	Adsorption	Entrapment	Cross-Linking	Covalence	Affinity
Preparation	Simple	Simple	Simple	Difficult	Difficult
Cost	Low	Moderate	Low	High	High
Binding	Weak bonds	Physical inclusion	Chemical bonds	Chemical bonds	Chemical bonds
Binding force	Weak	Weak	Strong	Strong	Very strong
Leakage	High	High	Moderate	Low	Low
Enzyme loading	Low	Low	High	High	High
Enzyme loss of activity	Negligible	Negligible	Moderate	Significant	Significant
Diffusional barriers	Low	Significant	Significant	Low	Low
Applicability	Wide	Wide	Selective	Selective	Very selective

The main strategies based on adsorption mechanisms to develop enzymatic biosensors are:

a. Physical adsorption: consisting of simple deposition of an enzyme onto a surface and its attachment through weak bonds.

b. Electrostatic interactions: enzymes can be electrostatically immobilized onto charged surfaces using layer-by-layer deposition, electrochemical doping, attachment on ion-exchanger beads, or retention onto Langmuir–Blodgett films (i.e., lamellar lipid films).

5.2.3 Entrapment/Encapsulation

Enzymes can be immobilized in 3D matrices using different approaches such as:

a. Electropolymerization: electrochemical polymerization is a simple approach for the controlled immobilization of enzymes on electrode surfaces. This one-step method involves the application of an appropriate potential or current to an electrode immersed in a solution containing both the enzyme and monomer molecules. During the electrochemical formation of the polymer, enzyme molecules are physically entrapped within the growing polymer network. Most of electropolymerized films used for biomolecule immobilization are conducting polymers such as polyaniline, polypyrrole, or polythiophene. Such conducting polymers can act as electron promoters effectively "wiring" the enzymes to the electrode surface [29,32].

b. Photopolymerization: polymers bearing photo-cross-linkable groups have largely been used to entrap enzymes during their polymerization by light exposition.

c. Sol–gel processes: sol–gel process involves hydrolysis of alkoxide precursors under acidic (or alkaline) conditions followed by condensation of the hydroxylated units, which leads to the formation of a porous gel.

d. Entrapment in a polysaccharide-based gel: enzymes can also be entrapped in a polysaccharide-based gel (e.g., alginate, chitosan, or agarose). These matrices are biocompatible and nontoxic, provide natural microenvironment to the enzyme, and also give sufficient accessibility to electrons to shuttle between the enzyme and the electrode.

e. Entrapment in a carbon paste: carbon paste is a mixture of graphite powder and an organic binder (pasting liquid) and is a popular electrode material used for the incorporation of biological component because it allows an intimate contact between incorporated enzymes, mediators, and transduction sites permitting a fast electron transfer.

f. Clay-modified electrodes: these are 2D layered inorganic solids with cavities that promote interactions with enzymes and intercalation of redox mediators.

5.2.4 Cross-Linking

In this approach, enzyme molecules are chemically bonded to solid supports or cross-linked to each other in the presence of glutaraldehyde or other bifunctional agents to significantly increase the attachment. It is a useful method to stabilize adsorbed biomaterials. This method is attractive due to its simplicity and the strong chemical binding achieved between biomolecules, but its drawbacks are that the cross-linking agents can interfere with the enzyme, there is increased risk of activity losses due to the distortion of the active enzyme conformation, and the chemical alterations of the active site of the enzyme during cross-linking.

5.2.5 Covalent Bonding

Covalent coupling of enzymes to polymeric supports is an attractive chemical immobilization method used to develop enzymatic biosensors. For this purpose, biocatalysts are bound to an inorganic material (e.g., controlled pore glass) and a natural (e.g., cellulose) or synthetic polymer (e.g., nylon) through functional groups that they contain and that are not essential for their catalytic activity. The binding of the enzymes to the solid support is generally carried out by initial activation of the surface using multifunctional reagents (e.g., glutaraldehyde or carbodiimide), followed by enzyme coupling to the activated support. It requires mild conditions under which reactions are performed, such as low temperature, low ionic strength, and pH in the physiological range. The main mechanisms of immobilization by covalent bonding are:

a. Binding between amine and carboxylic groups: the most used procedures to develop enzymatic biosensors are those based on using glutaraldehyde and carbodiimide. Carbodiimides allow the binding between the carboxyl groups of a support and the amino groups of an enzyme or vice versa; N-hydroxysuccinimide (NHS) can be associated to carbodiimide in order to improve immobilization efficiency. Enzyme immobilization can also be achieved using glutaraldehyde as the activating agent.

b. S–Au bonds: thiol-containing enzymes can be directly immobilized on gold surface because of the strong affinity between thiols and gold substrates.

c. Self-assembled monolayers (SAMs): enzymes can also be immobilized onto pre-formed SAMs via covalent binding. Alkanethiols containing carboxylic acid or amine groups are chemisorbed onto a gold surface, and binding with the respective groups of an enzyme is achieved using glutaraldehyde or carbodiimide activation.

5.2.6 Affinity

In an effort to design biosensors based on oriented and site-specific immobilization of enzymes, (bio)affinity bonds can be created between an activated support (e.g., with lectin, avidin, metal chelates) and a specific group (a tag) of the proteinic sequence of the enzyme (e.g., carbohydrate residue, biotin, histidine); an enzyme can contain affinity tags in its sequence (e.g., a sugar moiety), but in some cases, the affinity tag (e.g., biotin) needs to be artificially attached to the enzyme. This strategy allows to control the biomolecule orientation in order to avoid enzyme deactivation and/or active site blocking. Several affinity methods have been described to immobilize enzymes through streptavidin–biotin, lectin–carbohydrate, and metal cation–chelator interactions.

5.2.7 Nanomaterial-Based Enzymatic Biosensors

Over the last years, nanomaterials (metal and metal oxide nanoparticles, nanowires, carbon nanotubes, quantum dots), by virtue of their unique physical and chemical properties, have attracted considerable attention for the design of electrochemical and optical enzymatic biosensors that exhibit high sensitivity and stability. Biomolecules have been immobilized on nanostructure materials by traditional enzyme immobilization methods (adsorption, entrapment, or covalence). The rationale behind using nanoscale structures for immobilization is to reduce diffusion limitations, to enhance the electron transfer between the enzyme redox center and the electrode surface and to maximize the functional surface area in order to increase enzyme loading (and hence the catalytic activity of attached enzymes). Several review articles have been published discussing practical applications of different nanomaterials in enzymatic biosensing [27,29,33–36].

Carbon nanotubes possess remarkable electrical, mechanical, and structural properties [37]. Carbon nanotubes have an outstanding ability to mediate fast electron transfer kinetics for a wide range of electroactive species involved in enzymatic reactions, such as hydrogen peroxide or NADH. In addition, carbon nanotube chemical functionalization can be used to attach almost any desired chemical species to them. This has permitted the realization of composite electrodes comprising carbon nanotube-based electrodes with immobilized enzymes. In such electrodes, the carbon nanotubes mainly serve as transducers, communicating the signal effectively from the active enzyme centers to the substrate [29,31,38].

SiO_2 nanoparticles are also excellent matrices for enzyme immobilization due to their good biocompatibility and easy preparation [29].

Magnetic particles have also been used as bioimmobilization platforms and magnetic carriers of biomolecules. Advantages in terms of high enzyme loading, easy manipulation, and control of the bioassembly to a specific location on the transducer surface by application of a magnetic field and scope for regeneration and reusability can be underlined. Among magnetic nanoparticles, iron oxide nanoparticles are the most commonly used magnetic materials because of their good biocompatibility, strong magnetic properties, low toxicity, and easy preparation [29,38,39].

Enzymes exhibit nanoscale dimensions comparable to the dimensions of metal or semiconductor nanoparticles. Among these, selenium, zinc oxide, manganese oxide, gold, iron oxide, titanium oxide, nickel oxide, zirconium oxide, platinum, and silver nanoparticles are most commonly used [27,33,34,38]. Moreover, the direct adsorption of enzymes onto bulk metal surfaces frequently results in denaturation of the protein and loss of bioactivity, which can be avoided if enzymes are first adsorbed onto metal nanoparticles before being electrodeposited on the electrode surface [29]. Extensive research efforts were directed in the past few years toward the development of biomolecule–nanoparticle hybrid assemblies and their application to design biosensors. Metallic nanoparticles may be implanted into enzymes to act as nanoelectrodes that electrically "wire" the redox center of the enzyme with electrodes thus enhancing electron transfer rates [29]. Many reports exist of the use of metal oxide nanoparticles for the fabrication of enzymatic biosensors [38]. In addition, metal nanoparticles such as gold nanoparticles have drawn much attention due to their characteristic optical properties. When metal nanoparticles are forced to interact in close vicinity to each other, their absorption peak dramatically shifts to longer wavelengths.

This phenomenon can be exploited for the fabrication of enzyme-responsive nanoparticle networks when the assembly or disassembly of the nanoparticles is triggered by enzymatic activity [40].

Nanocrystals of semiconducting materials (CdS, CdSe), or quantum dots, are particularly interesting for sensing since their surface can be modified via enzyme attachment. Their fluorescence and electrochemiluminescence properties have been exploited for the fabrication of enzyme-amplified immunoassays [40–43].

5.3 Transducers in Enzymatic Biosensors

The vast majority of enzymatic biosensors existing today exploit either electrochemical (amperometric and potentiometric) or optical (absorbance, fluorescence, chemiluminescence) transduction to convert the action of the bioreceptor molecule into a measurable signal [44]. Other transducers (such as enzymatic thermistors) and detection modes (electrochemical impedance) do exist but are rather specialized and outside the scope of this work. Different enzymes commonly used in biosensors, their substrates, the species detected, and the detection modes used are listed in Table 5.4.

5.3.1 Electrochemical Transduction

Several key characteristics that make the electrochemical biosensors useful include their successful operation in turbid environments, satisfactory sensitivity, low power requirements, and the potential for miniaturization. The two most important classes of electrochemical transducers for biosensing include the amperometric and potentiometric/ion-selective field-effect transistor (ISFET) devices. More data on electrochemical biosensors can be found in [8,45].

Amperometric transduction is based on the application of a fixed suitable potential to a working electrode in reference to a Ag/AgCl reference electrode. As long as the potential is sufficient to initiate oxidation or reduction of an electroactive species, the current is proportional to concentration of, and is used to quantify, the target species. It is a convenient fact that enzymatic reactions often generate chemical species (H_2O_2, NADH, quinone) that can be measured by amperometry (Table 5.1). An example of an amperometric device is the aforementioned glucose oxidase–based biosensor for glucose, which relies on the amperometric detection of either the hydrogen peroxide formed or the oxygen consumed during the enzymatic reaction.

Depending on the type of contact and integration level of the enzyme and the transducer, amperometric enzymatic biosensors can be classified into

TABLE 5.4

Different Enzymes Commonly Used in Biosensors, Their Substrates, the Species Detected, and the Detection Mode Used

Substrate	Enzyme	Species Detected	Detection
Choline	Choline oxidase	H_2O_2	Amperometry
Ethanol	Alcohol oxidase	H_2O_2	Amperometry
Formaldehyde	Formaldehyde dehydrogenase	NADH	Amperometry
Glucose	Glucose oxidase	H_2O_2, O_2	Amperometry
Glutamine	Glutamine oxidase	H_2O_2	Amperometry
Glycerol	Glycerol dehydrogenase	NADH, O_2	Amperometry
Hypoxanthine	Xanthine oxidase	H_2O_2	Amperometry
Lactate	Lactate oxidase	H_2O_2	Amperometry
Oligosaccharides	Glucoamylase, glucose oxidase	H_2O_2	Amperometry
Phenol	Polyphenol oxidase	Quinone	Amperometry
Hormones	Alkaline phosphatase (as label)	*p*-nitrophenol	Amperometry
Aspartame	L-aspartase	NH_3	Potentiometry
Fats	Lipase	Fatty acids	Potentiometry
Glucose	Glucose oxidase	Gluconic acid	Potentiometry
Urea	Urease	NH_4^+, CO_2	Potentiometry
Nitrite	Nitrite reductase	NH_4^+	Potentiometry
Penicillin	Penicillinase	H^+	Potentiometry
Sulfate	Sulfate oxidase	HS^-	Potentiometry
Glucose	Glucose oxidase	Gluconic acid	ISFET
Urea	Urease	CO_2, NH_3	ISFET
Penicillin	Penicillinase	Penicillic acid	ISFET
Triolein	Lipase	Fatty acids	ISFET
Ethanol	Alcohol dehydrogenase	NADH	Absorbance
Ethanol	Alcohol dehydrogenase	NADH	Fluorescence
Glucose	Glucose oxidase	O_2	Absorbance
Urea	Urease	Ammonia	Absorbance
Lactate	Lactate monooxygenase	NADH	Absorbance
Penicillin	Penicillinase	Penicillic acid	Absorbance
Glutamate	Glutamate dehydrogenase	NADH	Absorbance
Pesticides	Acetylcholinesterase	Acetic acid	Absorbance
Pesticides	Alkaline phosphatase	*p*-nitrophenol	Absorbance
Chlorophenols	Horseradish peroxidase	H_2O_2	Chemiluminescence
Choline	Choline oxidase	H_2O_2	Chemiluminescence
Ethanol	Alcohol dehydrogenase	NADH	Electrochemiluminescence

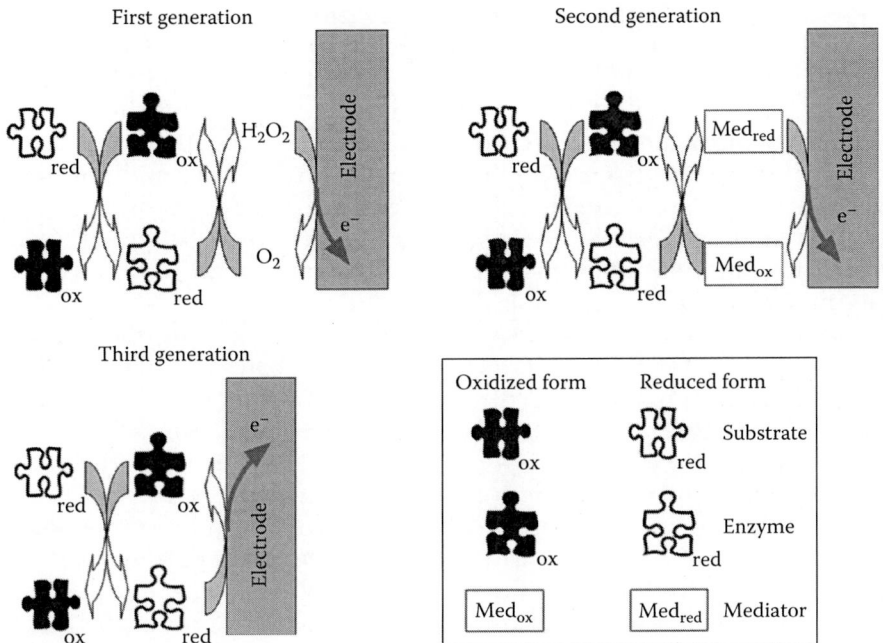

FIGURE 5.6
The three generations of amperometric biosensors.

three generations [15,46] (Figure 5.6). Biosensors of the *first generation* have the biorecognition element physically bound or entrapped into a membrane fixed on the transducer surface. The reactant or product of the enzymatic reaction diffuses to the transducer and causes the response. The original Clark enzyme electrode is a typical example of the first generation. Glucose oxidase is entrapped into a dialysis membrane attached to the electrode surface, and the reactants (glucose and oxygen) together with the products (gluconic acid and hydrogen peroxide) reach the electrode by diffusion. In the *second generation* of biosensors, the enzyme is directly adsorbed or covalently bound on the transducer surface without a semipermeable membrane. In this case, intermediate compounds, called *mediators*, are utilized to shuttle electrons between the enzyme and the transducer. Indeed, the active center of most of the redox enzymes is deeply buried inside the protein matrix hindering free electrical communication with the electrode material [35]. Mediators are defined as artificial electron donors or acceptors that can be readily oxidized or reduced at the reaction site of the enzyme with a low oxidation potential. Therefore, they should exhibit reversible and fast heterogeneous kinetics and ideally be pH independent. Common mediators are ferrocene and its derivates, ferrocyanide, methylene blue, benzoquinone, and N-methyl phenazine. Mediators can freely diffuse in the solution, but it is also possible to coimmobilize them on the surface of the transducer together with the enzyme or to use them as labels to the enzyme [35,45,47]. In the case of the *third generation* of biosensors, the redox enzyme is directly bound on the electrode surface in such a manner that direct electron transfer is possible between the active side of the enzyme and the transducer without the intervention of a mediator. The biosensors based on conducting polymers belong to this category. The design of biosensors based on direct electron transfer is the focus of significant research efforts over the last few years, and different strategies to immobilize redox enzymes on the surface of electrodes in order to achieve direct electron transfer between the enzyme and the transducer have been described [15,26,35,46].

Regarding transducer materials, electrodes are usually made of noble metals (gold or platinum) and different forms of carbon (carbon paste, glassy carbon, carbon fibers, nanotubes) [35]. Screen-printing technology is also very widely used due to simplicity, low cost, and disposability of the sensors [48].

In potentiometry, the potential developed across an ion-selective membrane forming part of an indicator electrode (hence the term ion-selective electrode) is measured against a reference electrode under conditions of zero current flow. The relationship between membrane potential and the concentration of the target species is governed by the Nernst equation [45]. Enzymatic potentiometric biosensors are conveniently fabricated by immobilizing the enzyme on the surface of the ion-selective membrane; upon contact with the analyte, the reaction product can diffuse to the membrane and cause change in the potential of the indicator electrode that is monitored.

Field-effect transistors (FETs) are a type of transistor that uses an electric field to control the conductivity of a channel (i.e., a region depleted of charge carriers) between two electrodes (i.e., the *source* and *drain*) in a semiconducting material. Control of the conductivity is achieved by varying the electric field potential, relative to the source and drain electrode, at a third electrode, known as the *gate*. Depending on the configuration and doping of the semiconducting material, the presence of a sufficient positive or negative potential at the gate electrode would either attract charge carriers (e.g., electrons) or repel charge carriers in the conduction channel. The current focus in biosensing applications is on ISFET and enzyme field-effect transistor (EnFET) devices. One of the most popular methods for the construction of FET-based biosensing devices is the immobilization of enzymes at the gate surface of ion-selective ISFET devices, creating an EnFET (Figure 5.7). The method of enzyme immobilization is a critical aspect of the device's performance and sensitivity. Biocatalytic reactions influence the presence of accumulated charge carriers at the gate surface in proportion to the original analyte concentration. This produces an electrical signal in the form of a measurable drain current. Examples of EnFET applications involve the analysis of penicillin, glucose, and urea [45,49,50].

The vast majority of potentiometric and EnFET enzymatic transducers rely on the measurement of H^+ or NH_4^+ ions produced in various enzymatic reactions by making use of pH- or NH_4^+-sensitive membranes, respectively (Table 5.4).

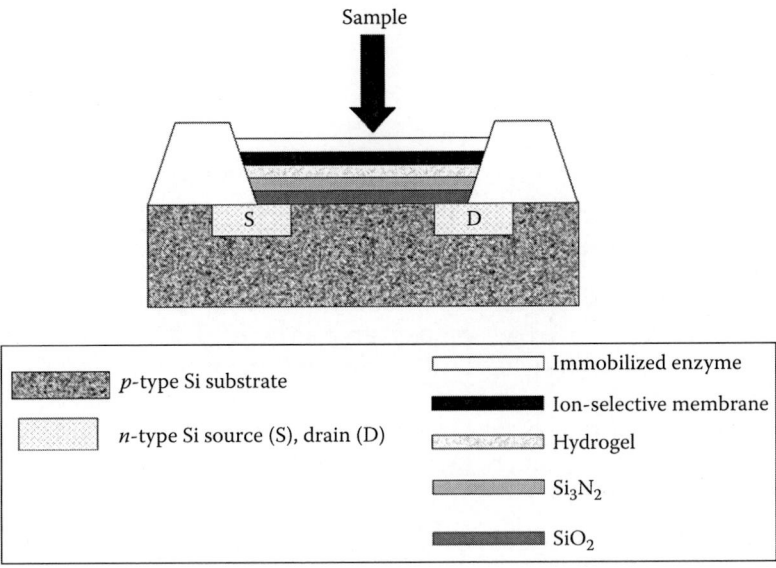

Sample

![p-type] *p*-type Si substrate	☐ Immobilized enzyme
☐ *n*-type Si source (S), drain (D)	■ Ion-selective membrane
	☐ Hydrogel
	Si_3N_2
	SiO_2

FIGURE 5.7
Schematic diagram of an EnFET.

5.3.2 Optical Transduction

Optical enzymatic biosensors can be used in combination with different types of spectroscopic techniques, for example, absorption, fluorescence, and chemiluminescence [16,30,51,52]. The enzymes commonly applied in optical biosensors are oxidases and oxidoreductases (which catalyze the oxidation of compounds using oxygen or NAD^+), esterases (which produce acids), decarboxylases (which produce CO_2), and deaminases (which produce NH_3). The biological receptor is most of the time immobilized close to, or directly on the surface of, an optical fiber. On interaction with the analyte, variation of the optical properties of the sensitive layer will occur and will be related to the concentration of the species analyzed. The optical sensor serving as the transducer for the enzyme-catalyzed reaction is often referred to as an *optrode*.

NADH has a strong absorbance at 340–360 nm, which can be used to monitor the concentration of substrates of dehydrogenase enzymes such as pyruvate or lactate. Other metabolites of clinical interest (e.g., penicillin, creatinine, glucose, urea) have been determined by use of multianalyte optical sensors based on pH transduction. The biosensors are fabricated by coimmobilization of the corresponding enzymes and pH-sensitive fluorophores (e.g., aminofluorescein) in a polymeric matrix. Many enzymatic sensors for the determination of organophosphorus pesticides are based on the use of acetylcholinesterase, an enzyme involved in neurochemical reactions that is inhibited by pesticides. Acetylcholinesterase can be immobilized in optical fiber biosensors, and detection is based on measuring the pH changes caused by release of acetic acid using pH-sensitive dyes [16,30].

Fluorescence is often applied in combination with oxidase enzymes; the decrease in the oxygen concentration is measured by the luminescence quenching of different transition metal complexes. In addition, several fiber-optic sensors have been based on detection of the fluorophore NADH at 455 nm produced as a result of the action on different substrates of dehydrogenase enzymes (e.g., glutamate or alcohol dehydrogenases) immobilized on optical fibers [16].

Chemiluminescence sensors, in which the analyte induces emission of light on interaction with a bioreceptor, can also be used. The popular chemiluminescence reaction between luminol and hydrogen peroxide catalyzed by horseradish peroxidase has served as the basis for different biosensors [16]. Electrochemiluminescence (electrogenerated chemiluminescence) biosensor systems mainly included detection of NADH using dehydrogenases and H_2O_2 using oxidases. Since many enzymes can produce H_2O_2 during their substrate-specific enzymatic reaction, electrochemiluminescence enzyme biosensors are made possible by coupling the luminol light-emitting reaction with enzyme-catalyzed reactions generating H_2O_2. $Ru(bpy)_2^{3+}$-based enzyme electrochemiluminescence biosensors operate on the $Ru(bpy)_2^{3+}$/ NADH co-reactant systems in conjunction with deoxygenases. Finally, many enzymes producing H_2O_2 during their substrate-specific enzymatic reaction can enhance quantum dot electrochemiluminescence, a property that has been exploited for the fabrication of "sandwich"-type immunosensors [41,42].

Chemiluminescence and electrochemiluminescence detection is widely used as a transduction technique in immunosensors and DNA sensors using horseradish peroxidase–labeled reported probes [42].

5.3.3 Enzymatic Biosensors for Multiplexed Assays

Enzyme biosensors also constitute an interesting alternative to carry out multidetection. They enable the analysis of samples containing analytes unable to be simultaneously detected at a conventional detector, by incorporation of different enzymes or coupling of several successive enzymatic reactions. The development of biosensor arrays, composed of miniaturized signal transducer elements, enables the real-time parallel monitoring of multiple species. Multianalyte systems have been used for the determination of analytes of interest in the food industry, for environmental analysis, and for clinical monitoring [46,53,54].

Regarding the food industry, detection of multiple analytes is particularly useful in order to monitor the concentrations of different reactants and products during fermentation processes. These compounds are mainly carbohydrates, alcohols, and organic acids. For instance, enzymatic amperometric biosensors have been developed to monitor simultaneously different compounds (glucose, ethanol, glycerol, lactate, and malate) during alcoholic fermentation and winemaking using oxidases and hehydrogenases. In addition, several commercial enzyme sensors are available in different forms, including autoanalyzers, manual laboratory instruments, and portable systems for use in the food industry [54].

For environmental applications, a multianalyte enzyme biosensor based on a disposable, thick film multielectrode was developed for the discrimination of binary mixtures of the pesticides paraoxon and carbofuran [53]. Another multisensor system comprised three EnFETs with three different enzymes (urease, butyrylcholinesterase, and acetylcholinesterase) inhibited at different degrees by different types of contaminants (organophosphorus pesticides, carbamates, and heavy metal ions) [55]. Biosensors were fabricated containing urease and acetylcholinesterase and organic chromophores whose optical properties are sensitive to changes in local pH caused by the enzyme reactions. The differential enzymatic inhibition caused by the presence of different heavy metal ions suggests that it is possible to design a sensor array for discrimination between different types of water pollutants [56].

Various clinical situations require the simultaneous monitoring of glucose and of other clinically important analytes. For example, the simultaneous monitoring of lactate and glucose is of considerable interest for patient monitoring during intensive care. Such coupling of multiple sensing elements requires that all the analytes are monitored independently at different levels and without cross talk. Different types of integrated miniaturized needle-type dual biosensors for amperometric intravascular glucose and lactate monitoring have been described. Similarly a needle-type sensor for the simultaneous continuous monitoring of glucose and insulin has been developed featuring enzyme-modified electrodes, which responded independently to insulin and glucose [46]. Multiparametric biochips have been developed for choline, glucose, and lactate based on the luminol/H_2O_2 electrochemiluminescence system using three different oxidases and luminol immobilized on an electrode. This biochip was extended to detect acetate using a tri-enzymatic sensing layer based on kinase oxidase activity. A reaction sequence using acetate kinase, pyruvate kinase, and pyruvate oxidase was shown to enable the production of H_2O_2 in response to acetate. Based on a similar entrapment concept of enzyme and luminol, a microarray of nine electrodes was used to develop multiparametric electrochemiluminescence biochips [41]. In view of the need to detect simultaneously multiple analytes in clinical diagnostics, commercial handheld analyzers have been marketed for the quantification of uric acid, lactate, and the activity of acetylcholine esterase; for cholesterol, triglycerides, and phospholipids [47]; and for glucose and β-ketone [44,57]. An optical array biosensor for simultaneous measurement of markers for renal disease (urea, creatinine, uric acid, glucose) was produced by immobilizing the appropriate enzymes and fluorescent dyes in a sol–gel matrix. For urea and creatinine, actions of the hydrolase were detected via an immobilized pH-sensitive fluorescent indicator. For glucose and uric acid determination, the glucose oxidase and uricase produced hydrogen peroxide, which was consumed by coimmobilized horseradish peroxidase to give a fluorescent signal at 590 nm as a result of reduction of Amplex red to resorufin. A handheld device for measurement of whole blood creatinine in point-of-care environments uses a known 3-enzyme cascade to generate hydrogen peroxide from creatinine and amperometric measurement of peroxide in a disposable strip format [58].

The simultaneous detection of the antibiotics penicillin and ampicillin was performed by immobilizing penicillinase and phenol red (a pH indicator) on an optrode. The pH changes during their hydrolysis of the analytes were monitored by absorbance measurements [59].

Finally, multianalyte enzymatic sandwich-type immunosensors (see next paragraph), operating in the multiplexed parallel mode (using different enzyme labels for each analyte on a single transducer) or in the spatially resolved mode (using a single enzyme label at many spatially resolved transducers), have been reported [42].

5.3.4 Enzymes Used as Labels in Biosensors

Enzyme-catalyzed reactions can produce exponential amplification effects when exploited for labeling of biomolecules (antibodies, aptamers, and DNA) that are used as reporter probes. One of the main reasons for this is that the biosensors based on reactions of the enzymatic labels provide high, steady, and reproducible signal amplification. Therefore, enzymatic amplification is a common strategy for the construction of immunosensors and DNA biosensors. Enzymatic reactions have been combined with an additional amplification process (e.g., redox cycling) or use of multienzyme labels per detection probe. In recent years, multienzyme label-based electrochemical biosensors that employ nanomaterials as tracer carriers have been developed. Using nanomaterials, the enzyme-to-carrier ratios enable very high signal amplification. The detection procedures are the same as those in conventional enzyme-based electrochemical and optical biosensors. Two enzymes (horseradish peroxidase and alkaline phosphatase) have been mainly used for labeling purposes, while glucose oxidase has been used in few reports [42,45,60,61].

The fabrication of immunosensors and DNA biosensors using enzymatic tracing is often based on a "sandwich" mode of operation. The target molecule is initially selectively immobilized on a primary capture biomolecule (antibody, DNA, or aptamer) attached to the surface of the transducer. A secondary probe biomolecule tagged with the selected enzyme is then selectively bound to the target species. Upon addition of the enzyme's substrate, the enzymatic product is measured (Figure 5.8). Due to the amplification cycle of enzymatic reactions, trace amount of the substrate molecule would result in significant enhancement of the concentration of the products [51,60,62]. "Sandwich" type of immunoassays for antigens

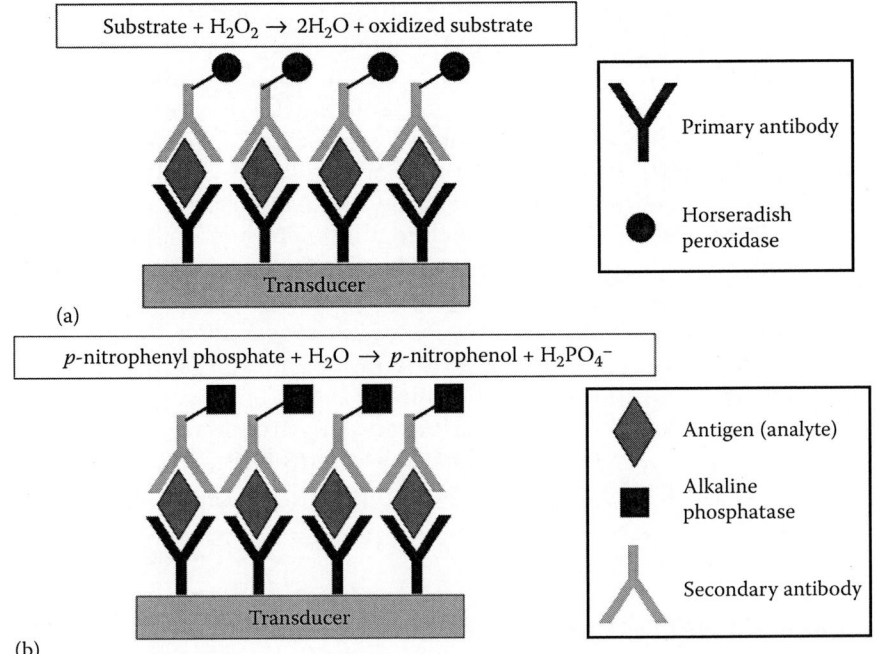

Substrate + $H_2O_2 \rightarrow 2H_2O$ + oxidized substrate

Primary antibody

Horseradish peroxidase

Transducer

(a)

p-nitrophenyl phosphate + $H_2O \rightarrow p$-nitrophenol + $H_2PO_4^-$

Antigen (analyte)

Alkaline phosphatase

Secondary antibody

Transducer

(b)

FIGURE 5.8
The principle of "sandwich" immunosensors using reporter probe antibodies labeled with (a) horseradish peroxidase and (b) alkaline phosphatase.

and bacteria utilizes antibodies or aptamers as reported probes. Enzyme-amplified electrochemical DNA assays have been developed in analogy to enzyme-labeled immunoassays. In this case, the capture and label biomolecules are oligonucleotides complimentary to the target [63].

For the quantitative and qualitative determination of human immunoglobulin (hIgG), pH–ISFETs were used as potentiometric transducers onto which anti-immunoglobulin antibodies labeled with urease were immobilized. Measurement was based on the pH changes as a function of concentration of IgG in the sample solution [49]. An optical fiber "sandwich" immunosensor for cholera antitoxin antibodies has been designed utilizing horseradish peroxidase–labeled goat anti-rabbit IgG immunoglobulin as a secondary antibody. The modified end of the optical fiber was placed in a solution of oxidizing reagent and luminol and the chemiluminescence was monitored [51]. Reporter antibodies labeled with horseradish peroxidase labels and linked to carbon nanotubes were used to construct a platform for sensitive amperometric detection of cancer biomarkers in serum and tissues [60].

In a "sandwich" assay suitable for *Cryptosporidium parvum* in water samples, the probe DNA was immobilized on a gold electrode. After hybridization with the target sequence, a secondary DNA probe tagged with alkaline phosphate was hybridized to the target. The enzyme reacted with its added substrate aminophenyl phosphate and generated p-aminophenol, which was detected electrochemically at the gold electrode [63]. A novel electrochemical DNA biosensor based on a "sandwich" sensing mode was constructed to detect PML/RARa fusion gene in acute promyelocytic leukemia. It involved a pair of LNA probes (capture probe immobilized at electrode surface and reporter probe tagged with HRP) operating in the amperometric mode [63]. Another amperometric immunosensor for ultrasensitive determination of cancer biomarkers and horseradish peroxide-encapsulated nanogold hollow microspheres as labels was reported [42].

Different aptamer-based electrochemical biosensors for thrombin have been reported by using two antithrombin aptamers in a sandwich format. The capture aptamer was immobilized on an electrode, and the probe aptamer was labeled with glucose dehydrogenase, pyrroquinoline quinone, glucose dehydrogenase, or alkaline phosphatase for the generation of electrochemical current upon addition of the suitable substrates [62].

5.4 Applications of Enzymatic Biosensors

5.4.1 Food Analysis

Current analytical methods in the food industry are time-consuming and require skilled labor and expensive instrumentation. However, food industry requires pocket-sized devices capable of online field measurements on undiluted samples of one or more parameters during the production or processing of foods. Most of these drawbacks can be conveniently addressed by enzymatic biosensors. The utility of enzymatic biosensors in food analysis has been reviewed in detail [54,64,65]. Representative applications are listed in Table 5.5.

Food additives, such as aspartame, sorbitol, benzoic acid, and sulfites, can be determined by exploiting their enzymatic conversion to redox-active species in the presence of oxidases, dehydrogenases, tyrosine, and other enzymes immobilized on electrode transducers.

Special attention has been drawn to the detection of contaminants, such as residues of pesticides, fertilizers, heavy metals, and organic compounds, given that the majority of these have a high level of toxicity and that the permitted levels of these compounds are regulated by national and international legislation. Biosensors are widely used to detect xenobiotic substances (such as additives, pesticides, heavy metals) as well as toxic compounds inherently contained in different foods [65].

The different pesticides used in food production can accumulate in the fatty tissue of animals. Biosensors used to detect these compounds are

TABLE 5.5

Examples of Enzymatic Biosensors in Food Analysis

Analyte	Enzyme	Detection
Aspartame	Carboxyl esterase, alcohol oxidase, carboxypeptidase, L-aspartase, peptidase, aspartate aminotransferase, glutamate oxidase and α-chymotrypsin	Amperometric
Sorbitol	Sorbitol dehydrogenase, NAD$^+$	Amperometric
Benzoic acid	Tyrosinase	Amperometric
Sulfites	Sulfite oxidase	Amperometric
Parathion	Parathion hydrolase	Amperometric
Propoxur and carbaryl	Acetylcholinesterase	Optical
Diazinon and dichlorvos	Tyrosinase	Amperometric
Paraoxon	Alkaline phosphatase	Optical
Nitrate	Nitrate reductase	Amperometric
	Nitrite reductase	Nitrite optical
Phosphate	Polyphenol oxidase and alkaline phosphatase, phosphorylase, phosphoglucomutase, and glucose-6-phosphate dehydrogenase	Amperometric
Arsenic, cadmium, bismuth	Cholinesterase	Electrochemical
Cadmium, copper, chrome, nickel, zinc	Urease	Optical
Copper and mercury	Glucose oxidase	Amperometric
Progesterone	Alkaline phosphatase-labeled progesterone (competitive immunoassay)	
Fructose	Fructose dehydrogenase	Amperometric
Lactose	β-Galactosidase	Amperometric
L-amino acids	D-amino acid oxidase	Amperometric
L-glutamate	L-glutamate oxidase	Amperometric
Ethanol	Alcohol oxidase, alcohol dehydrogenase	Amperometric
Cholesterol	Cholesterol oxidase and peroxidase	Amperometric
a-solanine	Butyrylcholinesterase	Potentiometric
Oxalate	Oxalate oxidase, catalase	Optical
Polyphenols	Tyrosinase, laccase	Amperometric
Fatty acids	Lipase	Amperometric
Biogenic amines	Amine oxidase and peroxidase	Amperometric
Hypoxanthine	Xanthine oxidase	Amperometric
Inosine	Xanthine oxidase	Amperometric
Lactic acid	Xanthine oxidase, diamine oxidase, polyamide oxidase	Amperometric

Sources: Turdean, G.L., *Int. J. Electrochem.*, Article ID 343125, 15, 2011; Monk, D.J. et al., *Anal. Bioanal. Chem.*, 379, 931, 2004; Barthelmebs, L. et al., Biosensors as analytical tools in food fermentation industry, *Bio-Farms for Nutraceuticals: Functional Food and Safety Control by Biosensors*, Giardi, M.T., Rea, G., Berra, B., (Eds.), Landes Bioscience and Springer Science & Business Media, New York, pp. 293–307, 2010; Prodromidis, M.I. and Karayannis, M.I., *Electroanalysis*, 14, 241, 2002; Cock, L.S. et al., *Chil. J. Agr. Res.*, 69, 270, 2009; Sharma, S.K. et al., *Curr. Appl. Phys.*, 3, 307, 2003.

mainly based on the inhibition of the enzymatic activity of enzymes such as cholinesterase, tyrosinase, or alkaline phosphatase [14,55,66]. In the optical detection mode, the substrate acetylcholine is added to the sample, and the enzyme activity is measured by monitoring the pH change (resulting from the hydrolysis of acetylcholine to acetic acid) by pH-sensitive indicators immobilized on an optical probe together with the enzyme [51,55,65].

In a similar manner, enzymatic devices with optical and electrochemical detection have been designed to determine the levels of highly toxic heavy metals such as arsenic, cadmium, and mercury in samples of water and soil, based on the inhibiting action of these target metals on enzymes such as urease, cholinesterase, and glucose oxidase [17,65,66].

Progesterone in cow's milk was determined by an electrochemical immunosensor operating in the competitive binding mode using alkaline phosphatase-labeled progesterone as the competitive agent and measurements of an amperometric signal in the presence of *p*-nitrophenyl phosphate using either amperometric or colorimetric assay [56].

Various food labeling regulations recognize the importance of determining "freshness" in terms of evaluating the appearance of compounds that produce disagreeable odors during periods of storage. Table 5.5 lists various biosensors developed to evaluate indicators of the freshness of foods (e.g., biogenic amines, hypoxanthine, inosine) mainly using oxidases [65,67].

Finally, thanks to advances in enzymatic biosensor technology, it is possible to determine and quantify online diverse compounds of importance in process control. Examples are amperometric enzymatic sensors for glucose (with glucose oxidase), lactulose (with β-galactosidase and fructose dehydrogenase), ethanol (with alcohol dehydrogenase), the amino acid lysine (with lysine oxidase), and lactic acid (with lactate oxidase) [54,64,65].

Commercial biosensors specifically intended for measurements in the food industry have been developed [54,65].

5.4.2 Environmental Monitoring

The extensive use of pesticides for agricultural purposes is the cause of their widespread presence in natural waters. Concerns about their toxicity and persistence in the environment has led the European Community to set limits on the concentration of pesticides in different environmental waters: the directive 98/83/EC on the quality of water for human consumption has set a limit of 0.1 μg/L for individual pesticides and of 0.5 μg/L for total pesticides. Enzymatic sensors, based on the inhibition of a selected enzyme, are the most common biosensors used for the determination of these compounds [14,55,66]. Based on the inhibition of acetylcholinesterase, choline oxidase, tyrosinase, and aldehyde dehydrogenase, various biosensors have been developed for the detection of organophosphorus and carbamate pesticides. Although sensitive, biosensors based on enzymatic inhibition are not selective [53,68]. One approach to solve the lack of specificity of acetylcholinesterase involves the genetic engineering of cholinesterase enzyme to obtain new specific enzymes for desired analytes or families [53]. Another strategy is to use enzymes more specific for particular classes of pesticides; as an example, the organophosphorus hydrolase enzyme is able to hydrolyze a number of organophosphorus pesticides such as paraoxon and parathion. Hydrolysis of these organophosphorus pesticides generates *p*-nitrophenol, which is an electroactive and chromophoric product. Thus, organophosphorus hydrolase could be combined with an optical transducer to measure the absorbance of *p*-nitrophenol or with an amperometric transducer to monitor the oxidation or reduction current of this product [53,55,56].

On the other hand, heavy metals are well known to inhibit the activity of enzymes, and application of this phenomenon to the determination of these hazardous toxic elements offers several advantages such as simplicity and sensitivity [17]. Urease is the most frequently utilized enzyme for inhibitive determination of metals. Optical, potentiometric, and ISFET transduction is normally used for the detection of pH changes in conjunction with urease biosensors [53]. However, as in the case of all biosensors relying on inhibition of the enzymatic activity, such devices lack selectivity [53,56,68].

Various enzymatic biosensors for phosphate and nitrite determination in natural waters with optical and electrochemical detection have appeared in the literature in recent years [53,56].

Phenolic compounds are widely distributed in the environment as organic pollutants. Amperometric biosensors with immobilized polyphenol oxidase, laccase, tyrosinase, and peroxidases are the most widely used for the detection of various phenols.

Table 5.6 lists some enzymatic biosensors applied to environmental monitoring.

5.4.3 Clinical Assays

There has been a great demand for rapid and reliable methods that can be used in biochemical laboratories for determination of substances in biological fluids such as blood, serum, and urine. There is also a demand to move clinical analysis from centralized laboratories to a doctor's clinic and patients self-testing at home. Most of the methods available in the market for rapid detection are based on enzyme electrodes [44,47]. Table 5.7 lists typical applications of enzyme sensors in the clinical field.

Currently, 85%–90% of the biosensor market is directly or indirectly associated with monitoring of glucose in clinical diagnostics. Diabetes is the fastest-growing chronic disease in the world, which afflicts around 2% of the world's population and 6% of the adult population of the Western world [69]. The number of diabetics is expected to reach 380 million by 2025. The history, challenges, and current trends in the development of glucose biosensors are reviewed in [46,69]. The fingerprick-type glucose biosensors (essentially operating on the principle first introduced in the

TABLE 5.6

Examples of Enzymatic Biosensors in Environmental Analysis

Analyte	Enzyme	Detection
Simazine	Peroxidase	Potentiometric
Parathion	Hydrolase	Amperometric
Paraoxon	Alkaline phosphatase	Optical
Carbaryl	Acetylcholinesterase	Amperometric
2,4-Dichlorophenoxiacetic acid	Acetylcholinesterase (immunoanalysis)	Amperometric
Atrazine	Glutathione S-transferase	Optical
Chlorpyrifos	Acetylcholinesterase	Amperometric
Malathion	Acetylcholinesterase	Amperometric
Mercury, cadmium, and arsenic	Urease	Potentiometric
Cadmium, copper, and lead	Urease	Potentiometric
Perchlorate	Perchlorate reductase	Amperometric
Nitrate	Cytochrome c nitrite reductase	Amperometric
Phosphate		
Phenols	Laccase and tyrosinase	Amperometric
Phenols	Polyphenol oxidase	Amperometric
Phenols	Horseradish peroxidase	Amperometric

Sources: Turdean, G.L., *Int. J. Electrochem.*, Article ID 343125, 15, 2011; Monk, D.J. et al., *Anal. Bioanal. Chem.*, 379, 931, 2004; Rodriguez-Mozaz, S. et al., *Pure Appl. Chem.*, 76, 723, 2004; Salgado, A.M. et al., Biosensor for environmental applications, *Environmental Biosensors*, Somerset, V., (Ed.), InTech, Rijeka, Croatia, pp. 3–16, 2011.

Clark–Lyons enzyme electrode) have enjoyed the greatest commercial success. These biosensors have also been extended to other key metabolites such as urea, creatinine, lactate, cholesterol, and uric acid. Comprehensive reviews of the current status of commercial clinical biosensors have been published [47,57,69].

Lactate measurements are useful in respiratory insufficiencies, shocks, heart failure, metabolic disorder, and monitoring of the physical condition of athletes. Lactate biosensors normally incorporate lactase oxidase or dehydrogenase, and the generated H_2O_2 and NADH are detected amperometrically or optically. Urea estimation is important in assessing kidney functions and disorders associated with it. Most of the urea biosensors are based on the potentiometric detection of NH_4^+ or HCO_3^- using ion-selective electrodes. Determination of cholesterol is clinically important since abnormal concentrations of cholesterol are related with hypertension, hyperthyroidism, anemia, and coronary artery diseases. Uric acid is one of the major products of purine breakdown in humans, and therefore its determination serves as a market for the detection of a range of disorders associated with

TABLE 5.7

Examples of Enzymatic Biosensors in Clinical Analysis

Analyte	Enzyme	Detection
Urea	Urease	Potentiometric, optical
Glucose	Glucose oxidase	Amperometric, optical
Glucose	Glucose oxidase	Potentiometric
Lipase	Triglycerides	Potentiometric
Cholesterol	Cholesterol oxidase	Amperometric
Lactate	Lactase oxidase	Amperometric
Lactate	Lactate dehydrogenase	Optical
Carnitine	Carnitine dehydrogenase and diaphorase	
Theophylline	Theophylline oxidase	
Creatine and creatinine	Creatininase, creatinase, and sarcosine oxidase	Potentiometric
Bilirubin	Bilirubin oxidase	Optical
Choline	Choline oxidase	Optical
Glutamate	Glutamate dehydrogenase	Optical
Mercury, cadmium, and arsenic	Urease	Potentiometric
Cadmium, copper, and lead	Urease	Potentiometric

Sources: Choi, M.M.F., *Microchim. Acta*, 148, 107, 2004; Malhotra, B.D. and Chaubey, A., *Sens. Actuat. B*, 91, 117, 2003; Pijanowska, D.G. and Torbicz, W., *Bull. Pol. Acad. Sci.: Tech. Sci.*, 53, 251, 2005.

altered purine metabolism while elevated levels of uric acid are observed in a wide range of a conditions such as leukemia, pneumonia, kidney injury, hypertension, and ischemia [47]. An increased level of triglycerides is an indirect coronary disease risk factor. The production of fatty acids by means of lipase results in pH changes that can be determined by potentiometry [49]. Optical biosensors for bilirubin (associated with gastric pathologies) and choline (an important neurotransmitter in mammals) have been developed incorporating their respective oxidases. As the target analytes are oxidized to biliverdin and betaine, respectively, the O_2 consumption in the enzymatic reaction can be quantitatively determined by measuring the fluorescence quenching of a luminescence dye immobilized on the biosensors [30]. The amino acid glutamate is another major neurotransmitter used in the nervous system for interneuronal communication. Since glutamate is released from neurons on a millisecond timescale into submicrometer spaces, the development of a glutamate biosensor with high temporal and spatial resolution is of great interest for the study of neurological function and disease. Several optical biosensors have been developed to monitor glutamate mainly based on the detection of NADH generated during glutamate oxidation in the presence of glutamate dehydrogenase [30].

Immunosensors for toxins (staphylococcal enterotoxin B) and pathogenic microorganisms (*S. aureus*) could also be fabricated by using enzyme-labeled antibodies [70].

Regarding point-of-care applications, multisensors, able to monitor multiple clinical analytes, have been developed [47,57].

5.4.4 Pharmaceutical Analysis

Since the analytical methods recommended by various pharmacopoeias for the detection of drugs are usually time-consuming, expensive, and laborious, biosensors are particularly suitable for pharmaceutical analysis. Amperometric enzymatic biosensors bearing enzymes such as oxidases, peroxidases, laccases, tyrosinase, and superoxide dismutase have been developed for oxidizable pharmaceutical compounds. Regarding drugs containing hydrolysable and labile groups and/or when there is H^+ liberation, the use of potentiometric biosensors is the most frequently used alternative (Table 5.8) [71].

Amperometric biosensors, fabricated by direct enzyme immobilization on an electrode surface, are the most widely used. Among the compounds with pharmaceutical significance that can be detected amperometrically are salicylates (using the enzyme salicylate hydroxylase in the presence of NADH and oxygen), paracetamol (acetaminophen) (using the oxidation of the compound by hydrogen peroxide catalyzed by horseradish peroxidase), and catecholamines (the catecholic group of which can be catalyzed by many oxidases). The use of vegetal tissues as a source of oxidases for biosensor

TABLE 5.8

Examples of Enzymatic Biosensors in Pharmaceutical Analysis

Analyte	Enzyme	Detection
Salicylates	Salicylate hydroxylase	Amperometric
Paracetamol (acetaminophen)	Horseradish peroxidase, tyrosinase, laccase	Amperometric
Epinephrine, dopamine, adrenaline	Polyphenol oxidase (from vegetable tissue and microbes)	Amperometric
Theophylline	Xanthine oxidase, theophylline oxidase	Amperometric
Phenothiazinic derivatives	Horseradish peroxidase	Amperometric
Clozapine	Horseradish peroxidase	Amperometric
Desipramine, pirlindole, and fluoxetine	Monoamine oxidases	Amperometric
Penicillin	Penicillinase	Potentiometric, optical
Ciprofloxacin	Horseradish peroxidase	Amperometric
Anthracyclines and sulfonamides	*ds*-DNA and horseradish peroxidase	Amperometric

Source: de Souza Gil, E. and de Melo, G.R., *Brazilian J. Pharm. Sci.*, 46, 375, 2010.

purposes represents a very promising alternative. The enzyme polyphenol oxidase, present in many kinds of plants, has been used as a source of enzymes for biosensor development focusing on the analysis of molecules containing catecholic and phenolic groups. The polyphenol oxidases catalyze oxidation of catechols or phenols to the respective quinones.

Methylxanthines (caffeine, theobromine, and theophylline) act as stimulants of the central nervous system and also as muscle relaxants. Xanthine oxidase and theophylline oxidase have been explored for the development of sensitive and selective enzyme biosensors applied to theophylline analysis. Since xanthines inhibit phosphodiesterase, phosphodiesterase-based biosensors have been developed for caffeine analysis.

Neuroleptics or antipsychotics (thioxanthenes and phenothiazines) as well as antidepressants (monoamine oxidase inhibitors and tricyclic compounds) are drugs used in the treatment of psychiatric disorders. The phenothiazinic derivatives were determined using amperometric biosensors containing horseradish peroxidase. Another horseradish peroxidase biosensor, constructed using enzyme immobilization in porous microparticles of silica magnetically linked to a glassy carbon electrode, was evaluated for the amperometric analysis of clozapine. The analytical signal is generated during clozapine oxidation in the presence of hydrogen peroxide. The most commonly used biosensor for monoamine oxidase inhibitors analysis is based on the very enzyme monoamine oxidases wired to amperometric transducers that detect the hydrogen peroxide liberated in the enzymatic reaction.

Biosensors for penicillin are based on hydrolysis by the enzyme penicillinase, which generates penicilloic acid detected by potentiometric or optical pH-sensitive transducers. The determination of anthracyclines and sulfonamides with glassy carbon *ds*DNA and horseradish peroxidase–modified electrodes was evaluated based on the oxidation of methylene blue serving as an electron mediator.

5.4.5 Future Perspectives of Enzymatic Biosensors

Multiplexed detection, the expanding use of nanomaterials, and efforts to achieve direct electron transfer in electrochemical transduction (all discussed in previous paragraphs) dominate current research on enzymatic biosensors. Future research topics include:

a. Implantable and wearable devices: miniaturized, implantable enzymatic biosensors are important since they are able to provide real-time data on the level of clinically important metabolite levels without the need for patient or medical intervention and regardless of the patient's physiological state (for instance, implantable biosensors are useful for diabetes management, which at present relies on measurements using test strips with blood drawn from fingerpricking). Challenges that have to be addressed are reliability (associated with biofouling and enzyme leaching/deactivation), sensor drift, biocompatibility, security/privacy, and miniaturization [15,25,72]. Over the last few years, integrated multidisciplinary approaches have been designed combining developments in biomimetic materials (to suppress the foreign body response), enzyme engineering (to promote enzyme stability), hybrid nanobiomaterials (to address endogenous interferences, to reduce the size of the sensor, and to maximize biocompatibility), and electronics (for self-calibration, to counter sensor drifts, for wireless communication, and to ensure privacy). As a result, it has been recently shown that successful continuous performance of implanted biosensors is possible for up to 4 months in rats and up to 1 year in pigs [43,72,73]. Current work is in progress with the view to develop wearable and implantable wireless sensor networks for healthcare monitoring [25,74] and textile-based wearable sensors [75].

b. Paper-based devices: a simple and low-cost but effective category of enzymatic bioanalytical sensors and microfluidic devices has been developed as paper-based assays. These disposable biosensors are suitable for use in developing countries and for field monitoring, where portability and affordability are important. The paper is patterned to create well-defined, millimeter-sized hydrophilic paper channels whose boundaries are defined by a hydrophobic polymer, and the enzyme is immobilized onto the sensing platform. Examples are the colorimetric assays of glucose, lactate, and uric acid in urine (based on the enzymatic conversion of glucose by glucose oxidase to form hydrogen peroxide that reacts with peroxidase and KI to form I_2 resulting in a brown color); bioactive paper "dipsticks" for detection of the acetylcholinesterase inhibitors paraoxon and aflatoxin B1; microfluidic paper-based platforms with integrated screen-printed electrodes for the rapid quantitative enzymatic detection of glucose, cholesterol, lactate, and alcohol in blood and urine and serum samples using the respective oxidases [43].

c. Lab-on-a-chip biosensing devices: integrated microfluidic chips are being developed to provide capabilities for automatic sample pretreatment and analysis of low amount of sample in a portable format with low reagent and power requirements. An example of lab-on-a-chip microfluidic device has been reported for automated detection of L-glutamate based on enzymatic detection. The enzymes (L-glutamate dehydrogenase and D-phenylglycine aminotransferase) were immobilized on the silicone surface of the chip [43]. A protein chip able to detect several biological warfare agents in an array format using a sandwich enzyme-linked immunosorbent assay has been reported [70].

d. Artificial enzymes: molecularly imprinted polymers could be used for the synthesis of enzyme mimics, by incorporating catalytically active groups into a polymeric matrix. In this case, the substrate, an intermediate or the product of an enzymatic reaction (or even their analogues), can be used as a template. Catalytic molecularly imprinted polymers that simulate the enzymatic activity of various enzymes have been reported, but only a few applications of peroxidase-mimicking polymers have found analytical use [76].

e. Noninvasive diagnostics: several studies suggest that saliva can be reliably used to identify protein and genetic markers. Due to the noninvasive manner of collecting samples, biosensors for saliva diagnostics are preferable to blood and serum analysis. Salivary α-amylase was detected on a biosensor consisting of a disposable test strip with immobilized 2-chloro-4-nitrophenyl-4-O-β-D-galactopyranosylmaltoside, which is hydrolyzed by amylase to form a yellow-colored product that was measured photometrically. A biosensor for phosphate was constructed by immobilizing pyruvate oxidase on a screen-printed electrode. The enzyme, in the presence of pyruvate and oxygen, generates H_2O_2, which is detected amperometrically [58].

References

1. K. Cammann, *Fresenius' Z. Anal. Chem.* 287 (1977) 1–9.
2. D.R. Thevenot, K. Toth, R.A. Durst, G.S. Wilson, *Anal. Lett.* 34 (2001) 635–659.
3. D.R. Thevenot, K. Toth, R.A. Durst, G.S. Wilson, *Biosens. Bioelectron.* 16 (2001) 121–131.
4. Biosensors WebBook, http://www.lsbu.ac.uk/water/enztech/index.html (assessed June 17, 2013).
5. Enzyme Technology, http://www.lsbu.ac.uk/water/enztech/index.html (assessed June 17, 2013).
6. Tran Minh Canh, *Biosensors*, Chapman and Hall, London, U.K. (1993).
7. F.G. Bănică, *Chemical Sensors and Biosensors: Fundamentals and Applications*, John Wiley & Sons, Chichester, U.K. (2012).
8. X. Zhang, H. Ju, J. Wang (Eds.), *Electrochemical Sensors, Biosensors and Their Biomedical Applications*, Elsevier, Amsterdam, the Netherlands (2008).
9. J. Cooper, T. Cass (Eds.), *Biosensors: A Practical Approach*, 2nd edn., Oxford University Press, Oxford, U.K. (2003).
10. B.R. Eggins, *Chemical Sensors and Biosensors*, Jonn Wiley & Sons, Chichester, U.K. (2002).
11. J.Y. Yoon, *Introduction to Biosensors*, Springer, New York (2013).
12. D.G. Buerk, *Biosensors: Theory and Applications*, Technomic Publishing Company, Inc., Lancaster, PA (1993).
13. Industry Experts, Biosensors—A global market overview, 2012. http://www.reportlinker.com/p0795991/Biosensors-A-Global-Market-Overview.html (assessed June 17, 2013).

14. R. Vargas-Bernal, E. Rodríguez-Miranda, G. Herrera-Pérez, Evolution and Expectations of enzymatic biosensors for pesticides, in: *Pesticides—Advances in Chemical and Botanical Pesticides*, R.P. Soundararajan (Ed.), InTech, Rijeka, Croatia (2012), pp. 329–356.
15. S. Borgmann, A. Schulte, S. Neugebauer, W. Schuhmann, Amperometric biosensors, in: *Advances in Electrochemical Science and Engineering*, R.C. Alkire, D.M. Kolb, J. Lipkowski (Eds.), Wiley VCH, Weinheim, Germany (2011), pp. 1–83.
16. M.D. Marazuela, M.C. Moreno-Bondi, *Anal. Bioanal. Chem.* 372 (2002) 664–682.
17. G.L. Turdean, *Int. J. Electrochem.* Article ID 343125, (2011) 15.
18. L.C. Clark, Jr., *Trans. Am. Soc. Artif. Int. Organs* 2 (1956) 41–48.
19. L.C. Clark, Jr., C. Lyons, *Ann. NY Acad. Sci.* 102 (1962) 29–45.
20. L. Clark, Jr., U.S. Patent #3, 539 (1970) 455.
21. S.J. Updike, J.P. Hicks, *Nature* 214 (1967) 986–988.
22. G. Guilbault, G. Lubrano, *Anal. Chim. Acta* 64 (1973) 439–455.
23. W.Z. Jia, K. Wang, X.H. Xia, *TrAC* 29 (2010) 306–318.
24. K.P. Voelkl, N. Opitz, D.W. Lubbers, *Fresenius'. Z. Anal. Chem.* 301 (1980) 162–163.
25. S. Vaddiraju, I. Tomazos, D.J. Burgess, F.C. Jain, F. Papadimitrakopoulos, *Biosens. Bioelectron.* 25 (2010) 1553–1565.
26. G. Preda, O. Bizerea, B. Vlad-Oros. Sol-gel technology in enzymatic electrochemical biosensors for clinical analysis, in: *Biosensors for Health, Environment and Biosecurity*, P.A. Serra (Ed.), InTech, Rijeka, Croatia (2011), pp. 363–388.
27. S.A. Ansari, Q. Husain, *Biotechnol. Adv.* 30 (2012) 512–523.
28. R.M. Twyman, Enzymes: Immobilized enzymes, in: *Encyclopedia of Analytical Science*, P. Worsfold, A. Townshend, C. Poole (Eds.), 2nd edn., Vol. 2, Elsevier, London, U.K. (2005), pp. 523–529.
29. A. Sassolas, L.J. Blum, B.D. Leca-Bouvierm, *Biotechnol. Adv.* 30 (2012) 489–511.
30. M.M.F. Choi, *Microchim. Acta* 148 (2004) 107–132.
31. K. Balasubramanian, M. Burghard, *Anal. Bioanal. Chem.* 385 (2006) 452–468.
32. D. Wei, A. Ivaska, *Chem. Anal. (Warsaw)* 51 (2006) 839–852.
33. Z. Zhao, H. Jiang, Enzyme-based electrochemical biosensors, in: *Biosensors*, P.A. Serra (Ed.), InTech, Rijeka, Croatia (2010), pp. 1–22.
34. H. Li, S. Liu, Z. Dai, J. Bao, X. Yang, *Sensors* 9 (2009) 8547–8561.
35. A.K. Sarma, P. Vatsyayan, P. Goswami, S.D. Minteer, *Biosens. Bioelectron.* 24 (2009) 2313–2322.
36. I. Willner, R. Baron, B. Willner, *Biosens. Bioelectron.* 22 (2007) 1841–1852.
37. A. Gustavo Rivas, M.D. Rubianes, M.C. Rodríguez, N.F. Ferreyra, G.L. Luque, M.L. Pedano, S.A. Miscoria, C. Parrado, *Talanta* 74 (2007) 291–307.
38. A.A. Ansari, M. Alhoshan, M.S. Alsalhi, A.S. Aldwayyan, Nanostructured metal oxides based enzymatic electrochemical biosensors, in: *Biosensors*, P.A. Serra (Ed.), InTech, Rijeka, Croatia (2010), pp. 23–46.
39. Y. Xu, E. Wang, *Electrochim. Acta* 84 (2012) 62–73.
40. R. de la Rica, D. Aili, M.M. Stevens, *Adv. Drug Delivery Rev.* 64 (2012) 967–978.
41. X.M. Chen, B.Y Su, X.H. Song, Q.A. Chen, X. Chen, X.R. Wang, *TrAC* 30 (2011) 665–676.
42. X. Pei, B. Zhang, J. Tang, B. Liu, W. Lai, D. Tang, *Anal. Chim. Acta* 758 (2013) 1–18.
43. C.R. Ispas, G. Crivat, S. Andreescu, *Anal. Lett.* 45 (2012) 168–186.
44. P. D'Orazio, *Clin. Chim. Acta* 334 (2003) 41–69.
45. D. Grieshaber, R. MacKenzie, J. Vörös, E. Reimhult, *Sensors* 8 (2008) 1400–1458.
46. J. Wang, *Chem. Rev.* 108 (2008) 814–825.

47. B.D. Malhotra, A. Chaubey, *Sens. Actuat. B* 91 (2003) 117–127.
48. M. Albareda–Sirvent, A. Merkoçi, S. Alegret, *Sens. Actuat. B* 69 (2000) 153–163.
49. D.G. Pijanowska, W. Torbicz, *Bull. Pol. Acad. Sci.: Tech. Sci.* 53 (2005) 251–260.
50. M. Pohanka, P. Skládal, *J. Appl. Biomed.* 6 (2008) 57–64.
51. D.J. Monk, Z. David, R. Walt, *Anal. Bioanal. Chem.* 379 (2004) 931–945.
52. A. Koyun, E. Ahlatcioğlu, Y.K. İpek, Biosensors and their principles, in: *A Roadmap of Biomedical Engineers and Milestones*, S. Kara (Ed.), InTech, Rijeka, Croatia (2012), pp. 115–142.
53. S. Rodriguez-Mozaz, M.P. Marco, M.J. Lopez de Alda, D. Barceló, *Pure Appl. Chem.* 76 (2004) 723–752.
54. L. Barthelmebs, C. Calas-Blanchard, G. Istamboulie, J.L. Marty, T. Noguer, Biosensors as analytical tools in food fermentation industry, in: *Bio-Farms for Nutraceuticals: Functional Food and Safety Control by Biosensors*, M.T. Giardi, G. Rea, B. Berra (Eds.), Landes Bioscience and Springer Science & Business Media, New York (2010), pp. 293–307.
55. N. Jaffrezic-Renault, *Sensors* 1 (2001) 60–74.
56. A.M. Salgado, L.M. Silva, A.F. Melo, Biosensor for environmental applications, in: *Environmental Biosensors*, V. Somerset (Ed.), InTech, Rijeka, Croatia (2011), pp. 3–16.
57. T. Ming-Hung Lee, *Sensors* 8 (2008) 5535–5559.
58. P. D'Orazio, *Clin. Chim. Acta* 412 (2011) 1749–1761.
59. T. Vo-Dinh, B. Cullum, *J. Anal. Chem.* 366 (2000) 540–551.
60. J. Li, S. Li, C.F. Yang, *Electroanalysis* 24 (2012) 2213–2229.
61. H. Yang, *Curr. Opin. Chem. Biol.* 16 (2012) 422–428.
62. Y. Xu, G. Cheng, P. He, Y. Fang, *Electroanalysis* 21 (2009) 1251–1259.
63. A. Liu, K. Wang, S. Weng, Y. Lei, L. Lin, W. Chen, X. Lin, Y. Chen, *TrAC* 37 (2012) 101–111.
64. M.I. Prodromidis, M.I. Karayannis, *Electroanalysis* 14 (2002) 241–261.
65. L.S. Cock, A.M. Zetty Arenas, A. Ayala Aponte, *Chil. J. Agr. Res.* 69 (2009) 270–280.
66. L.S.B. Upadhyay, N. Verma, *Anal. Lett.* 46 (2013) 225–241.
67. S.K. Sharma, N. Sehgal, A. Kumar, *Curr. Appl. Phys.* 3 (2003) 307–316.
68. M.D. Luque de Castro, M.C. Herrera, *Biosens. Bioelectron.* 18 (2003) 279–294.
69. https://www.imt.liu.se/edu/courses/TBMT32/pdfs/Biosensors%20-%20Essential%20Compnents.pdf (assessed February 8, 2013).
70. C. Bala, New Challenges in the design of bio(sensors) for biological warfare agents, in: *Portable Chemical Sensors: Weapons Against Bioterrorism*, D.P. Nikolelis (Ed.), NATO science for peace and security series A: Chemistry and biology, Springer Science & Business Media, Dordrecht, the Netherlands (2012), pp. 15–41.
71. E. de Souza Gil, G.R. de Melo, *Brazilian J. Pharm. Sci.* 46 (2010) 375–391.
72. S. Carrara et al., *Sensors* 12 (2012) 11013–11060.
73. C.N. Kotanen, F.G Moussy, S. Carrarac, A. Guiseppi-Elie, *Biosens. Bioelectron.* 35 (2012) 14–26.
74. A. Darwish, A.E. Hassanien, *Sensors* 11 (2011) 5561–5595.
75. J.R. Windmiller, J. Wang, *Electroanalysis* 25 (2013) 29–46.
76. G. Díaz-Díaz, D. Antuña-Jiménez, M.C. Blanco–López, M. Jesús Lobo-Castañón, A.J. Miranda-Ordieres, P. Tuñón-Blanco, *TrAC* 33 (2012) 68–80.

6

Antibody-Based Biosensors

Boris B. Dzantiev and Anatoly V. Zherdev

CONTENTS

6.1 Introduction

Antibodies as a recognizing (receptor) element have generated interest in
the development of different analytical systems, including biosensors, for
a long time. An article of Janata in 1975 is considered as the first development of immunosensors [1]. However, it actually describes a biosensor

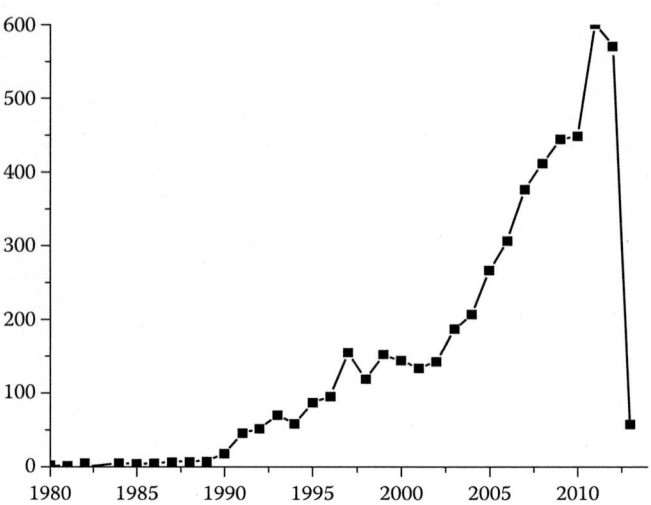

FIGURE 6.1
Dynamics of publications on immunosensors.

with receptor of non-antibody nature, concanavalin A, and its use for binding with mannan. Immunosensors are slightly younger: their first description was published in 1976 [2].

Over the years, thousands of works were issued in the field of immuno-sensation. Today, Thomson Reuters Web of Knowledge database features more than 5000 articles on the subject (information as of January 29, 2013). Bibliometric data (Figure 6.1) indicate that their number increases from year to year. But this growth is irregular: steady periods in latter halves of decades follow the surge of interest in the early 1990s and early 2000s.

Immunosensors are the subject matter of not only multiple experimental articles but also a number of reviews [3–15]. Taking into consideration the availability of these references as well as easy search of original experimental articles, we will not discuss the different types of immunosensors and immunosensoric control of compounds related to different classes.

This chapter aims to examine antibodies as the receptor elements for biosensors in comparison with alternative compounds. Recently, the antibodies have got a number of potential competitors. The articles in which the creators of biosensors using aptamers, molecularly imprinted polymers, and other new receptors write about their prospects and advantages in contrast with the "traditional" sensors based on antibodies are published regularly. Therefore, this review considers the following problems:

What antibodies and their derivatives as well as alternative receptor molecules are available to researchers? What differences of these compounds from each other should be considered when creating a biosensor?

What tasks have to be solved for the effective use of antibodies in biosensors? How successful is practice of immunosensors commercialization?

6.2 Antibodies Yesterday and Today: Their Diversity and Place in Relation to Other Bioreceptors

6.2.1 Polyclonal (Natural) Antibodies

Molecules of all antibodies have the general structural principle based on paired heavy and light polypeptide chains that allows to integrate them into the common chemical class of immunoglobulins. In most cases, the biosensors are used by immunoglobulins G (IgG) prevailing in serum. Y-shaped IgG molecule consists of three structural blocks, as shown in Figure 6.2: one branching constant (C) block and two identical variables (V). Disulfide bonds provide linkage of polypeptide chains of these blocks together.

Molecular diversity of immunoglobulins is provided by variable structural blocks through three mechanisms:

1. The multiplicity of genes encoding the variable domains; the one gene only is selected and expressed in each antibody-producing cell.
2. Additional somatic combination of segments by which these genes are divided.
3. Mutations in the course of B cell differentiation and their transformation into plasma cells.

The combined effect of these three processes provides high variability: several thousand kinds of heavy and light chains of immunoglobulins (their number varies slightly depending on animal specimen) can be formed in the body. Aggregation of light and heavy chains leads to the formation of the active antigen-binding antibody site that may have millions of variants (different authors estimate the range of $2 \cdot 10^7 - 10^8$). Contact with some antigen activates the line of antibody-producing cells that secrete immunoglobulins that can be bound with this antigen and neutralize it. Available diversity of antibodies provides the potential for the immune response to the compounds of all kinds.

The immune response varies from organism to organism and depends on the physiological characteristics of the animal as well as the kinds of antigens it contacts during its life [16]. Absolute replication of polyclonal antibody preparation properties during immunization of different animals is impossible. Ideally, to reproduce immunization results, it would be well to use linear and specific pathogen-free (SPF) animals. For the matter of that, both the first and especially the second solutions increase significantly the cost of antibody production process and are not prominent in the commercial (along with research) practice.

Even after successful immunization, a small fraction of serum immunoglobulins are the antibodies to immunogen; usually they amount to a few percent. Therefore, the preparation of antibodies without "ballast" nonspecific

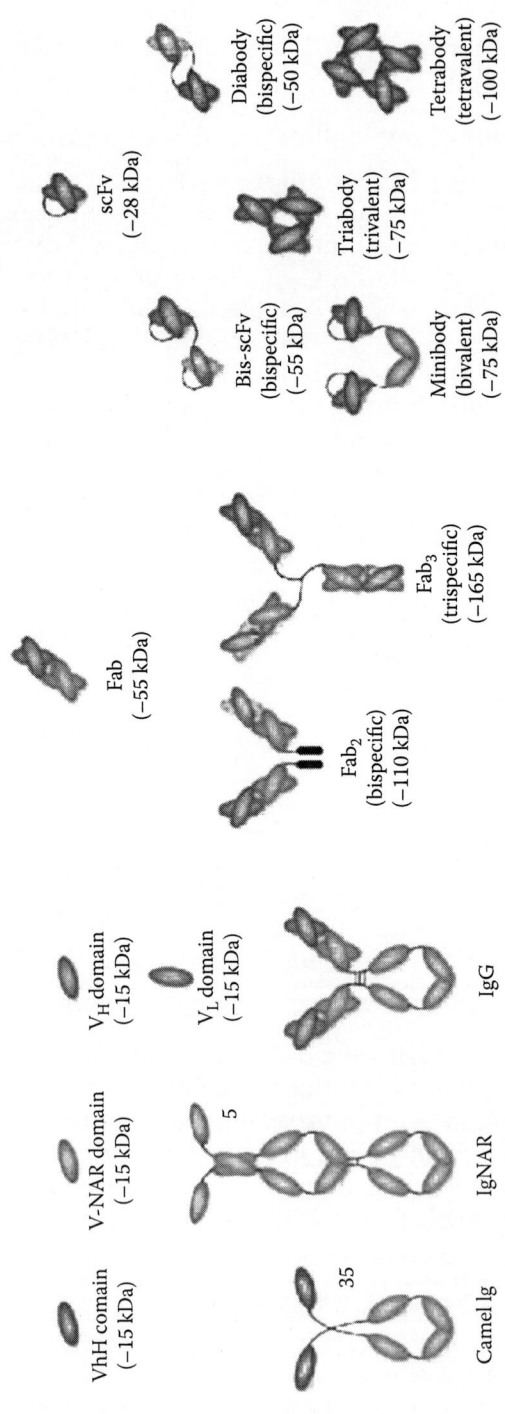

FIGURE 6.2
Diversity of antibodies, their fragments, and structures of recombinant antibodies. (Reprinted by permission from Macmillan Publishers Ltd. *Nat. Biotechnol.*, Holliger, P. and Hudson, P.J., Engineered antibody fragments and the rise of single, 23(9), 1126–1136, Copyright 2005.)

immunoglobulins is preferable to be used in biosensors. To clean the preparation of antibodies, it is usually passed through the antigen-containing sorbent and then the bound molecules are eluted. These obtained affine antibodies are specific for a particular antigen but anyway are the mixture of molecules with different structure of the antigen-binding sites that interact variously with the target antigen.

The history of antibodies to detect different compounds is much longer than the history of biosensors. By the time of biosensor development, the antisera were the only available source of antibodies, and all of the first publications on immunosensors describe the use of polyclonal antibodies specifically.

Although a number of researches described the intermolecular variability in preparations of monoclonal antibodies [17–24], it mainly depends on posttranslational modification or partial destruction and does not lead to the serious adverse effects. The most bioanalytical tasks provide consideration of monoclonal antibodies as the preparation with identical reactivity of all molecules.

The advantages and disadvantages of poly- and monoclonal antibodies as analytical reagents were discussed in many publications [25–31]. For example, it was suggested repeatedly that monoclonal antibodies have less affinity. However, the monoclonal technique replicates the polyclonal antibody pool that was formed during immunization. Therefore, the lower affinity of monoclonal antibodies observed in the comparative experiments is caused by their small sample from the polyclonal pool. (Working with polyclonal antisera, it is obvious that antibodies of the most affine clones make a dominating contribution to binding with the antigen.)

Regardless of obvious advantages—molecular identity and potential unlimited production—even for the moment, the monoclonal antibodies do not force out polyclonal antibodies as bioanalytical reagents. About a quarter of publications in the immunosensor-related bibliography of the recent years describes the works done using polyclonal antibodies. It should be noted that the cost plays the important, but not defining, role in the choice of reactants used. Converting to 1 mg of affine-purified antibodies, the polyclonal antibodies verge toward monoclonal antibodies by the price. It is more important that the sensor based on the polyclonal antibodies detects all molecular forms of antigen. Monoclonal analytical system is conversely very sensitive to the state of the epitope identified. Therefore, it may underestimate the concentration value that does not show the content of all molecular forms of the antigen.

Generally, it required several months to obtain poly- or monoclonal antibodies due to the generating processes of immune response *in vivo* and the growth rate of cultured cells. Currently available offers of services to produce antibodies against any antigen of the customers compete with each other, reducing immunization time, but do not change the situation radically. Depending on the intensity of immune response and dynamics of its development, it required 1.5–3 months for the *in vivo* part of the procedure [16], and then in case of monoclonal technology, you need about a month for obtaining, screening, and growing the selected clones.

In vitro lymphocyte immunization is a method to speed up the production of antibodies significantly. In this case, the standard immunization protocol is 10–12 days [32], and the technique that reduces it to 6 days was recently described [33].

6.2.2 Artificial Antibody Fragments and Natural Miniantibodies

Complicated geometry of immunoglobulin, a small molecule part of which only is involved in the formation of complex with antigen, is not optimal to solve many analytical problems. Due to this, the surface density of antigen-binding sites is low when full-sized antibodies occupy a large area during persorption on the sensor.

The traditional approach to the fragmentation of antibodies is their proteolysis using pepsin or papain enzymes. Their application allow only those fragments of antibodies (Fab regions) that are responsible for binding to the antigen (see Figure 6.2). The use of these preparations increases antigen-binding capacity of the sensor surfaces.

An interesting natural pseudo-fragment of immunoglobulins was found in *Tylopoda*. Camel and llama blood contains not only the usual full-sized immunoglobulins but also IgG molecules formed by two heavy chains only [34] (see Figure 6.2). The absence of variable light chains in those antibodies of *Tylopoda*, named as nanobodies, is balanced out by increase in size and area of the contact with antigen for variable areas of heavy chains [35]. The production technique for identical single-domain antibodies of *Tylopoda* was developed (using phage display, which will be discussed hereafter— see Section 6.2.4) [36]. For the moment, it is not clear how the differences in the structure of the antigen-binding nanobody sites affect their affinity and specificity to different classes of antigens, but the active work in this area is under way [37,38].

The similar structure of single-domain antibody, also being a diabody of heavy chains, is described for sharks [39]. This poorly studied "novel antigen receptor" (Ig-NAR) seems to be evolutionarily the most ancient natural miniantibody.

There were a number of successful applications of nanobodies in immunosensor described. For example, the research [40] specifies the detection of prostate-specific antigen in significant diagnostic concentration 10 ng/mL using a biosensor based on detection of surface plasmon resonance. Soon after, the authors managed to reduce prostate-specific antigen detection threshold in the same system by more than an order of magnitude via optimization of immobilization regime [41]. However, a systematic comparative study on analytical capabilities of nanobodies and conventional full-sized antibodies has not been conducted yet.

6.2.3 Recombinant Antibodies

Recently developed techniques for manufacturing recombinant proteins using microorganisms are actively applied to antibodies too [42–45]. The antibodies are the "churly" subject for the products of the most traditional systems of bacterial producers due to their large size, complex multichain structure, and necessity of posttranslational glycosylation. Although the affinity of the recombinant antibodies is not inferior to poly- and mono-clonal antibodies and even may exceed by site-directed mutagenesis (see Section 6.3), more difficult manufacturing limits their use with therapeutic tasks mainly. The formation of single-chain structures, maintaining the ability to interact with antigens, is the predominant trend in the design of recombinant antibodies. The most known forms of these artificial antibodies usually combining the elements of natural light and heavy chains are shown in Figure 6.2.

Artificial antibodies obtained upon microbiological synthesis are often aggregation-prone making them difficult to purify [44]. However, this problem is solved by selection of special conditions for refolding [46–48].

Successful application of recombinant antibodies in immunosensors is summarized in the recent review [49], analyzing several dozens of experimental publications. The immunosensor systems were developed to identify a number of pesticides and mycotoxins, prostate-specific antigen, receptor 2 of human epidermal growth factor, and bacterial surface antigens [41,50–53]. It is noted [49] that the advantage of using small molecules of recombinant antibodies (primarily monovalent miniantibodies scFvs) makes possible their genetically engineered modification to include amino acids interacting with metals (cysteine, histidine) or peptides providing oriented immobilization on sensor scaffolding with high surface density.

6.2.4 Display Antibody Technologies

Quick production of antibody to a "new" antigen, for which the commercial antibodies are not yet available, is of utmost interest for biosensors. In 1990, a new technology, dubbed as the display (application of preparations that are a mixture of antibody producers of different specificity), was implemented [54]. The producers' mixture that is used in display technologies is called library, despite its obvious difference—in real library you can immediately take the book in the definite place, not looking over the entire storage. There are various kinds of display technology—phage (most popular), cellular, ribosomal, and mRNA displays. High-throughput screening provides test of library containing 10^8–10^9 variants [55]. When the antibody that the researcher is interested in is selected, its producer is used to obtain the target antibody in necessary quantity.

The first successful application of phage display–selected miniantibodies in sensors is described in [56]. The authors demonstrate electrochemical detection of lactose, bacteria *Listeria monocytogenes,* and secreted mycobacterial enzyme MtKatG. The production by phage display technique and application of antibodies to C-reactive protein is to be noted recently in [57]. The authors of the latter work specify good affinity of three selected antibodies with binding constants in the range from 27 to 10 nM. Immobilization of the selected antibodies on quartz scaffolding provided gravimetric determination of the purified antigen in buffer with good reproducibility for 15 cycles.

In some recent works, the structure of antigen-binding peptide is determined on the basis of phage library screening, and it is used as a receptor of biosensor with no production of more complex antibody molecule. Thus, in some cases, a significant increase in sensitivity compared to conventional immunoassay is observed. For example, in [58], 2500-fold difference was reached for magneto-electrochemical determination of herbicide molinate. Regardless of obtained optimistic results, currently available data are still not enough for grounded conclusion about preference for biosensor analysis of preparations produced with the use of display technology and their competitiveness with conventional antibodies.

6.2.5 Alternative Scaffolds

Immunoglobulins are not the only group of proteins for which the combinatorial variability of specific sites provides production of molecules that specifically interact with a variety of compounds. Over the last 15 years, it described more than 50 groups of so-called protein scaffolds that also combine conservative basic structure with limited hypervariable segments providing receptor function [59–64]. The following scaffolds are best described:

- Affibodies [65]—the structures based on Z-domain of staphylococcal protein A
- Engineered Kunitz domains [66] that have variability within the disulfide-fixed structure of serine protease inhibitor
- Monobodies or adnectins [67], based on the 10th extracellular domain of human fibronectin III (10Fn3)
- Anticalins [68]—derived lipocalins, a large family of eight-chain proteins presented in various organisms
- DARPins [69]—ankyrin modified repeated domains
- Avimers [70], based on multimeric LDLR-A module
- Knottin-rich peptides [71]

Overall fold dimensions of these proteins typically range from 40–50 to 200 amino acids; the amount of amino acids in hypervariability sites is 6.5–16. This provides considerable diversity of binding sites and highly selective detection of various compounds. The molecular size of alternative scaffold proteins are usually less compared with immunoglobulins, and their production is associated with lesser difficulties and expenses. The alternative sensor proteins may be characterized by display screening to select molecules that are the most affine for target compound [72].

There is a number of commercial manufacturers for scaffold (affibody protein A (www.affibody.com; www.abcam.com), anticalin lipocalin (www.pieris-ag.com), adnectin fibronectin III (www.adnexustx.com), DARPin ankyrin (www.molecularpartners.com), Kunitz domain APPI (www.dyax.com)) that provide potential intensification of work in this area in the coming years.

Due to the scaffolds diversity, the crucial question in work with them is the choice of criteria to select the most advanced compounds is assumed.

Boersma and Pluckthun [73] suggest the following criteria:

- Minimal aggregation or available simple technique to prevent it
- High level of expression without further aggregation
- Modifiability of protein molecules to improve stability
- Absence of disulfide bonds leading to variation of assembly in the course of expression
- Available synthetic or recombinant formation of scaffold-based multivalent structures
- Simple chemical conjugation (e.g., via single cysteine residue in molecule)

In addition, Ruigrok et al. [74] emphasize the importance of protein stability in a wide range of temperature and pH, under the influence of detergents and organic solvents. That is especially important for their application in biosensors. Therefore, the efficiency of proteins from thermophilic microorganisms is expected.

At the beginning of this chapter, we mentioned several thousands of publications on application of antibodies in biosensors. Alternative protein scaffolds are used only in few biosensor researches. Thus, the works [75,76] describe fluorometric microarray biosensors with extremely high (fM) limits of detection. In [77], the scaffolds with binding constants (0.78–2.38 10^8 M^{-1}) were applied for the detection of C-reactive protein using fiber-optic sensor based on the principle of reflectometric interference spectroscopy. The protein is detected in concentrations up to 39 ng/mL; it fully satisfies the practical requirements.

Many of the alternative scaffolds are not inferior to antibodies by affinity making them potentially competitive receptors for biosensors. However, the technologies for integration of these alternative scaffolds in the intermolecular conjugates, the requirements for stabilization, and prevention of nonspecific blocking of active sites during analysis in complex matrices are elaborated very poorly for the moment.

6.2.6 Polypeptide and Peptide Receptors

Up-to-date technologies can not only use the natural variability of protein molecules but also create the libraries of peptide and polypeptide structures to select the receptor molecules, effectively detecting target compound. The present-day practice for biosensor application of peptide receptor includes two approaches that differ significantly.

The first approach is based on the formation of peptide libraries obtained due to random combination of amino acids and their screening for binding with target compound. The selected most affinitive preparation is characterized in composition and produced in preparative amounts—either using traditional method of solid-phase peptide synthesis or recombinant technologies [78]. Unfortunately, the affinity of peptides selected from these randomly generated libraries is not very high. Antibodies and alternative scaffolds overmatch them in binding constants by several orders and are considered primarily as preliminary affine concentration tools in this regard [79]. However, there are application examples of peptides selected from phage libraries in sensory systems: for example, gravimetric detection of *Edwardsiella tarda* in the range $8 \cdot 10^2$–$8 \cdot 10^6$ cells [80], troponin I by electrochemical and gravimetric sensor with detection limits 0.11–0.34 µg/mL [81], and staphylococcal enterotoxin B by the sensor based on surface plasmon resonance (detection limit was not determined, the binding constant is set equal to $4.2 \pm 0.7 \cdot 10^5$ M^{-1}) [82].

Variation of peptide structure provides detection of not only biopolymers but also various substrates of practical significance [82]—such as polystyrene and gold—for which the antibody production is almost impossible (available single publications leave open the question of immune response induction and applicability of such antibodies for analysis in multicomponent samples [83,84]). Thus, peptide YLTMPTP is taken to be effective for binding with polystyrene [85,86] (binding constant with syndiotactic polystyrene [$2 \cdot 10^{11}$ M^{-1}]) and VSGSSPDS [87] and LKAHLPPSRLPS [88] for gold. These interactions are not directly related to recognition receptor processes in biosensors, but they can be effectively used for specific immobilization before the analysis.

The second approach is based on a sensor with a large number of binding zones (array type) in which a variety of peptides are immobilized. Characteristics of the test sample are based on the whole data array on binding availability or unavailability in each zone [89]. Despite a number of obtained interesting results, this approach is far from practical use. We can confirm or confute the identification of individual proteins in purified form. More substantial analysis requires characteristics of binding of all other proteins and is difficult to interpret in the case of multicompound mixtures.

6.2.7 Other Competitors of Antibodies

The variety of receptor molecules is not limited to proteins. There are biosensors based on the selective interaction of lectins with carbohydrates [90–94], but such systems are suitable for a certain class of objects only. The general and very active developing areas are researches on application of molecular imprinted polymers (MIPs), in which the analyte binding sites are formed as its "fingerprints" during polymerization, and aptamers (the fragments of nucleic acids). This book contains chapters of Erdem, Palchetti, and Marti describing biosensoric applications of aptamers and MIPs. This obviates the need to repeat their structural features and mechanism of interaction with target compound in this chapter.

Also it should be noted that available publications include the reviews on analytical applications of aptamers [95–100] and MIPs [101–104]. If the MIPs were considered as a means of selective detection of low molecular substances only earlier, then a number of recent works demonstrate their successful application for high molecular compounds [105–107].

As for the aptamers, the potential and most promising direction for further development of their applications in biosensor is the transition from a set of nucleotides that are typical of DNA and RNA to a wide range of available nucleotides and, as a result, providing higher affinity and specificity of detection [108–112].

6.2.8 Summary: Antibodies or Other Receptor Elements?

The conclusions about characteristics of different receptor elements that are relevant to their application in sensors are summarized in Table 6.1.

Regardless of the discovery of new molecular forms of antibodies and alternative receptors, the major part of biosensor developments is still focused on full-sized antibodies from mammals. In general, it is related to availability of these reagents. The decades of immunological studies provide production and commercial availability of poly- and monoclonal antibodies against a very large number of compounds. The catalogs of leading suppliers offer dozens of thousands of antibody preparations. In addition, the application of antibodies in different conditions and different reaction media has been examined in detail. The synthesis of antibody-based reagents for one or another analytic system can also be carried out by standard methods referring to vast experience of predecessors.

Researchers working with alternative receptors emphasize the complexity of the molecular structure of immunoglobulins and their insufficient stability with respect to various physical and chemical factors. However, up-to-date methods of genetic engineering make it possible to improve the structure of antibodies for successful overcoming of the current limitations in affinity [113,114] and stability [115,116].

Characterization of molecular properties of alternative receptors is too limited for the moment, and unavailability of standard technologies for their upscaled production leads to rising costs of the reagents compared to conventional antibodies. A very small number of experiments on application of alternative receptors in biosensors do not provide reliable, universal conclusions about their relative advantages and disadvantages. The researchers

TABLE 6.1

Comparative Analysis of the Advantages and Disadvantages for Antibodies and Alternative Receptors

Receptor	Polyclonal Antibody	Monoclonal Antibody	Nanobody	scFv	Protein Scaffold	Peptide	MIP	Aptamer
Affinity and specificity	+	+	+	+	±	–	±	+
In vitro manufacturing	–	±	±	+	+	+	+	+
Cheapness	+	–	–	–	–	–	+	–
Simple conjugation	+	+	–	–	–	–	–	–
Stability in different media	±	±	±[a]	±[b]	±	+	+	+

[a] A very small amount of data; high thermal stability registered [39,195].
[b] Less stable than the full-sized antibodies, the tendency to aggregation.

have to solve a lot of problems that have been solved for antibodies during transition to meaningful quantitative determination of various compounds in complex biomatrices using alternative receptors. That is why selection of the major bioanalytical challenges currently and in the nearest future is justified in favor of antibodies as the effective recognition element of biosensors.

6.3 Problems Solved in the Development of Immunosensors

6.3.1 Initial Comments

Developing systems include the so-called direct and indirect immunosensors, depending on whether the sensor detects the act of interaction between antigen and antibody molecules directly or the additional marker (such as enzyme or fluorophore molecule) is used for detection of this event. The antibodies are successfully combined with different physical principles, implemented in present-day biosensors. The devices based on detection of electrons, photons, or weight change dominate among immunosensors (as well as among other types of biosensors). Thus, the electrochemical, optical, and piezoelectric immunosensors stand out. The features, capabilities, limitations, and new perspectives for each of these biosensor areas are discussed in the relevant chapters of this book and do not require duplication in relation to antibodies as receptor elements. In this section, we consider the issues that have to be solved during bioanalytical application of antibodies regardless of what method of signal recording is applied ultimately.

6.3.2 Immobilization of Antibodies

Majority of immunosensors use antibodies or antigen immobilized on the sensitive surface of the sensor. In the case of immobilized antigen, the recommendations for immobilization are not common and depend on its nature. If the chosen format of analysis is based on immobilization of antibodies (or their derivatives) at the surface of the biosensor and registration of its binding to the antigen in the test sample, the requirements for immobilization method are rather universal. The sensor should provide the maximum ratio of the response to the concentration of the analyte in the sample, thus achieving the lowest detection limits. The maximum compact immobilization of antibodies at the surface of the sensor, not leading, however, to structure change in the active site of the antibody or restriction of its availability to interact with antigen is preferable. Note that it is undesirable to improve the antigen-binding capacity using the procedures that reduce the detected signal—for example, to form additional layers that reduce conductivity of the electrochemical sensor. On the contrary, the spatial convergence of the label

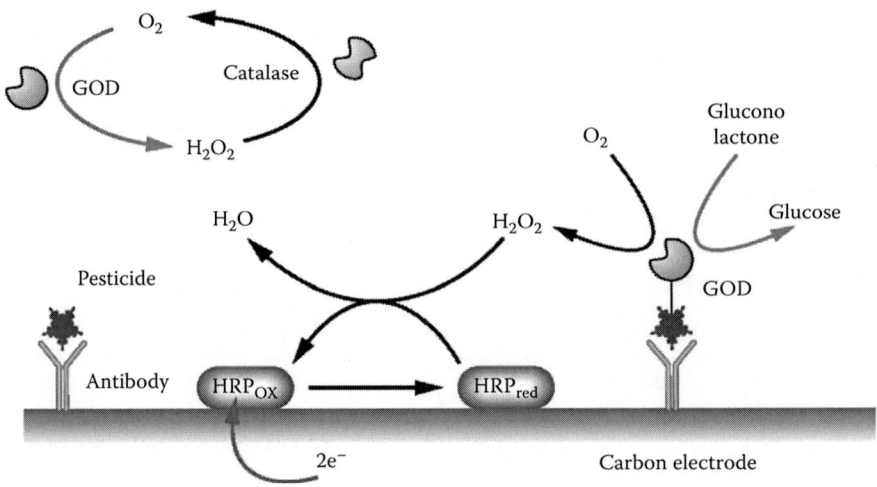

FIGURE 6.3
Bienzyme system for nonseparating electrochemical immunosensor. GOD, glucose oxidase; HRP, horseradish peroxidase.

and the sensitive surface allows in some cases to detect immune complexes without separation of bound and unbound label [117,118] (see Figure 6.3).

The simplest way to immobilize the antibody is the physical adsorption (Figure 6.4). However, there are a lot of systems described in which this process is accompanied by denaturation or significant conformational changes of antibodies [119]. In addition, according to the application of the sensor produced by noncovalent immobilization, especially in nonphysiological media, the number of antibodies dissociated from its surface increases. Clear alternative, which is

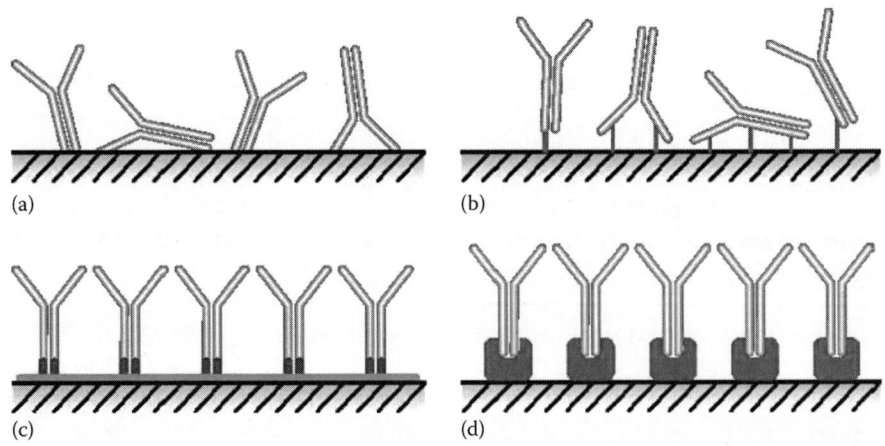

FIGURE 6.4
Main variants of antibody immobilization on sensor surface: (a) random noncovalent, (b) random covalent, (c) oriented via binding site of the antibody, (d) oriented via additional binding molecule.

also implemented in a large number of developments, is covalent coupling of anti-bodies to chemically activated surface of the sensor. Generally, the methods that are used randomly direct the coupled antibodies against the surface attachable—depending on which reactive group of antibodies is in contact with it.

The described problems could be potentially solved and are solved suc-cessfully in a number of cases by oriented immobilization of antibodies. It is important that unification of antibody orientation at such immobilization provide unification of antigen-binding properties and thus provide more reproducible characteristics of the sensor.

The antibodies can be immobilized with chosen orientation in different ways. For example, the availability of a limited number of cysteines that are considerably distant from the antigen-binding site in immunoglobulin mole-cule is widely used. Therefore, the immobilization method may include direct chemical conjugation of IgG with sensor surface coated with dextran [120] or thiol [121] layer. Fab fragment of antibody obtained via reduction of Fab2 frag-ment is immobilized at the surface of thiol even more easily than the native antibody [122]. C-terminal cysteine residue included into the recombinant antibody was also successfully used for oriented immobilization [123].

Oriented immobilization of antibodies can be also performed by their indirect binding to the sensor surface where the antibody-binding layer is preformed. Such natural proteins as staphylococcal protein A, streptococcal protein G [124,125], or (strept)avidin coupled to biotinylated antibodies [126] can perform the orienting function. For example, in [127], homogeneous antibody–antigen and antibody–protein A interactions were followed by a fast transfer of formed oriented immune complexes on the sensor surface by electrostatic interaction of oppositely charged polymers at the surface and in immune complexes. A pair of oppositely charged polyelectrolytes may be considered as efficient tools for such rapid separation of immuno-reactants from the reaction mixture [15,128] (Figure 6.5).

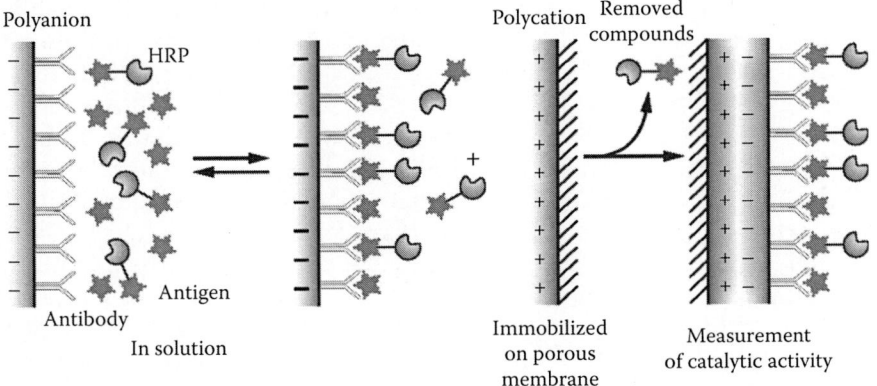

FIGURE 6.5
Principle of immunoreactants separation by the interaction of oppositely charged polyions.

Lowering the limit of detection, reached by oriented immobilization of antibodies, varies considerably for different immunosensors. In a recent paper [129], this parameter varied from 1–2 to 20–227 times for different antigens detected by immunosensor based on surface plasmon resonance. The mechanisms for these differences are unclear at present and require further detailed study for different kinds of immunosensors.

6.3.3 Affinity of Antibody and Limit of Detection for Immunosensor

Generally noted advantage of antibodies is their high affinity of complex formation with corresponding antigens, determinative capability of antigen detection in extremely low concentrations. However, the affinity of immune complexes is limited by the nature of immune response, and this factor determines the limit of detection for immunosensors.

The maximum association rate constant (ka) is determined by the diffusion rate of the protein molecules in solution and amounts to approximately $10^6\,M^{-1}c^{-1}$ [130]. The dissociation rate constant (kd) can be increased due to selection of antibodies *in vivo* [131,132] but up to a certain limit only. According to the "affinity ceiling" hypothesis [131,133,134], selection of antibodies with high affinity is less effective. This is due to the fact that for antibodies with kd = $10^{-4}\,c^{-1}$, the half-life of antigen complex with B cellular receptor is 30 min and is several times greater than endocytosis time of the antibody complex with the B cellular receptor (about 8.5 min). Further increase in half-life of the complex does not support proliferation of B cells [133–135]. Due to this, antibodies of IgG class against protein antigens during the secondary immune response are characterized by the values of dissociation constant (KD) in the range $10^{-7}–10^{-10}\,M^{-1}$.

It should be noted, however, that the foregoing limitation for selection of high-affine clones does not exclude the possibility that some antigens achieved significantly higher degree of complementarity with the antigen-binding site of the antibody and, accordingly, a higher binding constant. Non-dissociating antigen–antibody complexes with infinite affinity have been described in [136]. Similar effects have been described for the interaction between antigen and antibody–metal ion complex [137], as well as for alternative protein scaffold—affibodies [138].

It should be noted that the relationship between the binding constant of immunochemical reaction and limit of the target antigen detection in the immunoassay (immunosensor analysis) essentially depends on the kind of immunoassay format. Classifications of immunoassay formats are extremely various. However, for appraisal of their capabilities first of all, the formats are divided into noncompetitive and competitive ones (Figure 6.6). In case of the former, the immune complex formation process is identified directly similar to, for example, sandwich assay with formation of antibody–antigen–targeted antigen complexes. In the latter case, the competition between antigen in the sample and the second antigen preparation available in the system is registered for binding to antibodies with subsequent registration of the immune complexes formed.

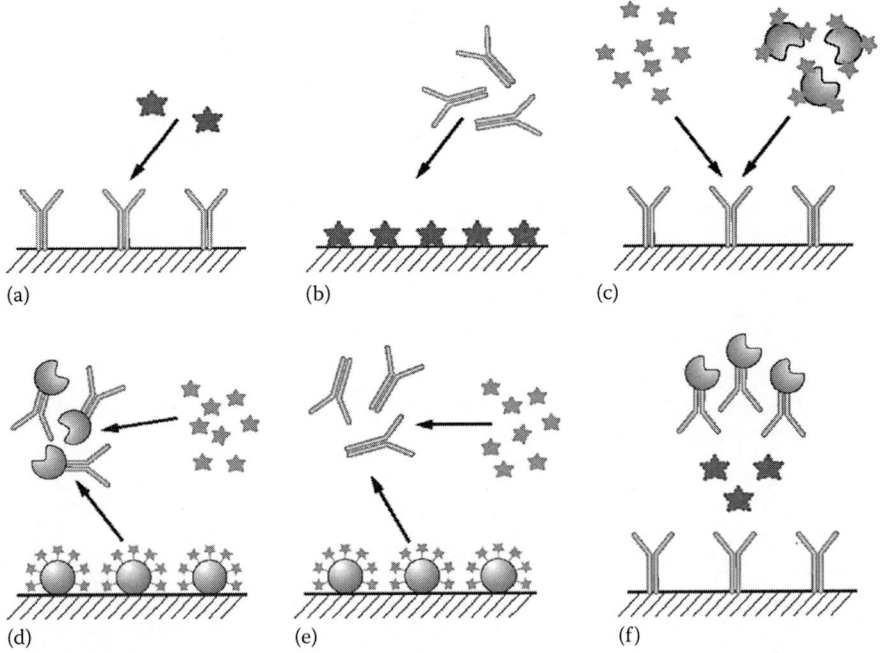

FIGURE 6.6
Main formats of immunosensor assays: (a) direct assay of antigen, (b) direct assay of antibodies, (c) competitive assay with immobilized antibodies and labeled antigen, (d) competitive assay with immobilized antigen and labeled antibodies, (e) competitive assay with immobilized antigen and unlabeled antibodies, (f) sandwich assay.

In 1986, Jackson and Ekins [139] carried out theoretical comparison of noncompetitive and competitive immunoassays in terms of sensitivity, precision, kinetics, and working range of analyte. The theoretical limits of detection for those methods and their relationship to the characteristics of antibodies were determined. After this, a number of developments in the immunoassay as a whole and in immunosensors appear with various methods of signal amplification in order to reach lower limit of detection [140–143]. However, these systems are also described by the general theory and differ only in the possibility of a more sensitive detection of the label due to its amplification.

It is expected that noncompetitive analysis would potentially identify very low concentrations of antigen, if it is admitted by sensitivity of label detection or direct detection of immune complexes as well as low background signal. There are several works that describe detection of individual antigen molecules using one or another means of signal amplification [144–146]. Metal nanoparticles [147,148] and formation of multilayer systems with antibodies and aptamers [149] are typical tools to amplify the signal in these sensors (Figures 6.7 and 6.8).

It is evident that the actual limit of valid analysis is higher since it requires not only signal registration but confirmation of its difference from the level of

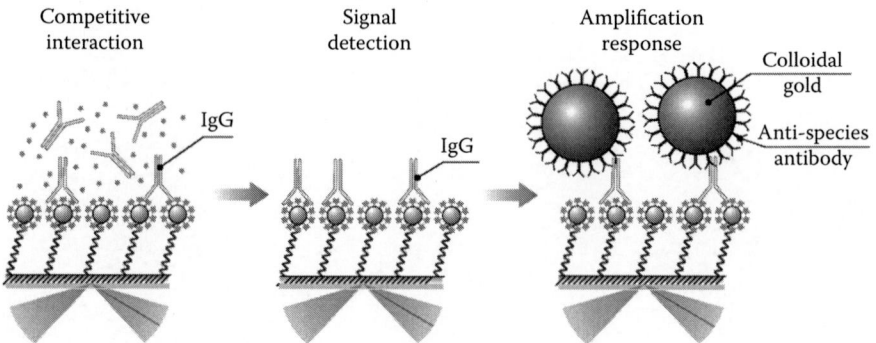

FIGURE 6.7
Principle of signal amplification in competitive surface plasmon immunosensor via indirect labeling of immune complexes by gold nanoparticles.

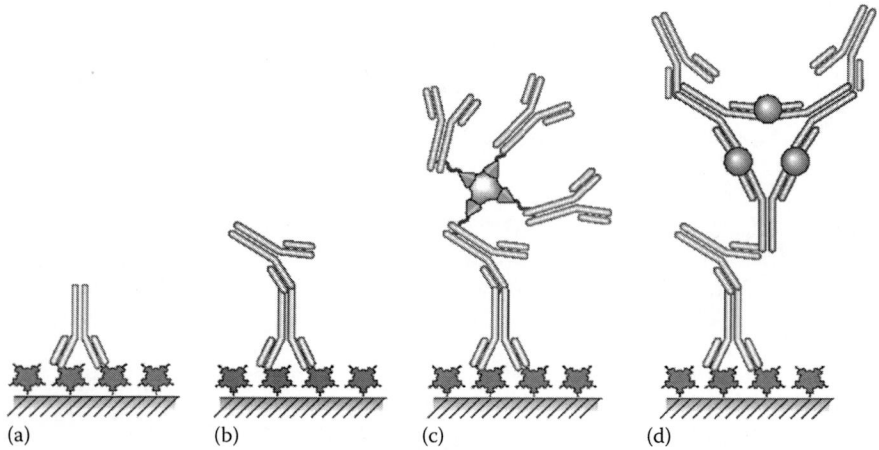

FIGURE 6.8
Standard method and three amplification methods for the detection of HBV antibodies using surface plasmon immunosensor: (a) direct assay, (b) amplification using sandwich assay, (c) avidin-biotinylated antibodies, and (d) peroxidase–anti-peroxidase (PAP) complex.

nonspecific interaction in the analytic system. Therefore, when considering such records, it is important to assess the accuracy and reproducibility of measurements aimed at detecting a few analyte molecules in the reaction medium [150].

In concluding this section, it should be emphasized that despite the emergence of new methodological solutions that reliably detect extremely low concentrations of immune complexes (and even single interaction), they generally require multistep measurements and additional complex instrumentation. Therefore, for highly sensitive immunoassay directed to widespread use, recommendations of Jackson and Ekins are still relevant and use of antibodies with maximum affinity is important.

6.3.4 Noncompetitive Immunoassay of Low Molecular Weight Antigens as a Way to Overcome Sensitivity Limitations

Both theoretical analysis [139] and practical implementation of various immunosensor systems demonstrate that two-site sandwich immunoassay enables reaching a lower detection limit as compared with the competitive formats. However, implementation of the sandwich assay in its traditional version provides for an antigen having two spatially separated antigenic determinants, which is not possible for low molecular weight compounds.

Can this contradiction be overcome to have a noncompetitive immunosensor detection of such compounds? Within the latest 2 dozen years, the researchers have put forth quite a few original approaches to address the issue based on the chemical modification of immunoreactants and their complexes, novel separation schemes, use of antibodies against special epitopes, etc. A detailed evaluation of these results has been presented in a recent review [151], which is why we shall confine ourselves to brief characteristics of several methodological solutions having maximal importance for immunosensors.

Ishikawa's method: The scheme proposed by the Ishikawa team provides conjugation of biotin with the compound to be detected prior to conducting the assay. As a result, a biotinylated analyte can be bound with the specific antibodies and with avidin, which allows implementing a two-site noncompetitive immunoassay [152]. The authors described extremely high sensitivity of the assay: up to 10 attomoles for angiotensin I and arginine vasopressin [153,154]. Nevertheless, a great number of stages and the associated long assay may limit application of the assay.

Solid-phase immobilized epitope immunoassay (SPIE-IA): In the assay scheme developed by Pradelles et al. [155], the analyte initially is bound to the immobilized antibody, and with the assistance of a cross-bridging reagent, the formed complex is fixed by covalent links. A dissociating treatment by HCl or methanol destroys links of the immune complex, yet the created covalent link is preserved during this process. As a result, the bound antigen molecule becomes accessible for immunochemical interaction, while a labeled antibody added to the system leads to the formation of a triple detectable complex. Subsequently, for a series of systems, 10–300-fold fall of the analyte detection limit using this method as compared to the traditional competitive immunoassay was described [156–158].

Idiometric assay: According to the immune network theory, the anti-idiotypic antibodies that are specific to variable section of initial antibody may be subdivided into two basic types: those nonsensitive to antigen presence at binding site (alphatype) and antibodies whose binding with variable section is competitively inhibited by the antigen (betatype) [159]. Based on this concept, Barnard and Kohen [160] have proposed an original noncompetitive assay for estradiol detection in serum. After forming (immobilized antibody + antigen contained in the sample) complexes on a surface, anti-idiotypic betatype antibodies are added to block the antibodies that have not reacted with the antigen. A subsequent addition of labeled anti-idiotypic alphatype antibodies allows to determine the antigen content in the sample. Later on, the authors described applications of the approach to solve various analytical tasks [161–163]. Thus, for instance, a recent work [164] has demonstrated a threefold reduction in the limit of cortisol detection, which allows an efficient characterization of the physiologically significant concentration range of this compound in biosamples.

Noncompetitive immunoassay using anti-metatype antibodies: According to Voss et al. [165], the anti-metatype antibodies are bound to the immune complex but are featured by an extremely low affinity in respect of the single antigen or antibody molecules. The property has been used by Self et al. [166] to detect digoxin via registering the formation of "anti-digoxin antibody–digoxin-labeled anti-metatype antibody" complexes.

Open sandwich immunoassay (OSIA): The method is based on the association effect of the separated VH and VL chains from the variable antibody domain that is greatly reinforced in antigen presence. By implementing the approach, Ueda et al. [167] have immobilized a VH fragment of anti-lysozyme antibody on surface and then added in a lysozyme-containing sample and a labeled VH fragment. The label bounding degree is lysozyme concentration dependent and enables a correct quantification of the antigen. Although the lysozyme is not a hapten and may be detected by noncompetitive immunoassays, OSIA is potentially acceptable to determine various antigens. Afterward, the authors have also described other applications of the given immunoassay [168,169].

6.3.5 Reaction Capability of Antibodies under Various Conditions

Antibodies as rather large biological molecules are quite stable to variations of conditions within the "physiological limits," but on borders of the stability area, they may reduce their affinity in a reversible way. Immunoassay conducted directly in the tested sample matrix could cause differences in comparison with the optimal antigen–antibody reaction conditions under such parameters as temperature, pH, and ionic strength, which is extremely important to be taken into account when developing any antibody-based sensors. The problem is addressed in part by diluting the sample with the reaction buffer that, however, also reduced the analyte detection limit in the initial sample.

The situation around the optimum temperature is quite nontrivial. The antigen–antibody reaction that is an exothermic one has a higher efficiency at lower temperatures. Therefore, it would be logical to conduct immunoassays at the temperature around 4°C. In reality, a considerable interaction deceleration is observed under these conditions, while the antibody conformation modifications at low temperatures affect adversely the affinity. Therefore, carrying out immunodetection at these temperatures is only justified in case of a low analyte stability. Nonetheless, for a fair part of the compounds, abandoning the physiological 37°C or higher and choosing to conduct all the assay stages at the ambient room temperatures have no negative consequences.

The need for immunodetecting hydrophobic substances dictated the choice of conducting immune reactions in water–organic solvent mixtures, which enables testing without a preliminary dilution of extract (which is important for determination of low concentrations of the analyte), as well as a methodological assay streamlining.

The major reasons for inactivation of protein molecule, when incubated with the organic solvents, are the changes in the properties of the medium surrounding the protein (at high solvent concentrations) and hydrogen bounds and hydrophobic interactions (at low solvent concentrations) [170,171].

The immunochemical interactions in nonaqueous media have been studied at the phenomenological level. The description has been provided for the solvent effect (methanol, ethanol, methyl fluoroform, hexane, acetonitrile, dimethylformamide (DMFA), and dimethylsulfoxide) on the antibodies' antigen-binding capability [53,172,173]. A work [53] has established that the presence of 20% of ethanol leads to a fall of the monoclonal antibody F_{ab} fragment capability to bind with its antigen (human ferritin) 2.4- or 25.8-fold, depending on its structure. Melnikova et al. [174] demonstrated that minor changes of antibody affinity may be already observed at organic solvent concentration of 5%. After incubation of anti-ferritin monoclonal antibodies with DMFA, a two- to sixfold increase in the binding constant values was observed, while the maximum activation effect was observed at DMFA concentrations of 5%–13% and 11%–40%, depending on the antibodies. A work [175] has shown that *E. coli*-expressed monoclonal IgG fragment against atrazine containing a mobile bridge and a disulfide link between H- and L-chains has the highest stability to the denaturing effect of polar and nonpolar solvents.

There are examples of a successful use of the organic solvents in the immunosensors. Thus, for instance, an article [175] has demonstrated that at a 10% methanol content it is possible to quantify the atrazine extracted from soil. An increase in the methanol content up to 30% leads to the assay's sensitivity loss. A further methanol addition up to 50% causes a substantial signal decrease that excludes conducting an assay under these conditions. Quite interesting is an approach that combines methanol use with a low-temperature assay [176]. It has been shown that an optimum condition for the detection of etofenprox insecticide is provided under the incubation temperature of 4°C in the presence of 50% methanol.

Triazine and phenylurea pesticides have been detected in undiluted (up to 99%) ethanol or hexane by solid-phase ELISA and immunofluorescent assay, with covalently immobilized antibodies and a labeled antigen [177]. Low-temperature experiments have demonstrated that the immunoagents are stable at a wide range of ethanol concentrations (from 0% to 99% by volume).

Atrazine ELISA at 0°C in the presence of hexane has led to a 74-fold IC_{50} increase as compared with the Tween–phosphate buffer. Determination of N-(2-N-chloroacetyl-aminobenzyl)-N'-4-chlorphenylurea in hexane, by contrast, has enabled a threefold IC_{50} reduction versus the Tween–phosphate buffer [177]. The ELISA sensitivity and selectivity in various solvents (water, ethanol, and hexane) have correlated with the antigen solubility. The obtained findings represent a significant interest also for developing immunosensor determination of water-insoluble compounds.

Some recent publications have presented data on an efficient immunosensor detection of triazine herbicides in a 50% chloroform hexane [178] and aflatoxin B1 and ochratoxin A in methanol extracts [148,179,180]. Apparently, a high antibody stability in these sensor systems that is not characteristic of pure antibody solutions is related to the immobilization effect, as well as complexing with a hydrophobic analyte [180], yet these issues have still not been studied in detail.

Some specific antigens generate own demands to the reaction media. For example, in immunoassay of surfactants, such common tools preventing nonspecific binding as Tween and Triton should be excluded [181].

6.3.6 Molecular Design of Antibodies to Control Their Affinity, Specificity, and Stability

The natural techniques of antibody structure variation (see Section 6.2.1) have their limitations. The recombinant variability is restrained by a set of segments used to create an antibody gene, while a somatic hypermutation is, predominantly, about single nucleotide replacements at short sections of amino acid chains. Therefore, for a directed modification of antibody affinity in respect of the target antigen and its structural analogues, it is justifiable to apply modern technologies that enable implementing many more molecular versions of antibodies. An artificial mutagenesis of the antibody-producing lines provides a large number of versions, out of which the researchers can select an optimum preparation to address their tasks. Whereas on the basis of theoretical spatial structure calculations, one might propose specific structural modifications of antigen-binding site to be implemented by the genetic engineering methods. An ideal solution would be provided by a combination of several mutation changes, the high-performance screening techniques, as well as the phage display SELEX technology to be used for characterization and optimum selection of antibodies [182].

Ensuring an efficient folding of the recombinant antibody preparations obtained, as well as improving their thermodynamic stability, are seen as important assignments of the molecular design of recombinant derivatives of antibodies. Of particular significance is the issue of miniantibody stability, so far as, when designing the Fv and scFv fragments, the variable domains find themselves beyond their usual surroundings, where they are stabilized by the constant domains of the light (CL) and heavy (CH1) chains. It results in the miniantibodies with no additional structural modifications losing their stability [183,184]. The frequently used fusion of VL and VH domains leads to the creation of a helically stabilized Fv fragment hsFv [185]. Alternative solutions are represented by disulfide stabilization of the Fv fragment accompanied by formation of the dsFv miniantibody [186], disulfide link introduction into scFv accompanied by formation of ds-scFv [187], and reformatting scFv into a structure similar to the natural Fab fragment [188].

However, the basis objective pursued when developing new recombinant antibodies is their affinity increment and, consequently, a reduction in the detection limit of the immunosensor implemented on the basis of these antibodies. The key parameter of antigen–antibody reaction is the value of kinetic association constant. Reaching the values of the variable of the size of order 10^5 M^{-1} s^{-1} allows an immunochemical interaction and its outcome registration within approximately 15 min [119]. The value of kinetic KD is a less critical parameter for the immunosensor measurements that are overwhelmingly based on the registration of immune compound formation rather than dissociation. The kinetic KD values of the order 10^{-3} s^{-1} enable minimizing the contribution of the dissociation process, when taking measurements. Proceeding from these considerations, the molecular design task should involve above all ensuring a greater accessibility of antigen-binding site and a growing probability of efficient contacts with the antigen molecule. Nonetheless, the developments known today solve the problem of modifying the two constant kinetic values, while the constant dissociation value variations may modify significantly the antibody specificity *vis-a-vis* the structural analogues of the antigen to be determined [113].

Some examples of modified antibody affinity using the genetic engineering are considered later.

Razai et al. [189] describe an increased affinity of the single-chain miniantibodies that neutralize the botulinic toxin BoNT of serotype A (BoNT/A) achieved using a nonspecific mutagenesis and a subsequent yeast display screening. For two miniantibodies, a 45-fold (from $9.43 \cdot 10^{-10}$ to $2.1 \cdot 10^{-11}$ M) and 37-fold (from $8.44 \cdot 10^{-10}$ to $2.26 \cdot 10^{-11}$ M) constant dissociation value reduction was achieved. At the same time, in respect of the full-sized antibodies, comparable and even somewhat greater changes, from $6.07 \cdot 10^{-11}$ to $1.71 \cdot 10^{-12}$ M and from $4.51 \cdot 10^{-11}$ to $5.54 \cdot 10^{-13}$ M, respectively, were observed. The authors indicate that growing affinity was related to the changing values of association and dissociation constants. Due to a higher affinity, the authors have succeeded in ensuring the detection of botulinic toxin with an extremely low detection level of up to $1 \cdot 10^{-13}$ M.

The techniques of molecular design based on computer-aided electrostatic interaction at the antigen-binding antibody site analysis ensure on average the same gain in the antibody affinity and immunodetection sensitivity. Thus, for instance, Lippow et al. [190] describe a 10-fold increase of affinity for antibody against epidermal growth factor receptor (up to 52 pM) and 140-fold increase (up to 30 pM) for antibody against lysozyme. The approach versatility has been confirmed by their successful use for the antibodies and to other protein antigens, as well as to fluorescein, a low molecular compound.

The present-day status in the area of molecular antibody design has been summarized in the recent reviews [113,191]. Their authors indicate the importance of addressing two tasks: selection of complementary amino acid chains in the hypervariable sections and calculation of a sterically optimum configuration of the antigen-binding site formed after combining heavy and light chains. At the same time, the available data enable the statement that the key role in ensuring the affinity is played by one of the six hypervariable loops, namely, the CDR-H3 loop.

A massive pool of experience has been accumulated in using the recombinant antibodies obtained by the molecular design techniques in immunosensor assays. Thus, for instance, one of the works [192] contains a description of a 17-fold increase in the affinity of antibody versus atrazine, which has ensured similar sensitivity gains in the immunoenzyme assay and immunosensor assay based on the surface plasmon resonance registration.

We should further emphasize that the genetic engineering methods enable collecting on the basis of the miniantibody blocks quantivalent structures featured by a significantly higher affinity. Thus, for instance, a tetravalent multimer of antibody to disialoganglioside GD2, which is a marker of a row of tumor cells, has been successfully used in the immunosensor for a direct detection of the target compound with the help of the multigravimetric method [193].

6.4 Conclusions

The discussion presented previously enables formulating the most promising directions for the improvement of antibody-based sensors:

- Increasing the antigen-binding capacity of the sensitive sensor surface attributed to the use alongside the full-size antibodies of their synthetic or recombinant fragments, oriented immobilization of antibodies or their fragments, and search for new surfaces and labels that ensure the most efficient generation and transfer of the detected signal

- Molecular antibody design: increasing its affinity, specificity regulation, and reaching higher stability under the sensor operation conditions, including regeneration cycles

- Search for new solutions enabling the combination of high immunosensor sensitivity with kinetic mode, as well as for new signal amplification systems (both for direct immunosensors and the immunosensors using the labeled compounds)

At present, in spite of great many research works in the sphere of the bio- and immunosensors, their range as commercially available products is not so broad. Today's biosensor market, in general, is formed by quite a limited product list. Approximately 50% of the sales volume is accounted for the glucose sensors, which jointly with the assay systems for blood-solved gases and blood electrolytes content make 90% of the world market [194]. Although the immunosensors are in demand for important practical tasks and despite the evident success of the R&D developments to achieve a high sensitivity, rapidity, miniaturization, and a higher performance, immunosensor integration with the common practices is changing very slowly.

However, some success is undeniable. Generally, these are serial-production immunosensors based on the registration of surface plasmon resonance and versions of flow immunochips in array format with a fluorescent detection. One may venture saying that a further extension of the commercially available immunosensor list is to be expected shortly, which immunosensors will implement the scientific information accumulated in this area.

The authors thank A.E. Urusov for help in figure preparation.

References

1. Janata, J., Immunoelectrode. *Journal of the American Chemical Society*, 1975; **97**(10): 2914–2916.
2. Aizawa, M. et al., An enzyme immunosensor for IgG. *Journal of Solid-Phase Biochemistry*, 1976; **1**(4): 319–328.
3. Ricci, F. et al., A review on novel developments and applications of immunosensors in food analysis. *Analytica Chimica Acta*, 2007; **605**(2): 111–129.
4. Liu, G. and Y. Lin, Nanomaterial labels in electrochemical immunosensors and immunoassays. *Talanta*, 2007; **74**(3): 308–317.
5. Skottrup, P.D., M. Nicolaisen, and A.F. Justesen, Towards on-site pathogen detection using antibody-based sensors. *Biosensors and Bioelectronics*, 2008; **24**(3): 339–348.
6. Hirst, E.R. et al., Bond-rupture immunosensors-a review. *Biosensors and Bioelectronics*, 2008; **23**(12): 1759–1768.
7. Hu, W.H. and C.M. Li, Nanomaterial-based advanced immunoassays. *Wiley Interdisciplinary Reviews-Nanomedicine and Nanobiotechnology*, 2011; **3**(2): 119–133.
8. Ramirez, N.B., A.M. Salgado, and B. Valdman, The evolution and developments of immunosensors for health and environmental monitoring: Problems and perspectives. *Brazilian Journal of Chemical Engineering*, 2009; **26**(2): 227–249.
9. Van Dorst, B. et al., Recent advances in recognition elements of food and environmental biosensors: A review. *Biosensors and Bioelectronics*, 2010; **26**(4): 1178–1194.
10. Luppa, P.B., L.J. Sokoll, and D.W. Chan, Immunosensors—Principles and applications to clinical chemistry. *Clinica Chimica Acta*, 2001; **314**(1–2): 1–26.

11. Dzwolak, W., R. Koncki, and S. Glab, Immunosensors in analytical chemistry. *Chemia Analityczna*, 1996; **41**(5): 715–736.
12. Holford, T.R.J., F. Davis, and S.P.J. Higson, Recent trends in antibody based sensors. *Biosensors and Bioelectronics*, 2012; **34**(1): 12–24.
13. Hock, B., Antibodies for immunosensors—A review. *Analytica Chimica Acta*, 1997; **347**(1–2): 177–186.
14. Stefan, R.I., J.F. van Staden, and H.Y. Aboul-Enein, Immunosensors in clinical analysis. *Fresenius Journal of Analytical Chemistry*, 2000; **366**(6–7): 659–668.
15. Dzantiev, B.B., A.V. Zherdev, and N.A. Byzova, Immunochemical approaches for rapid detection of biologically active compounds. *Defense against Bioterror: Detection Technologies, Implementation Strategies and Commercial Opportunities*, eds. D. Morrison et al., Vol. 1. Springer: Dordrecht, the Netherlands, 2005, pp. 291–301.
16. Leenaars, P. et al., The production of polyclonal antibodies in laboratory animals—The report and recommendations of ECVAM Workshop 35. *Atla-Alternatives to Laboratory Animals*, 1999; **27**(1): 79–102.
17. Mierendorf, R.C. and R.L. Dimond, Functional-heterogeneity of monoclonal-antibodies obtained using different screening assays. *Analytical Biochemistry*, 1983; **135**(1): 221–229.
18. Serrato, J.A. et al., Differences in the glycosylation profile of a monoclonal antibody produced by hybridomas cultured in serum-supplemented, serum-free or chemically defined media. *Biotechnology and Applied Biochemistry*, 2007; **47**: 113–124.
19. Kunkel, J.P. et al., Comparisons of the glycosylation of a monoclonal antibody produced under nominally identical cell culture conditions in two different bioreactors. *Biotechnology Progress*, 2000; **16**(3): 462–470.
20. Muthing, J. et al., Effects of buffering conditions and culture pH on production rates and glycosylation of clinical phase I anti-melanoma mouse IgG3 monoclonal antibody R24. *Biotechnology and Bioengineering*, 2003; **83**(3): 321–334.
21. Vlasak, J. and R. Ionescu, Heterogeneity of monoclonal antibodies revealed by charge-sensitive methods. *Current Pharmaceutical Biotechnology*, 2008; **9**(6): 468–481.
22. Storch, G.A. and C.S. Park, Monoclonal-antibodies demonstrate heterogeneity in the G-glycoprotein of prototype strains and clinical isolates of respiratory syncytial virus. *Journal of Medical Virology*, 1987; **22**(4): 345–356.
23. Liu, H.C. et al., Heterogeneity of monoclonal antibodies. *Journal of Pharmaceutical Sciences*, 2008; **97**(7): 2426–2447.
24. Dengl, S. et al., Aggregation and chemical modification of monoclonal antibodies under upstream processing conditions. *Pharmaceutical Research*, 2013; **30**: 1–20.
25. Tracy, R.P. et al., Comparison of monoclonal and polyclonal antibody-based immunoassays for osteocalcin—A study of sources of variation in assay results. *Journal of Bone and Mineral Research*, 1990; **5**(5): 451–461.
26. Shankaran, D.R. et al., Evaluation of the molecular recognition of monoclonal and polyclonal antibodies for sensitive detection of 2,4,6-trinitrotoluene (TNT) by indirect competitive surface plasmon resonance immunoassay. *Analytical and Bioanalytical Chemistry*, 2006; **386**(5): 1313–1320.
27. Tommasi, M. and S. Raspanti, Comparison of calcitonin determinations by polyclonal and monoclonal IRMAs. *Clinical Chemistry*, 2007; **53**(4): 798–799.

28. Borowitz, D., R. Lin, and S.S. Baker, Comparison of monoclonal and polyclonal ELISAs for fecal elastase in patients with cystic fibrosis and pancreatic insufficiency. *Journal of Pediatric Gastroenterology and Nutrition*, 2007; **44**(2): 219–223.

29. Dominguez, J. et al., Comparison of a monoclonal with a polyclonal antibody-based enzyme immunoassay stool test in diagnosing *Helicobacter pylori* infection before and after eradication therapy. *Alimentary Pharmacology & Therapeutics*, 2006; **23**(12): 1735–1740.

30. Prontera, C. et al., Comparison between analytical performances of polyclonal and monoclonal electrochemiluminescence immunoassays for NT-proBNP. *Clinica Chimica Acta*, 2009; **400**(1–2): 70–73.

31. Le Berre, M. and M. Kane, Biosensor-based assay for domoic acid: Comparison of performance using polyclonal, monoclonal, and recombinant antibodies. *Analytical Letters*, 2006; **39**(8): 1587–1598.

32. Zafiropoulos, A. et al., Induction of antigen-specific isotype switching by *in vitro* immunization of human naive B lymphocytes. *Journal of Immunological Methods*, 1997; **200**(1–2): 181–190.

33. Kato, M. et al., A method for inducing antigen-specific IgG production by *in vitro* immunization. *Journal of Immunological Methods*, 2012; **386**(1–2): 60–69.

34. Hamerscasterman, C. et al., Naturally-occurring antibodies devoid of light-chains. *Nature*, 1993; **363**(6428): 446–448.

35. Muyldermans, S., C. Cambillau, and L. Wyns, Recognition of antigens by single domain antibody fragments: The superfluous luxury of paired domains. *Trends in Biochemical Sciences*, 2001; **26**(4): 230–235.

36. Desmyter, A. et al., Antigen specificity and high affinity binding provided by one single loop of a camel single-domain antibody. *Journal of Biological Chemistry*, 2001; **276**(28): 26285–26290.

37. Tillib, S.V., "Camel nanoantibody" is an efficient tool for research, diagnostics and therapy. *Molecular Biology*, 2011; **45**(1): 66–73.

38. Muyldermans, S. et al., Camelid immunoglobulins and nanobody technology. *Veterinary Immunology and Immunopathology*, 2009; **128**(1–3): 178–183.

39. Dooley, H., M.F. Flajnik, and A.J. Porter, Selection and characterization of naturally occurring single-domain (IgNAR) antibody fragments from immunized sharks by phage display. *Molecular Immunology*, 2003; **40**(1): 25–33.

40. Huang, L. et al., Prostate-specific antigen immunosensing based on mixed self-assembled monolayers, camel antibodies and colloidal gold enhanced sandwich assays. *Biosensors and Bioelectronics*, 2005; **21**(3): 483–490.

41. Saerens, D. et al., Engineering camel single-domain antibodies and immobilization chemistry for human prostate-specific antigen sensing. *Analytical Chemistry*, 2005; **77**(23): 7547–7555.

42. Wesolowski, J. et al., Single domain antibodies: Promising experimental and therapeutic tools in infection and immunity. *Medical Microbiology and Immunology*, 2009; **198**(3): 157–174.

43. Weisser, N.E. and J.C. Hall, Applications of single-chain variable fragment antibodies in therapeutics and diagnostics. *Biotechnology Advances*, 2009; **27**(4): 502–520.

44. Holliger, P. and P.J. Hudson, Engineered antibody fragments and the rise of single domains. *Nature Biotechnology*, 2005; **23**(9): 1126–1136.

45. Carter, P.J., Potent antibody therapeutics by design. *Nature Reviews Immunology*, 2006; **6**(5): 343–357.

46. Berdichevsky, Y. et al., Matrix-assisted refolding of single-chain Fv-cellulose binding domain fusion proteins. *Protein Expression and Purification*, 1999; **17**(2): 249–259.

47. Suttnar, J. et al., Procedure for refolding and purification of recombinant proteins from escherichia-coli inclusion-bodies using a strong anion-exchanger. *Journal of Chromatography B-Biomedical Applications*, 1994; **656**(1): 123–126.

48. Lee, M.H. et al., Bacterial expression and *in vitro* refolding of a single-chain Fv antibody specific for human plasma apolipoprotein B-100. *Protein Expression and Purification*, 2002; **25**(1): 166–173.

49. Zeng, X.Q., Z.H. Shen, and R. Mernaugh, Recombinant antibodies and their use in biosensors. *Analytical and Bioanalytical Chemistry*, 2012; **402**(10): 3027–3038.

50. Lowe, J. et al., Development of a novel homogenous electrochemiluminescence assay for quantitation of ranibizumab in human serum. *Journal of Pharmaceutical and Biomedical Analysis*, 2010; **52**(5): 680–686.

51. Halamek, J. et al., Highly sensitive detection of cocaine using a piezoelectric immunosensor. *Biosensors and Bioelectronics*, 2002; **17**(11–12): 1045–1050.

52. Harris, R.D. et al., Integrated optical surface plasmon resonance immunoprobe for simazine detection. *Biosensors and Bioelectronics*, 1999; **14**(4): 377–386.

53. Stocklein, W.F.M. et al., Diphenylurea hapten sensing with a monoclonal antibody and its Fab fragment: Kinetic and thermodynamic investigations. *Analytica Chimica Acta*, 1998; **362**(1): 101–111.

54. McCafferty, J. et al., Phage antibodies—Filamentous phage displaying antibody variable domains. *Nature*, 1990; **348**(6301): 552–554.

55. de Wildt, R.M.T. et al., Antibody arrays for high-throughput screening of antibody–antigen interactions. *Nature Biotechnology*, 2000; **18**(9): 989–994.

56. Benhar, I. et al., Recombinant single chain antibodies in bioelectrochemical sensors. *Talanta*, 2001; **55**(5): 899–907.

57. Al-Halabi, L. et al., Recombinant antibody fragments allow repeated measurements of C-reactive protein with a quartz crystal microbalance immunosensor. *mAbs*, 2013; **5**(1): 140–149.

58. Arevalo, F.J. et al., Ultra-sensitive electrochemical immunosensor using analyte peptidomimetics selected from phage display peptide libraries. *Biosensors and Bioelectronics*, 2012; **32**(1): 231–237.

59. Skerra, A., Engineered protein scaffolds for molecular recognition. *Journal of Molecular Recognition*, 2000; **13**(4): 167–187.

60. Binz, H.K., P. Amstutz, and A. Pluckthun, Engineering novel binding proteins from nonimmunoglobulin domains. *Nature Biotechnology*, 2005; **23**(10): 1257–1268.

61. Nygren, P.A. and A. Skerra, Binding proteins from alternative scaffolds. *Journal of Immunological Methods*, 2004; **290**(1–2): 3–28.

62. Lofblom, J., F.Y. Frejd, and S. Stahl, Non-immunoglobulin based protein scaffolds. *Current Opinion in Biotechnology*, 2011; **22**(6): 843–848.

63. Gill, D.S. and N.K. Damle, Biopharmaceutical drug discovery using novel protein scaffolds. *Current Opinion in Biotechnology*, 2006; **17**(6): 653–658.

64. Skerra, A., Alternative non-antibody scaffolds for molecular recognition. *Current Opinion in Biotechnology*, 2007; **18**(4): 295–304.

65. Nygren, P.A., Alternative binding proteins: Affibody binding proteins developed from a small three-helix bundle scaffold. *FEBS Journal*, 2008; **275**(11): 2668–2676.

66. Nixon, A.E. and C.R. Wood, Engineered protein inhibitors of proteases. *Current Opinion in Drug Discovery & Development*, 2006; **9**(2): 261–268.
67. Lipovsek, D., Adnectins: Engineered target-binding protein therapeutics. *Protein Engineering Design & Selection*, 2011; **24**(1–2): 3–9.
68. Skerra, A., Alternative binding proteins: Anticalins—Harnessing the structural plasticity of the lipocalin ligand pocket to engineer novel binding activities. *FEBS Journal*, 2008; **275**(11): 2677–2683.
69. Stumpp, M.T., H.K. Binz, and P. Amstutz, DARPins: A new generation of protein therapeutics. *Drug Discovery Today*, 2008; **13**(15–16): 695–701.
70. Silverman, J. et al., Multivalent avimer proteins evolved by exon shuffling of a family of human receptor domains. *Nature Biotechnology*, 2005; **23**(12): 1556–1561.
71. Kolmar, H., Alternative binding proteins: Biological activity and therapeutic potential of cystine-knot miniproteins. *FEBS Journal*, 2008; **275**(11): 2684–2690.
72. Enander, K. et al., A versatile polypeptide platform for integrated recognition and reporting: Affinity arrays for protein-ligand interaction analysis. *Chemistry-a European Journal*, 2004; **10**(10): 2375–2385.
73. Boersma, Y.L. and A. Pluckthun, DARPins and other repeat protein scaffolds: Advances in engineering and applications. *Current Opinion in Biotechnology*, 2011; **22**(6): 849–857.
74. Ruigrok, V.J.B. et al., Alternative affinity tools: More attractive than antibodies? *Biochemical Journal*, 2011; **436**: 1–13.
75. Renberg, B. et al., Affibody protein capture microarrays: Synthesis and evaluation of random and directed immobilization of affibody molecules. *Analytical Biochemistry*, 2005; **341**(2): 334–343.
76. Renberg, B. et al., Affibody molecules in protein capture microarrays: Evaluation of multidomain ligands and different detection formats. *Journal of Proteome Research*, 2007; **6**(1): 171–179.
77. Albrecht, C. et al., A new assay design for clinical diagnostics based on alternative recognition elements. *Biosensors and Bioelectronics*, 2010; **25**(10): 2302–2308.
78. Pande, J., M.M. Szewczyk, and A.K. Grover, Phage display: Concept, innovations, applications and future. *Biotechnology Advances*, 2010; **28**(6): 849–858.
79. Giraudi, G. et al., Solid-phase extraction of ochratoxin A from wine based on a binding hexapeptide prepared by combinatorial synthesis. *Journal of Chromatography A*, 2007; **1175**(2): 174–180.
80. Choi, H. and S.J. Choi, Detection of Edwardsiella tarda by fluorometric or biosensor methods using a peptide ligand. *Analytical Biochemistry*, 2012; **421**(1): 152–157.
81. Wu, J. et al., Development of a troponin I biosensor using a peptide obtained through phage display. *Analytical Chemistry*, 2010; **82**(19): 8235–8243.
82. Soykut, E.A., F.C. Dudak, and I.H. Boyaci, Selection of staphylococcal enterotoxin B (SEB) binding peptide using phage display technology. *Biochemical and Biophysical Research Communications*, 2008; **370**(1): 104–108.
83. Huang, G.S., Y.S. Chen, and H.W. Yeh, Measuring the flexibility of immunoglobulin by gold nanoparticles. *Nano Letters*, 2006; **6**(11): 2467–2471.
84. Park, T.J. et al., Protein nanopatterns and biosensors using gold binding polypeptide as a fusion partner. *Analytical Chemistry*, 2006; **78**(20): 7197–7205.
85. Serizawa, T., P. Techawanitchai, and H. Matsuno, Isolation of peptides that can recognize syndiotactic polystyrene. *ChemBiochem*, 2007; **8**(9): 989–993.

86. Serizawa, T., T. Sawada, and T. Kitayama, Peptide motifs that recognize differences in polymer-film surfaces. *Angewandte Chemie-International Edition*, 2007; **46**(5): 723–726.
87. Huang, Y. et al., Programmable assembly of nanoarchitectures using genetically engineered viruses. *Nano Letters*, 2005; **5**(7): 1429–1434.
88. Nam, K.T. et al., Virus-enabled synthesis and assembly of nanowires for lithium ion battery electrodes. *Science*, 2006; **312**(5775): 885–888.
89. Margulies, D. and A.D. Hamilton, Combinatorial protein recognition as an alternative approach to antibody-mimetics. *Current Opinion in Chemical Biology*, 2010; **14**(6): 705–712.
90. Zeng, X.Q. et al., Carbohydrate-protein interactions and their biosensing applications. *Analytical and Bioanalytical Chemistry*, 2012; **402**(10): 3161–3176.
91. Gorityala, B.K. et al., Design of a "Turn-Off/Turn-On" biosensor: Understanding carbohydrate-lectin interactions for use in noncovalent drug delivery. *Journal of the American Chemical Society*, 2012; **134**(37): 15229–15232.
92. Pei, Z.C. et al., Real-time analysis of the carbohydrates on cell surfaces using a QCM biosensor: A lectin-based approach. *Biosensors and Bioelectronics*, 2012; **35**(1): 200–205.
93. Sota, H. et al., Quantitative lectin-carbohydrate interaction analysis on solid-phase surfaces using biosensor based on surface plasmon resonance. *Recognition of Carbohydrates in Biological Systems Pt A: General Procedures*, 2003; **362**: 330–340.
94. Geo, J.Q., D.J. Liu, and Z.X. Wang, Screening lectin-binding specificity of bacterium by lectin microarray with gold nanoparticle probes. *Analytical Chemistry*, 2010; **82**(22): 9240–9247.
95. Wang, Y.X. et al., Application of aptamer based biosensors for detection of pathogenic microorganisms. *Chinese Journal of Analytical Chemistry*, 2012; **40**(4): 634–642.
96. Hianik, T. and J. Wang, Electrochemical aptasensors—Recent achievements and perspectives. *Electroanalysis*, 2009; **21**(11): 1223–1235.
97. Strehlitz, B., N. Nikolaus, and R. Stoltenburg, Protein detection with aptamer biosensors. *Sensors*, 2008; **8**(7): 4296–4307.
98. Han, K., Z.Q. Liang, and N.D. Zhou, Design strategies for aptamer-based biosensors. *Sensors*, 2010; **10**(5): 4541–4557.
99. Sassolas, A., L.J. Blum, and B.D. Leca-Bouvier, Electrochemical aptasensors. *Electroanalysis*, 2009; **21**(11): 1237–1250.
100. Iliuk, A.B., L.H. Hu, and W.A. Tao, Aptamer in bioanalytical applications. *Analytical Chemistry*, 2011; **83**(12): 4440–4452.
101. Fodey, T. et al., Developments in the production of biological and synthetic binders for immunoassay and sensor-based detection of small molecules. *Trac-Trends in Analytical Chemistry*, 2011; **30**(2): 254–269.
102. Walcarius, A. and M.M. Collinson, Analytical chemistry with silica sol-gels: Traditional routes to new materials for chemical analysis, in *Annual Review of Analytical Chemistry*. Annual Reviews: Palo Alto, CA, 2009, pp. 121–143.
103. Haupt, K. et al., Molecularly imprinted polymers, in *Molecular Imprinting*, ed. K. Haupt. Springer-Verlag Berlin: Berlin, Germany, 2012, pp. 1–28.
104. Moreno-Bondi, M.C. et al., Immuno-like assays and biomimetic microchips, in *Molecular Imprinting*, ed. K. Haupt. Springer-Verlag Berlin: Berlin, Germany, 2012, pp. 111–164.

105. Yang, K.G. et al., Protein-imprinted materials: rational design, application and challenges. *Analytical and Bioanalytical Chemistry*, 2012; **403**(8): 2173–2183.
106. Kryscio, D.R. and N.A. Peppas, Critical review and perspective of macromolecularly imprinted polymers. *Acta Biomaterialia*, 2012; **8**(2): 461–473.
107. Verheyen, E. et al., Challenges for the effective molecular imprinting of proteins. *Biomaterials*, 2011; **32**(11): 3008–3020.
108. Imaizumi, Y. et al., Nucleotide modification and polymerase engineering for creating a novel class of artificial nucleic acid aptamers, in *5th European Conference of the International Federation for Medical and Biological Engineering, Parts 1 and 2*, ed. A. Jobbagy. Springer: New York, 2012, pp. 1023–1026.
109. He, W.G. et al., X-aptamers: A bead-based selection method for random incorporation of druglike moieties onto next-generation aptamers for enhanced binding. *Biochemistry*, 2012; **51**(42): 8321–8323.
110. Eaton, B.E., The joys of *in vitro* selection: Chemically dressing oligonucleotides to satiate protein targets. *Current Opinion in Chemical Biology*, 1997; **1**(1): 10–16.
111. Vaught, J.D. et al., Expanding the chemistry of DNA for *in vitro* selection. *Journal of the American Chemical Society*, 2010; **132**(12): 4141–4151.
112. Kasahara, Y. et al., Effect of 3'-end capping of aptamer with various 2',4'-bridged nucleotides: Enzymatic post-modification toward a practical use of polyclonal aptamers. *Bioorganic & Medicinal Chemistry Letters*, 2010; **20**(5): 1626–1629.
113. Altshuler, E.P., D.V. Serebryanaya, and A.G. Katrukha, Generation of recombinant antibodies and means for increasing their affinity. *Biochemistry-Moscow*, 2010; **75**(13): 1584–1605.
114. Wark, K.L. and P.J. Hudson, Latest technologies for the enhancement of antibody affinity. *Advanced Drug Delivery Reviews*, 2006; **58**(5–6): 657–670.
115. Demarest, S.J. and S.M. Glaser, Antibody therapeutics, antibody engineering, and the merits of protein stability. *Current Opinion in Drug Discovery & Development*, 2008; **11**(5): 675–687.
116. Traxlmayr, M.W. and C. Obinger, Directed evolution of proteins for increased stability and expression using yeast display. *Archives of Biochemistry and Biophysics*, 2012; **526**(2): 174–180.
117. Dzantiev, B.B. et al., Determination of the herbicide chlorsulfuron by amperometric sensor based on separation-free bienzyme immunoassay. *Sensors and Actuators B-Chemical*, 2004; **98**(2–3): 254–261.
118. Ivnitski, D. and J. Rishpon, A one-step, separation-free amperometric enzyme immunosensor. *Biosensors and Bioelectronics*, 1996; **11**(4): 409–417.
119. Saerens, D. et al., Antibody fragments as probe in biosensor development. *Sensors*, 2008; **8**(8): 4669–4686.
120. Sun, Y. et al., Design and performances of immunoassay based on SPR biosensor with magnetic microbeads. *Biosensors and Bioelectronics*, 2007; **23**(4): 473–478.
121. Wink, T. et al., Self-assembled monolayers for biosensors. *Analyst*, 1997; **122**(4): R43–R50.
122. Lee, W. et al., Immobilization of antibody fragment for immunosensor application based on surface plasmon resonance. *Colloids and Surfaces B-Biointerfaces*, 2005; **40**(3–4): 143–148.
123. Torrance, L. et al., Oriented immobilisation of engineered single-chain antibodies to develop biosensors for virus detection. *Journal of Virological Methods*, 2006; **134**(1–2): 164–170.

124. Lee, K.G. et al., The investigation of protein A and Salmonella antibody adsorption onto biosensor surfaces by atomic force microscopy. *Biotechnology and Bioengineering*, 2008; **99**(4): 949–959.

125. Starodub, N.F. et al., Immunosensor for the determination of the herbicide simazine based on an ion-selective field-effect transistor. *Analytica Chimica Acta*, 2000; **424**(1): 37–43.

126. Shen, Z.H. et al., Recombinant antibody piezoimmunosensors for the detection of cytochrome P4501B1. *Analytical Chemistry*, 2007; **79**(4): 1283–1289.

127. Plekhanova, Y.V. et al., A new assay format for electrochemical immunosensors: Polyelectrolyte-based separation on membrane carriers combined with detection of peroxidase activity by pH-sensitive field-effect transistor. *Biosensors and Bioelectronics*, 2003; **19**(2): 109–114.

128. Yazynina, E.V. et al., Immunoassay techniques for detection of the herbicide simazine based on use of oppositely charged water-soluble polyelectrolytes. *Analytical Chemistry*, 1999; **71**(16): 3538–3543.

129. Trilling, A.K. et al., The effect of uniform capture molecule orientation on biosensor sensitivity: Dependence on analyte properties. *Biosensors and Bioelectronics*, 2013; **40**(1): 219–226.

130. Schlosshauer, M. and D. Baker, Realistic protein–protein association rates from a simple diffusional model neglecting long–range interactions, free energy barriers, and landscape ruggedness. *Protein Science*, 2004; **13**(6): 1660–1669.

131. Foote, J. and H.N. Eisen, Kinetic and affinity limits on antibodies produced during immune-responses. *Proceedings of the National Academy of Sciences of the United States of America*, 1995; **92**(5): 1254–1256.

132. Maynard, J. and G. Georgiou, Antibody engineering. *Annual Review of Biomedical Engineering*, 2000; **2**: 339–376.

133. Batista, F.D. and M.S. Neuberger, Affinity dependence of the B cell response to antigen: A threshold, a ceiling, and the importance of off-rate. *Immunity*, 1998; **8**(6): 751–759.

134. Cauerhff, A., F.A. Goldbaum, and B.C. Braden, Structural mechanism for affinity maturation of an anti-lysozyme antibody. *Proceedings of the National Academy of Sciences of the United States of America*, 2004; **101**(10): 3539–3544.

135. Ho, M. et al., *In vitro* antibody evolution targeting germline hot spots to increase activity of an anti-CD22 immunotoxin. *Journal of Biological Chemistry*, 2005; **280**(1): 607–617.

136. Chmura, A.J., M.S. Orton, and C.F. Meares, Antibodies with infinite affinity. *Proceedings of the National Academy of Sciences of the United States of America*, 2001; **98**(15): 8480–8484.

137. Trisler, K. et al., A metalloantibody that irreversibly binds a protein antigen. *Journal of Biological Chemistry*, 2007; **282**(36): 26344–26353.

138. Holm, L., P. Moody, and M. Howarth, Electrophilic affibodies forming covalent bonds to protein targets. *Journal of Biological Chemistry*, 2009; **284**(47): 32906–32913.

139. Jackson, T.M. and R.P. Ekins, Theoretical limitations on immunoassay sensitivity—Current practice and potential advantages of fluorescent eu-3+ chelates as nonradioisotopic tracers. *Journal of Immunological Methods*, 1986; **87**(1): 13–20.

140. Sun, Z.F. et al., Sensitive immunosensor for tumor necrosis factor alpha based on dual signal amplification of ferrocene modified self-assembled peptide nanowire and glucose oxidase functionalized gold nanorod. *Biosensors and Bioelectronics*, 2013; **39**(1): 215–219.

141. Maragos, C.M., Signal amplification using colloidal gold in a biolayer interferometry-based immunosensor for the mycotoxin deoxynivalenol. *Food Additives and Contaminants Part a-Chemistry Analysis Control Exposure & Risk Assessment*, 2012; **29**(7): 1108–1117.
142. Ge, S.G., X.L. Jiao, and D.R. Chen, Ultrasensitive electrochemical immunosensor for CA 15-3 using thionine-nanoporous gold-graphene as a platform and horseradish peroxidase-encapsulated liposomes as signal amplification. *Analyst*, 2012; **137**(19): 4440–4447.
143. Lee, T.H. et al., Signal amplification by enzymatic reaction in an immunosensor based on localized surface plasmon resonance (LSPR). *Sensors*, 2010; **10**(3): 2045–2053.
144. Schopf, E. and Y. Chen, Attomole DNA detection assay via rolling circle amplification and single molecule detection. *Analytical Biochemistry*, 2010; **397**(1): 115–117.
145. Huang, S.X. and Y. Chen, Ultrasensitive fluorescence detection of single protein molecules manipulated electrically on Au nanowire. *Nano Letters*, 2008; **8**(9): 2829–2833.
146. Hohlbein, J. et al., Surfing on a new wave of single-molecule fluorescence methods. *Physical Biology*, 2010; **7**(3): 031001.
147. Mayer, K.M. et al., A single molecule immunoassay by localized surface plasmon resonance. *Nanotechnology*, 2010; **21**(25): 255503.
148. Urusov, A.E. et al., Ochratoxin A immunoassay with surface plasmon resonance registration: Lowering limit of detection by the use of colloidal gold immunoconjugates. *Sensors and Actuators B-Chemical*, 2011; **156**(1): 343–349.
149. Guo, L.H. and D.H. Kim, LSPR biomolecular assay with high sensitivity induced by aptamer-antigen-antibody sandwich complex. *Biosensors and Bioelectronics*, 2012; **31**(1): 567–570.
150. de la Rica, R. and M.M. Stevens, Plasmonic ELISA for the ultrasensitive detection of disease biomarkers with the naked eye. *Nature Nanotechnology*, 2012; **7**(12): 821–824.
151. Fan, M. and J. He, Recent progress in noncompetitive hapten immunoassays: A review. In: *Trends in Immunolabelled and Related Techniques*, ed. E. Abuelzein. InTech: Shanghai, China, 2012.
152. Tanaka, K. et al., Novel and sensitive noncompetitive (2-site) enzyme-immunoassay for haptens with amino-groups. *Journal of Clinical Laboratory Analysis*, 1990; **4**(3): 208–212.
153. Hashida, S. et al., Novel and sensitive noncompetitive enzyme-immunoassay (hetero-2-site enzyme-immunoassay) for arginine vasopressin in plasma. *Analytical Letters*, 1991; **24**(7): 1109–1123.
154. Ishikawa, E., K. Tanaka, and S. Hashida, Novel and sensitive noncompetitive (2-site) immunoassay for haptens with emphasis on peptides. *Clinical Biochemistry*, 1990; **23**(5): 445–453.
155. Pradelles, P. et al., Immunometric assay of low-molecular-weight haptens containing primary amino-groups. *Analytical Chemistry*, 1994; **66**(1): 16–22.
156. Volland, H. et al., Recent developments for SPIE-IA, a new sandwich immunoassay format for very small molecules. *Journal of Pharmaceutical and Biomedical Analysis*, 2004; **34**(4): 737–752.
157. Lebeau, L. et al., Synthesis of haptens for the development of a solid-phase immobilized epitope-immunoassay (SPIE-IA) of AZT-TP. *Tetrahedron Letters*, 1999; **40**(23): 4323–4326.

158. Marnet, P.G. et al., Enzyme immunometric assay (SPIE-IA) for oxytocin, in *Oxytocin: Cellular and Molecular Approaches in Medicine and Research*, eds. R. Ivell and J.A. Russell. Plenum Press Div Plenum Publishing Corp: New York, 1995, pp. 613–614.

159. Urbain, J., M. Slaoui, and O. Leo, Idiotypes, recurrent idiotypes and internal images. *Annales D Immunologie*, 1982; **D133**(2): 179–189.

160. Barnard, G. and F. Kohen, Idiometric assay—Noncompetitive immunoassay for small molecules typified by the measurement of estradiol in serum. *Clinical Chemistry*, 1990; **36**(11): 1945–1950.

161. Barnard, G. et al., The measurement of estrone-3 glucuronide in urine by non-competitive idiometric assay. *Journal of Steroid Biochemistry and Molecular Biology*, 1995; **55**(1): 107–114.

162. Barnard, G., H. Karsiliyan, and F. Kohen, Idiometric assay, the 3rd way—A non-competitive immunoassay for small molecules. *American Journal of Obstetrics and Gynecology*, 1991; **165**(6): 1997–2000.

163. Barnard, G. et al., The measurement of progesterone in serum by a noncompetitive idiometric assay. *Steroids*, 1995; **60**(12): 824–829.

164. Niwa, T. et al., An enzyme-linked immunometric assay for cortisol based on idiotype-anti-idiotype reactions. *Analytica Chimica Acta*, 2009; **638**(1): 94–100.

165. Voss, E.W. et al., Polyclonal antibodies specific for liganded active-site (metatype) of a high-affinity anti-hapten monoclonal-antibody. *Molecular Immunology*, 1988; **25**(8): 751–759.

166. Self, C.H., J.L. Dessi, and L.A. Winger, High-performance assays of small molecules—Enhanced sensitivity, rapidity, and convenience demonstrated with a noncompetitive immunometric anti-immune complex assay system for digoxin. *Clinical Chemistry*, 1994; **40**(11): 2035–2041.

167. Ueda, H. et al., Open sandwich ELISA: A novel immunoassay based on the interchain interaction of antibody variable region. *Nature Biotechnology*, 1996; **14**(13): 1714–1718.

168. Ueda, H. et al., Homogeneous noncompetitive immunoassay based on the energy transfer between fluorolabeled antibody variable domains (open sandwich fluoroimmunoassay). *Biotechniques*, 1999; **27**(4): 738–742.

169. Ueda, H. et al., An optimized homogeneous noncompetitive immunoassay based on the antigen-driven enzymatic complementation. *Journal of Immunological Methods*, 2003; **279**(1–2): 209–218.

170. Narhi, L.O. et al., Induction of alpha-helix in the beta-sheet protein tumor necrosis factor-alpha: Thermal- and trifluoroethanol-induced denaturation at neutral pH. *Biochemistry*, 1996; **35**(35): 11447–11453.

171. Avdulov, N.A. et al., Direct binding of ethanol to bovine serum albumin: A fluorescent and C-13 NMR multiplet relaxation study. *Biochemistry*, 1996; **35**(1): 340–347.

172. Dooley, H. et al., Stabilization of antibody fragments in adverse environments. *Biotechnology and Applied Biochemistry*, 1998; **28**: 77–83.

173. Wasacz, F.M., J.M. Olinger, and R.J. Jakobsen, Fourier-transform infrared studies of proteins using nonaqueous solvents—Effects of methanol and ethyleneglycol on albumin and immunoglobulin-G. *Biochemistry*, 1987; **26**(5): 1464–1470.

174. Melnikova, Y.I. et al., Antigen-binding activity of monoclonal antibodies after incubation with organic solvents. *Biochemistry-Moscow*, 2000; **65**(11): 1256–1265.

175. Skladal, P., Effect of methanol on the interaction of monoclonal antibody with free and immobilized atrazine studied using the resonant mirror-based biosensor. *Biosensors and Bioelectronics*, 1999; **14**(3): 257–263.

176. Katagiri, M. et al., Effects of methanol and temperature on enzyme immunoassay with monoclonal antibodies specific to the insecticide etofenprox. *Bioscience Biotechnology and Biochemistry*, 1999; **63**(11): 1988–1990.

177. Stocklein, W.F.M. et al., Sensitive detection of triazine and phenylurea pesticides in pure organic solvent by enzyme linked immunosorbent assay (ELISA): Stabilities, solubilities and sensitivities. *Analytica Chimica Acta*, 2000; **405**(1–2): 255–265.

178. Tomassetti, M., E. Martini, and L. Campanella, New immunosensors operating in organic phase (OPIEs) for analysis of triazinic pesticides in olive oil. *Electroanalysis*, 2012; **24**(4): 842–856.

179. Ben Rejeb, I. et al., Development of a bio-electrochemical assay for AFB(1) detection in olive oil. *Biosensors and Bioelectronics*, 2009; **24**(7): 1962–1968.

180. Nabok, A.V. et al., Registration of T-2 mycotoxin with total internal reflection ellipsometry and QCM impedance methods. *Biosensors and Bioelectronics*, 2007; **22**(6): 885–890.

181. Mart'ianov, A.A. et al., Immunoenzyme assay of nonylphenol: Study of selectivity and detection of alkylphenolic non-ionic surfactants in water samples. *Talanta*, 2005; **65**(2): 367–374.

182. Yan, X.H. and Z.R. Xu, Production of human single-chain variable fragment (scFv) antibody specific for digoxin by ribosome display. *Indian Journal of Biochemistry & Biophysics*, 2005; **42**(6): 350–357.

183. Backmann, N. et al., A label-free immunosensor array using single-chain antibody fragments. *Proceedings of the National Academy of Sciences of the United States of America*, 2005; **102**(41): 14587–14592.

184. Harrison, J.S., A. Gill, and M. Hoare, Stability of a single-chain Fv antibody fragment when exposed to a high shear environment combined with air-liquid interfaces. *Biotechnology and Bioengineering*, 1998; **59**(4): 517–519.

185. Arndt, K.M., K.M. Muller, and A. Pluckthun, Helix-stabilized Fv (hsFv) antibody fragments: Substituting the constant domains of a Fab fragment for a heterodimeric coiled-coil domain. *Journal of Molecular Biology*, 2001; **312**(1): 221–228.

186. Webber, K.O. et al., Preparation and characterization of a disulfide-stabilized Fv fragment of the anti-tac antibody—Comparison with its single-chain analog. *Molecular Immunology*, 1995; **32**(4): 249–258.

187. Young, N.M. et al., Thermal stabilization of a single-chain Fv antibody fragment by introduction of a disulphide bond. *FEBS Letters*, 1995; **377**(2): 135–139.

188. Quintero-Hernandez, V. et al., The change of the scFv into the Fab format improves the stability and *in vivo* toxin neutralization capacity of recombinant antibodies. *Molecular Immunology*, 2007; **44**(6): 1307–1315.

189. Razai, A. et al., Molecular evolution of antibody affinity for sensitive detection of botulinum neurotoxin type A. *Journal of Molecular Biology*, 2005; **351**(1): 158–169.

190. Lippow, S.M., K.D. Wittrup, and B. Tidor, Computational design of antibody-affinity improvement beyond *in vivo* maturation. *Nature Biotechnology*, 2007; **25**(10): 1171–1176.

191. Kuroda, D. et al., Computer-aided antibody design. *Protein Engineering Design & Selection*, 2012; **25**(10): 507–521.

192. Kramer, K., D. Rau, and B. Hock, Comparison of affinity ranking and immuno-chemical key data as measure for molecular antibody evolution. *Spectroscopy-an International Journal*, 2003; **17**(2–3): 355–365.

193. Cheung, N.K.V. et al., Single-chain Fv-streptavidin substantially improved therapeutic index in multistep targeting directed at disialoganglioside GD2. *Journal of Nuclear Medicine*, 2004; **45**(5): 867–877.

194. Babineau, B., M. Best, and S. Farrell, *Commercially Available Biosensors. Applications, Availability, and Marketability.* http://faculty.uml.edu/xwang/16.541/2011/presentation1/Commercially%20Available%20Biosensors%20group10.pptx, 2011, p. 26.

195. Dumoulin, M. et al., Single-domain antibody fragments with high conformational stability. *Protein Science*, 2002; **11**(3): 500–515.

7

Ion Channel Switch- and Lipid Film-Based Biosensors

Tibor Hianik, Dimitrios P. Nikolelis, and Georgia-Paraskevi Nikoleli

CONTENTS

7.1 Introduction

Lipids are amphiphilic molecules, which possess both hydrophilic head groups and hydrophobic chains. Depending on the concentration and the ratio of the sizes of the head groups and tails, they can spontaneously form different structures such as *bilayer or multilayer sheets*, *micelles*, and *liposomes* in aqueous media. In these structures, the hydrophobic chains are hidden from water and the hydrophilic groups remain in the solution. Natural cell membranes consist mainly from different lipids such as phospholipids, glycolipids, cholesterol, and proteins. The protein fraction varies from 20% to 80% depending on the type of membrane. The peripheral proteins such as cytochrome c (cyt c) are embedded to the surface of the membrane, while integral proteins, for example, bacteriorhodopsin, span the hydrophobic part of the lipid bilayer.

The biomembrane is the basic cell structure that not only serves as the barrier that separates intracellular and extracellular environment but contains biomacromolecules that provide various functions important for living organisms, such as transport of ions and other compounds, receptor functions, and immunity response. The lipid environment protects the biopolymers against degradation. This extremely thin (~5 nm) lipid matrix with incorporated bioreceptors is a unique biosensor developed by nature. Since the discovery of model bilayer lipid membranes (BLMs) by Mueller et al. [1], there were attempts to use them in biosensing applications. However, these free-standing BLMs were fragile and not suitable for long-term use. The low mechanical and electrical stability of BLM was the obstacle to their practical applications. Recent advances in the stabilization of lipid bilayer have resulted in the preparation of lipid bilayer-based biosensors for detection of a variety of real samples. Substantial progress has been made after 2000 in the development of stabilized lipid films that have started substantial interest in the application of BLM in biosensing. Since our previous review [2], additional progresses were made in this direction. More recently, the number of publications directed to the practical use of lipid films as biosensors has increased considerably. Lipid membranes represent an appropriate biocompatible structure for the development of new types of biosensors with fast response (on the order of a few seconds) and high sensitivity (i.e., nanomolar detection limits) that may eventually be used in health diagnosis and in-field applications for food analysis and environmental monitoring.

Most of these biosensors are cost efficient, easy to use, fast responding, and are good alternative to mostly expensive, time-consuming standard analytical and screening methods (i.e., chromatographic procedures or mass spectroscopy). These devices will be able to be either regenerated for uses multiple times or be used as disposable sensors in a single format.

Studies are currently focused on the development and evaluation of methods for the incorporation of ion channels, enzymes, antibodies, or receptors in electrochemical, optical, or mass-sensitive detectors with lipid film as a recognition element or as an ion channel switch (ICS)-based device. An example of the development of lipid film-based sensors is glass fiber filter-supported BLMs with incorporated artificial receptor that has been recently developed by our group for the rapid detection of carbofuran in fruits and vegetables [3], and for the rapid detection of doping materials in the urine of athletes [4]. The same principle will be fully extended for the detection of heavy metals, insecticides, and other toxicants in foods. Novel trends include nanofabricated surfaces composed of carbon nanotubes or nanoporous alumina. Substantial attention was made also to the formation of hybrid supported BLM (sBLM) on micro- and nanoporous substrates. The signal generation, especially those connected with redox reactions, has been enhanced by the modification of lipid layer by nanoparticles. This chapter therefore reflects these novel trends in biosensors based on sBLM. The methods of preparation of stabilized lipid films and recent achievements in the development of sBLM-based biosensors are described.

7.2 Methods for Preparation of Biosensors Based on Lipid Films

Freely suspended BLMs have been used for the last five decades for investigations of transduction of electrochemical signal through lipid membranes and their preparation has been extensively described elsewhere [5]. However, these membranes are highly unstable and they are prone to break under electrical or mechanical forces. The preparation of stabilized lipid membranes on solid support is described herein as this technology represents a route for applications as practical biosensors. The solid support depends on the applications and may consist of a metallic surface or ultrafiltration membrane. The former type is mainly used for the construction of disposable biosensors, whereas the latter is for continuous monitoring of an analyte.

7.2.1 Metal-Supported Lipid Layers

The metal electrodes (gold, silver, platinum, and copper) are frequently used in electrochemical biosensors due to their rich surface chemistry and possibility of use in various sensing applications. The surface of these materials can be covered by lipid films and those modified by bioreceptors and mediators. These lipid film-coated metal electrodes have been proved as suitable electrochemical biosensors [6]. Gold surface as a support for lipid films has been used also in surface plasmon resonance (SPR) [7] and acoustic quartz crystal microbalance (QCM) [8] biosensors.

Several methods have been reported in the formation of lipid films at metal surface. These methods are based on typical self-assembly properties of amphiphilic molecules in water and in spontaneous chemisorption of thiolated compounds on the surface of noble metals (mostly gold and platinum). Substantial interest to self-assembled monolayers (SAM) appeared since the pioneering work by Nuzzo and Allara [9] who prepared SAM by chemisorptions of alkanethiolates on a gold surface. These systems are most frequently used to date. The preparation of the chemisorbed monolayer requires smoothing and careful cleaning of gold surface. The electrode should be then immersed into the alkanethiol solution for at least 24 h. It has been shown that formation of the alkanethiol SAM is a two-step process. First step is rather fast, (few minutes) consisting approximately 80%–90% coverage of the surface. Second step is much slower (up to 20 h) and resulted in complete surface coverage [10]. The process of chemisorptions can be, however, accelerated by the application of positive voltage (approximately 0.6 V) on a metal electrode relative to the reference Ag/AgCl electrode [11]. In addition to chemisorptions technique, the method of preparation of SAM by immersion of amphiphilic molecules to a clean gold surface with subsequent drying and immersing in buffer resulted in molecular layers of comparable properties such as chemisorbed SAM in respect to the degree of surface coverage and decrease of double-layer capacitance [10]. The preparation of SAM on a metal surface has been recently reviewed by Chen and Li [6]. Preparation of BLM requires immobilization of a second monolayer on top of alkanethiol SAM. Monolayer can be prepared with various techniques, for example, by simple immersion of the gold plate with alkylthiol monolayer into the lipid solution (in n-hexane), by vesicle fusion [12], or by Langmuir–Blodgett (LB) method (Figure 7.1A) [13]. These systems can serve as enzyme, immuno-, or genetic biosensors [14]. In an ICS biosensor, the so-called tethered sBLM was developed. These membranes contain special hydrophilic spacers between lipids, ion channels, and the gold support providing water filed volume that is necessary for monitoring the ion flux across the channels incorporated in lipid bilayer (Figure 7.1B) [15]. The principles of biosensors based on ICS are explained in Section 7.3.

The tethered membranes have been successfully used also for investigating the mechanisms of interaction of short peptides with lipid bilayers [16], pore-forming segment of nicotinic acetylcholine receptor [17], transmembrane peptide, porin [18], the behavior of integral protein cyt c oxidase [19] and H⁺-ATP synthases in lipid bilayer [20], α-hemolysin [17], and cyt c

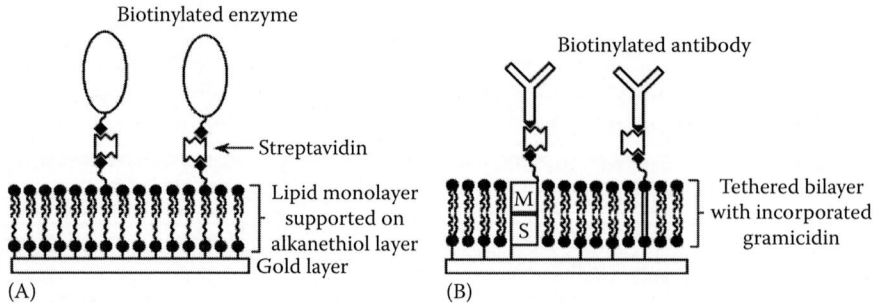

FIGURE 7.1
The scheme of (A) lipid monolayer supported on chemisorbed alkanethiol layer and (B) tethered BLM with incorporated gramicidin channel on a gold substrate. (Partially adapted from Moradi-Monfared, S. et al., *Biosens. Bioelectr.*, 34, 261, 2012.) The enzymes or antibodies were immobilized on the outer layer of supported BLM through biotinylated linker and streptavidin.

oxidase [21]. The formation of tethered membranes and those formed on nanopores (see later text) has been reviewed recently [22,23].

A rather simple system of preparation of sBLM at the freshly cut tip of Teflon-coated metallic wire was reported by Tien and Salamon [24]. The technique of formation of sBLMs is based on the interaction of an amphipathic lipid molecule with a nascent metallic surface. The procedure consists of two steps. First, one end of a Teflon-coated stainless steel metal wire (diameter of the wire is usually 0.1–0.5 mm) is immersed in a lipid solution (this can be, e.g., 2% phospholipid in n-hexadecane or other organic solvent) and then, while still immersed, the tip is cut off with a miniature guillotine (good reproducibility of sBLMs requires reproducible cutting of the wire). Second, the fresh tip of the wire, having become coated with lipid, was placed in the electrolyte (usually 0.1–1 M KCl, buffered with usual buffers, e.g., Tris-HCl), whereupon the lipid film spontaneously thinned, forming a self-assembled lipid bilayer.

Except stainless steel, also gold, platinum, or silver wire can be used. Teflon-coated wires are commercially available. In addition, the coating of the wire by electrochemical polymerization can be used [25]. The disadvantage of this method is the low reproducibility of sBLM electrical properties and uncertainty of lipid layer structure at the metal surface. Except the bilayer, also monolayers and multilayers could coexist [26]. However, defects in lipid bilayers structure could be eliminated by the application of a negative DC potential during the formation of sBLM on the silver wire tip [27]. The electrical properties of these membranes were studied in detail by electrochemical impedance spectroscopy (EIS) [28].

Puu et al. [29] reported the method of formation of lipid bilayer on smooth platinum support by means of LB technique and with a liposome fusion. They succeed to incorporate into these supported membranes various functional proteins and enzymes. Supported lipid bilayers can be further stabilized by bacterial S-layers. The regular S-protein array serves also as a convenient matrix for immobilization of biopolymers [30].

Lipid monolayers and bilayers can be formed also on a mercury surface. The advantage is the relatively easy formation of the film. Using mercury drop electrode, the fresh surface is available for preparation of lipid films and their investigation by electrochemical methods. Disadvantage is the toxicity of mercury, which makes this method not useful for practical applications (see Ref. [31] for recent review).

7.2.2 BLM Formed on a Surface of Glassy Carbon Electrode

Due to the fact that glassy carbon is inert and has wide potential window in electrochemistry, it is rather a suitable support for the formation of sBLM [32]. Initially, Deinhammer et al. [33] reported a method of modification of glassy carbon electrode (GCE) by primary amines, allowing preparation of hydrophobic surface [34]. However, even without this modification, the sBLM can be formed at GCE surface by simple dispersion of lipids in hexane. It has been demonstrated by Siontorou et al. [35], who also showed the successful incorporation of DNA into lipid bilayers. GCE has been used also in work by Wu et al. [36]. They formed BLM by immersing the drop (5 μL) of dimyristoylphosphatidylcholine (DMPC) in chloroform (2 mg mL^{-1}) onto the polished surface of GCE. The electrode has been immediately immersed in phosphate-buffered saline (PBS). The formation of BLM was approved by impedance spectroscopy. The specific capacitance obtained 0.37 μF cm^{-2} suggests the thickness of hydrophobic part of BLM 4.8 nm. Tong et al. [37] reported similar parameters for sBLM composed of DMPC and supported on GCE. Huang et al. [38] used negatively charged dipalmitoylphosphatidylglycerol (DPPG) for preparation of sBLM on a GCE support. The specific capacitance determined by impedance spectroscopy has been 0.26 μF cm^{-2}. They used these membranes to study the interaction of antimicrobial peptide nisin with lipid bilayer. At relatively low concentration (above 150 μmol L^{-1}), nisin induced conductivity of sBLM; however, at higher concentrations (above 750 μmol L^{-1}), it disrupted the lipid layer acting like surfactant. Karabaliev [39] studied the effect of divalent cations on the properties of sBLM formed on GCE. He observed that the addition of divalent cations to the electrolyte prior to the formation of the lipid film improves the bilayer integrity. The integrity of lipid films on a GCE support can be improved also by the application of positive voltage 0.4 V versus SCE, while high voltage (1.8 V) or negative voltage disintegrates the membrane [40].

7.2.3 Stabilized Lipid Films Formed on a Glass Fiber Filter and on an Agar Support

Another form of stabilized lipid films that allowed their use in the development of practical biosensors was based on using ultrafiltration membranes [41]. The use of ultrafiltration membrane-supported lipid films has allowed the practical application of these detectors in real samples for the determination of aflatoxin M$_1$ in milk and milk preparations [42]. The lipid film is formed on a microporous filter glass fiber disk (diameter of ca. 0.9 cm) [41,42]. Briefly, the setup for the formation of stabilized BLMs consisted of two Plexiglas chambers separated by a Saran Wrap (PVDC; DowBrands L.P., Indianapolis, IN) partition of a thickness of approximately 10 μm. This plastic sheet was cut to more than twice the size of the contact area of the faces of the chambers and folded in half; a hole of 0.32 mm diameter was made through the double-layer of the plastic film by punching with a perforation tool [43]. A microporous glass fiber disk (diameter of ca. 0.9 cm) was placed between the two plastic layers and centered on the 0.32 mm orifice. The partition with the filter membrane in place was then clamped between the Plexiglas chambers. One of the chambers was machined to contain an electrochemical cell with a circular shape (diameter 1.0 cm and depth 0.5 cm) connected with plastic tubing for the flow of the carrier solution; an Ag/AgCl reference electrode was immersed in the waste of the carrier electrolyte solution. The second chamber was machined to contain a cylindrical cell having its longitudinal axis perpendicular to the flow of the carrier solution. The upper hole of this cell was circular (surface area of about 0.2 cm²) and the lower was elliptical (with diameters 0.5 and 1.4 cm parallel and vertical to the flow of the carrier electrolyte solution, respectively) facing the opposing cell. An Ag/AgCl reference electrode was placed into the cylindrical cell and an external 25 mV DC voltage was applied across the filter membrane between the two reference electrodes. A digital electrometer (Model 614, Keithley Instruments, Cleveland, OH) was used as a current-to-voltage converter. A peristaltic pump (Gilson Model MINIPULS 3) was used for the flow of the carrier electrolyte. Injections of the samples tested were made with a Hamilton repeating dispenser with a disposable tip (Hamilton Co., Nevada). The electrochemical cell and electronic equipment were isolated in a grounded Faraday cage. A simple scheme of the apparatus used is presented in Figure 7.2. The process of stabilized BLM formation for flow injection experiments was previously described [41,42]. Lipid solution in chloroform (10 µL) was added dropwise from a microliter syringe to the water surface in the cylindrical cell near the partition. The level of the electrolyte solution was then dropped below the aperture and then raised again within a few seconds. The formation of "solvent-free"

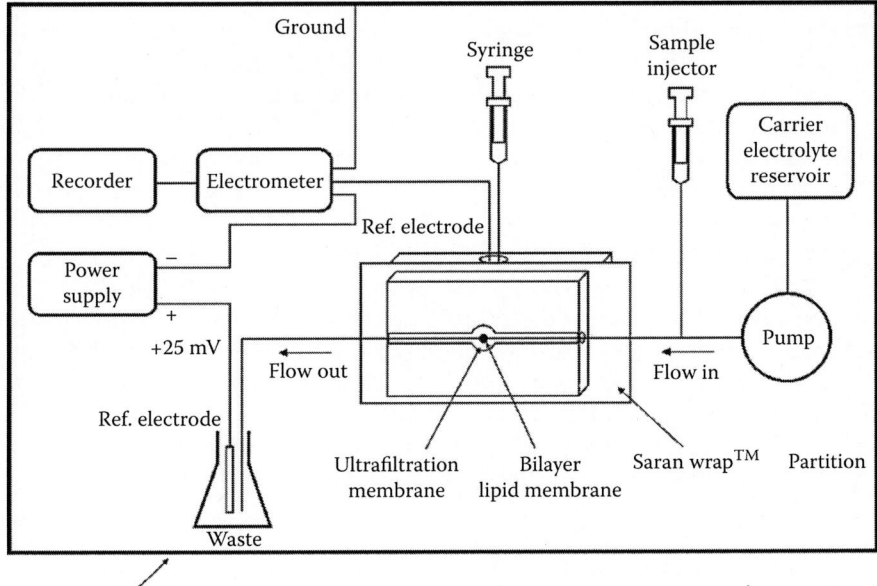

FIGURE 7.2
A simplified version of the apparatus used for flow-through experiments using filter-supported BLMs.

BLMs was verified by the magnitude of the transmembrane ion current obtained and by electrochemical characterization using gramicidin D.

A simple method of BLM formation on an agar support has been proposed simultaneously by Ottova et al. [44] and Ziegler et al. [45]. In this method, first, the Teflon tube of an inner diameter between 0.5 and 1 mm is filed by 2% agar in 1 M KCl. The BLM is formed at freshly cut end of the tube immersed in a buffer by painting method [1]. The advantage of this method is the possibility to study the mechanisms of ionic transport through ion channels, because the ions can diffuse into the agar gel.

7.2.4 Bilayers Stabilized by Polymerization and Formed on Semiconductor and Carbon Nanotube Surfaces

7.2.4.1 Polymer-Supported Bilayer Lipid Membranes

There are three main reasons why attachments of thin films to the transducer surface are essential for the fabrication of chemo- and biosensors. Although biomolecules can be directly deposited on the transducer surface, the resulting assembly is generally only a monolayer thick and the surface concentration is not very high. To provide the signal as high as possible (to enhance the sensitivity), it is desirable to increase the surface concentration of biomolecules. This can be achieved by incorporating biomolecules in, for example, a polymer coating, with a thickness corresponding to several monolayers. A larger number of biomolecules can thus be incorporated. The film material chosen must allow the diffusion of substrate to the active site of the biomolecules (enzyme or antibody) and this requires a relatively flexible structure in which the substrate binding and regeneration of an enzyme or antibody should not be substantially depressed. The principle of transduction must also be maintained, so that a long-range electron transport from a redox enzyme in a polymer to an electrode surface should be possible.

A second reason for using polymer matrices on the transducer surface is to provide additional selectivity to the reaction. Some potentially disturbing molecules can be excluded by using permselective membrane barriers. Examples include the use of ion exchange polymers such as Nafion® that effectively block transport of anionic species by Donnan exclusion and cellulose acetate that functions as a size exclusion barrier.

Third, the film must also possess properties that allow the biomolecules to function properly. Several biological compounds, in particular transport proteins that have great potential as recognition element in biosensors, function satisfactorily only in a lipid layer environment. Artificial membrane-like structures are therefore required to maintain biospecific reactivity.

Stabilized lipid films were prepared by polymerization that was previously described in the literature [46–49]. Stabilized lipid films were initially prepared by polymerization by heating at 60°C [46]. In recent works, however, the polymerization took place by using ultraviolet (UV) irradiation instead of thermal polymerization [47]. The latter method had the advantage that a recognition element, for example, enzyme, could be incorporated prior to polymerization without loss of its activity. Briefly, 0.8 mL of a mixture containing 4% w/v egg phosphatidylcholine and partially oxidized cholesterol in n-hexane was mixed with 0.07 mL of methacrylic acid, 0.8 mL of ethylene glycol dimethacrylate, 8 mg of 2,2'-azobis-(2-methylpropionitrile), and 1.0 mL of acetonitrile. The mixture was sparged with nitrogen for about 1 min and sonicated for 30 min. For the preparation of the stabilized lipid films, 0.15 mL of this mixture was spread on the microfilter. The filter with the mixture was then irradiated using the UV deuterium lamp. Raman spectrometry was used to monitor the kinetics of the process of polymerization. The measuring setup was similar to that presented in Figure 7.2. These membranes were stable in air storage for repetitive uses.

7.2.4.2 Lipid Films Supported on Semiconductors

The semiconductors, such as indium–tin oxide (ITO) have been used for preparing sBLM with photoelectric response [50]. ITO surface after careful cleaning in methanol is hydrophilic. When a drop of phospholipid solution is added to the surface and then immersed in a buffer, the lipid bilayer is formed spontaneously within approximately 10 min. The impedance spectroscopy allowed determining specific resistance, R_m, and specific capacitance, C_m, of sBLM, and those doped by C_{60} fullerene. The value of C_m equal to 0.56 ± 0.02 µF cm^{-2} suggested the formation of lipid bilayer at the surface. Due to photoelectric properties of ITO, the authors observed clear photoelectric response of sBLM. The current that appeared by illumination of sBLM linearly increased with increasing light intensity. This sBLM can be used as a tool for the study of the light-induced properties of biomembranes.

7.2.4.3 Lipid Films Supported on Carbon Nanotubes

Supported BLMs were formed also on the surface of multiwall carbon nanotubes (MWCNTs) grown on Ta plate covered by 8–50 nm of the cobalt layer [51]. The drop of lipid solution without or with saturated C_{60} (5% phosphatidylcholine in n-decane) has been added onto the surface of the MWCNTs and then immersed into PBS. The formation of BLM monitored by CV method has been completed spontaneously within 10 min. The static water contact angles on the surface of MWCNTs were of 6°–7°, indicating a hydrophilic property of this surface. The thickness of the BLM (4.38 nm), determined from CV data, suggests bilayer structure of the lipid film with phospholipid head group contacted with MWCNTs. Similar conclusion was made by Kanyo et al. [52]. However, depending on the method of preparation of MWCNTs, the surface can be also hydrophobic that allows contact with lipid hydrophobic chains [53]. The incorporation of C_{60} into these BLMs resulted in increase of the current up to 1 µA that was much higher in comparison with those of unmodified BLM (1 nA) [51]. Interesting behavior of C_{60}-doped BLM was observed by CV at the presence of

10 mmol L^{-1} redox probe $Fe(CN)_6^{3-/4-}$. While peak separation potential ΔE_p has been 59 mV for bar MWCNTs and did not depend on scan rate, this value for MWCNTs covered by BLM-doped with C_{60} increased from 59 to 94 mV with the increase of the scan rates from 0.4 to 1.00 V s^{-1}, which indicate a quasi-reversible redox reaction at C_{60}-BLM/MWCNTs electrode. The authors suggested that the transport of electrons by C_{60} through BLM is not fast enough to match the redox reaction rate at the MWCNTs covered by BLM. It has been shown that BLM doped by C_{60} formed on MWCNT support is a useful tool for the study of photo-electric properties of lipid membranes. The naked MWCNTs revealed reversible photocurrent response following illumination by white light (150 W halogen light source). Coverage by BLM without C_{60} resulted in disappearance of the photocurrent, but presence of C_{60} caused recovery of the photocurrent, although the current density was lower in comparison with naked MWCNTs layer [51].

The stability of sBLM and their sensitivity toward various analytes can be improved by modification of lipid layer using nanoparticles [51,54,55]. For example, Liu et al. [55] has detected cyt c at Au nanoparticle-modified sBLM. The interaction of cyt c with these sBLMs resulted in increase membrane capacitance and resistance. Au nanoparticle-modified sBLMs were rather use-ful for preparation of biosensor for the detection of estrogen 17β-estradiol [56].

7.2.5 BLM Formed on Microporous Materials

The main drawback of sBLM is difficulties in applications requiring measure-ments of the ionic conductivity. These difficulties can be overcome by using micro- or nanoporous materials. It is known that the stability of BLM depends on the size and properties of the aperture. Stability of BLM is substantially increased with decreasing the aperture diameter and the annular–septum contact angle [55]. So far the most traditional material used for the formation of BLM was Teflon. However, it is rather difficult to obtain in Teflon the aper-ture with diameter less than 100 μm. But polycarbonate filters have pores of μm or lower size. An interesting method of formation of polycarbonate filter-supported BLM has been reported by Favero et al. [57]. They used filters with the pore size of 1 μm. First, one side of the filter has been covered by gold layer of 50 nm thickness. Special care in sputtering the gold provided that the pores were not blocked. This has been tested by scanning electron microscopy (SEM) and by atomic force microscopy (AFM) methods. Then the filter has been immersed into the octadecanethiol (ODT) dissolved in ethanol overnight. This resulted in the formation of ODT SAM. The filter with ODT was placed into the specially constructed Teflon cell and the BLM at the cis site of the cell was formed by convenient painting method by addition of small amount of lecithin dissolved in n-hexane/isobutanol 10:1 mixture. These hybrid BLMs have been modified by valinomycin or gramicidin D. The ionic conductivity and selectivity has been similar to those of conventional free-standing BLM. Agarose-supported sBLMs that formed on microporous holes (100 μm in diam-eter) have been also reported. These BLMs were formed by simple painting

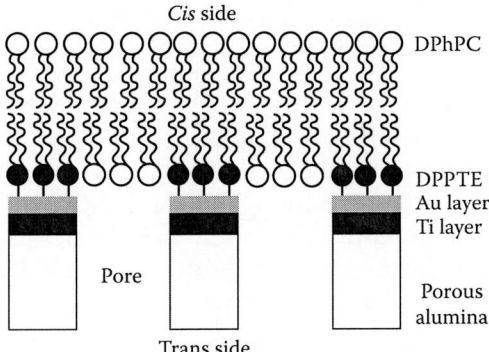

FIGURE 7.3
Scheme of nano-BLM composed of DPhPC formed on porous alumina covered by gold layer with chemisorbed DPPTE monolayer. (Adapted from Römer, W. and Steinem, C., *Biophys. J.*, 86, 955, 2004.)

method using n-decane as a solvent. Channel former molecules such as alamethicin were able to diffuse through the agarose and form ionic channels in bilayer [56]. Romer and Steinem [58] reported the formation of nano-BLM using similar methods like those described earlier; however, instead of polycarbonate filter, they used porous alumina substrate with pore diameter of 55 and 280 nm (Figure 7.3). One side of the porous material was coated with a thin gold layer and then by chemisorption of 1,2-dipalmitoyl-sn-glycero-3-phosphothioethanol (DPPTE). The hybrid BLM on the hydrophobic surface has been formed from 1,2-diphytanoyl-sn-glycero-3-phosphocholin (DPhPC). First, the surface was pretreated with DPhPC dissolved in pentane and dried in nitrogen. Then the BLM was formed by painting method. As a solvent the n-decane (1% solution) was used. The formation process and electrical parameters of BLM were monitored by impedance spectroscopy. The specific capacitance (0.65 ± 0.2) $\mu F\,cm^{-2}$ and specific resistance $1.6 \times 10^8\,\Omega\,cm^2$ were comparable to those of conventional BLM. The bilayer nature of BLM has been confirmed also by the incorporation of gramicidin and alamethicin into BLMs. The channel formers exhibited typical conductance states. In addition to polycarbonate filters and porous alumina, also photoresistive material can be used for the preparation of micropores as a support for the formation of BLM.

Liu et al. [55] used photoresist for the formation of apertures in the range of 10–40 μm that serves for the formation of BLM from phosphatidylcholine. They even succeeded in the formation of nanoaperture in silicon with a diameter of approximately 660 nm. The BLMs were modified by ionophore dibenzo-18-crown-6; thus, this system served as ion-selective electrode for detection of K^+ ions with sensitivity of 1 μmol L^{-1} and linear range 1 μmol L^{-1}–0.1 mol L^{-1}. The evidence on BLM formation was based on the measurement of the current induced by gramicidin. The BLMs were stable for several tens of hours.

A novel robust platform providing formation of solvent-free lipid bilayers over arrays of cylindrical nanopores has been reported in paper by Kresak et al. [59]. The nanopores were milled through thin Si_3N_4 diaphragms using a focused ion beam. Nanopore-spanning BLMs (npsBLMs) were formed reproducibly by directed fusion of giant unilamellar vesicles (GUVs) to the pore-containing diaphragms. The arrays of npsBLMs exhibit electrical resistances in the GΩ range, lifetimes of up to several days, and breakdown voltages above 250 mV. The npsBLMs revealed typical conducting properties for gramicidin and alamethicin like classical free-standing BLM.

The use of spark-assisted chemical engraving (SACE) to produce glass apertures that are suitable for the formation of artificial BLMs has been described [60]. Prior to use, the glass apertures were rendered hydrophobic by a silanization process and were then incorporated into a simple microfluidic device. Successful BLM formation and the subsequent acquisition of single-channel recordings are demonstrated. Due to the simplicity and rapidity of the SACE process, these glass apertures could be easily integrated into an all-glass microfluidic system for BLM formation.

7.3 Biosensors Based on Ion Channel Switch

The biosensors based on ICS utilize the conducting properties of gramicidin that form the ion channel in lipid bilayer by association of its two monomers. The monomer of gramicidin is composed of 15 mostly hydrophobic amino acids. In the hydrophobic part of the membrane, it folds into the helix with an inner hydrophilic pore of diameter 0.4 nm. This pore is well suited for conducting monovalent cations. The conducting properties and the kinetics of formation of gramicidin channel have been studied in detail using free-standing BLMs [61]. It has been shown that the conductivity of single-channel fluctuates between open and closed state with discrete conducting jumps. It has been suggested that this behavior is due to association and dissociation of gramicidin monomers that diffuse laterally in both BLM monolayers. Due to the thickness fluctuations and forces between monomers, they associate into the conducting dimers. The lifetime of the gramicidin channel (dimer) depends on the thickness of the hydrophobic part of the BLM and increases with decreasing the thickness. While for relatively thick solvent-containing membrane the lifetime is in the order of a second or less, a substantial increase in lifetime takes place for thin, and solvent-free membranes. This lifetime further increases in supported membranes due to the less fluctuation in the membrane thickness in comparison with free-standing BLMs. Moreover, if the monomers at the monolayer close to the solid support are fixed by hydrophilic linker to the metal support, the lifetime further increases and provides stable

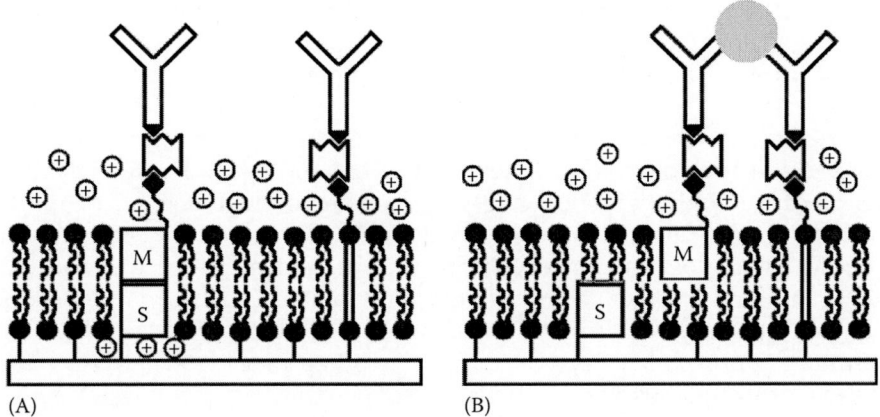

(A)   (B)

FIGURE 7.4
The scheme explaining the principle of ICS immunosensor. Without the target, the gramicidin dimer composed of supported (S) and mobile (M) monomers is in conducting state and the ion flux is observed (A). Addition of target resulted in dissociation of the dimer and decrease of the BLM conductance (B). The target-specific antibodies are attached to the streptavidin and biotinyl-ated linker to the mobile gramicidin monomer and to the tethered lipid with linked hydrophobic chains, respectively. (Adapted from Moradi-Monfared, S. et al., *Biosens. Bioelectr.*, 34, 261, 2012.)

ion flux through the membrane. The conductance of gramicidin channel increases with decreasing the radius of ion-hydrated shell. In particular the conductivity is higher for K^+ ions in comparison with Na^+ ions.

Based on the peculiarities of association and dissociation of gramicidin monomers in supported lipid membranes, Cornell et al. [15] developed an immunobiosensor based on the ICS. The principle of biosensor functioning is schematically presented in Figure 7.4.

The tethered BLM is formed on a gold support. Some of the lipids and gramicidin monomers in the monolayer close to the support are anchored to a gold layer by hydrophilic linker. In addition to conventional phospholipids, also those with connected hydrophobic chains are immobilized to a gold layer by the linker. The polar part of these lipids is modified by biotin-containing linker. The same modification is performed also for gramicidin monomers diffusing in outer monolayer. For this purpose the biotin is attached to the C- terminus of gramicidin via a number of aminocaproyl linker groups [62]. Addition of streptavidin and biotinylated antibodies resulted in the creation of sensing surface (Figure 7.4A). Without target (protein, virus, bacteria, etc.), the BLM is characterized by conductance due to the conductivity of stabi-lized gramicidin dimers. However, addition of a target resulted in disso-ciation of dimers due to affinity of the antibody to the target and due to the formation of complex between the target and two antibodies—bound to freely diffusing gramicidin monomer and to that attached at fixed lipid. This situation is shown in Figure 7.4B. As a result the conductivity decreases. The conductivity is measured usually by impedance method at fixed AC voltage (50 mV) and relatively low frequency (20–100 Hz). The plot of admittance as a function of time represents typical record of binding event at the sensor surface. This method has been used for detection of influenza A virus in clinical samples [63]. Recently the kinetic model of the sensor has been reported and successfully compared with the real system in monitoring the complexation of biotinylated linkers with streptavidin [64]. The model allowing prediction of the kinetics changes in sensor conductance following the binding event. The structural studies using NMR showed that biotinylated gramicidin analogues are very similar to the native gramicidin and that the gramicidin analogues with longer biotinylated linker are more accessible for binding streptavidin or avidin [65].

Similar approach has been used also for ICS genosensor [66]. In this case, instead of antibodies the biotinylated single-stranded DNA (ssDNA) are immobilized onto the freely diffusing gramicidin A monomers and fixed lipids, respectively. Addition of complementary strand containing two complementary chains separated by 10–15 nucleotides caused hybridization and dissociation of gramicidin dimmer. This is observed as a decrease in BLM conductance. The results were correlated based on measurement DNA hybridization by SPR method. Unfortunately more deep analysis such as effect of nucleotide mismatch was not performed. In general the biosensors based on ICS are rather elegant systems. However, this system has not been so far substantially widespread.

Another interesting approach consisting in development of artificial ion channels that can be switched to the opening or closing state by certain chemical groups. Husaru et al. [67] reported artificial ion channel based on calix [4] resorcinarene. Opening and closing of the channel has been achieved by the azo group that changes its conformation by light irradiation.

7.4 Biosensors Based on Supported Lipid Films and Their Analytical Applications

The development of a device based on lipid membranes for analytical applications may include the following steps:

1. Development of stabilized lipid membranes such as those supported on metal, on ultrafiltration membranes, or on a polymer such as polymethylacrylate.
2. Determination of the optimal method to convert the lipid membranes into systems sensitive to desired analytes. This can be achieved by the incorporation of a receptor.

3. Performance of experimental studies to find out whether the permeability of the lipid membranes is altered by an analyte/lipid membrane interaction and to what extent. Investigations were performed to define the optimum experimental conditions for maximized response.

4. Development of the techniques for the direct selective and automated monitoring or rapid screening (in a single format) of compounds of biomedical, pharmaceutical, environmental, and industrial interest, such as insecticides, herbicides, toxins, and gas pollutants, based on the results of the studies mentioned earlier. Therefore, a considerable improvement in terms of cost efficiency, availability, and speed of biomedical, pharmaceutical, environmental, and industrial monitoring procedures can be made.

The specificity of lipid film devices is established with incorporation of a "receptor" (i.e., enzyme, antibody, or natural or artificial receptor). However, even in these cases, proteins and lipids may interfere with the determination of an analyte. In such a case, a flow-through system should be used to avoid protein adsorption and interferences. For example, in order to use lipid film devices for the direct detection of an analyte in dairy products and other foods that may contain proteins and lipids, the effect of lactoglobulin (a common protein present in milk and other dairy products) was investigated in our studies using variable flow rates [42,68]. The effect of this protein was investigated by injecting solutions containing variable concentrations of lactoglobulin into the flowing carrier electrolyte solution at variable flow rates. Our results indicated that concentrations of lactoglobulin smaller than 0.15% (w/v) provided no ion current transients at any flow rates (i.e., including stopped flow). Concentrations of lactoglobulin in the range of 0.15%–0.3% (w/v) produced transient electrochemical signals that could be eliminated by using flow rates above 2 mL/min. Similar results were obtained with albumin [46]. Control experiments using a buffered bovine serum albumin (BSA) solution of concentration similar to that found in human biofluids (6%–8% w/v in human serum and 0.5–1.4 mg mL^{-1} in urine) were performed [46]. The results have shown that no interference is observed for concentrations of albumin up to 3.2 g L^{-1}. For larger concentrations, interference from albumin in the form of random ion current transients of no discrete pulse height due to adsorption on BLMs occurred after 2.5 min from the injection of albumin in the electrolyte solution. Therefore, lipid film-based sensors can be used for the rapid detection of an analyte in human fluids.

The investigation of interferences of samples containing ascorbic acid (AA), glucose, leucine, glycine, tartrate, citrate, bicarbonate, caffeine, and lactoglobulin at concentrations up to 3.2 g L^{-1} has shown no interferences [68].

Lipids that are present in dairy products or human fluids may also interfere with the lipid molecules that constitute the BLM. However, our recent electrochemical experiments have shown that when milk samples or human serum samples were injected in the flow of the carrier electrolyte solution, no alterations of the background membrane conductivity occurred. The lipids are probably not able to fuse with the BLM due to the protective effect exerted by proteins, since those two constituents are known to form lipoproteins, with the proteins protruding into the aqueous phase [69].

7.4.1 Ion-Selective sBLM Modified by Carriers and Ion Channels

In the first studies of sBLM formed according to the method developed by Tien and Salamon [24,25], there was an attempt to check, whether would be possible, their modification by ion carriers, for example, valinomycin (VAL), in order to use this system as an ion-selective electrode. The problem however remains with the presence of metal support that does not allow further diffusion of the ions. This problem has been solved in the paper by Ziegler et al. [45,70] and Ottova et al. [44], who used agar-supported lipid films and successfully demonstrated ionic channel characteristics similar to that observed for free-standing membranes (see also Section 7.2.3.). These agar-supported lipid films have been used also for studying the interaction of short peptides with lipid membranes [71]. The advantage of agar-supported BLM in addition to their stability is that they require minimal volume of buffer—around 50–100 μL—which is at least 10 times lower than that used in typical experiments with free-standing BLM. Unfortunately, the main disadvantage of agar-supported films is their lower stability in comparison with metal-supported films. Therefore, further attempts were focused on development of the supported lipid systems that contain hydrophilic spacer between metal and lipid film, which can be filed by electrolyte and thus represent more suitable model of biomembrane. Such a system was proposed by Cornell et al. [15] that developed sophisticated sBLM at a gold layer and contained hydrophilic spacers between gold and bilayer. As we described in the previous section, these tethered BLMs were used in the development of ICS biosensors.

However, in potentiometric mode, even SAM formed by alkanethiols and phospholipids modified by carriers can be used as ion-selective electrodes. Ding et al. [72] showed that using the potentiometry, the sBLM-based biosensor with incorporated VAL allowed to detect K^+ in the linear range of 10 μmol L^{-1}– 0.1 mol L^{-1} with limit of detection (LOD) 1 μmol L^{-1}. The coefficients of selectivity $K_{K+, Na+}$, $K_{K+, Li+}$, $K_{K+, Ca2+}$, and K_{Mg+2} were 10^{-4}, 10^{-4}, 2×10^{-5}, and 3×10^{-5}, respectively. The stability of the sensor was rather high and reached up to 2 months if stored at −10°C. Similar behavior has been observed for monensin incorporated for alkanethiol/phospholipid sBLM [73].

Cheng et al. [74] proposed another approach for the determination of K^+. They studied capacitive response toward increased concentrations of K^+ of sBLM with incorporated VAL. An increase of sBLM capacitance with increasing K^+ ion concentrations has been observed. The increase of capacitance has been due to increase of dielectric permittivity of sBLM caused by increase of concentration of VAL–K$^+$ complexes. The sensitivity of the sensor was rather high (LOD 0.1 µmol L^{-1}) in comparison with other electrochemical sensors (typically 1 µmol L^{-1}).

An ammonium ion minisensor based on self-assembled BLMs on a metal support and gramicidin D as a channel-forming ionophore was also developed [75]. Semisynthetic platelet-activating factor (PAF) was found to enhance the transport of ammonium ions and to reduce potassium interference. This microfabricated ammonium sensor provides advantages of extremely fast response times, high selectivity to ammonium ions in the presence of volatile amines, and capability of analyzing small volumes of samples with detection limits of 1 mmol L^{-1} of ammonium ions. The device can be simply and reliably fabricated at low cost and can be used as a disposable sensor.

7.4.2 Enzyme Biosensors

Lipid film-coated electrodes represent a unique tool for preparation of enzyme biosensors and for study of the mechanisms of enzymatic reactions at the surfaces [14,76]. The enzyme can be incorporated into the lipid films by means of dissolution of enzyme molecules in lipid solution from which the film is prepared or by immobilization of the enzyme at the lipid film surface. In this respect the works by Snejdarkova et al. are of great significance for further progress in this field. In these works either streptavidin [77] or avidin [78] has been used for the immobilization of glucose oxidase (GOX) to the lipid film surface prepared on a tip of a freshly cut stainless-steel wire coated by insulating polymer or Teflon. Further we have used this approach also for the immobilization of other enzymes, for example, urease on a polypyrrole film [79] or bienzyme system that contained acetylcholinesterase and choline oxidase (CHO) [80].

Let us show peculiarities of lipid-coated electrodes modified by enzymes using an example of the most detailed studied system composed of GOX attached to the supported lipid film using avidin–biotin technology [78]. Teflon-coated stainless steel [78] or platinum wires [81] have been used for the preparation of lipid film with immobilized GOX. The formation of the film is rather simple and is based on the technique developed by Tien and Salamon [24]. The clean wire is immersed into the drop of lipid solution dissolved in n-decane–butanol mixture (8:1 v/v). Then the tip is cut by sharp scalpel and immersed in a buffer (typically 0.1 mol L^{-1} KCl + 10 mmol L^{-1} Tris-HCl, pH 7.0). Electrolyte pH affects GOX activity; therefore, it is important to work in optimal pH (approximately pH 7). In a buffer the self-assembled formation of the lipid film takes place. Phospholipids isolated from crude ox brain extract (COB) are convenient for the formation of these films. The lipids can be modified by D-biotin–N-hydroxy-succinimide ester [77]. In order to use avidin–biotin technology, it is also important to modify the enzyme by avidin or streptavidin. For this purpose the method based on formation of avidin–GOX (A–GOX) conjugates by cross-linking with glutaraldehyde is convenient [82]. The modification of the film surface containing biotinylated phospholipids consists in immersion of lipid-coated wire into the A–GOX buffer solution. The kinetics of the binding of A–GOX onto the film surface can be studied by impedance spectroscopy or using electrostriction method. It is expected that immobilization of A–GOX to the lipid film surface will result in strong binding of avidin to the biotin sides. This binding is noncovalent but very robust having a dissociation constant 10^{-15} mol L^{-1}. In addition there exists a strong interaction between avidin molecules at the surface of the film. All these processes result in the stabilization of film structure. This stabilization partially causes restriction of the mobility of the lipids that in turn affect the mechanical properties of the lipid film. Substantial influence of binding process on the mechanical properties of the film can be therefore expected. In addition GOX is negatively charged. Therefore, the binding process should result in changes of the surface potential of the film. Attachment of A–GOX to the film surface should also change electrical capacitance of the system. These phenomena have been observed experimentally [78]. The interaction of A–GOX with surface of biotinylated lipid film resulted in the increase of elastic modulus, decrease of membrane capacitance, and increase of surface potential. The saturation of membrane capacitance and surface potential starts already at A–GOX concentration of 30 nmol L^{-1}. The restriction of the mobility of phospholipids following the adsorption of the conjugate A–GOX resulted also in the increase of relaxation time of reorientation of the dipole moments [83].

The detection of glucose in a buffer can be performed amperometrically [84]. One of the frequently used methods is based on anodic reoxidation of enzyme usually at a potential of approximately +0.6 V (positive terminal on a working electrode):

$$GOX\,(FAD) + glucose \rightarrow GOX\,(FADH) + gluconolactone \qquad (7.1)$$

$$GOX\,(FADH)_2 + O_2 \rightarrow GOX\,(FAD) + 2H^+ + 2e^- \qquad (7.2)$$

Thus, the changes of the current in a system A–GOX electrode–reference electrode is a measure of the concentration of glucose degraded by enzymes at the surface of lipid film. As we mentioned earlier, the lipid film is poorly permeable by charged particles, such as ions or electrons. Therefore, to increase the sensitivity, it is necessary to modify the lipid film by electron carrier, for example, tetracyanoquinodimethane (TCNQ) [4], tetrathiafulvalene (TTF) [36,85], or platinum nanoparticles [51]. The sensor response following addition of glucose is in the order of 1 min. Wu et al. [36] reported an even faster (20 s) response. The plot of the current as a function of glucose concentration has a typical shape expected from enzymatic reactions following Michaelis–Menten kinetics, that is, it is a curve with saturation (Figure 7.5A).

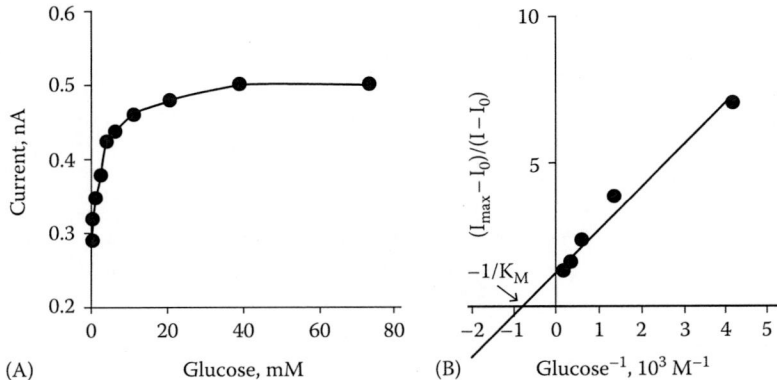

FIGURE 7.5
(A) Dependence of membrane current (I) on concentration of glucose for sBLM containing biotinylated phospholipids with immobilized A–GOX. (B) Plot of $(I_{max} - I_0)/(I - I_0)$ versus reciprocal concentration of glucose for results presented in (Figure 7.5A). I_0 is the current in absence of glucose, I is the current at certain glucose concentration, and I_{max} is the current at saturation, that is, at high glucose concentration. The intersection of this line with x axis is equal to $-1/K_M$. (Reproduced from *Bioelectrochem. Bionerg.*, 41, Hianik, T., Snejdarkova, M., Passechnik, V.I., Rehak, M., and Babincova, M., 221, Copyright 1996, with permission from Elsevier.)

The sensitivity of the sensor response depends on the potential applied as well as on modification of the supporting layer by the mediator. The dependence of the current versus concentration of glucose can be linearized using the Lineweaver–Burk equation:

$$\frac{1}{v} = \frac{1}{v_{max}}\left(1 + \frac{K_M}{c}\right) \tag{7.3}$$

where
   v is velocity of enzyme reaction at certain concentration, c, of the substrate
   v_{max} is maximal velocity of the reaction (at high concentration of substrate when saturation of v as a function of c takes place)
   K_M is Michaelis constant

For amperometric enzyme sensor, Equation 7.3 can be transformed to

$$\frac{(I_{max} - I_0)}{(I - I_0)} = 1 + \frac{K_M}{c} \tag{7.4}$$

where
   I_0 is the current in absence of glucose
   I is the current at certain glucose concentration
   I_{max} is the current at saturation, that is, at high glucose concentration

The plot $(I_{max} - I_0)/(I - I_0)$ as a function of $1/c$ should be a straight line. The intersection of this line with x axis is equal to $-1/K_M$. The example of analysis using Lineweaver–Burk plot for lipid film-based enzyme electrode is shown in Figure 7.5B [78]. The obtained value $K_M = 0.66 \pm 0.18$ mmol L^{-1} is close to the value reported by Bartlett et al. [84] ($K_M = 1$ mmol L^{-1}) for the enzyme electrode with GOX modified by ferrocene derivatives and immobilized on GCE, that is, no lipid film was present.

The analysis of the properties of GOX sensor based on lipid films prepared on various supports, including stainless steel, platinum, polypyrrole, and Nafion films modified by ferrocene, showed that the best sensitivity and best resistance to various interferences (ascorbic acid [AA], paracetamol, uric acid [UA]) have been obtained by GOX sensor formed on the base of lipid films supported on Nafion film with incorporated ferrocene (Fc). Nafion film was supported on the tip of a platinum wire coated by insulating polymer [81]. The presence of the mediator—Fc—entrapped into Nafion film allowed substantial increase in sensitivity of the sensor (17.7 µA mmol L^{-1} cm^{-2}), which is almost 1000 times more in comparison with the sensor prepared on the base of lipid film on a stainless steel support. In addition, the presence of mediator allowed to decrease the potential for oxidation H_2O_2 to +0.4 V instead of +0.6 V for films without mediator, similar to the systems modified by TCNQ [78]. This allowed considerably reduced action of various interferences. Despite the fact that at +0.4 V during the first 3 days of use the sensitivity decreased almost by 40%, during subsequent 3 weeks, the sensor was stable. These unique properties open possibilities also for practical application of the lipid film-based sensors in complex biological liquids. Using anionic lipids, such as dimyristoyl phosphatidylglycerol (DMPG), it is possible to substantially minimize the interaction with sensor surface of negatively charged interfering compounds such as AAs, UAs, or p-acetamidophenol [36].

A technique for the stabilization of lipid membrane-based biosensors with incorporated acetylcholinesterase (AChE) that retains its activity for repetitive uses appeared in the literature [47,86]. Microporous filters composed of glass fibers were used as supports for the stabilization of these sensors. The lipid film is formed on the filter by polymerization by heating at 60°C [86] or using UV radiation prior its use [47]. Methacrylic acid was the functional monomer, ethylene glycol dimethacrylate was the cross-linker, and 2,2'-azobis-(2-methylpropionitrile) was the initiator. AChE is incorporated within this mixture prior to polymerization. This method for preparation of stabilized lipid membranes was investigated using Raman spectroscopy [47]. The results have indicated that the kinetics of polymerization is completed within 4 h. The retain in activity of the enzyme was studied using electrochemical experiments, which have shown that this mild technique of polymerization can now be used to incorporate a protein in the lipid membranes without loss of their activity. The analyte was injected into flowing streams of a carrier electrolyte solution. Hydronium ions produced by the enzymatic reaction at the lipid film surface caused dynamic alterations of the electrostatic fields and phase structure of membranes, and as a result ion current transients were obtained; the magnitude of these signals was correlated to the substrate concentration, which could be determined at the micromolar level. The response times were approximately 10 s. This will allow the practical use of the techniques for chemical sensing based on lipid membrane technology and commercialization of these devices.

The stabilized supported lipid membranes were used for the flow injection analysis of pesticides [86]. The principle of the method is based on the degree of inhibition of the enzyme by the pesticide. Carbofuran was chosen as a typical pesticide. The novelty of the method is based on the fact that an "air-segmented" flow injection technique can be used as compared to previous systems that are based on lipid film technology but are not stable in air. This allows determination of the pesticide using the degree of inhibition and reactivation of enzyme by injections of substrate. Carbofuran was determined at concentration levels of 10^{-7}–10^{-9} mol L^{-1}. The investigation of the effect of potent interferences included a wide range of compounds usually found in foods and also proteins and lipids. The technique was applied in various real samples of fruits, vegetables, and dairy products. The recovery ranged between ca. 96% and 106%, which shows no interferences from the matrix effects.

Another approach of stabilization lipid layer has been proposed by Zheng et al. [87], who reported horseradish peroxidase (HRP)-based sBLM sensor for the detection of hydrogen peroxide. The lipid bilayer has been formed at the surface of GCE by addition of the small amount of DMPC dissolved in chloroform and in subsequent electrode immersing in PBS. After the addition of HRP, sBLM was stabilized by electropolymerized polydopamine film. The hydrogen peroxide was detected by measurement of electrocatalytic reduction current in the presence of hydroquinone as a mediator. The lipid layer substantially stabilized the activity of HRP, which retain, 84% of the initial response after storage at 4°C and using the sensor once a day over the period of 3 weeks. High sensitivity (LOD = 0.1 μmol L^{-1}) and linear range 0.27 μmol L^{-1} – 3.1 mmol L^{-1} of H_2O_2 detection is promising for practical applications.

An original approach for detection of the activity of CHO embedded at the surface of supported lipid membrane has been reported by Jiao et al. [88]. The detection was based on measuring chemiluminescence of luminol induced by hydrogen peroxide, the product of choline oxidation. The supported films were prepared by LB deposition method. The lipid monolayer at the PBS subphase was prepared by spreading of multilamellar vesicles composed of phospholipids and contained luminal modified by hydrophobic chain, and IgG that specifically bind CHO. This work successfully demonstrated the reagentless detection of CHO activity immobilized in the natural environment of lipid film. The LB method has been used also for deposition of the mixed monolayers composed of GOX and stearic acid (SA) or octadecylamine (ODA) on a platinum support. The amperometric detection of GOX and the sensor stability was better for ODA layers. The good linearity between

0.1 and 5 mmol L^{-1} glucose was obtained [89]. The deposition on a solid support of lipid multilayers contained tyrosinase and lutetium bisphthalocyanine has been effective in amperometric detection of phenol derivative-based antioxidants [90].

Serious problem in fabrication of sBLM enzyme biosensor consists in poor solubility of most enzymes in hydrophobic environment of the lipid bilayer. Certainly the GOX, HRP, and AChE enzymes are globular, water-soluble molecules. Therefore, they are usually immobilized on a surface of lipid layer. However, layer-by-layer (LbL) technology of the formation of biosensor especially those based on liposomes can solve the problem with enzyme solubility. Most recently Guan et al. [91] reported preparation of AChE and lipid film-based amperometric biosensor for detection of organophosphate pesticide dichlorvos. The AChE has been encapsulated into the large unilamellar liposomes of the diameter 7.3 ± 0.85 μm composed of L-α-phosphatidylcholine. The lipid bilayer was modified by porins that form pores, which allow diffusion of the pesticides and the AChE substrate, the acetylthiocholine chloride (ATCl), inside the liposomes from the surrounding buffer. The GCE surface of the sensor was first covered by positively charged chitosan following adsorption of liposomes. This process was repeated several times to form multilayer sensing system. The detection of the pesticide was based on its inhibition of the AChE activity. With increased concentration of dichlorvos, the AChE activity decreased, which resulted in decreasing of transformation of the substrate ATCl into the electroactive thiocholine. Thus, the decrease of the redox signal of thiocholine was used as a sensor response. The biosensor revealed LOD = 0.86 ± 0.098 μg L^{-1} and linear range of detection of dichlorvos at concentrations 0.25–1.5 and 1.5–10 μmol L^{-1}. Also the stability of the sensor was rather high. The sensor retained its activity even after 15-days storage at 4°C. No interferences were observed at presence of 0.1 mmol L^{-1} urea or 0.1 mmol L^{-1} glucose or 0.1 mmol L^{-1} fructose.

7.4.3 DNA Biosensors

The potentiality of BLMs to monitor DNA hybridization and to develop biosensors for rapid nucleic acid detection has been investigated [92]. Planar free-suspended and self-assembled metal-supported BLMs were also used to monitor *in situ* the incorporation of single-stranded deoxyribonucleic acid (ssDNA) in these membranes. The results have shown that the adsorption of ssDNA at lipid membranes (as a medium for DNA incorporation on an electrode surface) can occur much faster, using milder conditions, and smaller amounts of DNA than other techniques described in the literature.

The use of sBLMs as transducers for the direct rapid electrochemical monitoring of DNA hybridization was reported in [92]. Characterized oligomers based on ssDNA thymidylic acid icosanucleotides that were terminated with C16 (dT20-C16) and deoxyadenylic acid icosanucleotides (dA20) were used for the hybridization procedure at the lipid membrane surface.

The incorporation of nucleic acids into the lipid matrix was investigated by differential scanning calorimetric (DSC) experiments. The sensor was regenerable for multiple cycles of application. A comparison of the present DNA biosensor with other standard methods of monitoring DNA hybridization (i.e., voltammetric and spectrophotometric) has revealed that the sBLM-based biosensor has a number of advantages such as regenerability for multiple cycles of application, low detection limits for DNA (a few hundreds of fmoles), fast response times (few minutes), mild conditions of hybridization, and capability of analyzing small volumes of sample.

The simple impedimetric genosensor based on DNA hybridization and surfactant-based supported film has been reported in our work [93,94]. The oligonucleotide dT_{15} modified by hydrophobic chain has been incorporated in the hydrophobic region of surfactant Brij-52 bilayer. Addition of complementary oligonucleotide dA_{15} resulted in changes of impedance spectrum. In particular, decreasing of real part of the high frequencies was noticed. At the same time, imaginary part does not change drastically. Therefore, the hybridization of DNA leads to the decreasing of membrane resistance. Probably it can be caused by local violations of inner bilayer order due to the hydrocarbon chains of modified oligonucleotides. The background spectra for the Brij-52 bilayer system with immobilized DNA probes was same as the spectra measured in the presence of noncomplementary oligonucleotide $AT_8C_6G_7$. Thus, the response generation has been caused by specific complement coupling with DNA probes immobilized on the surfactant membrane surface. To develop the biosensor based on the affinity interaction detection, it is not necessary to measure whole impedance spectrum in wide frequency range of applied potential. The most sufficient changes were observed in real part of impedance in the frequency range 1–10 kHz. The highest response of the system is observed at the frequency of 2 kHz. The change of the response was approximately 40% from background level. It is two times higher than in similar system reported earlier [95].

An interesting approach for detection of the DNA hybridization has been proposed by Aoki et al. [96]. The detection was based on the reduction of the ionic conductivity after DNA hybridization, through a peptide nucleic acid (PNA) probe bounded to sBLM on a gold surface. Recently the agar-supported BLM modified by ssDNA, but without hydrophobic chain, has been demonstrated as a suitable genosensor. This sensor has been used for amperometric detection of DNA hybridization [97].

The sBLM modified by ssDNA can enhance the sensitivity of the lipid layer towards the detection of various toxins and drugs. In papers by Siontorou et al. [98,99], the sBLM were formed from phosphatidylcholine in n-hexane on the tip of Teflon-coated silver wire of the diameter 0.5–1 mm. The BLMs were modified by 5′-hexadecyl-deoxythymidylic acid icosanucleotide (dT_{20}-C_{16}). It has been shown that the presence of ssDNA improved the detecting limit of hydrazines and aflatoxin. Similar approach has been used for detection atenolol, the β-adrenoreceptor blocking agent. The sBLM sensor modified by ssDNA provided fast response (few seconds) to alterations of atenolol concentration (20–200 μmol L^{-1}) in electrolyte solution. The sensor was stable for long periods of time (over 48 h) [100,101].

Supported sBLM formed on GCE with incorporated double-stranded DNA (dsDNA) from calf thymus revealed properties similar to ionic channels. It has been shown that the conductivity of the sBLM can be controlled by quinacrine ions, while other ions such as K$^+$ or Na$^+$ had no effect on membrane conductivity. Thus, supported BLM modified by dsDNA serves as a receptor for quinacrine ions [37]. Novel trends in DNA biosensors have been recently reviewed also by Teles and Fonseca [102]. The peculiarities of LB lipid films with incorporated short ssDNA containing poly(butadiene) chains have been reported in Ref. [103]. The mixed films formed stable monolayers at an air–water interphase and can be potentially used for deposition on a solid surface as a basis for hybridization sensor.

7.4.4 Affinity Biosensors Based on Artificial and Natural Receptors

Supported BLM allows incorporation of natural or artificial receptors and thus could serve as a biosensor for detection of hormones or other ligands. However, even using lipids with specific head groups, it is possible to discriminate small molecules, for example, the neurotransmitter dopamine (DA). DA plays a significant role in the central nervous system. The loss of DA may cause serious illness such as schizophrenia or Parkinson's disease [104]. Therefore, control of DA concentration in biological fluids is important for neurochemistry and medicine. So far the major attention in DA detection was concentrated on the application of electrochemical methods. This is due to DA electroactivity that can be easily oxidized at various surfaces (gold, platinum, carbon, glassy carbon, and those modified by various compounds including conducting polymers, nanoparticles, etc.). The physiological concentration of DA in human plasma is 0.5 nmol L^{-1}. However, the concentration of the major interfering compounds in DA detection— AA and UA—is much higher, that is, 100 and 15 μmol L^{-1}, respectively [105]. AA and UA are also electroactive and their oxidation potentials are similar to those of DA. This causes substantial problems in electrochemical detection of DA. Despite several sophisticated detection systems, the sensitivity of electrochemical detection is typically above 20 nmol L^{-1} of DA (see [106] and references herein). Supported BLM can be advantageously used for DA detection and other hormones, like epinephrine (EP, known as adrenaline). The lipid bilayer blocks the access of electroactive species to a metal surface. It has been shown that sBLM composed of DMPC-doped with mediator 5,5-ditetradecyl-2-(2-trimethyl-ammonioethyl)-1,3-dioxane bromide (DTDB) on a GCE is favorable for electrochemical reactions of cationic species like DA and related catecholamines. This sensor exhibited high electrocatalytic activity towards the oxidation of DA and AA, but the catalytic potential of AA oxidation was by 135 and 160 mV more negative than those of EP and DA, respectively. The sensitivity of DTDB-doped BLM is due to the dioxane group as well as the suitable length of DTDB molecule. The current response towards flow injection addition of EP was linear in a range of 10^{-8}–10^{-4} mol L^{-1} [107].

Platinum-supported BLM, modified by anthraquinone-2-sulphonic acid (AQS), has been used for nonenzyme detection of glucose using CV method at the presence of redox probe potassium ferricyanide. It has been shown that both the oxidation and the reduction current peaks decreased at the presence of glucose in concentration range varying from 10 to 320 mmol L^{-1}. The pretreatment of membrane with physiological concentration of insulin (0.13 nmol L^{-1}) evoked slight increase of the sBLM conductance at the presence of glucose [108].

The high affinity of nitric oxide (NO) to lipids initiated the application of sBLM for detection of NO [51]. It has been shown that increase in NO concentration since 0–16 µmol L^{-1} resulted in gradual increase of membrane capacitance. This has been attributed to the increase of dielectric permittivity of the inner membrane part. Karabaliev and Kochev [109] reported the possibility of using sBLM-containing cholesterol as an impedimetric sensor for detection of saponin. Saponins are complex glycosides synthesized in plants. They are known as highly active substances that cause hemolysis of the red blood cells with release of hemoglobin. sBLMs were formed at the surface of GCE from a mixture of natural lecithin and cholesterol dissolved in n-hexane. The changes in sBLM resistance have been observed already at 0.01% saponin. sBLM formed by the method of Tien and Salamon [24] and modified by fullerene C_{60} has been shown as a sensitive tool for amperometric detection of smell compounds—various neutral odorants [110]. The detection has been based on blocking the mediated electron transport due to the more compact lipid film following incorporation of odorants. The sensitivity of lipid films to the detergents has been used for amperometric detection of the nonionic detergent Triton X-100. The lipid films were formed inside the nanoporous alumina using liposomes. The sensing system was able to detect the alteration in electrical properties of the lipid films by detergent [111].

However, the most efficient biosensors for selective detection of various compounds are based on artificial or natural receptors incorporated into lipid films. The supported lipid films modified by calixarenes allowed sensitive and rapid detection of doping compounds such as DA in the urine of athletes [112] and insecticides in fruits and vegetables [113]. Recent applications in this field include the construction of a disposable chemosensor for the selective rapid detection of carbofuran [2] and food hormones (i.e., naphthalene acetic acid (NAA)) in fruits and vegetables [114] and zinc in water [115].

We also developed the sBLM-based biosensor with incorporated calix [6] arene containing six phenol units that form a cavity specific to cyt c [116]. Cyt c is a small protein that plays an important role in electron transport at the inner membrane of the mitochondria, but it is responsible also for cell apoptosis. This protein is released from the mitochondria and its concentration can be considered as an indicator of normal functioning of the cell. Therefore, there is high interest in the development of an efficient method of cyt c detection. The sensor has been prepared by the formation of lipid monolayer at the chemisorbed ODT layer at the gold surface of the electrode (diameter 2 mm) using liposome fusion. The small unilamellar liposomes that contained calixarenes in various molar ratios have been used. The scheme of the sensing layer containing the calixarenes is presented in Figure 7.6A. The sensor response following the addition of cyt c was monitored by EIS either by changes in electrical capacitance or by measuring the changes in electron transfer resistance at the presence of redox couple 5 mM $K_4[Fe(CN)_6]$ and $K_3[Fe(CN)_6]$ (1:1). Increased concentration of cyt c resulted in the decrease of capacitance due to the increase of the thickness of the sensing layer (Figure 7.6B). At the same time, the charge transfer resistance decreased. This is the consequence of the increase in the positive charge at the surface due to the incorporation of protonated amino groups of lysine

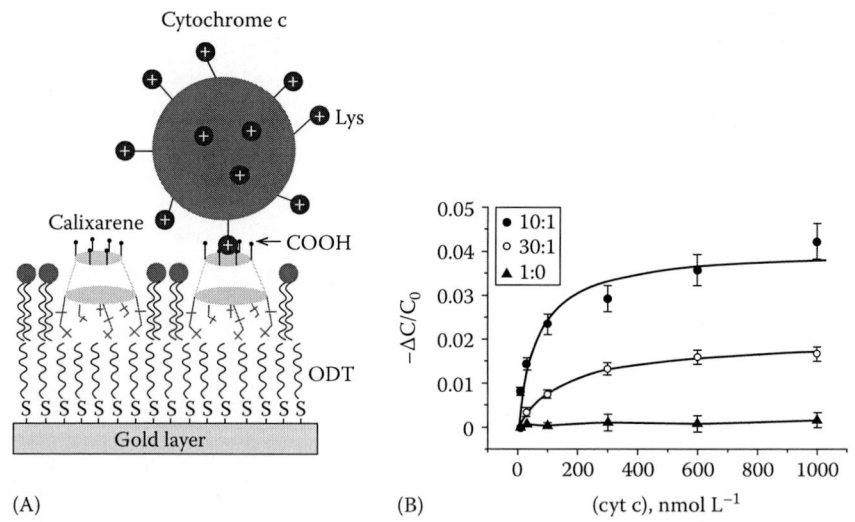

(A)   (B)   (cyt c), nmol L^{-1}

FIGURE 7.6

(A) The scheme of biosensor based on cyt c-sensitive calixarenes incorporated into the phospholipid monolayer supported on gold layer with chemisorbed ODT. The cyt c is incorporated by lysine (Lys) residue into the cavity of calixarene contained at the upper part of carboxyl groups. The lower part of calixarene contained hydrophobic chains allowing its incorporation into the hydrophobic part of the lipid monolayer. (B) The plot of the relative changes of the sensor capacitance ($\Delta C = C/C_0$; C_0 and C are the capacitance in the absence and in the presence of the analyte, respectively) as a function of cyt c concentration for the pure sensors without (1:0) and with calixarenes at lipid–calixarene mole ratios 30:1 and 10:1, respectively. The full lines are fit according to Langmuir isotherm. (From Mohsin, M.A., Banica, F.-G., Oshima, T., and Hianik, T.: *Electroanalysis*. 23. 1229, 2011. Copyright Wiley-VCH Verlag GmbH & Co. KGaA. Adapted and reproduced with permission.)

residues at the surface of cyt c into the cavity of calixarene. However, most of the positively charged lysine residues were presented at the cyt c surface. Thus, this behavior was in good agreement with the results of experiments on the measurement zeta potential [8]. The sensitivity of detection of cyt c was as low as 10 nmol L^{-1}. However, a much higher concentration of free lysine (>30 mmol L^{-1}) was necessary to induce a similar effect. Nonspecific interactions of cyt c with sBLM without calixarenes were negligible.

The investigations of electrochemical interactions of NAA with stabilized lipid films supported on a methacrylate polymer on a glass fiber filter with incorporated auxin-binding protein 1 receptor for the development of a biosensor for the rapid detection of this compound in fruits were described in Ref. [114]. NAA was injected into the flowing streams of a carrier electrolyte solution, the flow of the electrolyte solution stopped, and an ion current transient was obtained; the magnitude of the signal was correlated to NAA concentration, which could be determined at the micromolar level. NAA preconcentrates at the lipid membrane surface, which causes dynamic alterations of the electrostatic fields and phase structure of membranes. The response time was ca. 5 min and NAA was determined at concentration levels of μM. The effect of potent interferences included a wide range of compounds. The results showed no interferences from these compounds in concentration levels usually found in real samples. The method was applied for the determination of NAA in fruits and the reproducibility of the method was checked in samples of various fruits. A quantitative method for the detection of NAA in fruits that can be complimentary to HPLC methods was also provided. These lipid films can be used as portable sensors for the rapid detection of NAA in fruits by non-skilled personnel.

The air stable lipid films supported on a methacrylate polymer on a glass fiber filter with incorporated artificial receptor have been used for detection of zinc [115]. This minisensor was constructed for the electrochemical flow injection analysis of zinc. The device can sense the analyte in a drop (75 μL) of sample. Zinc was injected into flowing streams of carrier electrolyte solution. A complex formation between the calix [4] arene phosphoryl receptor and zinc took place. This enhanced the preconcentration of zinc at the lipid membrane surface, which in turn caused structural changes of membranes. As a result, ion current transients were obtained and the magnitude of these signals was correlated to the substrate concentration. The response time was approximately 5 s and zinc was determined at concentration levels of nmol L^{-1}. The analytical curve was linear in the concentration range 0.1–1.2 μmol L^{-1} with detection limit of 0.05 μmol L^{-1} and a relative standard deviation lower than 4%. The effect of potent interferences included a wide range of other metals. As an analytical demonstration, trace concentrations of Zn(II) were successfully detected in real samples of waters without any laborious and time-consuming treatment.

Voltage-activated sodium (Nav) channels are essential in generating and propagating nerve impulses, placing them among the most widely targeted ion channels by toxins from venomous organisms. An increasing number of spider toxins have been shown to interfere with the voltage-driven activation process of mammalian Nav channels, possibly by interacting with one or more of their voltage sensors. A paper appeared in the literature that focuses on the mechanism by which spider toxins affect Nav channel gating and the possible applications of these toxins in the drug discovery process [117].

A novel system of integrating artificial lipid bilayer (biomimetic membrane) with single-walled carbon nanotube networks (SWNT net)-based field-effect transistor (FET) was developed and demonstrated that such hybrid nano-electronic biosensors can specifically and electronically detect the presence and dynamic activities of ionophores (specifically, gramicidin and calcimycin) in their native lipid environment [118]. This technique can potentially be used to examine other membrane proteins (e.g., ligand-gated ion channels, receptors, membrane insertion toxins, and antibacterial peptides) for the purposes of biosensing, fundamental studies, or high-throughput drug screening.

A nanostructure electrochemical biosensor was developed to directly detect and screen estrogenic substances based on estrogen receptor (ER) binding without the use of radio- or enzyme-labeled compounds [119]. The biosensor was fabricated by immobilization of ERs in sBLM modified with Au nanoparticles, and the properties of the modified electrodes were characterized by cyclic voltammetry and impedance spectroscopy. The results indicated that the biosensor was able to detect the natural estrogen 17β-estradiol with an acceptable linear correlation ranging from 5 to 150 ng L^{-1} and a detection limit of 1 ng L^{-1}. The biosensor could also detect other known xenoestrogens such as bisphenol A and 4-nonylphenol with satisfied sensitivity and quantitative results. The biosensor showed good reliability and repeatability, and the Au nanoparticles greatly enhanced the sensitivity and stability of the biosensor. Moreover, estrogenic activity of water samples determined by this biosensor was in good agreement with that determined by MCF-7 cell proliferation assay.

Advances in the field of electrochemical biosensors based on lipid films with incorporated synthetic receptor were recently reported [120]. This report included the design and characterization of a novel bioelectric-sensing platform engineered by coupling an ion channel, which serves as the electrical probe, to G-protein-coupled receptors (GPCRs), a family of receptors that detect molecules outside the cell. These ion channel-coupled receptors may potentially detect a wide range of ligands recognized by natural or altered GPCRs, which are known to be major pharmaceutical targets. This could form a unique platform for label-free drug screening.

We already mentioned the advantage of micro- and nanoporous materials for the formation of stabilized lipid films (Section 2.5). In the most recent work by Lazzara et al. [121], it has been demonstrated that anodic aluminum oxide (AAO) can be advantageously used for the formation of stable lipid films containing NTA(Ni) head group to extract two (His)-tagged proteins, namely, PIGEA14 (small protein composed of 126 amino acids, M.w.14 kDa) and ezrin (larger protein composed of 586 amino acids, M.w. = 70 kDa). Earlier it has been shown by QCM and SPR methods that these proteins adsorb to a gold surface covered with phospholipids modified by NTA(Ni) [122,123]. However, due to the large surface area of AAO, which is 80 times larger in comparison with the flat surface, a high effectivity of protein extraction was achieved. Moreover, the used approach consisting in formation supported films by small unilamellar vesicles with diameter 20–50 nm that diffuse into the small AAO pores (diameter 75 nm) allowed to apply time-resolved optical waveguide spectroscopy and confocal laser scanning fluorescence microscopy to monitor protein adsorption to the lipid layer. Thus, the AAO-supported lipid films could serve as efficient tools for protein extraction.

Modification of supported lipid membranes and liposomes with the receptors allowing substantial amplification of the signal in detection of target species using QCM method. The method of signal amplification in QCM is crucial especially when target compounds have a relatively small mass that does not induce substantial changes in resonant frequency of the quartz transducer. An interesting approach has been proposed by Chen et al. [124], who reported amplified QCM detection of cholera toxin (CT) based on surface agglutination of ganglioside-bearing liposomes. The sensor surface was composed of lipid monolayer adsorbed at ODT monolayer chemisorbed at the gold layer of QCM transducer. This supported lipid film was modified by monosialoganglioside (GM1), which specifically interacted with CT. The unilamellar liposomes (diameter approximately 150 nm) composed of DPPC and GM1 (mole ratio 40:1) were used for signal amplification. Addition of CT to a surface of lipids modified by GMI at the presence of the GM1-containing liposomes resulted in substantial changes of the resonant frequency, which was considerably larger in comparison with that when only CT was added to the sensor surface (without liposomes). Thus, the agglutination of the liposomes at the sensor surface at the presence of CT enormously amplified the signal that resulted in substantial increase of the sensitivity of toxin detection. The detection limit achieved was as low as 25 µg L^{-1} of CT. At the same time, due to biocompatibility of sBLM, the nonspecific interactions with other compounds such as BSA, IgG, IgE, HSA, thrombin, and lysozyme were rather low in comparison with specific effect of CT. Additional advantage of the sensor was the possibility of its fast regeneration. By rinsing the sBLM with ethanol, it was possible to remove the lipid layer without damaging the ODT monolayer. Addition of GM1-modified liposomes to the ODT surface allowed to prepare fresh sBLM within 30 min.

7.5 Conclusions

The present chapter describes the preparation and analytical applications of biosensors based on lipid film technology. Recent technological advances include the construction of stabilized supported lipid film-based devices on a polymer or another support with incorporated enzyme or a receptor (natural or artificial) stable in air. These sensors reveal detection limits in the nanomolar range. The most important aspect of the present effort is to provide a commercial portable sensor that can be used for in-field and market applications. Present technology starts to offer chemo- or biosensors that can be portable and be used by nonspecialized personnel for rapid *in situ* detections of various compounds.

The results have shown that a diversity of lipid film-based biosensors can be reused after storage in air even after a period of couple of months and can be reproducibly fabricated with simplicity and low cost. These analytical methods are faster and have a lower cost than those based on chromatographic techniques and can be used as rapid portable detectors complimentary to these methods for in-field and market measurements in foods, and for environmental monitoring.

Recently advanced lithography has enabled the fabrication of nanopores and the insertion of nanoparticles, nanostructuration of porous supports in thin membranes where artificial bilayers can be assembled. This process accommodates a wide range of lipid composition in stable form, with the inclusion of membrane proteins. Therefore, the application of nanotechnology to this field of lipid film technology has allowed miniaturization and will result in mass production of sensors. Producing the smaller patterns will enable sensors to respond faster, with a higher degree of sensitivity, and at lower production costs. Development of sensors using the present technologies will offer improved sensitivity for detection with high specificity at the molecular level, with an increment of several orders of magnitude over currently available techniques with applications in medical diagnostics, environmental monitoring, pharmaceutical screening, and food processing.

Acknowledgments

The authors gratefully acknowledge the financial support by the Agency for Promotion Research and Development (project Nos. LPP-0250-09, SK-GR-0006-11). The authors also express their gratitude for the financial help of the Greek Ministry of Development, General Secretariat for Research and Technology, and the European Commission (European Regional Development Fund and Natural Resources), which cofunded the present Greece–Slovakia project (contract 12SLO_ET30_1036).

References

1. P. Mueller, D.O. Rudin, H.T. Tien, and W.C. Wescott, *Nature* 194 (1962) 979.
2. D.P. Nikolelis, T. Hianik, and G.-P. Nikoleli, *Electroanalysis* 22 (2010) 2747.
3. D.P. Nikolelis, G. Raftopoulou, N. Psaroudakis, and G.-P. Nikoleli, *Electroanalysis* 20 (2008) 1574.
4. D.P. Nikolelis, N. Psaroudakis, and N. Ferderigos, *Anal. Chem.* 77 (2005) 3217.
5. D.P. Nikolelis and U.J. Krull, *Electroanalysis* 5 (1993) 539.
6. D. Chen and J. Li, *Surf. Sci. Rep.* 61 (2006) 445.
7. A. Abdennour, M.J. Linman, and Q. Cheng, *Sens. Actuators B* 156 (2011) 169.
8. Z. Garaiová, M.A. Mohsin, V. Vargová, F.-G. Banica, and T. Hianik, *Bioelectrochemistry* 87 (2012) 220.
9. R.G. Nuzzo and D.L. Allara, *J. Amer. Chem. Soc.* 105 (1983) 4481.
10. W. Pan, C.J. Durning, and N.J. Turro, *Langmuir* 12 (1996) 4469.
11. C.M.A. Brett, S. Kresák, T. Hianik, and A.M. Oliveira-Brett, *Electroanalysis* 15 (2003) 557.
12. M. Akram, M.C. Stuart, and D.K.Y. Wong, *Anal. Chim. Acta* 504 (2004) 243.
13. H. Lang, C. Duschl, M. Grätzel, and H. Vogel, *Thin Solid Films* 210–211 (1992) 818.
14. T. Hianik, *Bioelectrochemistry: Fundamentals, Experimental Techniques and Applications* (Ed. P. N. Bartlett), Wiley: Chichester, U.K., 2008, pp. 87–156.
15. B.A. Cornell, V.L.B. Braach-Maksvytis, L. King, B.D.J. Raguse, I. Wieczorek, and R.J. Pace, *Nature* 387 (1997) 580.
16. P. Vitovič, S. Kresák, R. Naumann, S. Schiller, R.N.A.H. Ruthven, R.N. Mc Elhaney, and T. Hianik, *Bioelectrochemistry* 63 (2004) 169.
17. I.K. Vockenroth, D. Fine, A. Dodobalapur, A.T.A. Jenkins, and I. Köper, *Electrochem. Commun.* 10 (2008) 323.
18. S.R. Jadhav, Y. Zheng, R.M. Garavito, R.M. Worden, *Biosens. Bioelecton.* 24 (2008) 831.
19. R. Naumann, E.K. Schmidt, A. Jonczyk, K. Fendler, B. Kadenbach, T. Liebermann, A. Offenhäusser, and W. Knoll, *Biosens. Bioelectron.* 14 (1999) 651.
20. R. Naumann, T. Baumgart, P. Gräber, A. Jonczyk, A. Offenhäusser, and W. Knoll, *Biosens. Bioelectron.* 17 (2002) 25.
21. M.G. Friedrich, M.A. Plum, M.G. Santonicola, V.U. Kirste, W. Knoll, B. Ludwig, and R.L.C. Naumann, *Biophys. J.* 95 (2008) 1500.
22. C. Danelon, S. Terrettaz, O. Guenat, M. Koudelka, and H. Vogel, *Methods* 46 (2008) 104.
23. W. Knoll, I. Köoper, R. Naumann, E.-K. Sinner, *Electrochim. Acta* 53 (2008) 6680.
24. H.T. Tien and Z. Salamon, *Bioelectrochem. Bioenerg.* 22 (1989) 211.
25. M. Rehák, M. Snejdárková, and M. Oto, *Biosens. Bioelectron.* 9 (1994) 337.
26. V.I. Passechnik, T. Hianik, S.A. Ivanov, and B. Sivak, *Electroanalysis* 10 (1998) 295.
27. H. Haas, G. Lamura, and A. Gliozzi, *Bioelectrochemistry* 54 (2001) 1.
28. F. Bordi, C. Cametti, and A. Gliozzi, *Bioelectrochemistry* 57 (2002) 39.
29. G. Puu, I. Gustafson, E. Artursson, and P.A. Ohlsson, *Biosens. Bioelectr.* 10 (1995) 463.
30. D. Pum, J.L. Toca-Herrera, and U.B. Sleytr, *Int. J. Mol. Sci.* 14 (2013) 2484.
31. A. Nelson, *Curr. Opin. Coll. Interf. Sci.* 15 (2010) 455.
32. E. Wang and X. Han, *Advances in Planar Lipid Bilayers and Liposomes* (Eds. H. T. Tien and A. Ottova-Leitmannova), Elsevier: Amsterdam, the Netherlands, Vol. 2, 2005, pp. 261–303.

33. R.S. Deinhammer, M. Ho, J.W. Anderegg, and M.D. Porter, *Langmuir* 10 (1994) 1306.
34. J. Liu and S. Dong, *Electrochem. Commun.* 2 (2000) 707.
35. C.G. Siontorou, A.M.O. Brett, and D.P. Nikolelis, *Talanta* 43 (1996) 1137.
36. Z. Wu, B. Wang, S. Dong, and E. Wang, *Biosens. Bioelectron.* 15 (2000) 143.
37. Y. Tong, X. Han, Y. Song, J. Jiang, and E. Wang, *Biophys. Chem.* 105 (2003) 1.
38. W. Huang, Z. Zhang, X. Han, J. Wang, J. Tang, S. Dong, and E. Wang, *Biophys. Chem.* 99 (2002) 271.
39. M. Karabaliev, *Bioelectrochemistry* 71 (2007) 54.
40. H. Zhang, Z. Zhang, J. Li, and S. Cai, *Electrochem. Commun.* 9 (2007) 605.
41. D.P. Nikolelis, C.G. Siontorou, V.G. Andreou, and U.J. Krull, *Electroanalysis* 7 (1995) 531.
42. V.G. Andreou and D.P. Nikolelis, *Anal. Chem.* 70 (1998) 2366.
43. D.P. Nikolelis and U.J. Krull, *Talanta* 39 (1992) 1045.
44. X.D. Lu, A. Ottova-Leitmannova, and H.T. Tien, *Bioelectrochem. Bioenerg.* 39 (1996) 285.
45 W. Ziegler, M. Gaburjakova, J. Gaburjakova, V. Tvarozek, and T. Hianik, *Biologia* 51 (1996) 683.
46. D.P. Nikolelis and M. Mitrokotsa, *Biosens. Bioelectr.* 17 (2002) 565.
47. D.P. Nikolelis, G. Raftopoulou, G.-P. Nikoleli, and M. Simantiraki, *Electroanalysis* 18 (2006) 2467.
48. D.P. Nikolelis, G. Raftopoulou, P. Chatzigeorgiou, G.-P. Nikoleli, and K. Viras, *Sens. Actuator B* 130 (2008) 577.
49. D.P. Nikolelis, D.A. Drivelos, M.G. Simantiraki, and S. Koinis, *Anal. Chem.* 76 (2004) 2174.
50. H. Gao, G.-A. Luo, J. Feng, A.L. Ottova, and H.T. Tien, *J. Photochem. Photobiol. B* 59 (2000) 87.
51. J.-S. Ye, A. Ottova, H.T. Tien, and F.-S. Sheu, *Bioelectrochemistry* 59 (2003) 65.
52. T. Kanyo, Z. Konya, A. Kukovecz, F. Berger, I. Dekany, and I. Kiricsi, *Langmuir* 20 (2004) 1656.
53. C. Richard, F. Balavoine, P. Schultz, T.W. Ebbesen, and C. Mioskowski, *Science* 300 (2003) 775.
54. J.F. Hicks, F.P. Zamborini, A.J. Osisek, and R.W. Murray, *J. Am. Chem. Soc.* 123 (2001) 7048.
55. B.R.D. Liu, D. Rieck, B.J. Van Wie, G.J. Cheng, D.F. Moffett, and D.A. Kidwell, *Biosens. Bioelectron.* 24 (2009) 1843.
56. W. Xia, Y. Li, Y. Wan, T. Chen, J. Wei, Y. Lin, and S. Xu, *Biosens. Bioelectron.* 25 (2010) 2253.
57. G. Favero, A. D'Annibale, L. Campanella, R. Santucci, and T. Ferri, *Anal. Chim. Acta* 460 (2002) 23.
58. W. Römer and C. Steinem, *Biophys. J.* 86 (2004) 955.
59. S. Kresak, T. Hianik, and R.L.C. Naumann, *Soft Matter.* 5 (2009) 4021.
60. M.E. Sandison, M. Zagnoni, M. Abu-Hantash, and H. Morgan, *J. Micromech. Microengin.* 17 (2007) 189.
61. S.R. Hladky and D.A. Haydon, *Biochim. Biophys. Acta* 174 (1972) 294.
62. C.J. Morton, G.H. Tabo, R.E. Koeppe, and F. Separovic, *Int. J. Peptide Res. Ther.* 12 (2006) 243.
63. S.Y. Oh a, B. Cornell, D. Smith, G. Higgins, C.J. Burrell, and T.W. Koka, *Biosens. Bioelectr.* 23 (2008) 1161.
64. S. Moradi-Monfared, V. Krishnamurthy, and B. Cornell, *Biosens. Bioelectr.* 34 (2012) 261.

65. F. Separovic, *Austral. J. Chem.* 56 (2003) 163.
66. S.W. Lucas and M.M. Harding, *Anal. Biochem.* 282 (2000) 70.
67. L. Husaru, R. Schulye, G. Steiner, T. Wolf, W.D. Habicher, and R. Salyer, *Anal. Bioanal. Chem.* 382 (2005) 1882.
68. G.-P. Nikoleli, D.P. Nikolelis, and C. Methenitis, *Anal. Chim. Acta*, 675 (2010) 58.
69. A. Townshed, *Encyclopaedia of Analytical Chemistry*, Academic Press Limited: London, U.K., Vol. 4, 1995, p. 2520 and Vol. 9, p. 5.
70. W. Ziegler, M. Gaburjakova, J. Gaburjakova, B. Sivak, V. Rehacek, V. Tvarozek, and T. Hianik, *Coll. Surf. A* 140 (1998) 357.
71. T. Hianik, U. Kaatze, D.F. Sargent, R. Krivanek, S. Halstenberg, W. Pieper, J. Gaburjakova, M. Gaburjakova, M. Pooga, and U. Langel, *Bioelectrochem. Bioenerg.* 42 (1997) 123.
72. L. Ding, J. Li, E. Wang, and S. Dong, *Thin Solid Films* 293 (1997) 153.
73. J.H. Li, L. Ding, E.K. Wang, and S.J. Dong, *J. Electroanal. Chem.* 414 (1996) 17.
74. Z. Cheng, L. Luo, Z. Wu, E. Wang, and X. Yang, *Electroanalysis* 13 (2001) 68.
75. D.P. Nikolelis, C.G. Siontorou, U.J. Krull, and P.L. Katrivanos, *Anal. Chem.* 68 (1996) 1735.
76. M. Trojanowicz, *Planar Lipid Bilayers (BLMs) and Their Applications* (Eds. H.T. Tien and A. Ottova-Leitmannova), Elsevier: Amsterdam, the Netherlands, 2003, pp. 807–845.
77. M. Snejdarkova, M. Rehak, and M. Otto, *Anal. Chem.* 65 (1993) 665.
78. T. Hianik, M. Snejdarkova, V.I. Passechnik, M. Rehak, and M. Babincova, *Bioelectrochem. Bioenerg.* 41 (1996) 221.
79. T. Hianik, Z. Cervenanska, T. Krawczynsky vel Krawczyk, and M. Snejdarkova, *Mater. Sci. Eng. C* 5 (1998) 301.
80. M. Rehák, M. Snejdárková, and T. Hianik, *Electroanalysis* 9 (1997) 1072.
81. M. Trojanowicz and A. Miernik, *Electrochim. Acta* 46 (2001) 1053.
82. M. Wilchek and E.A. Bayer, *Methods Enzymol.* 184 (1990) 746.
83. T. Hianik, *Rev. Molec. Biotechnol.* 74 (2000) 189.
84. P.N. Bartlett, V.Q. Bradford, and R.G. Whitaker, *Talanta* 38 (1991) 57.
85. S. Campuzano, B. Serra, M. Pedrero, F.J. Manuel de Villena, and J.M. Pingarrón, *Anal. Chim. Acta* 494 (2003) 187.
86. D.P. Nikolelis, M. Simantiraki, G.C. Siontorou, and K. Toth, *Anal. Chim. Acta* 537 (2005) 169.
87. L. Zheng, L. Xiong, D. Zheng, Y. Li, Q. Liu, K. Han, W. Liu, K. Tao, S. Yang, and J. Xia, *Talanta* 85 (2011) 43.
88. T. Jiao, B.D. Leca-Bouvier, P. Boullanger, L.J. Blum, and A.P. Girard-Egrot, *Coll. Surf. A* 354 (2010) 284.
89. K.-H. Wang, M.-J. Syu, C.-H. Chang, and Y.-L. Lee, *Sens. Actuators B* 164 (2012) 29.
90. C. Apetrei, P. Alessio, C.J.L. Constantino, J.A. de Saja, M.L. Rodriguez-Mendez, F.J. Pavinatto, E. Giuliani Ramos Fernandes, V. Zucolotto, and O.N. Oliveira Jr., *Biosens. Bioelectr.* 226 (2011) 2513.
91. H. Guan, F. Zhang, J. Yu, and D. Chi, *Food Res. Int.* 49 (2012) 15.
92. C.G. Siontorou, D.P. Nikolelis, P.A.E. Piunno, and U.J. Krull, *Electroanalysis* 9 (1997) 1067.
93. M.Y. Vagin, A.A. Karyakin, and T. Hianik, *Bioelectrochemistry* 56 (2002) 91.
94. M.Y. Vagin, E.E. Karyakina, T. Hianik, and A.A. Karyakin, *Biosens. Biolectron.* 18 (2003) 1031.
95. K. Hashimoto, K. Ito, and Y. Ishimori, *Sens. Actuators B* 46 (1998) 220.

96. H. Aoki, P. Bulmann, Y. Umezawa, *Electroanalysis* 12 (2000) 1272.
97. H. Zhou, N. Liu, Z. Gao, H. Wang, and Y. Fang, *Electrochem. Commun.* 10 (2008) 787.
98. C.G. Siontorou, D.P. Nikolelis, and U.J. Krull, *Anal. Chem.* 72 (2000) 180.
99. C.G. Siontorou, V.G. Andreou, D.P. Nikolelis, and U.J. Krull, *Electroanalysis* 12 (2000) 747.
100. D.P. Nikolelis, S.-S.E. Petropoulou, and M.V. Mitrokotsa, *Bioelectrochemistry* 58 (2002) 107.
101. U.J. Krull, D.P. Nikolelis, and J. Zeng, *Planar Lipid Bilayers (BLMs) and Their Applications* (Eds. H. T. Tien and A. Ottova-Leitmannova), Elsevier: Amsterdam, the Netherlands, 2003, pp. 767–787.
102. F.R.R. Teles and L.P. Fonseca, *Talanta* 77 (2008) 606.
103. L. Caseli, C.P. Pascholati, F. Teixeira Jr., S. Nosov, C. Vebert, A.H.E. Müeller, and O.N. Oliveira Jr., *J. Coll. Int. Sci.* 347 (2010) 56.
104. A.M. Sardat, C. Czudek, and G.P. Reynolds, *Neuroreport* 7 (1997) 910.
105. L. Lin, J. Chen, H. Yao, Y. Chen, Y. Zheng, and X. Lin, *Bioelectrochemistry* 73 (2008) 11.
106. E. Shams, A. Balbei, A.R. Taheri, and M. Kooshki, *Bioelectrochemistry* 75 (2009) 83.
107. V. Sacks-Granek and J. Rishpon, *Advances in Planar Lipid Bilayers and Liposomes* (Eds. H.T. Tien and A. Ottova-Leitmannova), Elsevier: Amsterdam, the Netherlands, Vol. 3, 2006, pp. 1–35.
108. G. Laputkova and J. Szabo, *Bioelectrochemistry* 56 (2002) 185.
109. M. Karabaliev and V. Kochev, *Sens. Actuators B* 88 (2003) 101.
110. I. Szymanska, H. Radecka, J. Radecki, and D. Kikut-Ligaj, *Biosens. Bioelectron.* 16 (2001) 911.
111. J.-B. Largueze, K. El Kirat, and S. Morandat, *Coll. Surf. B.* 79 (2010) 33.
112. D.P. Nikolelis, G. Raftopoulou, and C.G. Siontorou, *Electroanalysis* 20 (2005) 1870.
113. D.P. Nikolelis, G. Raftopoulou, M. Simantiraki, N. Psaroudakis, G.-P. Nikoleli, and T. Hianik, *Anal. Chim. Acta* 620 (2008) 134.
114. D.P. Nikolelis, N. Ntanos, G.-P. Nikoleli, and K. Tampouris, *Protein Peptide Lett.* 15 (2008) 789.
115. D.P. Nikolelis, G. Raftopoulou, N. Psaroudakis, and G.-P. Nikoleli, *Int. J. Environ. Anal. Chem.* 89 (2009) 211.
116. M.A. Mohsin, F.-G. Banica, T. Oshima, and T. Hianik, *Electroanalysis* 23 (2011) 1229.
117. F. Bosmans and K.J. Swartz, *Trends Pharmacol. Sci.* 31 (2010) 175.
118. Y. Huang, P.V. Palkar, L.J. Li, H. Zhang, and P. Chen, *Biosens. Bioelectron.* 25 (2010) 1834.
119. W. Xia, Y. Li, Y. Wan, T. Chen, J. Wei, Y. Lin, and S. Xu, *Biosens. Bioelectron.* 25 (2010) 1253.
120. C.J. Moreau, J.P. Dupuis, J. Revilloud, K. Arumugam, and M. Vivaudou, *Nat. Nanotechnol.* 3 (2008) 620.
121. T.D. Lazzara, D. Behn, T.-T. Kliesch, A. Janshoff, and C. Steinem, *J. Coll. Interf. Sci.* 366 (2012) 57.
122. D. Behn, S. Bosk, H. Hoffmeister, A. Janshoff, R. Witzgall, and C. Steinem, *Biophys. Chem.* 150 (2010) 47.
123. S. Bosk, J.A. Braunger, V. Gerke, and C. Steinem, *Biophys. J.* 100 (2011) 1708.
124. H. Chen, Q.-Y. Hu, J.-H. Jiang, G.-L. Shen, and R.-Q. Yu, *Anal. Chim. Acta* 657 (2010) 204.

8

Oligonucleotide and Nucleic Acid–Based Biosensors

Ilaria Palchetti

CONTENTS

8.1 Introduction

Natural nucleic acids (deoxyribonucleic acid [DNA], and ribonucleic acid [RNA]) are macromolecules composed of nucleotide chains. Nucleotides consist of a sugar ring (deoxyribose in DNA and ribose in RNA), a phosphate group, and a nucleic acid ring (which contains a basic nitrogen group and hence is the base in the nucleotide). These nucleotides form phosphodiester bonds between the 3′ group of one sugar molecule and 5′ group of another. The sugar–phosphate backbone does not change, but the base group can be different throughout the chain. There are four nucleic acid bases in DNA nucleotides, to give four different nucleotides: adenine, cytosine, guanine, and thymine. Similarly, there are four nucleic acid bases in RNA, to give four different nucleotides: adenine, cytosine, guanine, and uracil. The bases can be either a purine (adenine or guanine) or a pyrimidine (thymine, cytosine, or uracil). DNA is a double helix and has two complementary antiparallel strands, whereas RNA is, in many of its biological roles, single stranded. The role of nucleic acids (NAs) in nature is of tremendous importance since they are responsible for carrying genetic information from one generation to another. Passage of the genetic information is done by Watson–Crick base pair recognition of complementary strands. Each adenine in one strand makes two hydrogen bonds with thymine (or uracil) in the complementary strand,

and each guanine makes three hydrogen bonds with cytosine. Additionally, the DNA sequence is matched with its complementary sequence in RNA, which is further used to make a protein that satisfies the genetic code. The process of building the RNA from the DNA is called transcription, whereas the process of building the protein based on the RNA is called translation.

In recent years, NAs have been incorporated into a wide range of biosensors and bioanalytical assays, due to their wide range of physical, chemical, and biological activities [1,2]. A biosensor is defined as a compact analytical device incorporating a biological or biologically derived sensing element either integrated within or intimately associated with a physicochemical transducer. An NA-based biosensor employs as the sensing element an oligonucleotide, with a known sequence of bases, or a complex structure of DNA or RNA. Some examples are reported in the following sections.

8.2 Nucleic Acid Hybridization Biosensors

NA hybridization biosensors (also termed genosensors) result from the integration of a sequence-specific probe (usually a short synthetic oligonucleotide) and a signal transducer. The probe, immobilized onto the transducer surface, acts as the biorecognition molecule and recognizes, through hybridization, the target DNA or RNA (Figure 8.1). End labels, such as thiols, disulfides, amines, and biotin are commonly incorporated to immobilize the probes to transducer surfaces. Selection of the probe nucleotide sequence depends very much on the target sequence and certain specific applications require the right choice of probe length [1]. Oligodeoxyribonucleotides (ODNs) are less susceptible to hydrolysis than oligonucleotides with RNA backbone and thus are commonly used. Recently, probes produced by chemical changes to the backbone of naturally occurring DNA or RNA are used more and more in NA sensing techniques. Among these, locked nucleic acid (LNA) and peptide nucleic acid (PNA) are the most used. LNA is a bicyclic NA where a ribonucleoside is linked between the 2'-oxygen and 4'-carbon atoms with a methylene unit. General properties of LNA oligonucleotides include highly stable base pairing with DNA and RNA, exceptionally high thermal stability, compatibility with most enzymes, and predictable melting behavior. PNA is a synthetic NA that has an achiral neutral polyamide backbone formed by repetitive units of N-(2-aminoethyl)glycine linked to the bases. The PNA molecule, being resistive to attack by nucleases, provides an extra edge over the use of conventional or naturally existing NAs.

Several variables affect the hybridization event at the transducer–solution interface and should be controlled carefully. These include salt concentration, temperature, and the presence of accelerating agents, contact time, and length of probe sequence. Then, once the target NA has been captured onto

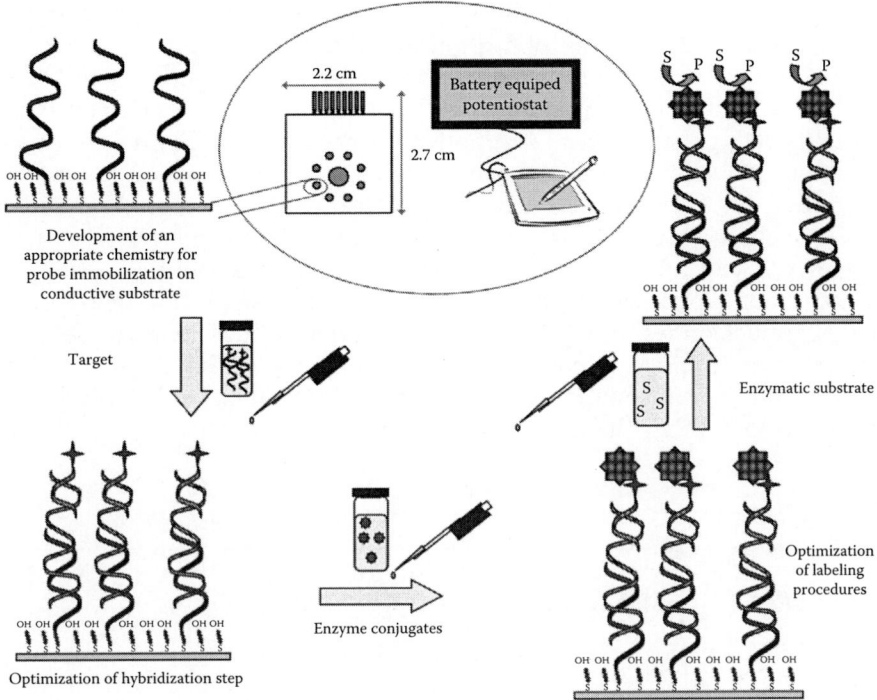

FIGURE 8.1
Example of an electrochemical genosensor-based assay: The voltammetric transducers are modified with the probes. The target is captured at the electrode surface and labeled (i.e., with an enzyme). Voltammetric methods can be used to detect the hybrids. A portable instrumentation can be used.

the sensor surface, a range of different approaches can be used for transducing the biorecognition event; these can be broadly divided into label-free and label-based schemes. Label-free approaches typically rely on the measurement of changes in the characteristics of sensing layer, before and after the hybridization reaction. Piezoelectric and surface plasmon resonance (SPR) transductions are essentially label-free techniques. By contrast, when organic and organometallic electroactive compounds, nanoparticles, and enzymes are permanently bound (e.g., covalently or via (strept)avidin–biotin interactions) to one of the constituents of the surface-tethered duplex, the method will be considered as label-based. Sensitivity and reliability of label-based approaches are often still unrivalled.

Along with the use of capture probes with improved selectivity and hybridization efficiency, specific pretreatments of the samples can be used to greatly enhance the yield of the heterogeneous hybridization events, and consequently, the sensitivity of detection.

Detection thus depends on a preamplification of the target sequence. This is true for most genoassays and especially for DNA detection since RNA-based assays benefit from the typically high number of ribosomes per cell. The most commonly used preamplification techniques are the polymerase chain reaction (PCR) and nucleic acid sequence-based amplification (NASBA) [2]. PCR is a technique widely used in molecular biology. It derives its name from one of its key components, a DNA polymerase used to amplify (i.e., replicate) a piece of DNA by in vitro enzymatic replication. As PCR progresses, the DNA thus generated is itself used as a template for replication. This sets in motion a chain reaction in which the DNA template is exponentially amplified. NASBA technology is based on simultaneous enzymatic activity of reverse transcriptase, T7 RNA polymerase, and RNase in combination with two oligonucleotides. It depends on the selective primer–template recognition to drive a cyclical, and exponential amplification of the target sequence.

More detailed information regarding hybridization-based NA biosensors are reported in other chapters of this book.

8.3 Aptamer-Based Biosensor

About 25 years ago, NA began to find a new role in the field of materials science and biotechnology [2]. Aptamers are single stranded DNA or RNA ligands that can be selected for different targets starting from a huge library of molecules containing randomly created sequences [3,4]. Aptamers with affinity for a large variety of molecules, including virtually any class of proteins (enzymes, membrane proteins, viral proteins, etc.), peptides, drugs, toxins, low-molar-mass ligands, and ions have been selected. The selection process is called systematic evolution of ligands by exponential enrichment (SELEX). The main advantage of aptamers is overcoming the use of animals or cell lines for the production of molecules. Antibodies against molecules that are not immunogenic are difficult to generate. On the contrary, aptamers are isolated by in vitro methods that are independent of animals: an in vitro combinatorial library can be generated against any target. In addition, the generation of antibodies in vivo means that it is the animal immune system that selects the sites on the target protein to which the antibodies bind. The in vivo parameters restrict the identification of antibodies that can recognize targets only under physiological conditions limiting the extension to which the antibodies can be functionalized and applied. Moreover, the aptamer selection process can be manipulated to obtain aptamers that bind a specific region of the target and with specific binding properties in different binding conditions. After the selection, aptamers are produced by chemical synthesis and purified to a very high degree by eliminating the batch-to-batch variation found when using antibodies. By chemical synthesis, modifications in the aptamer can be introduced enhancing the stability, affinity, and specificity of the molecules. Often the kinetic parameters of aptamer-target complex can be changed for higher affinity or specificity. Another advantage over antibodies can be seen in the higher temperature stability of aptamers and they can recover their native active conformation after denaturation; in fact antibodies are large proteins sensitive to temperature and they can undergo irreversible denaturation.

More information on aptamer-based biosensors can be found in other chapters of this book.

8.4 DNAzyme- and Aptazyme-Based Biosensor

Thus, from what's reported earlier, the conclusions are that NAs can be employed to develop affinity biosensors (genosensors and aptasensors) that allow the detection of affinity-based interactions between biomolecules, for example, DNA hybridization or antibody–antigen (target–aptamer) recognition. But this is only partial truth. In recent years, NAs have been used also to create, miming natural ribozyme, a system capable of catalytic activity. The term "nucleic acid enzyme" is used to identify these NA structures that have catalytic activity. Ribozymes (literally enzymes made of RNA) are found in nature and mediate phosphodiester bond cleavage and formation and peptide bond formation. Artificial ribozymes have been obtained by means of combinatorial chemistry approaches, such as in vitro selection and in vitro evolution, and have been shown to catalyze quite a broad array of other chemical reactions. DNAzyme catalysts have not been found in nature, and all known DNAzymes were isolated by in vitro selection. So far, most of their substrates have been found to be NAs. Among the many classes of DNAzymes, RNA-cleaving DNAzymes are the most widely used, mainly because of their simple reaction conditions, fast turnover rates, and significant modifications on their substrate lengths [5]. DNAzymes can perform chemical modifications on NAs, while aptamers can bind a broad range of molecules. A combination of the two has generated a new class of functional NAs known as allosteric DNAzymes or aptazymes. Both DNAzymes and aptazymes have been employed to develop biosensors [2].

Liu and Lu were the first to employ DNAzyme sensitive to chemical stimuli [6]. Their work has been focused on the use of Pb^{2+}-specific DNAzyme (Figure 8.2A). The catalytic DNA used is "17E," a variant of the "8–17" deoxyribozyme. In the presence of the analyte, the enzyme carries out a catalytic reaction such as hydrolytic cleavage (Figure 8.2B).

The fluorosensor described in [6] was constructed by labeling the 5′-end of the substrate with a fluorophore and the 3′-end of the enzyme strand with a fluorescence quencher. In the absence of lead ions, 17E exhibits a very weak activity towards the fluorogenic RNA-containing substrate. However, when lead is present, 17E rapidly cleaves the substrate, leading to a strong

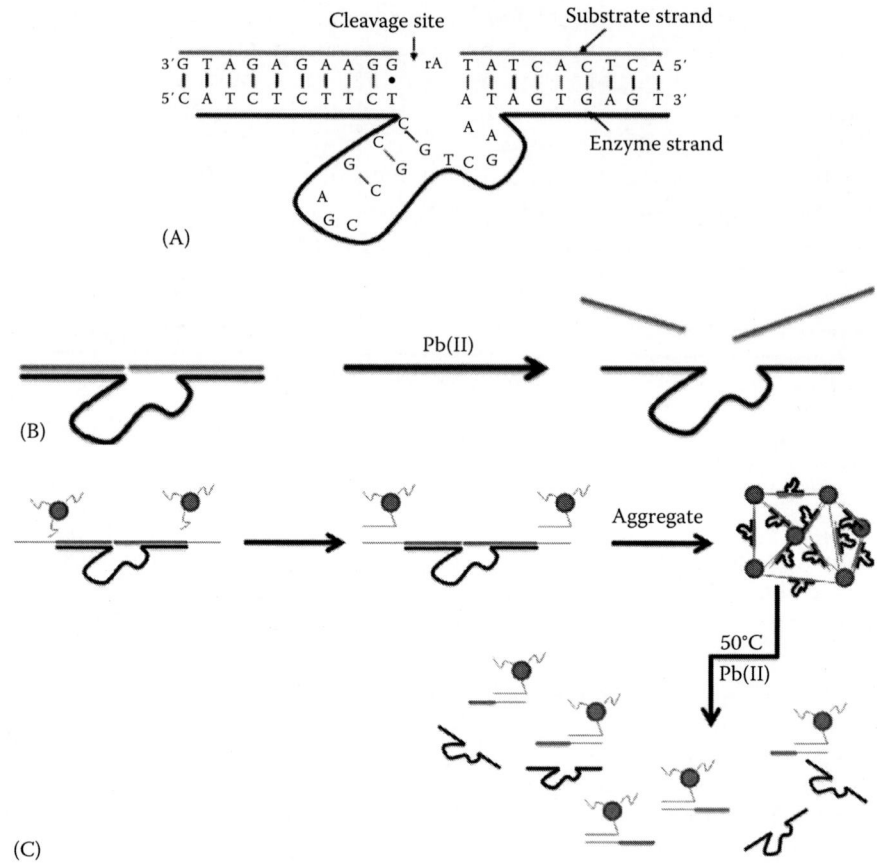

FIGURE 8.2
(A) Secondary structure of the 8–17 DNAzyme system that consists of an enzyme strand and a substrate strand. The cleavage site is indicated by a black arrow. Except for a ribonucleoside adenosine at the cleavage site r(A), all other nucleosides are deoxyribonucleosides. (B) Cleavage in the presence of Pb(II). (C) Scheme of DNAzyme directed assembly of gold nanoparticles and their application for Pb(II) sensing. In this system the substrate stand has been extended on both 3′ and 5′ ends for 12 bases, which are complementary to the 12 mer DNA attached to the 13 nm gold nanoparticles. (Reprinted from Palchetti, I. and Mascini, M., *Analyst*, 133(7), 846, 2008. With permission.)

fluorescence signal. Lu's group has also used 17E with gold nanoparticles for the design of a colorimetric assay [7]; in this case, gold nanoparticles were coated with a 12 mer oligonucleotide (Figure 8.2C). In the absence of lead, the substrate hybridizes with the oligonucleotide on the particle causing aggregation of the gold nanoparticles and the production of blue color. In the presence of Pb^{2+}, the DNAzyme catalyzes the hydrolytic cleavage of the substrate, which does not bind to the oligonucleotide on the particles anymore and prevents the formation of the particle aggregate causing the formation of red color.

UV–Vis spectroscopy was used to detect the changes in color and for the quantitative detection of Pb^{2+}.

Several other DNAzymes (or deoxyribozymes) or aptazymes that have been selected for metal ions and for organic molecules such as adenosine or cocaine have been reported in literature [8].

Another kind of most commonly used colorimetric probe is G-quadruplex DNAzyme. Under certain conditions, some G-quadruplexes can bind hemin to form DNAzymes that possess peroxidase-like activity. The hemin/G-quadruplex acts as a catalytic nanostructure that mimics the function of horseradish peroxidase. The DNAzyme catalyzes the oxidation of ABTS2—(2,2'-azinobis (3-ethylbenzthiazoline)-6-sulfonate) to the colored ABTS—or catalyzes the oxidation of luminal or 3,3,5,5-tetramethylbenzidine sulfate (TMB) by hydrogen peroxide (H_2O_2) [9,10] with the concomitant generation of chemiluminescence. These properties were used to develop colorimetric, and chemiluminescence detection schemes for DNA or aptamer-substrate complexes.

8.5 DNA-Based Biosensor for Toxicity Evaluation

Finally, short DNA oligonucleotide chains and genomic DNA have been employed to develop biosensor for toxicity evaluation. The elaboration of new highly sensitive, specific, and rapid methods of determination of environmentally toxic compounds has always been a challenge to analytical chemists. This is due to the complex ecological situation and the demand for diagnosing the consequences of the impact of biologically active contaminants upon the human organism. For these reasons, the development and characterization of rapid, sensitive, and inexpensive assays for detection of chemically induced damage to DNA caused by pollutants is an important issue in environmental monitoring. A significant number of short-term tests for genotoxicity/ mutagenicity have been developed to determine the extent of environmental hazards in polluted water and sediments. These bioassays have been used in well-constructed batteries of cytotoxicity and genotoxicity tests to complement traditional approaches. These tests typically fall into one of the several classes depending on their mechanism and end point. These classes include bacterial mutagenesis, cultured mammalian cell mutagenesis, chromosomal damage, and DNA damage. Despite the description of short-term, many of these assays are expensive to run, that require sophisticated technical expertise, and are not well suited to be adapted to screening applications. In recent years, there has been an enormous increase in the use of NAs as a tool in the recognition or monitoring of chemical compounds of environmental interest that may affect the genome. In this context, DNA-based biosensors can be a valid approach to detect DNA damage and can be designed to be rapid, inexpensive, and user friendly. Similarly, the effect of drugs or natural organic substances on NAs can be exploited by using DNA damage biosensors. Different kinds of transducers have been employed; each of them has their proper advantages and disadvantages. In particular, electrochemical DNA biosensors can be used: (1) for the detection of DNA strand breaks and base damage; and (2) for the detection of electroactive substances that specifically interact with DNA (covalently and/or non-covalently). The measurement of DNA damage using electrochemical biosensors has been demonstrated using the direct measurement of oxidation–reduction properties of the bases or indirectly using electrochemical probes.

Guanine is the most easily oxidized of the four DNA bases and has an oxidation potential of 1.06 V versus SCE at pH 7. The three remaining nucleobases have more positive oxidation potentials. Guanine and adenine bases in DNA can be oxidized on solid electrodes such as mercury, glassy carbon, and pyrolytic graphite [11–13].

In [13] disposable carbon screen-printed electrodes were used. In this approach genomic calf thymus DNA was used as receptor and square wave voltammetry as detection technique. The analytical signal was the decrease in the magnitude of the oxidation current peak of guanine after interaction with the sample solution (Figure 8.3).

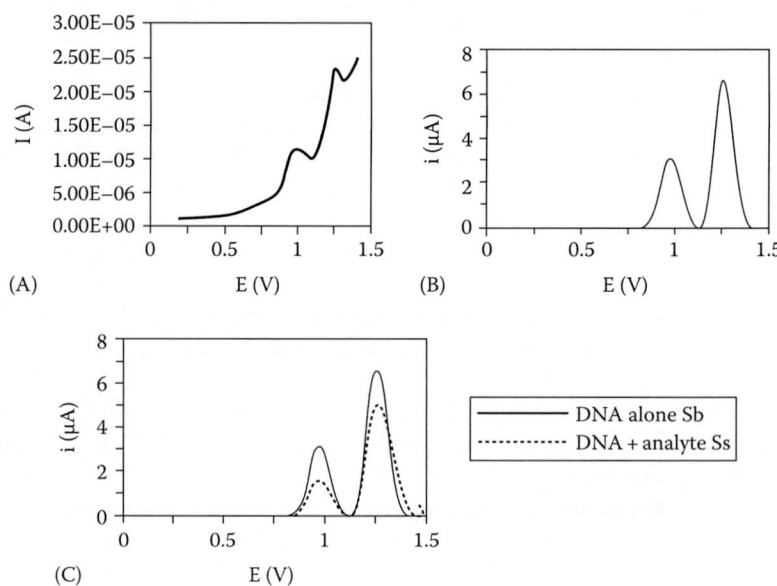

FIGURE 8.3
Redox behavior of guanine (+1.0 V) and adenine (+1.25 V) bases after a square wave voltammetric scan carried out with a graphite screen-printed working electrode. Shown are the normal signal (A), the signal after the baseline correction (B), and the decrease of the DNA peaks after the interaction with a pollutant agent (C). Sb, blank signal; Ss, sample signal. (Reprinted from Palchetti, I. and Mascini, M., *Analyst*, 133(7), 846, 2008. With permission.)

Rusling's group showed that $Ru(bpy)_3^{2+}$ and derivatives can be used to detect DNA damage. When double stranded (ds)DNA in ultrathin films was damaged, guanines became more available to react rapidly with the catalyst, providing an increased current in the square wave oxidation peak. $[Ru(bpy)_2\text{-}PVP]^{2+}$ and $[Ru(bpy)_2poly(4\text{-}vinylpyridine)_{10}Cl)]Cl$ [14] were also used by Rusling's group to develop a self-contained, reagentless sensor for toxicity screening based on the detection of DNA damage.

These biosensors were also used to measure toxic aromatic amines, oxidative damage, bioactivated benzo(a)pyrene, hydrazine, triazine-based herbicides, and fluorine derivatives [13–20].

Also some heavy metals are known to have great affinity with DNA and to cause mutagenesis and carcinogenesis. Babkina and Ulakhovich [21] proposed a method for the determination of heavy metals based on biospecific preconcentration of metal ions on an electrochemical DNA biosensor followed by the destruction of DNA metal complexes with ethylenediaminetetraacetate and voltammogram recording.

DNA damage has also been measured using fluorescence-based biosensors. Time-resolved fluorescence measurements were used to detect radiation-induced changes in DNA unwinding behavior from DNA isolated from white blood cells [22]. DNA adducts of benzo(a)pyrene were measured using low-temperature fluorescence on a gold biosensor chip [23].

Changes in melting–annealing behavior that were observed in real time using a double strand selective fluorescent indicator dye have also been used to measure DNA damage induced by radiation and chemical mutagens such as styrene oxide, glutaraldehyde, acrolein, allylamine, chloroacetone, acrylonitrile, bromoethane, crotonaldehyde, and benzo(a)pyrene [24–26]. This screening assay was sensitive to various forms of DNA damage including strand breaks, cross-links, and adduct formation. The assay was also demonstrated with respect to the effects of well-characterized genotoxins such as mitomycin C as well as cytotoxic, and nongenotoxic compounds such as phenol, cyclohexane, and toluene. Figure 8.4 presents the concentration dependence for crotonaldehyde [26].

A fiber-optic capillary fluorescence system, which employs the long-wavelength intercalating fluorophore TO-PRO-3 (TP3), was used to analyze toxic aromatic amines, antibiotics, and several kinds of antitumor drugs. Compounds that interact with the TP3–DNA complex were indirectly detected by a decrease in the fluorescence intensity [27].

DNA damage generated by organophosphate pesticides was studied by Hianik et al. [28] using the sensitivity of the acoustic shear wave technique to the interaction of DNA strands with damaged DNA containing apurinic or apyrimidinic (AP) sites.

Recently, DNA damage biosensor has been applied to the detection of the antioxidant effect of apple and orange juices [29].

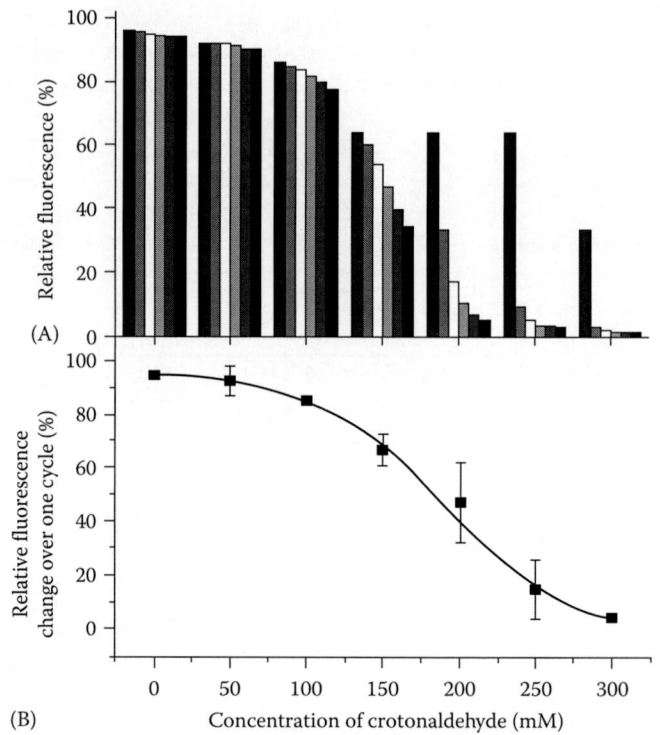

FIGURE 8.4

Concentration dependence for crotonaldehyde. (A) Melting–annealing profiles. The successive bars represent maximum fluorescence measured relative to the control for each cycle. (B) Analysis using one cycle. (Reprinted from Kailasam, S. and Rogers, K.R., *Chemosphere*, 66, 165, 2007. With permission.)

8.6 Conclusion

In summary, NAs are undoubtedly excellent molecules for the development of smart and innovative biosensors. The integration of nanotechnology, microfluidics, and bioanalytical systems clearly represents one of the future directions of all biosensor research. When these different areas are combined, the possibilities seem endless, and there are high hopes of solving many current problems such as sample preparation, real portability, single-molecule detection, analytical speed, and reliability. Many obstacles will have to be overcome in order to fulfill all of these hopes; however, some exciting examples, typically of subsystems rather than complete bioanalytical sensors, can already be found in the current literature. Miniaturization of PCR devices, for instance, offers several advantages such as short assay time, low reagent consumption, and rapid heating/cooling rates, as well as great potential for integrating multiple processing modules to reduce size and power consumption. This will undoubtedly benefit areas such as genosensor assays, since multiple processes, including sample collection and pretreatment, DNA extraction, amplification, hybridization, and detection, can be performed on a single, self-contained microfluidic platform. Such miniaturization of the analytical instrumentation will enable the transportation of laboratory to the sample source, as required for point-of-care testing. In conclusion, in the near future, NA sensing and biosensor technology itself will undoubtedly benefit from nanotechnology and μTAS technology.

References

1. J. Labuda, A. M. O. Brett, G. Evtugyn, M. Fojta, M. Mascini, M. Ozsoz, I. Palchetti, E. Paleček, and J. Wang, Electrochemical nucleic acid-based biosensors: Concepts, terms, and methodology (IUPAC technical report), *Pure Appl. Chem.* (2010), 82(5), 1161–1187.
2. I. Palchetti and M. Mascini, Nucleic acid biosensors for environmental pollution monitoring, *Analyst* (2008), 133(7), 846–854.
3. M. Mascini, I. Palchetti, and S. Tombelli, Nucleic acid and peptide aptamers: Fundamentals and bioanalytical aspects, *Angew. Chem. Int. Ed.* (2011), 51(6), 1316–1332.
4. I. Palchetti and M. Mascini, Electrochemical nanomaterial-based nucleic acid aptasensors, *Anal. Bioanal. Chem.* (2012), 402(10), 3103–3114.
5. Y. Lu and J. Liu, Functional DNA nanotechnology: Emerging applications of DNAzymes and aptamers, *Curr. Opin. Biotechnol.* (2006), 17, 580–588.
6. J. Li and Y. Lu, A general method to convert DNAzymes into fluorescent sensors using catalytic beacon, *J. Am. Chem. Soc.* (2000), 122, 10466–10467.
7. J. Liu and Y. Lu, A colorimetric lead biosensor using DNAzyme-directed assembly of gold nanoparticles, *J. Am. Chem. Soc.* (2003), 125, 6642–6643.
8. M. Mascini and I. Palchetti (Eds.), *Nucleic Acid Biosensors for Environmental Pollution Monitoring*, ISBN: 978-1-84973-131-7, RSC Publishing, 2011.
9. Z. Zhang, E. Sharon, R. Freeman, X. Liu, and I. Willner, Fluorescence detection of DNA, adenosine-5′-triphosphate (ATP), and telomerase activity by zinc (II)-protoporphyrin IX/G-quadruplex labels, *Anal. Chem.* (2012), 84, 4789–4797.
10. Y. Du, B. Li, S. Guo, Z. Zhou, M. Zhou, E. Wang, and S. Dong, Quadruplex-based DNAzyme for colorimetric detection of cocaine: Using magnetic nanoparticles as the separation and amplification element, *Analyst* (2011), 136, 493–497.
11. J. F. Rusting, Sensors for toxicity of chemicals and oxidative stress based on electrochemical catalytic DNA oxidation, *Biosens. Bioelectron.* (2004), 20, 1022–1028.
12. E. Palecek and M. Fojta, Detecting DNA hybridization and damage, *Anal. Chem.* (2001), 73, 74A–83A.
13. F. Lucareli, I. Palchetti, G. Marrazza, and M. Mascini, Electrochemical DNA biosensor as a screening tool for the detection of toxicants in water and wastewater samples, *Talanta* (2002), 56, 949–957.

14. B. Wang and J. F. Rusling, Voltammetric sensor for chemical toxicity using [Ru(bpy)2poly(4-vinylpyridine)10Cl)]⁺ as catalyst in ultrathin films. DNA damage from methylating agents and an enzyme-generated epoxide, *Anal. Chem.* (2003), 75, 4229–4235.

15. M. Fojta, T. Kubicarova, and E. Palecek, Electrode potential-modulated cleavage of surface-confined DNA by hydroxyl radicals detected by an electrochemical biosensor, *Biosens. Bioelectron.* (2000), 15, 107.

16. J. Wang, G. Rivas, M. Ozsoz, D. H. Grant, X. Cai, and C. Parrado, Microfabricated electrochemical sensor for the detection of radiation-induced DNA damage, *Anal. Chem.* (1997), 69, 1457–1460.

17. B. Wang, I. Jansson, J. B. Schenkman, and J. F. Rusling, Evaluating enzymes that generate genotoxic benzo[a]pyrene metabolites using sensor arrays, *Anal. Chem.* (2005), 77, 1361.

18. J. Wang, M. Chicharro, G. Rivas, X. Cai, N. D. Percio, A. M. Farias, and H. Shiraishi, DNA biosensor for the detection of hydrazines, *Anal.Chem.* (1996), 68, 2251–2254.

19. A. M. Oliveira-Brett and L. A. da Silva, A DNA-electrochemical biosensor for screening environmental damage caused by s-triazine derivatives, *Anal. Bioanal. Chem.* (2002), 373, 717–723.

20. V. Vlyskoiil, J. Labuda, and J. Barek, Voltammetric detection of damage to DNA caused by nitro derivatives of fluorene, *Anal. Bioanal. Chem.* (2010), 397, 233–241.

21. S. S. Babkina and N. A. Ulakhovich, Complexing of heavy metals with DNA and new bioaffinity method of their determination based on amperometric DNA-based biosensor, *Anal.Chem.* (2005), 77, 5678–5685.

22. G. Cosa, A. L. Vinette, J. R. N. McLean, and J. C. Scaiano, DNA damage detection technique applying time-resolved fluorescence measurements, *Anal. Chem.* (2002), 74, 6163.

23. N. M. Grubor, R. Shinar, R. Jankowiak, M. D. Porter, and G. J. Small, Novel biosensor chip for simultaneous detection of DNA carcinogen adducts with low temperature fluorescence, *Biosens. Bioelectron.* (2004), 19, 547.

24. K. Ramanathan, R. K. Gary, A. Apostol, and K. R. Rogers, A fluorescence-based assay for DNA damage induced by radiation, chemical mutagens and enzymes, *Curr. Appl. Phys.* (2003), 3, 99.

25. K. Ramanathan and K. Rogers, A fluorescence based assay for DNA damage induced by styrene oxide, *Sens. Actuat. B* (2003), 91, 205.

26. S. Kailasam and K. R. Rogers, A fluorescence-based screening assay for DNA damage induced by genotoxic industrial chemicals, *Chemosphere* (2007), 66, 165–171.

27. Y. Liu and B. Danielsson, Fluorometric broad-range screening of compounds with affinity for nucleic acids, *Anal. Chem.* (2005), 77, 2450–2454.

28. T. Hianik, X. Wang, S. Andreev, N. Dolinnaya, T. Oretskaya, and M. Thompson, DNA-duplexes containing abasic sites: Correlation between thermostability and acoustic wave properties, *Analyst* (2006), 131, 1161–1166.

29. G. Ziyatdinova and J. Labuda, Biosensor with protective membrane for the detection of DNA damage and antioxidant properties of fruit juices, *Electroanalysis* (2012), 24, 2333–2340.

9

De Novo DNA Synthesis and Its Biosensor Detection

Anatoly N. Reshetilov, Nadezhda V. Zyrina, and Lyudmila A. Zheleznaya

CONTENTS

9.1 Introduction

Initiation of specific synthesis by deoxyribonucleic acid (DNA) polymerases is known to require a template—a single-stranded DNA—and a primer, a short oligonucleotide with the free 3′-OH end complementary to a template chain segment. After *E. coli* DNA polymerase I had been isolated, partially purified preparations of this enzyme were found to be capable of synthesizing long deoxyadenylate–deoxythymidylate (dAT) copolymers without primers and template added. This synthesis (named *de novo* or nonprimed DNA synthesis) is given considerable attention by investigators at present. Detailed analysis of the effect has revealed that thermophilic DNA polymerases synthesize high-molecular DNA from deoxynucleotide triphosphate (dNTP) in the absence of any initiating nucleic acid. This chapter briefly reviews the molecular biological bases of *de novo* DNA synthesis and presents data in a novel approach to its analysis based on semiconductor transducers. Application of nonprimed synthesis techniques and novel DNA synthesis/sequencing detection methods is a valuable approach relating molecular biology on the whole and primed/non-primed DNA synthesis in particular to green synthesis of nanoparticles.

9.2 Molecular Biological Bases of *De Novo* DNA Synthesis

Initiation of DNA synthesis by DNA polymerases usually requires a template DNA (a single-stranded DNA) and a primer, a short oligonucleotide with the free 3'-OH end complementary to a segment of the template DNA (Kornberg and Baker 1992). Soon after the isolation of *E. coli* DNA polymerase I, partially purified preparations of this enzyme were found to be capable of synthesizing long dAT copolymers without primers and template DNA added (Schachman et al. 1960). This synthesis named *de novo* or nonprimed DNA synthesis was studied in sufficient detail (Radding and Kornberg 1962; Kornberg et al. 2004). However, the enzyme purification techniques available at that time failed to yield highly purified preparations, and observed synthesis was attributed to an insufficient elimination of DNA impurities from the preparation.

Much later, Ogata and Miura returned to this synthesis to find that thermophilic DNA polymerase from the Archaea *Thermococcus litoralis* (Tli) could synthesize DNA of up to 50,000 bp from dNTP in the absence of any initiating nucleic acid (Ogata and Miura 1997). After a 1 h lag period, the reaction continued almost linearly, and in 4 h all dNTP were included into the high-molecular product. The possible impurities of nucleic acids, which could serve as a template or primer in the reaction of synthesis, were ruled out, as the authors thoroughly pretreated all reagents. Additional analysis showed that a linear double-stranded DNA was the synthesis product. The authors named the revealed synthesis "creative" or *ab initio* DNA synthesis and suggested that genetic information could potentially be created directly by protein.

In their follow-up works, Ogata and Miura studied the effect of temperature, ionic strength, and pH on *ab initio* DNA synthesis by Tli DNA polymerase (Ogata and Miura 1998). When the reaction temperature was changed, the DNA sequence of the synthesized product changed significantly. Thus, at 69°C, (TAAT)n repeats were the reaction products; at 84°C, (TATCCGGA)n; and at 89°C, (TATCGCGATAGCGATCGC)n. Ionic strength of the reaction conditions also affected the character of the sequence: the sequence (TATCTAGA)n was synthesized at 0 mM KCl; (TATATACG)n at 50 mM KCl; and (TATAGTTATAAC)n, at 100 mM KCl. When pH of the reaction was changed from 6.8 to 10.8, the maximum size of synthesized DNA decreased from 50,000 down to 3,000 bp, but the character of the sequence did not change. These results showed that the composition of synthesized DNA was to a significant extent affected by the reaction conditions.

Ogata and Miura suggested that genetic information, which could have been created by protein under primeval conditions, strongly depended on environmental factors.

They found that thermophilic DNA polymerase from the bacterium *Thermus thermophilus* (Tth) also synthesized long DNA repeats in the complete absence of exogenous template and primer DNA (Ogata and Miura 1998). Similar to DNA synthesis by Tli DNA polymerase, this synthesis was characterized by a 1 h lag period. After that, the reaction proceeded almost linearly, and 2.2% of substrates were incorporated into the high-molecular product in 4 h. Synthesized DNA had a 25% GC content and the sequences were mainly represented by (CATGTATA)n, (TGTATGTATACATACATA)n, and (TATACGTA)n tandem repeats.

At the same time, Hanaki et al. (1997, 1998) conducted similar studies of primer- and template-independent DNA synthesis by various thermophilic DNA polymerases. These authors showed that Taq DNA polymerase and some others formed polymer poly-d(AT) from deoxyadenosine triphosphate (dATP) and deoxythymidine triphosphate (dTTP), whereas Tth, Vent, Vent(exo-), Pfu, Ultma, BcaBEST, and KOD DNA polymerases did not perform this synthesis. Besides, Hanaki and coauthors note that primer-/template-independent synthesis of poly-d(AT) is performed by DNA polymerases that possess the 5'-3' exonuclease and deoxynucleotidyl transferase (TdT) activities.

Hanaki and coauthors showed that template-independent synthesis of poly-d(AT) by thermophilic DNA polymerases involved two reactions: slow formation of 16–19-nucleotide d(AT) without the participation of primed matrix and rapid elongation of oligonucleotide d(AT) due to self-priming. These authors also noted that in the absence of substrates, Taq DNA polymerase, due to the 5'-3' exonuclease activity, degraded high-molecular polymer d(AT) to polymers of less than 75 bp in size, which were stable to further cleavage. This fact enabled them to make a conclusion of the simultaneous elongation and degradation reactions in matrix- and primer-independent synthesis.

A new type of template-/primer-independent DNA synthesis was found by Liang et al. (2004). They showed that a sufficiently slow DNA synthesis described by Ogata and Miura was strongly stimulated by adding restriction endonuclease to the reaction mixture with DNA polymerase. In the presence of restriction endonuclease, the lag period was reduced to 4 min, whereas in the experiments by Ogata and Miura, the lag period in the absence of the endonuclease was 1 h. After the lag period, DNA synthesis became linear and ended in 2 h, when all dNTP were used. The yield of synthesized DNA was more than 90% of the theoretically possible amount, whereas in the absence of restriction endonucleases, only 2.2% of substrates were included in the high-molecular product even after a 4 h synthesis.

The pattern of DNA synthesis strongly depended on the amount of endonuclease introduced. First, as the amount of restriction endonuclease increased, the synthesized high-molecular DNA went up too. However, the pattern changed with further increase in the amount of restriction endonuclease: a broad range of low-molecular DNA products were registered up to the absence of electrophoretically detected products. Intensive DNA synthesis was also observed using various DNA polymerases and restriction endonucleases. Synthesized DNA consisted of a multiple repeated sequence, the one recognized by the particular restriction endonuclease participating in the synthesis. Sequences similar to those synthesized in the reaction were found in genomic DNA of many organisms.

We found one more type of *de novo* DNA synthesis (Zyrina et al. 2007). Intensive synthesis was observed in the presence of a nicking endonuclease. Similar to restriction endonucleases, nicking endonucleases recognize a short specific sequence in double-stranded DNA and cleave DNA at a fixed position relative to the recognized sequence. However, unlike restriction endonucleases, nicking endonucleases make a nick in only one, predetermined DNA strand. Nicking endonucleases are designated by letter N before the abbreviated name of the genus of a bacterium from which the enzyme was isolated. Depending on which (top or bottom) strand the nicking endonuclease cleaves, it is designated as Nt or Nb, respectively.

Though nicking endonucleases were discovered comparatively recently (Abdurashitov et al. 1996), they have already found use in the development of various methods, including those for medical diagnostics (Tan et al. 2007; Kuhn and Frank-Kamenetskii 2008). A possibility of using nickase to transfer a cargo representing an oligonucleotide was demonstrated (Bath et al. 2005). It is assumed that nicking endonucleases can be used to solve some problems of DNA nanotechnologies and DNA-based computers (Wang et al. 2001; Zhang et al. 2002).

We operated with the nicking endonuclease Nt.BspD6I (Zheleznaya et al. 2001). It recognizes the 5'-GAGTC-3'/5'-GACTC-3' sequence (site) in double-stranded DNA and cleaves only the strand containing the GAGTC sequence downstream of the recognition site towards the 3' end. The enzyme exhibits an activity in the presence of Mg^{2+} ions at 55°C. The DNA polymerase Bst is compatible with it in respect to temperature; we used a large fragment of the polymerase, in which there is no 5'–3' endonuclease activity. Synthesis in the presence of nicking endonuclease depends on its amount and incubation time. As the amount of Nt.BspD6I increases (incubation time, 1 h), high-molecular DNA is first synthesized and its amount increases, then it starts to be degraded (Figure 9.1). A similar pattern is observed, depending on time, if the amount of Nt.BspD6I in the reaction mixture corresponds to that at which high-molecular DNA is synthesized. Analysis of the synthesized DNA showed it to consist of numerous repeats of the sequence of the site recognized by the nicking endonuclease Nt.BspD6I.

Thus, the phenomenon of *de novo* DNA synthesis—the ability of DNA polymerase to synthesize DNA in the presence of template DNA and a primer—has become an obvious fact. What is more, the synthesis is strongly stimulated in the presence of restriction endonucleases and nicking

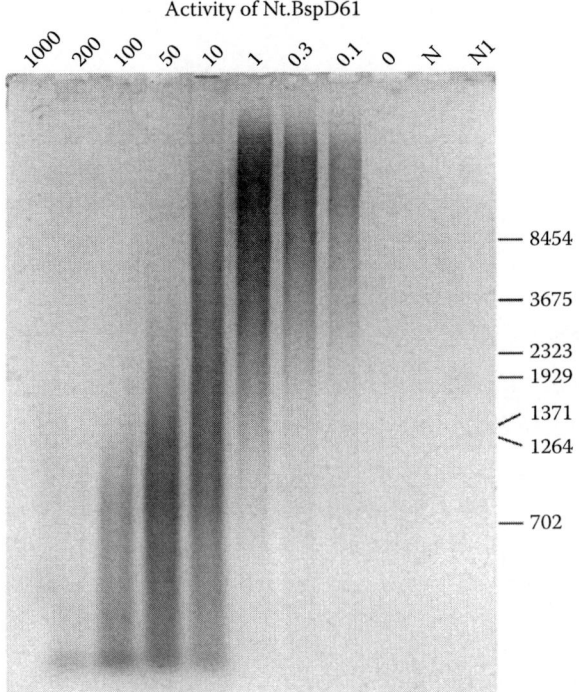

FIGURE 9.1
Template-/primer-independent DNA synthesis stimulated by nicking endonuclease Nt.BspD6I after 1 h incubation. Reaction products in 1% agarose gel. Top numbers are activities of Nt.BspD6I in reaction. N, Nt.BspD6I inactivated by heating; N1, incubation without Bst DNA polymerase in the presence of 2.3 units of Nt.BspD6I. The DNA size marker is shown in bp.

endonucleases. The most impressive result of works on this synthesis is the diverse sequence of repeats synthesized in the presence of only DNA polymerase, DNA polymerase and restriction endonuclease, and DNA polymerase and nicking endonuclease. Already Ogata and Miura interpreted this synthesis as a creation of new genetic information by protein. In their experiments, information was set out and created by DNA polymerase. It is evident that in experiments with restriction endonuclease and nicking endonuclease, information was set out by endonucleases, because DNA consists of repeats of sequences recognized by endonucleases. In this context, a decrease of the lag period in the presence of endonucleases is explainable by the fact that these enzymes are highly specific to recognized sequences, and respectively, they rapidly pick up recognized triphosphates from the mixture of nucleoside triphosphates. A higher efficiency of synthesis in the presence of endonucleases is associated with the ability of these enzymes to cleave DNA, and DNA cleavage is a necessary stage of synthesis. High-molecular DNA in solution represents a random coil, whose diameter is proportional to the length of DNA. At a DNA length of about 100,000 bp, the diameter of the coil is about several hundred nanometers. Thus, such a DNA is a model of the nanoparticle.

Establishing the phenomenon of *de novo* DNA synthesis would, probably, change the view on the mechanisms of the increase of the number of repeats in some genes of genomic DNA, which are causes of serious neurological disorders. Simplicity of the reaction and a high yield would enable a rapid and sufficiently cheap production of DNA for practical use.

9.3 Electrochemical Approach to the Registration of Molecular Effects in DNA

9.3.1 Direct Electrochemical Registration of DNA Synthesis

Direct label-free detection of DNA synthesis is a novel and currently intensively developed approach. Proposed new ideas and devices can greatly simplify the sequencing technique and contribute to its sooner commercial use for various, primarily diagnostic, purposes.

Label-free detection of DNA synthesis based on the electrochemical registration of the process was first considered by Pourmand et al. (2006). That study can be considered to be seminal in this respect. Those authors formulated the detection principle and showed that the process of DNA synthesis can be performed directly using the electronic circuits and techniques known in electrochemical measurements.

The principle of label-free detection of synthesis can be presented as follows. Self-priming single-stranded DNA molecules (template molecules) are immobilized on the surface of a gold electrode through a thiol-reactive self-assembled monolayer. The electrode is treated with the Klenow (exo⁻) fragment of DNA polymerase. An electric signal, indicating the incorporation of a complementary nucleotide into the template molecule, is generated in response to separate additions of dNTP solution. The signal is characterized by high amplitude, reaches a peak value of 400 pA, and emerges essentially without delay. Its time of development is about 50 ms; the first phase is followed by the second one, which characterizes its further transformation and it is called the "fall–rise" phase. It is observed within the time range of 200–400 ms, after which the signal decays. The integral of the measured current is about 90 pC at a density of about 6.0×10^{11} immobilized DNA molecules/cm^2 of the electrode. If a solution containing a noncomplementary dNTP is added at some point of time, no current signal is observed. No current is produced, either if a complementary dNTP is added in the absence of DNA polymerase or in the absence of DNA template or if the template is not in an immobilized state on the electrode surface. The lack of signal in the control experiments indicates that its emergence is due only to the reaction determined by the presence of a complementary nucleotide and the simultaneous presence of DNA polymerase and immobilized DNA. The current emerging in this case can be due only to processes caused by the incorporation of the nucleotide into the primer strand.

Incorporation of each nucleotide leads to increase in the total negative charge of the DNA molecule per one electron. The charge emerges as a result of the removal of a proton from the 3′-OH group of primer DNA in the catalytic reaction. The overall process is shown schematically in Figure 9.2. Due to the electroneutrality, an increase of the total negative charge of the DNA molecule is compensated by an equivalent increase of the total positive charge in solution owing to the evolution of a proton. Both electrical charges—positive and negative—induce surface charges of unlike signs on the electrode, which is electrically isolated from the solution with a layer of thiol-reactive molecules. For the considered electrode, the value of induced charge is in fact invariable within the range of inducer distances from 1 nm to 30 μm. A significant drop

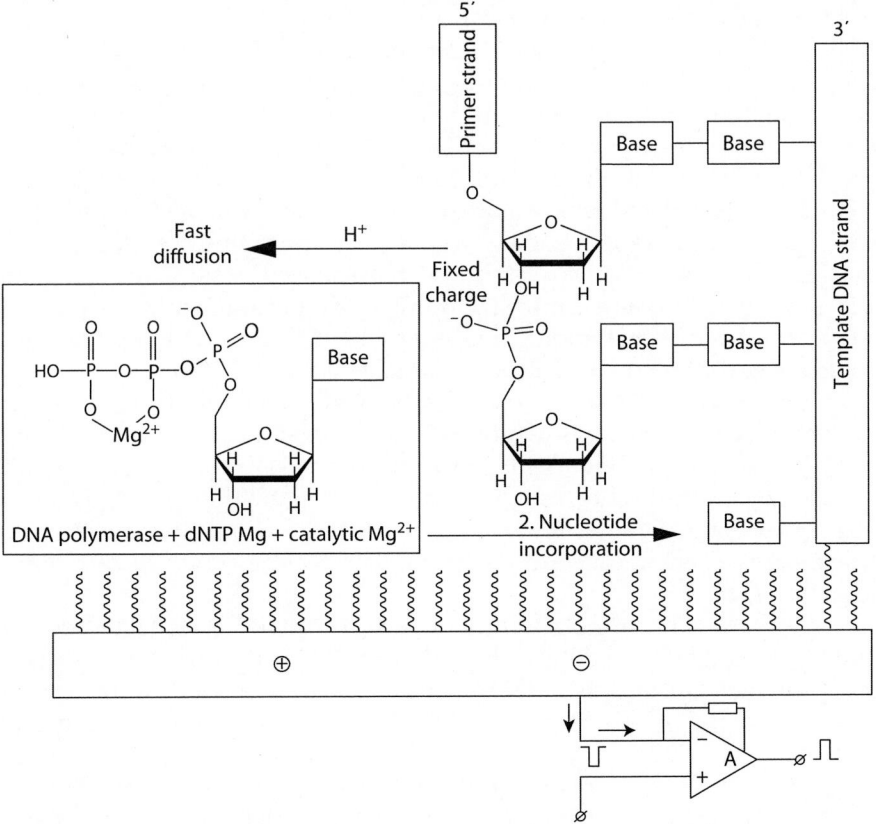

FIGURE 9.2
Mechanism of DNA synthesis detection. (After Pourmand, N. et al., *Proc. Natl. Acad. Sci. USA*, 103(17), 6466, 2006.)

of induced signal occurs only if the inducers are removed at a distance of 30 μm, which is the zone of a reliable detection of the synthesis. The charge of the DNA molecule immobilized on the electrode is, thus, fixed at the closest distance from the electrode surface. This distance is about 100 nm. At the same time, the evolved/released proton is not bound to the DNA molecule and freely, at a high rate, diffuses into solution. Over the measurement time of a single cycle of nucleotide incorporation, that is, over the time of the general development of the signal, the distance to which the proton can diffuse is about 140 μm. The induction effect makes it possible to pick up the signal due to the incorporation of a single nucleotide into the template DNA molecule. In actual fact, what is really detected is the sum of local current pulses caused by the reaction on all DNA molecules immobilized on the electrode. This leads to the emergence of a large current detected by a voltage-clamp amplifier. In essence, the theoretical analysis of this process and conclusions made testify to the proposed approach to the implementation of the label-free detection of DNA synthesis or, at a certain setup of the measurement circuit, to the implementation of label-free sequencing (Pourmand et al. 2006).

The measured current is equal to the quotient of the division of the flown charge by the reaction time, that is, $I(t) = dQ(t)/dt$, where Q is charge, I is current, and t is time. To assess the real value of current, the surface density of DNA molecules was measured. The data showed it to be about 1 fmol DNA immobilized on a surface of 0.0009 cm^2, which corresponds to about 6.7×10^{11} DNA molecules per square centimeter. These data testify to a high efficiency of signal transduction by a gold electrode of this design. Herewith, the signal-to-noise ratio was 200 units. If, based on the signal-to-noise ratio of 50 units, we estimate the minimal number of molecules whose sequencing can be done by this technique, it will be 1.2×10^{10}/cm^2. At a standard electrode size of 0.0009 cm^2, the number of molecules will be 1.8×10^7.

Let us present data on the sensitivity of the method to measurement preparation conditions. It is mentioned by Pourmand et al. (2006) that about 70% of experiments were unsuccessful, which was explained by insignificant deviations in the electrode fabrication technique that led to significant changes in the generation of current. It should be noted that the electrode functionality checkup method consists of conducting standard capacitance measurements.

In summary, we can note the following. Pourmand et al. (2006) demonstrated direct label-free detection of electric charge perturbations during a DNA polymerase-catalyzed DNA synthesis. DNA molecules were immobilized on a coated, electrically isolated gold electrode. To detect the signal, the electrode was switched into the circuit containing a voltage-clamp amplifier. The effect was designated by the term "charge perturbations detection" in DNA synthesis. The concept of a perturbed charged state is based on the principle of electroneutrality and the induction of surface charge in a polarizable electrode. When a fixed negative charge emerges from a DNA molecule, it can be detected by the current flowing in the external circuit.

This induction effect is enhanced by the high rate diffusion of a positive charge, a proton, and from the reaction region. In principle, DNA synthesis associated with the emergence of an induced charge can be detected without DNA immobilization due to the high velocity of proton diffusion. This velocity is, at least, higher than the diffusion velocity of a nonimmobilized DNA molecule.

To date, several approaches to DNA sequencing by synthesis, based on enzymatic or photochemical reactions and making use of labels, have been proposed. These approaches are briefly described in Zubov (2010). Pourmand et al. (2006) demonstrated that the detection of sequencing can be implemented by the registration of electrochemical reactions that accompany this process. The label-free detection technique demonstrates one of the most remarkable aspects of this approach—reduced costs and increased rates of assay. The potential of multichannel label-free detection should also be mentioned. Thus, Pourmand et al. (2006) showed the possibility of a simultaneous use of two amplifiers performing the registration of an induced charge placed at a distance of 300 μm from each other. This fact broadens the potential of the described technique and implies the simultaneous use of several amplifiers operated in parallel. At present, we are aware of the commercial production of detection systems including up to 16 V-clamp amplifiers in one module. The use of such systems can be considered as a tool for multichannel detection intended to solve the problem of the detection of biological pathogens, to reveal genetic mutations, and to identify unknown DNA sequences. On the whole, this approach can be used for detecting look-alike enzyme reactions, exhibited in pH changes. Thus, it can be concluded that electrodes enabling the detection of an induced charge are a reliable and efficient basis for developing biosensors with the view of registering diverse molecular and diagnostic processes (Pourmand et al. 2006).

9.3.2 Application of Semiconductor Instruments for DNA Sequencing

Recently, R&D of semiconductor devices for DNA sequencing has been greatly intensified. In particular, there are examples of using semiconductor devices for sequencing without making use of labels and optical methods (Rothberg et al. 2009, 2010, 2011). The point of attraction in such developments is, on the whole, the significance of information contained in the genetic code of the DNA molecule. This information is important for man as an absolute value, but at the same time, it is of paramount importance for applications such as biotechnology and medicine. To obtain it, inexpensive devices should be developed, which could be manufactured in sizeable amounts for practical applications.

Let us trace the main points related to the advancement of this approach by using a series of works on the development of novel principles of semiconductor sequencing as an example (Rothberg et al. 2011). As a brief summary preceding the analysis of this issue, we should note a number of significant points characterizing this trend. Thus, the possibility of implementing DNA sequencing by means of semiconductor technology—which is characterized by low-cost devices—has been described. The registration technique is based on direct detection using integrated circuits and does not make use of complex methods involving optical labels. Sequence data are obtained by directly sensing the ions produced by template-directed DNA polymerase synthesis using all-natural nucleotides on a massively parallel semiconductor sensing device or ion chip. The sensing element is an ion-sensitive (pH-sensitive) field-effect transistor (FET). The sensing elements are incorporated into the supporting chip (supporting template). The total number of sensing elements in the template is 1.2 million cells, using which the parallel sensing of the results of independent reactions are done. The device based on this principle of detection is cheap to fabricate on a large scale and can be readily modified (Ion torrent: Semiconductor sequencing; Ion torrent: The chip is the machine). This system was shown to be efficient in sequencing three bacterial genomes and a human genome. An important direction for modification is the possibility to increase the number of detecting cells (unit sensors). The system was shown to remain highly reliable when the number of unit sensors was increased tenfold.

It should be noted that the interest in DNA sequencing is expressed both on the part of scientific research and on the part of applied use, first and foremost medicine. A decrease in the cost of analysis and sequencing time can solve a number of diverse problems. They are associated first of all with the treatment of cancer, search for human genetic disorders, treatment of infectious diseases, and studies of individual genomes. Important fields of application are ecology and studies of ancient DNA. Although *de novo* sequencing costs have noticeably dropped, there is a natural desire to continue this tendency. Researchers believe that, when developing sequencers, one should remember Moore's law formulated for semiconductor devices and implied the exponential increase of the density of components in an integrated circuit. There are estimates to show that the desirable cost of genome sequencing should not exceed U.S. $1000. Let us dwell in more detail on the features of genome sequencing by means of semiconductor transducers (Rothberg et al. 2011).

Until recently, the genomic DNA sequencing process was sufficiently expensive, because costly detection tools and related expendable materials had to be used in the respective instruments. The new trend in sequencing consisted in the development of a label-free principle of analysis and the use for this purpose of microelectronics technology of detection based on complementary metal–oxide–semiconductor (CMOS) sensors. At the first stage of the development of sequencing detection system, it was proposed to use a multisensor system containing 1.2 million single sensors. Each sensor is a FET registering the pH change of solution. All transistors are combined with CMOS sensors into an integrated system for signal detection and data presentation for further processing. CMOS systems are not costly and are broadly used in computers, digital cameras, and mobile phones. Their low cost considerably reduces the overall cost of sequencers on their basis. The measurement principle is based on the change of pH during the attachment of a nucleotide to the single-stranded DNA template, which is a fragment of the whole DNA molecule. As we mentioned earlier, the synthesis is performed by DNA polymerase and is accompanied with the release of a proton (Pourmand et al. 2006). A single-stranded DNA is immobilized on the gate of an FET. Its DNA fragment is immobilized on each transistor. In total, the clones make up a whole DNA molecule. The release of protons, summed up with respect to the multitude of DNA clones present in a unit cell, can be reliably registered from each transistor. As the proton release is unambiguously related to the type of nucleotides, which are supplied in a synchronous succession, the instrument constructs the proton release pattern with respect to the entire set of sensors. The data obtained are sent to be processed. This system appears to be an alternative to the optical system, in which for generation of signal each attached nucleotide is equipped with an optical label. Despite the fact that there exists a considerable number of various electrochemical systems for the registration of protons, the FET was chosen as the sensing element or the primary pickup. This is due to its high chemical-sensitivity characteristics in the detection of the proton concentration in its technological combination with CMOS technology. Efforts by Ion Torrent Systems Inc., which was the first to propose this approach, were focused on the development of an electronic chip with a large number of fast, similar-parameter pH-sensitive transistors (Rothberg et al. 2009, 2010).

Consider some quantitative characteristics of the approach. The total duration of the complete measurement procedure is about 2 h. The chip contains 1.2 million sensors with 25 million base pairs arranged on their surface. The integral circuit consists of a chip containing the sensor elements, each of which has a floating gate. The authors developed a cheap and scalable simple electronic architecture of the sensor part based on CMOS technology. A schematic representation of the measuring cell or the primary sensor element is given in Figure 9.3.

A 3.5 μm in diameter well is used for sequencing. The well is enclosed by the 3 μm high dielectric wall. A tantalum pentoxide layer is used as an ion-selective material. The sensor element has a high chemical sensitivity to protons (58 mV/pH). High-speed addressing and readout are used for the registration of signal. When a nucleotide (dNTP) is incorporated into a DNA strand, a proton is released; this event changes the pH of the solution in the well (ΔpH). The pH change leads to a change of the charge state of the input–transistor gate. Owing to this, the surface potential of the pH-sensitive layer and potential (ΔV) of the FET gate also changes. Single sensor elements are arranged on the chip in a 2-D array. A row select register enables one row of the horizontal readout register. The column select register selects one of the columns for output to external electronics. The single sensor coupled with

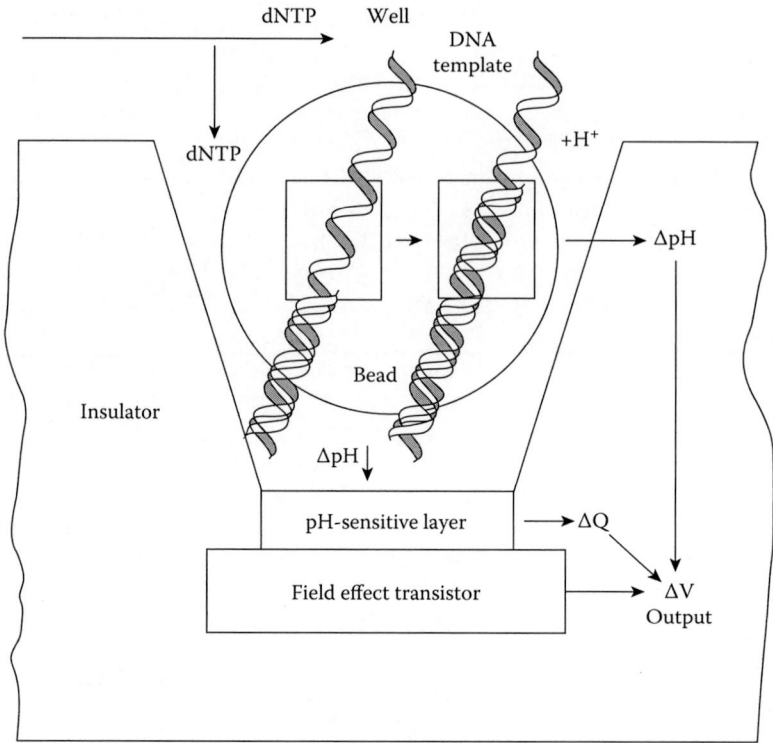

FIGURE 9.3
A simplified view of the measuring cell. (From Rothberg, J.M. et al., *Nature*, 475(7356), 348, 2011.)

processing chips provides for the direct transformation of the signal emerging in biological material into an electronic signal. In contrast to the optical readout, this approach does not use the accumulation of photons to amplify signals. Instead, a direct readout of the effect of proton release during the incorporation of nucleotides is used.

Sensor chips are packaged into disposable polycarbonate bodies, which form flow cells. This packaging also serves to isolate the supporting electronics from the fluids and makes it possible to provide convenient sample loading. Besides that, flow cells serve to provide electrical and fluidic interfaces to the sequencing instrument. Chips are designed and fabricated with 1.5, 7.2, and 13 M FETs. On the basis of the placement of the flow cell on the sensor array, 1.2, 6.1, and 11 M wells and sensors are exposed to fluids, with 99.9% of the sensors sensitive to pH and usable for DNA sequencing. Increasing the numbers of sensors per chip was first achieved by increasing the die area, from 10.6 mm × 10.9 mm to 17.5 mm × 17.5 mm. The next step was to increase the density of the sensors by reducing the number of transistors per sensor from three to two. Chip density is determined by the number of transistors in the node, which contains the measuring and auxiliary CMOS transistors.

Using a 0.35 μm CMOS node, the minimum spacing for a three-transistor sensor is 5.1 and 3.8 μm for a two-transistor sensor. The diameters of the measuring wells in this case are equal to 3.5 and 1.3 μm respectively. The search for variants to increase the density of the sensors showed that it was possible to form wells containing two transistors and having a diameter of 1.3 μm with a spacing of 1.68 μm between the wells. Such an increase of density became possible using a 110 nm node.

There are no costly optical components in semiconductor sequencing technology. This enabled a simplification of the sequencer design and a significant reduction of its cost. The readout of signals from the pH-sensitive chip and the supply of reagents to the flow cell of the instrument are monitored by the electronic control system. The nucleotide sequence is read out owing to a decrease of pH at the incorporation of one of the four nucleotides into the DNA. Before sequencing, DNA is fragmented and ligated to adaptors, and adaptor-ligated libraries are clonally amplified by emulsion PCR onto beads of 2 μm in diameter. Then a primer and DNA polymerase are added to the beads, and the beads are placed into wells on the chip. The size of the beads is slightly smaller than that of the wells, so only one DNA fragment fits into each well. The surface of the chip contains more than a million of wells; at the bottom of each well, there is a pH-sensitive FET reacting even to small changes in the concentration of protons (H^+).

For DNA sequencing, solutions of four deoxynucleoside triphosphates are passed in turn through a flow cell on the chip. If a nucleotide in solution is complementary to the template-strand base adjoining the primer, then DNA polymerase extends the primer by one link, which is accompanied with the release of pyrophosphate and a decrease of pH. If there is no complementary base, pH does not change. Then the cell is washed and a solution of another deoxynucleoside triphosphate is fed to it. Their multiple passing makes it possible to determine the nucleotide sequence of more than 200 bp long in each of the DNA clones placed into wells in only 2–3 h.

The performance of this method depends on the number of pH-sensitive wells on the chip, the efficiency of their filling with bead carriers of DNA fragments, purity of deoxynucleoside triphosphates, and a number of other optimizable factors. By the end of this year, Ion Torrent Systems Inc. plans to bring the performance of the semiconductor sequencer to 1 billion base pairs (1 Gb).

In summary, we can note that Rothberg et al. (2011) first showed the possibility of genomic sequencing of a bacterial genome and a human genome by a semiconductor device. The device makes use of FETs and standard CMOS transducers. All in all, expenses for the annual production of CMOS integrated circuits exceed 50 billion dollars. The aim of the development was to show a new way for the use of highly integrated systems. Larger and denser arrays should be produced. The available CMOS nodes should enable the production of 10 million sensor ion chips suitable for routine human genome sequencing.

9.3.3 Use of Semiconductor pH-Sensitive Transducers for Registration of Nonspecific DNA Synthesis

The results of studies (Pourmand et al. 2006; Rothberg et al. 2011) and the published results of *de novo* DNA synthesis were a stimulus for conducting a series of initial experiments and elucidating the possibility of using the electrochemical approach for the detection of DNA synthesis. As a biological model, we chose nonprimed DNA, which makes use of dNTP and two enzymes—nickase and DNA polymerases—as starting elements. Our aim was to understand if DNA synthesis can be detected by means of a FET at all, as well as under the measurement conditions when elements of the reaction are not in an immobilized state. It should be noted that this novel statement of the problem has not been described earlier. By the estimates given by Pourmand et al. (2006), this reaction is not optimized for such a detection.

We used FETs with a tantalum pentoxide gate dielectric layer. The chemical sensitivity of the transistors was 55–58 mV/pH. The measurements were carried out in a transistor's microcuvette, whose volume was 3 μL. A 150 mM NaCl solution containing 1 mM MgCl$_2$, and pH 8.0 was used as a measuring medium. No buffer component was used. Bst DNA polymerase (large fragment) and Nt.BspD6I nickase were used as biological components. The nickase, a 70.8 kDa protein, recognized the 5′-GAGTC-3′/5′-GACTC-3′ site on the double-stranded DNA molecule and cleaved only the strand containing the sequence GAGTC at a distance of 4 bp from the recognition site towards the 3′ end. The experiment consisted of two stages. At the first stage, DNA was synthesized in a medium containing 150 mM NaCl, dNTP (150 μM), nickase (5 activity units), and DNA polymerase (2 activity units). Duration of the synthesis was 60 min; temperature of the reaction medium, 55°C. At the second stage, we assessed the pH value of the resulting solution in comparison with pH of the solution containing no DNA polymerase and which was not subjected to heating. The reaction of synthesis was controlled by chromatography for the presence of high-molecular products.

The reaction of nonspecific DNA synthesis was found to lead to a shift of pH to an acidic region. Under chosen measurement conditions, an average pH shift is assessed to be 1.5 units at a measurement error of about 20%. The presence of only nickase in the reaction medium did not lead to the formation of either low-molecular reaction products (double-stranded DNA molecules with a small number of base pairs, less than 200) or high-molecular reaction products (double-stranded DNA molecules with a large number of base pairs, more than 1000). Thus, we succeeded in showing that nonspecific DNA synthesis was accompanied with a pH shift. This pH shift occurs under conditions, which differ from those described by Pourmand et al. (2006), that is, representing new measurement conditions. Application of FETs enabled detection in small volumes, of the order of 3–5 μL, thus reducing unproductive expenses for reagents. Therefore, in this series of experiments, we succeeded in showing a DNA synthesis running within the framework of the considered protocol; the synthesis can be detected using an electrochemical transducer—a pH-sensitive FET.

9.4 Conclusion

In a brief conclusion, we would note that the reviewed issues of the molecular biological bases for nonprimed DNA synthesis and the emerging methods of semiconductor detection are directly related to the advancement of nanotechnologies. An understanding of these processes is of great significance for progress in green synthesis of nanoparticles. Electrodes enabling the detection of induced charge in DNA synthesis are a reliable and efficient starting point in the development of biosensors for registration of diverse molecular and diagnostic processes. Application of non-primed synthesis techniques and novel DNA synthesis/sequencing detection methods relate molecular biology as a whole and nonprimed synthesis in particular to green synthesis of nanoparticles.

Acknowledgments

The work was supported by the Federal Goal-oriented Programs "Scientific and Scientific-educational Personnel of Innovative Russia" for 2009–2013, State contract No 14.740.11.0370, and "R&D for the Priority Directions of Russia's Science and Technology Complex for 2007–2012," State contract No 16.512.11.2126.

References

Abdurashitov, M. A., O. A. Belichenko, A. B. Shevchenko, and C. X. Degtyarev. 1996. N.BstSE—site-specific nuclease from *Bacillus stearothermophilus* SE-589. *Molecular Biology (Moscow)* 30:1261–1267.

Bath, J., S. J. Green, and A. J. Turberfield. 2005. A free-running DNA motor powered by a nicking enzyme. *Angewandte Chemie (International Edition)* 44:4358–4361.

Hanaki, K., T. Odawara, T. Muramatsu, Y. Kuchino, M. Masuda et al. 1997. Primer/template independent synthesis of poly d(A-T) by Taq polymerase. *Biochemical and Biophysical Research Communications* 238:113–118.

Hanaki, K., T. Odawara, N. Nakajima, Y. K. Schimizu, C. Nozaki et al. 1998. Two different reactions involved in the primer/template independent polymerization of dATP and dITP by Taq DNA polymerase. *Biochemical and Biophysical Research Communications* 244:210–219.

Ion Torrent: Semiconductor sequencing. http://www.iontorrent.com/lib/images/PDFs/ion_prod_a.pdf (accessed on December 1, 2012).

Ion Torrent: The chip is the machine. http://www.invitrogen.com/site/us/en/home/brands/Ion-Torrent.html (accessed on December 1, 2012).

Kornberg, A. and T. A. Baker. 1992. *DNA Replication*, 2nd edn. W. H. Freeman and Company, New York.

Kornberg, A., L. L. Bertsch, J. F. Jackson, and H. G. Khorana. 2004. Enzymatic synthesis of deoxyribonucleic acid. XVI. Oligonucleotides as templates and the mechanism of their replication. *Proceedings of the National Academy of Sciences of the United States of America* 51:315–323.

Kuhn, H. and M. D. Frank-Kamenetskii. 2008. Labeling of unique sequences in double-stranded DNA at sites of vicinal nicks generated by nicking endonucleases. *Nucleic Acids Research* 36:e40.

Liang, X., K. Jensen, and M. D. Frank-Kamenetskii. 2004. Very efficient template/primer-independent DNA synthesis by thermophilic DNA polymerase in the presence of a thermophilic restriction endonuclease. *Biochemistry* 43:13459–13466.

Ogata, N. and T. Miura. 1997. Genetic information "created" by archaebacterial DNA polymerase. *Biochemical Journal* 324:667–671.

Ogata, N. and T. Miura. 1998. Creation of genetic information by DNA polymerase by archaeon *Thermococcus litoralis*: Influences of temperature and ionic strength. *Nucleic Acids Research* 26:4652–4656.

Pourmand, N., M. Karhanek, H. H. J. Persson, C. D. Webb, T. H. Lee et al. 2006. Direct electrical detection of DNA synthesis. *Proceedings of the National Academy of Sciences of the United States of America* 103(17):6466–6470.

Radding, C. and A. Kornberg. 1962. Enzymatic synthesis of deoxyribonucleic acid. XIII. Kinetics of primed and *de novo* synthesis of deoxynucleotide polymers. *The Journal of Biological Chemistry* 237:2877–2882.

Rothberg, J. M., W. Hinz, K. L. Johnson, and J. Bustillo. 2009. *Methods and Apparatus for Measuring Analytes Using Large Scale FET Arrays*. Ion Torrent Systems Incorporated, Guilford, CT, p. 116.

Rothberg, J. M., W. Hinz, T. M. Rearick, J. Schultz, W. Mileski et al. 2011. An integrated semiconductor device enabling non-optical genome sequencing. *Nature* 475(7356):348–352.

Rothberg, J. M., J. C. Schultz, K. L. Johnson, T. M. Rearick, M. J. Milgrem et al. 2010. *Integrated Sensor Arrays for Biological and Chemical Analysis*. Ion Torrent Systems Incorporated, Guilford, CT, p. 63.

Schachman, H. K., J. Adler, C. M. Radding, I. R. Lehman, and A. Kornberg. 1960. Enzymatic synthesis of deoxyribonucleic acid. VII. Synthesis of a polymer of deoxyadenylate and deoxythymidylate. *The Journal of Biological Chemistry* 235:3242–3249.

Tan, E., B. Erwin, S. Dames, K. Voelkerding, and A. Niemz. 2007. Isothermal DNA amplification with gold nanosphere-based visual colorimetric readout for herpes simplex virus detection. *Clinical Chemistry* 53:2017–2020.

Wang, L., J. G. Hall, M. Lu, Q. Liu, and L. M. Smith. 2001. A DNA computing readout operation based on structure-specific cleavage. *Nature Biotechnology* 19:1053–1059.

Zhang, X., H. Yan, Z. Shen, and N. C. Seeman. 2002. Paranemic cohesion of topologically-closed DNA molecules. *Journal of the American Chemical Society* 124:12940–12941.

Zheleznaya, L. A., T. A., Perevyazova, D. V. Alzhanova, and N. I. Matvienko. 2001. Site-specific nickase from Bacillus species strain D6. *Biochemistry (Moscow)* 66:989–993.

Zubov, V. V. 2010. Devices for reading DNA. *Khimiya i Zhizn* 7:4–7.

Zyrina, N. V., L. A. Zheleznaya, E. V. Dvoretsky, V. D. Vasiliev, A. Chernov et al. 2007. N.BspD6I DNA nickase synthesis of non-palindromic repetitive DNA by Bst DNA polymerase. *Biological Chemistry* 388:367–372.

10

Aptasensors: The New Trends

Audrey Sassolas and Jean-Louis Marty

CONTENTS

10.1 Introduction

In 1990, Ellington's group [1], Gold's group [2], and Robertson's group [3] independently reported the development of an *in vitro* selection technique that allowed the discovery of specific nucleic acid sequences that bind nonnucleic acid targets with high affinity and specificity. The technique by which these molecules are obtained is called systematic evolution of ligands by exponential enrichment (SELEX), and the resulting deoxyribonucleic acid (DNA) or ribonucleic acid (RNA) oligonucleotides are referred to as aptamers.

Aptamers, often called "synthetic antibodies," can rival antibodies in both therapeutic and diagnostic applications. Aptamers show very high affinity to a wide range of targets such as proteins, metal ions, and pathogenic microorganisms. Aptamers possess several competitive advantages over antibodies such as their accurate and reproducible chemical production [4]. Thus, immunization and animal hosts are not necessary to produce aptamers. Moreover, the selected nucleic acids bind their targets with affinity and specificity comparable to those of antibodies. Aptamers are more stable than Ab and they can be selected in extreme conditions, whereas antibodies are only stable in physiological conditions. Aptamers can also undergo reversible denaturation, and they can be easily modified with new functional groups without affecting their activity. Due to its many advantages, numerous aptamer-based detection systems have been developed for the detection of a wide range of targets from protein to small molecules.

Aptamers can be used as sensing elements in biosensors. A biosensor is made of two closely associated elements: a biological recognition element (e.g., aptamer) and a transducer (electrochemical, mass, optical, or thermal). The biological recognition element has to be immobilized on the transducer surface. Biosensors based on aptamers as biorecognition elements have been coined "aptasensors." These new recognition molecules can be applied for specific detection of various analytes ranging from small ions to large proteins.

Aptasensors can be classified according to three configurations [5]:

1. The direct mode or target-induced conformation change mode

   The targets directly bind to their aptamers and this event subsequently leads to the changes of detectable properties. For instance, the conformational change of aptamers to specific pattern can modify the position of a signal moiety covalently bound to the end of the aptamer or the weight of the sensing layer.

2. Complementary strand displacement mode

This strategy is related to the unique ability of nucleic acids to form duplex structures with their complementary sequences. In the absence of target, based-pairing between the aptamer and the complementary sequences is favored. When the target binds to the aptamer, the hybridization is disrupted leading to the release of the complementary strand. Therefore, via adequate labeling of the complementary strand or the aptamer, the molecular recognition event can be effectively converted into a detectable signal. Because of its great simplicity and its very broad applicability, the duplex-to-complex change concept has been successfully exploited in the design of various aptasensors.

3. Sandwich mode

Some protein targets, such as platelet-derived growth factor (PDGF) and thrombin, have dual binding sites that enable them to bind two recognition molecules and form sandwich-like complexes. Since these proteins could specifically bind either aptamers or antibodies, the sandwich structures have three basic formats: aptamer–target–aptamer, aptamer–target–antibody, and antibody–target–antibody. The first two formats have been widely used in the design of aptasensors.

To develop aptasensors, different detection modes can be employed, among which electrochemical and optical ones are the most used. Recently, several reviews [6–11] presented aptasensors. In this chapter, aptasensors are described according to their transduction principle.

10.2 Optical Aptasensors

Some targets can be detected by using several optical methods: chemiluminescence (CL), fluorescence surface plasmon resonance (SPR), surface-enhanced Raman scattering spectroscopy (SERS), optical resonance, diffraction grating, Brewster angle straddle interferometry, evanescent-field-coupled (EFC) waveguide-mode, or surface plasmon-coupled directional emission (SPCDE).

10.2.1 Fluorescence Detection

Fluorescence represents the main detection mode described for most aptasensors. The first aptasensor was based on fluorescence [12]. Since that time, numerous fluorescence-based aptasensors have been described. There are a large number of well-characterized fluorophores and quenchers that can be used to modify nucleic acids during the automated synthesis or simple postsynthesis chemistry.

A fluorescent aptasensor [13] using a 58 mer RNA aptamer selective to L-adenosine immobilized onto the core of a multimode fiber using avidin–biotin interactions was developed. The detection was based on competitive binding of fluorescein isothiocyanate (FITC)-labeled L-adenosine with unlabeled analyte. This device allowed to detect L-adenosine in the submicromolar range.

A 15 mer antithrombin DNA aptamer was labeled with FITC [14]. Then, the labeled aptamers were immobilized on a glass slide via a flexible linker. Nonlabeled thrombin was selectively detected by following changes in the evanescent-wave-induced fluorescence anisotropy of the immobilized aptamer. Interaction with the target caused an increase in the size of fluorescent complex and a change in the rotational diffusion rate, and thus in the measured fluorescence anisotropy. The dynamic range of this biosensor extended from nanomolar to micromolar concentrations of thrombin, and the detection limit was 5×10^{-9} M.

A fiber-optic biosensor using an aptamer receptor for thrombin detection has been developed [15]. Antithrombin DNA aptamers were immobilized on the surface of silica microspheres that were distributed in microwells on the distal tip of an imaging fiber coupled to a modified epifluorescence microscope system. A competitive binding system between thrombin and fluorescein-labeled thrombin showed a detection limit of 10^{-9} M for the free target.

10.2.2 Chemiluminescence Detection

Highly sensitive CL detection has often been used in aptasensors.

A thiolated antithrombin aptamer was immobilized onto a gold-coated glass slide (Figure 10.1) [16]. After interaction with thrombin, the sensor surface reacted with microperoxidase 11 (MP-11)—aptamer conjugates that acted as catalytic reporting labels. The MP-11 is a heme-containing oligopeptide that catalyzes, like horseradish peroxidase (HRP), the biocatalytic generation of luminol CL in the presence of H_2O_2. The emitted light intensity, which was measured with a photon-counting spectrometer equipped with a cooled photomultiplier detection system, increased with the concentration of thrombin. The detection limit was 10^{-8} M.

Some authors demonstrated that metallic nanoparticles (NPs) can be used to generate the CL of luminol in the presence of H_2O_2 ([17,18]). Willner and coworkers used PtNPs functionalized with aptamers as labels to detect thrombin [19]. A gold-coated glass slide was modified with thiolated thrombin aptamer. After interaction with the target and subsequently with thrombin aptamer-functionalized PtNPs, NPs could catalyze the CL reaction in the presence of luminol and H_2O_2. In this case, the detection limit was 10^{-9} M of thrombin.

Quantum dots (QDs) could also generate efficient and stable electrochemiluminescence (ECL) signal. A QD-based ECL aptasensor was developed

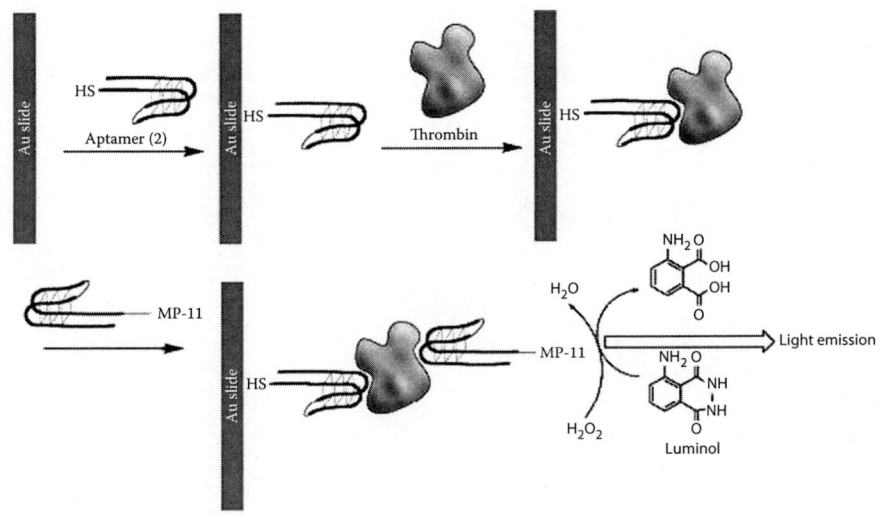

FIGURE 10.1
Thrombin analysis with the use of the aptamer–MP-11 hybrid label and using CL as a readout signal. (From Shlyahovsky, B. et al., *Biosens. Bioelectron.*, 22, 2570, 2007.)

for thrombin detection [20]. A thiolated aptamer was first immobilized on a gold electrode. A sandwich structure was formed between immobilized aptamer, thrombin, and a biotinylated aptamer. Then, streptavidin-modified QDs were bound to the sandwich-type structure via affinity. The ECL intensity was proportional to the logarithm of thrombin concentration from 2.7 to 55×10^{-8} M.

An ECL aptasensor was described to detect thrombin using aptamer-based sandwich format and ruthenium complex as ECL label [21]. Aptamers were immobilized on gold NPs bound to an indium tin oxide (ITO) electrode. In the presence of thrombin, a complex was formed between immobilized aptamer, the target, and a $Ru(bpy)_3^{2+}$-labeled secondary aptamer. As the modified electrode was immersed in a tri-*n*-propylamine electrolyte solution, ECL response emerged under an applied cyclic potential. The detection limit was 10^{-8} M.

An electrogenerated chemiluminescent aptasensor using ruthenium complex as an ECL label was also developed for the determination of cocaine [22]. A $Ru(bpy)_2(dcbpy)$-labeled aptamer was immobilized onto a gold electrode surface via thiol–Au interaction. Then, the electrode was placed in a cell facing photomultiplier tube. An enhanced ECL signal was generated upon recognition of the target cocaine, attributed to a change in the conformation of the aptamer from a random coil-like configuration to a three-way junction structure, the tag being in close proximity to the sensor surface. The detection limit was 10^{-9} M. $Ru(bpy)_2(dcbpy)$ label was also used to develop a displacement ECL aptasensor for lysozyme detection [23].

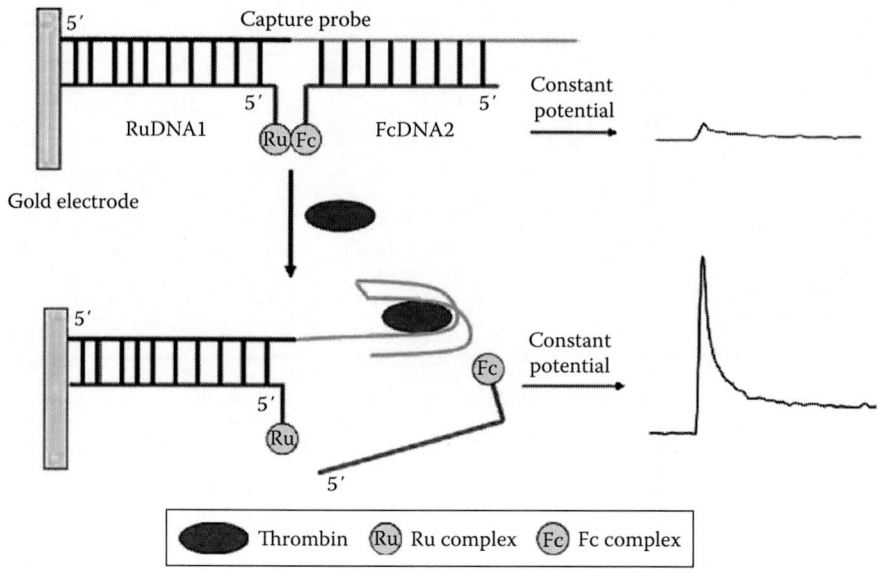

FIGURE 10.2
Schematic diagram of the ECL biosensing method for the detection of thrombin using quenching electrogenerated CL of tris(2,2′-bipyridine)ruthenium(II) by ferrocene. (From Li, Y. et al., *Electrochem. Commun.*, 10, 1322, 2008.)

Ru(bpy)$_3^{2+}$-doped silica NPs have also been used as labels to develop electrochemiluminescent aptasensors for thrombin detection [24,25]. Ru(bpy)$_3^{2+}$ molecules were incorporated in a silica matrix [26]. ECL intensity can be enhanced due to the increase of the number of Ru(bpy)$_3^{2+}$ molecules doped per NP.

An optical aptasensor based on a structure-switching ECL-dequenching mechanism was developed for the detection of thrombin (Figure 10.2) [27]. A thiolated DNA sequence constituted of a stem complementary to an oligonucleotide labeled with a ruthenium derivative and the thrombin-binding aptamer sequence complementary to an oligonucleotide tagged with ferrocene as an ECL quencher was immobilized on gold electrode. In the absence of thrombin, the three DNA oligonucleotides assembled into a tripartite duplex structure in which ferrocene and the ruthenium derivative were in close proximity, causing an ECL quenching. In the presence of the target, the aptamer sequence formed an aptamer–thrombin complex and the DNA sequence labeled with ferrocene was released. Thus, an ECL signal proportional to thrombin concentration in the range from 2×10^{-10} to 2×10^{-7} M was observed. The detection limit was 6×10^{-11} M.

10.2.3 SPR Detection

SPR is a label-free optical technique frequently used to develop biosensors especially to detect proteins using aptamers [7]. In order to detect a molecule, the aptamer is immobilized onto the surface. Since the analyte binds to the ligand, the mass and the refractive index increase.

Most of the aptasensors are based on direct mode. An SPR-based biosensor for the detection of human immunodeficiency virus 1 (HIV) trans-activator of transcription (Tat) protein was developed [28]. In this work, biotinylated RNA aptamer, specific for HIV-1 Tat protein, was immobilized on the dextran surface of a chip that was modified with streptavidin. This biosensor showed a linear range up to 2.5 mg L^{-1}. This system could be regenerated using a solution of 12 mM NaOH with 1.2% EtOH, and more than 50 cycles could be performed on the same chip without loss in sensitivity. Poly(dimethylsiloxane) (PDMS) multichannels were used for studies of DNA aptamer–human immunoglobulin E (IgE) interactions by SPR imaging [29]. The sensing surface was prepared with thiol-terminated aptamers through a self-assembling process in PDMS channels defined on a gold substrate. A linear relationship was obtained between the signal intensity and the IgE concentration between 8.4×10^{-9} and 8.4×10^{-8} M. The detection limit was 2×10^{-9} M of IgE. A single-stranded DNA (ssDNA) aptamer-based SPR biosensor was also developed to detect retinol-binding protein 4 (RBP4), a biomarker in the diagnosis of type 2 diabetes [30]. A biotinylated aptamer was immobilized on a streptavidin-modified gold surface. This detection system showed a detection limit of 7.5×10^{-8} M. Recently, an aptasensor for the detection of thrombin based on the use of gold surface modified with 4-mercaptobenzoic acid was developed [31]. The biosensor demonstrated to be highly selective of human thrombin even in the presence of large excess of bovine thrombin, bovine serum albumin, cytochrome c, lysozyme, and myoglobin. By using the SPR, the detection limit was 1.2×10^{-10} M.

Sandwich configurations were also described for the development of aptasensors. An enzymatically amplified SPR imaging methodology was also described for the detection of thrombin [32]. In the first step, a surface-immobilized aptamer was used to capture thrombin. Then, an HRP-conjugated antibody was introduced forming an aptamer–thrombin–antibody sandwich structure. This surface was then exposed to an HRP substrate, the 3,3′,5,5′-tetramethylbenzidine (TMB), that created a dark blue precipitate on the surface, was detected by SPR imaging. This method was also used to detect vascular endothelial growth factor (VEGF). In this instance, a biotinylated VEGF antibody and then an antibiotin-conjugated HRP were used. In this case, 10^{-12} M of VEGF could be detected [33].

10.2.4 Other Optical Detection Modes

10.2.4.1 Surface Plasmon-Coupled Directional Emission

Because fluorophores near metallic surfaces or particles have unique characteristics, including decreases in fluorescence intensity, quantum yield, and photostability, some research has recently focused on the near-field interaction between fluorophores and metals, especially the phenomenon of SPCDE [34–37]. Xie and coworkers developed the first SPCDE-based aptasensor for thrombin detection [38]. Aptamer labeled with Texas Red was immobilized on a gold surface via avidin–biotin interactions. In the absence of the target, the fluorophore was close to the metal surface resulting in the quenching of fluorescence, and no signal was detected by SPCDE. When the target thrombin (10^{-6} M) was added, the fluorophore was displaced, and the distance between the fluorophore and gold increased causing fluorescence enhancement.

10.2.4.2 Surface-Enhanced Raman Scattering Spectroscopy

This detection technique presents several advantages compared to fluorescence. A Raman dye can be either fluorescent or nonfluorescent, and a minor chemical modification of the dye molecule can lead to a new dye with a different Raman spectrum even if the two dyes exhibit virtually indistinguishable fluorescence spectra [39]. Moreover, the spectral specificity of a SERS probe is excellent in comparison to that of the fluorescence method.

A reagentless aptameric biosensor based on SERS spectroscopy was also developed to detect cocaine [40]. Tetramethylrhodamine (TMR)-labeled aptamer was immobilized on a SERS substrate. In the absence of the target, the aptamer was partially unfolded. So the TMR moiety remained away from the substrate and yielded a weak SERS signal. In the presence of cocaine, the aptamer folded into stable three-way junction, in which the TMR moiety came in close proximity to the SERS substrate, generating an enhanced SERS signal. Detection limit was 10^{-6} M.

A SERS aptasensor was also described to detect thrombin [41]. Thiolated thrombin-binding aptamers were immobilized onto a gold substrate (Figure 10.3). A sandwich structure was formed between the immobilized

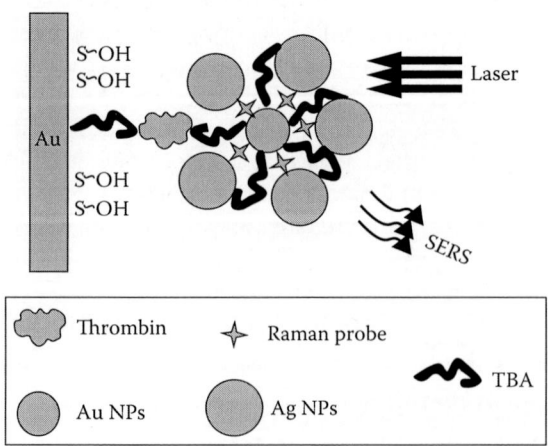

FIGURE 10.3
SERS aptasensor for protein recognition. (From Wang, Y. et al., *Chem. Commun.*, 28, 5220, 2007.)

aptamer, the protein target, and a secondary aptamer bound to AuNPs, which were labeled by a Raman reporter (R6G). Then, silver NPs aggregated on AuNPs yielding electromagnetic hot spots. Thus, the Raman signal of the R6G was greatly enhanced due to the large electromagnetic coupling effect produced by the hot spots between AgNPs and AuNPs. The detection limit of this SERS aptasensor was 5×10^{-10} M.

10.2.4.3 Optical Resonance

Optical microspheres are excellent candidates to develop label-free biosensors [42]. The light propagates in microspheres in the form of whispering gallery modes (WGMs). WGM occurs when light travels in a dielectric medium of circular geometry (e.g., a microsphere). After repeated total internal reflections at the curved boundary, the electromagnetic field can close on itself, giving rise to resonances [43]. The WGM's resonant wavelength is extremely sensitive to changes in refractive index near the sphere's surface when molecules bind to or are removed from the surface. A microsphere resonator biosensor was developed using an aptamer as the receptor molecule on the surface [44]. Thrombin-binding aptamers were bound to a functionalized sphere surface. The aptamer-modified spheres were immersed in distilled water and were put in physical contact with a fiber prism. Different concentrations of thrombin were gradually injected in a fluidic cell. Upon each injection, the WGM shifted to a higher wavelength, indicating binding between proteins and aptamers on the sphere surface. The detection limit of thrombin was estimated to be 10^{-6} M.

10.2.4.4 Diffraction Grating

The principle of diffraction-based sensing is simple: specific capture molecules are patterned on a surface so that the pattern will diffract light in a predictable manner. The simplest example is diffraction grating, which is a pattern of repeating parallel lines, yielding a diffraction image consisting of a row of equally spaced spots [45–47]. The initial pattern of immobilized capture molecules will produce a diffraction image of weak intensity upon illumination. When the surface with patterned probes is placed in contact with a solution of complementary target molecules, binding of target molecules to the capture molecules occurs, increasing the height and/or refractive index of the existing diffraction pattern, and the resulting diffraction image intensity will increase.

Diffraction grating was adopted to detect S-adenosyl homocysteine (SAH), an analyte consisting of the nucleoside adenine joined to the amino acid homocysteine by a 5' thioether linkage, using an adenosine-specific RNA aptamer [48]. Biotinylated RNA sequence was immobilized on streptavidin-modified micropatterns. In the presence of the target, a sandwich complex was formed between SAH, immobilized adenosine aptamers, and antiSAH-antibody-coupled magnetic beads. Diffraction gratings formed by the SAH-antibody beads binding to the adenosine aptamer-modified surface produced a diffraction pattern upon illumination by laser light. The detection limit was 6.4×10^{-11} M.

A sensitive self-assembled optical diffraction aptasensor was developed for the detection of platelet-derived growth factor B-chain (PDGF-BB) using a microbead-based rolling circle amplification (RCA) as a signal enhancement method. RCA has been proven to enhance signals for detecting a variety of analytes due to its sensitivity [49]. RCA requires a circular template and a primer with a free 3′ end that is then extended via DNA or RNA polymerase with strand displacement abilities to result in a single-stranded concatemeric product consisting of tandem repeats of the complement of the circular template. A biotinylated antiPDGF-BB-specific aptamer is immobilized on streptavidin-coated periodic patterns. PDGF-BB is then introduced and captured by the aptamer. A second aptamer with an additional primer sequence bound to the protein. A padlock probe hybridized to the primer tail of the aptamer is ligated and RCA is initiated. The diffraction grating is then formed by introducing streptavidin-labeled beads that conjugated to biotinylated probes, which are bound to RCA-amplified concatemers on the periodic patterns. Illuminating the pattern with a laser light produced diffraction modes with varying intensities that depended on the number of beads present on the patterns (Figure 10.4).

10.2.4.5 Evanescent-Field-Coupled Waveguide-Mode Biosensor

An EFC waveguide-mode aptasensor was developed to detect vitamin B12 and human coagulation factor IXa [50]. In this device, the excitation of waveguide-modes changed the reflectivity of the incident light over a narrow angular region. The waveguide-modes are sensitive to the dielectric environment near the surfaces of the waveguides, and variations in this dielectric environment result in changes in the reflectivity that can be used as a means of detecting small concentrations of molecules [51].

EFC waveguide sensor was first used to detect interactions between aptamer and vitamin B12 [50]. Waveguide surface perforated by nanoholes was modified by streptavidin that bound biotinylated (dT_{24}) oligonucleotide. This oligonucleotide was used to immobilize RNA aptamer having 24 nucleotides of an "A" residue at its 3′ end. In the presence of 10^{-6} M of the target, a shift in the resonance position and a change in the reflectivity were clearly observed. The EFC waveguide sensor was also used to detect 5×10^{-7} M of human factor IXa.

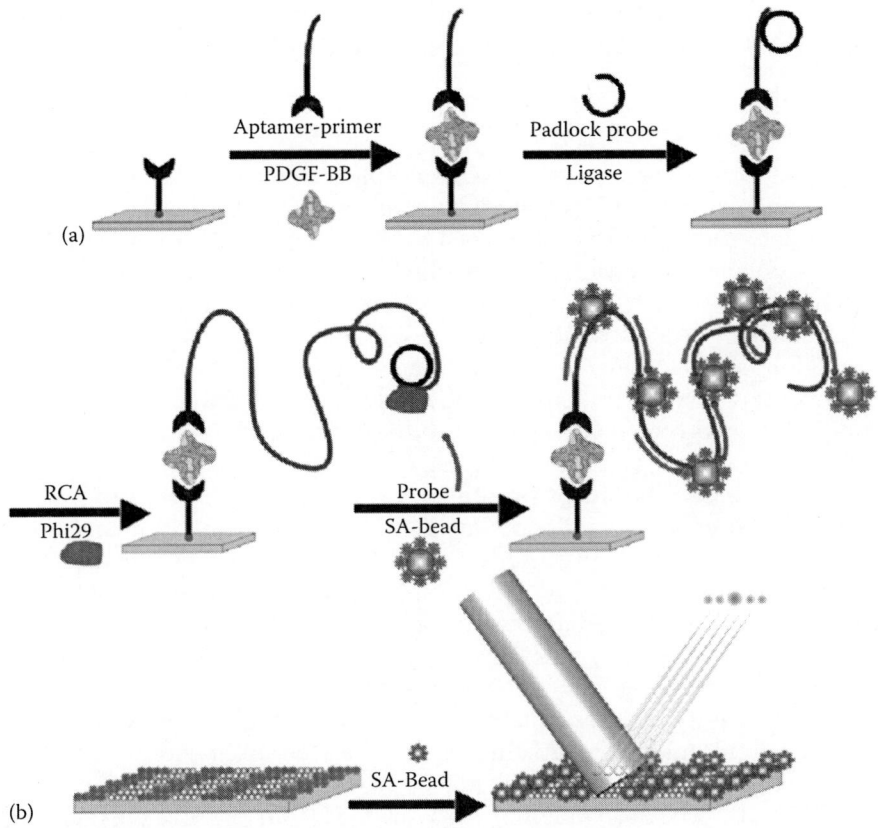

FIGURE 10.4

(a) Schematic of RCA-based microbead detection assay in combination with aptamers. A biotinylated antiPDGF-BB-specific aptamer is immobilized on streptavidin coated periodic patterns. PDGF-BB is introduced and captured by the aptamer. An aptamer–primer complex, with an additional primer sequence binds to the protein. A padlock probe hybridized to the primer tail of the aptamer is ligated and RCA is initiated. Streptavidin conjugated beads bind to elongated concatemers via hybridized biotinylated probes. (b) Self-assembled streptavidin (SA)-coated beads on the RCA-based micropattern form diffraction gratings that yield diffraction modes upon illumination with a laser.

10.3 Mass-Sensitive Aptasensors

Mass-sensitive detection measures mass changes on the sensor surface, requiring no additional reagents for the labeling. Mass-sensitive aptasensors are normally more applicable for the detection of large analytes like proteins or cells but typically not for small molecules. Frequently, mass-sensitive aptasensors include acoustic wave-based sensors (quartz crystal microbalance (QCM) and Love-wave devices) and micromechanical cantilever-based sensors.

10.3.1 QCM-Based Biosensors

A QCM sensor is a mass-sensitive sensor capable of measuring mass changes on a quartz crystal surface [52–54]. It consists of a thin quartz disk sandwiched between a pair of electrodes. Quartz is a piezoelectric material that deforms when an electric field is applied across the electrode. The quartz crystal has a resonant frequency dependent on the total oscillating mass. This frequency decreases with an increase in material on the QCM surface.

An IgE-specific aptamer and conventional antiIgE antibodies were used to develop quartz crystal biosensors [55]. These ligands were covalently immobilized on a gold-coated QCM. The immunosensor and the aptasensor were both allowed to detect a minimum IgE concentration of 5×10^{-10} M. However, the aptamer-based biosensor presented a linear measurement range up to 5×10^{-8} M, whereas the antibody-based receptor layer could linearly detect up to 5×10^{-9} M. Liss et al. (2002) tried to regenerate the antibody-coated chips with an acid buffer (0.2 M glycine-HCl, pH 2.2) and the aptamer-coated surface with 50 mM ethylenediaminetetraacetic acid (EDTA) or the same glycine buffer. The antibody layer was irreversibly damaged, whereas the binding of IgE to the aptamers was completely reversible, allowing the biosensor reuse.

A QCM-based biosensor obtained by immobilizing a biotinylated RNA aptamer, specific for the Tat protein of HIV-1, onto the streptavidin-coated gold electrode of the crystal was described [56]. Tat protein can be linearly detected up to 2.5 mg L^{-1}. The detection limit was 0.65 mg L^{-1}. To improve the reproducibility of both the aptamer immobilization and the interaction between the aptamer and the target, the biotinylated aptamer was heated at 90°C for 1 min to unfold the RNA strand. After this thermal treatment, the biotin label at the 5′ end was available for the interaction with streptavidin on the crystal surface. In this case, the target could be detected up to 1.25 mg L^{-1} but with a lower detection limit of 0.25 mg L^{-1}. Moreover, this treatment induced a better reproducibility for the interaction between the aptamer and Tat.

The QCM method was also used to analyze the binding of thrombin to a linear aptamer and to a molecular beacon aptamer [57]. Both biotinylated aptamers were immobilized to a surface of AT-cut crystal covered by neutravidin. Addition of thrombin (5×10^{-9} to 10^{-7} M) to the surface resulted in the decrease of the resonance frequency for both linear aptamer and molecular beacon aptamer-based biosensors. The sensor based on the linear aptamer revealed an approximately 1.5 times higher sensitivity in comparison with the molecular beacon aptamer.

A QCM aptasensor for human thrombin detection was developed on the basis of polymeric forms of phenothiazine dyes [58]. Two phenothiazine dyes, methylene blue (MB) and methylene green, were polymerized onto a gold electrode at one side of the crystal. Then, avidin was electrostatically precipitated or cross-linked with human serum albumin by glutaraldehyde. Biotinylated aptamers were immobilized via avidin–biotin binding. The developed aptasensors were allowed to detect 10^{-8}–10^{-7} M of thrombin.

10.3.2 Love-Wave Biosensors

Love-wave biosensors are a special type of surface acoustic wave sensors, which use shear horizontal waves guided in a layer on the surface of the sensor to reduce energy dissipation of the acoustic wave to the fluid and to increase the surface sensitivity.

Thrombin or Rev DNA aptamers were covalently immobilized onto a gold surface [59]. The sensitivity of these sensors to proteins was determined by binding fluorescently labeled proteins to the aptamer-derivatized Love-wave sensor. A calibration curve was obtained by pipetting and drying a fluorescent protein of known concentration to an Au-covered Si substrate. Thus, the mass of protein bound to the sensor could be evaluated. Detection limits of approximately 75 pg/cm^2 were obtained for thrombin and Rev peptide. In the same way, RNA or DNA thrombin aptamers were covalently immobilized on a Love-wave sensor surface covered with 11-mercaptoundecanoic using carbodiimide coupling [60]. The interaction of thrombin with the specific immobilized aptamer was studied. An RNA antithrombin aptamer was also immobilized on a Love-wave surface using glutaraldehyde [61]. This sensor could be regenerated by use of 0.1 N NaOH.

10.3.3 Cantilever-Based Biosensors

Microcantilever sensors have recently emerged as promising tools to detect biomolecular interactions [62–64]. These systems offer many advantages compared to conventional sensors: label-free detection, high precision, reliability, reduced size, easy fabrication of multielement sensor arrays, and a small thermal mass. Microcantilever sensors are based on responses due to either surface stress variation or mass loading. Interaction between an immobilized ligand (e.g., aptamer) and an analyte (e.g., protein) causes a change of the surface stress of the cantilever and can be detected as changes in the cantilever deflection (Δx).

A label-free protein detection method was developed using a cantilever-based biosensing [65]. This sensor used two adjacent cantilevers that constitute a sensor/reference pair. The sensor cantilever was functionalized with specific thiolated aptamer for *Thermus aquaticus* (Taq) DNA polymerase, whereas the reference cantilever was blocked with nonspecific ssDNA. The Taq-aptamer binding induced a change in surface stress causing a differential cantilever bending that depended on the target concentration. The injection of 5×10^{-10} M Taq DNA polymerase into the fluidic chamber resulted in 32 nm of differential cantilever bending.

A nanomechanical aptamer-based biosensor was developed for the detection of hepatitis C virus (HCV) helicase [66]. Amine-modified RNA aptamers were immobilized on the functionalized surface of microcantilever. In the presence of the target, HCV helicase interacted with the aptamer causing a resonant frequency shift.

Recently, a novel DNA aptamer-based sensing system using atomic force microscopy (AFM) was developed for the detection of thrombin [67]. A DNA aptamer is immobilized on a gold chip surface. The target molecules are added to the DNA aptamer immobilized on the chip for a reaction between the DNA aptamer and the target molecules. After the reaction, the gold chip is applied to AFM with the target molecules on a modified cantilever. Since DNA aptamers that are already bound with target molecules are not able to bind to the target on the cantilever, it is assumed that the lower affinity force between the gold chip and the cantilever is due to the higher thrombin concentration added to the gold chip.

10.4 Electrochemical Aptasensors

Compared to other detection systems, electrochemistry offers many advantages: high sensitivity, robustness, fast response, low cost, possibility of miniaturization, and mass production. This detection method was used for the first time 8 years ago. Two main approaches of detection have been used for electrochemical aptasensors: the use of aptamer labels (enzymes, metal NPs, and redox compounds), covalently or noncovalently bound to aptamers, and aptamer label-free detection systems. Measurement of changes of electrochemical features after target-binding could then be correlated to target concentration. "Signal-on" (positive readout signal) and less sensitive "signal-off" (negative readout signal) aptasensors have been described, based on target binding-induced conformational change of aptamers or on target binding-induced strand displacement or even on both processes [8].

10.4.1 Covalent Labels

Redox labels are commonly covalently linked to terminal groups of aptamers. The labeling position has to be carefully selected so as not to interfere with the folding of the aptamer or to modify an essential binding group. Labels such as enzymes, metal NPs, MB, ferrocene (Fc), or anthraquinone have been used.

10.4.1.1 Enzyme Labels

Ikebukuro et al. reported the first electrochemical aptasensor that was also the first sandwich-based aptasensor. Two different aptamers for thrombin detection were used: a thiolated aptamer was immobilized onto the gold electrode, and another aptamer was labeled with glucose dehydrogenase from *Burkholderia cepacia* (GlcDH) [68] or with oxygen-insensitive pyrrolo-quinoline quinone glucose dehydrogenase from *Acinetobacter calcoaceticus* (PQQ-GlcDH) [69]. Thrombin was detected in a sandwich manner with the aptamer immobilized onto the gold electrode; the enzyme-labeled aptamer was added, and the electric current generated by glucose addition was measured at 0.1 V vs. Ag/AgCl, in a 1-methoxyphenazine methosulfate (m-PMS) containing buffer. With GlcDH, 1 μM of thrombin was selectively detected. With PQQ-GlcDH, the linear range extended from 4×10^{-8} M to 10^{-7} M, with a detection limit of 10^{-8} M. This one-shot sensor presented a surface that was difficult to thoroughly regenerate contrary to an immunosensor.

Electrochemical aptasensors using HRP as an enzyme label have been developed. A sandwich format has been designed [70]. Thiolated aptamers were immobilized on a gold electrode and incubated with thrombin before the reaction with HRP-labeled aptamers. The electrochemical measurement was performed in the presence of hydrogen peroxide and a diffusional osmium-based mediator ($[Os(bpy)_2(pyr–CH_2–NH_2)]Cl$). The assay suffered from significant nonspecific adsorption of HRP-labeled aptamers. Recently, an electrochemical aptasensor for the detection of cocaine was developed [71]. The aptasensor was constructed by cleaving anticocaine aptamer into two fragments: One was assembled on a gold electrode surface, while the other was modified with biotin at 3′-end, which could be further labeled with streptavidin-horseradish peroxidase (SA-HRP). Upon binding with cocaine, the HRP-labeled aptamer fragment/cocaine complex formed on the electrode increased the reduction current of hydroquinone (HQ) in the presence of H_2O_2. Differential pulse voltammetry (DPV) was allowed to sensitively detect cocaine with the dynamic range from 10^{-7} M to 5×10^{-5} M and the detection limit down to 2×10^{-8} M. Marty's group developed an electrochemical aptasensor for the detection of ochratoxin A (OTA) [72]. In this case, free OTA competed with labeled alkaline phosphatase–ochratoxin A (ALP–OTA) for the binding to the DNA aptamer immobilized on magnetic beads. The electrochemical detection was thus achieved through a suitable substrate for the enzyme ALP, by DPV. The aptasensor obtained using this novel approach allowed a detection limit of 0.11 μg L^{-1} and was also used for the analysis of wine samples. An electrochemical aptasensor for the detection of IgE was also developed [73]. This system was based on a competitive format using a biotinylated aptamer. ExtrAvidin–ALP conjugate was used for labeling and electrochemical detection by DPV.

10.4.1.2 Methylene Blue

Heeger and coworkers developed a signal-off electrochemical aptasensor for the thrombin detection [74]. Before the binding of thrombin, the electrochemical active redox moiety, MB, which was covalently labeled at the end of aptamer, could transfer electron with the electrode surface due to the flexible conformation of the aptamer. However, once thrombin was captured by the

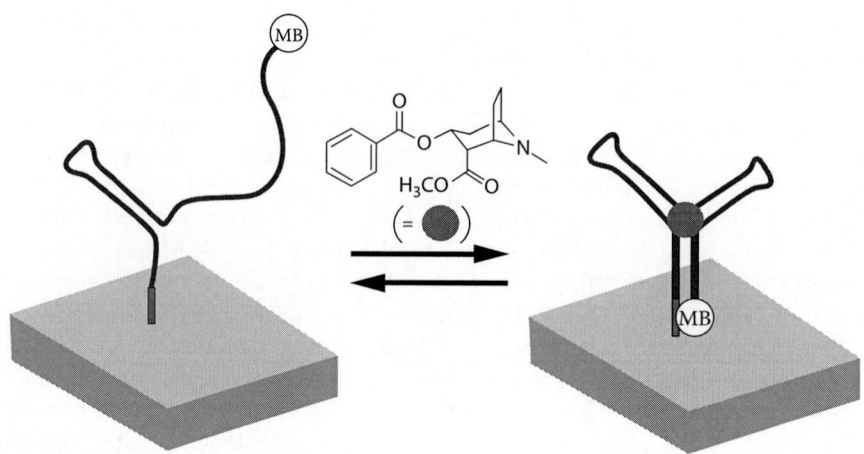

FIGURE 10.5
The electronic aptamer-based cocaine biosensor. MB: methylene blue. (From Baker, B.R. et al., *J. Am. Chem. Soc.*, 128, 3138, 2006.)

aptamer and the G-quadruplex structure was formed, the MB moiety was kept away from the electrode surface, resulting in electrochemical signal-off. The main disadvantage of such design is negative signal.

To circumvent this problem, signal-on electrochemical aptasensor can be developed. A signal-on electrochemical aptasensor was also described for the electrochemical detection of cocaine [75]. This sensor was fabricated by the self-assembly of an MB-tagged aptamer on a 1 mm² gold electrode via the alkanethiol group. In the absence of the target, the aptamer was partially unfolded, with only one of its three double-stranded stems intact (Figure 10.5). In the presence of the target, the aptamer folded into the cocaine-binding three-way junction, increasing the observed MB reduction peak. The same strategy was used to develop an electrochemical sensor for the detection of PDGF directly in blood serum using MB-modified PDGF-binding aptamer. The sensor showed a detection limit of 5×10^{-11} M in 50% blood serum [76].

10.4.1.3 Ferrocene

A reusable signal-off aptasensor based on target-binding-induced Fc-labeled aptamer displacement has also been developed for adenosine detection [77]. Thiolated capture DNA probes were self-assembled on a gold NP-covered gold surface before partial hybridization with Fc-labeled aptamer probes. Upon adenosine-binding, dissociation of the signaling aptamers led to a decrease in the Fc oxidation peak current measured by alternating current voltammetry (ACV). A low detection limit of 2×10^{-8} M and a wide linear range from 10^{-7} M to 10^{-5} M were obtained. Regeneration could be performed by thermal dehybridization in hot water.

Different signal-on electrochemical aptasensors have been described. For instance, Radi and coworkers reported a signal-on aptasensor for potassium ion detection [78]. Ferrocene-labeled thiolated DNA aptamers containing multiple guanine-rich segments were self-assembled on gold. In the presence of potassium ions, immobilized aptamers were converted from a random coil structure to a potassium-specific compact G-quadruplex structure, thus holding the Fc label in close proximity to the electrode surface. The ferrocene oxidation peak current measured by square-wave voltammetry (SWV) could be linearly related to potassium ion concentration in a range varying from 0.1 to 0.8×10^{-3} M. Regeneration of the sensing interface could be performed in 0.1 M HCl. The same group also reported reusable signal-on aptasensors based on thrombin-binding-induced conformational change of immobilized Fc-labeled aptamers [79]. Signal enhancement by self-assembling gold NPs onto a gold electrode has recently been demonstrated with a signal-on electrochemical aptasensor based on immobilization of ferrocene-labeled aptamers [80]. Thiolated redox-active aptamers were self-assembled on gold NPs. Cocaine-binding led to conformational change of the aptamer, thus holding the redox moieties in close proximity to the electrode. The Fc oxidation peak current measured by SWV increased with cocaine concentration, giving a linear range from 1 to 15×10^{-6} M and a detection limit of 5×10^{-7} M. Sensitivity was shown to be 10-fold higher when gold NPs were self-assembled on the gold surface. Regeneration was possible and good storage stability under a 15-day period was mentioned. An amperometric aptasensor for theophylline detection based on an immobilized ferrocene-labeled RNA aptamer was reported [81]. In the presence of theophylline, a structural rearrangement of the aptamer occurred, so that the ferrocene redox probe was close to the gold electrode surface and could induce an increase in its peak current measured by DPV. A dynamic range from 0.2×10^{-6} M (detection limit) to 10^{-7} M of theophylline was reported. Detection could also be performed in serum. An electrochemical aptasensor was also developed by coimmobilizing a Fc-labeled thrombin-binding aptamer and MP-11 on a gold electrode [82]. Thrombin addition led to the aptamer conformational change so that the Fc label approached MP-11, thus giving rise to a reduction current in the presence of hydrogen peroxide.

A carbon nanotube (CNT)-enhanced electrochemical aptasensor was also developed for the detection of thrombin [83]. Multiwalled carbon nanotubes (MWCNTs) were used as carriers of the electrochemical capture probe to improve the sensitivity of the aptasensor. In the absence of the target, an aptamer labeled with Fc was hybridized with immobilized DNA strand. Then, a strong current signal was observed when an appropriate potential was applied. In the presence of thrombin, the aptamer preferred to form an aptamer–target complex, which was thermodynamically more stable. As a result, the aptamer labeled with Fc left from the electrode surface, thus decreasing the electrochemical signal. The detection limit was 5×10^{-13} M.

A reusable and sensitive aptasensor for PDGF-BB detection based on proximity-dependent surface hybridization has also been reported

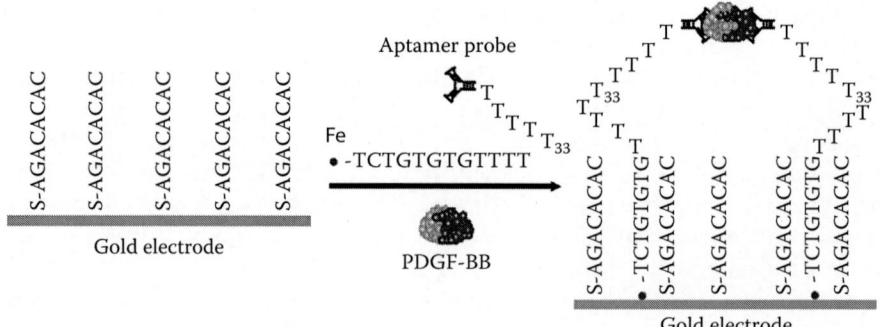

FIGURE 10.6

An electrochemical aptasensor for protein detection based on proximity-dependent surface hybridization. Simultaneous recognition of target brings two proximate aptamer probes, promoting the flanking Fc-labeled tail sequences to hybridize together with immobilized strands and triggering redox current. The absence of the target only allows independent annealing of the tail sequences, disabling the hybridization of the short sequences due to predesigned low melting temperature. (From Zhang, Y.-L. et al., *J. Am. Chem. Soc.*, 129, 15448, 2007.)

(Figure 10.6) [84]. The system was based on the simultaneous recognition of the target molecule by a pair of ferrocene-labeled aptamer probes, followed by their hybridization with immobilized oligonucleotide strands. In the presence of PDGF-BB, the ferrocene labels of the aptamer tail sequences were located close to the electrode surface, thus giving rise to a measurable redox current. The detection limit was 1 ng L^{-1}. Regeneration of the sensor interface was performed by washing with 1 M NaOH at 50°C, thus allowing its reuse over eight times without loss of sensitivity.

10.4.1.4 Anthraquinone

Based on signal-off architecture, target-induced strand release of a redox-modified aptamer from the aptamer–capture DNA duplex has been proposed as a general strategy to design electrochemical aptasensors for the detection of a wide range of molecules [85]. Detection has been reported for a large protein like thrombin and for a small molecule like ATP. In the duplex configuration, the anthraquinone redox tag was confined near the gold electrode surface, thus giving an intense electrochemical signal. Upon target-binding, release of the redox-labeled aptamer strand from the electrode into solution led to a significant reduction of the electrochemical signal that could be correlated to the target concentration. Concentrations of 10^{-5} M of ATP and as low as 10^{-8} M of thrombin have been detected.

10.4.1.5 Metal Nanoparticles

Polsky et al. (2006) [86] reported the first amplified aptasensor using electrocatalytic NPs. A sandwich configuration based on self-assembled thiolated aptamers on gold surface reacting with thrombin and then with PtNP-labeled signaling aptamers was designed. PtNPs were used as catalysts showing higher stability than enzymes that were traditionally used as redox labels. PtNPs catalyzed reduction of H_2O_2 was added in the working medium before linear sweep voltammetry (LSV) measurements. Electrocatalytic reduction of hydrogen peroxide led to a cathodic current that could be related to thrombin concentration. Detection limit was 10^{-9} M.

Gold NPs have also been used for the development of amplified electrochemical aptasensors. For instance, an electrochemical aptasensor for the detection of PDGF was reported [87]. A sandwich structure was designed, based on self-assembled thiolated aptamers on gold reacting with PDGF and then with gold NP-labeled aptamers. $[Ru(NH_3)_5Cl]^{2+}$ was used as the electrochemical probe that could electrostatically interact with aptamers, especially with numerous strands linked to NPs. Indeed, gold NPs were used to load more aptamers and they gave rise to larger peak currents. $[Ru(NH_3)_5Cl]^{2+}$ peak currents measured by cyclic voltammetry could be used to detect as low as 10^{-14} M of PDGF. In undiluted blood serum, 10^{-12} M could also be detected. Regeneration in 10% SDS solution had been performed once. An amplified impedimetric aptasensor for thrombin detection was also reported, based on aptamers and rhodamine 6G (R6G)-modified AuNPs [88]. R6G was used to overcome the repulsion between the NPs and to stabilize them. A thrombin-binding aptamer was initially immobilized on a gold electrode to capture the target. In the presence of thrombin, aptamer-functionalized AuNPs could further bind to thrombin, forming a sandwich sensing system on the electrode. The AuNPs binding to target led to a significant increase of the electron transfer impedance. Thrombin was linearly detected from 5×10^{-11} to 1.8×10^{-8} M. The detection limit was 2×10^{-11} M.

10.4.2 Non-Covalent Labels

Aptasensors using labels that are noncovalently bound to aptamers have also been reported. Redox species can intercalate into double-stranded DNA (dsDNA) or interact with nucleic acid sequence by electrostatic interactions.

10.4.2.1 Intercalated Redox Species

Some protocols used redox-active MB that intercalates into dsDNA rather than being covalently tethered to the aptamer. An electrochemical detection method of a beacon-type aptamer biosensor was developed using MB intercalation [89]. A beacon aptamer was covalently immobilized on a gold surface. MB was intercalated into the beacon sequence. In the presence of the target, the hairpin-forming beacon aptamer was conformationally changed, releasing the intercalated MB. In the presence of thrombin, the cathodic peak of MB decreased. The linear range extended up to 5.1×10^{-8} M, with a detection limit of 1.1×10^{-8} M.

A reusable aptasensor for adenosine detection has recently been reported [90]. The binding of the target induced aptamer displacement from the aptamer–capture probe duplex structure that was immobilized on a gold electrode surface covered by a gold NP film. An external electroactive indicator, MB, interacted with DNA, so that target binding led to a decreased amount of adsorbed MB and a corresponding decreased redox current of the indicator that was measured by DPV. Linear range extended from 5×10^{-9} M to 10^{-6} adenosine and the detection limit was of 10^{-9} M.

10.4.2.2 Cationic Redox Species

A signal-off aptasensor for the detection of thrombin was developed using a cationic polythiophene bearing a Fc substituent as an electrochemical mediator [91]. The human α-thrombin aptamer was immobilized on a gold electrode. Without interaction between the aptamer and the target, the polythiophene bound electrostatically to the negatively charged surface and a high current peak was observed. On the contrary, if the interaction between the aptamer and the target was achieved, polymer could not bind to the surface and no current peak was measured. A significant decrease of the signal was obtained for concentrations higher than 10^{-6} M.

$[Ru(NH_3)_6]^{3+}$ can also be used as the signaling redox molecule for the development of electrochemical aptasensor. A signal-off aptasensor was developed for the detection of lysozyme [92]. DNA aptamers were self-assembled on a gold electrode. The redox cations were electrostatically bound to the negatively charged aptamer phosphate backbone. Upon positively charged lysozyme-binding, the overall density of negative charges decreased on the surface, thus leading to a decrease in the integrated charge of $[Ru(NH_3)_6]^{3+}$ reduction peak that was measured by cyclic voltammetry (CV). Lysozyme could be detected at physiological concentrations, in the range from 0.5 to 50 mg L^{-1}. A chronocoulometric aptasensor based on the use of $[Ru(NH_3)_6]^{3+}$ as the signaling redox molecule and complementary strand displacement strategy was also developed for adenosine monophosphate (AMP) detection [93]. Negatively charged duplexes of self-assembled aptamers and short complementary DNA strands were electrostatically bound with $[Ru(NH_3)_6]^{3+}$. Upon AMP-binding, aptamer conformational change led to the release of the short complementary DNA and to the decrease of the amount of surface charges. With such a signal-off aptasensor, a moderate sensitivity was obtained, with a linear range extending from 10^{-7} to 10^{-3} M of AMP.

10.5 Aptamer Label-Free Detection

Aptamer label-free devices eliminate additional aptamer labeling procedure, but in some cases, they use other labeled molecules such as labeled aptamer-complementary DNA strand.

10.5.1 Use of Labeled Aptamer-Complementary DNA Strand

This strategy is based on the formation of a duplex structure between label-free aptamer and labeled complementary sequence. Based on this principle, signal-off and signal-on aptasensors have been developed.

Signal-off aptasensor based on target-binding-induced dissociation of an electroactive aptamer-complementary DNA oligonucleotide was developed for the detection of thrombin [94]. dsDNA duplexes were formed by hybridization of ferrocene-labeled short aptamer-complementary DNA oligonucleotides with aptamers self-assembled on a gold electrode. Dissociation of the labeled oligonucleotide led to a decrease in the DPV current signal due to the Fc moiety. Responses could be measured from 2×10^{-9} M of thrombin and the linear range extended up to 10^{-8} M. The aptasensor was shown to be highly specific and reusable. Recently, a signal-off electrochemical aptasensor based on target-induced strand release was developed for highly sensitive detection of chloramphenicol in honey [95]. Aptamer was immobilized on electrode and then hybridized with the complementary biotinylated strand to form aptamer/DNA duplex. In the presence of chloramphenicol, the biotinylated DNA strand was released from the electrode. Then the binding of streptavidin–ALP to the remaining biotinylated DNA strand led to enzyme-amplified electrochemical signal, which decreased with the increase in target concentration. Under optimal conditions, the electrochemical signal was linear with the logarithm of chloramphenicol concentrations in the range from 10^{-9} to 10^{-6} M with the detection limit of 0.29×10^{-9} M of chloramphenicol.

A reagentless signal-on aptasensor based on target-induced strand displacement has been described [96] in Figure 10.7. A thiolated DNA aptamer was self-assembled on a gold electrode before hybridization with a MB-labeled partially complementary DNA sequence. The rigid duplex regions prevented MB from approaching the electrode surface. Upon

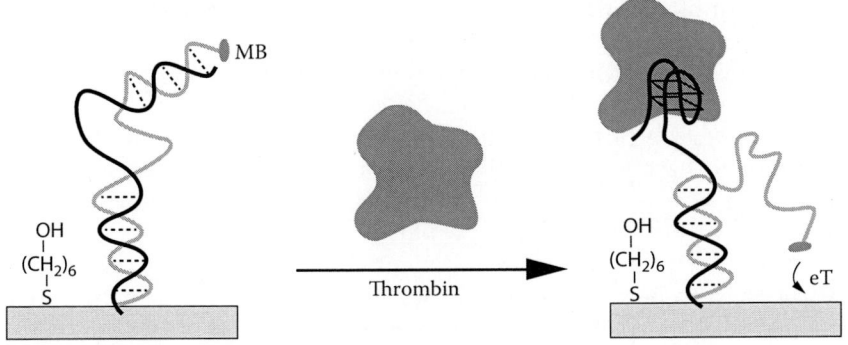

FIGURE 10.7
Reagentless "signal-on" aptasensor based on target-induced strand displacement. (From Xiao, Y. et al., *J. Am. Chem. Soc.*, 127, 17990, 2005.)

thrombin-binding, the aptamer adopted a G-quadruplex conformation, and a flexible and labeled single-stranded element led MB close to the gold surface, thus giving rise to a detectable current measured by ACV. Detection of thrombin concentrations as low as 3×10^{-9} M could be performed, and a physiologically relevant linear range was observed up to 8×10^{-8} M.

10.5.2 Label-Free Aptasensors

10.5.2.1 Amperometric Aptasensors

Numerous label-free aptasensors that need the presence of a redox probe such as $[Fe(CN)_6]^{3-}/^{4-}$ in solution have been reported. A reusable label-free signal-off aptasensor for adenosine detection based on small target-binding-induced structure-switching of the immobilized aptamer probes has been developed [97]. Amino-modified aptamers were covalently immobilized on an o-phenylenediamine-modified gold electrode via glutaraldehyde coupling. Adenosine-binding led to a structure change of the aptamer probes that formed adenosine–aptamer complexes. Decrease of the peak current was measured by ACV in the presence of $[Fe(CN)_6]^{3-}/^{4-}$ as the redox probe. Adenosine could be detected in a linear range extending from 10^{-7} M to 10^{-4} M, with a detection limit of 10^{-8} M. Regeneration could be performed by thermal dehybridization in hot water. A label-free signal-off aptasensor for 17β-estradiol detection has also been reported [98]. DNA aptamer immobilization on a streptavidin-modified gold electrode chip was performed via avidin–biotin interaction. After binding of the endocrine-disrupting chemical compound, current decrease was measured by SWV in the presence of $[Fe(CN)_6]^{3-}/^{4-}$. A linear range was obtained from 0.01 to 1×10^{-9} M. Recently, a novel label-free aptasensor for dopamine determination was reported. The aptamer was immobilized on a graphene–polyaniline nanocomposite film-modified electrode [99]. To quantify the amount of target, the peaks of SWV were monitored using the redox couple $[Fe(CN)_6]^{3-}/^{4-}$. The electrochemical aptasensor showed a linear response in the linear range 7×10^{-12} M–9×10^{-8} M and a detection limit of 1.98×10^{-12} M. The electrochemical aptasensor was successfully tested on human serum samples. A label-free and reagentless aptasensor allowing direct detection of thrombin without using additional electroactive label and based on gold NP redox properties has been very recently described [100]. Aptamers were immobilized by avidin–biotin interaction on a screen-printed electrode modified with gold NPs. Stripping voltammetry detection allowed to detect thrombin concentration as low as 10^{-9} M.

A label-free amperometric aptasensor based on thrombin acting both as a protein to be detected and as an enzyme catalyst has also been described [70]. Thiolated aptamers were immobilized on a gold electrode and incubated with thrombin. The electrochemical measurement was performed in the presence of β-Ala-Gly-Arg-pNA as the thrombin substrate, thus leading to p-nitroaniline production that could be detected. The electrochemical detection was faster and more sensitive than the optical reference one.

A label-free and reagentless signal-off aptasensor was developed for the detection of ghrelin using Spiegelmers, a new generation of aptamers with increased stability [101]. Spiegelmers are enantiomeric aptamers composed of L-nucleotides, resistant to nucleases, showing an extraordinary stability in biological fluids. The target was detected using the oxidation signal of guanine. The electroactivity of this base led to a significantly enhanced voltammetric signal. When ghrelin interacted with the aptamer, a decrease of the guanine oxidation peak was measured by SWV. Ghrelin was linearly detected between 0.01 and 1 mg L^{-1}.

10.5.2.2 Impedimetric Aptasensors

Electrochemical impedance spectroscopy (EIS) sensors are widely used for the detection of biomolecular interactions because they do not need labels [102,103]. Label-free EIS aptasensors are based on the monitoring of changes in the electron transfer resistance at the electrodes as a result of target-binding at proper pH. Experiments are performed in the presence of $[Fe(CN)_6]^{3-/4-}$ as the redox-active probe. Increase of the interfacial electron transfer resistance can be due to the formation of an insulating layer or to hydrated radius of a denatured protein blocking the electrode surface. On the contrary, decrease of the electron transfer resistance can be due to the screening of the negative charge of the aptamers by a positively charged protein.

Numerous label-free impedimetric aptasensors have been described for the detection of a wide range of molecules. A label-free impedimetric aptasensor for thrombin detection has been described [104]. DNA aptamers were self-assembled on a microfabricated thin gold surface. The formation of aptamer–thrombin complexes on the electrodes gave rise to an increase of electron transfer resistance in the presence of $[Fe(CN)_6]^{3-/4-}$ as the redox probe. The detection limit was 10^{-8} M. A label-free impedimetric aptasensor based on signal amplification by denaturation of the captured protein to be detected has also been described [105]. Aptamers were self-assembled onto a gold electrode. After specific binding, the captured thrombin was unfolded into a denatured conformation by using guanidine hydrochloride. The denaturation of the protein into its primary structure with a larger hydrated radius efficiently blocked the $[Fe(CN)_6]^{3-/4-}$ electron transfer to the gold electrode, thus amplifying the electron transfer resistance and improving the sensitivity. A detection limit as low as 10^{-14} M of thrombin was obtained and the linear range extended from 0.01 to 0.1 × 10^{-12} M. Recently, an impedimetric aptasensor was developed for the detection of OTA [106]. The aptamer was immobilized onto organized mixed layers of diazonium salts via click chemistry. The interaction between OTA and the aptamer induced an increase of the electron transfer resistance. The dynamic range of this impedimetric aptasensor ranges from 1.25 to 500 ng L^{-1} OTA with a detection limit of

0.25 ng L^{-1}. Recently, an impedimetric aptasensor was also developed for the sensitive and selective detection of acetamiprid, an insecticide [107]. To improve the sensitivity of the aptasensor, gold NPs were electrodeposited on the bare gold electrode surface that was employed as a platform for aptamer immobilization. With the addition of acetamiprid, the formation of acetamiprid–aptamer complex on the AuNP-deposited electrode surface resulted in an increase of electron transfer resistance (ETR). A wide linear range was obtained from 5 to 600 × 10^{-9} M with a low detection limit of 10^{-9} M. An impedimetric aptasensor was also developed for the label-free detection of human IgE [108]. Amino-terminated IgE aptamers were covalently attached to carboxyl-modified nanocrystalline diamond surfaces using carbodiimide chemistry. EIS was applied to measure the changes in interfacial electrical properties that arise when the aptamer-functionalized diamond surface was exposed to IgE solutions. During incubation, the formation of aptamer–IgE complexes caused a significant change in the capacitance of double layer, in good correspondence with the IgE concentration. The linear dynamic range of IgE detection was from 0.03 to 42.8 mg L^{-1}. Recently, a label-free impedimetric aptasensor was also developed for the detection of *Plasmodium* lactate dehydrogenase (LDH), a biomarker for malaria [109]. Both *Plasmodium vivax* LDH and *Plasmodium falciparum* LDH were selectively detected with a detection limit of 10^{-12} M. Furthermore, the aptasensor clearly distinguished between malaria-positive blood samples of two major species (*P. vivax* and *P. falciparum*) and a negative control, indicating that it may be a useful tool for the diagnosis, monitoring, and surveillance of malaria. An impedimetric aptasensor was also developed for the detection of lysozyme [110]. In this case, the amino-linked DNA aptamer was immobilized onto single-walled carbon nanotube (SWCNT)-modified screen-printed electrode. CNT presents several advantages such as high surface area and fast heterogeneous electron transfer. The aptasensor allowed selective detection of lysozyme with a detection limit of 8.62 × 10^{-7} M.

10.5.2.3 ISFET-Based Aptasensors

Aptamers can be advantageously used in field-effect transistor (FET)-type formats because of their small sizes. Aptamers are smaller in size than the Debye length of the double layer (ca. 3 nm at 10 mM ionic concentrations), so that the binding event with a protein can occur within the electrical double layer in buffer solution. Changes in the charge distribution can thus easily be detected by FETs.

So et al. (2005) developed an SWCNT-FET aptasensor [111]. An antithrombin DNA aptamer was covalently linked to the side wall of CNTs that were assembled between source and drain electrodes. Target-binding led to measurable decrease in conductance. Detection limit was around 10^{-8} M. The same group applied SWCNT-FETs for the detection of *Escherichia coli* [112].

Aptamer-modified CNT-FETs have been designed for IgE detection [113]. These label-free biosensors are based on the covalent immobilization of aptamers on CNT channels. Target addition led to a sharp decrease in the source–drain current, related to the screening effect of IgE molecules on the device. The nonspecific binding of IgE was successfully suppressed. The detection limit was reported as 2.5×10^{-10} M and the linear dynamic range as 2.5×10^{-10} to 2×10^{-8} M. The performances of the aptamer-based CNT-FETs were better than those obtained with antibody-based CNT-FETs under similar conditions.

Thrombin detection was also reported by using a label-free ion-gated apta-FET sensor based on the charge transport properties of polypyrrole NTs [114]. Aptamers were covalently linked to NT networks onto interdigitated microelectrode surfaces. Intermolecular interactions occurring during the aptamer–thrombin complex formation led to a decrease in the source–drain current.

A selective label-free and reagentless aptasensor based on an ion-selective field-effect transistor (ISFET) device has been described for the detection of a small molecule, AMP [115]. In this case, the covalently immobilized and gate-confined aptamer formed a duplex with a short nucleic acid sequence. Target binding led to the separation of the aptamer/nucleic acid duplex. The negatively charged oligonucleotide release affects the ionization of the gate and, thus, the gate-to-source potential of the ISFET device.

Parallel analysis of two analytes, cocaine and AMP, has been described using a bifunctional aptamer [123]. The bifunctional aptamer includes two aptamer units for cocaine and AMP. A nucleic acid sequence was immobilized on the gate of a FET device and was then hybridized with the bifunctional aptamer. In the presence of cocaine or AMP, the bifunctional aptamer was released. The change in the charge of the gate gave rise to changes in the gate-to-source potential that could be related to the analyte concentration. Cocaine and AMP could be detected at concentrations ranging from 10^{-6} to 10^{-4} M and from 10^{-6} to 10^{-4} M, respectively.

10.5.2.4 *Potentiometric Aptasensor*

A potentiometric aptasensor was developed for the detection of thrombin. This simple and cheap device was based on poly(phenothiazine) conducting polymers electropolymerized on a glassy carbon electrode. Avidin-modified polymer surfaces obtained by direct electrostatic precipitation have then been used to immobilize biotinylated antithrombin DNA aptamers. Measurement of the difference in the potential of the sensor in pH 3 and in pH 7.6 media gave rise to potentiometric thrombin detection in the concentration range from 10^{-9} to 10^{-6} M.

CNT properties have been exploited for the development of efficient potentiometric aptasensors. A potentiometric CNT aptasensor was developed for the detection of a variable surface glycoprotein (VSG) from African trypanosomes [116]. An amine-modified aptamer was immobilized on a carboxylated SWCNT-modified electrode. SWCNTs are highly sensitive to changes in the ionic environment at their interface as well as to redox conditions in the solution. Variations in the electromotive force are measured in real time upon the direct addition of diluted real blood samples containing the target protein thus eliminating the need of preliminary matrix removal. The same group developed a potentiometric SWCNT aptasensor for the detection of *E. coli* [117]. The selective aptamer–target interaction significantly changed the electrical potential, thus allowing for both interspecies and interstrain selectivity and enabling the direct detection of the target. A potentiometric aptasensor based on aptamer-functionalized CNTs was also developed for the detection of thrombin [118].

10.6 Conclusion

Aptamers that bind nonnucleic acid targets (e.g., protein, ions, amino acids, antibiotics, vitamins, cocaine, organic dyes, whole cells, and microorganisms) were first chemically produced in 1990 using the SELEX process. Aptamers show a very high affinity for their targets, with dissociation constants typically varying from the micromolar to low picomolar range, comparable to some of those monoclonal antibodies, sometimes even better. Aptamers have also been selected with a high binding specificity [119]. Compared to antibodies, aptamers present many advantages, especially the fact of avoiding the use of animals for their production.

The application of aptamers as recognition elements in biosensors offers several advantages such as [120,121]

- Aptamers can be selected in conditions that are similar to that of real media, making them ideal for food applications.
- Aptamers can be modified during immobilization or labeled without effect on their affinity.
- Aptamers can be subjected to repeated cycles of denaturation and regeneration.
- Aptamers, once selected, can be synthesized with high reproducibility and purity from commercial sources.
- Aptamers often undergo significant conformational changes upon target-binding. This offers great flexibility in the design of novel biosensors with high detection sensitivity and selectivity.

Different detection techniques have been contemplated. Frequently, the detection methods are based on a tagging approach. Among them, the fluorescence-based technique is often used, due to its high sensitivity, the ease of labeling aptamers with fluorescent dyes, the availability of many different fluorophores and quenchers, and the inherent capability for real-time detection [5]. But the entire optical system including laser diode, photodiode, and filter is not very suited for miniaturization and appears very costly. Electrochemistry is a promising tool often employed to develop aptasensors. Labels, such as enzymes, ferrocene, MB, redox cations, or NPs, can be used, but label-free electrochemical detection of hybridization is also a very attractive approach.

Nanotechnology is playing an important role in the development of efficient aptasensors. Different types of nanomaterials with different properties have been used, and they offer exciting new opportunities to improve the performance of biosensors for the development of aptasensors [86,87,111,116].

Aptamer-based biosystems are still immature compared to immunoassays, which reflects the limited availability of aptamer types and the relatively poor knowledge of surface-immobilization technologies for aptamers [5]. However, important advances have been realized since 1996 [12], and aptamer-based detection systems appear as highly efficient devices with enormous potential.

Future direction will probably see growth of microchips and nanochips. The development of microarrays based on DNA aptamers used as receptors can be seen as a logical continuation of the DNA-chip technology development [122], although the principle of target recognition is not based on hybridization but is analogous to the immunochemical assay.

References

1. Ellington, A. D. and Szostak, J. W. (1990). In vitro selection of RNA molecules that bind specific ligands. *Nature*, **346**, 818–822.
2. Tuerk, C. and Gold, L. (1990). Systematic evolution of ligands by exponential enrichment: RNA ligands to bacteriophage T4 DNA polymerase. *Science*, **249**, 505–510.
3. Robertson, D. L. and Joyce, G. F. (1990). Selection in vitro of an RNA enzyme that specifically cleaves single-stranded DNA. *Nature*, **344**, 467.
4. Jayasena, S. D. (1999). Aptamers: An emerging class of molecules that rival antibodies in diagnostics. *Clinical Chemistry*, **45**, 1628–1650.
5. Song, S., Wang, L., Li, J., Fan, C., and Zhao, J. (2008). Aptamer-based biosensors. *Trends in Analytical Chemistry*, **27**, 108–117.
6. Sett, A., Das, S., Sharma, P., and Bora, U. (2012). Aptasensors in health, environment and food safety monitoring. *Open Journal of Applied Biosensor*, **1**, 9–19.
7. Sassolas, A., Blum, L. J., and Leca-Bouvier, B. D. (2011). Optical detection systems using immobilized aptamers. *Biosensors and Bioelectronics*, **26**, 3725–3736.

8. Sassolas, A., Blum, L. J., and Leca-Bouvier, B. D. (2009). Electrochemical aptasensors. *Electroanalysis*, **21**, 1237–1250.

9. Hianik, T. and Wang, J. (2009). Electrochemical aptasensors—Recent achievements and perspectives. *Electroanalysis*, **21**, 1223–1235.

10. Hong, P., Li, W., and Li, J. (2012). Applications of aptasensors in clinical diagnostics. *Sensors*, **12**, 1181–1193.

11. Velasco-Garcia, M. N. and Missailidis, S. (2009). New trends in aptamer-based electrochemical biosensors. *Gene Therapy and Molecular Biology*, **13**, 1–9.

12. Davis, K. A., Abrams, B., Lin, Y., and Jayasena, S. D. (1996). Use of a high affinity DNA ligand in flow cytometry. *Nucleic Acids Research*, **24**, 702–706.

13. Kleinjung, F., Klussmann, S., Erdmann, V. A., Scheller, F. W., Fürste, J. P., and Bier, F. F. (1998). High-affinity RNA as a recognition element in a biosensor. *Analytical Chemistry*, **70**, 328–331.

14. Potyrailo, R. A., Conrad, R. C., Ellington, A. D., and Hiefte, G. M. (1998). Adapting selected nucleic acid ligands (aptamers) to biosensors. *Analytical Chemistry*, **70**, 3419–3425.

15. Lee, M. and Walt, D. R. (2000). A fiber-optic microarray biosensor using aptamers as receptors. *Analytical Biochemistry*, **282**, 142–146.

16. Shlyahovsky, B., Li, D., Katz, E., and Willner, I. (2007). Proteins modified with DNAzymes or aptamers act as biosensors or biosensor labels. *Biosensors and Bioelectronics*, **22**, 2570–2576.

17. Li, Z.-P., Wang, Y.-C., Liu, C.-H., and Li, Y.-K. (2005). Development of chemiluminescence detection of gold nanoparticles in biological conjugates for immunoassay. *Analytica Chimica Acta*, **551**, 85–91.

18. Zhang, Z.-F., Cui, H., Lai, C.-Z., and Liu, L.-J. (2005). Gold nanoparticle-catalyzed luminol chemiluminescence and its analytical applications. *Analytical Chemistry*, **77**, 3324–3329.

19. Gill, R., Polsky, R., and Willner, I. (2006). Pt nanoparticle functionalized with nucleic acid as catalytic labels for the chemiluminescent detection of DNA and proteins. *Small*, **3**, 1037–1041.

20. Huang, H. and Zhu, J.-J. (2009). DNA aptamer-based QDs electrochemiluminescence biosensor for the detection of thrombin. *Biosensors and Bioelectronics*, **25**, 927–930.

21. Fang, L., Lü, Z., Wei, H., and Wang, E. A electrochemiluminescence aptasensor for detection of thrombin incorporating the capture aptamer labeled with gold nanoparticles immobilized onto the thio-silanized ITO electrode. *Analytica Chimica Acta*, **628**, 80–86.

22. Li, Y., Qi, H., Peng, Y., Yang, J., and Zhang, C. (2007). Electrogenerated chemiluminescence aptamer-based biosensor for the determination of cocaine. *Electrochemistry Communications*, **9**, 2571–2575.

23. Bai, J., Wei, H., Li, B., Song, L., Fang, L., Lv, Z., Zhou, W., and Wang, E. (2008). [Ru(bpy)2(dcbpy)NHS] labeling/aptamer-based biosensor for the detection of lysozyme by increasing sensitivity with gold nanoparticle amplification. *Chemistry–An Asian Journal*, **3**, 1935–1941.

24. Wang, W., Zhou, J., Yun, W., Xiao, S., Chang, Z., He, P., and Fang, Y. (2007). Detection of thrombin using electrogenerated chemiluminescence based on $Ru(bpy)_3^{2+}$-doped silica nanoparticle aptasensor *via* target protein-induced strand displacement. *Analytica Chimica Acta*, **598**, 242–248.

25. Wang, X.-Y., Yun, W., Zhou, J.-M., Dong, P., He, P.-G., and Fang, Y.-Z. (2008). Ru(bpy)$_3$$^{2+}$-doped silica nanoparticle aptasensor for detection of thrombin based on electrogenerated chemiluminescence. *Chinese Journal of Chemistry*, **26**, 315–320.

26. Wang, L., Yang, C., and Tan, W. (2005). Dual-luminophore-doped silica nanoparticles for multiplexed signaling. *Nano Letters*, **5**, 37–43.

27. Li, Y., Qi, H., Peng, Y., Gao, Q., and Zhang, C. (2008). Electrogenerated chemiluminescence aptamer-based method for the determination of thrombin incorporating quenching of tris(2,2'-bipyridine)ruthenium by ferrocene. *Electrochemistry Communications*, **10**, 1322–1325.

28. Tombelli, S., Minunni, A., Luzi, E., and Mascini, M. (2005). Aptamer-based biosensors for the detection of HIV-1 Tat protein. *Bioelectrochemistry*, **67**, 135–141.

29. Wang, Z. Z., Wilkop, T., Xu, D. K., Dong, Y., Ma, G. Y., and Cheng, Q. (2007). Surface plasmon resonance imaging for affinity analysis of aptamer - protein interactions with PDMS microfluidic chips. *Analytical and Bioanalytical Chemistry*, **389**, 819–825.

30. Lee, S. J., Youn, B. S., Park, J. W., Niazi, J. H., Kim, Y. S., and Gu, M. B. (2008). ssDNA aptamer-based surface plasmon resonance biosensor for the detection of retinol binding protein 4 for the early diagnosis of type 2 diabetes. *Analytical Chemistry*, **80**, 2867–2873.

31. Jalit, Y., Gutierez, F. A., Dubacheva, G., Goyer, C., Coche-Guerente, L., Defrancq, E., Labbé, P., Rivas, G. A., and Rodriguez, M. C. (2013). Characterization of a modified gold platform for the development of a label-free anti-thrombin aptasensor. *Biosensors and Bioelectronics*, **41**, 424–429.

32. Li, Y., Lee, H. J., and Corn, R. M. (2006). Fabrication and characterization of RNA aptamer microarrays for the study of protein-aptamer interactions with SPR imaging. *Nucleic Acids Research*, **34**, 6416–6424.

33. Li, Y., Lee, H. J., and Corn, R. M. (2007). Detection of protein biomarkers using RNA aptamer microarrays and enzymatically amplified surface plasmon resonance imaging. *Analytical Chemistry*, **79**, 1082–1088.

34. Geddes, C., Gryczynski, I., Malicka, J., Gryczynski, Z., and Lakowicz, J. (2004). Directional surface plasmon coupled emission. *Journal of Fluorescence*, **14**, 119–123.

35. Lakowicz, J. R. (2004). Radiative decay engineering 3. Surface plasmon-coupled directional emission. *Analytical Biochemistry*, **324**, 153–169.

36. Lakowicz, J. R. (2005). Radiative decay engineering 5: Metal-enhanced fluorescence and plasmon emission. *Analytical Biochemistry*, **337**, 171–194.

37. Matveeva, E., Gryczynski, Z., Malicka, J., Lukomska, J., Makowiec, S., Berndt, K., Lakowicz, J., and Gryczynski, I. (2005). Directional surface plasmon-coupled emission: Application for an immunoassay in whole blood. *Analytical Biochemistry*, **344**, 161–167.

38. Xie, T., Liu, Q., Cai, W., Chen, Z., and Li, Y. (2009). Surface plasmon-coupled directional emission based on a conformational-switching signaling aptamer. *Chemical Communications*, **22**, 3190–3192.

39. Kneipp, K., Kneipp, H., Itzkan, I., Dasari, R. R., and Feld, M. S. (1999). Ultrasensitive chemical analysis by Raman spectroscopy. *Chemical Reviews*, **99**, 2957–2976.

40. Chen, J., Jiang, J., Gao, G., Shen, G., and Yu, R. (2008). A new aptameric biosensor for cocaine based on surface-enhanced Raman scattering spectroscopy. *Chemistry–A European Journal*, **14**, 8374–8382.

41. Wang, Y., Wei, H., Li, B., Ren, W., Guo, S., Dong, S., and Wang, E. (2007). SERS opens a new way in aptasensor for protein recognition with high sensitivity and selectivity. *Chemical Communications*, **28**, 5220–5222.
42. White, I. M., Hanumegowda, N. M., and Fan, X. D. (2005). Subfemtomole detection of small molecules with microsphere sensors. *Optics Letters*, **30**, 3189–3191.
43. Quan, H. and Guo, Z. (2005). Simulation of whispering-gallery-mode resonance shifts for optical miniature biosensors. *Journal of Quantitative Spectroscopy and Radiative Transfer*, **93**, 231–243.
44. Zhu, H. Y., Suter, J. D., White, I. M., and Fan, X. D. (2006). Aptamer based microsphere biosensor for thrombin detection. *Sensors*, **6**, 785–795.
45. Goh, J. B., Loo, R. W., and Goh, M. C. (2005). Label-free monitoring of multiple biomolecular binding interactions in real-time with diffraction-based sensing. *Sensors and Actuators B-Chemical*, **106**, 243–248.
46. Goh, J. B., Loo, R. W., McAloney, R. A., and Goh, M. C. (2002). Diffraction-based assay for detecting multiple analytes. *Analytical and Bioanalytical Chemistry*, **374**, 54–56.
47. Goh, J. B., Tam, P. L., Loo, R. W., and Goh, M. C. (2003). A quantitative diffraction-based sandwich immunoassay. *Analytical Biochemistry*, **313**, 262–266.
48. Acharya, G., Chang, C.-L., Holland, D. P., Thompson, D. H., and Savran, C. A. (2007). Rapid detection of *S*-adenosyl homocysteine using self-assembled optical diffraction gratings. *Angewandte Chemie-International Edition*, **46**, 1–4.
49. Lee, J., Icoz, K., Roberts, A., Ellington, A. D., and Savran, C. A. (2010). Diffractometric detection of proteins using microbead-based rolling circle amplification. *Analytical Chemistry*, **82**, 197–202.
50. Gopinath, S. C. B., Awazu, K., Fujimaki, M., Sugimoto, K., Ohki, Y., Komatsubara, T., Tominaga, J., Gupta, K. C., and Kumar, P. K. R. (2008). Influence of nanometric holes on the sensitivity of a waveguide-mode sensor: Label-free nanosensor for the analysis of RNA aptamer-ligand interactions. *Analytical Chemistry*, **80**, 6602–6609.
51. Fujimaki, M., Rockstulh, C., Wang, X., Awazu, K., Tominaga, J., Koganezawa, Y., Ohki, Y., and Komatsubara, T. (2008). Silica-based monolithic sensing plates for waveguide-mode sensors. *Optic Express*, **16**, 6408–6416.
52. Janshoff, A., Galla, H. J., and Steinem, C. (2000). Piezoelectric mass-sensing devices as biosensors–An alternative to optical biosensors? *Angewandte Chemie-International Edition*, **39**, 4004–4032.
53. O'Sullivan, C. K. and Guilbault, G. G. (1999). Commercial quartz crystal microbalances–theory and applications. *Biosensors and Bioelectronics*, **14**, 663–670.
54. Janshoff, A. and Steinem, C. (2001). Quartz crystal microbalance for bioanalytical applications. *Sensors Update*, **9**, 313–354.
55. Liss, M., Petersen, B., Wolf, H., and Prohaska, E. (2002). An aptamer-based quartz crystal protein biosensor. *Analytical Chemistry*, **74**, 4488–4495.
56. Minunni, M., Tombelli, S., Gullotto, A., Luzi, E., and Mascini, M. (2004). Development of biosensors with aptamers as bio-recognition element: The case of HIV-1 Tat protein. *Biosensors and Bioelectronics*, **20**, 1149–1156.
57. Hianik, T., Ostatna, V., Sonlajtnerova, M., and Grman, I. (2007). Influence of ionic strength, pH and aptamer configuration for binding affinity to thrombin. *Bioelectrochemistry*, **70**, 127–133.

58. Porfirieva, A., Evtugyn, G., and Hianik, T. (2007). Polyphenothiazine modified electrochemical aptasensor for detection of human α-thrombin. *Electroanalysis*, **19**, 1915–1920.

59. Schlensog, M. D., Gronewold, T. M. A., Tewes, M., Famulok, M., and Quandt, E. (2004). A Love-wave biosensor using nucleic acids as ligands. *Sensors and Actuators B-Chemical*, **101**, 308–315.

60. Gronewold, T. M. A., Glass, S., Quandt, E., and Famulok, A. (2005). Monitoring complex formation in the blood-coagulation cascade using aptamer-coated SAW sensors. *Biosensors and Bioelectronics*, **20**, 2044–2052.

61. Jung, A., Gronewold, T. M. A., Tewes, M., Quandt, E., and Berlin, P. (2007). Biofunctional structural design of SAW sensor chip surfaces in a microfluidic sensor system. *Sensors and Actuators B-Chemical*, **124**, 46–52.

62. Carrascosa, L. G., Moreno, M., Alvarez, M., and Lechuga, L. M. (2006). Nanomechanical biosensors: A new sensing tool. *TrAC-Trends in Analytical Chemistry*, **25**, 196–206.

63. Sepaniak, M., Datskos, P., Lavrik, N., and Tipple, C. (2002). Microcantilever transducers: A new approach to sensor technology. *Analytical Chemistry*, **74**, 568A–575A.

64. Lavrik, N. V., Sepaniak, M. J., and Datskos, P. G. (2004). Cantilever transducers as a platform for chemical and biological sensors. *Review of Scientific Instruments*, **75**, 2229–2253.

65. Savran, C. A., Knudsen, S. M., Ellington, A. D., and Manalis, S. R. (2004). Micromechanical detection of proteins using aptamer-based receptor molecules. *Analytical Chemistry*, **76**, 3194–3198.

66. Hwang, K. S., Lee, S.-M., Eom, K., Lee, J. H., Lee, Y.-S., Park, J. H., Yoon, D. S., and Kim, T. S. (2007). Nanomechanical microcantilever operated in vibration modes with use of RNA aptamer as receptor molecules for label-free detection of HCV helicase. *Biosensors and Bioelectronics*, **23**, 459–465.

67. Miyachi, Y., Ogino, C., Amino, T., and Kondo, A. (2011). Development of a novel aptamer-based sensing system using atomic force microscopy. *Journal of Bioscience and Bioengineering*, **112**, 511–514.

68. Ikebukuro, K., Kiyohara, C., and Sode, K. (2004). Electrochemical detection of protein using a double aptamer sandwich. *Analytical Letters*, **37**, 2901–2909.

69. Ikebukuro, K., Kiyohara, C., and Sode, K. (2005). Novel electrochemical sensor system for protein using the aptamers in sandwich manner. *Biosensors and Bioelectronics*, **20**, 2168–2172.

70. Mir, M., Vreeke, M., and Katakis, L. (2006). Different strategies to develop an electrochemical thrombin aptasensor. *Electrochemistry Communications*, **8**, 505–511.

71. Zhang, D.-W., Sun, C.-J., Zhang, F.-T., Xu, L., Zhou, Y.-L., and Zhang, X.-X. (2012). An electrochemical aptasensor based on enzyme linked aptamer assay. *Biosensors and Bioelectronics*, **31**, 363–368.

72. Barthelmebs, L., Hayat, A., Limiadi, A. W., Marty, J. L., and Noguer, T. (2011). Electrochemical DNA aptamer-based biosensor for OTA detection using superparamagnetic nanoparticles. *Sensors and Actuators B-Chemical*, **156**, 932–937.

73. Papamichael, K. I., Kreuzer, M. P., and Guilbault, G. G. (2007). Viability of allergy (IgE) detection using an alternative aptamer receptor and electrochemical means. *Sensors and Actuators B-Chemical*, **121**, 178–186.

74. Xiao, Y., Lubin, A. A., Heeger, A. J., and Plaxco, K. W. (2005). Label-free electronic detection of thrombin in blood serum by using an aptamer-based sensor. *Angewandte Chemie-International Edition*, **44**, 5456–5459.

75. Baker, B. R., Lai, R. Y., Wood, M. S., Doctor, E. H., Heeger, A. J., and Plaxco, K. W. (2006). An electronic, aptamer-based small-molecule sensor for the rapid, label-free detection of cocaine in adulterated samples and biological fluids. *Journal of the American Chemical Society*, **128**, 3138–3139.

76. Lai, R. Y., Plaxco, K. W., and Heeger, A. J. (2007). Aptamer-based electrochemical detection of picomolar platelet-derived growth factor directly in blood serum. *Analytical Chemistry*, **79**, 229–233.

77. Wu, Z. S., Guo, M. M., Zhang, S. B., Chen, C. R., Jiang, J. H., Shen, G. L., and Yu, R. Q. (2007). Reusable electrochemical sensing platform for highly sensitive detection of small molecules based on structure-switching signaling aptamers. *Analytical Chemistry*, **79**, 2933–2939.

78. Radi, A. E. and O'Sullivan, C. K. (2006). Aptamer conformational switch as sensitive electrochemical biosensor for potassium ion recognition. *Chemical Communications*, **2006**, 3432–3434.

79. Sanchez, J. L. A., Baldrich, E., Radi, A. E. G., Dondapati, S., Sanchez, P. L., Katakis, I., and O'Sullivan, C. K. (2006). Electronic 'off-on' molecular switch for rapid detection of thrombin. *Electroanalysis*, **18**, 1957–1962.

80. Li, X., Qi, H., Shen, L., Gao, Q., and Zhang, C. (2008). Electrochemical aptasensor for the determination of cocaine incorporating gold nanoparticles modification. *Electroanalysis*, **20**, 1475–1482.

81. Ferapontova, E. E. and Gothelf, K. V. (2009). Optimization of the electrochemical RNA-aptamer based biosensor for theophylline by using a methylene blue redox label. *Electroanalysis*, **21**, 1261–1266.

82. Mir, M., Jenkins, T. A., and Katakis, I. (2008). Ultrasensitive detection based on aptamer beacon electron transfer chain. *Electrochemistry Communications*, **10**, 1533–1536.

83. Liu, X., Li, Y., Zheng, J., Zhang, J., and Sheng, Q. (2010). Carbon nanotube-enhanced electrochemical aptasensor for the detection of thrombin. *Talanta*, **81**, 1619–1624.

84. Zhang, Y.-L., Huang, Y., Jiang, J.-H., Shen, G.-L., and Yu, R.-Q. (2007). Electrochemical aptasensor based on proximity-dependent surface hybridization assay for single-step, reusable, sensitive protein detection. *Journal of the American Chemical Society*, **129**, 15448–15449.

85. Yoshizumi, J., Kumamoto, S., Nakamura, M., and Yamana, K. (2008). Target-induced strand release (TISR) from aptamer-DNA duplex: a general strategy for electronic detection of biomolecules ranging from a small molecule to a large protein. *The Analyst*, **133**, 323–325.

86. Polsky, R., Gill, R., Kaganovsky, L., and Willner, I. (2006). Nucleic acid-functionalized Pt nanoparticles: Catalytic labels for the amplified electrochemical detection of biomolecules. *Analytical Chemistry*, **78**, 2268–2271.

87. Wang, J., Meng, W., Zheng, X., Liu, S., and Li, G. (2009). Combination of aptamer with gold nanoparticles for electrochemical signal amplification: Application to sensitive detection of platelet-derived growth factor. *Biosensors and Bioelectronics*, **24**, 1598–1602.

88. Li, B., Wang, Y., Wei, H., and Dong, S. (2008). Amplified electrochemical aptasensor taking AuNPs based sandwich sensing platform as a model. *Biosensors and Bioelectronics*, **23**, 965–970.

89. Bang, G. S., Cho, S., and Kim, B. G. (2005). A novel electrochemical detection method for aptamer biosensors. *Biosensors and Bioelectronics*, **21**, 863–870.

90. Feng, K., Sun, C., Kang, Y., Chen, J., Jiang, J. H., Shen, G.-L., and Yu, R.-Q. (2008). Label-free electrochemical detection of nanomolar adenosine based on target-induced aptamer displacement. *Electrochemistry Communications*, **10**, 531–535.

91. Le Floch, F., Ho, H. A., and Leclerc, M. (2006). Label-free electrochemical detection of protein based on a ferrocene-bearing cationic polythiophene and aptamer. *Analytical Chemistry*, **78**, 4727–4731.

92. Cheng, A. K. H. and Yu, H.-Z. (2007). Aptamer-based biosensors for label-free voltammetric detection of lysozyme. *Analytical Chemistry*, **79**, 5158–5164.

93. Shen, L., Chen, Z., Li, Y., Jing, P., Xie, S., He, S., He, P., and Shao, Y. (2007). A chronocoulometric aptamer sensor for adenosine monophosphate. *Chemical Communications*, **21**, 2169–2171.

94. Lu, Y., Zhu, N., and Mao, L. (2008). Aptamer-based electrochemical sensors that are not based on the target binding-induced conformational change of aptamers. *Analyst*, **133**, 1256–1260.

95. Yan, L., Luo, C., Cheng, W., Mao, W., Zhang, D.-W., and Ding, S. (2012). A simple and sensitive electrochemical aptasensor for determination of Chloramphenicol in honey based on target-induced strand release. *Journal of Electroanalytical Chemistry*, **687**, 89–94.

96. Xiao, Y., Piorek, B. D., Plaxco, K. W., and Heeger, A. J. (2005). A reagentless signal-on architecture for electronic, aptamer-based sensors via target-induced strand displacement. *Journal of the American Chemical Society*, **127**, 17990–17991.

97. Zheng, F., Wu, Z. S., Zhang, S.-B., Guo, M.-M., Chen, C.-R., Shen, G.-L., and Yu, R.-Q. (2008). Aptamer-based electrochemical biosensors for highly selective and quantitative detection of adenosine. *Chemical Research in Chinese Universities*, **24**, 138–142.

98. Kim, Y. S., Jung, H. S., Matsuura, T., Lee, H. Y., Kawai, T., and Gu, M. B. (2007). Electrochemical detection of 17β-estradiol using DNA aptamer immobilized gold electrode chip. *Biosensors and Bioelectronics*, **22**, 2525–2531.

99. Liu, S., Xing, X., Yu, J., Lian, W., Li, J., Cui, M., and Huang, J. (2012). A novel label-free electrochemical aptasensor based on graphene-polyaniline composite film for dopamine determination. *Biosensors and Bioelectronics*, **36**, 186–191.

100. Suprun, E., Shumyantseva, V., Bulko, T., Rachmetova, S., Rad'ko, S., Bodoev, N., and Archakov, A. (2008). Au-nanoparticles as an electrochemical sensing platform for aptamer-thrombin interaction. *Biosensors and Bioelectronics*, **24**, 825–830.

101. Mascini, M., Papamichael, K. I., Mevola, I., Pravda, M., and Guibault, G. G. (2007). Ghrelin detection using spiegelmer-capture molecules. *Analytical Letters*, **40**, 403–430.

102. Daniels, J. S. and Pourmand, N. (2007). Label-free impedance biosensors: Opportunities and challenges. *Electroanalysis*, **17**, 1239–1257.

103. Pejcic, B. and De Marco, R. (2006). Impedance spectroscopy: Over 35 years of electrochemical sensor optimization. *Electrochimica Acta*, **51**, 6217–6229.

104. Cai, H., Lee, T. M.-H., and Hsing, I. M. (2006). Label-free protein recognition using an aptamer-based impedance measurement assay. *Sensors and Actuators B-Chemical*, **114**, 433–437.

105. Xu, Y., Yang, L., Ye, X., He, P., and Fang, Y. (2006). An aptamer-based protein biosensor by detecting the amplified impedance signal. *Electroanalysis*, **18**, 1449–1456.

106. Hayat, A., Sassolas, A., Marty, J. L., and Radi, A. E. (2013). Highly sensitive ochratoxin A impedimetric aptasensor based on the immobilization of azido-aptamer onto electrografted binary film via click chemistry. *Talanta*, **103**, 14–19.

107. Fan, L., Zhao, G., Shi, H., Liu, M., and Li, Z. (2013). A highly selective electro-chemical impedance spectroscopy-based aptasensor for sensitive detection of acetamiprid. *Biosensors and Bioelectronics*, **43**, 12–18.

108. Tran, D. T., Vermeeren, V., Grieten, L., Wenmackers, S., Wagner, P., Pollet, J., Janssen, K. P. F., Michiels, L., and Lammertyn, J. (2011). Nanocrystalline dia-mond impedimetric aptasensor for the label-free detection of human IgE. *Biosensors and Bioelectronics*, **26**, 2987–2993.

109. Lee, S. J., Song, K.-M., Jeon, W., Jo, H., Shim, Y.-B., and Ban, C. (2012). A highly sensitive aptasensor towards Plasmodium lactate dehydrogenase for the diag-nosis of malaria. *Biosensors and Bioelectronics*, **35**, 291–296.

110. Rohrbach, F., Karadeniz, H., Erdem, A., Famulok, M., and Mayer, G. (2012). Label-free impedimetric aptasensor for lysozyme detection based on carbon nanotube-modified screen-printed electrodes. *Analytical Biochemistry*, **421**, 454–459.

111. So, H.-M., Won, K., Kim, Y. H., Kim, B.-K., Ryu, B. H., Na, P. S., Kim, H., and Lee, J.-O. (2005). Single-walled carbon nanotube biosensors using aptamers as molecu-lar recognition elements. *Journal of the American Chemical Society*, **127**, 11906–11907.

112. So, H.-M., Park, D.-W., Jeon, E.-K., Kim, Y.-H., Kim, B. S., Lee, C.-K., Choi, S. Y., Kim, S.-C., Chang, H., and Lee, J.-O. (2008). Detection and titer estimation of *Escherichia coli* suing aptamer-functionalized single-walled carbon-nanotube field-effect transistors. *Small*, **4**, 197–201.

113. Maehashi, K., Katsura, T., Kerman, K., Takamura, Y., Matsumoto, K., and Tamiya, E. (2007). Label-free protein biosensor based on aptamer-modified car-bon nanotube field-effect transistors. *Analytical Chemistry*, **79**, 782–789.

114. Yoon, H., Kim, J.-H., Lee, N., Kim, B.-G., and Jang, J. (2008). A novel sensor plat-form based on aptamer-conjugated polypyrrole nanotubes for label-free electro-chemical protein detection. *ChemBioChem*, **9**, 634–641.

115. Zayats, M., Huang, Y., Gill, R., Ma, C.-A., and Willner, I. (2006). Label-free and reagentless aptamer-based sensors for small molecules. *Journal of the American Chemical Society*, **128**, 13666–13667.

116. Zelada-Guillen, G. A., Tweed-Kent, A., Niemann, M., Göringer, H. U., Riu, J., and Rius, F. X. (2013). Ultrasensitive and real-time detection of proteins in blood using a potentiometric carbon-nanotube aptasensor. *Biosensors and Bioelectronics*, **41**, 366–371.

117. Zelada-Guillen, G. A., Bhosale, S. V., Riu, J., and Rius, F. X. (2010). Real-time potentio-metric detection of bacteria in complex samples. *Analytical Chemistry*, **82**, 9254–9260.

118. Düzgün, A., Maroto, A., Mairal, T., O'Sullivan, C., and Rius, F. X. (2010). Solid-contact potentiometric aptasensor based on aptamer functionalized carbon nanotubes for the direct determination of proteins. *Analyst*, **135**, 1037–1041.

119. Jenison, R. D., Gill, S. C., Pardi, A., and Polisky, B. (1994). High-resolution molecular discrimination by RNA. *Science*, **263**, 1425–1429.

120. Deisingh, A. K. (2006). Aptamer-based biosensors: Biomedical applications. *RNA Towards Medicine*, **173**, 341.

121. O'Sullivan, C. K. (2002). Aptasensors–the future of biosensing. *Analytical and Bioanalytical Chemistry*, **372**, 44–48.

122. Sassolas, A., Leca-Bouvier, B. D., and Blum, L. J. (2008). DNA biosensors and microarrays. *Chemical Reviews*, **108**, 109–139.

123. Elbaz, J., Shlyahovsky, B., Li, D., and Willner, I. (2008). Parallel analysis of two analytes in solutions or on surfaces by using a bifunctional aptamer: Applications for biosensing and logic gate operations. *ChemBioChem.* **9**, 232–239.

11

Tissue, Microorganisms, Organelles, and Cell-Based Biosensors

Theodoros Varzakas, Georgia-Paraskevi Nikoleli, and Dimitrios P. Nikolelis

CONTENTS

11.1 Introduction

This chapter describes the use of different sensor technologies that can identify cells, tissues, microorganisms, and organelles.

Surface plasmon resonance (SPR) biosensors, electrochemical impedance spectroscopy (EIS) biosensor, whole-cell and cell array biosensors, antibody-immobilized microcantilever (MC) resonators, enzyme-based biosensors and nanoparticles (NPs), aptamers, luminescent biosensors, photonic crystal microcavity biosensors, nanomaterials and biosensors, microwire sensors, and oligoaziridine biosensors are described with applications on living cells or microorganisms. Their principles have been explained along with their mode of action.

11.2 AlGaN/GaN Sensor Technology

Aluminum gallium nitride (AlGaN)/gallium nitride (GaN)sensor technology is an excellent potential successor to silicon-based field-effect transistor (Si-FET) sensors because of their high thermal and chemical stability in liquids and excellent compatibility with living cells [1,2].

The conductive channel using the AlGaN/GaN sensor technology forms spontaneously and contains a high-electron density and mobility 2-D electron gas (2-DEG). The 2-DEG is located close to the surface, which makes it extremely sensitive to changes in surface charge density [3]. Thus, highly sensitive, reference electrode-free, AlGaN/GaN-based sensors can be fabricated [4,5]. By growing cells on the gate area of AlGaN/GaN field-effect transistors (FETs), complex biophysical properties of cells, associated with changes in the activity of ion channels, can be studied label-free and in real-time. As biosensors, AlGaN/GaN devices with live cells are not only capable of detecting field potential changes generated by ion transport through the cell membrane, but because of the relationship between gate and channel charge, the change is amplified. Such an approach has important applications for fundamental biomedical research as well as drug discovery and assessment.

Overall this technology has exciting potential for producing cheap, portable, and even personalized sensors capable of multiple measurements on living cells, without destroying them, in complex solutions such as cell culture medium, urine, blood, or saliva.

In the last decade, AlGaN/GaN FET sensors have been used to record cell action potentials, noninvasive cell electrophysiological measurements, and electrical stimulation of cells in vitro. However, the strength of the recorded signal was only in the range of tens to hundreds of microvolts. Moreover, these devices employed a reference electrode as part of the measurement set up.

Podolska et al. [6] reported the detection of live cell biological activity for label-free AlGaN/GaN-based biosensors. Live cells were seeded in solution onto the exposed gate region of packaged AlGaN/GaN FET devices and subsequently subjected to various chemical changes in the solution.

Cell depolarization with KCl was used to induce biological signals. Significant optimization of experimental parameters was performed, including environmental conditions and cell seeding concentration. Selectivity towards biological activity was improved by buffering with 30 mM 4-(2-hydroxyethyl)-1-piperazineethanesulfonic acid (HEPES). Reproducible calcium concentration-dependent responses were observed with saturation at the normal physiological level of around 2.5 mM. Cells in a calcium-free solution were pretreated in separate experiments with calcium ion channel inhibitors like mibefradil, nisoldipine, and HC-030031 and the calcium channel activator S-BayK(–)8644 before calcium dosing. The resulting responses were compared to the depolarization-induced calcium intake and demonstrated physiologically relevant behavior of calcium ion channels.

11.3 Bacterial Sensors

Adanyi et al. [7] reported on the optimization of the immobilization of silicatein-modified *E. coli* BL21AI cells onto the SiO_2-type chips (buffer concentration, pH, temperature, reaction time, etc.) and the study of the biological properties, in particular the inhibitory effect of stressors/environmental pollutants on the novel bacterial sensor in real-time.

Silicateins have been identified and characterized, which form the axial filaments of the spicules of the siliceous sponges, consisting of not only amorphous silica, among others. These enzymes are able to catalyze the polycondensation and deposition of silica at mild conditions. Silicateins can be expressed in *E. coli*. The recombinant proteins are expressed on the surface of the cell wall and are able to catalyze the formation of a polysilicate net around the bacterial cells providing the possibility for further attachment to the surface of SiO_2 containing sensor chips. With this mild immobilization process, it is now possible to prepare novel microbial sensors based on optical waveguide lightmode spectroscopy (OWLS).

The effect of oxidative stress was investigated by exposing the sensors containing biosilica-immobilized *E. coli* BL21AI cells to various concentrations of hydrogen peroxide (H_2O_2). The effect of antibiotics was tested using chloramphenicol (CAP), which is effective against a variety of Gram-positive and Gram-negative bacteria, and penicillin G, which destroys the bacterial cell wall. In addition, the inhibition by carbofuran (CF) pesticide was also tested. CF is a highly toxic compound that inhibits cholinesterase activity. They concluded that the novel bacterial sensor consisting of the silicatein-modified *E. coli*

BL21AI cells immobilized on OWLS sensor surface could be an effective tool to detect the presence of different types of pollutants in real-time measurement.

Living bacteria can be immobilized by physicochemical methods onto the measuring surface of the sensors. The method used for immobilization is very important since it may influence bioactivity and cell viability. Horváth et al. [8] tested a grating-coupled planar optical waveguide sensor by monitoring the adhesion of *E. coli* K12 cells to the sensor surface. Yeh and Ramsden [9] investigated a simple method for determining the number of bacteria adsorbed on a planar optical waveguide. Covalent binding methods that rely on the formation of a stable covalent bond cannot be applied because exposure to harmful chemicals and harsh reaction conditions may damage the bacterial cell membrane and decrease the biological activity [10]. At milder conditions, cells can be entrapped into various membranes/coatings, for example, sol/gel-derived composite materials composed of silica and a grafting copolymer [11]. Premkumar et al. [12] demonstrated the encapsulation of genetically engineered bioluminescent *E. coli* in thick silicate films, while Odaci et al. [13] entrapped the cells together with a chitosan (CS) matrix.

11.4 Surface Plasmon Resonance Biosensors

SPR biosensors are one of the most sensitive optical biosensors widely applied for chemical sensing and biosensing characterizations as reported by Homola et al. [14]. SPR biosensors do not need reagents or labels for the detection of a target analyte; the immobilized biological recognition element (bioreceptor) can be regenerated and reused for continuous or multiple detection, and because the bioreceptor and transducer are integrated into one single sensor, on-site detection can be easily achieved [15]. Although various biological recognition elements are available as receptors (such as enzymes, antibodies, microbes, and organelles), antibodies are widely used as effective binding partners in SPR biosensors.

Therefore, improvements in the immunoaffinity of antibody–antigen reactions can enhance the sensitivity and specificity of SPR biosensors [16].

Wei et al. [17] developed a biosensor based on SPR for the rapid identification of *C. jejuni* in broiler samples. They examined the specificity and sensitivity of commercial antibodies against *C. jejuni* with six *Campylobacter* strains and six non-*Campylobacter* bacterial strains. Antigen–antibody interactions were studied using enzyme-linked immunosorbent assay (ELISA) and a commercially available SPR biosensor platform (Spreeta™). *Campylobacter* cells killed with 0.5% formalin had significant lower antibody reactivity when compared to live cells or cells inactivated with 0.5% thimerosal or heat (70°C for 3 min) using ELISA. The SPR biosensor showed good sensitivity with commercial antibodies against *C. jejuni* at 10^3 CFU mL^{-1} and a low cross-reactivity with *Salmonella* serotype *typhimurium*. The sensitivity of the SPR was similar when testing spiked broiler meat samples. However, research is still needed to reduce the high background observed when sampling meat products.

A novel SPR biosensor using lectin as bioreceptor was developed by Wang et al. [18] for the rapid detection of *Escherichia coli* (*E. coli*) O157:H7. The selective interaction of lectins with carbohydrate components from bacterial cell surface was used as the recognition principle for the detection. Five types of lectins from *Triticum vulgaris, Canavalia ensiformis, Ulex europaeus, Arachis hypogaea, and Maackia amurensis* were employed to evaluate the selectivity of the approach for binding *E. coli* O157:H7 effectively. A detection limit of 3×10^3 CFU mL^{-1} was obtained for the determination of *E. coli* O157:H7 when using the lectin from *T. vulgaris* as the binding molecule. Furthermore, the proposed biosensor was used to detect *E. coli* O157:H7 in real food samples. Results showed that the lectin-based SPR biosensor was sensitive, reliable, and effective for the detection of *E. coli* O157:H7, which holds a great promise in food safety analysis.

11.5 Electrochemical Impedance Spectroscopy (EIS) Biosensor

EIS, which is based on the change of the charge transfer resistance (Rct) using a redox couple, has received much attention due to its high-sensitivity and label-free characteristics.

A novel label-free EIS biosensor for direct cancer cell detection based on the interaction between carbohydrate and lectin has been developed by Hu et al. [19] with good sensitivity and selectivity. In the present work, concanavalin A (ConA), a mannose-specific lectin, was immobilized on a gold disk electrode to fabricate the ConA sensor. This sensor was incubated with the cancer cell sample, and the binding of cancer cells with ConA resulted in the change of Rct.

EIS measurement was employed to measure the impedance change that reveals the concentration of cancer cells. This method has been successfully applied in human liver cancer cell Bel-7404 for direct and sensitive detection with a detection limit of 234 cells mL^{-1}. This method could be extended to carry out multicomponent diagnosis applications, thus providing enormous potential for applications of cancer monitoring and therapy.

11.6 Whole-Cell and Cell Array Biosensors

Whole-cell biocatalysts offer several advantages over free enzymes, including high stability, reduced purification requirements, low preparation cost, and efficient cofactor regeneration [20]. Whole-cell-based biosensors are used for high-throughput drug discovery, clinical diagnosis, and environmental monitoring of chemicals and heavy metals [21–25]. As a key factor in the development of a whole-cell biosensor, the surface immobilization technique can affect both sensitivity and stability [21,26–31]. Physical (entrapment or adsorption) and chemical (cross-linking) methods have been widely used for surface immobilization of cells [32].

A whole-cell array biosensor for the efficient detection of neurotoxic organophosphate compounds (OPs) was developed through the immobilization of recombinant *Escherichia coli* cells containing periplasmic-expressing organophosphorus hydrolase (OPH) onto the surface of a 96-well microplate using mussel adhesive protein (MAP) as a microbial cell-immobilizing linker [33]. Both the paraoxon-hydrolyzing activity and fluorescence microscopy analyses demonstrated that the use of MAP in a whole-cell biosensor increased the cell-immobilizing efficiency and enhanced the stability of immobilized cells compared to a simple physical adsorption-based whole-cell system. Scanning electron microscopic analyses also showed that the *E. coli* cells were effectively immobilized on the MAP-coated surface without any pretreatment steps. The whole-cell array biosensor system, prepared using optimal MAP coating (50 µg cm^{-2}) and cell loading (4 OD$_{600}$), detected paraoxon levels as low as 5 µM with high reproducibility, and its quantitative detection range was 5–320 µM. The MAP-based whole-cell array biosensor showed a good long-term stability for 28 days with 80% retained activity and a reusability of up to 20 times. In addition, paraoxon in tap water was also successfully detected without a reduction in sensitivity. Their results indicate that the proposed MAP-based whole-cell array system could be used as a potential platform for a stable and reusable whole-cell biosensor.

Cell-based sensors represent a subclass of sensor technology systems with an overwhelming development for the measurement of substances in liquid environments over the last two decades.

Cell-based assays are currently considered central to toxicity testing and environmental exposure testing [34–37]. One of the main advantages of cell-based approaches is their possibility not only to detect specific substances but to supply functional information, that is, information about general cytotoxicity. Up to now, sensors incorporating living cells face the drawback of a steady and high risk of contamination and cell death, caused by changes in the cellular environment.

Cell-based sensors for the detection of gases have long been underrepresented, due to the cellular requirement of being cultured in a liquid environment. Bohrn et al. [38] established a cell-based gas biosensor for the detection of toxic substances in air by adapting a commercial sensor chip (Bionas®), previously used for the measurement of pollutants in liquids. Cells of the respiratory tract (A549, RPMI 2650, V79) that survive at a gas phase in a natural context are used as biological receptors. The physiological cell parameters acidification, respiration, and morphology are continuously monitored in parallel. Ammonia was used as a highly water-soluble model gas to test the feasibility of the sensor system. Infrared measurements confirmed the sufficiency of the medium draining method. This sensor system provides the basis for many sensor applications such as environmental monitoring, building technology, and public security.

Hofmann et al. [39] presented the time-resolved detection of chemically induced stress upon intracellular signaling cascades by using genetically modified sensor cells based on the human keratinocyte cell line HaCaT. The cells were stably transfected with a HSP72-GFP reporter gene constructed to create an optical sensor cell line expressing a stress-inducible reporter protein. The time- and dose-dependent performance of the sensor cells is demonstrated and discussed in comparison to a label-free impedimetric monitoring approach (electric cell-substrate impedance sensing (ECIS)). Moreover, a microfluidic platform was established based on µSlidesI^{0,4} Luer to allow for a convenient, sterile, and incubator-independent time-lapse microscopic observation of the sensor cells. Cell growth was successfully achieved in this microfluidic setup, and the cellular response to a cytotoxic substance could be followed in real-time and in a noninvasive, sensitive manner. This study paved the way for the development of micrototal analysis systems that combine optical and impedimetric readouts to enable an overall quantitative characterization of changes in cell metabolism and morphology as a response to toxin exposure. By recording multiple parameters, a detailed discrimination between competing stress- or growth-related mechanisms is possible, thereby presenting an entirely new in vitro alternative to skin irritation tests.

There is a demand in cell culture assay applications (for instance, drug toxicity testing) for biosensors that can monitor metabolite levels continuously, even in unstirred solutions. These microbiosensors need to be of sufficiently small dimension that they exhibit radial diffusional behavior and therefore produce a sustained steady-state current response in the presence of a given analyte concentration; also they do not perturb the system by consuming glucose in the culture medium.

A water-based carbon screen-printing ink formulation, containing the redox mediator cobalt phthalocyanine (CoPC) and the enzyme glucose oxidase (GOx), was investigated for its suitability to fabricate glucose microbiosensors in a 96-well microplate format [40]: (1) the biosensor ink was dip-coated onto a platinum (Pt) wire electrode, leading to a satisfactory amperometric performance; (2) the ink was deposited onto the surface of a series of Pt microelectrodes (10–500 mm diameter) fabricated on a silicon substrate using MEMS (microelectromechanical systems) microfabrication techniques (capillary deposition proved to be successful); a Pt microdisk electrode of Z100 mm was required for optimum biosensor performance; and (3) MEMS processing was used to fabricate suitably sized metal (Pt) tracks and pads onto a silicon 96-well format base chip, and the glucose biosensor ink was screen-printed onto these pads to create glucose microbiosensors. When formed into microwells, using a 340 µL volume of buffer, the microbiosensors produced steady-state amperometric responses that showed linearity up to 5 mM glucose (CV = 6% for n = 5 biosensors). When coated, using an optimized protocol, with collagen in order to aid cell adhesion, the biosensors continued to show satisfactory performance in culture medium (linear range to 2mM, dynamic range to 7 mM, CV = 5.7% for n = 4 biosensors). Finally, the operation of these collagen-coated microbiosensors, in 5-well 96-well format microwells, was tested using a 5-channel multipotentiostat. A relationship between amperometric response due to glucose and cell number in the microwells was observed. These results indicate that microphotolithography and screen-printing techniques can be combined successfully to produce microbiosensors capable of monitoring glucose metabolism in 96-well format cell cultures. The potential application areas for these microbiosensors are discussed.

11.7 Antibody-Immobilized Microcantilever (MC) Resonators

Microcantilever-based mass resonators, with selective coatings for target immobilization, represent one of the most promising classes of label-free biosensor platforms [41]. The change in resonance frequency of the sensor is a function of mass binding to the cantilever surface: the obtained mass resolution is in the nano- to zeptogram range, when vibrational curves are monitored directly in liquid environment (for real-time measurements) or in vacuum conditions (to minimize dumping effects and therefore enhancing mass sensitivity), respectively [42]. Cantilever sensors represent a technological platform adaptable to different molecular recognition approaches and to various diverse biosensing applications, such as identification and quantification of microorganisms, biomolecules, and nucleic acids [43]. Furthermore, such technique was recently demonstrated to be successfully integrable as a diagnostic tool on a Lab-On-Chip (LOC) platform to reduce assay time and limit sampling and/or sample preparation, providing compact and portable objects for biomolecule [44,45] or microorganism identification [46].

Probably in consideration of the high mass sensitivity and precision needed to detect small targets with low molecular weight, works on the successful detection of residues of immunogenic small molecules such as pesticides [47–49], toxins [50], hormones, and antibiotics [51] were carried out using the static approach, where the deflection of the cantilever due to surface stress variation generated by antigen/antibody binding is monitored. However, static mode is often subjected to restrictions, such as stabilization problems due to thermal drift [52], and difficulty to link the deflection change due to specific molecular adsorption to the amount of the adsorbed molecules [53]. This implies that immunodetection by MC resonators is preferred when a quantitative response on molecular interactions is required.

To date, control strategies in detecting anabolic agents for promoting growth of food-producing animals are mainly related to screening techniques based on immunochemical and physiochemical methods, whose major limit is represented by relatively low analytical sensitivity. As a consequence, consumers are currently exposed to molecules with potential carcinogenic effects such as 17b-estradiol, the most powerful substance with estrogenic effect. Therefore, high analytical sensitivity screening and confirmatory methods are required, coupled easiness of use and efficiency. Ricciardi et al. [54] reported on the immunodetection of 17b-estradiol in serum by antibody-immobilized MC resonators, an innovative biosensing platform able to quantify an adsorbed target mass (such as cells, nucleic acids, and biomolecules), thanks to a shift in resonance frequency. Their tool based on MC resonator arrays has shown to be capable of discriminating treated and untreated animals, showing the ability of detecting traces of 17b-estradiol in serum at concentrations lower than the present accepted physiological serum concentration threshold value (40 ppt), and commercial ELISA tests (25 ppt). The method exhibits a limit of detection of 20 ppt and a limited cross-reactivity with high concentrations (10 ppb) of similar molecules (testosterone).

11.8 Enzyme-Based Biosensors and Nanoparticles

The performance of an enzyme-based biosensor relies mostly on the properties of the supporting materials. They should provide a good environment for enzyme immobilization and should maintain their biological activity [55]. Lots of materials have been used for enzyme immobilization including inorganic materials [56], organic materials [57], and biomaterials [58,59]. Biomaterials are more ideal as an enzyme immobilization platform since they are more biocompatible with enzymes. Among these biomaterials, eggshell membrane (ESM) has proved to be an effective and stable enzyme immobilization bioplatform because it does not only maintain the enzyme activity but also extends the shelf-life of the immobilized enzyme [60–62].

Currently the use of NPs in biosensors has become one of the hottest research areas. NPs can offer advantages, such as large surface-to-volume ratio, high surface reaction activity, and strong adsorption ability to immobilize the desired biomolecules [63,64]. Many metal and semiconductor NPs have been applied to prepare modified electrodes [65]. Among these metal NPs, gold nanoparticles (AuNPs) are better conductors and offer good microenvironment for retaining the activity of enzyme [66]. They can bind directly with enzymes without disrupting their biological recognition properties [67]. Nowadays, it is revealed that AuNPs also exhibit excellent catalytic effects on many important chemical reactions [68]. In addition, AuNPs are able to reduce the insulating effect of the protein shell and thus enhance electron transfer in the reaction processes.

The catalytic effect of AuNPs is highly size-dependent [69]. The unique active sites [70] and electronic states [71] of AuNPs can lead to their anomalous catalytic activity although the mechanism is still not fully understood.

Zheng et al. [72] reported on the fabrication and application of a glucose biosensor based on the catalytic effect of AuNPs on enzymatic reaction for blood glucose determination. AuNPs were initially in situ synthesized on the surface of an ESM that was subsequently immobilized with GOx to produce a GOx-AuNPs/ESM. The GOx-AuNPs/ESM was positioned on the surface of an oxygen (O_2) electrode to form a GOx-AuNPs/ESM glucose biosensor. The effects of pH, concentration of phosphate buffer solution, and amount of GOx on the response of the GOx-AuNPs/ESM glucose biosensor were studied in detail. AuNPs on GOx/ESM can improve the calibration sensitivity (30% higher than GOx/ESM without AuNPs) and stability (87.3% of its initial response to glucose after 10-week storage), and shorten the response time (<30 s) of the glucose biosensor. The linear working range for the GOx-AuNPs/ESM glucose biosensor is 8.33 µM–0.966 mM glucose with a detection limit of 3.50 µM (S/N = 3).

The biosensor has been successfully applied to determine the glucose in human blood serum samples and the results compared well to a standard spectrophotometric method commonly used in hospitals. They demonstrated that the developed GOx-AuNPs/ESM glucose biosensor has potential in biomedical analysis.

Unnikrishnan et al. [73] reported a simple electrochemical approach for the immobilization of GOx on reduced graphene oxide (RGO). The immobilization of GOx was achieved in a single step without any cross-linking agents or modifiers. A simple solution phase approach was used to prepare exfoliated graphene oxide (GO), followed by electrochemical reduction to get RGO–GOx biocomposite. The direct electrochemistry of GOx was revealed at the RGO–GOx modified glassy carbon electrode (GCE). The electrocatalytic and electroanalytical applications of the proposed film were studied by cyclic voltammetry (CV) and amperometry. It is notable that the glucose determination has been achieved in mediator-free conditions. RGO–GOx film showed very good stability, reproducibility, and high selectivity. The developed biosensor exhibits excellent catalytic activity towards glucose over a wide linear range of 0.1–27 mM with a sensitivity of 1.85 µA mM^{-1} cm^{-2}. The facile and easy electrochemical approach used for the preparation of RGO–GOx may open up new horizons in the production of cost-effective biosensors and biofuel cells.

Qi et al. [74] presented the synthesis of bacteria-mediated bioimprinted films for selective bacterial detection. Marine pathogen sulfate-reducing bacteria (SRB) were chosen as the template bacteria. CS doped with reduced graphene sheets (RGSs) was electrodeposited on an indium tin oxide electrode, and the resulting RGSs–CS hybrid film served as a platform for bacterial attachment. The electrodeposition conditions were optimized to obtain RGSs–CS hybrid films with excellent electrochemical performance. A layer of nonconductive CS film was deposited to embed the pathogen, and acetone was used to wash away the bacterial templates. EIS was performed to characterize the stepwise modification process and monitor the SRB population. Faradic impedance measurements revealed that the Rct increased with increased SRB concentration. A linear relationship between DRct and the logarithm of SRB concentration was obtained within the concentration range of 1.0×10^4–1.0×10^8 cfu mL^{-1}. The impedimetric sensor showed good selectivity towards SRB based on size and shape. Hence, the selectivity for bacterial detection can be improved if the bioimprinting technique is combined with other bio-recognition elements.

High level of oxidative stress is involved in the formation of incipient tumor and carcinomatous cells. Chang et al. [75] has explored a facile strategy to assess the oxidative stress elicited by H_2O_2 in cells with amperometric current–time technique in vitro. An electrochemical biosensor exhibiting high-sensitivity and selectivity to H_2O_2 is fabricated by the integration of grapheme with AuNPs and poly(toluidine blue O) films. The efflux of H_2O_2 from several representative tumor cells and normal cells on exposure to ascorbic acid could be detected by using the graphene-based nanocomposite films. The results indicate that tumor cells release much more H_2O_2 than do the normal cells. The novel sensor raises the possibility for clinical diagnostic application to evaluate the higher level of intracellular oxidative stress of tumor cells in comparison with normal cells.

A novel biocompatible film assembled by combining GO and poly-L-lysine (PLL) for adhesion and electrochemical impedance detection of leukemia K562 cells was proposed by Zhang et al. [76]. The biocompatible film showed an improved immobilization capacity for living cells and a good biocompatibility for preserving the activity of the immobilized living cells. The immobilized K562 cells on the biocompatible film-modified electrode can be directly monitored with EIS in the presence of $[Fe(CN)6]^{3-/4-}$ as redox probes. A highly sensitive electrochemical impedance method for the detection of leukemia K562 cancer cells was developed. Under the optimized conditions, the increased electron-transfer resistance (Ret) with a good correlation to the logarithmic value of concentration of K562 cells ranging from 10^2 to 10^7 cells mL^{-1}, and with the detection limit of 30 cells mL^{-1} (S/N = 3). Additionally, the proposed method was used to describe the viability of cells and to evaluate the effectiveness of the antitumor drug nilotinib on K562 cells. The obtained results of nilotinib cytotoxicity are well agreed with those from WST-1 assays. Furthermore, the work demonstrates that a highly biocompatible film of PLL/GO assembled is also expected to be an appropriate matrix for the electrochemical investigation of adhesion, proliferation, apoptosis of other relevant mammalian cells that is not limited to adherent cells, and the study of cell-based biosensors.

11.9 Aptamers

The discovery of aptamer, especially those selected for binding tumor markers or cancer cells, may provide a potential solution for early diagnosis and therapy for cancers [77,78]. Aptamer-based detection may not only exhibit desirable selectivity and specificity but also show some distinguished advantages, such as ease in molecular design and modification, and the capability of passing through some barriers due to its low molecular weight [79,80]. Therefore, based on the discovery of deoxyribonucleic acid (DNA) aptamers targeting mucin 1 (MUC1) [81], which is a glycoprotein expressed on the apical surface of epithelial cells and whose overexpression is often associated with some kinds of cancers, Gendler (2001) and coworkers [82] and some others have developed a few methods to detect MUC1-overexpressed breast cancer cells, in which MUC1-binding aptamers and different measuring techniques have been well applied [83–86].

Breast cancer is one of the most critical threats to the health of women, and the development of new methods for early diagnosis is urgently required, so Zhu et al. [87] reported a method to detect Michigan cancer foundation-7 (MCF-7) human breast cancer cells with considerable sensitivity and selectivity by using electrochemical technique. In this method, a MUC1-binding aptamer is adopted to recognize MCF-7 human breast cancer cells, while enzyme labeling is employed to produce amplified catalytic signals. The molecular recognition and the signal amplification are elaborately integrated by fabricating an aptamer–cell–aptamer sandwich architecture on an electrode surface, thus a biosensor for the detection of MCF-7 is fabricated based on the architecture. The detection range can be from 100 to 1×10^7 cells, and the detection limit can be as low as 100 cells. The method is also cost-effective and conveniently operated, implying potential help for the development of early diagnosis of breast cancer.

There has been great progress in the development of functional DNA-based sensors for the detection of metal ions. However, many functional DNAs are vulnerable to hydrolysis by nucleases in human blood. In addition, the detection methods that are based on DNA often exhibit interference due to the high blood concentrations of other ions, such as K^+ and Na^+. Therefore, Chung et al. [88] selected highly Pb^{++}-specific DNA-aptamer sequences based on circular dichroism (CD) spectroscopy of 4G-rich DNA sequences and Hg^{2+}-specific T-rich DNA sequences and immobilized them on AuNPs for the simultaneous detection of Pb^{2+} and Hg^{2+} in human serum. They used AuNPs because these have a superior fluorescence-quenching efficiency over a broad range of wavelengths compared with other organic quenchers. In addition, AuNPs have a stabilizing effect on the immobilized DNA, which makes it more resistant to degradation by nucleases than free DNA. As a result, even in the presence of deoxyribonuclease (DNase), they were able to simultaneously detect Pb^{2+} and Hg^{2+} in serum at concentrations as low as 128 and 121 pM, respectively, within 10 min. These detection limits for Pb^{2+} and Hg^{2+} were 39-fold and 26.4-fold lower, respectively, than the detection limits that were obtained using free DNAs. Given the multicolor-fluorescence-quenching capability of the AuNPs and the possibility of developing functional nucleic acids for the detection of other metal ions, this study extends the application of oligonucleotides to a point-of-care detection system for the detection of multiple harmful metal ions in body fluids.

Sensitive, reliable, and simple detection of sequence-specific DNA-binding proteins (DBPs) is of paramount importance in the area of proteomics, genomics, and biomedicine. Liu et al. [89] describes a novel fluorescent-amplified strategy for ultrasensitive, visual, quantitative, and "turn-on" detection of DBP. A Forster resonance energy transfer (FRET) assay utilizing a cationic conjugated polymer (CCP) and an intercalating dye was designed to detect a key transcription factor, nuclear factor-kappa B (NF-kB), the model target. A series of label-free DNA probes bearing one or two protein-binding sites (PBS) were used to identify the target protein specifically. The binding DBP protects the probe from digestion by exonuclease III, resulting in high-efficient FRET due to the high affinity between the intercalating dye and duplex DNA, as well as strong electrostatic interactions between the CCP and DNA probe. By using label-free hairpin DNA or double-stranded DNA containing two PBS as probe, they could detect as low as 1 pg μL^{-1} of NF-kB in HeLa nuclear extracts, which is 10,000-fold more sensitive than the previously reported methods. The approach also allows naked-eye detection by observing the fluorescent color of solutions with the assistance of a handheld UV lamp. Additionally, a less than 10% relative standard deviation was obtained, which offers a new platform for superior precision and low-cost and simple detection of DBP. The features of this optical biosensor show promising potential for early diagnosis of many diseases and high-throughput screening of new drugs targeted to DBPs.

11.10 Luminescent Biosensors

The development of methods for illuminating key components of the cellular networks requires the application of sophisticated tools. One novel approach for creating such powerful tools relies on structurally engineered reporter proteins to create intracellular sensors. The signal emitted by these sensors is regulated by their interaction with cellular components [90]. One of the important procedures that has been developed by several groups in the past decade is the use of split-protein reassembly (also called protein-fragment complementation), wherein initially nonfunctional fragments of split-protein are induced to reassemble through the specific interactions [91–93]. For the first time, split-protein reassembly was reported on ubiquitin, which showed reassembly of fragments upon interaction with potential partners [94].

Firefly luciferase is one of the most promising reporter protein that has been validated for split-protein strategies because of its applicability in imaging of living animals. Moreover, it requires no high-energy excitation source because light emission results from an endogenous chemical reaction and thus does not suffer from background noise and tissue damage in signal measurements [95].

Inositol 1,4,5-trisphosphate (IP3) is a crucial second messenger that regulates complicated signaling processes in various physiological events. Alteration in its content has been observed in many diseases. Hence, the development of a high-throughput screening system to monitor temporal changes of IP3 is essential for screening new potential therapeutic compounds. Toward a simple, sensitive, and rapid method for measuring IP3, Ataei et al. [96] describe the development and application of a novel biosensor based on luciferase fragment-assisted complementation strategy, which converts the ligand-induced conformational changes to light. Designed sensor comprising the IP3-binding core domain of IP3-receptor fused between complementary nonfunctional fragments of firefly luciferase allows direct detection of IP3 in the presence of luciferin substrate both in cell lysate and in living cells. According to the results presented, the screening time was very fast and maximum response was obtained up to 11-fold higher than untreated cells. Moreover, the designed biosensor was able to monitor the release of IP3 upon the induction by different inducers like bradykinin and ATP. The current biosensor not only provides a specific IP3 detector in vitro but also facilitates monitoring of the response of IP3 in living organisms.

Rhodamine is one of the most attractive fluorochromes because of its photo physical properties [97]. Recently, much effort has been focused on the development of rhodamine-based chemosensors [98–102] and polymeric chemosensors [103,104] for the detection of heavy metal ions.

Since Fe^{3+} is a fluorescence quencher because of its paramagnetic nature [105], the development of chemosensors that exhibit fluorescence enhancement upon binding with Fe^{3+} would be very much attractive. According to the theory of hard and soft acids and bases (HSAB), a stable complex is formed between a hard base and a hard acid such as Fe^{3+} ions. This theory offers a possibility to develop ligands with improved selectivity toward a particular metal ion depending upon the strengths of HSAB. The nature and number of the external chelating moieties incorporated with the rhodamine [106,107] play an important role for tuning metal ion selectivity.

The number and nature of coordinating entities as well as the size of chelating cavity in rhodamine-based chemosensors were tuned to enhance the selectivity and sensitivity of Fe^{3+} ions. An intense pink color development and enhancement in fluorescence emission intensity of chemosensor 5 upon complex formation at pH 7.4 enabled the detection of Fe^{3+} ions in the presence of other competitive metal ions like Li^+, Na^+, K^+, Cs^+, Mg^{2+}, Ca^{2+}, Sr^{2+}, Cr^{3+}, Mn^{2+}, Fe^{2+}, Cu^{2+}, Co^{2+}, Ni^{2+}, Zn^{2+}, Cd^{2+}, Hg^{2+}, and Pb^{2+}. A plausible application of chemosensor 5 in the imaging of live fibroblast cells exposed to Fe^{3+} ions is also demonstrated [108].

11.11 Photonic Crystal Microcavity Biosensors

Chakravarty et al. [109] experimentally demonstrated label-free photonic crystal (PC) microcavity biosensors in silicon-on-insulator (SOI) to detect the epithelial–mesenchymal transition (EMT) transcription factor, ZEB1, in minute volumes of sample. Multiplexed specific detection of ZEB1 in lysates from NCI-H358 lung cancer cells down to an estimated concentration of 2 cells μL^{-1} is demonstrated. L13 photonic crystal microcavities, coupled to W1 photonic crystal waveguides, are employed in which resonances show high Q in the bio-ambient phosphate-buffered saline (PBS). When the sensor surface is derivatized with a specific antibody, the binding of the corresponding antigen from a complex whole-cell lysate generates a change in refractive index in the vicinity of the photonic crystal microcavity, leading to a change in the resonance wavelength of the resonance modes of the photonic crystal microcavity. The shift in the resonance wavelength reveals the presence of antigen. The sensor cavity has a surface area of ~11 mm². Multiplexed sensors permit simultaneous detection of many binding interactions with specific immobilized antibodies from the same biosample at the same instant. Specificity was demonstrated using a sandwich assay that further amplifies the detection sensitivity at low concentrations. The device represents a proof-of-concept demonstration of label-free, high-throughput and, multiplexed detection of cancer cells with specificity and sensitivity on a silicon chip platform.

11.12 Nanomaterials and Biosensors

In order to apply biosensors to physiological research, miniaturization of sensors is required for a high spatial resolution, because the sensors are often operated at cell or tissue level. Miniaturization increases the resistance of the sensor, which significantly decreases the maximum attainable sensitivity [110]. The sensitivity issue affects not only the limit of detection but also the capability of measuring very small changes in concentration over time [111], while the small changes can be key to exploring important physiological phenomena, such as β cell glucose consumption during insulin secretion [112]. One effective way to solve the low-sensitivity problem is to enhance the electrochemical transduction via incorporating nanomaterials.

Physiological studies require sensitive tools to directly quantify transport kinetics in the cell/tissue spatial domain under physiological conditions. Although biosensors are capable of measuring concentration, their applications in physiological studies are limited due to the relatively low-sensitivity, excessive drift/noise, and inability to quantify analyte transport. Nanomaterials significantly improve the electrochemical transduction of microelectrodes and make the construction of highly sensitive microbiosensors possible. Furthermore, a novel biosensor modality, self-referencing (SR), enables the direct measurement of real-time flux and drift/noise subtraction. SR microbiosensors based on nanomaterials have been used to measure the real-time analyte transport in several cell/tissue studies coupled with various stimulators/inhibitors. These studies include glucose uptake in pancreatic B cells, cancer cells, muscle tissues, intestinal tissues, and *P. aeruginosa* biofilms; glutamate flux near neuronal cells; and endogenous indole-3-acetic acid flux near the surface of *Zea mays* roots. Results from the SR studies provide important insights into cancer, diabetes, nutrition, neurophysiology, environmental, and plant physiology studies under dynamic physiological conditions, demonstrating that the SR microbiosensors are an extremely valuable tool for physiology research [113].

11.13 Microwire Sensors

Lu et al. [114] developed a label-free biosensor based on electrochemical impedance measurement followed by dielectrophoretic force and antibody–antigen interaction for detection and quantification of foodborne pathogenic bacteria. In their previous work, gold–tungsten wires (25 µm in diameter) were functionalized by coating with polyethyleneimine–streptavidin–anti-*Escherichia coli* antibodies to improve the sensing specificity, and fluorescence intensity measurement was employed to quantify bacteria captured by the sensor. The focus of this research was to evaluate the performance of the developed biosensor by monitoring the changes of electron-transfer resistance (DRet) of the microwire after the bioaffinity reaction between bacterial cells and antibodies on its surface as an alternative quantification technique to fluorescence microscopy. EIS has been used to detect and validate the resistance changes in a conventional three-electrode system in which $[Fe(CN)_6^{3-}] / [Fe(CN)_6^{4-}]$ served as the redox probe. The impedance data demonstrated a linear relationship between the increments of DRet and the logarithmic concentrations of *E. coli* suspension in the range of 10^3–10^8 CFU mL^{-1}. In addition, there were little changes of DRet when the sensor worked with *Salmonella*, which clearly evidenced the sensing specificity to *E. coli*. EIS was proven to be an ideal alternative to fluorescence microscopy for enumeration of captured cells.

11.14 Oligoaziridine Biosensors

Since polyaziridines are known to be good metal chelating agents (Rivas et al. [115]) and can be easily synthesized using supercritical carbon dioxide (scCO2) as a clean and sustainable technology, a fully integrated sensing system comprising both pyrene and polyaziridine's potential was envisaged.

A water-soluble biocompatible aziridine-based biosensor with pendant anthracene units was synthesized by radical polymerization of N-substituted aziridines in scCO2 [116]. The binding ability of the sensor towards a series of metal ions was examined by comparing the fluorescence intensities of the solutions before and after the addition of 100 equivalents of a solution of the metal ion chloride salt. A fast, simple, and highly optical sensitive dual behavior, "off–on" and "on–off" response, was observed after the biosensor was exposed to the metal cations in aqueous solution. Zinc presented the highest fluorescence enhancement (turn on), and copper presented the highest fluorescence quenching (turn off). The response time was found to be instantaneous, and the detection limit was achieved even in the presence of excess metal cation competitors. By using immunofluorescence microscopy, it was also shown that oligoaziridine acts as an "on–off" probe through highly sensitive (detection limit of 1.6 nM), selective, and reversible binding to copper anions under physiologic conditions using live human fibroblast cells. The stoichiometry for the action of the biosensor with Cu^{2+} was determined by a job plot and it indicates the formation of an oligoaziridine–Cu^{2+} 1:2 adduct.

11.15 Use of PDIF-CN2 Molecules in the Development of n-Type Organic Field-Effect Transistors for Biosensing Applications

Organic electronics may represent an important new tool for the analysis of structures ranging from single molecules to cellular events. Specifically, organic field-effect transistors (OFET) are potentially powerful devices for the real-time detection/transduction of biosignals. Despite this interest up-to-date, the experimental data useful to support the development of OFET-based biosensors are still few, and in particular, n-type (electron-transporting) devices, being fundamental to develop highly performing circuits, have been scarcely investigated.

Barra et al. [117] made films of N,N'-1H,1H-perfluorobutyldicyanoperylene-carboxydi-imide (PDIF-CN2) molecules, a recently introduced and very promising n-type semiconductor, and these have been evaporated on glass and silicon dioxide substrates to test the biocompatibility of this compound and its capability to stay electrically active even in liquid environments. They found that PDIF-CN2 transistors can work steadily in water for several hours. Biocompatibility tests, based on in vitro cell cultivation, remarked the need to functionalize the PDIF-CN2 hydrophobic surface by extra-coating layers (i.e., PLL) to favor the growth of confluent cellular populations.

They demonstrated that PDIF-CN2 compound is an interesting organic semiconductor to develop electronic devices to be used in the biological field. This work contributes to define a possible strategy for the fabrication of low-cost and flexible biosensors, based on complex organic complementary metal-oxide-semiconductor (CMOS) circuitry including both p- (hole-transporting) and n-type transistors.

11.16 Giant Magnetoresistance (GMR) Biosensor

Cell-surface interactions such as cell spreading and phagocytosis represent important aspects in biology and are of special interest for biomedical applications. Adherent cells like fibroblasts continually probe their environment, and they need to attach to and spread on an underlying surface in order to perform numerous biological functions.

Apart from conventional optical end-point detection methods, other optical techniques such as internal reflection microscopy (IRM) or total internal reflection fluorescence microscopy (TRIFM) are employed to follow the process of cell spreading with high spatial and temporal resolution [118–120]. In addition to the visualized cell/surface contact area, the separation distance between the cell and the substrate surface can also be quantified. Shoshi et al. [121] investigated the phagocytic behavior of human fibroblast cells during their spreading process on particle-immobilized sensor surfaces. Special focus is put on the susceptibility of cells during their adhesion process to the cell membrane's competitive mechanism of phagocytosis. To monitor the influence of particle uptake on the spreading characteristics in real-time, they employed a previously introduced magnetic approach based on giant magnetoresistance (GMR) sensors, magnetic particles, and microfluidics [122]. Research on real-time monitoring of cell-surface interactions at nanoscale level is of high importance in cellular biophysics, material science, and the development of future biomaterials for biomedical applications [123].

11.17 Rolling Circle Amplification Biosensors– DNA Hybridization Biosensors

In recent years, an alternative nucleic acid amplification technique called rolling circle amplification (RCA), was adapted to the detection of DNA instead. This amplification method is achievable at constant temperature (e.g., 60°C) simply by mixing circular single-stranded DNA (ssDNA) probe, DNA polymerase, and nicking enzyme. Unlike conventional nucleic acid amplification reactions such as polymerase chain reaction (PCR), this reaction does not require exogenous primer, which often causes primer dimerization or nonspecific amplification [124]. RCA is a simple isothermal enzymatic process that can be used to generate very long ssDNA molecules with tandem repeats [125,126].

The robustness, high potential, and simplicity of RCA soon made it a powerful DNA diagnostic technology among other isothermal amplification techniques for probe/signal amplification.

An electrochemical biosensor based on RCA and NP aggregates for highly sensitive identification of DNA and cancer cells was established by Ding et al. [127]. First, a "sandwich-type" DNA complexes containing target DNA were constructed on the surface of the magnetic beads. Second, one part of the primer in the "sandwich-type" DNA complexes induced the RCA in the system. Then the long RCA products were digested to construct another "sandwich-type" DNA complex for the electrochemical detection. Differential pulse voltammetry (DPV) peaks with high signal intensity were obtained, and the signal intensities were found to be dependent on the amount of Fc, which is related to the concentration of target DNA. Under optimum conditions, the electrochemical signal intensity increased with the increase in the concentration of target DNA. A detection limit of 2.8×10^{-18} M of target DNA was achieved. Combined with aptamer technology, the proposed signal amplification strategy was also used for the identification of cancer cells with a detection limit of 100 Ramos cells mL^{-1}.

A sensitive, fast, and inexpensive method for direct electrochemical detection of target DNA sequences in nonamplified genomic DNA samples is described by Alipour et al. [128]. Hybridization detection relies on the alteration in guanine oxidation signal following hybridization of the probe with complementary genomic DNA. Initially, the method was tested to detect target DNA on low cycle number PCR amplicons. Having obtained promising detection results from only a five-cycle product, the feasibility of target sequence detection in extracted genomic DNA without PCR amplification, but with the vortex-mediated fragmentation of the large genomic DNA into small pieces, was examined. Experimental variables affecting the efficiency of sensor were investigated. Detection experiments with various noncomplementary genomic

DNAs as well as a proper probe, nonspecific with respect to all genomic samples, confirmed the excellent selectivity of the approach. The sensitivity of the method for analyzing the vortex-mediated fragmentized genomic DNA samples is estimated to be approximately 0.58 ng μL^{-1}.

11.18 Fluorescence Quenching

A powerful antibody-based fluorescent biosensor strategy called Quenchbody has been described by Abe et al. [129]. The Quenchbody works on the principle of antigen-dependent removal of quenching effect on a fluorophore that had been quenched by intrinsic tryptophan (Trp) residues of a single-chain antibody variable region (scFv). This Quenchbody-based assay is superior than previous methods. First, it does not require any additional reagents to detect the target antigen but only needs Quenchbody itself, with some nonessential ingredients such as bovine serum albumin (BSA) and nonionic surfactant to protect the Quenchbody from nonspecific adhesion to the cuvette. Second, the method is markedly rapid and simple, since it does not need either immobilization or washing steps but only needs adding the sample directly to the Quenchbody reagent to measure its fluorescence. Finally, the technology can be applied to the detection of a range of antigens from small haptens to large proteins with the same assay principle.

Jeong et al. [130] described the detection of vimentin phosphorylation at Ser71 (PS71) and Ser82 (PS82) based on the Quenchbody technology. Protein phosphorylation is a key event in intracellular signal transduction, and fluorescent biosensor for the specific phosphorylation event in a target protein is considered highly useful as a tool of cellular biology and drug screening. PS71 and PS82 are important vimentin phosphorylation targets that occur during anaphase and metaphase of mitosis, respectively. First, they investigated new fluorescent dyes that can be incorporated for making Quenchbodies. In addition to carboxytetramethylrhodamine (TAMRA) used previously, two rhodamine dyes, rhodamine 6G (R6G) and ATTO 520 that share similar molecular structure, were successfully incorporated to anti-BGP Quenchbody. The fluorescence intensity of the resultant Quenchbodies showed marked increases in an antigen concentration-dependent way, suggesting their utility as a label of Quenchbody. Then they tried to detect PS71 and PS82 of vimentin with the corresponding Quenchbodies and optimized reaction conditions as well as the Quenchbody structure itself. This study presented the first detection of protein serine phosphorylation with an antibody-based fluorescent biosensor.

Due to its simplicity, the Quenchbody-based phosphorylation biosensors will be widely applicable to in vitro diagnostics, drug screening, and imaging in a rapid, simple, and highly sensitive manner.

11.19 Membrane Engineering

An alternative approach to cell transfection is the recently developed technology of membrane engineering as reported by Kintzios [131]. It is a generic methodology of artificially inserting (usually by electroinsertion) tens of thousands of receptor molecules on the cell surface, thus rendering the cell a selective responder against analyte binding to the inserted receptors. Receptor molecules can vary from antibodies to enzymes to polysaccharides. The working assumption of the method is that attachment of the target molecule to its respective receptor causes a change in the cell membrane structure, which is measurable as a change in the cell membrane potential.

Kintzios previously reported [132] the first application of this technology for the construction of an ultrasensitive electrophysiological superoxide sensor, which was based on membrane engineered mammalian cells immobilized in an alginate matrix. The membrane engineering process involved the electroinsertion of superoxide dismutase (SOD) molecules in the membranes of Vero fibroblast cells, which acted as catalytic units able to convert O_2^- to H_2O_2. Superoxide dismutation triggered changes to the cell membrane potential (partly effected by a rapid decrease of intracellular Ca^{2+} concentration) that were measured by appropriate microelectrodes, according to the principle of the bioelectric recognition assay (BERA). The sensor instantly responded to picomole concentrations of O_2^- with a detection limit (S/N = 3) of 100 pM. This sensor was further applied to the measurement of superoxide concentrations during the developmental cycle of duckweed (*Spirodela polyrhiza*) [133].

Membrane engineering is a generic methodology for increasing the selectivity of a cell biosensor against a target molecule by electroinserting target specific receptor molecules on the cell surface. Moschopoulou et al. [134] have previously reported the construction of an ultrasensitive superoxide anion $\left(O_2^-\right)$ sensor based on immobilized cells, which have been membrane engineered with SOD. In the present study, they provided evidence that superoxide dismutation triggered changes to the membrane potential of membrane engineered fibroblast cells, as confirmed by electrophysiological and fluorescence assays. These changes were associated with changes in $[Ca^{2+}]_{cyt}$, as revealed by the selective inhibition of intracellular calcium ion traffic. In addition, by conducting selective inhibition assays, they showed that electroinserted SOD molecules retained their characteristic catalytic properties. They also investigated the effect of the concentration of electroinserted SOD molecules on the performance of the superoxide assay. Finally, they increased the sensitivity of the sensor by 100-fold to a detection limit of 1 pM O_2^- by changing the concentration of immobilized cells on the performance of the biosensor.

11.20 Sulfur-Oxidizing Bacteria (SOB) Biosensors

Recently, sulfur-oxidizing bacteria (SOB) is used for the continuous monitoring of toxicity and their use circumvents many of the obstacles associated with conventional assays [135–137]. SOB, first identified by Sergei Winogradsky in 1885, are chemoautotrophic bacteria that oxidize reduced sulfur compounds to sulfuric acid in the presence of O_2.

The environmental risk assessment of toxic chemicals in stream water requires the use of a low-cost standardized toxicity bioassay. Hassan et al. [138] studied a biosensor for the detection of toxic chemicals in stream water using SOB in continuous mode. The biosensor depends on the ability of SOB to oxidize sulfur particles under aerobic conditions to produce sulfuric acid. The reaction results in an increase in electrical conductivity (EC) and a decrease in pH. The biosensor is based on the inhibition of SOB in the presence of toxic chemicals by measuring changes in EC and pH. They found that the SOB biosensor can detect Cr^{6+} at a low concentration (50 ppb), which is lower than many whole-cell biosensors. The effect of organic material in real stream water on SOB activity was studied. Due to the presence of mixotrophic SOB, they found that the presence of organic matter increases the SOB activity, which decreases the biosensor start-up period. Low alkalinity (22 mg L^{-1} $CaCO_3$), increased effluent EC, and decreased effluent pH, which are optimal for biosensor operation. While at high alkalinity (820 mg L^{-1} $CaCO_3$), the activity of SOB decreased a little. They found that the system can detect 50 ppb of Cr^{6+} at low alkalinity (22 mg L^{-1} $CaCO_3$) in a few hours, while complete inhibition was observed after 35 h of operation at high alkalinity (820 mg L^{-1} $CaCO_3$).

References

1. I. Cimalla, F. Will, K. Tonisch, M. Niebelschótz, V. Cimalla, V. Lebedev et al., AlGaN/GaN biosensor—Effect of device processing steps on the surface properties and biocompatibility, *Sensors and Actuators B: Chemical* 123 (2007) 740–748.
2. A. Podolska, S. Tham, R.D. Hart, R.M. Seeber, M. Kocan, M. Kocan, U.K. Mishra, K.D.G. Pfleger, G. Parish, and B.D. Nener, Biocompatibility of semiconducting AlGaN/GaN material with living cells, *Sensors and Actuators B: Chemical* 169 (2012) 401–406.
3. O. Ambacher, B. Foutz, J. Smart, J.R. Shealy, N.G. Weimann, K. Chu et al., Two dimensional electron gases induced by spontaneous and piezoelectric polarization in undoped and doped AlGaN/GaN heterostructures, *Journal of Applied Physics* 87 (2000) 334.
4. A.B. Encabo, J. Howgate, M. Stutzmann, M. Eickhoff, and M.A. Sanchez-Garcva, Ultrathin GaN/AlN/GaN solution-gate field effect transistor with enhanced resolution at low source-gate voltage, *Sensors and Actuators B: Chemical* 142 (2009) 304–307.

5. A. Podolska, M. Kocan, A.M.G. Cabezas, T.D. Wilson, G.A. Umana-Membreno, B.D. Nener, G. Parish, S. Keller, and U.K. Mishra, Ion versus pH sensitivity of ungated AlGaN/GaN heterostructure-based devices, *Applied Physics Letters* 97 (2010) 012108-1–012108-3.

6. A. Podolska, L.C. Hool, K.D.G. Pfleger, U.K. Mishra, G. Parish, and B.D. Nener, AlGaN/GaN-based biosensor for label-free detection of biological activity, *Sensors and Actuators B: Chemical* 177 (2013) 577–582.

7. N. Adanyi, Z. Bori, I. Szendro, K. Erdelyi, X. Wang, H.C. Schroder, and W.E.G. Muller, Bacterial sensors based on biosilica immobilization for label-free owls detection, *New Biotechnology* (2013) doi:10.1016/j.nbt.2013.01.006.

8. R. Horváth, H.C. Pedersen, N. Skivesen, D. Selmeczi, and N.B. Larsen, Optical waveguide sensor for on-line monitoring of bacteria, *Optics Letters* 28 (2003) 1233–1235.

9. Y.P. Yeh and J.J. Ramsden, Quantification of the number of adsorbed bacteria on an optical waveguide, *The Journal of Physical Chemistry* 10 (2010) 53–54.

10. Y. Lei, W. Chen, and A. Mulchandani, Microbial biosensors, *Analytica Chimica Acta* 568 (2006) 200–210.

11. J. Jia, M. Tang, X. Chen, L. Qi, and S. Dong, Co-immobilized microbial biosensor for BOD estimation based on sol–gel derived composite material, *Biosensors and Bioelectronics* 18 (2003) 1023–1029.

12. J.R. Premkumar, R. Rosen, S. Belkin, and O. Lev, Sol–gel luminescence biosensors: Encapsulation of recombinant *E. coli* reporters in thick silicate films, *Analytica Chimica Acta* 462 (2002) 11–23.

13. D. Odaci, S. Timur, and A. Telefoncu, Bacterial sensors based on chitosan matrices, *Sensors and Actuators B: Chemical* 134(1) (2008) 89–94.

14. J. Homola, S.S. Yee, and G. Gauglitz, Surface plasmon resonance sensor: Review, *Sensors and Actuators B: Chemical* 54 (1999) 3–15.

15. J.G. Quinn, S. O'Neill, A. Doyle, C. McAtamney, D. Diamond, B.D. MacCraith, and R. O'Kennedy, Development and application of surface plasmon resonance-based biosensors for the detection of cell–ligand interactions, *Analytical Biochemistry* 281 (2000) 135–143.

16. R.E. Mandrell and M.R. Wachtel, Novel detection techniques for human pathogens that contaminate poultry, *Current Opinion in Biotechnology* 10 (1999) 273–278.

17. D. Wei, O.A. Oyarzabal, T.-S. Huang, S. Balasubramanian, S. Sista, and A.L. Simonian, Development of a surface plasmon resonance biosensor for the identification of *Campylobacter jejuni*, *Journal of Microbiological Methods* 69 (2007) 78–85.

18. Y. Wang, Z. Ye, C. Si, and Y. Ying, Monitoring of *Escherichia coli* O157:H7 in food samples using lectin based surface plasmon resonance biosensor, *Food Chemistry* 136 (2013) 1303–1308.

19. Y. Hu, P. Zuo, and B.-C. Ye, Label-free electrochemical impedance spectroscopy biosensor for direct detection of cancer cells based on the interaction between carbohydrate and lectin, *Biosensors and Bioelectronics* 43 (2013) 79–83.

20. C. de Carvalho, Enzymatic and whole cell catalysis: Finding new strategies for old processes, *Biotechnology Advances* 29 (2011) 75–83.

21. J. Kumar and S.F. D'Souza, Microbial biosensor for detection of methyl parathion using screen printed carbon electrode and cyclic voltammetry, *Biosensors and Bioelectronics* 26 (2011) 4289–4293.

22. Q. Liu, H. Cai, Y. Xu, L. Xiao, M. Yang, and P. Wang, Detection of heavy metal toxicity using cardiac cell-based biosensor, *Biosensors and Bioelectronics* 22 (2007) 3224–3229.

23. Q. Liu, H. Yu, Z. Tan, H. Cai, W. Ye, M. Zhang, and P. Wang, In vitro assessing the risk of drug-induced cardiotoxicity by embryonic stem cell-based biosensor, *Sensors and Actuators B: Chemical* 155 (2011) 214–219.
24. P. Mulchandani, W. Chen, A. Mulchandani, J. Wang, and L. Chen, Amperometric microbial biosensor for direct determination of organophosphate nerve agents using recombinant *Moraxella sp.* with surface expressed organophosphorus hydrolase, *Biosensors and Bioelectronics* 16 (2001) 433–437.
25. A. Perdikaris, N. Vassilakos, I. Yiakoumettis, O. Kektsidou, and S. Kintzios, Development of a portable, high throughput biosensor system for rapid plant virus detection, *Journal of Virological Methods* 177 (2011) 94–99.
26. G.S. Alvarez, M.L. Foglia, G.J. Copello, M.F. Desimone, and L.E. Diaz, Effect of various parameters on viability and growth of bacteria immobililzed in sol-gel-derived silica matrices, *Applied Microbiology and Biotechnology* 82 (2009) 639–646.
27. T. Braschler, R. Johann, M. Heule, L. Metref, and P. Renaud, Gentle cell trapping and release on a microfluidic chip by in situ alginate hydrogel formation, *Lab on a Chip* 5 (2005) 553–559.
28. A.C. Jen, M.C. Wake, and A.G. Mikos, Hydrogels for cell immobilization, *Biotechnology and Bioengineering* 50 (1996) 357–364.
29. S.K. Jha, M. Kanungo, A. Nath, and S.F. D'Souza, Entrapment of live microbial cells in electropolymerized polyaniline and their use as urea biosensor, *Biosensors and Bioelectronics* 24 (2009) 2637–2642.
30. J. Kumar and S.F. D'Souza, An optical microbial biosensor for detection of methyl parathion using Sphingomonas sp. immobilized on microplate as a reusable biocomponent, *Biosensors and Bioelectronics* 26 (2010) 1292–1296.
31. J. Kumar and S.F. D'Souza, Immobilization of microbial cells on inner epidermis of onion bulb scale for biosensor application, *Biosensors and Bioelectronics* 26 (2011) 4399–4404.
32. S.F. D'Souza, Microbial biosensors, *Biosensors and Bioelectronics* 16 (2001) 337–353.
33. C.S. Kim, B.-H. Choi, J.H. Seo, G. Lim, and H.J. Cha, Mussel adhesive protein-based whole cell array biosensor for detection of organophosphorus compounds, *Biosensors and Bioelectronics* 41 (2013) 199–204.
34. F. Zucco, I. De Angelis, E. Testai, and A. Stammati, Toxicology investigations with cell culture systems: 20 years after, *Toxicology in vitro* 18 (2004) 153–163.
35. N. Bhogal, C. Grindon, R. Combes, and M. Balls, Toxicity testing: Creating a revolution based on new technologies, *Trends in Biotechnology* 23 (2005) 299–307.
36. A. Bentley, A. Atkinson, J. Jezek, and D.M. Rawson, Whole cell biosensors— Electrochemical and optical approaches to ecotoxicity testing, *Toxicology in vitro* 15 (2001) 469–475.
37. Regulation (EC) No 1907/2006 of the European Parliament and of the Council available online: http://eur-lex.europa.eu/LexUriServ/LexUriServ.do?uri = OJ:L:2006:396:0001:0849:EN:PDF; (accessed: January 19, 2013).
38. U. Bohrn, E. Stutz, K. Fuchs, M. Fleischer, M.J. Schoning, and P. Wagner, Monitoring of irritant gas using a whole-cell-based sensor system, *Sensors and Actuators B: Chemical* 175 (2012) 208–217.
39. U. Hofmann, S. Michaelis, T. Winckler, J. Wegener, and K.-H. Feller, A whole-cell biosensor as in vitro alternative to skin irritation tests, *Biosensors and Bioelectronics* 39 (2013) 156–162.

40. R.M. Pemberton, T. Cox, R. Tuffin, I. Sage, G.A. Drago, N. Biddle et al., Microfabricated glucose biosensor for culture well operation, *Biosensors and Bioelectronics* (2012): http://dx.doi.org/10.1016/j.bios.2012.11.032i.

41. P.S. Waggoner and H.G. Craighead, Micro-and nanomechanical sensors for environmental, chemical, and biological detection, *Lab on a Chip* 7 (2007) 1238–1255.

42. J.L. Arlett, E.B. Myers, and M.L. Roukes, Comparative advantages of mechanical biosensors, *Nature Nanotechnology* 6 (2011) 203–215.

43. B.N. Johnson and R. Mutharasan, Biosensing using dynamic-mode cantilever sensors: A review, *Biosensors and Bioelectronics* 32 (2012) 1–18.

44. C. Ricciardi, G. Canavese, R. Castagna, I. Ferrante, A. Ricci, S.L. Marasso, L. Napione, and F. Bussolino, Integration of microfluidic and cantilever technology for biosensing application in liquid environment, *Biosensors and Bioelectronics* 26 (2010) 1565–1570.

45. P.S. Waggoner, C.P. Tan, and H.G. Craighead, Microfluidic integration of nanomechanical resonators for protein analysis in serum, *Sensors and Actuators B: Chemical* 150 (2010) 550–555.

46. C. Ricciardi, G. Canavese, R. Castagna, G. Digregorio, I. Ferrante, S.L. Marasso, A. Ricci, V. Alessandria, K. Rantsiou, and L.S. Cocolin,) Online portable microcantilever biosensors for Salmonella enterica serotype Enteritidis detection, *Food and Bioprocess Technology* 3 (2010) 956–960.

47. M. Alvarez, A. Calle, J. Tamayo, L.M. Lechuga, A. Abad, and A. Montoya, Development of nanomechanical biosensors for detection of the pesticide DDT, *Biosensors and Bioelectronics* 18 (2003) 649–653.

48. R. Raiteri, G. Nelles, H.J. Butt, W. Knoll, and P. Skladal, Sensing of biological substances based on the bending of microfabricated cantilevers, *Sensors and Actuators B: Chemical* 61 (1999) 213–217.

49. C.R. Suri, J. Kaur, S. Gandhi, and G.S. Shekhawat, Label-free ultra-sensitive detection of atrazine based on nanomechanics, *Nanotechnology* 19 (2008), doi: 10.1088/0957-4484/19/23/235502.

50. S.H. Tark, A. Das, S. Sligar, and V.P. Dravid, Nanomechanical detection of cholera toxin using microcantilevers functionalized with ganglioside nanodiscs, *Nanotechnology* 21 (2010) 435502.

51. W. Tan, Y. Huang, T. Nan, C. Xue, Z. Li, Q. Zhang, and B. Wang, Development of protein A functionalized microcantilever immunosensors for the analyses of small molecules at parts per trillion levels, *Analytical Chemistry* 82 (2010) 615–620.

52. F. Lochon, I. Dufour, and D. Rebiere, A microcantilever chemical sensors optimization by taking into account losses, *Sensors and Actuators B: Chemical* 118 (2006) 292–296.

53. K. Eom, H.S. Park, D.S. Yoon, and T. Kwon, Nanomechanical resonators and their applications in biological/chemical detection: Nanomechanics principles, *Physics Reports* 503 (2011) 115–163.

54. C. Ricciardi, I. Ferrante, R. Castagna, F. Frascella, S.L. Marasso, K. Santoro, M. Gili, D. Pitardi, M. Pezzolato, and E. Bozzetta, Immunodetection of 17b-estradiol in serum at ppt level by microcantilever resonators, *Biosensors and Bioelectronics* 40 (2013) 407–411.

55. T.H. Wink, S.J. van Zuilen, A. Bult, and W.P. van Bennekom, Self-assembled monolayers for biosensors, *Analyst* 122 (1997) 43R–50R.

56. X.L. Luo, J.J. Xu, W. Zhao, and H.Y. Chen, Glucose biosensor based on ENFET doped with SiO2 nanoparticles, *Sensors and Actuators B: Chemical* 97 (2004) 249–255.

57. H.Y. Wang and S.L. Mu, Bioelectrochemical characteristics of cholesterol oxidase immobilized in a polyaniline film, *Sensors and Actuators B: Chemical* 56 (1999) 22–30.
58. H. Han, Y. Li, H. Yue, Z.D. Zhou, D. Xiao, and M.M.F. Choi, Clinical determination of glucose in human serum by a tomato skin biosensor, *Clinica Chimica Acta* 395 (2008) 155–158.
59. X.F. Yang, Z.D. Zhou, D. Xiao, and M.M.F. Choi, A fluorescent glucose biosensor based on immobilized glucose oxidase on bamboo inner shell membrane, *Biosensors and Bioelectronics* 21 (2006) 1613–1620.
60. M.M.F. Choi, W.S.H. Pang, D. Xiao, and X.J. Wu, An optical glucose biosensor with eggshell membrane as an enzyme immobilisation platform, *Analyst* 126 (2001) 1558–1563.
61. M.M.F. Choi, M.M.K. Liang, and A.W.M. Lee, A biosensing method with enzyme immobilized eggshell membranes for determination of total glucosinolates in vegetables, *Enzyme and Microbial Technology* 36 (2005) 91–99.
62. D. Xiao and M.M.F. Choi, Aspartame optical biosensor with bienzyme-immobilized eggshell membrane and oxygen-sensitive optode membrane, *Analytical Chemistry* 74 (2002) 863–870.
63. W.Y. Cai, Q. Xu, X.N. Zhao, J.J. Zhu, and H.Y. Chen, Porous gold-nanoparticle-$CaCO_3$ hybrid material: Preparation, characterization, and application for horseradish peroxidase assembly and direct electrochemistry, *Chemistry of Materials* 18 (2006) 279–284.
64. L.Q. Yang, X.L. Ren, F.Q. Tang, and L. Zhang, A practical glucose biosensor based on Fe3O4 nanoparticles and chitosan/nafion composite film, *Biosensors and Bioelectronics* 25 (2009) 889–895.
65. R. Baron, B. Willner, and I. Willner, Biomolecule–nanoparticle hybrids as functional units for nanobiotechnology, *Chemical Communications* 4 (2007) 323–332.
66. D.X. Li, Q. He, Y. Cui, L. Duan, and J.B. Li, Immobilization of glucose oxidase onto gold nanoparticles with enhanced thermostability, *Biochemical and Biophysical Research Communications* 355 (2007) 488–493.
67. A.A. Vertegel, R.W. Siegel, and J.S. Dordick, Silica nanoparticle size influences the structure and enzymatic activity of adsorbed lysozyme, *Langmuir* 20 (2004) 6800–6807.
68. M.D. Hughes, Y.J. Xu, P. Jenkins, P. McMorn, P. Landon, D.I. Enache et al., Tunable gold catalysts for selective hydrocarbon oxidation under mild conditions, *Nature* 437 (2005) 1132–1135.
69. T. Kiyonaga, T. Akita, and H. Tada, Au nanoparticle electrocatalysis in a photoelectrochemical solar cell using CdS quantum dot-sensitized TiO2 photoelectrodes, *Chemical Communications* 15 (2009) 2011–2013.
70. A. Wieckowski, E. Savinova, and C.G. Vayenas, *Catalysis and Electrocatalysis at Nanoparticle Surfaces*, Marcel Dekker, Inc., New York, 2003.
71. M. Valden, X. Lai, and D.W. Goodman, Onset of catalytic activity of gold clusters on titania with the appearance of nonmetallic properties, *Science* 281 (1998) 1647–1650.
72. B. Zheng, S. Xie, L. Qian, H. Yuan, D. Xiao, and M.M.F. Choi, Gold nanoparticles-coated eggshell membrane with immobilized glucose oxidase for fabrication of glucose biosensor, *Sensors and Actuators B: Chemical* 152 (2011) 49–55.
73. B. Unnikrishnan, S. Palanisamy, and S.-M. Chen, A simple electrochemical approach to fabricate a glucose biosensor based on graphene–glucose oxidase biocomposite, *Biosensors and Bioelectronics* 39 (2013) 70–75.

74. P. Qi, Y. Wan, and D. Zhang, Impedimetric biosensor based on cell-mediated bioimprinted films for bacterial detection, *Biosensors and Bioelectronics* 39 (2013) 282–288.
75. H. Chang, X. Wang, K.-K. Shiu, Y. Zhu, J. Wang, Q. Li, B. Chen, and H. Jiang, Layer-by-layer assembly of graphene, Au and poly (toluidine blue O) films sensor for evaluation of oxidative stress of tumor cells elicited by hydrogen peroxide, *Biosensors and Bioelectronics* 41 (2013) 789–794.
76. D. Zhang, Y. Zhang, L. Zheng, Y. Zhan, and L. He, Graphene oxide/poly-L-lysine assembled layer for adhesion and electrochemical impedance detection of leukemia K562 cancer cells, *Biosensors and Bioelectronics* 42 (2013) 112–118.
77. B. Soontornworajit, Y. Wang, Nucleic acid aptamers for clinical diagnosis: Cell detection and molecular imaging, *Analytical and Bioanalytical Chemistry* 399 (2011) 1591–1599.
78. P. Ying, Z. Zhu, H. Liu, J. Zhang, J. Liu, and W. Tan, *Analytical and Bioanalytical Chemistry* 397 (2010) 3225–3233.
79. M. Famulok, G. Mayer, and M. Blind, Nucleic acid aptamers—From selection in vitro to applications in vivo, *Accounts of Chemical Research* 33 (2000) 591–599.
80. X. Chen, Y. Huang, and W. Tan, Using aptamer-nanoparticle conjugates for cancer cells detection, *Journal of Biomedical Nanotechnology* 4 (2008) 400–409.
81. C.S.M. Ferreira, C.S. Matthews, and S. Missailidis, DNA aptamers that bind to MUC1 tumour marker: Design and characterization of MUC1-binding single-stranded DNA aptamers, *Tumor Biology* 27 (2006) 289–301.
82. S.J. Gendler and J. Mammary, MUCI, the renaissance molecule, *Journal of Mammary Gland Biology and Neoplasia* 6 (2001) 339–353.
83. T. Li, Q. Fan, T. Liu, X. Zhu, J. Zhao, and G. Li, Detection of breast cancer cells specially and accurately by an electrochemical method, *Biosensors and Bioelectronics* 25 (2010) 2686–2689.
84. J. Zhao, X. He, B. Bo, X. Liu, Y. Yin, and G. Li, A 'signal-on' electrochemical aptasensor for simultaneous detection of two tumor markers, *Biosensors and Bioelectronics* 34 (2012) 249–252.
85. W. Wei, D. Li, X. Pan, and S. Liu, Electrochemiluminescent detection of Mucin 1 protein and MCF-7 cancer cells based on the resonance energy transfer, *Analyst* 137 (2012) 2101–2106.
86. J. Li, X. Zhong, F. Cheng, J. Zhang, L. Jiang, and J. Zhu, One-pot synthesis of aptamer-functionalized silver nanoclusters for cell-type-specific imaging, *Analytical Chemistry* 84 (2012) 4140–4146.
87. X. Zhu, J. Yang, M. Liu, Y. Wu, Z. Shen, and G. Li, Sensitive detection of human breast cancer cells based on aptamer–cell–aptamer sandwich architecture, *Analytica Chimica Acta* 764 (2013) 59–63.
88. C.H. Chung, J.H. Kim, J. Jung, and B.H. Chung, Nuclease-resistant DNA aptamer on gold nanoparticles for the simultaneous detection of Pb^{2+} and Hg^{2+} in human serum, *Biosensors and Bioelectronics* 41 (2013) 827–832.
89. X. Liu, L. Ouyang, X. Cai, Y. Huang, X. Feng, Q. Fan, and W. Huang, An ultrasensitive label-free biosensor for assaying of sequence-specific DNA-binding protein based on amplifying fluorescent conjugated polymer, *Biosensors and Bioelectronics* 41 (2013) 218–224.
90. B. Binkowski, F. Fan, and K. Wood, Engineered luciferases for molecular sensing in living cells, *Current Opinion in Biotechnology* 20 (2009) 14–18.

91. K.E. Luker, M.C. Smith, G.D. Luker, S.T. Gammon, H. Piwnica-Worms, and D. Piwnica-Worms, Kinetics of regulated protein–protein interactions revealed with firefly luciferase complementation imaging in cells and living animals, *Proceedings of the National Academy of Sciences of the United States of America* 101 (2004) 12288–12293.

92. T. Ozawa, Designing split reporter proteins for analytical tools, *Analytica Chimica Acta* 556 (2006) 58–68.

93. M. Torkzadeh-Mahani, F. Ataei, M. Nikkhah, and S. Hosseinkhani, Design and development of a whole-cell luminescent biosensor for detection of early-stage of apoptosis, *Biosensors and Bioelectronics* 38 (2012) 362–368.

94. N. Johnsson and A. Varshavsky, Split ubiquitin as a sensor of protein interactions in vivo, *Proceedings of the National Academy of Sciences of the United States of America* 91 (1994) 10340–10344.

95. C.H. Contag and M.H. Bachmann, Advances in in vivo bioluminescence imaging of gene expression, *Annual Review of Biomedical Engineering* 4 (2002) 235–260.

96. F. Ataei, M. Torkzadeh-Mahani, and S. Hosseinkhani, A novel luminescent biosensor for rapid monitoring of IP3 by split-luciferase complementary assay, *Biosensors and Bioelectronics* 41 (2013) 642–648.

97. R.W. Ramette and E.B. Sandell, Rhodamine B equilibria, *Journal of the American Chemical Society* 78 (1956) 4872–4878.

98. L. Prodi, F. Bolletta, M. Montalti, and N. Zaccheroni, Luminescent chemosensors for transition metal ions, *Coordination Chemistry Reviews* 205 (2000) 59–83.

99. B. Valeur and I. Leray, Design principles of fluorescent molecular sensors for cation recognition, *Coordination Chemistry Reviews* 205 (2000) 3–40.

100. X. Chen, T. Pradhan, F. Wang, J.S. Kim, and J. Yoon, Fluorescent chemosensors based on spiroring-opening of xanthenes and related derivatives, *Chemical Reviews* 112 (2012) 1910–1956.

101. M. Beija, C.A.M. Afonso, and J.M.G. Martinho, Synthesis and applications of rhodamine derivatives as fluorescent probes, *Chemical Society Reviews* 38 (2009) 2410–2433.

102. H.N. Kim, M.H. Lee, H.J. Kim, J.S. Kim, and J. Yoon, A new trend in rhodamine-based chemosensors: Application of spirolactam ring-opening to sensing ions, *Chemical Society Reviews* 37 (2008) 1465–1472.

103. Q. Zou, X. Li, J. Zhang, J. Zhou, B. Sun, and H. Tian, Unsymmetrical diarylethenes as molecular keypad locks with tunable photochromism and fluorescence via Cu^{2+} and CN coordinations, *Chemical Communications* 48 (2012) 2095–2097.

104. H.N. Kim, Z. Guo, W. Zhu, J. Yoon, and H. Tian, Recent progress on polymer-based fluorescent and colorimetric chemosensors, *Chemical Society Reviews* 40 (2011) 79–93.

105. A.P. De Silva, H.Q.N. Gunarante, T. Gunnlaugsson, A.J.M. Huxley, C.P. McCoy, J.T. Rademacher et al., Signaling recognition events with fluorescent sensors and switches, *Chemical Reviews* 97 (1997) 1515–1566.

106. J. Mao, L. Wang, W. Dou, X. Tang, Y. Yan, and W. Liu, Tuning the selectivity of two chemosensors to Fe(III) and Cr(III), *Organic Letters* 9 (2007) 4567–4570.

107. Y. Zhao, X.-B. Zhang, Z.-X. Han, L. Qiao, C.-Y. Li, L.-X. Jian et al., Highly sensitive and selective colorimetric and off-on fluorescent chemosensor for Cu^{2+} in aqueous solution and living cells, *Analytical Chemistry* 81 (2009) 7022–7030.

108. N.R. Chereddy, K. Suman, P.S. Korrapati, S. Thennarasu, and A.B. Mandal, Design and synthesis of rhodamine based chemosensors for the detection of Fe^{3+} ions, *Dyes and Pigments* 95 (2012) 606–613.

109. S. Chakravarty, W.-C. Lai, Y. Zou, H.A. Drabkin, R.M. Gemmill, G.R. Simon, S.H. Chin, and R.T. Chen, Multiplexed specific label-free detection of NCI-H358 lung cancer cell line lysates with silicon based photonic crystal microcavity biosensors, *Biosensors and Bioelectronics* 43 (2013) 50–55.

110. A.J. Bard and L.R. Faulkner, *Electrochemical Methods: Fundamentals and Applications*, 2nd edn., Wiley, New York, 2000.

111. J. Shi, T. Cha, J. Claussen, A. Diggs, J.H. Choi, and D.M. Porterfield, Microbiosensors based on DNA modified single-walled carbon nanotube and Pt black nanocomposites, *Analyst*, 136 (2011) 4916–4924.

112. S.-K. Jung, L.M. Kauri, W.-J. Qian, and R.T. Kennedy, *The Journal of Biological Chemistry* 275(9) (2000) 6642–6650.

113. J. Shi, E.S. McLamore, and D.M. Porterfield, Nanomaterial based self-referencing microbiosensors for cell and tissue physiology research, *Biosensors and Bioelectronics* 40 (2013) 127–134.

114. L. Lu, G. Chee, K. Yamada, and S. Jun, Electrochemical impedance spectroscopic technique with a functionalized microwire sensor for rapid detection of foodborne pathogens, *Biosensors and Bioelectronics* 42 (2013) 492–495.

115. B.L. Rivas, E. Pereira, and A. Maureira, *Polymer International* 58 (2009) 1093–1114.

116. V.P. Raje, P.I. Morgado, M.P. Ribeiro, I.J. Correia, V.D.B. Bonifacio, P.S. Branco, and A.A. Ricardo, Dual on–off and off–on switchable oligoaziridine biosensor, *Biosensors and Bioelectronics* 39 (2013) 64–69.

117. M. Barra, D. Viggiano, P. Ambrosino, F. Bloisi, F.V. Di Girolamo, M.V. Soldovieri, M. Taglialatela, and A. Cassinese, Addressing the use of PDIF-CN2 molecules in the development of n-type organic field-effect transistors for biosensing applications, *Biochimica et Biophysica Acta* (2012): http://dx.doi.org/10.1016/j.bbagen.2012.11.025.

118. J.S. Burmeister, L.A. Olivier, W.M. Reichert, and G.A. Truskey, Application of total internal reflection fluorescence microscopy to study cell adhesion to biomaterials, *Biomaterials* 19 (1998) 307–325.

119. E. Cretel, D. Touchard, A.M. Benoliel, P. Bongrand, and A. Pierres, Early contacts between T lymphocytes and activating surfaces, *Journal of Physics: Condensed Matter* 22 (2010) 194107.

120. P. Ryzhkov, M. Prass, M. Gummich, J.-S. Kuhn, C. Oettmeier, and H.-G. Dobereiner, Adhesion patterns in early cell spreading, *Journal of Physics: Condensed Matter* 22 (2010) 194106.

121. A. Shoshi, J. Schotter, P. Schroeder, M. Milnera, P. Ertl, R. Heer, G. Reiss, and H. Brueckl, Contemporaneous cell spreading and phagocytosis: Magnetoresistive real-time monitoring of membrane competing processes, *Biosensors and Bioelectronics* 40 (2013) 82–88.

122. A.I. Shoshi, J. Schotter, P. Schroeder, M. Milnera, P. Ertl, V. Charwat et al., Magnetoresistive-based real-time cell phagocytosis monitoring, *Biosensors and Bioelectronics* 36(1) (2012) 116–122.

123. M. Gardel and U. Schwarz, Cell-substrate interactions, *Journal of Physics: Condensed Matter* 22 (2010) 190301.

124. T. Murakami, J. Sumaoka, and M. Komiyama, Sensitive isothermal detection of nucleic-acid sequence by primer generation–rolling circle amplification, *Nucleic Acids Research* 37 (2009) e19.

125. B. Schweitzer, S. Roberts, B. Grimwade, W. Shao, M. Wang, Q. Fu et al., Multiplexed protein profiling on microarrays by rolling-circle amplification, *Nature Biotechnology* 20 (2002) 359–365.

126. B. Schweitzer, S. Wiltshire, J. Lambert, S. O'Malley, K. Kukanskis, Z. Zhu, S.F. Kingsmore, P.M. Lizardi, and D.C. Ward, Immunoassays with rolling circle DNA amplification: A versatile platform for ultrasensitive antigen detection, *Proceedings of the National Academy of Sciences of the United States of America* 97 (2000) 10113–10119.

127. C. Ding, N. Wang, J. Zhang, and Z. Wang, Rolling circle amplification combined with nanoparticle aggregates for highly sensitive identification of DNA and cancer cells, *Biosensors and Bioelectronics* 42 (2013) 486–491.

128. E. Alipour, M.H. Pournaghi-Azar, M. Parvizi, S.M. Golabi, and M.S. Hejazi, Electrochemical detection and discrimination of single copy gene target DNA in non-amplified genomic DNA, *Electrochimica Acta* 56 (2011) 1925–1931.

129. R. Abe, H. Ohashi, I. Iijima, M. Ihara, H. Takagi, T. Hohsaka, and H. Ueda, *Journal of the American Chemical Society* 133 (2011) 17386–17394.

130. H.-J. Jeong, Y. Ohmuro-Matsuyama, H. Ohashi, F. Ohsawa, Y. Tatsu, M. Inagaki, and H. Ueda, Detection of vimentin serine phosphorylation by multicolor Quenchbodies, *Biosensors and Bioelectronics* 40 (2013) 17–23.

131. S. Kintzios, *Molecular Identification through Membrane-Engineered Cells*, USPTO Application #20090170068 (2009). http://www.freshpatents.com/Chemistry--molecular-biology-and-microbiology-dt200907ntc435.php

132. G. Moschopoulou and S. Kintzios, Application of membrane-engineering to bioelectric recognition cell sensors for the ultra-sensitive detection of superoxide radical: A novel biosensor principle, *Analytica Chimica Acta* 573 (2006) 90–96.

133. G. Moschopoulou, I. Papanastasiou, O. Makri, N. Lambrou, G. Economou, K. Soukouli, and S.E. Kintzios, Cellular redox-status is associated with regulation of frond division in Spirodela polyrrhiza, *Plant Cell Reports* 26 (2007) 2063–2069.

134. G. Moschopoulou, T. Valero, and S. Kintzios, Superoxide determination using membrane-engineered cells: An example of a novel concept for the construction of cell sensors with customized target recognition properties, *Sensors and Actuators B: Chemical* 175 (2012) 78–84.

135. S.H.A. Hassan and S.E. Oh, Improved detection of toxic chemicals by *Photobacterium phosphoreum* using modified Boss medium, *Journal of Photochemistry and Photobiology B: Biology* 101(1) (2010) 16–21.

136. S.W. Van Ginkel, S.H.A. Hassan, and S.E. Oh, Detecting endocrine disrupting compounds in water using sulfur-oxidizing bacteria, *Chemosphere* 81 (2010) 294–297.

137. S.E. Oh, S.H.A. Hassan, and S.W. Van Ginkel, A novel biosensor for detecting toxicity in water using sulfur-oxidizing bacteria, *Sensors and Actuators B: Chemical* 154(1) (2011) 17–21.

138. S.H.A. Hassan, S.W. Van Ginkel, and S.-E. Oh, Effect of organics and alkalinity on the sulfur oxidizing bacteria (SOB) biosensor, *Chemosphere* 90 (2013) 965–970.

12

Nanotechnology and Nanofabrication Applications in Chemical Sensing

Magnus Willander and Zafar H. Ibupoto

CONTENTS

12.1 Introduction

Biosensors are analytical devices with a sensing biological material including microorganisms, antibodies, enzymes, cell receptors, tissues, derived biomaterials, and finally a transducing microsystem. The transducer is the main component in a biosensor that converts biological signal into a measurable electrical signal. The transducer component is comprised of electrochemical, optical, thermometric, magnetic, or piezoelectric material. The biosensors have gotten significant importance in medical diagnosis, environmental monitoring, genetics, food industry, and defense applications. The high importance of biosensors in the previously mentioned applications can be correlated to their simplicity, high sensitivity, and practical usability for real sample analysis and on spot analysis [1–4]. Leland Clark Jr. known as the pioneer of sensor field, developed the oxygen electrode in 1956 [1]. Clark provided a solid platform for the engineering of biosensors based on his concept that has been considered significantly since the opening of oxygen electrode. In 1962, the Clark concept was further extended experimentally by trammelling glucose oxidase on an oxygen electrode through a dialysis membrane, and the glucose concentration was determined with decreased concentration of oxygen [2]. Later in 1967, an independent and an operational enzyme-based electrode for the detection of glucose was developed [3]. A potentiometric urease that immobilized an ammonium-selective liquid membrane-based electrode for the detection of urea was for the first time introduced in 1969 [4] and some other electrodes such as pH electrodes [5], and pH ion-selective field-effect transistors (ISFETs) as transducers were designed [6]. In 1970, the first ISFET was constructed [7]; afterwards, several researchers followed it as the way until the end of 1980. These sensor devices are based on silicon technology; therefore, they can be raised in number at cheap price. Although this technology is acceptable, there are many production hurdles, and it can take a long time to solve these problems. Although a lot of research in the sensor technology has been done [8,9], more work is still needed in this research field [10].

Recently, with the revolutionary advent of nanotechnology, a number of new nanomaterials have been synthesized and their properties have been investigated. Moreover, these novel nanomaterials are being used strongly in the biosensor technology. The use of nanomaterials in biosensing describes the consolidation of different sciences such as material science, molecular engineering, chemistry, and biotechnology. The nanomaterials have great impact on the working performance of biosensors by improving the sensitivity and recognition of biomolecules at small volumes. The biosensors based on nanomaterials have a great practical usability in biomolecular recognition, pathogenic diagnosis, and environmental applications [11–13]. The most advantageous properties of nanomaterials for the biosensing applications include larger surface to volume ratio, different physical and chemical properties from micro surface, and easy to detect target analyte in small volumes. Moreover, nanomaterials have a great diagnosing ability inside the biological cell. The use of nanomaterials inside the biological cell makes a clear difference of nanodevices from the bulk device that has been used for the biosensing purpose.

Due to the presence of highly sensitive instruments that provide simplicity in the characterization, a great degree of novel materials can be synthesized with excellent properties at nanoscale level. Having materials at nanoscale clearly demonstrates different performances than its bulk form; therefore, nanodevices are superior in response than bulk devices of the same material. The material science is showing a very close relation with biostuff due to the favorable properties exhibited by synthesized materials. In order to have a highly sensitive and specific biosensor, the choice of using substrate for disseminating the sensing stuff is very important for evaluating the performance of the biosensor. Several materials at nanoscale have been utilized for the development of biosensors due to their distinctive and typical physical, mechanical, chemical, optical, and magnetic characteristics that significantly improve the sensitivity and recognition of biomolecule sensing. Such nanomaterials include gold nanoparticles [14], carbon nanotubes [15], magnetic nanoparticles [16], and quantum dots [17].

Also nanostructures including nanowires [7,18–23] and nanocrystals [24–28] are capable of working efficiently for the detection of biological or chemical substance with more or less similar sensitivity and selectivity. The high selectivity and sensitivity of these nanostructures can be attributed to their comparable dimensions to that of biological or chemical substances that are to be sensed, and also these nanostructures behave as excellent principal transducers for generating strong electrical signals. Among other nanostructures, inorganic nanowires and nanocrystals possess both distinctive electrical [19–21,29–44] and optical [24–28] characteristics at high magnitude; therefore, one can make good use of these for sensing at large scale.

Today, optical biosensors based on semiconductor nanocrystals exhibit several advantages over conventional organic molecular dyes due to their remarkable emission properties and simplicity in tuning [24–28]. Moreover, recently the stabilized lipid films [45,46], the graphene [47], and the chitosan composite with multiwalled carbon nanotube-based electrodes [48] have also been used for the sensing of different bioactive compounds. The nanowires composed of semiconducting materials exhibit permutable electrical properties that make them suitable models for sensor technology, and also label-free electrical display in readable form is highly appealing for various applications [49–57]. The electronic devices based on the nanoscale approach are easily incorporated into miniaturized forms, and electrical signals can be transferred to the outdoor world. The extraordinary properties exhibited by nanowires and their high sensitivity indicate that the devices based on nanowires can bring radical change at a wide range in sensor technology and also in biological and medicinal fields [58].

Among several materials, ZnO nanowires/nanorods are considered potential candidates for biosensing applications because of their excellent properties including biosafety, biocompatibility, piezoelectric behavior, nontoxicity, high electron transfer communication, better analytical working activity, high sensitivity, and cheap and easy synthesis [59–71]. Due to the high isoelectric point (IEP) of ZnO approximately 9.5 that provides the more attractive environment for the immobilization of low IEP protein molecules such as enzymes, deoxyribonucleic acid (DNA), and antibodies through electrostatic interactions with strong mechanical stability [72]. Moreover, ZnO exhibits high ionic properties around 60% and it is relatively highly stable at physiological pH values. These advantages can be exploited in generating stable and reproducible signals from different analyte concentrations. The electrochemical nanosensors based on the ZnO nanowires and nanorods have good selectivity for real sample analysis, which could be more appropriate for diagnostic purposes with increased sensitivity. The diameter of nanowires/nanorods is relatively similar to the size of chemical and biological substances to be sensed, and this makes nanowires/nanorods excellent transducers with strong electrical signals. Our group has carried out intracellular measurements in describing different aspects close to the cell membrane using ZnO nanowires/nanorods since 2006 till today [66,67,70,71,73,74].

Our work was further followed by Lieber et al. [75] in measuring the active potential inside the cell. In addition to nanowires/nanorods, ZnO nanotubes are also considered highly efficient for the development of biosensors due to their high surface to volume ratio and high porosity. Due to this, a large quantity of proteinaceous molecules (enzymes, antibodies, etc.) can be immobilized, which results in high sensitivity and fast response time for the fabricated biosensor. ZnO nanotube-based glucose, uric acid, and C-reactive protein (CRP) biosensors have been reported by our group [76–78].

12.2 Results and Discussions

ZnO nanorods and nanotubes were successfully fabricated on gold-coated glass substrate and consequently used for the chemical detection of ascorbic acid and CRP [78,79].

12.2.1 ZnO Nanorod-Based Selective L-Ascorbic Biosensor

ZnO nanorod-based ascorbic acid biosensor was developed by immobilizing the nanorods with ascorbate oxidase in conjunction with 3-glycidoxypropyltrimethoxysilane (GPTS). This process of functionalization of nanorods was followed in two steps. Initially, 20% GPTS solution was put on the ZnO nanorods dropwise and consequently left at 70°C for 1 h. Next, the unnecessary GPTS molecules were taken away by phosphate buffer solution, and then nanorods were dipped in 2 mg/mL prepared ascorbate oxidase solution for 3 min.

Afterward, ascorbate oxidase-immobilized nanorods were dried at room temperature, and then sensor electrodes were left at 4°C for 22 h for the proper immobilization of ZnO nanorods [80]. The electrochemical response of ascorbate oxidase-immobilized ZnO nanorod-based sensor electrode for various ascorbic acid concentrations is shown in Figure 12.1. It can be observed from

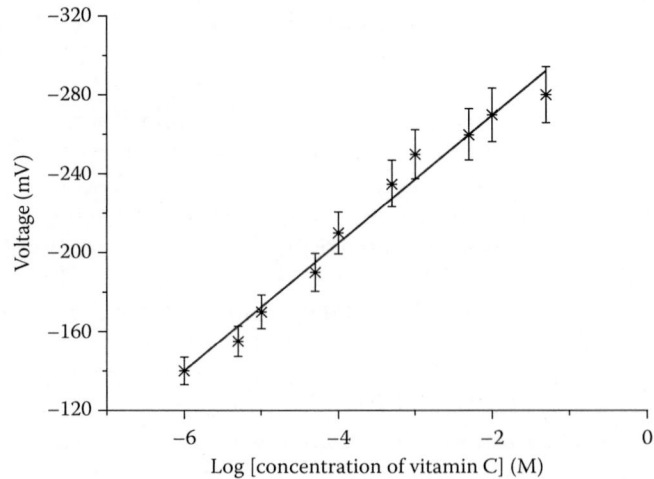

FIGURE 12.1
Calibration curve using enzyme-immobilized ZnO nanorod-based biosensor electrode representing the electrochemical response at different l-ascorbic acid concentrations from 1.0×10^{-6} to 5.0×10^{-2} M with Ag/AgCl reference electrode.

Figure 12.1 that the response shown by the proposed response is according to Nernst equation. The sensing mechanism for ZnO nanorod-based ascorbic acid can be described as follows: during the insertion of sensor electrode in the electrolytic solution of ascorbic acid, the presence of ascorbate oxidase on ZnO nanorods has enhanced catalytic efficiency for the oxidation of ascorbic acid into dehydroascorbic (DAA) and it resulted in the generation of hydronium ions and two electrons inside the testing solution, and this phenomenon of charge creation is responsible for the output response on the screen of the reading device. The potential difference electromotive force (EMF) is mainly related to the number of charges present on the surface of ZnO nanorods, and its value is different for different concentrations of analyte. Moreover, for the ascorbic acid biosensor, the electrochemical response is also dependent on the catalytic efficiency of ascorbate oxidase for the oxidation of ascorbic acid. The proposed ZnO nanorod-based ascorbic acid biosensor has demonstrated a wide range of detection of ascorbic acid from 1.0×10^{-6} to 5.0×10^{-2} with high linearity, sensitivity, stability, and selectivity. The present biosensor has shown a fast time response of less than 10 s as shown in Figure 12.2. All the obtained results have indicated the potential applicability of the proposed ascorbic acid biosensor for the analysis of ascorbic acid in food and real-time samples. The performance evaluation of the proposed ascorbic acid was also monitored in terms of reproducibility, repeatability, and selectivity. Reproducibility is one of the parameters for the evaluation of performance, and for this experiment, five independent biosensor electrodes were prepared under a similar set of conditions, and the observed results have shown the acceptable reproducibility with relative standard deviation of less than 5% in 0.05 mM as shown in Figure 12.3.

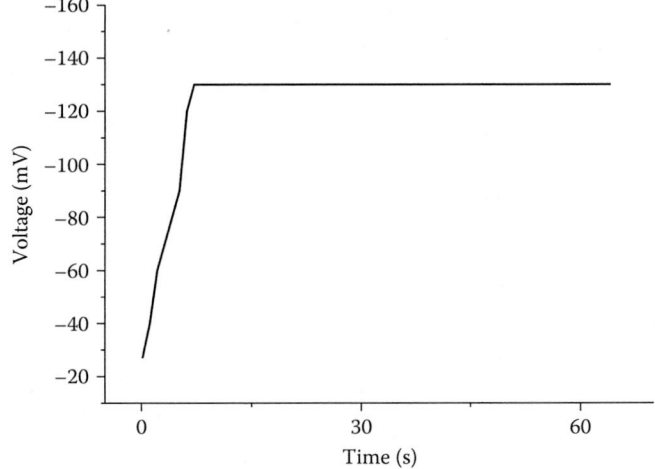

FIGURE 12.2
Calibration curve representing the time response of the proposed sensor in 1.0×10^{-4} M ascorbic acid test solution.

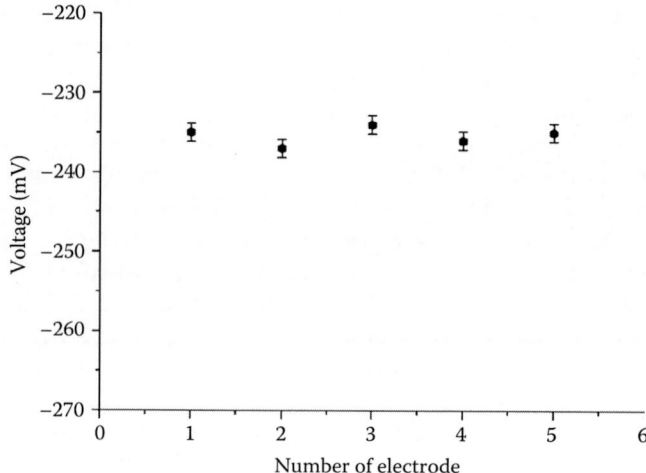

FIGURE 12.3
Calibration curve representing the biosensor-to-biosensor reproducibility in 5.0×10^{-5} M ascorbic acid test solution.

The repeatability was monitored in a single fabricated biosensor for three consecutive days, and the observed results are shown in Figure 12.4. It can be inferred from Figure 12.4 that the biosensor maintained its detection range, linearity, and sensitivity. For each experiment, the biosensor electrode was dipped in phosphate buffer solution in order to obtain the desired results with relatively

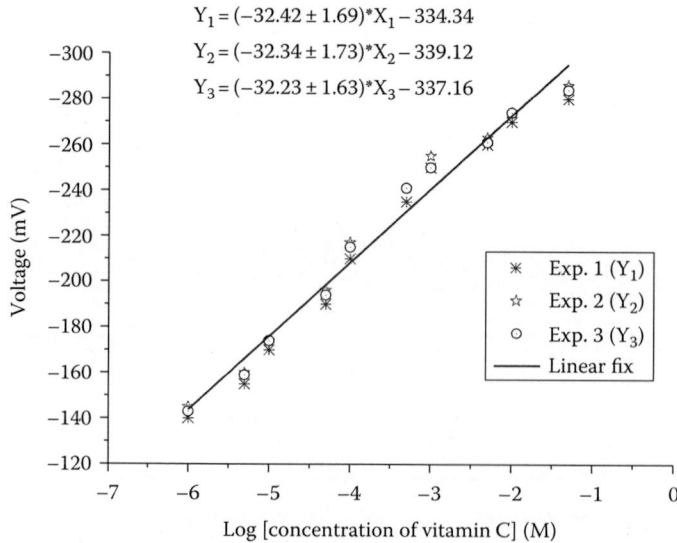

$$Y_1 = (-32.42 \pm 1.69)^*X_1 - 334.34$$
$$Y_2 = (-32.34 \pm 1.73)^*X_2 - 339.12$$
$$Y_3 = (-32.23 \pm 1.63)^*X_3 - 337.16$$

FIGURE 12.4
Calibration curve representing the proposed biosensor reusability at room temperature after 2–3 h span in the ascorbic acid test solution from 1.0×10^{-6} to 5.0×10^{-2} M concentration range.

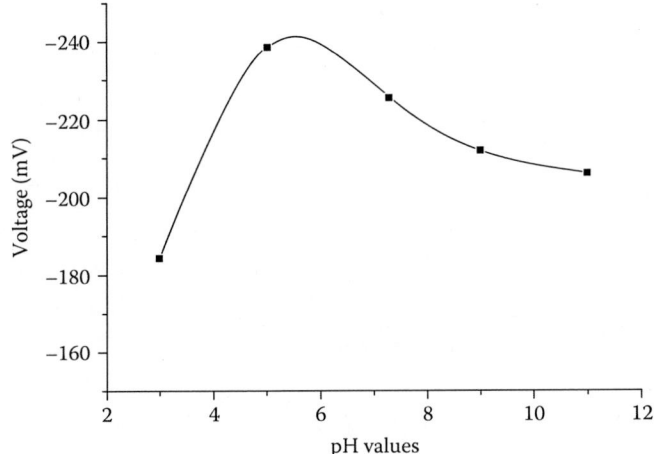

FIGURE 12.5
Calibration curve representing the study of EMF response with the variation of pH values.

similar response for the next experiment. When the biosensor was not in use, it was kept at 4°C. Selectivity is the primary element for the evaluation of performance of the proposed biosensor in the presence of common interferents. The mixed method was used for this study, and the response of ascorbic acid biosensor for ascorbic acid was monitored in the presence of potassium ions, sodium ions, zinc ions, calcium ions, glucose, fructose, and copper ions in the interfering amount of each interferent. For the detected range of ascorbic acid, we added a significant volume of each interferent from the concentration range 1.0×10^{-6} to 5.0×10^{-2} M. The biosensor remained very selective towards the ascorbic acid except that copper ions have shown slight interference with ascorbate oxidase as shown in Figure 12.5. Furthermore, the pH influence on the electrochemical response of ascorbic acid biosensor was monitored in order to confirm the working pH of the proposed biosensor for long-term stability and usability. For this experiment, a pH range of 3–12 was selected and the observed electrochemical response against this pH range is shown in Figure 12.6. It can be seen from Figure 12.6 that the output response of the biosensor is maximum at pH 5 [81], and the response is shown in decreasing order due to the instability of ascorbic acid electrolyte at neutral or alkaline pH [82]. However, we carried out all experiments around neutral pH in order to avoid the possible dissolution of ZnO nanorods in either very acidic or alkaline pH values.

12.2.2 ZnO Nanotube-Based Disposable C-Reactive Protein Biosensor

ZnO nanorods were chemically etched into nanotubes using 3–5 M concentrations of potassium chloride (KCl). The chemical etching of ZnO nanorods depends on several parameters such as duration of etching, concentration of KCl used, and temperature. ZnO nanotubes exhibit several attractive

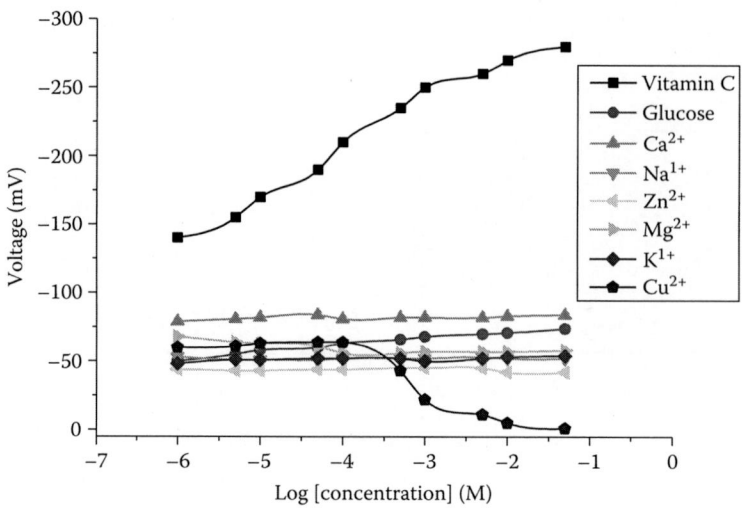

FIGURE 12.6
Representing the effect of different interfering ions on the output response of the proposed biosensor.

features, but for biosensing application, their porosity and hollow nature gives them excellent contributions towards the development of biosensors with enhanced performance. By exploiting these advantages, we have developed for the first time disposable potentiometric biosensor for the detection of CRP using the antibody-immobilized ZnO nanotube sensor electrode. In the development of CRP biosensor based on ZnO nanotubes, we concluded that the two properties of this material significantly contributed to the generation of output potential including high ionic property and IEP of approximately 9.5. The piezoelectric features of ZnO nanomaterial have assisted in the output potential in addition to the charge atmosphere given by the CRP. The high IEP of ZnO helped in the provision of strong binding of immobilized antibody. The ZnO nanotube-based CRP biosensor has demonstrated acceptable sensitivity; high linearity, reproducibility, repeatability, and selectivity; and fast response time. The fabrication of biosensor was as follows: Initially, ZnO nanorods were prepared on the gold-coated glass substrate using hydrothermal growth method, and then nanorods were chemically etched in 3–5 M KCl for 16–18 h at 80°C. Afterwards, ZnO nanotubes were functionalized with antiCRP using the electrostatic physical adsorption method. The antibody solution of 1.5×10^2 mg/L was prepared in 1 mM phosphate buffer solution, and also 2.5% glutaraldehyde was added in this solution, and consequently, ZnO nanotubes were dipped in this solution for 5 min. The antibody-immobilized ZnO nanotube-based biosensor electrodes were dried at room temperature and kept at 4°C overnight. The electrochemical response of antibody-immobilized ZnO nanotube-based biosensor for the detection of CRP was measured by the potentiometric technique using pH meter model 728. Keithley 2400, an electrical tool, was used for the measurement of response time. The biosensors were left at 4°C, when not in use. The sensing mechanism of antibody-immobilized ZnO nanotube-based biosensor for the detection CRP can be described in terms of rapid complex formation between antibody and CRP molecules on the surface of ZnO. In the biosensing system of the present case, antibody-functionalized ZnO nanotubes were used as working electrode and silver–silver chloride (Ag/AgCl) as reference electrode, and both electrodes were connected to a pH meter.

ZnO nanotubes behaved as excellent transducers and strongly bound the antibody molecules and later a fast complex formation took place between the immobilized antibody and CRP molecules inside the testing solution. The better performance of the proposed biosensor can be attributed to the high IEP value of ZnO that firmly attracted the low IEP protein molecules such as the antibody in the present case. In this study, six antibody clones and (CRP-8)-immobilized ZnO nanotube-based biosensor electrodes were used in different concentrations of CRP. The potentiometric response of these biosensor electrodes was measured for 1.0×10^{-6} to 1.0×10^{0} mg/L concentration of CRP, and the observed linear response was found for 1.0×10^{-5} mg/L to 1.0×10^{0} mg/L. All concentrations of CRP were made in 1 mM phosphate buffer solution of pH 7.3. Similarly, ZnO nanotubes were also used in the same concentration range of CRP and the observed response was found negligible as shown in Figure 12.7a. However, antibody-immobilized ZnO nanotube

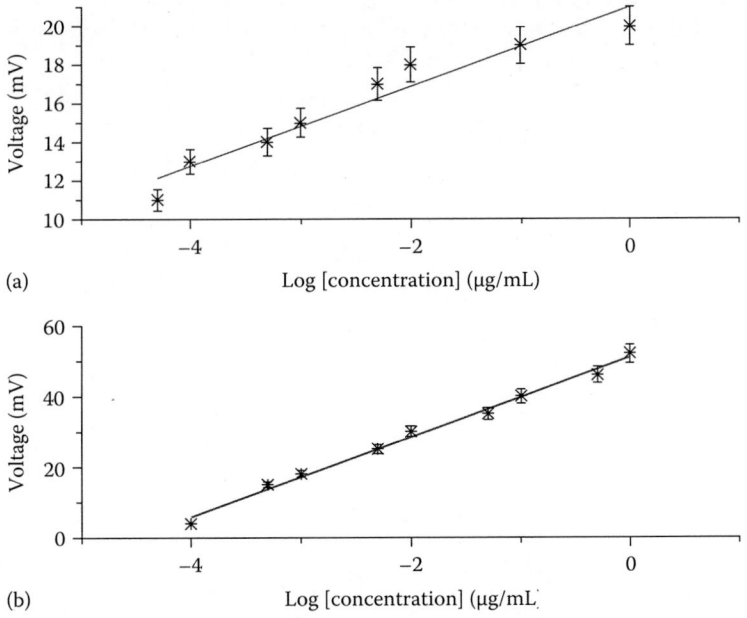

(a)

(b)

FIGURE 12.7
(a) The calibration curve of bare ZnO nanotubes for CRP antigen and (b) the calibration curve of antibody-immobilized ZnO nanotubes for CRP.

biosensor electrode has shown fast, and stable output potential, and linear and selective response for the detection of CRP as shown in Figure 12.7b. This obvious output response for antibody-immobilized ZnO nanotube biosensor can be assigned to fast and favorable complex formation between immobilized antibody and CRP on the surface of ZnO. Moreover, this response of biosensor can also be illustrated in terms of the physical property of ZnO itself such as piezoelectric that might be involved in the production of stable electrical signal in addition to rapid complex formation between immobilized antibody and CRP molecules. This strange and advantageous response of the antibody-immobilized ZnO nanotubes in the presence of a nonparticular ionic environment is making a novel approach for the design of a new biosensor for protein detection on ZnO and also other materials that exhibit similar properties as ZnO possesses at nanoscale. Figure 12.7b shows the detection range of fabricated biosensor with good linearity, acceptable sensitivity of 13.17 ± 0.42 mV/decade, and regression coefficient value of 0.99. The proposed antibody-immobilized ZnO nanotube-based biosensor electrode was also tested in nonspecific protein such as bovine serum albumin (BSA) in order to confirm the response of biosensor for other proteins, and we observed no significant response in this experiment. This experiment clearly demonstrated the selectivity of fabricated biosensor for CRP detection. The biosensor has shown a fast response time of less than 10 s when used in the detected range of CRP using fast addition method as shown in Figure 12.8.

Furthermore in this study, effects of pH and temperature were examined for the proposed biosensor. In pH effect, the pH of 1×10^{-2} mg/L concentration was adjusted by adding 0.1 M HCl and NaOH solutions, and the observed response for different pH values is shown in Figure 12.9. The biosensor has shown maximum response at pH 7 that is quite close to optimum

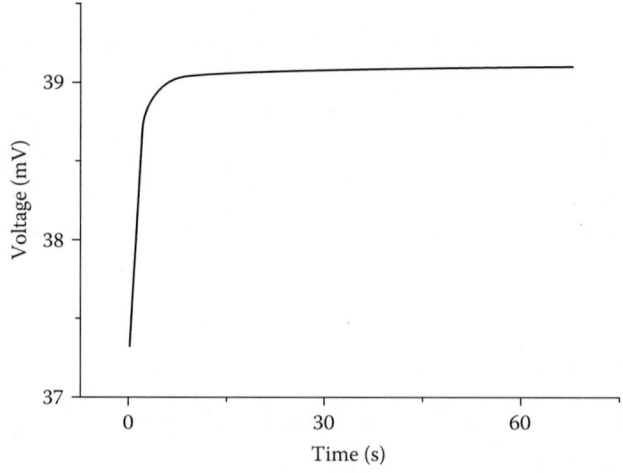

FIGURE 12.8
The response time of the sensor.

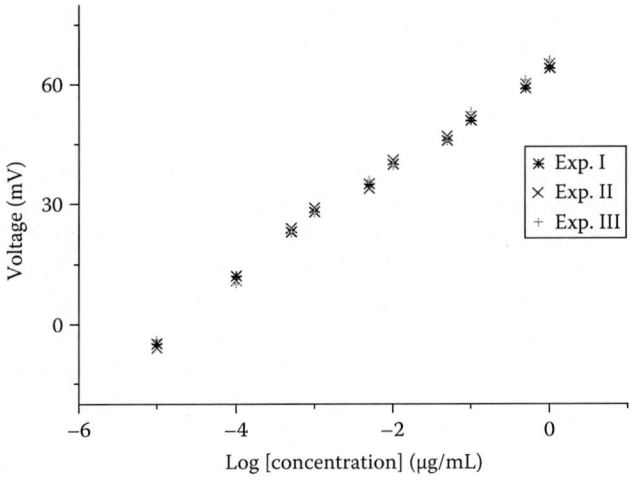

FIGURE 12.9
The repeatability of the sensor.

pH of antibody around 7.2. However, the response of the biosensor in higher pH values of the CRP solution was observed in descending order, which could be due to the slow response of the antibody for the attraction of CRP molecules. In acidic medium, the decreasing response of the biosensor might be attributed to the instability of ZnO due to its amphoteric nature. Besides pH effect, the influence of temperature on the performance of biosensor was monitored in order to know the change in the output potential at different temperatures due to the change in the mobility of charge produced close to the surface of ZnO nanotubes. Figure 12.10 shows the response of

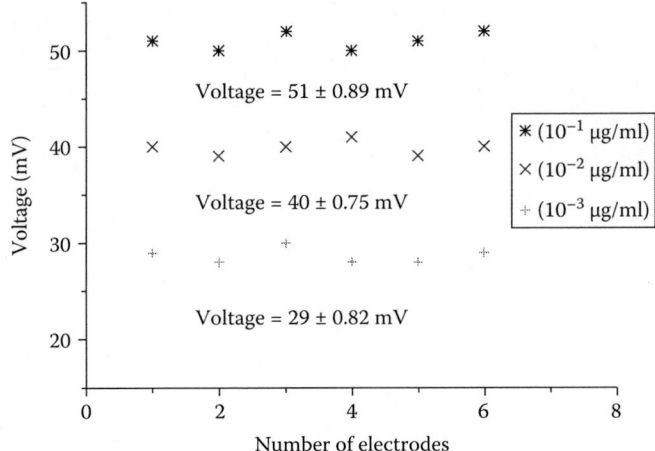

FIGURE 12.10
The reproducibility of the sensor.

the biosensor for different temperatures, and it can be seen that the biosensor has shown maximum response at 55°C. Afterward, the response of the biosensor can be observed in decreasing order, which could be due to the retardation of activity of proteins such as the antibody in the present study. Apart from this experiment, all other measurements were carried out at room temperature because of the simplicity in doing the experiments. The proposed antibody-immobilized ZnO nanotube-based biosensor has demonstrated good repeatability, reproducibility, selectivity, and stability, fast response time, and low limit of detection. The repeatability for a biosensor is studied to confirm the usability of the same sensor electrode for a certain period of time under the same set of conditions. In this study, one of the biosensor electrodes was examined for three consecutive days, and the biosensor revealed the same output response for this period of time as shown in Figure 12.11. The biosensor electrode was left at 4°C when not in use, and after each experiment using the CRP solution, the electrode was cleaned with phosphate buffer solution and dried respectively. In order to evaluate the biosensor-to-biosensor response, six biosensor electrodes were designed under the same set of conditions for the evaluation of reproducibility. In this experiment, these biosensor electrodes were used in different concentrations of CRP such as 1.0×10^{-1}, 1.0×10^{-2} and 1.0×10^{-3} mg/L, and the biosensor revealed good excellent reproducibility with relative standard deviation less than 5% as shown in Figure 12.12. In addition to repeatability and reproducibility, the selectivity of the proposed biosensor was examined using the separation solution method [83]. The response of antibody-immobilized ZnO nanotube-based biosensor was supervised in

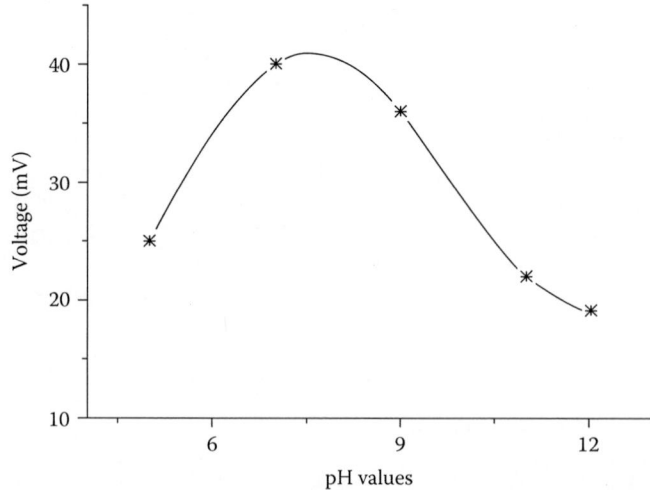

FIGURE 12.11
The effect of pH of solution on the response of the sensor.

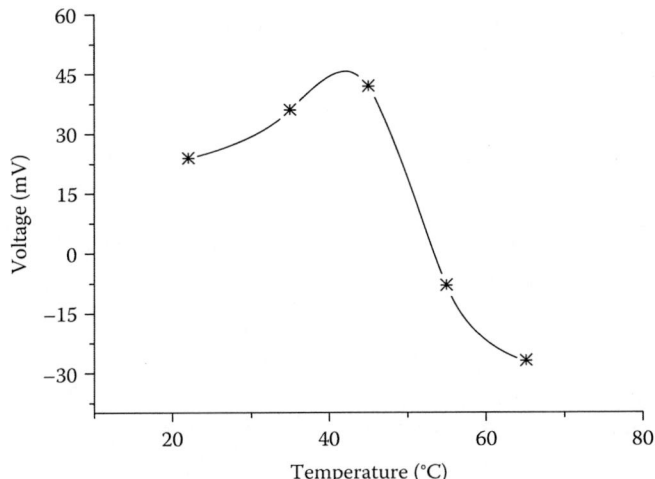

FIGURE 12.12
The effect of temperature on the response of the sensor.

the presence of common species in the human blood serum such as glucose, urea, uric acid, sodium, potassium, and iron ions. During monitoring, we observed the negligible response of biosensor for these species, and also it was observed that the biosensor could be used as a disposable biosensor for the detection of CRP.

12.3 Conclusion

The previous examples show the use of nanomaterials in biosensing due to their high surface to volume ratio, ability to detect the analyte from small volumes, and large signal to noise ratio, especially the wireless nanosensors that transfer message through the mobile. Moreover, nanosensors exhibit fast response time due to their high sensitivity toward the target substance. The trend towards the use of nanomaterials for biosensing applications in the present time describes the possible outcome of nanochips in the near future containing different nanostructures that will be used for the detection of various analytes at the same time. This is possible by using the same nanomaterial containing different independent single nanocrystals on the same chip, and these can be functionalized separately with different biosensitive materials, detecting their respective target analytes simultaneously.

References

1. L.C. Clark Jr., *Transactions of the American Society for Artificial Internal Organs* 2 (1956) 41–48.
2. L.C. Clark Jr., C. Lyons, *Annals of the New York Academy of Sciences* 102 (1962) 29–45.
3. S.J. Updike, J.P. Hicks, *Nature* 214 (1967) 986–988.
4. G.G. Guilbault, J. Montalvo, *Journal of the American Chemical Society* 91 (1969) 2164–2569.
5. M.M. Yanina, V.I. Sorochinskii, B.I. Kurganov, R.S. Balugyan, *Journal of the Analytical Chemistry of the USSR* 42 (1987) 1208–1213.
6. S. Arisawa, R. Yamamoto, *Thin Solid Films* 210 (1992) 443–445.
7. P. Bergveld, *IEEE Transactions on Biomedical Engineering* 17 (1970) 70–77.
8. P. Bergveld, *Sensors and Actuators B-Chemical* 88 (2003) 1–20.
9. J.D. Newman, A.P.F. Turner, R.S. Marks, D.C. Cullen, I. Karube, C.R. Lowe, and H.H. Weetall (Eds.), *Handbook of Biosensors and Biochips*, John Wiley & Sons, Ltd., New York (2007), pp. 1–16.
10. X.L. Luo, J.J. Xu, H.Y. Chen, *Chinese Journal of Analytical Chemistry* 32 (2004) 1395–1400.
11. Y. Zhang, M. Yang, N.G. Portney, D. Cui, G. Budak, E. Ozbay, M. Ozkan, C.S. Ozkan, *Biomedical Microdevices* 10 (2008) 321–328.
12. D. Cui, J. Nanosci, *Nanotechnology* 7 (2007) 1298–1314.
13. B. Pan, D. Cui, C.S. Ozkan, M. Ozkan, P. Xu, T. Huang, F. Liu, H. Chen, Q. Li, R. He, F. Gao, *The Journal of Physical Chemistry* C 112 (2008) 939–944.
14. B. Pan, D. Cui, P. Xu, Q. Li, T. Huang, R. He, F. Gao, *Colloids Surface A* 295 (2007) 217–222.
15. D. Cui, F. Tian, S.R. Coyer, J. Wang, B. Pan, F. Gao, R. He, Y. Zhang, *Journal of Nanoscience and Nanotechnology* 7 (2007) 1639–1646.
16. B. Pan, D. Cui, Y. Sheng, C. Ozkan, F. Gao, R. He, Q. Li, P. Xu, T. Huang, *Cancer Research* 67 (2007) 8156–8163.
17. X. You, R. He, F. Gao, J. Shao, B. Pan, D. Cui, *Nanotechnology* 18 (2007) 035701:1–035701:5.
18. A.M. Morales, C.M. Lieber, *Science* 279 (1998) 208.
19. J. Hu, T.W. Odom, C.M. Lieber, *Accounts of Chemical Research* 32 (1999) 435–445.
20. C.M. Lieber, *MRS Bulletin* 28 (2003) 486.
21. Y. Cui, X. Duan, Y. Huang, C.M. Lieber, and Z.L. Wang (eds.), *Nanowires and Nanobelts*, Springer Science and Business Media, New York (2003), pp. 3–68.
22. Y. Xia, P. Yang, Y. Sun, Y. Wu, B. Mayers, B. Gates, Y. Yin, F. Kim, H. Yan, *Advanced Materials* 15 (2003) 353–389.
23. Z.L. Wang, *Materials Today* 7 (2004) 26–33.
24. C.M. Niemeyer, *Angewandte Chemie (International Edition)* 40 (2001) 4128–4158.
25. W.C.W. Chan, D.J. Maxwell, X. Gao, R.E. Bailey, M. Han, S. Nie, *Current Opinion in Biotechnology* 13 (2002) 40–46.
26. J.L. West, N.J. Halas, *Annual Review of Biomedical Engineering* 5 (2003) 285–292.
27. P. Alivisatos, *Nature Biotechnology* 22 (2004) 47–52.
28. P. Gould, *Materials Today* 7 (2004) 36–43.

29. Y. Cui, X. Duan, J. Hu, C.M. Lieber, *The Journal of Physical Chemistry B* 104 (2000) 5213–5216.
30. X. Duan, Y. Huang, Y. Cui, J. Wang, C.M. Lieber, *Nature* 409 (2001) 66–69.
31. Y. Cui, C.M. Lieber, *Science* 291 (2001) 851–853.
32. Y. Huang, X.F. Duan, Y. Cui, L.Y. Lauhon, K.H. Kim, C.M. Lieber, *Science* 294 (2001) 1313–1317.
33. M.S. Gudiksen, L.J. Lauhon, J. Wang, D.C. Smith, C.M. Lieber, *Nature* 415 (2002) 617–620.
34. Y. Huang, X. Duan, Y. Cui, C.M. Lieber, *Nano Letters* 2 (2002) 101–104.
35. L.J. Lauhon et al., *Nature* 420 (2002) 57–61.
36. Y. Cui, Z. Zhong, D. Wang, W.U. Wang, C.M. Lieber, *Nano Letters* 3 (2003) 149–152.
37. M.C. McAlpine, R.S. Friedman, S. Jin, K.H. Lin, W.U. Wang, C.M. Lieber, *Nano Letters* 3 (2003) 1531–1535.
38. Z. Zhong, D. Wang, Y. Cui, M.W. Bockrath, C.M. Lieber, *Science* 302 (2003) 1377–1379.
39. N. Panev, A.I. Persson, N. Sköld, L. Samuelson, *Applied Physics Letters* 83 (2003) 2238.
40. S. Jin, D. Whang, M.C. McAlpine, R.S. Friedman, Y. Wu, C.M. Lieber, *Nano Letters* 4 (2004) 915–919.
41. A.B. Greytak, L.J. Lauhon, M.S. Gudiksen, C.M. Lieber, *Applied Physics Letters* 84 (2004) 4176–4178.
42. Y. Wu, J. Xiang, C. Yang, W. Lu, C.M. Lieber, *Nature* 430 (2004) 61–65.
43. G. Zheng, W. Lu, S. Jin, C.M. Lieber, *Advanced Materials* 16 (2004) 1890–1893.
44. M.T. Bjork, C. Thelander, A.E. Hansen, L.E. Jensen, M.W. Larsson, L.R. Wallenberg, L. Samuelson, *Nano Letters* 4 (2004) 1621–1625.
45. D.P. Nikolelis, M. Mitrokotsa, *Biosensors and Bioelectronics* 17 (2002) 565–572.
46. G.P. Nikoleli, M.Q. Israr, N. Tzamtzis, D.P. Nikolelis, M. Willander, N. Psaroudakis, *Electroanalysis* 24 (2012) 1285–1295.
47. M.Q. Israr, K. ul Hasan, J.R. Sadaf, I. Engquist, O. Nur, M. Willander, B. Danielsson, *Journal of Biosensors and Bioelectronics* 2 (2011) 1000109
48. X. Che, R. Yuan, Y. Chai, J. Li, Z. Song, W. Li, X. Zhong, *Colloids and Surfaces B: Biointerfaces* 84 (2011) 454–461.
49. Y. Cui, Q. Wei, H. Park, C.M. Lieber, *Science* 293 (2001) 1289–1292.
50. E. Comini, G. Faglia, G. Sberveglieri, Z. Pan, Z.L. Wang, *Applied Physics Letters* 81 (2002) 1869–1871.
51. X.T. Zhou, J.Q. Hu, C.P. Li, D.D.D. Ma, C.S. Lee, S.T. Lee, *Chemical Physics Letters* 369 (2003) 220–224.
52. C. Li, D. Zhang, X. Liu, S. Han, T. Tang, J. Han, C. Zhou, *Applied Physics Letters* 82 (2003) 1613.
53. J. Hahm, C.M. Lieber, *Nano Letters* 4 (2004) 51–54.
54. W.U. Wang, C. Chen, K. Lin, Y. Fang, C.M. Lieber, *Proceedings of the National Academy of Science United States of America* 102 (2005) 3208–3212.
55. F. Patolsky, G. Zheng, O. Hayden, M. Lakadamyali, X. Zhuang, C.M. Lieber, *Proceedings of the National Academy of Science United States of America* 101 (2004) 14017–14022.
56. A. Kolmakov, M. Moskovits, *Annual Review Materials Research* 34 (2004) 151–180.
57. Q. Wan, Q.H. Li, Y.J. Chen, T.H. Wang, X.L. He, J.P. Li, C.L. Lin, *Applied Physics Letters* 84 (2004) 3654–3656.

58. F. Patolsky, C.M. Lieber, *Materials Today* 8 (2005) 20–28.
59. S.P. Singh, S.K. Arya, P. Pandey, B.D. Malhotra, S. Saha, K. Sreenivas, V. Gupta, *Applied Physics Letters* 91 (2007) 063901–063903.
60. P.H. Yeh, Z. Li, Z.L. Wang, *Advanced Materials* 21 (2009) 4975–4978.
61. T.Y. Wei, P.H. Yeh, S.Y. Lu, Z.L. Wang, *Journal of the American Chemical Society* 131 (2009) 17690–17695.
62. N. Kumar, A. Dorfman, J.I. Hahm, *Nanotechnology* 17 (2006) 2875–2881.
63. S.M.U. Ali, O. Nur, M. Willander, B. Danielsson, *IEEE Transactions on Nanotechnology* 8 (2009) 678–683.
64. S.M.U. Ali, O. Nur, M. Willander, B. Danielsson, *Sensors and Actuators B* 145 (2010) 869–874.
65. M.H. Asif, S.M.U. Ali, O. Nur, M. Willander, C. Brännmark, P. Strålfors, U. Englund, F. Elinder, B. Danielsson, *Biosensors and Bioelectronics* 25 (2010) 2205–2211.
66. S.M.U. Ali, M.H. Asif, A. Fulati, O. Nur, M. Willander, C. Brännmark, P. Strålfors, U.H. Englund, F. Elinder, B. Danielsson, *IEEE Transaction on Nanotechnology* 10 (2011) 913–919.
67. S.M.U. Ali, N.H. Alvi, Z.H. Ibupoto, O. Nur, M. Willander, B. Danielsson, *Sensors & Actuators: B Chemical* 2 (2011) 241–247.
68. E. Topoglidis, E. Palomares, Y. Astuti, A. Green, C.J. Campbell, J.R. Durrant, *Electroanalysis* 17 (2005) 1035–1041.
69. S.M. Al-Hilli, R.T. Al-Mofarji, M. Willander, *Applied Physics Letters* 89 (2006) 073119-1–073119-3.
70. S.M. Al Hilli, M. Willander, A. Öst, P. Strålfors, *Journal of Applied Physics* 102 (2007) 084304.
71. S.M.U. Ali, O. Nur, M. Willander, B. Danielsson, *IEEE Transactions on Nanotechnology* 8 (2009) 678–683.
72. J.X. Wang, X.W. Sun, A. Wei, Y. Lei, X.P. Cai, C.M. Li, Z.L. Dong, *Applied Physics Letters* 88 (2006) 233106–233109.
73. S.M. Al-Hilli, M. Willander, *Nanotechnology* 20 (2009) 175103.
74. S.M. Al-Hilli, M. Willander, *Sensors* 9 (2009) 7445–7480.
75. X. Duan, R. Gao, P. Xie, T. Cohen-Karni, Q. Qing, H.S. Choe, B. Tian, X. Jiang, C.M. Lieber, *Nature Nanotechnology* 7 (2011) 174–179.
76. S.M.U. Ali, M. Kashif, Z.H. Ibupoto, M. Fakhar-e-Alam, U. Hashim, M. Willander, *Micro & Nano Letters* 6 (2011) 609–613.
77. S.M.U. Ali, Z.H. Ibupoto, C.O. Chey, O. Nur, M. Willander, *Chemical Sensors* 1 (2011) 19.
78. Z.H. Ibupoto, S.M.U. Ali, K. Khun, M. Willander, *Journal of Biosensors and Bioelectronics* 2 (2011) 3.
79. Z.H. Ibupoto, N. Jamal, K. Khun, M. Willander, *Sensors and Actuators B* 166–167 (2012) 809–814.
80. C.C. Jung, H.T. Yen, C.C. Chien, *IEEE Sensors Journal* 8 (2008) 1571–1577.
81. J.C.B. Fernandes, L.T. Kubota, G.D.O. Neto, *Analytica Chemica Acta* 385 (1999) 3–12.
82. R.H. Garrett, C.M. Grisham, *Biochemistry*, 2nd edn., Saunders College Publishing, Orlando, FL (1999).
83. P.R. Buck, E. Lindneri, *Pure and Applied Chemistry* 66 (1994) 2527–2536.

13

Bioelectronic Tongues

Manel del Valle, Xavier Cetó, and Manuel Gutiérrez-Capitán

CONTENTS

13.1 Introduction to BioET

There is unpaired activity in the design of new sensors and biosensors, normally directed to the implementation of new concepts, designs, or configurations, in all cases heading to improved biodevices showing perfect selectivity. Opposite to the search of biosensors with better specificity, there is a different approach that appeared in the late 1990s that proposes the use of arrays of sensors in order to obtain some added value in the generation of analytical information. Some variants of this approach are the use of extra sensors for the detection of malfunctioning episodes, the design of parallel multidetermination schemes to accelerate sample throughput, or the assembly of sensor arrays with cross-sensitivity features, which has led to what is referred to as electronic noses and tongues. According to the agreed International Union of Pure and Applied Chemistry (IUPAC) definition [1], an electronic tongue (ET) is "a multisensor

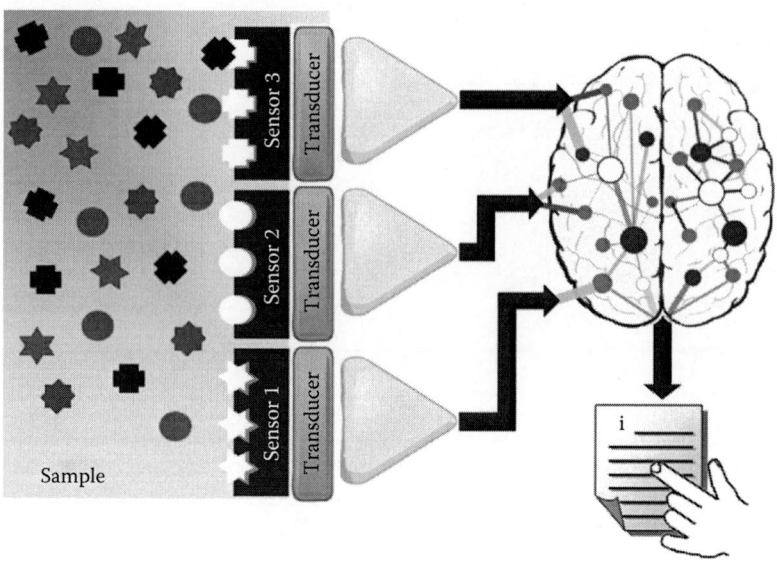

FIGURE 13.1
The use of the sensor array as key concept in the development of an ET analysis system.

system, consisting of a number of low-selective sensors using advanced mathematical procedures for signal processing based on pattern recognition and/or multivariate data analysis—artificial neural networks (ANNs), principal component analysis (PCA), etc." The underlying motivation of ETs is different from the general trend, that is, to use low-selectivity sensors or with cross-response features, a prerequisite for the development of these biomimetic systems, outlined in Figure 13.1. In these ET systems, each sensor uses certain recognition element to differentiate the response. This can be different ionophores responsible for the potentiometric response, different metal electrodes responsible for different redox behavior, or different catalysts responsible for different voltammetric response. Many research papers are found along the ca. one-decade history of ETs that use sensors of different types, but a new class is emerging lately, which has received the name bioelectronic tongue (BioET); this specific type is characterized by including one or several biosensors into the sensor array. To be considered a biosensor, the sensor's recognition element must be of biological origin, namely, enzymes, antigens, antibodies, nucleic acids, receptor proteins, cells, or even tissues. Our research group was pioneer in this variant, developing a BioET with voltammetric biosensors for glucose [2], another BioET with voltammetric biosensors for phenolic compounds [3], or also, a BioET employing potentiometric biosensors for the determination of urea [4].

13.1.1 Building of a BioET

An ET can be defined as an analytical system applied to liquid analysis formed by a sensor array in order to generate multidimensional analytical data, plus a chemometric processing tool to obtain the sought information from these complex data. Recent reviews can be consulted in the literature to check the extension of variants that have been employed up to now to develop these systems [5]. Although there are ETs employing optical sensors or piezoelectric (mass) sensors, most of the described works employ sensors of the electrochemical type [6]. Among these, many exploit sensors of potentiometric [7,8] or voltammetric type [9], although systems employing impedimetric sensors are also significant [10]. To respond to the expectation in the sensor community, recent special issues of significant journals in the analytical chemistry field have appeared dealing with the topic [11,12].

With respect to the chemometric tools that may be used with (Bio)ET systems, although this is one of their chief parts, a detailed description is out of the scope of this chapter and they will only be mentioned as used; a proper exposition can be consulted in recent reviews, well for ET systems [13], or for electronic nose systems [14], that is, the equivalent analytical systems but using gas sensors and applied to gas media, given data processing details are mostly equivalent between both.

Basically, the lay reader needs to know that certain data processing algorithms are more devised for the identification/classification of samples, such as PCA; therefore, they are used for qualitative applications. Other procedures, for example, regression using partial least squares (PLS), are specially conceived for quantification purposes, consequently mainly used for multi-determination applications [15]. The mentioned chemometric tools are conventional pattern recognition techniques, in essence of linear nature, which can be somehow limited if the sensors considered have clear nonlinear behavior. To improve results, researchers have suggested the use of ANN, which is a massively parallel computing technique, especially suited to nonlinear sensor responses and very much related to human pattern recognition [16]. Concerning the experience in our laboratory, we are most in favor of the uses of ANNs, as these are very powerful modeling tools, amenable to both qualitative and quantitative applications [17].

Figure 13.2 sketches the operation of an ET system based on an ANN model. In this case, it has developed a multidetermination application, that is, to simultaneously yield the concentrations of two species A and B from the readings provided by a sensor array. Measures are directly fed to an input layer of elements that distribute information to a number of processing elements in the hidden layer. The operation is directly inspired by the biological brain, as each previous element acts as a neuron; then, it is activated if enough information is introduced into it. The sought information (in this case concentration values of the two species A and B) is obtained from the output layer, also by processing its entering information as just specified. To develop the proper response model, one must obtain the proper weight connections distributing the information, in a process known as backpropagation; for the model, the way to activate the output of each neuron is also important, that is, the transfer function. To obtain the weights, a mathematical optimization procedure is followed, in which known samples are first

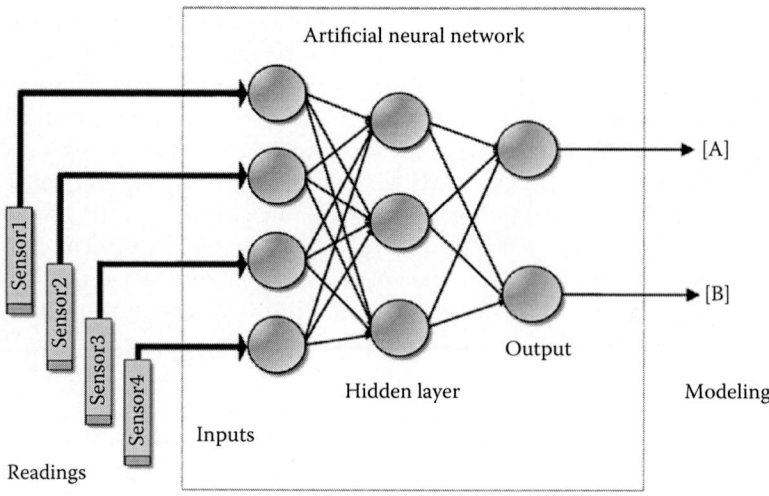

FIGURE 13.2
The ET concept applied for the multidetermination of several species employing a neural network model.

presented to the system, and a learning algorithm is applied (this is the training phase of the model building). Other details of the final configuration are the number and layers of neurons used, also to be obtained by trial and error. Once the response has been built, the recommended method is to use new data to verify its performance, as any other validation should be biased by previously known information [16]. For this purpose, new samples are presented to the system, and the obtained responses are compared to those expected (testing phase). A usual procedure to check these data is to build obtained versus expected comparison graphs, and use the correlation coefficient, slope, and intercept of the linear regression of the data as indicators of the goodness of modeling. For each modeled species, a sufficiently significant correlation should be obtained, and the general trend should be absent from any bias. As it has been said, the use of ANNs will normally provide better results in comparison with other tools, especially if there are nonlinearities in the biosensor response. But the drawback is that one needs to provide it with enough number of samples, so the weight coefficients (the training) can be obtained with enough confidence; it should also be remembered that the final results are not to be conditioned by which samples are designated for the training phase and which others for testing. These conflicts have been counterbalanced with the use of sequential flow systems capable of preparing in automated operation a sufficiently large set of samples [18]. For a detailed description of the use of ANNs to build multidetermination response models, the reader can refer to [19]. In [20], the use of the shareware software EasyNN-plus (http://www.easynn.com/dlennp.htm) has been shown for that purpose.

One would expect that biosensors don't need to be used in array mode, as they usually involve very selective, almost specific determinations.

But although of the general thought, biosensor use can present response to a group of specific substances, for example, sugars or pesticides, or may show interfering effects by tertiary substances. If one creates a set of biosensors, where each biosensor recognizes one or a group of analytes or compensates for the presence of interferents, then with the help of advanced statistical methods, very accurate characterization of the sample can be obtained. Also the qualitative applications are reported, where the analytes discriminated by the biosensors help in the classification or identification of sample types. A third possibility is still listed, which is the inhibitory effect; BioETs have been devised also to distinguish between inhibitors, for example, between the different pesticides that alter the response of acetylcholinesterase biosensors.

One of the precautions to be taken for the correct operation of the BioET is that colinearity must be absent; that is, every element forming the array must originate a signal that should not be linearly dependent on the rest of the biosensors. This can be accomplished in the following different ways: one is through the use of different recognition elements or through the use of biological elements of equivalent function but different origin (e.g., the use of equivalent enzymes but from different animals to show different responses, or in more advanced way, through the use of genetically engineered elements); the other is by affecting the response of the biological element, for example, through the use of chemical modifiers, that is, the use of redox mediators to modify the voltammetric responses or also with the use of physical modifiers, for example, the use of metallic catalysts or new carbon forms (e.g., carbon nanotubes).

13.1.2 BioET in the Literature

Several designs of BioETs can be found in the literature. For the multisensory approach to raise interest, it must bring some added value, that is, it should provide additional information different to the individual concentrations yielded by each single sensor. This can be the imaging of a concentration profile from a multichannel portrait or some error prevention ability gained in the approach. The potential of these new tools lies in new aspects of analytical information that may be obtained, different from the simple presence/absence of analytes; examples of these are the correction of a matrix effect, the obtaining of some characteristic not directly attainable through a chemical composition, like the identification of a sample variety or the correlation versus a perception from a panel of sensory experts.

The type of applications devised with BioETs can be grouped, first, in qualitative type, like identification of species, classification of sample varieties, or recognition of adulteration episodes; conversely, the application can be of quantitative type, normally the multidetermination of a set of chemical species, an interesting objective in process control. A different, more bioinspired trend is the artificial taste concept [21] devised with the goal of performing automated taste perception, especially in the industrial field. This idea, initiated by the works of Prof. Toko in Japan [22], tries to develop a sensor array inspired in the tongue's papillae and/or responding to the basic taste types (sweet, sour, salty, and bitter, plus the far-east concept *umami*, which means delicious); the ultimate goal is to mimic the human taste assessment in cases where a human expert is not feasible, 24 h continuous automatic control, poisonous/extreme conditions, etc. Recent attempts, completely biomimetic, have been described with immobilization of taste receptors on microelectrodes and recording of its nerve signals [23].

The first analytical system employing an array of biosensors was described by Magalhães and Machado in an application to determine creatinine in urine samples [24]. The approach used two identical creatinine potentiometric biosensors employing the enzyme creatinine amidohydrolase and four PVC-membrane ion-selective electrodes (ISEs) for K^+, Na^+, NH_4^+, and Ca^{2+}, intended to compensate any interference from urine matrix. The setup was calibrated employing a multivariate response model based on PLS, and the results validated against the standard Jaffé method.

In our laboratory, the first potentiometric BioET was designed employing an array formed by two urea biosensors based on the enzyme urease, covalently linked to carboxylated PVC (PVC-COOH), plus ISEs for H^+, K^+, Na^+, NH_4^+, and cationic generic response [4]. Response models were built employing PLS and ANN, with slightly better performance for the latter. With this array, it was possible to determine urea in clinical samples without the need to separate endogenous ammonium or the interfering ions sodium and potassium. In the same progress line, the system was enlarged with an additional potentiometric biosensor employing creatinine amidohydrolase to determine urea, creatinine, and alkaline interferents [25]. Such a BioET is ready at the moment for its clinical application in studies of renal function or the monitoring of hemodialysis processes.

Next, significant examples of the use of voltammetric biosensors must be presented. The experience of our research group in the development of biosensors has allowed the immobilization of the biological component on the surface of the electrochemical transducers and also to incorporate it in the bulk of biocomposite materials, showing a large versatility, especially for enzyme-based biosensors.

In a significant contribution from our laboratory [2], Gutés devised a voltammetric BioET working in a flow system using different glucose biosensors, formed by epoxy–graphite biocomposites with glucose oxidase enzyme and different metal catalysts (Pt, Pd, and Au–Pd) to promote a differentiated response. The system was successfully applied to the multidetermination of glucose and ascorbic acid in an attempt to determine simultaneously glucose plus its typical interfering species in biological fluids without any separation stage or barrier membrane. In this case, the sensor array used the same enzyme, but the response in the array could be differentiated through the use of different metal catalysts. The metal catalysts improved the biosensor response by decreasing the oxidation potential for hydrogen peroxide from the enzymatic reaction. This automated ET was applied in juice samples.

A more conventional approach is the use of separate sensing channels for each analyte. For example, there is the work of Moser et al. that describes a flow system with an enzymatic array to determine glutamine, glucose, lactate, and glutamate in food products [26]. However, this case did not obtain any profit from the cross-response to counterbalance interfering species or matrix effects. In another work [27], a screen-printed enzyme array employing tyrosinase, horseradish peroxidase, acetyl cholinesterase, and butyrylcholinesterase was applied in environmental analysis to identify PCA in different water qualities like untreated, alarm, alert, and normal. A pioneering work in the literature used an array employing enzymes monoamine oxidase, tyramine oxidase, and diamine oxidase plus a blank electrode to resolve mixtures of biogenic amines histamine, tyramine, and putrescine [28]. The amperometric array system worked at a fixed potential of 700 mV on screen-printed substrates, to perform multiple determination on fish, meat products, sauerkraut, beer, dairy products, wine, and further fermented foods; biogenic amine concentrations were predicted with the help of an ANN system and compared with the conventional high performance liquid chromatography (HPLC) determination.

There are several works in the literature related to the determination of phenols and polyphenols. In an interesting case study [27], Solná and Skládal attempted the determination of various phenolic compounds combining the responses of a multiple screen-printed sensor array, which employed amperometric biosensors made using laccase, tyrosinase, and peroxidase enzymes. The amperometric flow-injection determination of phenolic compounds at discrete applied potentials using the multichannel biosensor was demonstrated at the nM level, although this work did not use the multivariate approach with the existing cross-responses. A contribution more in consonance with the BioET approach was the determination of mixtures of phenolic compounds (phenol, catechol, and 3-cresol) from the overlapping voltammogram obtained with the tyrosinase enzyme biosensor [3]. ANN was the tool used to resolve the mixture at the mM level. Torrecilla and coworkers derived a similar work shortly afterwards to determine the phenolic content in olive oil mill wastewater, in which the main difference was the use of the laccase enzyme for building the biosensor [29].

In an interesting approach, a BioET was devised employing three enzymatic biosensors based on tyrosinase and three different rare earth phthalocyanines as electron mediators to evaluate the changes that occur along the aging of beers. For this purpose, alcoholic and nonalcoholic beers were analyzed using cyclic voltammetry, showing significant changes during the aging process. The features extracted from the cyclic voltammograms were used to perform PCA and linear discriminant analysis (LDA) in order to discriminate between their types and age [30]. Advantages of the approach for the flavor sensory evaluation during the elaboration of nonalcoholic beer were also reviewed recently [31]. In a further attempt to use the phenol-responsive enzymes, the coimmobilization of laccase and tyrosinase on the same electrode using glutaraldehyde as a cross-linker and Nafion-ion exchanger as a protective layer provided a biosensor with the integrated response of two enzymes to five individual polyphenols. This system showed a good agreement between the estimated total phenol index in beer samples and that obtained using the Folin–Ciocalteu reagent [32].

In our laboratory, we developed a voltammetric BioET formed by an array of epoxy–graphite biosensors, bulk modified with redox enzymes laccase and tyrosinase, aimed to the simultaneous determination of different polyphenols in the wine matrix. Departure information was the set of voltammograms generated with the biosensor array, and ANNs were used for the extraction and quantification of phenolic compounds relevant in wine. The system was applied to resolve ternary mixtures of catechol, caffeic acid, and catechin [33]. In a subsequent work, the Folin–Ciocalteu index as a global content of polyphenolic compounds in wine was also correctly predicted [34].

An interesting variant of BioETs is to use inhibition-based biosensors, such as those employed for the determination of pesticides; the principle here is to use enzymes from different biological origins, which may show different degrees of inhibition to a set of substances. A first biosensor array was constructed employing acetylcholinesterase enzymes from different origins in order to obtain a differentiated inhibitory effect from different pesticides [35]. The array employed the wild-enzyme type from electric eel (EE) and two different genetically modified enzymes (B1 and B394). In this way, an inhibitory BioET was constructed to resolve dichlorvos and carbofuran mixtures at the nM level. As pesticide inhibition was irreversible and biosensors cannot be reused, responses were obtained from single-use, disposable, screen-printed amperometric electrodes. Recently, a similar strategy was attempted but employing an automated flow injection analysis (FIA) system, in which dichlorvos and methyl paraoxon pesticide mixtures were resolved [36]. These initial results have been extended to new pesticides, in this case the ternary mixture of chlorpyrifos oxon, chlorfenvinfos, and azinphos-methyl oxon [37].

Similar to the acetylcholinesterase inhibition, the system of sarcosine oxidase was explored for the determination of carboxylic acids [38]. This enzyme was inhibited by citric, malic, succinic, acetic, and formic acids, while tartaric and lactic acid did not provide any inhibition effect. This response feature might be used as a starting point to monitor mixtures of carboxylic acids, for example, in wine production.

BioET principles might be extended to multiplex detection with arrays of biosensors based on antibodies or DNA, with the interesting advantage that data treatment could be used to solve a limited specificity [39]. In any case, it is evident that there is a trend in the food and clinical diagnostic field to adopt multichannel schemes as inspired in the DNA chips, which now evaluate the electrochemical transduction for the sake of simplicity and portability [40]. From the works found in the literature, there is the interesting work of Wilson and Nie, who described a multiplexed electrochemical immunosensor for the determination of seven tumor markers using a single chip [41]. The simultaneous determination of carcinoembryonic antigens, α-fetoprotein, β-human chorionic gonadotropin, ferritin, or carbohydrate antigens 15-3, 125, or 19-9, were demonstrated at the ng/mL level, although the system did not employ any cross-response feature. A more advanced system [42], which is able to correct the cross talk between assays employing the cross-reactivity profiles of eight antibodies for androgenic anabolic steroids, permitted the determination of four of these (androstenedione, methyl boldenone, progesterone, and stanozolol) in the 0.1–300 nM range.

Analogously, it is fascinating that the design of electrochemically addressed systems for multiplex sensing of DNA sequences [43], a tool that may be the basis for systems aimed at clinical diagnostics or food control [44]. In our laboratory, a pioneer work to detect two model genes from a single electrode was described [45]. Multiplex biosensing idea was shown with an impedimetric genosensor scheme in which two DNA probes were immobilized on an electrode surface, and hybridization experiments were monitored by following its electrochemical impedance spectrum. The direct observation did not allow discernment of which DNA targets were present in the sample. The application of the ET concept, where an ANN response model was built from the complex impedance spectra, permitted the correct identification of all the training cases in a leave-one-out scheme. Equivalent schemes were also developed for protein detection by an array of aptamer-modified electrodes [46], demonstrated in the model work with the detection of human IgE.

From the scope of the use of BioETs, this chapter will now present two basic applications, one employing potentiometric sensors and the other with voltammetric sensors. The first example will employ urea and creatinine biosensors based on urease and creatinine deiminase, respectively, covalently immobilized onto ammonium-selective electrodes and its coupling in an array with ISEs sensitive to ammonium, potassium, and sodium. Generic sensors to alkaline ions will be also included. The response model based on ANNs will attempt the simultaneous determination of urea and creatinine plus ammonium, potassium, and sodium in clinical samples.

In the second example, a BioET formed by an array of four voltammetric enzyme-modified biosensors will be applied in the analysis of phenolic compounds found in beers. One blank electrode, a laccase biosensor, a tyrosinase biosensor, and one electrode bulk modified with copper nanoparticles will form the array; these modifiers will be selected in order to incorporate differentiated or catalytic response toward phenolic compounds and aimed to their simultaneous resolution. In this case, the highly complex electrochemical responses obtained from the biosensors will need special preprocessing employing the discrete wavelet transform in order to extract the significant components and compress the departure information. Then, the obtained wavelet coefficients will feed an ANN model specially trained to predict major phenolic compounds found in beer: ferulic, gallic, and sinapic acids.

To finish this introduction, Table 13.1 summarizes a selection of the applications, where the different research in the clinical, food, and environmental field has been depicted.

TABLE 13.1

Table Summarizing Relevant Applications Employing the BioET Concept

Application	Field	Detection Principle	Description	Data Processing	References
Resolution of mixtures of biogenic amines: histamine, tyramine, and putrescine	Food	Amperometry at fixed potential	Array employing enzymes monoamine oxidase, tyramine oxidase, and diamine oxidase plus a blank electrode	ANN	[28]
Determination of creatinine	Clinical	Potentiometry	Array employing creatinine amidohydrolase biosensor plus ISEs for alkaline ions	PLS	[24]
Determination of urea and potassium	Clinical	Potentiometry	Array employing two urease biosensors plus sodium and potassium ISEs	PCR, PLS	[47]
Determination of urea and interfering ions	Clinical	Potentiometry	Array employing urease biosensor plus ISEs for alkaline ions	ANN	[4]
Determination of urea, creatinine, and interfering ions	Clinical	Potentiometry	Array employing urease biosensor, creatinine amidohydrolase biosensor, plus ISEs for alkaline ions	ANN	[25]
Phenols	Environmental	Amperometry at fixed potential	Screen-printed enzyme electrode array employing laccase, tyrosinase, and HRP	Do not use cross-response	[48]
Glucose + interfering species	Food, clinical	Voltammetry, LS	Automated flow system using an array formed by different GOD enzyme sensors with different metal catalysts: Pt, Pd, and Au–Pd	ANN	[2]
Phenol index	Environmental	Voltammetry, LS	Laccase biosensor overlapping response is used to determine global phenolic content in olive oil mill wastewater.	ANN	[29]

Aging of beer	Food	Voltammetry, CV	Sensor array made from three enzymatic biosensors employing laccase and three different phthalocyanines as redox mediators	PCA, LDA, and ANN	[30]
Resolution of mixtures of catechol, caffeic acid, and catechin in wine	Food	Voltammetry, CV	Sensor array made from enzymatic biosensors using tyrosinase, laccase, Cu nanoparticles, and a blank electrode	Feature extraction + ANN	[33]
Total polyphenol content (Folin–Ciocalteu index) in wine	Food	Voltammetry, CV	Sensor array made from enzymatic biosensors using tyrosinase, laccase, Cu nanoparticles, and a blank electrode	FFT pretreatment + ANN	[34]
Resolution of mixtures of ferulic, gallic, and sinapic acids in beer	Food	Voltammetry, CV	Sensor array made from enzymatic biosensors using tyrosinase, laccase, Cu nanoparticles, and a blank electrode	Slicing integral + ANN	[49]
Resolution of insecticides dichlorvos and methyl paraoxon	Environmental	Chronoamperometry	Automated flow system with a sensor array formed by different acetylcholinesterases, from EE and two different genetically modified enzymes B1 and 3394 from Drosophila	ANN	[36]
Resolution of insecticides chlorpyrifos oxon, chlorfenvinfos, and azinphos-methyl oxon	Environmental	Chronoamperometry	Flow system to record inhibition time transient caused by pesticides on a genetically modified acetylcholinesterase from Drosophila B131 enzyme electrode	ANN	[37]

(continued)

TABLE 13.1 (continued)

Table Summarizing Relevant Applications Employing the BioET Concept

Application	Field	Detection Principle	Description	Data Processing	References
Determination of the sum of citric, succinic, malic, acetic, and formic carboxylic acids in wine	Food	Chronoamperometry	Flow system to record inhibition time transient caused by carboxylic acids on a sarcosine oxidase enzyme electrode	Do not use cross-response	[38]
Determination of mixtures of 4 anabolic androgenic steroids	Clinical	Spectrophotometry at fixed wavelength	Multiple competitive ELISA immunoassays with different cross-reactivity profiles	K-NN	[42]
Multiplex determination of IgE	Clinical	Impedance	Aptamer probes immobilized by electrical addressing	Do not use cross-response	[46]
Multiplex determination of proteins	Clinical	Fluorescence at fixed wavelength	Aptamer probes immobilized on beads over a micromachined chip to perform capture and sandwich assay formats	Do not use cross-response	[50]

GOD, glucose oxidase; HRP, horseradish peroxidase; LS, linear sweep voltammetry; CV, cyclic voltammetry; K-NN, K-nearest neighbor classifier; FFT, fast Fourier transform.

13.2 BioET Employing Potentiometric Sensors

Urea is a typical analyte to be determined employing biosensors [51]. For this purpose, urease enzyme is used, both with employment of potentiometric biosensors and also with optical ones. The system reported here is a BioET, aimed for the determination of urea and creatinine as analytes of interest in hemodialysis processes. As urea is the main and final product of protein metabolism, its level often provides information on the nutritional status in a body system. On its side, creatinine is a marker of an amino acid pool, which high level indicates slow turnover of muscle protein. These two final metabolites are normally employed as indicators of waste products accumulated in patients with diminished renal function. Both molecules are transported by blood to the kidneys, where they are further filtered and eliminated in the urine. Therefore, it is quite common to determine these two species in serum samples [52]. However, an advanced renal failure would occur before a significant increase of these two compounds in blood is detected. For early detection of the different nephropathies, it is also necessary to quantify the amount of these two substances eliminated in the urine [53]. Equivalent determinations are also performed in the evaluation of performance of hemodialysis. Checking the literature, few biparametric systems exist allowing the simultaneous determination of urea and creatinine in clinical samples. The different systems found are based on enzymatic methods [54–58], and they need to remove any interference present before the analysis.

In this study, we present a BioET to analyze urine samples based on two types of potentiometric biosensors for urea and creatinine determination plus selective and generic ISEs for ammonium, potassium, and sodium [25]. These enzymatic electrodes and ISEs, based on PVC membranes, provide a cross-response towards the different species considered. The chemometric tool used to build the multivariate response model is an ANN, which is fed with the readings originated from the biosensor array. The major advantage of this system is not only to determine simultaneously urea and creatinine without any pretreatment step, but also to quantify their most severe interferences, ammonium, potassium, and sodium ions, these being also of clinical interest.

13.2.1 Fabrication of the Potentiometric Biosensors

Potentiometric sensors used were all solid-state ISEs with a solid contact made of a conductive composite. This is the habitual configuration in our laboratories [59]. Figure 13.3 is a photograph of the electrodes used. PVC and PVC-COOH membranes were formed by solvent casting the sensor cocktail onto the solid contact. PVC-COOH was employed as an anchoring matrix for enzymes needed in the urea and creatinine biosensors. The formulation of the different membranes used is outlined in Table 13.2.

FIGURE 13.3

Picture of the potentiometric sensors; beside on top, a mounting gland for watertight assembly in a vessel or conduction.

TABLE 13.2

Formulation of the Different Polymeric Membranes Used

Electrode	PVC	PVC-COOH	Plasticizer	Ionophore	Additive
Ammonium 1	33.0%	—	66.0% BPA	1.0% nonactin	—
Ammonium 2	—	33.0%	66.0% BPA	1.0% nonactin	—
Potassium	31.7%	—	66.7% DOS	1.0% valinomycin	0.6% KpClPB
Sodium	21.8%	—	70.0% NPOE	6.0% bis(12-crown-4)	2.2% KpClPB
Generic 1	29.0%	—	67.0% DOS	4.0% dibenzo-18-crown-6	—
Generic 2	27.0%	—	70.0% DBS	3.0% lasalocid	—

BPA, bis(1-butylpentyl) adipate; DOS, dioctyl sebacate; NPOE, o-nitrophenyl octyl ether; DBS, dibutyl sebacate; KpClPB, potassium tetrakis(4-chlorophenyl) borate.

The sensor array used consisted of 12 electrodes altogether: two urea biosensors, two creatinine biosensors, two ISEs for ammonium, two for potassium, two for sodium, and two electrodes of generic response toward alkaline ions (Generic 1 and 2). For the fabrication of urea and creatinine biosensors, the enzymes urease (EC 3.5.1.5) and creatinine deiminase (EC 3.5.4.21) were utilized, respectively. Both enzymes catalyze the hydrolysis of their respective biomolecules, obtaining ammonium (NH_4^+) ions as products. The concentration of the ammonium generated increased with the concentration of the metabolite. Therefore, in this study, an ammonium sensor was used as the transducer for both types of biosensors.

Immobilization of enzymes on the NH_4^+-selective PVC-COOH membrane (Ammonium 2, Table 13.2) was performed by the carbodiimide reaction, using N-(3-dimethylaminopropyl)-N'-ethylcarbodiimide hydrochloride. This reaction facilitated the formation of covalent amide bonds between the polymeric membrane and the enzyme. This immobilization method produces devices with good sensitivity and lifetime, given that it provides sensor surfaces more stable toward changes in pH, temperature, and ionic strength. Once washed in 50 mM Tris-HCl buffer (pH 7.5), biosensors were stored in the refrigerator with the same buffer to preserve the enzyme activity [60].

13.2.2 Potentiometric Characterization of the Biosensors

All the sensors were calibrated to check that they presented the correct behavior to be included in the array. In the case of the biosensors for urea and creatinine, the calibration curves were obtained by means of the method of analyte addition: the variation of potential originated by the addition of accumulated microvolumes of stock solutions in 50 mM Tris-HCl buffer pH 7.5 was determined. Tris-HCl buffer was chosen because it has a lower interference effect than other buffer solutions that contain alkaline ions. These interfering ions would increase the limit of detection (LD) considerably; thus, their concentration has to be controlled. In addition, pH 7.5 has been considered the most appropriate one, both for the catalytic activity of the enzymes and for the detection of ammonium ions. The calibration curves obtained for each species (urea and creatinine) are depicted in Figure 13.4.

As it can be observed, while urea biosensors show a sensitivity value around +30 mV/decade due to their enzyme kinetics, creatinine biosensors show a slope close to the Nernstian behavior, about +50 mV/decade. On the other hand, the LD is quite similar for both analytes, around 1.0×10^{-4} M. The linear range is around two decades of concentration, up to 1.0×10^{-2} M, for the two species, which is appropriate for the proposed application.

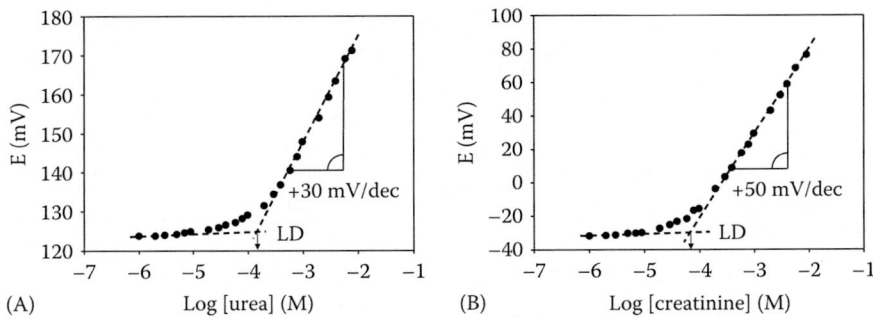

FIGURE 13.4
Typical calibration curves obtained for (A) urea and (B) creatinine using the prepared potentiometric biosensors.

In order to estimate the lifetime of the prepared biosensors, frequent calibrations in the Tris-HCl buffer were performed. In the case of urea biosensors, the results showed that in terms of sensitivity and LD, the loss of response was practically null along the 20 days of study. In the case of creatinine biosensors, the variation of these two parameters was less than 15% during 27 days. Therefore, the stability of the biosensor response was appropriate to be included in the array. Finally, to check the response time of these biosensors, some analyte additions were performed, while the potentials in mV were recorded every second. The results of these experiments showed that 95% of the final response was in the range 1–3 min for both cases.

13.2.3 Response Modeling of the BioET

The training of the potentiometric BioET was performed based on the emf readings of the biosensor array corresponding to a set of synthetic samples defined from a statistical experimental design. They were prepared in the laboratory using a Tris-HCl buffer as background solution and their details are summarized in Table 13.3. The composition of the 27 solutions used to build the five analyte response model was designed according to a fractional factorial design with three levels of concentration and five factors (3^{5-2}) [61]. The external validation set was formed by 13 additional synthetic solutions (test set) with their concentration values randomly generated. As it can be seen in Table 13.3, the concentration of the five analytes ranged over two orders of magnitude (urea and creatinine, from 0.1 to 10 mM; ammonium, from 0.005 to 0.5 mM; and potassium and sodium, from 0.05 to 5 mM), as defined by possible physiological levels and which supposes an extensive training space. Besides, the system was also applied to real urine samples collected from three volunteers and the results compared with reference determination procedures. To obtain a greater variability of the concentrations, eight new spiked urine samples were prepared by adding different concentrations of the five analytes.

Once the readings of potential were acquired for each sample, the next step was the construction of ANN model. Selecting the optimum topology of an ANN represented a hurdle, because of the difficulty to predict the optimum configuration in advance. In fact, this configuration was obtained by a trial and error procedure [16]. Fortunately, some parameters were defined by the system itself: the number of input neurons, which was 12 (one for each sensor from the array), and the number of output neurons, 5 (one for each considered analyte). Besides, our previous experience on potentiometric ETs and some preliminary tests allowed us to fix other variables: the transfer function for the output layer, as a linear one (*purelin*), the use of a single hidden layer, and the use of the Bayesian regularization as a learning algorithm with a learning rate of 0.1 and momentum of 0.4 [19].

Therefore, we studied the 20 possible configurations that result from combining the transfer functions *tansig* and *logsig* and a number of neurons in

TABLE 13.3

Composition of the Synthetic Solutions Used to Build the Multivariate Response Model

Solution	[Urea] (mM)	[Creatinine] (mM)	[NH$_4^+$] (mM)	[K$^+$] (mM)	[Na$^+$] (mM)
1	0.100	3.140	0.043	1.580	1.573
2	0.481	2.760	0.100	3.287	5.000
3	0.860	2.380	0.157	5.000	3.287
4	1.242	6.580	0.214	1.383	4.810
5	1.623	6.200	0.272	3.096	3.096
6	2.004	5.820	0.329	4.810	1.383
7	2.385	10.00	0.386	1.192	2.906
8	2.765	9.620	0.440	2.906	1.192
9	3.146	9.240	0.500	4.600	4.600
10	3.527	5.440	0.024	1.002	4.400
11	3.908	5.040	0.081	2.715	2.715
12	4.300	4.680	0.138	4.400	1.002
13	4.700	8.860	0.195	0.810	2.525
14	5.100	8.480	0.253	2.525	0.810
15	5.400	8.100	0.310	4.200	4.200
16	5.800	2.000	0.367	0.620	0.620
17	6.192	1.620	0.420	2.335	4.000
18	6.600	1.240	0.481	4.000	2.335
19	7.000	7.720	0.005	0.431	2.144
20	7.300	7.340	0.062	2.144	0.431
21	7.700	6.960	0.119	3.858	3.858
22	8.096	0.860	0.176	0.240	0.240
23	8.500	0.481	0.233	1.954	3.667
24	8.900	0.100	0.291	3.667	1.954
25	9.200	4.280	0.348	0.050	3.477
26	9.600	3.900	0.400	1.763	1.763
27	10.00	3.520	0.460	3.477	0.050
28[a]	8.600	2.581	0.129	0.123	4.500
29[a]	0.620	5.140	0.257	0.519	4.800
30[a]	1.000	5.461	0.100	2.732	1.410
31[a]	8.600	2.000	0.332	4.900	0.499
32[a]	7.400	8.460	0.420	2.610	3.570
33[a]	7.900	6.380	0.319	4.600	3.705
34[a]	7.400	1.340	0.067	1.574	2.854
35[a]	4.900	6.160	0.308	3.616	1.026
36[a]	2.229	0.530	0.026	2.440	4.497
37[a]	2.227	2.660	0.133	3.296	0.790
38[a]	2.550	1.920	0.430	2.100	2.218
39[a]	6.400	8.040	0.096	3.361	0.342
40[a]	7.909	2.280	0.400	4.603	0.900

[a] Samples that formed the external validation set.

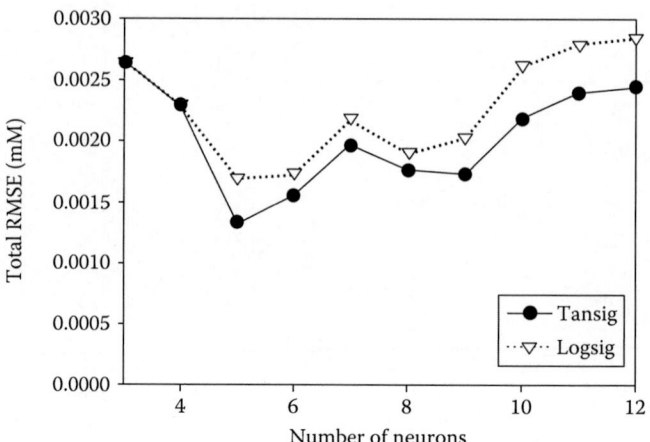

FIGURE 13.5
Plot of total RMSE (mM) versus the number of neurons in the hidden layer.

the hidden layer between 3 and 12. In order to choose the best transfer function, the root-mean-square error (RMSE, for each analyte and the total value, Equation 13.1) was examined for each ANN configuration. The representation of the total RMSE versus the number of neurons in the hidden layer is shown in Figure 13.5. These RMSE values are calculated as the mean of three replicates of the same configuration. As can be observed, although there is not a big difference between the RMSE of the two functions, generally smaller errors are obtained when using the *tansig* function:

$$RMSE = \sqrt{\frac{\sum_{ij}(c_{ij} - \hat{c}_{ij})^2}{5n - 1}} \tag{13.1}$$

After selecting the transfer function, the next step was to define the number of neurons in the hidden layer. Therefore, the slope, intercept, and correlation coefficient obtained from the found versus expected comparison graphs (for external validation set) were represented along the number of neurons employed in the hidden layer. These values were also obtained from the mean of three replicates. Thus, the choice of the best configuration was done considering the closeness to the ideal values of 1.0 for slope, 0.0 for intercept, and 1.0 for the correlation coefficient. As can be observed in Figure 13.6, this was achieved when six neurons were employed in the hidden layer.

13.2.4 Application

Figure 13.7 illustrates the performance of the optimized model for the external test set and for the real samples, comparing found versus expected values for urea, ammonium, potassium, sodium, and creatinine.

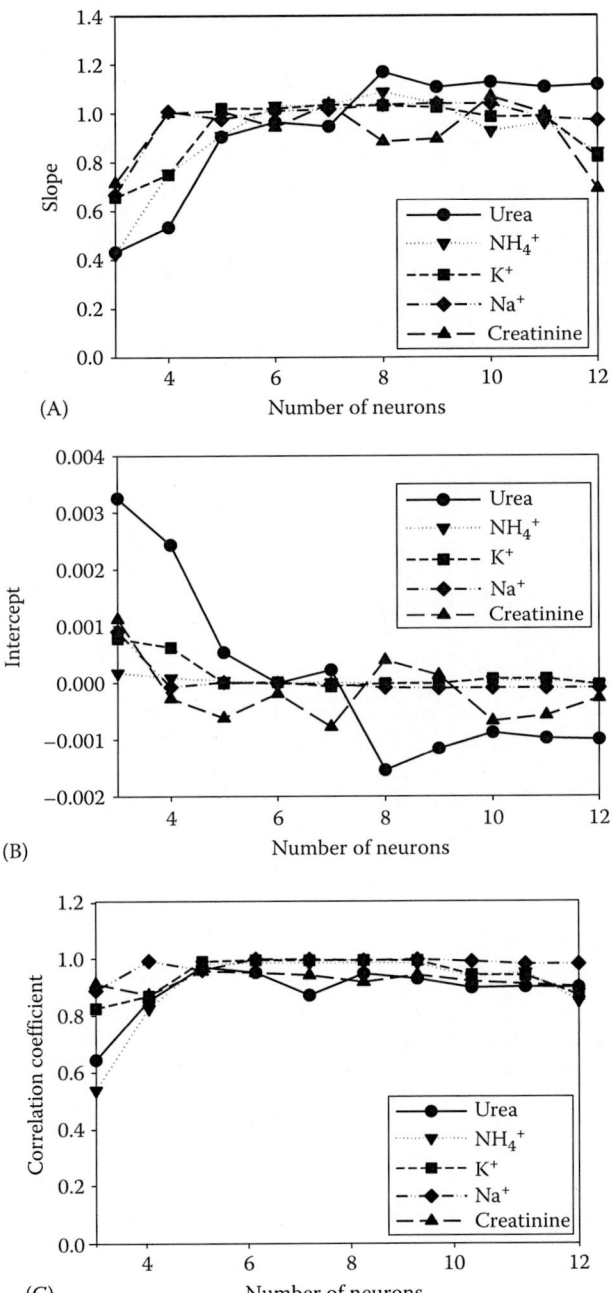

FIGURE 13.6
Plot of (A) slope, (B) intercept, and (C) correlation coefficients versus number of neurons in the hidden layer for the external test set.

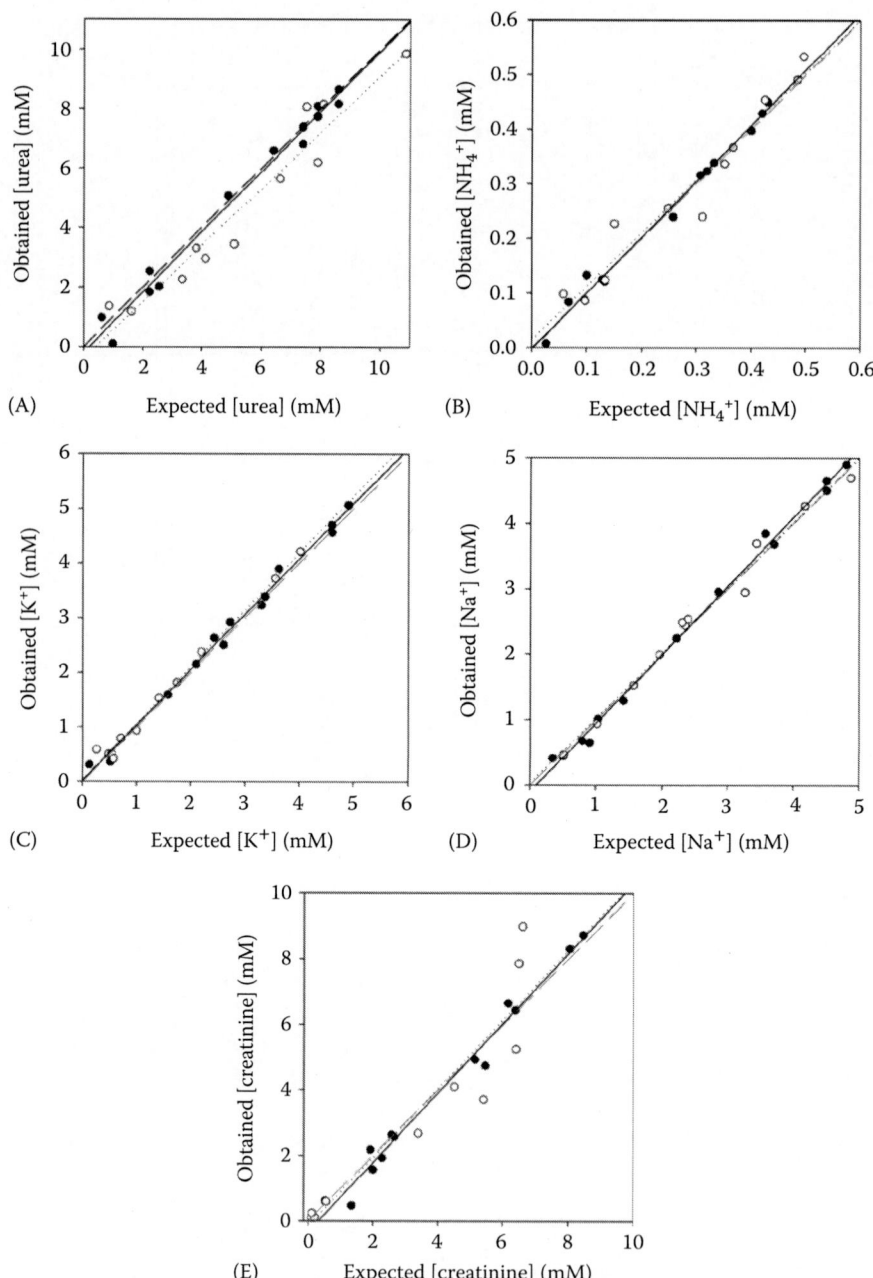

FIGURE 13.7
Performance of the optimized model for the samples in the test set (•) and the real urine samples (o) for (A) urea, (B) ammonium, (C) potassium, (D) sodium, and (E) creatinine.

TABLE 13.4

Regression Data for the Comparison of Results Provided by the Proposed BioET for the Five Considered Species, Considering the External Test Set and the Real Samples (Intervals Calculated at the 95% Confidence Level)

Analyte	Test Subset			Real and Spiked Samples		
	Correlation	Slope	Intercept	Correlation	Slope	Intercept
Urea	0.992	1.01 ± 0.08	-0.20 ± 0.51	0.967	0.95 ± 0.19	-0.40 ± 1.17
Creatinine	0.991	1.06 ± 0.09	-0.35 ± 0.45	0.940	1.05 ± 0.29	-0.17 ± 1.19
Ammonium	0.995	1.02 ± 0.07	0.00 ± 0.02	0.970	0.98 ± 0.18	0.01 ± 0.06
Potassium	0.996	1.02 ± 0.06	0.02 ± 0.18	0.995	1.04 ± 0.07	0.02 ± 0.14
Sodium	0.998	1.04 ± 0.04	-0.10 ± 0.12	0.991	0.99 ± 0.10	0.04 ± 0.27

Predictions were good for the individual ions and for their mixtures. The analysis of the three urine samples plus the eight spiked samples was performed just by its simple direct measurement with the biosensor array, followed by interpolation using the ANN model previously optimized. This states the simplicity of the developed procedure. Obtained results were compared with those obtained by established reference methods. Table 13.4 includes the regression data for the five considered species. In all cases, the uncertainty intervals (calculated at the 95% confidence level) included the ideal slopes of 1.0 as well as 0.0 intercepts. Dispersion between real and spiked samples was higher than that obtained with the test set; nevertheless, the correlation coefficients were highly significant in all cases.

13.3 Voltammetric BioET

Phenolic compounds are a group of naturally occurring compounds that are found in fruits and vegetables or beverages like tea, beer, or wine. Although individual content values for each of the single ones are rather low, a wide variety of them are present with a quite high global content, being their content of clear interest given their antioxidant properties that provide great health benefits and their effect in beverages' sensorial features (viz., in their color, body, astringency, and stability) [62].

In this context, the applicability of electrochemical biosensors in the analysis of antioxidant compounds, including phenolics, is promising and there is a growing interest in the development of such devices [63]. Thus, they represent an attractive alternative to traditional laboratory methods given their low cost and their ease of use to carry out on-field analyses. Nevertheless, it is deemed that further work is still required to avoid and/or take into account the interference problem. The combination of biosensors with chemometric tools such as PCA or ANNs may represent an alternative to conventional biosensing methods and may take benefit of the advantages of both parts. On the one hand, we have the specificity and selectivity of biosensors, and on the other, the use of ANNs modeling to derive meaning from complex or imprecise data, for example, in multivariate calibration or pattern recognition. This coupling, known as BioET, represents the most recent variant of ETs and may be postulated as a tool combining chemometrics to solve interference problems from biosensors and biosensors as the tool that solves the selectivity problem from the ETs.

Nevertheless, a known problem when voltammetric sensors are used is the large dimensionality of the generated data that hinders their treatment [64], that is, when a complete voltammogram is recorded for each sensor from the array. This is, perhaps, the main reason why this approach has been rarely used in the literature, especially if ANNs are to be used, in which case the departure information needs to be preprocessed.

A direct solution is the use of multiway processing methods (samples × sensors × polarization potential) like multiway PLS (nPLS), but the intricacy of the theoretical background and complexity of the technique is also critical [65]. One solution when dealing with a set of voltammograms is to employ a preprocessing stage for data reduction. The main objective of this step is to reduce the complexity of the input signal preserving the relevant information while making it compatible with ANNs, which in addition allows to gain advantages in training time, to avoid redundancy in input data, and to obtain a model with better generalization ability [64]. This compression stage may be achieved through the use of methods such as feature extraction [33], PCA [66], kernels [67], windowed slicing integral [49], discrete wavelet transform (DWT) [68], or even fast Fourier transform (FFT) [34].

As an illustrative example, we reported the application of a voltammetric BioET towards the simultaneous quantification of three major phenolic compounds found in beers. Electrochemical responses were obtained from a set of four bulk modified voltammetric biosensors, and then those were preprocessed employing DWT in order to extract the significant information and compress the departure data. Finally, the obtained coefficients were fed in an ANN model specially trained to predict individual phenolic compounds present in beer (Figure 13.8).

13.3.1 Construction of the BioET Array

A crucial step in the development of biosensors is the proper selection of the biomolecule, which will act as the recognition element, and its type of interaction with the target molecule, that is, biological stability, reaction type and transduction process, and generation of possible interfering by-products. Also, similar considerations need to be taken into account when selecting the electrochemical modifiers for the modified sensors.

In our case, based on previous studies with phenolic compounds [33] and following the conventional methodology previously established by our research

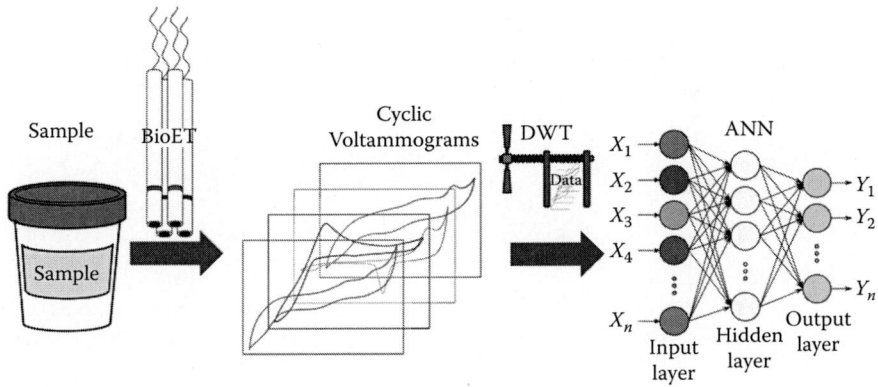

FIGURE 13.8

Processing scheme of the voltammetric BioET approach. After measuring the samples with the BioET array, signals are preprocessed employing DWT. Then, the obtained scores are fed in an ANN model that carries out the quantification of the analytes. Finally, appropriate weights and biases are applied by the learning algorithm until the targets are reached within the established error.

group [69], an array of four different graphite–epoxy voltammetric biosensors was prepared using bare graphite C and adding different modifiers such as tyrosinase, laccase, and copper nanoparticles to the bulk mixture—one component per electrode plus a blank electrode without any modifier (Table 13.5). Electrode fabrication begins with the soldering of a copper disk to an electric connector. Then, the connector is introduced to a PVC tube, where the composite paste will be deposited. These fabrication steps are illustrated in Figure 13.9.

On one hand, tyrosinase and laccase were chosen as those are usually employed for the development of amperometric biosensors toward phenolic compounds. On the other hand, copper nanoparticles were chosen given that both tyrosinase and laccase are copper-containing enzymes. Then, it was thought that some catalytic effect could be derived, a fact finally confirm in the sensor response.

TABLE 13.5

Formulation of the Biocomposites Used for Preparing the BioET

Sensor	EPO-TEK H77[a] (%)	Graphite (%)	Modifier
GEC	85	15	—
Tyr	83	15	2% tyrosinase from mushroom (EC 1.14.18.1, 4276 U/mg)
Lac	83	15	2% laccase from *Trametes versicolor* (EC 1.10.3.2, 21 U/mg)
Cu	83	15	2% copper nanoparticles (50 nm)

[a] Previously mixing the resin (part A) with its corresponding hardener (part B) in a ratio 20:3 (w/w)

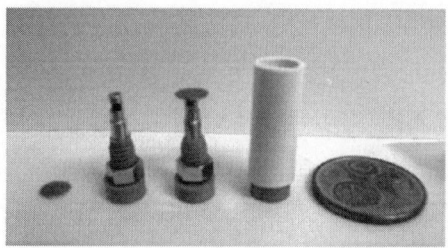

FIGURE 13.9
Steps in the construction of graphite–epoxy composite electrodes. From left to right: fixing a copper disk to a connector, assembly into the PVC tube, and incorporation of the graphite–epoxy mixture together with the modifiers.

These enzymes belong to the class of copper-containing oxidases, which catalyze the reduction of molecular oxygen by different electron donors. In those reactions, the oxygen is reduced directly to water without the intermediate formation of hydrogen peroxide. To be precise, tyrosinase catalyzes the hydroxylation of monophenols to catechols, which in turn are further oxidized to o-quinones, both using molecular oxygen; then o-quinones produced in the enzymatic reaction are electrochemically reduced to o-diphenols, resulting in the recycling of the diphenols at the applied negative potential. Laccase catalyzes the oxidation of phenols giving phenoxy radical species that are converted to quinones in the second stage of the oxidation; then as in tyrosinase, these quinones can be electrochemically reduced again to phenols.

For biosensor preparation, the resin EPO-TEK H77 (Epoxy Technology, Billerica, MA, United States) and its corresponding hardener compound were mixed in the ratio 20:3 (w/w). Afterwards, 15% (w/w) of graphite (50 mm BDH Laboratory Supplies) and 2% (w/w) of the modifier (either the enzyme or the catalyst) were added to the previous mixture before hardening. Then, it was manually homogenized for 60 min, and afterwards the paste was allowed to harden for 7 days at 40°C. Finally, the electrode surface was polished with different sandpapers of decreasing grain size, with a final electrode area of 28 mm².

13.3.2 BioET Characterization

Initially, prior to its application to the resolution of phenolic mixtures, biosensors responses were characterized towards the standard solutions of three of the major phenolic compounds found in beer, that is, ferulic, gallic, and sinapic acids. This was done in order to confirm that differentiated signals were observed for the different electrodes, generating enough rich data that can be a useful departure point for the multivariate calibration model.

As an example, voltammetric responses of copper-modified sensor are illustrated in Figure 13.10A. As a general trend, two processes are observed for

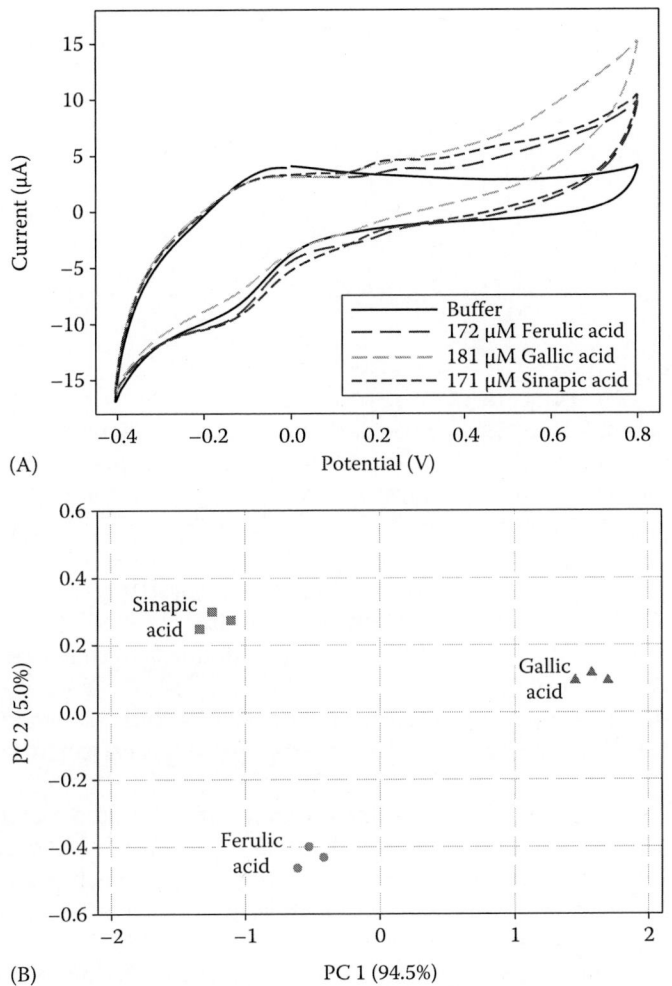

FIGURE 13.10
(A) Example of the voltammetric responses obtained with Cu sensor toward compounds under study. (B) Score plot of the first two components obtained after PCA of the different phenolic compound standards.

all the sensors corresponding to the oxidation of the corresponding phenol to its quinone form and the reduction of the quinone to the phenolic form. Moreover, some peaks that could be attributed to the oxidation of the methoxy groups of the phenolic compounds are observed. Thus, as expected, two peaks are obtained for sinapic acid, one for ferulic acid, and no peaks are observed in the case of gallic acid, which does not have any methoxy group (Figure 13.11).

To confirm this differentiated behavior, and as a previous step prior to the construction of the quantitative model, raw voltammetric responses of those standards were analyzed employing PCA.

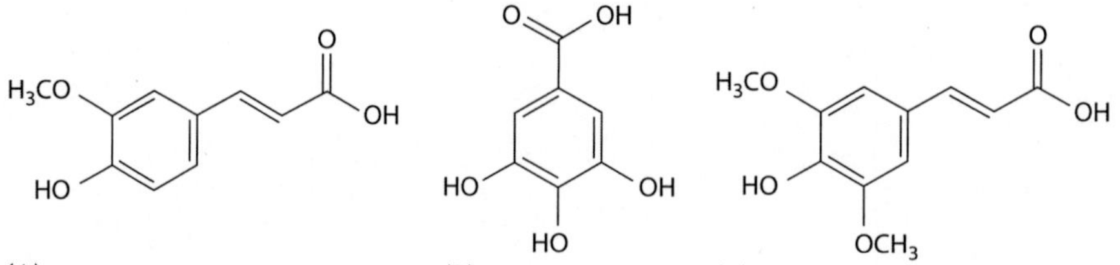

FIGURE 13.11
Chemical structures of the polyphenols under study: (A) ferulic acid, (B) gallic acid, and (C) sinapic acid.

PCA allows the projection of the information carried by the original variables onto a smaller number of underlying ("latent") variables called principal components (PCs) with new coordinates called scores, obtained after data transformation. Then by plotting the PCs, one can view interrelationships between different variables and detect and interpret sample patterns, groupings, similarities, or differences [70].

In this manner, raw data from voltammetric BioET were analyzed by means of PCA, achieving clear discrimination for the three compounds as could be seen in Figure 13.10B. Moreover, it could be seen that samples were sorted along PC1 depending on their number of methoxy groups, with gallic acid on the right far from the others and ferulic and sinapic acids close to each other (these containing one and two groups, respectively), mainly separated by PC2.

The next step, once it was confirmed that differentiated behavior was obtained by the BioET array towards the different compounds under study, was to evaluate more deeply the electrode response for each phenolic compound. Thus, standard solutions of increasing concentration of the three polyphenols were measured (Figure 13.12).

As expected, the intensity of the peaks associated with phenols increased with the concentration of the phenolic compounds. Thus, in order to characterize the electrode sensitivity for each compound, the peak currents associated to phenols were plotted against the concentration of each species. Although maximum response differed slightly for each compound, the same potential for each sensor was chosen when building the calibration plot, like would be done in a conventional amperometry (current measured at a single potential).

In this sense, Table 13.6 summarized the obtained calibration curve parameters for each of the different biosensors from the BioET array. As expected, marked mix-response for the different involved compounds was obtained; also, differentiated response between the different biosensors used was observable, being this a highly desirable departure point for studies with BioET systems.

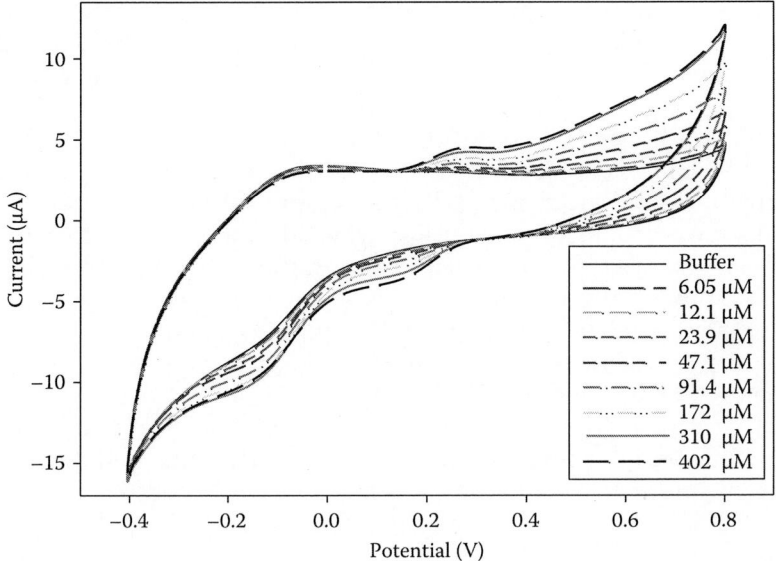

FIGURE 13.12

Voltammetric response of "Cu" sensor toward standard solutions of increasing ferulic acid concentration.

TABLE 13.6

Summary of the Responses of the BioET Array toward the Studied Phenolic Compounds

Sensor		Ferulic Acid		Gallic Acid		Sinapic Acid	
		an	*cat*	*an*	*cat*	*an*	*cat*
GEC	E (V)	0.28	0.08	0.28	0.08	0.28	0.08
	Sensitivity (µA/M)	1756	−2104	4918	—	2428	−2108
Tyr	E (V)	0.40	−0.01	0.40	−0.01	0.40	−0.01
	Sensitivity (µA/M)	1276	1227	4633	—	1780	−954
Lac	E (V)	0.40	−0.01	0.40	−0.01	0.40	−0.01
	Sensitivity (µA/M)	1114	1104	3630	—	1692	−790
Cu	E (V)	0.31	0.08	0.31	0.08	0.31	0.08
	Sensitivity (µA/M)	3381	−3891	8555	—	5228	−4234

13.3.3 BioET Multidetermination of Phenolic Compounds

In order to prove the capabilities of the BioET to achieve the simultaneous quantification of different phenolic compound mixtures, a total set of 42 samples were manually prepared with a concentration range for the three species from 0 to 200 µM for each phenolic compound. The set of samples was divided into two data subsets: a training subset formed by 27 samples (64%), which were distributed in a cubic design and used to build the response model [33], plus 15 additional samples (36%) for the testing subset, distributed randomly along the experimental domain and used to evaluate the models, predictive ability.

As discussed, when dealing with voltammetric sensor arrays, the main problem is the huge dimensionality of the recorded data. Even if differentiated response in terms of specificity can be found in a conventional amperometric manner (intensity measured at a single potential), huge improvements in results can be obtained when using the whole curve instead of particular peaks; only, this requires a preprocessing step in order to reduce the high dimensionality of the recorded signals (samples × sensors × potentials), while, at the same time, keeping the relevant information.

In this report, reduction of the large data record generated for each sample was achieved by means of DWT [68], employing Daubechies wavelet mother function and a fourth decomposition level, which allowed the reduction of signals from each voltammogram down to 23 coefficients without any loss of relevant information, attaining a compression ratio of 91.4%. Then, the obtained coefficients were used to build a model that allows the prediction of the desired parameters (Figure 13.8).

The first step in building the ANN model is selecting the topology of the neural network used. Given the difficulties to predict the optimum configuration in advance, this consists of a trial-and-error process where several parameters (training algorithm, number of hidden layers, number of neurons, transfer functions, etc.) are fine-tuned in order to find the best configuration that optimizes the performance of the neural network model [16].

For this proposal, a systematic study of the number of neurons in the hidden layer and combinations of functions in both hidden and output layers were tested. In our case, we varied the number of neurons in the hidden layer between 1 and 12 and evaluated the use of combinations of four different transfer functions (i.e., *logsig, purelin, tansig,* and *satlins*) in both the hidden and output layers.

For the selection of the optimal topology, DWT–ANN model was trained with 64% of the data, using the remaining 36% (testing subset) to characterize the accuracy of the quantification model with unbiased data. Subsequently, comparison graphs of predicted versus expected concentrations for the three determined phenols were built to easily check the performance of the ANN model. After this step, the best configuration was chosen taking into account, the topology which gave better slope, intercept, and correlation coefficient values (i.e., close to ideal values of 1, 0, and 1, respectively).

For easier interpretation of the results and to simplify the selection of the best model, the difference of the optimal values for the correlation coefficients (Δr), the slopes (ΔSlope), and the intercept (ΔIntercept) from the comparison graphs of the three analytes was calculated according to Equations 13.2 and 13.3, where x represents each parameter. Hence, given ideal values for the slope and correlation coefficient are 1, they will be ranged between 3 and 0 (Equation 13.2); they value 3 when there is no prediction capability by the model and decrease as model behavior improves [49]. Similarly, as ideal intercept value is 0,

Δ Intercept will be closer to 0 as modeling performance improves (Equation 13.3). Additionally, as an extra evaluator of the model performance, the RMSE for each of the configurations is also calculated (Equation 13.4). Thus, the optimum topology will be the one that also gives the lowest *RMSE* value:

$$\Delta x = \sum_{j} \left| 1 - x_j \right| \tag{13.2}$$

$$\Delta x = \sum_{j} \left| x_j \right| \tag{13.3}$$

$$RMSE = \sqrt{\frac{\sum_{ij} (c_{ij} - \hat{c}_{ij})^2}{3n - 1}} \tag{13.4}$$

To better visualize the performance indicators of the different evaluated topologies, each of the parameters is plotted against the number of neurons and for each of the transfer function couples. As can be seen in Figure 13.13, very bad models are obtained when employing only one or two neurons

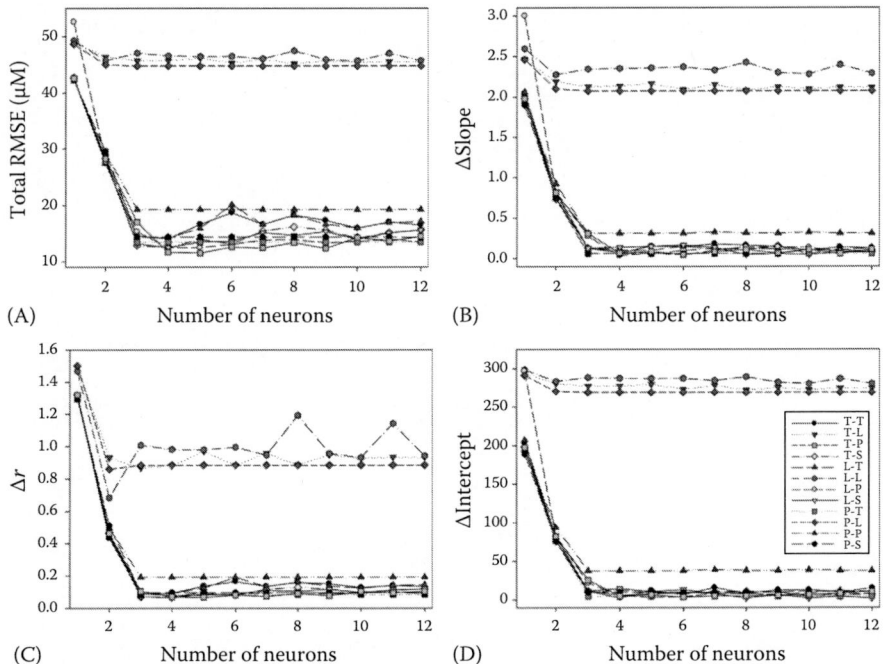

(A) (B) (C) (D)

FIGURE 13.13
Detailed results of the ANN optimization. Obtained (A) RMSE, (B) slopes, (C) correlation coefficient values, and (D) intercept values of obtained versus expected comparison graphs for the testing subsets are plotted against different numbers of neurons in the hidden layer.

in the hidden layer, with a clear improvement from then on; moreover, for higher values, there is also a decrease in model performance that may be due to its overfitting. Also poor models are obtained when using *logsig* transfer function in the output layer. However, as shown in Figure 13.13A, it can be seen how the combination of *tansig* and *purelin* functions led to the lowest RMSE values, thus selecting this configuration as the optimal one.

Upon completion of an extensive study varying its configuration, the final ANN architecture had 92 neurons (corresponding to the coefficients obtained from the DWT analysis) in the input layer, five neurons and *tansig* transfer function in the unique hidden layer, and three neurons and *purelin* transfer function in the output layer (one for each compound).

As can be observed in Figure 13.14, a satisfactory trend is obtained for the three compounds, with regression lines almost indistinguishable from the

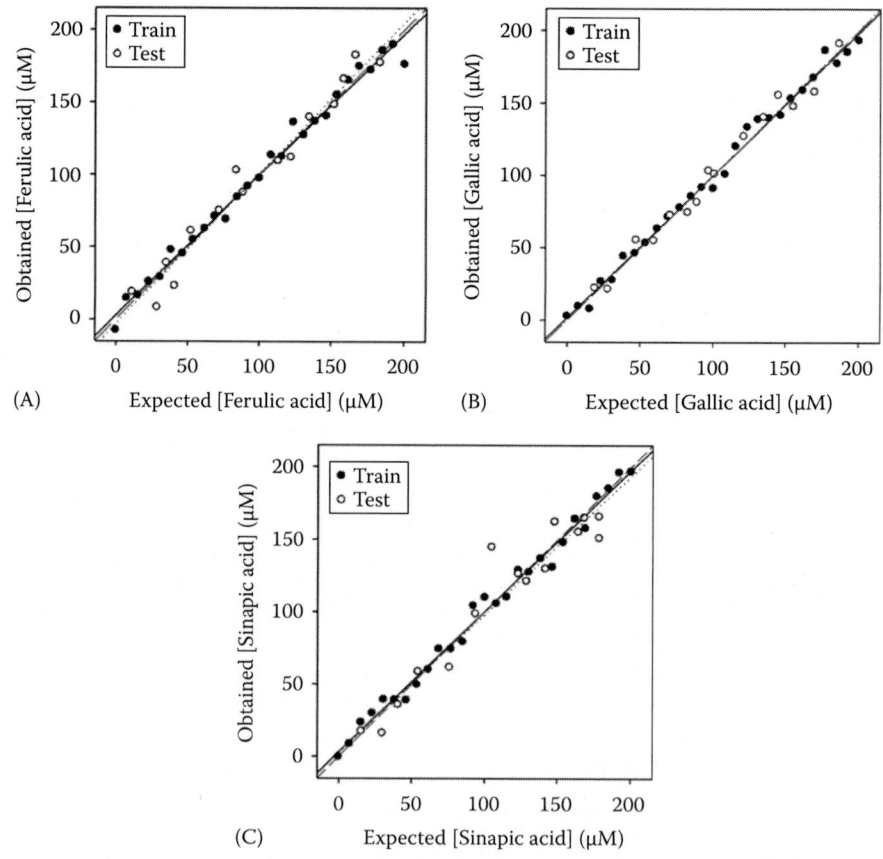

FIGURE 13.14
Modeling ability of the optimized DWT-ANN. Set adjustments of obtained versus expected concentrations for (A) ferulic, (B) gallic, and (C) sinapic acids. Dashed line corresponds to theoretical diagonal line.

TABLE 13.7

Fitted Regression Lines of the Comparison between Obtained versus Expected Results Provided by the Proposed BioET for the Samples of Training and Testing Subsets, and the Three Considered Phenols (Intervals Calculated at the 95% Confidence Level)

	Training Subset			Testing Subset		
Phenol	Correlation	Slope	Intercept (μM)	Correlation	Slope	Intercept (μM)
Ferulic acid	0.994	0.971 ± 0.044	2.7 ± 5.1	0.982	1.039 ± 0.119	-2.9 ± 13.1
Gallic acid	0.996	0.985 ± 0.036	1.6 ± 4.2	0.991	1.003 ± 0.083	0.2 ± 9.3
Sinapic acid	0.996	0.972 ± 0.035	2.9 ± 4.1	0.961	0.953 ± 0.165	2.8 ± 20.2

theoretical ones. Additionally, regression parameters are calculated, and as expected from the graphs, a good linear trend is attained for all the cases, but as usual, with improved behavior for the training subsets due to the lower dispersion. Obtained values are summarized in Table 13.7. As it could be expected, taking into account that the external test subset data are not employed at all for modeling, better performance was attained for the training subset; however, its good fit is a measure of the accomplished modeling performance. Despite this, the results obtained for both subsets are close to the ideal values, with intercepts close to 0 and slopes and correlation coefficients close to 1.

13.4 Conclusions

There is a clear demand for simpler and more efficient procedures to obtain biochemical information on different fields. Along evolution in new analytical technologies, the use of biosensors is a clear option for a fast, simple, and cheap gathering of information. In the last decade, a new trend in the sensor field has arisen, which is to couple multidimensional sensor information with advanced computer processing strategies; when this approach uses biosensors, the developed devices, known as BioETs, may provide further advances, deriving applications of qualitative and quantitative nature, even custom adapted to specific requirements. This chapter, apart from introducing general ideas, has shown two BioET developments, in this case with departure point on different sensor transduction techniques.

Urea and creatinine biosensors based on covalent immobilization of enzymes urease and creatinine deiminase, respectively, were derived from ammonium ISEs to be included in the sensor array. Their characterization showed adequate cross-response characteristics to complement ammonium, potassium, sodium, and generic ISEs. ANNs were applied as processing tool to obtain a calibration model from the measured potentials. The developed potentiometric BioET provided an extraordinarily simple procedure, with direct measurement, to quantify concentrations of urea and creatinine in real urine samples without the necessity of eliminating alkaline interferences or endogenous ammonium. Such a system would facilitate clinical studies of renal function or the monitoring of hemodialysis processes in real time.

In the second example, qualitative discrimination and quantitative resolution at low concentration level of mixtures of typical phenolic compounds present in beers were achieved by means of a voltammetric BioET. To this aim, the response from different enzymatic biosensors was combined to enrich the analytical departure information, which, being more complex in this case, needed a feature extraction process employing DWT. The resolution of signal overlapping by ANN allowed obtaining a good response model for three important phenols in the beverage field (ferulic, gallic, and sinapic acids), in an application comparable to more complex analytical techniques such as HPLC.

Acknowledgments

Financial support for this work was provided by the Spanish Ministry of Science and Innovation, MCINN (Madrid), through the project CTQ2010–17099, and the program ICREA Academia. X. Cetó thanks the support of the Dept. d'Innovació, Universitats i Empresa de la Generalitat de Catalunya, for the predoctoral grant.

References

1. Y. Vlasov, A. Legin, A. Rudnitskaya, C. Di Natale, A. D'Amico, *Pure Appl. Chem.*, 77 (2005) 1965–1983.
2. A. Gutés, A. B. Ibañez, M. del Valle, F. Cespedes, *Electroanalysis*, 18 (2006) 82–88.
3. A. Gutés, F. Céspedes, S. Alegret, M. del Valle, *Biosens. Bioelectron.*, 20 (2005) 1668–1673.
4. M. Gutiérrez, S. Alegret, M. del Valle, *Biosens. Bioelectron.*, 22 (2007) 2171–2178.
5. P. Ciosek, W. Wroblewski, *Analyst*, 132 (2007) 963–978.
6. M. del Valle, *Electroanalysis*, 22 (2010) 1539–1555.
7. Y. G. Vlasov, A. V. Legin, A. M. Rudnitskaya, *Russ. J. Gen. Chem.*, 78 (2008) 2532–2544.
8. A. Bratov, N. Abramova, A. Ipatov, *Anal. Chim. Acta*, 678 (2010) 149–159.
9. F. Winquist, *Microchim. Acta*, 163 (2008) 3–10.
10. A. Riul, Jr., C. A. R. Dantas, C. M. Miyazaki, O. N. Oliveira, Jr., *Analyst*, 135 (2010) 2481–2495.
11. M. del Valle, *Microchim. Acta*, 163 (2008) 1–2.

12. M. del Valle, *Sensors*, 11 (2011) 10180–10186.
13. E. Richards, C. Bessant, S. Saini, *Electroanalysis*, 14 (2002) 1533.
14. S. M. Scott, D. James, Z. Ali, *Microchim. Acta*, 156 (2006) 183–207.
15. P. Ciosek, W. Wroblewski, *Sens. Actuators B-Chem.*, 114 (2006) 85–93.
16. F. Despagne, D. L. Massart, *Analyst*, 123 (1998) 157R–178R.
17. M. del Valle, *Int. J. Electrochem.*, 2012 (2012) Article ID 986025.
18. A. Durán, M. Cortina, L. Velasco, J. A. Rodriguez, S. Alegret, M. del Valle, *Sensors*, 6 (2006) 19–29.
19. M. del Valle, in *Comprehensive Analytical Chemistry* (Eds. S. Alegret, A. Merkoçi), Elsevier, Amsterdam, the Netherlands, (2007), pp. 721–753.
20. M. Gutiérrez, D. Calvo, M. del Valle, in *Comprehensive Analytical Chemistry* (Eds. S. Alegret, A. Merkoçi), Elsevier, Amsterdam, the Netherlands, (2007), pp. e311–e330.
21. K. Toko, *Sens. Actuators B-Chem.*, 64 (2000) 205–215.
22. K. Toko, *Biomimetic Sensor Technology*, Cambridge University Press, Cambridge, U.K., (2000), p. 1.
23. Q. Liu, F. Zhang, D. Zhang, N. Hu, H. Wang, K. J. Hsia, P. Wang, *Biosens. Bioelectron.*, 40 (2013) 115–120.
24. J. M. C. S. Magalhães, A. A. S. C. Machado, *Analyst*, 127 (2002) 1069–1075.
25. M. Gutiérrez, S. Alegret, M. del Valle, *Biosens. Bioelectron.*, 23 (2008) 795–802.
26. I. Moser, G. Jobst, G. A. Urban, *Biosens. Bioelectron.*, 17 (2002) 297–302.
27. E. Tønning, S. Sapelnikova, J. Christensen, C. Carlsson, M. Winther-Nielsen, E. Dock, R. Solna et al., *Biosens. Bioelectron.*, 21 (2005) 608–617.
28. J. Lange, C. Wittmann, *Anal. Bioanal. Chem.*, 372 (2002) 276–283.
29. J. S. Torrecilla, M. L. Mena, P. Yáñez-Sedeño, J. García, *J. Agric. Food. Chem.*, 55 (2007) 7418–7426.
30. M. Ghasemi-Varnamkhasti, M. L. Rodríguez-Méndez, S. S. Mohtasebi, C. Apetrei, J. Lozano, H. Ahmadi, S. H. Razavi, J. Antonio de Saja, *Food Control*, 25 (2012) 216–224.
31. M. Ghasemi-Varnamkhasti, S. S. Mohtasebi, M. L. Rodríguez-Méndez, M. Siadat, H. Ahmadi, S. H. Razavi, *Trends Food Sci. Tech.*, 22 (2011) 245–251.
32. M. ElKaoutit, I. Naranjo-Rodriguez, K. R. Temsamani, M. Dominguez de la Vega, J. L. H. H. de Cisneros, *J. Agric. Food Chem.*, 55 (2008) 8011–8018.
33. X. Cetó, F. Céspedes, M. I. Pividori, J. M. Gutiérrez, M. del Valle, *Analyst*, 137 (2012) 349–356.
34. X. Cetó, F. Céspedes, M. del Valle, *Talanta*, 99 (2012) 544–551.
35. M. Cortina, M. del Valle, J.-L. Marty, *Electroanalysis*, 20 (2008) 54–60.
36. G. Valdés-Ramírez, M. Gutiérrez, M. del Valle, M. T. Ramírez-Silva, D. Fournier, J. L. Marty, *Biosens. Bioelectron.*, 24 (2009) 1103–1108.
37. G. A. Alonso, R. B. Dominguez, J. L. Marty, R. Muñoz, *Sensors*, 11 (2011) 3791–3802.
38. J. Zeravik, K. Lacina, M. Jilek, J. Vlcek, P. Skládal, *Microchim. Acta*, 170 (2010) 251–256.
39. C. A. Spinks, *Trends Food Sci. Tech.*, 11 (2000) 210–217.
40. A. C. Mak, S. J. Osterfeld, H. Yu, S. X. Wang, R. W. Davis, O. A. Jejelowo, N. Pourmand, *Biosens. Bioelectron.*, 25 (2010) 1635–1639.
41. M. S. Wilson, W. Nie, *Anal. Chem.*, 78 (2006) 6476–6483.
42. D. Calvo, N. Tort, J. P. Salvador, M. P. Marco, F. Centi, S. Marco, *Analyst*, 136 (2011) 4045–4052.

43. K. M. Roth, K. Peyvan, K. R. Schwarzkopf, A. Ghindilis, *Electroanalysis*, 18 (2006) 1982–1988.
44. S. Ingebrandt, C.-K. Yeung, M. Krause, A. Offenhäusser, *Biosens. Bioelectron.*, 16 (2001) 565–570.
45. A. Bonanni, D. Calvo, M. del Valle, *Electroanalysis*, 20 (2008) 941–948.
46. D. Xu, H. Han, W. He, Z. Liu, D. Xu, X. Liu, *Electroanalysis*, 18 (2006) 1815–1820.
47. D. P. A. Correia, J. M. C. S. Magalhães, A. A. S. C. Machado, *Microchim. Acta*, 163 (2008) 131–137.
48. R. Solná, P. Skládal, *Electroanalysis*, 17 (2005) 2137–2146.
49. X. Cetó, F. Céspedes, M. del Valle, *Electroanalysis*, 25 (2012) 68–76.
50. R. Kirby, E. J. Cho, B. Gehrke, T. Bayer, Y. S. Park, D. P. Neikirk, J. T. McDevitt, A. D. Ellington, *Anal. Chem.*, 76 (2004) 4066–4075.
51. M. Singh, N. Verma, A. K. Garg, N. Redhu, *Sens. Actuators B-Chem.*, 134 (2008) 345–351.
52. E. H. Taylor, in *Chemical Analysis* (Ed. J. D. Winefordner), John Wiley & Sons, New York, (1989).
53. J. D. Burtis, E. R. Ashwood, in *Tietz Textbook of Clinical Chemistry*, W.B. Saunders, Philadelphia, PA, (1994).
54. M. Jurkiewicz, S. Alegret, J. Almirall, M. Garcia, E. Fabregas, *Analyst*, 123 (1998) 1321–1327.
55. I. Karube, H. Matsuoka, S. Suzuki, E. Watanabe, K. Toyama, *J. Agric. Food. Chem.*, 32 (1984) 314–319.
56. K. Matsumoto, H. Kamikado, H. Matsubara, Y. Osajima, *Anal. Chem.*, 60 (1988) 147–151.
57. A. Radomska, R. Koncki, K. Pyrzynska, S. Glab, *Anal. Chim. Acta*, 523 (2004) 193–200.
58. C. S. Rui, K. Sonomoto, Y. Kato, *Anal. Sci.*, 8 (1992) 845–850.
59. J. Gallardo, S. Alegret, M. A. de Roman, R. Munoz, P. R. Hernandez, L. Leija, M. del Valle, *Anal. Lett.*, 36 (2003) 2893–2908.
60. R. Koncki, A. Radomska, S. Glab, *Anal. Chim. Acta*, 418 (2000) 213–224.
61. L. Zhang, Y. Z. Liang, J. H. Jiang, R. Q. Yu, K. T. Fang, *Anal. Chim. Acta*, 370 (1998) 65–77.
62. A. S. Arribas, M. Martínez-Fernández, M. Chicharro, *TrAC Trends Anal. Chem.*, 34 (2012) 78–96.
63. D. M. A. Gil, M. J. F. Rebelo, *Eur. Food Res. Technol.*, 231 (2010) 303–308.
64. X. Cetó, F. Céspedes, M. del Valle, *Microchim. Acta*, 180 (2013) 319–330.
65. R. Bro, *Crit. Rev. Anal. Chem.*, 36 (2006) 279–293.
66. R. M. de Carvalho, C. Mello, L. T. Kubota, *Anal. Chim. Acta*, 420 (2000) 109–121.
67. R. Gutiérrez-Osuna, H. T. Nagle, *IEEE Trans. Syst. Man Cybern. B Cybern.*, 29 (1999) 626–632.
68. L. Moreno-Barón, R. Cartas, A. Merkoçi, S. Alegret, M. del Valle, L. Leija, P. R. Hernandez, R. Muñoz, *Sens. Actuators B-Chem.*, 113 (2006) 487–499.
69. F. Céspedes, E. Martínez-Fàbregas, S. Alegret, *TrAC Trends Anal. Chem.*, 15 (1996) 296–304.
70. I. T. Jolliffe, *Principal Component Analysis*, Springer, New York, (2002), p. 488.

14

Molecularly Imprinted Polymer-Based Biosensors

Mihrican Muti and Arzum Erdem

CONTENTS

14.1 Introduction

Molecular imprinting is a versatile technique for the preparation of synthetic receptors, on the basis of complexes between a template molecule and polymerizable monomers. After polymerization process, the products are called molecularly imprinted polymers (MIPs). These polymers are prepared by using cross-linked polymers containing cavities specific to an analyte. These cavities are created by copolymerization of cross-linking monomers and functional monomers along with an imprinting molecule or template. In the molecular imprinting technique, the first important step is the interactions of template molecule with functional monomer before polymerization. The monomer-template interactions may occur in two different ways: covalent or noncovalent interactions. The second important step is to remove the template molecule from the resulting polymer to generate the vacant recognition sites for the template molecule in terms of both functional groups and size [1–3]. The MIP then selectively rebinds to the analyte compound [4,5]. By this way, MIP technologies imitate biological antibody systems that function on the basis of specific binding sites where an antigen binds strongly to an antibody [5].

MIPs are attracting widespread attention due to their potential to deliver robust molecular recognition elements targeted toward essentially any guest present in any environment for various analytical applications, such as solid-phase extraction (SPE) [6], chromatography [7], capillary electrophoresis [8], sensors [9,10], and catalysis [11].

The presence of active pharmaceuticals in waste and in surface waters has become a major concern in environment pollution. This could be due to excretion from human and veterinary use, an inappropriate disposal of unused pharmaceuticals, and an incomplete elimination in wastewater treatment plants [12–14].

Molecular imprinting is a technique for the creation of materials with tailor-made recognition sites. This method has garnered significant attention in the field of environmental science and technology owing to its high selectivity for target molecules [15].

In the treatment of trace contaminants, there is a large usage of MIP because they can be specifically designed to remove one or a group of target compounds. This is an advantage over nonspecific technologies, such as activated carbon, which may be consumed while removing large amounts of nontrace contaminants from water [16–47].

MIP is conventionally used as an SPE media for analytical chemistry. During most SPE procedures, the wide range of contaminants contained in water samples for environmental analysis are coextracted, and complex extraction procedures may be required to isolate specific contaminants. The preconcentration and extraction method can perform simultaneously with MIP technologies because the binding sites on the MIPs are designed to remove a specific contaminant from the water system [4]. Nonimprinted polymers (NIPs), on the other hand, are cross-linked polymeric materials that have macropores containing adsorption sites for organic molecules. They are synthesized using the same procedure as MIP, but in the absence of a template. Hence, they have the same chemical properties as MIP; however, they do not contain any specific cavities [48]. NIP exhibits strong nonspecific binding, which is attributed to hydrophobic interactions between organic compounds and polymers. In the MIP studies, NIP particles should be studied as a control against MIP particles to compare nonspecific binding to template-specific binding.

14.1.1 Synthesis of Molecularly Imprinted Polymers

For molecularly imprinting, the common approach involves first the complexation in solution of a template molecule with functional monomers, through noncovalent bonds, followed by polymerization of these monomers around the template with the help of a cross-linker in the presence of an initiator. After polymerization, template molecules are removed by extensive washing steps to disrupt the interactions between the template and the monomers; thus, it makes the binding sites available, that is, cavities, complementary to the template in size, shape, and position of the functional group. The choice of the chemical reagents making up the MIP judicious in order to create highly specific cavities designed for the template molecule.

The synthesis of an MIP results in the production of a polymer containing cavities similar in size and shape, and incorporating functionalities complementary to the template molecule used during its production. Once the MIP is obtained, the cavities generated inside are a frozen print of the original template molecule. If the template molecules can adopt to many different conformations, they cannot be efficiently extracted. The template molecule, functional monomer, and the cross-linking agent are the main components involved in the production of an MIP. To obtain the highest affinity of the MIP towards the target analyte, monomer composition used in the production of MIP should be selected carefully. Some exhaustive studies have been made of this selection in the literature [49]. The choice of the right functional monomer is very important for MIP construction because functional monomer determines, on one hand, the stability of the complex formed before and during the polymerization process and, on the other hand, determines the ability of the MIP to interact selectively with the target molecule. For most functional monomers, the selective retention of the analytes onto the polymer can be established by H bond, or ionic interaction, depending upon the solvent and the pH of the sample being percolated.

There are many different commercially available functional monomers selected for the molecular imprinting polymerization according to the binding properties to the template molecule. The most widely used monomers are methacrylic acid (MAA) and 4-vinylpyridine (4-VP), as can be seen in Tables 14.1 and 14.2 (summarizing some of their most recent molecularly imprinted solid-phase extraction [MISPE] applications). MAA is the preferred monomer for interacting with basic compounds, due to its nature, whereas 4-VP is preferred for acidic compounds. Strong hydrogen binding can be established for both compounds, and they have been used for extracting either acidic or basic compounds. For example, MAA has been used to extract diverse substances such as antibiotics [50], antiepileptic drugs [51], or doping agents [52], whereas 4-VP has also been used for extracting 2,4-nitrophenol [53] from river water and enrofloxacin from milk samples [54]. To increase further the selectivity of an MIP, and to synthesize the functional monomer according to the functionalities present in the target molecule, it is the most convenient way. In such cases, the functional monomer synthesized generally has several interaction points with the template molecule. There are specially designed functional monomers synthesized for the selective extraction of penicillin [55] or riboflavin [56] from aqueous matrices, among other compounds in the literature.

Hydrogen binding is the most likely interaction between the functional monomer and the template molecule. This is a very effective interaction between these two molecules. The third party involved in the synthesis of an MIP is the

TABLE 14.1

Food Toxicants Detected by Using Different Techniques in Combination with MIP

Analyte (Template)	Functional Monomer	Detection Limit	Method	References
4-Nonylphenol	2-Amino thiophenol	3.20×10^{-7} mol/L	Amperometry	[16]
Ametryn	MAA	—	HPLC	[17]
Parathion	MAA	$5.0 \cdot 10^{-10}$ M	SWV	[18]
Pirimicarb	MAA	—	Spectrometry	[19]
Paraoxon	PTMOS, APTES, TEOS, and TMOS	—	Colorimetry	[20]
Methidathion	N,N'-MBAA	0.02 mg/L	HPLC	[21]
DEP	MAA	—	Chromatography	[22]
Chloramphenicol	MAA	2.0×10^{-9} M	DPV	[23]
Vomitoxin	Polypyrrole	>1 ng/mL	SPR	[24]
Patulin	4,4'-Azobis(4-cyanopentanoic acid) ACPA	10 µg/kg	HPLC	[25]
MPA	4-VP	0.17 µg/kg	LC/MS	[26]
Zearalenone	1-Allypiperazine	—	LC/MS	[27]
Fumonisin B	DEAEM	22 µg/kg	HPLC	[30]
Moniliformin	MAA	—	HPLC	[31]

Abbreviations: Analyte: DEP, Diethyl(3-methylureido) (phenyl)methylphosphonate; Functional monomers: PTMOS, phenyl trimethoxysilane; APTES, aminopropyltriethoxysilane; TMOS, tetramethylorthosilicate; TEOS, tetraethyl orthosilicate; ACPA, 4-4'Azobis(4-cyanopentanoic acid); DEAEM, 2-(diethylamino)ethyl methacrylate; Measurement techniques: HPLC, high-performance liquid chromatography; SWV, square wave voltammetry; DPV, differential pulse voltammetry; SPR, surface plasmon resonance; LC/MS, liquid chromatography/mass spectrometry.

cross-linking agent, whose function is to deliver mechanical stability, stabilize the molecular recognition site, and control the porosity of the polymer. There are many different cross-linking agents used for the polymerization process, the most widely used is ethylene glycol dimethacrylate (EGDMA). The cross-linking agent is always used in excess to both the template and the functional monomer. However, it is not as important as the functional monomer; choosing the right one is advisable. Choice of the cross-linking agent can affect the hydrophobicity, especially when the sample of interest is a water-based matrix.

Since MIPs are polymeric sorbents obtained by polymerizing suitable monomers in the presence of a template molecule, one way to classify them is in terms of the interaction between the functional monomer and the template molecule during polymerization. Depending on the nature of this interaction, MIPs can be classified according to whether they are obtained using the covalent, non-covalent, or semicovalent approach. In the covalent approach, all of the interaction process before and after polymerization occur through by the

TABLE 14.2

Environmental Pollutants Detected by Using Different Techniques
in Combination with MIP

Analyte (Template)	Functional Monomer	Detection Limit	Method	References
3,3'-Dichlorobenzidine	Functional organosilanes	—	HPLC	[32]
BPA	4-VP	5 µg/L	HPLC	[33]
Triclosan	PTMOS, APTES, TEOS	—	HPLC	[34]
PAH	4-VP	0.5 ng/L	GC–MS	[35]
2,4,6-TCP	MAA	1.9 µg/L	HPLC	[36]
2,4-DMP	4-VP	2.2 ng/mL	HPLC	[37]
2-CP	MAA	0.05 ng/L	HPLC	[38]
BPA	BPA–terthiophene and carbazole	—	EIS	[39]
4-Nitrophenol	MAA	—	LC	[41]
Chloroguaiacol	4-VP	27 nmol/L	Amperometry	[42]
TBBPA	APTES	2 ng/mL	LC	[43]
4-Aminophenol	MAA	3 µmol/L	Amperometry	[44]
Parathion	MAA	49.0 ng/L	SWV	[45]
Ciprofloxacin	MAA	5.5 ng/L	LC–MS/MS	[46]
TNT	MAA	1.5×10^{-9} mol/L	SWV	[47]

Abbreviations: Analyte: PAH, polycyclic aromatic hydrocarbon; TBBPA, tetrabromobisphenol A; TNT, trinitrotoluene; Functional monomers: PTMOS, phenyl trimethoxysilane; APTES, aminopropyltriethoxysilane; TEOS, tetraethyl orthosilicate; BPA, bisphenol A; Measurement techniques: HPLC, high-performance liquid chromatography; SWV, square wave voltammetry; GC–MS, gas chromatography–mass spectrometry; EIS, electrochemical impedance spectrometry; LC/MS, liquid chromatography/mass spectrometry.

covalent bonds. Before polymerization, the interaction of template molecule with functional monomer, and also after polymerization the interaction of template molecule with MIP occur according to this interaction procedure. The covalent approach is not used to synthesize MIPs and apply them in MISPE applications, because this approach has slow kinetics involved in establishing a covalent bond during the extraction of analyte from the sample and the cleavage of this covalent bond during the elution of analyte from the MIP. The most widely used approach to synthesizing MIPs for use in MISPE applications has traditionally been the noncovalent approach. In this approach the interaction between the functional monomer and the template molecule before the polymerization process and the interaction of the target analyte onto the MIP during the extraction process are through noncovalent interactions. Another synthetic protocol for obtaining MIPs is known as the semicovalent approach. In this case, the interaction between the functional monomer and the template molecule before the polymerization process is covalent, whereas the interaction of the target analyte and the MIP once the polymer is in use is through noncovalent interactions. In recent years, several studies have reported the use of the semicovalent approach [57,58]. An interesting comparison between the noncovalent and the semicovalent approach can be established from two different studies reported in the literature. In the study of Cacho et al. [59], MIP is obtained using the semicovalent approach [59] and it is compared with those previously obtained using the noncovalent approach [60] to determine their respective effectiveness in the selective extraction of triazines from real samples. From this comparison, it is concluded that the MIP obtained using the semicovalent approach enabled a better cleanup of the sample than the MIP obtained using the noncovalent approach, at least for this specific application. Due to the different interactions established between the template and the functional monomer before the polymerization process in the noncovalent or semicovalent approach, a different ratio of template molecule to functional monomer is needed to ensure the best imprinting of polymer. The ratio normally used in the semicovalent approach is rather low and is generally 1:1 or 1:2 [57,59], whereas for the non-covalent approach, the ratios typically range from 1:4 to 1:8, depending upon the complexity of the template and the affinity of the functional monomer to the template.

14.1.1.1 Synthesis Protocols

Several synthetic protocols have been developed to obtain these polymers; some of them are mentioned later.

14.1.1.1.1 Traditional (Bulk) Polymerization

The most widely used synthetic protocol has been occurring by traditional polymerization (TP). In this polymerization technique all components are dissolved in the low volume of a suitable solvent, also known as porogen, and left to polymerize, thus creating a monolithic polymer. MIPs are being increasingly used in these two extraction techniques: solid-phase microextraction (SPME) [61] and stir-bar sorptive extraction (SBSE) [62]. TP has been the technique of choice for synthesizing homemade fibers for this extraction technique. A polymer obtained by TP must be crushed, ground, and sieved to obtain useful particles for MISPE applications. This process is laborious and time-consuming and produces low yields of the desired range of particles, which are also irregular in size and shape. Despite all these drawbacks, TP is still the most widely used synthetic protocol, because it is the simplest method and requires limited organic chemistry and no specialized equipment. TP has been successfully used to synthesize many different MIPs to extract selective antibiotics from plasma [63], or human urine samples [64,65], or even nerve-gas agents from soil sample extracts [66]. This polymerization has also been the choice for synthesizing the previously mentioned MIPs using a homemade functional monomer [55,56]. Several polymerization techniques have been developed for MISPE applications in order to overcome the irregularities in size and shape of the imprinted particles and to reduce the time needed to obtain them. The aim of these techniques is to obtain spherical particles and a higher polymerization yield than TP. The spherical shape also improves both packing of the cartridge and mass transfer of the analyte.

14.1.1.1.2 Precipitation Polymerization

Precipitation polymerization (PP) is one of the polymerization techniques to overcome the drawbacks of TP, and this technique is the most widely used technique for obtaining spherical particles. By using this polymerization technique spherical particles of a size suitable for MISPE applications, in a single preparative step, can be obtained. The basic principle of the PP approach is that when the polymeric chains growing in solution reach a certain critical mass, they precipitate from the solution. The particle size thus obtained is generally smaller than 10 μm [67] and is sometimes even in the sub-μm range; this size is proper for MISPE applications. For this technique a further drawback is that PP is not as robust as TP in terms of the possibility of imprinting any given template molecule. In order to succeed with this polymerization technique, the polymerization conditions must be carefully chosen to give good-quality products, and this places some restrictions in the monomers and porogens that can be used. The template molecule can also influence the outcome of the polymeric particles. Several studies have shown that the presence of a template molecule during the polymerization process has very different effects on the MIP and the NIP because the presence of the template molecule can decrease the polymerization yield [68], increase the particle size [69], or have no effect at all [70]. Nevertheless, there are several studies where MIPs obtained using the PP protocol are used as sorbents in MISPE applications. It is reported on the extraction of tebuconazole from food samples [49], and it is reported extensively on the extraction of several pesticides from vegetables [71]. When TP is compared with PP, it is found that PP outperformed TP regarding not only synthetic issues (polymerization yield and particle uniformity) but also sorbent capacities when sorbents are applied to MISPE protocols [51,69].

14.1.1.1.3 Multistep Swelling Polymerization

Multistep swelling polymerization (MSS) is another approach for obtaining spherical particles. With this technique uniformly sized particles are suspended in water, and after several additions of suitable organic solvents, the initial particles swell to a final size in the range of 5–10 μm. All polymerization components are added to the solution of swollen particles, which are in desired size and afterwards, the polymerization is induced. Compared to PP, this approach is more robust in molecular imprinting than any given molecule because none of the compounds influence the process of obtaining spherical particles. This synthesized protocol has greater complexity compared with TP and PP polymerization technique. Thus, this type of polymerization has not been widely used as either TP or PP.

14.1.1.1.4 Suspension Polymerization

Suspension polymerization (SP) is a different approach used to obtain spherical particles. In this case, all the components involved in the polymerization process are dissolved together in an appropriate organic solvent, and this solution is further added to a larger volume of an immiscible solvent. In this polymerization protocol it is important to stir the system vigorously in order to form droplets (typically in the μm range), and then the polymerization reaction is induced. A certain amount of emulsifier like polyvinyl alcohol is added to the polymerization process to overcome the aggregation. The particles obtained under this protocol are also spherical in shape, and their size typically is in the range of 10–100 μm. In both SP and MSS polymerization techniques, the common feature is that any given molecule can be used to deliver imprinted materials and the most commonly used dispersing agent in both cases is water. However, in both cases, the presence of water during the polymerization process may jeopardize proper interactions between the template and functional monomer. Although this kind of polymerization technique has not been used as commonly as other polymerization techniques in recent years, different MIPs synthesized according to the SP were used in the literature [72–77].

14.1.1.1.5 Grafting

Another polymerization technique aimed at delivering spherical particles is grafting. But this technique is also used to produce composite materials [78]. Silica particles are used as starting material, and all the components involved in the polymerization process are adsorbed within these particles before the polymerization process starts. Once the polymer is formed, the silica is etched away to reveal final product of spherical particles [79]. These particles are in the specular image of the original silica particle. In contrast to all the previously reported techniques, grafting polymerization requires more synthetic skill and the use of corrosive solvents. Nevertheless, over the past few years, several studies have focused on the use of this technique to synthesize MIPs and applied them for extraction [78]. These techniques previously mentioned are the most widely used synthesis protocols that have been described in the literature for obtaining MIPs and applied in MISPE applications over the past few years. MIPs have wide acceptance in performing selective extractions, however, these sorbents have also become commercially available for performing routine selective extractions. A schematic illustration of molecular imprinting is given in Scheme 14.1.

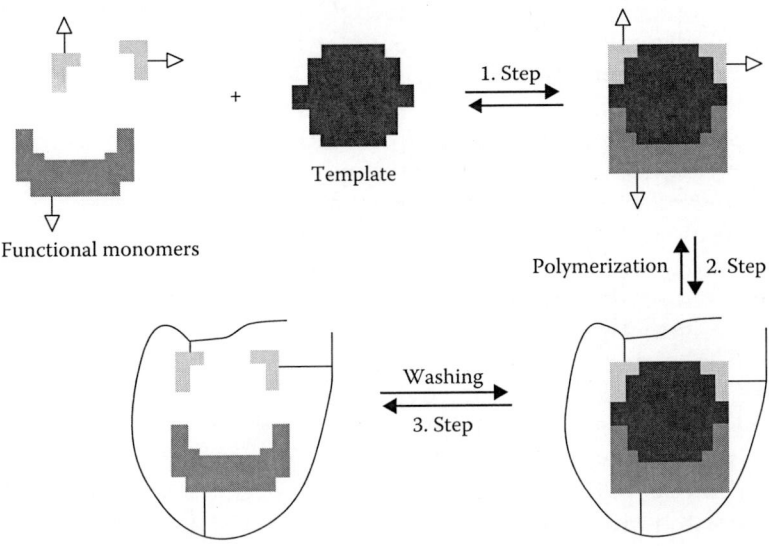

SCHEME 14.1
Schematic illustration of molecular imprinting.

14.2 MIP-Based Biosensor Developed for Monitoring of Food Toxicants and Environmental Pollutants

14.2.1 MIP-Based Biosensor for the Detection of Food Toxicants

Food contamination due to chemicals from natural or anthropogenic sources is a significant source of food-borne illness, and it poses severe risks to human health, although effects are often difficult to link with a particular food. The dangerous substances in food may include natural toxicants, such as mycotoxins, phycotoxins, and phytotoxins [80–82]; environmental contaminants, such as polychlorinated dioxins [83] and polycyclic aromatic hydrocarbons (PAHs) [84]; and chemicals, such as pesticides and veterinary drugs deliberately used to increase the food supply, whose residues can be present in the processed food and potentially affect human health [85]. It is now largely accepted that chemical contamination of food can affect human health not only after a single massive exposure but, more often, after continuous exposure to low doses of toxic chemicals that can be related to several chronic diseases, including some types of cancers and serious hormonal dysfunctions. Public awareness about chemicals in food is relatively high, and the consumers continue to express concern about the risks to their health due to the deliberate addition of chemicals to food. Thus, efficient analytical protocols based on efficient analytical processes—sensitive, selective, fast, inexpensive, and suitable for sample mass screenings—are required by legislation, health authorities, and companies operating in the food market. Contemporary analytical methods have the sensitivity required for contamination detection and quantification, but direct application of these methods on food samples can be rarely performed. Usually, contaminants are present in food at low concentration (ng-μg/g) levels, dispersed in highly complex (thousand of different components) and morphologically structured matrices, with an elevated degree of sample-to-sample variability. Thus, such a type of matrix introduces severe disturbances, and analysis can be performed only after some cleanup and preconcentration steps [86–88]. Current sample pretreatment methods, mostly based on the SPE technique, are very fast and economical but show a lack of selectivity, while methods based on immunoaffinity extraction are very selective but expensive and not suitable for harsh environments [89,90]. As economical, rapid, and selective cleanup methods (relying on "intelligent" materials) are needed, SPE and cleanup methods based on MIPs (MISPE) seem to represent natural candidates to circumvent the drawbacks typical of more traditional SPE techniques [91–93]. Recent years have seen a significant increase of the MISPE technique in the food contaminant analysis. In fact, this technique seems to be particularly suitable for extractive applications where analyte selectivity in the presence of very complex samples represents the main problem. The molecular structure of some food toxicants is illustrated in Figure 14.1.

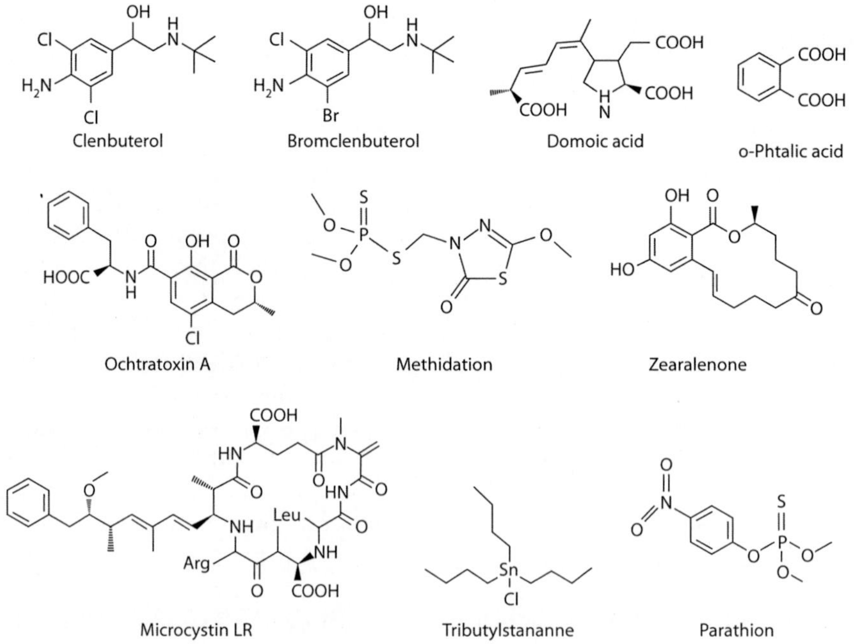

FIGURE 14.1
Molecular structure of some food toxicants.

4-nonylphenol is ubiquitous in foodstuffs, including fresh fruits and vegetables, human breast milk, livestock products, and rice [94]. For the accurate and rapid determination of 4-nonylphenol, it is important to develop new methods. Huang et al. [16] developed a novel electrochemical sensor for sensitive and fast determination of 4-nonylphenol (4-NoP). This group modified gold electrode using MIP supported with TiO_2 nanoparticles and gold nanoparticles (AuNPs) to fabricate a sensor for the detection of 4-NoP. In the present study, high binding capacity and selectivity for 4-NoP with the limit of detection (LOD) (S/N = 3) as 3.20×10^{-7} mol/L at the linear relationship of the current of 4-NoP-imprinted sensor from 4.80×10^{-4} to 9.50×10^{-7} mol/L (r = 0.998) in a good repeatability and stability was reported. This group also reported that this MIP electrochemical sensor has presented a great potential in the real sample analysis [16].

Koohpaei et al. used MIPs for the detection of triazinic herbicide ametryn by using high-performance liquid chromatography (HPLC) technique with the chemometric approach [17]. Parathion is an organophosphate compound that has high toxicity and widespread uses in agricultural areas [95,96]. These organophosphate compounds are used as pesticides, insecticides, and chemical war agents. The design and construction of an extra high-selective voltammetric sensor for parathion by using an acrylic-based MIP was developed by Alizadeh with the detection limit as $5.0 \cdot 10^{-10}$ M (S/N = 3) at the linear relationship over parathion concentration in the range of 1.7×10^{-9}–9.0×10^{-7} M. It was reported that this square wave voltammetric sensor has very high selectivity for parathion determination even at very low concentration [18].

Pirimicarb is a food toxicant that is a kind of carbamate pesticide. This pesticide is widely used in agriculture. Gao et al. [19] developed MIP for the determination of the pirimicarb. Poly(methacrylic acid) (PMAA) was grafted on the surface of silica gel particles ($PMAA/SiO_2$) and pirimicarb was used here as the template. Next, the molecular imprinting was carried out and pirimicarb-imprinted material MIP-$PMAA/SiO_2$ was obtained. The selectivity of the MIP was tested by using propoxur and atrazine whose structures were approximate to pirimicarb. It was reported that MIP-$PMAA/SiO_2$ possesses excellent recognition ability and combining selectivity for pirimicarb [19].

A different toxic organophosphate compound is paraoxon that is widely used as agricultural pesticide and herbicide. Paraoxon was imprinted as the template in a thin film of sol–gel polymer, and colorimetric assay was performed for the detection of paraoxon [20]. It was reported that this method is very sensitive for the detection of pmole amounts of bound paraoxon into the thin film in relatively short assay time, and nonspecific binding was negligible in comparison with the specific binding that could be monitored in the broad range of concentrations [20].

HPLC method by using MIP to detect methidathion from olive oil was developed by Bakas et al. N,N'-methylene bisacrylamide (MBAA) was chosen as the functional monomer that has strong binding activity towards methidathion. The selectivity of methidathion-imprinted polymers was evaluated by using three different pesticide solutions including fenthion, malathion, and dimethoate. These polymers demonstrated highest affinity for methidathion and integration of all important parameters was the crucial factor for successful development and optimization of the protocol for extraction of methidathion from such complex matrix like olive oil. The proposed method proved to be highly accurate, quick, and inexpensive with the LOD as 0.02 mg/L for concentrations ranging from 0.1 to 9 mg/L ($r^2 = 0.996$) [21].

MIPs were used in different studies for the recognition of organophosphate pesticide (OP) analogs. A novel compound, diethyl(3-methylureido) (phenyl) methylphosphonate (DEP), was used as the template molecule. IR, NMR, and HPLC methods were performed in this study. DEP-imprinted MIPs showed the specific recognition ability for DEP and OPs possessing alkaline groups through selective binding experiments and chromatographic experiments reported [22].

Chloramphenicol is a broad-spectrum antibiotic, previously used in veterinary medicine. Therefore, the detection of CAP in animal food samples is highly important. High-selective chloramphenicol sensor was proposed by Alizadeh et al. It was reported that this sensor showed interesting resistance against the interference effects of the electroactive nitro aromatic compounds even in the trace amount of milk samples. This method can be used as an alternative method for reliable CAP determination in complex real samples like milk with the detection limit as 2.0×10^{-9} M (S/N = 3), and this sensor showed a linear response range of $8.0 \times 10^{-9}–1.0 \times 10^{-6}$ M [23].

Mycotoxins are toxic secondary metabolites produced by fungi occurring on agricultural commodities. Mycotoxins exhibit a broad range of toxic effects (carcinogenic, neurotoxic, nephrotoxic, immunosuppressive, and estrogenic) implying a potential threat to human and animal health. Choi et al. [24] developed molecularly imprinted polypyrrole (MIPPy) film for the detection of the mycotoxin deoxynivalenol (DON) using a surface plasmon resonance (SPR) transducer. It was reported in their study that MIPPy–SPR sensor exhibited a linear response for the detection of DON in the range of 0.1–100 ng/mL ($R^2 = 0.988$), with a low detection limit of >1 ng/mL. According to their results, it was concluded that this sensor has also high binding affinity for DON [24].

MIPs were synthesized with grafting polymerization technique and characterized for the detection of patulin in different studies. Adsorption selectivity and binding capacity of these polymers were investigated, and the detection limit was obtained as 10 μg/L, ($R^2 = 0.9998$) [25].

Smet et al. [26] developed MIP to determine the mycophenolic acid (MPA) in maize. In this study it was reported that analysis of naturally contaminated samples by LC–MS/MS demonstrated that imprinted cleanup is a suitable technique for the detection of MPA in maize. Detection limit was estimated as 0.17 μg/kg. This report also indicated that purification of maize samples using the developed imprinted polymers is not costly because of the inexpensive synthesis and the multiple use of one MIP cartridge [26].

Mausia et al. synthesized MIPs as specific adsorbents for zearalenone by PP. Specificity, selectivity, and enrichment ability of these polymers were investigated in this study [27].

Lucci et al. [28] developed a method for the cleanup and preconcentration of zearalenone from corn and wheat samples employing MIP as the selective sorbent for SPE. The precision and accuracy of this technique the satisfactory with a repeatability standard deviation (RSD) between 1.1 and 1.9 µg/kg in wheat samples and between 1.4 and 2.6 µg/kg in maize samples, and it was reported that the RSD ranges from 2.5% to 6.2% in wheat samples and from 0.9% to 6.8% in maize samples [28].

T-2 toxin is one of the most important fusarium toxins [97] that are natural contaminants in food, especially in grains (corn, wheat, oat, and barley), and they are an important food safety issue because of their toxic effects and worldwide occurrence in food and feed [98–100]. The synthesis of a T-2 toxin imprinted polymer and its application in food analysis are reported in the present study. The synthesis of a T-2-selective imprinted polymer and its implementation in the cleanup MISPE of maize, barley and oat samples containing frequently this toxin. The detection limits were estimated as 0.4, 0.5, and 0.6 µg/kg in maize, barley, and oat samples, respectively [29].

Fumonisins are secondary metabolites that are produced by fusarium molds, primarily *F. verticillioides* and *F. proliferatum* [99]. Toxicological studies of fumonisins revealed that they could cause several animal diseases such as pulmonary edema in swines, equine leukoencephalomalacia in horses, and esophageal cancer and hepatic cancer in horses and rats [101–103]. Fumonisin-selective imprinted polymer was developed by Smet et al. for the detection of fumonisin in bell pepper, rice, and corn flakes. It was reported that the analysis of fumonisin in complex food matrices requires selective and sensitive detection of this carcinogenic mycotoxins [30].

Moniliformin is a very important mycotoxin produced with low molecular weight that has worldwide potential to contaminate maize and other cereal grains [31]. Imprinted polymers to bind moniliformin were synthesized using two different templates by Appell et al. Binding interaction, template effects, reproducibility, and activities of these MIPs were also investigated, and it was reported that the moniliformin-binding polymers show potential as solid sorbents to remove moniliformin from liquids and as packing in MISPE columns [31].

14.2.2 MIP-Based Biosensor for the Detection of Environmental Pollutants

Environmental pollutants are contaminating substances introduced to the environment by humans that can endanger public health and inflict damage to living resources and ecological systems [104]. Current activities and advances in industrialization and technological processes have consequently introduced harmful chemicals into water, air, and land. The number and level of hazardous substances such as environmental pollutants, agrochemicals, and sewage wastes have dramatically increased in the recent years [105]. Water pollution occurs by adding contaminating substances into the pure water resources directly or indirectly, and this makes the water inappropriate for drinking or bathing [106]. Good-quality drinking water is being polluted by industrial/environmental wastes, and moreover, development and survival of aquatic and wildlife is being disturbed by this water pollution that was reported [107]. Toxic metals and solvents that are part of effluent wastewater of industries are major components of water pollution [108].

Phenolic compounds are considered as priority pollutants in the environment that are present in water resources at very low concentrations. Therefore, the determination of low concentration of phenol requires the development of selective SPE methods, which is one of the preconcentration techniques prior to their instrumental determination [109]. A different approach to determine phenolic compound is to use molecularly imprinted SPE. Various phenolic compounds have been used as templates in the preparation of selective MISPE sorbents. For example, pentachlorophenol [110], 4-chlorophenol [111], 2,4-dichlorophenol (2,4-DCP) [52], 2,4,6-trichlorophenol (2,4,6-TCP) [36], 4-nitrophenol [41], chloramphenicol [112], catechol [113], bisphenol A (BPA) [33], 4-nonylphenol [16], p-acetamidophenol [114], and caffeic acid and p-hydroxybenzoic acid [115] are used as templates. Some of these templates are rich in functional groups so that MIPs exhibit high recognition abilities and selectivities toward the same structure of analogs.

3,3'-Dichlorobenzidine is an environmental pollutant that is widely employed in the manufacture of dyes for cloth, paper, leather, and other related products [116–118]. 3,3'-Dichlorobenzidine and degradation products, 3-chlorobenzidine and benzidine, are of environmental concern due to their carcinogenic nature. Environmental samples containing 3,3'-dichlorobenzidine are very complex, and the concentration of 3,3'-dichlorobenzidine is usually very low. Therefore, current analysis of 3,3'-dichlorobenzidine in environmental matrices often requires cleanup and preconcentration steps that are complex and time-consuming [32]. Hu et al. designed an improved analytical technique for the determination of 3,3'-dichlorobenzidine by using molecularly imprinted polysiloxane microspheres (MIPS) that were synthesized by covalent imprinting. Some parameters such as adsorption capacity and recognition selectivity were investigated in this study, and it was reported that MIPS exhibits a high adsorption capacity and good recognition selectivity towards 3,3'-dichlorobenzidine [32].

BPA is one of the phenolic compounds used in the production of polycarbonate plastics and epoxy resins. BPA can cause adverse health effects on ecosystems and on human beings by mimicking or interfering hormonal activities to affect growth, development, and reproduction. The effects of BPA on the health may be cumulative and are irreversible, endangering the sustainable development of humans [119]. Even trace BPA produces adverse effects on aquatic life. Therefore, the determination of BPA in very low concentration is important.

Molecularly imprinted polymeric microspheres (MIPMs) were developed by Xie et al. for the determination of BPA in environmental water, which exist in low concentration. The interference properties were investigated in the presence of heavy metals and humic acid with the detection limit of BPA as 5 µg/L [33].

Triclosan (5-chloro-2-(2,4-dichlorophenoxy)phenol) is an important antibacterial ingredient found in personal care products (PPCPs) such as soaps, detergents, toothpastes, disinfectants, cosmetics, and pharmaceuticals [120]. PPCPs and their metabolites are continuously introduced into the environment [13,121], which may affect water quality and potentially impact the ecosystem and human health. Synthetic core–shell MIPs were prepared for the extraction of trace triclosan in environmental water samples by Gao et al. The surface molecular imprinting technique combined with a sol–gel process using the carbon nanotubes (CNTs) coated with silica was developed [34]. It was reported that imprinted polymers exhibit fast kinetics, high capacity, and favorable selectivity and these MIPs developed could be used for the selective adsorption and determination of triclosan from environmental matrices. According to this study MIPs showed good repeatability and reproducibility even though different batches of MIPs were used [34].

PAHs are one of the most important groups of environmental pollutants that can be transferred into the human food chain [122] and they exist in air dust. Some organic species includes alkanes, alkenes, carboxylic acids, carbonyl compounds, and aromatic compounds such as PAHs. These compounds in the environment are well justified, and many of them have proven carcinogenic and/or mutagenic properties. SPE with an MIP has been developed to determine five probable human carcinogenic PAHs in ambient air dust by using gas chromatography–mass spectrometry (GC–MS) technique [35]. Multiple-template imprinting on a single polymer trap was performed and used as SPE material for environmental analysis that reduced analysis time, labor, and cost by performing simultaneous extraction of five carcinogenic PAHs. It was reported that efficient SPE extraction using MIP was successfully applied to the off-line GC–MS ion-trap detection of PAHs in ambient air dust samples. It was shown that undesirable matrix interference in the samples could be removed by MISPE and this pretreatment increased the sensitivity of the MS ion-trap detector in analysis of air samples. Besides easy handling and high stability of MIP, enabled reliable and efficient pretreatment in determination of PAHs in air dust at ultra-trace concentrations was reported [35].

2,4,6-TCP is the other phenolic compound that causes environmental pollution. Molecularly imprinted micro-solid-phase extraction (MIMSPE) is used by Feng et al. for the selective preconcentration of 2,4,6-TCP as phenolic compounds from environmental water samples. To determine the extraction efficiency of the MIMSPE, various parameters have been evaluated. It was reported that MAA-co-divinylbenzene (DVB) polymer showed good selectivity and enrichment efficiency as the MIMSPE coupled with HPLC for enrichment and analysis of phenolic compounds in environmental water. It was also reported that the high recoveries and satisfied precision for all the analytes proved that the method was valid for the analysis of some phenolic compounds in different matrices (tap water, river water, and sewage water) [36].

A different study to determine 2,4-dimethylphenol (2,4-DMP) from environmental water, and MIP was used as a class-selective sorbent in MISPE for preconcentration and determination of phenolic compounds from the environmental water. The variables affecting the extraction efficiency of MISPE procedure were systematically investigated to facilitate the class-selective extraction of phenols from spiked water samples. It was reported that imprinted polymer showed good selectivity for 2,4-DMP with 3.7 ng/mL detection limit in the linear range of 0.04–10.0 µg/mL [37].

MIP was proposed to improve the recognition ability of the MISPE of 2-chlorophenol (2-CP) from environmental waters by El-Sheikh et al. Two MISPE methods for 2-CP detection were developed. One of these methods is based on the polymer imprinted with 2-CP that was used as the extracting sorbent. But this method suffered from low selectivity and high detection limit of 2-CP (7.10 ng/L). In the other method, a polymer imprinted with 4-amino-antipyrine (4-AAP) derivatized 2-CP (2-CP-4-AAP). According to the report, the second method showed high recognition ability/selectivity towards 2-CP-4-AAP with lower detection limit of 0.05 ng/L for 2-CP-4-AAP compared to the first one [38].

Magnetic molecularly imprinted polymers (MMIPs) were synthesized for the selective recognition of 2,4-DCP in different studies. The prepared MMIPs exhibited excellent specific recognition, thermal stability, and saturation magnetization, and they could be easily separated from the suspension by an external magnetic field from aqueous solutions reported in this study. It was reported that it allows for the leading to a fast and selective recognition of 2,4-DCP [40].

Noncovalent and semicovalent approaches were used for the selective extraction of 4-nitrophenol (4-NP) from water samples by using MIP. It was reported that noncovalent MIP was more selective but the semicovalent one showed slightly higher recoveries of 4-NP [41].

Amperometric detection in chloroguaiacol (4-chloro-2-methoxyphenol) at submicromolar levels is detected using MIP by Tarley et al. The satisfactory chloroguaiacol determination was easily achieved even in the equimolar presence of investigated analogous phenolic compounds, and the MIP exhibited good performance by application of method in river water that contains humic substances. It was reported as sensitive, selective, and precise, and obviously fast results were obtained with this method [42].

Highly selective MIPs on silica gel particles were developed for tetrabromobisphenol A (TBBPA) in water samples by Yin et al. Diphenolic acid (DPA) and BPA were used for the detection of selectivity that has similar structure as TBBPA. It was reported that this method has high selectivity and sufficient capacity for enrichment and rapid analysis of TBBPA residues in environmental water samples [43].

MIP-modified glassy carbon electrode (GCE) was developed by Neto et al. for the determination of 4-aminophenol by using amperometry. It was reported that this modified electrodes exhibit high selectivity and the preparation procedures were simple, fast, and reproducible with detection limit as 3.0 µmol/L. This sensor provided a lower detection limit, with similar sensitivity and linear range compared with other sensors and biosensors for phenol [123–125] described in the literature [44].

OPs have high toxicity and widespread uses in agricultural areas. To develop a selective and sensitive method for the detection of OPs in water, plants, soil, and foodstuffs are very important. MIP-modified GCE was developed by Alizadeh et al. for the determination of OP parathion by using square wave voltammetry (SWV). It was reported that parathion-imprinted polymer exhibited a high selectivity as absorber material for parathion extraction and determination in aqueous samples. The detection limit of this method was described as 49.0 ng/L with RSD of 5.7% (n = 5) in the linear dynamic range of 0.20–467.4 µg/L [45].

Trace amounts of pharmaceutical residues in water, such as antibiotics, cause environmental pollution, so it is necessary to monitor them [1]. Fluoroquinolones (FQs) are an important group of antibiotics. A simple method based on magnetic separation for selective extraction of FQs from environmental water samples using MMIP was developed by Chen et al. This proposed method was successfully applied to determine FQs in the presence of ciprofloxacin, enrofloxacin, lomefloxacin, levofloxacin, fleroxacin, and sparfloxacin in different water samples, such as lake water, river water, and primary and final sewage effluent. MMIP showed great analytical potential in the separation, purification, and concentration of analytes in large volumes of real water samples at low concentration level [46].

A highly selective voltammetric sensor for 2,4,6-trinitrotoluene (TNT) was introduced by Alizadeh et al. This method was applied for design and preparation of the highly selective and sensitive electrochemical sensor for TNT determination. The carbon paste electrode modified with MIP having recognition sites for TNT and square wave voltammetry was used for the detection of TNT. The detection limit was obtained as 1.5×10^{-9} mol/L with a dynamic linear range of 5×10^{-9}–1×10^{-6} mol/L. It was reported that these TNT-imprinted polymer-modified electrodes exhibited very high selectivity even at low concentration [47].

14.3 Conclusion

The applications of MIP for detection of analytes in the biological samples present some advantages, such as mechanical and chemical stability, easy preparation, low cost, and high selectivity. Furthermore, these MIP particles can be regenerated and reused repeatedly. These particles would need to be removed after their application, and it can be achieved through physical, chemical, or magnetic separation techniques. MIP-based biosensor used in food toxicant detection and environmental analysis brings low detection limit and good reproducibility due to specific binding between analyte and polymer matrix [18,24,35,37,38,42,43,46].

Acknowledgments

A.E. would like to express her gratitude to the Turkish Academy of Sciences (TUBA) as the associate member of TUBA for its partial support.

References

1. B. Sellergren (Eds.), *Molecularly Imprinted Polymers, Man Made Mimics of Antibodies and Their Applications in Analytical Chemistry: Techniques and Instrumentation in Analytical Chemistry*, Vol. 23, Elsevier, Amsterdam, the Netherlands, 2001.
2. M. Yan and O. Ramström (Eds.), *Molecularly Imprinted Materials: Science and Technology*, Marcel Dekker, New York, 2005.
3. C. Alexander, H.S. Andersson, L.I. Andersson, R.J. Ansell, N. Kirsch, I.A. Nicholls, J. O'Mahony, and M.J. Whitcombe, *J. Mol. Recognit.* 19 (2006) 106–180.
4. N. Masqué, *TrAC, Trend. Anal. Chem.* 20 (2001) 477–486.
5. P.S. Sharma, F.D. Souza, and W. Kutner, *TrAC, Trend. Anal. Chem.* 34 (2012) 59–77.
6. B. Sellergren, *TrAC, Trend. Anal. Chem.* 18 (1999) 164–174.
7. F. Barahona, E. Turiel, P.A.G. Cormack, and A. Martin-Esteban, *J. Sep. Sci.* 34 (2011) 217–224.
8. L. Schweitz, L.I. Andersson, and S. Nilsson, *Anal. Chim. Acta* 435 (2001) 43–47.
9. S.E. Diltemiz, A. Denizli, A. Ersöz, and R. Say, *Sens. Actuat. B* 133 (2008) 484–488.
10. T.A. Sergeyeva, O.A. Slinchenko, L.A. Gorbach, V.F. Matyushov, O.O. Brovko, S.A. Piletsky, L.M. Sergeeva, and G.V. Elska, *Anal. Chim. Acta* 659 (2010) 274–279.
11. J. Liu and G. Wulff, *J. Am. Chem. Soc.* 126 (2004) 7452–7453.
12. F. Sacher, F.T. Lange, H.J. Brauch, and I. Blankenhorn, *J. Chromatogr. A* 938 (2001) 199–210.
13. T. Heberer, *Toxicol. Lett.* 131 (2002) 5–17.
14. J. Fick, H. Söderström, R.H. Lindberg, C. Phan, M. Tysklind, and D.G.J. Lasson, *Environ. Toxicol. Chem.* 28 (2009) 2522–2527.
15. M. Khajeh and E. Sanchooli, *Environ. Chem. Lett.* 9 (2011) 177–183.
16. J. Huang, X. Zhang, S. Liu, Q. Lin, X. He, X. Xing, W. Lian, and D. Tang, *Sens. Actuators B* 152 (2011) 292–298.
17. A.R. Koohpaei, S.J. Shahtaheri, M.R. Ganjali, A. Rahimi Forushani, and F. Golbabaei, *Talanta* 75 (2008) 978–986.
18. T. Alizadeh, *Electroanalysis* 21 (2009) 1490–1498.
19. B. Gao, J. Wang, F. An, and Q. Liu, *Polymer* 49 (2008) 1230–1238.

20. S. Marx and A. Zaltsman, *Int. J. Environ. Anal. Chem.* 83 (2003) 671–680.
21. I. Bakas, N.B. Oujji, E. Moczko, G. Istamboulie, S. Piletsky, E. Piletska, I.A. Ichou, E. Ait-Addi, T. Noguer, and R. Rouillon, *Anal. Chim. Acta* 734 (2012) 99–105.
22. S. Kang, Y. Xu, L. Zhou, and C. Pan, *J. Appl. Polym. Sci.* 124 (2012) 3737–3743.
23. T. Alizadeh, M.R. Ganjali, M. Zare, and P. Norouzi, *Food Chem.* 130 (2012) 1108–1114.
24. S.W. Choi, H.J. Chang, N. Lee, and H.S. Chun, *Sensors* 11 (2011) 8654–8664.
25. D. Zhao, J. Jia, X. Yu, and X. Sun, *Anal. Bioanal. Chem.* 401 (2011) 2259–2273.
26. D.D. Smet, V. Kodeck, P. Dubruel, C.V. Peteghem, E. Schacht, and S.D. Saeger, *J. Chromatogr.* A1218 (2011) 1122–1130.
27. T. Mausia, D.D. Smet, Q. Guorun, C.V. Peteghem, D. Zhang, A. Wu, and S.D. Saeger, *Anal. Lett.* 44 (2011) 2633–2643.
28. P. Lucci, D. Derrien, F. Alix, C. Pérollier, and S. Bayoudh, *Anal. Chim. Acta* 672 (2010) 15–19.
29. D.D. Smet, S. Monbaliu, P. Dubruel, C.V. Peteghem, E. Schacht, and S.D. Saeger, *J. Chromatogr.* A1217 (2010) 2879–2886.
30. D.D. Smet, P. Dubruel, C.V. Peteghem, E. Schacht, and S.D. Saeger, *Food Addit. Contam. Part A* 26 (2009) 874–884.
31. M. Appell, D.F. Kendra, E.K. Kim, and C.M. Maragos, *Food Addit. Contam. Part A* 24 (2007) 43–52.
32. Y. Hu, R. Hu, Q. Zhu, J. Zhan, H. Liu, and B. Yao, *Environ. Chem. Lett.* 10 (2012) 275–280.
33. Y. Xie, H. Li, L. Wang, Q. Liu, Y. Shi, H. Zheng, M. Zhang, Y. Wu, and B. Lu, *Water Res.* 45 (2011) 1189–1198.
34. R. Gao, X. Kong, F. Su, X. He, L. Chen, and Y. Zhang, *J. Chromatogr. A* 1217 (2010) 8095–8102.
35. R.J. Krupadam, B. Bhagat, and M.S. Khan, *Anal. Bioanal. Chem.* 397 (2010) 3097–3106.
36. Q. Feng, L. Zhao, and J.M. Lin, *Anal. Chim. Acta* 650 (2009) 70–76.
37. P. Qi, J. Wang, J. Jin, F. Su, and J. Chen, *Talanta* 81 (2010) 1630–1635.
38. A.H. El-Sheikh, R.W. Al-Quse, M.I. El-Barghouthi, and F.S. Al-Masri, *Talanta* 83 (2010) 667–673.
39. D.C. Apodaca, R.B. Pernites, R. Ponnapati, F.R. Del Mundo, and R.C. Advincula, *Macromolecules* 44 (2011) 6669–6682.
40. J. Pan, L. Xu, J. Dai, X. Li, H. Hang, P. Huo, C. Li, and Y. Yan, *Chem. Eng. J.* 174 (2011) 68–75.
41. E. Caro, N. Masque, R.M. Marce, F. Borrull, P.A.G. Cormack, and D.C. Sherrington, *J. Chromatogr. A* 963 (2002) 169–178.
42. C.R.T. Tarley, M.G. Segatelli, and L.T. Kubota, *Talanta* 69 (2006) 259–266.
43. Y.M. Yin, Y.P. Chen, X.F. Wang, Y. Liu, H.L. Liu, and M.X. Xie, *J. Chromatogr. A* 1220 (2012) 7–13.
44. J.R.M. Neto, W.J.R. Santos, P.R. Lima, S.M.C.N. Tanaka, A.A. Tanaka, and L.T. Kubota, *Sens. Actuators. B* 152 (2011) 220–225.
45. T. Alizadeh, M.R. Ganjali, P. Nourozi, and M. Zare, *Anal. Chim. Acta* 638 (2009) 154–116.
46. L. Chen, X. Zhang, Y. Xu, X. Du, X. Sun, L. Sun, H. Wang, Q. Zhao, A. Yu, H. Zhang, and L. Ding, *Anal. Chim. Acta* 662 (2010) 31–38.
47. T. Alizadeh, M. Zare, M.R. Ganjali, P. Norouzi, and B. Tavana, *Biosens. Bioelectron.* 25 (2010) 1166–1172.

48. V. Pichon and F. Chapuis-Hugon, *Anal. Chim. Acta* 622 (2008) 48–61.
49. M.L. Hu, M. Jiang, P. Wang, S.R. Mei, Y.F. Lin, X.Z. Hu, Y. Shi, B. Lu, and K. Dai, *Anal. Bioanal. Chem.* 387 (2007) 1007–1016.
50. E. Caro, R.M. Marcé, P.A.G. Cormack, D.C. Sherrington, and F. Borrull, *Anal. Chim. Acta* 562 (2006) 145–151.
51. A. Beltran, E. Caro, R.M. Marcé, P.A.G. Cormack, D.C. Sherrington, and F. Borrull, *Anal. Chim. Acta* 597 (2007) 6–11.
52. B. Claude, P. Morin, S. Bayoudh, and J. Ceaurriz, *J. Chromatogr. A* 81 (2008) 1196–1197.
53. Q.Z. Feng, L.X. Zhao, W. Yan, F. Ji, Y.L. Wei, and J.M. Lin, *Anal. Bioanal. Chem.* 391 (2008) 1073–1079.
54. G. Qu, A. Wu, X. Shi, Z. Niu, W. Xie, and D. Zhang, *Anal. Lett.* 41 (2008) 1443–1458.
55. J.L. Urraca, M.C. Moreno-Bondi, A.J. Hall, and B. Sellergren, *Anal. Chem.* 79 (2007) 695–701.
56. P. Manesiotis, A.J. Hall, J. Courtois, K. Irgum, and B. Sellergren, *Angew. Chem., Int. Ed.* 44 (2005) 3902–3906.
57. D.K. Alexiadou, N.C. Maragou, N.S. Thomaidis, G.A. Theodoridis, and M.A. Koupparis, *J. Sep. Sci.* 31 (2008) 2272–2282.
58. Y.Q. Xia, T.Y. Guo, H.L. Zhao, M.D. Song, B.H. Zhang, and B.L. Zhang, *J. Sep. Sci.* 30 (2007) 1300–1306.
59. C. Cacho, E. Turiel, A. Martín-Esteban, D. Ayala, and C. Pérez-Conde, *J. Chromatogr. A* 1114 (2006) 255–262.
60. C. Cacho, E. Turiel, A. Martín-Esteban, C. Pérez-Conde, and C. Cámara, *Anal. Bioanal. Chem.* 376 (2003) 491–496.
61. D. Djozan and B. Ebrahimi, *Anal. Chim. Acta* 616 (2008) 152–159.
62. X. Zhu, J. Cai, J. Yang, Q. Su, and Y. Gao, *J. Chromatogr. A* 1131 (2006) 37–44.
63. Y. Tang, Z. Huang, T. Yang, X. Hu, and X. Jiang, *Anal. Lett.* 38 (2005) 219–226.
64. A. Beltran, R.M. Marcé, P.A.G. Cormack, D.C. Sherrington, and F. Borrull, *J. Sep. Sci.* 31 (2008) 2868–2874.
65. A. Beltran, N. Fontanals, R.M. Marcé, P.A.G. Cormack, and F. Borrull, *J. Sep. Sci.* 32 (2009) 3319–3326.
66. S. Le Moullec, A. Bégos, V. Pichon, and B. Bellier, *J. Chromatogr. A* 1108 (2006) 7–13.
67. C. Cacho, E. Turiel, and C. Pérez-Conde, *Talanta* 78 (2009) 1029–1035.
68. H. Sambe, K. Hoshina, R. Moaddel, I.W. Wainer, and J. Haginaka, *J. Chromatogr. A* 1134 (2006) 88–94.
69. A. Beltran, R.M. Marcé, P.A.G. Cormack, and F. Borrull, *J. Chromatogr. A* 1216 (2009) 2248–2253.
70. J. Wang, P.A.G. Cormack, D.C. Sherrington, and E. Khoshdel, *Angew. Chem., Int. Ed.* 42 (2003) 5336–5338.
71. E. Turiel, J.L. Tadeo, P.A.G. Cormack, and A. Martín-Esteban, *Analyst* (Cambridge, U.K.) 130 (2005) 1601–1607.
72. X. Liu and J. Lei, *Polym. Eng. Sci.* 52 (2012) 2099–2105.
73. X. Wang, Q. Fang, S. Liu, and L. Chen, *Anal. Bioanal. Chem.* 404 (2012) 1555–1564.
74. L. Geng, X. Kou, J. Lei, H. Su, G. Ma, and Z. Su, *J. Chem. Technol. Biotechnol.* 87 (2012) 635–642.
75. J. Zhang, J. Ma, X. Yue, X. Bu, and Y. Han, *J. Appl. Polym. Sci.* 124 (2012) 723–728.
76. J. Kong, Y. Wang, C. Nie, D. Ran, and X. Jia, *Anal. Methods* 4 (2012) 1005–1011.

77. J. Qiao, H. Yan, H. Wang, and Y. Lv, *J. Sep. Sci.* 34 (2011) 2668–2673.
78. F.G. Tamayo and A. Martín-Esteban, *J. Chromatogr. A* 1098 (2005) 116.
79. E. Yilmaz, T. Adali, O. Yilmaz, and M. Bengisu, *React. Funct. Polym.* 67 (2007) 10–18.
80. J.A. Lewis and G.R. Fenwick, *Food Contaminants: Sources and Surveillance*, C.S. Creaser and R. Purchase (Eds.), The Royal Society of Chemistry, Cambridge, U.K., 1991, pp. 1–20.
81. J.W. DeVries, M.W. Trucksess, and L.S. Jackson (Eds.), *Mycotoxins and Food Safety*, Kluwer, New York, 2002.
82. H.P. van Egmond, *Anal. Bioanal. Chem.* 378 (2004) 1152–1160.
83. S.M. Hays and L.L. Aylward, *Regul. Toxicol. Pharm.* 37 (2003) 202–217.
84. S. Moret and L.S. Conte, *J. Chromatogr. A* 882 (2000) 245–253.
85. J.D.G. McEvoy, *Anal. Chim. Acta* 473 (2002) 3–26.
86. M. Careri, F. Bianchi, and C. Corradini, *J. Chromatogr. A* 970 (2002) 3–64.
87. P.L. Buldini, L. Ricci, and J.L. Sharma, *J. Chromatogr. A* 975 (2002) 47–70.
88. A. Juan-Garcia, G. Font, and Y. Pico, *J. Sep. Sci.* 28 (2005) 787–790.
89. V. Pichon, M. Delaunay-Bertoncini, and M.C. Hennion, *Sampling and Sample Preparation for Field and Laboratory*, J. Pawliszyn (Ed.), Comprehensive analytical chemistry, Elsevier Science, Amsterdam, the Netherlands, 2002, pp. 1081–1100.
90. M.C. Hennion and V. Pichon, *J. Chromatogr. A* 1000 (2003) 29–52.
91. B. Sellergren, F. Lanza, *Molecularly Imprinted Polymers: Man Made Mimics of Antibodies and Their Applications in Analytical Chemistry*, B. Sellergren (Ed.), Elsevier Science, Amsterdam, the Netherlands, 2001, pp. 355–375.
92. F. Lanza and B. Sellergren, *Adv. Chromatogr.* 41 (2001) 138–173.
93. E. Caro, R.M. Marće, F. Borrull, P.A.G. Cormack, and D.C. Sherrington, *TrAC, Trend. Anal. Chem.* 25 (2006) 143–154.
94. C. Uguz, O. Varisli, C. Agca, and Y. Agca, *Reprod. Toxicol.* 28 (2009) 542–549.
95. J.P. Hsu, H.G. Wheeler, D.E. Camann, H.J. Schattenberg, R.G. Lewis, and A.E. Bond, *J. Chromatogr. Sci.* 26 (1988) 181–189.
96. A.A. Ciucu, C. Negulescu, and R.P. Baldwin, *Biosens. Bioelectron.* 18 (2003) 303–310.
97. I. Sospedra, J. Blesa, J.M. Soriano, and J. Mañes, *J. Chromatogr. A* 1217 (2010) 1437–1440.
98. B. Bakan, D. Melcion, D. Richard-Molard, and B. Cahagnier, *J. Agric. Food Chem.* 50 (2002) 728–731.
99. R. Krska, E. Welzig, and H. Boudra, *Anim Feed. Sci. Tech.* 137 (2007) 241–264.
100. M. Zachariasova, O. Lacina, A. Malachova, M. Kostelanska, J. Poustka, M. Godula, and J. Hajslova, *Anal. Chim. Acta* 662 (2019) 51–61.
101. L. Harrison, B. Colvin, J. Greene, L. Newman, and J. Cole, *J. Vet. Diagn. Invest.* 2 (1990) 217–221.
102. W. Gelderblom, N. Kriek, W. Marasas, and P. Thiel, *Carcinogenesis* 12 (1991) 1247–1125.
103. J. Rheeder, W. Marasas, P. Thiel, E. Sydenham, G. Shephard, and D. Schalkwijk, *Phytopathology* 82 (1992) 353–357.
104. P.Y. Li and H. Qian, *Iran J. Environ. Health* 8 (2011) 41–48.
105. J.H. Rodriguez, S.B. Weller, E.D. Wannaz, A. Klumpp, and M.L. Pignata, *Ecol. Indic.* 11 (2011) 1673–1680.
106. A. Clark, T. Turner, K.P. Dorothy, J. Goutham, C. Kalavati, and B. Rajanna, *Ecotoxicol. Environ. Saf.* 56 (2003) 390–397.

107. S. Jobling and C.R. Tyler, *Pure Appl. Chem.* 75 (2003) 2219–2234.
108. M. Muneer, I.A. Bhatti, and S. Adeel, *Asian J. Chem.* 22 (2010) 7453–7459.
109. R. Cela, M.P. Llompart, and I. Rodriguez, *J. Chromatogr. A* 885 (2000) 291–304.
110. D. Han, G. Fang, and X. Yan, *J. Chromatogr. A* 1100 (2005) 131–136.
111. E. Caro, R. Marce, P. Cormack, D. Sherrington, and F. Borrull, *J. Chromatogr. A* 995 (2003) 233–238.
112. M.L. Mena, L. Agui, P. Martinez-Ruiz, P. Yanez-Sedeno, A.J. Reviejo, and J.M. Pingarron, *Anal. Bioanal. Chem.* 376 (2003) 18–25.
113. C.R.T. Tarley and L.T. Kubota, *Anal. Chim. Acta* 548 (2005) 11–19.
114. M.L. Yang and Y.Z. Li, *Anal. Lett.* 37 (2004) 2043–2052.
115. C. Michailof, P. Manesiotis, and C. Panayiotou, *J. Chromatogr. A* 1182 (2008) 25–33.
116. T.J. Haley, *Clin. Toxicol.* 8 (1975) 13–42.
117. G. Choudhary, *Chemosphere* 32 (1996) 267–291.
118. L. Wang, J. Yan, W. Hardy, C. Mosley, S. Wang, and H.T. Yu, *Toxicology* 207 (2005) 411–418.
119. C.A. Staples, P.B. Dorn, G.M. Klecka, S.T. O'Block, and L.R. Hariis, *Chemosphere* 36 (1998) 2149–2173.
120. H.N. Bhargava, A. Patricia, and B.S. Leonard, *Am. J. Infect. Control* 24 (1996) 209–218.
121. D.W. Kolpin, E.T. Furlong, M.T. Meyer, E.M. Thurman, S.D. Zaugg, L.B. Barber, and H.T. Buxton, *Environ. Sci. Technol.* 36 (2002) 1202–1211.
122. National Research Council, *Polycyclic Aromatic Hydrocarbons: Evaluation and Effects*, Committee on Pyrene and Selected Analogues, Board on Toxicology and Environmental Health Hazard, Commission of Life Sciences, National Academy of Press, Washington, DC, 1983.
123. X. Zhang, S. Wang, and Q. Shen, *Microchim. Acta* 149 (2005) 37–48.
124. W. Suna, K. Jiao, S. Zhang, C. Zhang, and Z. Zhang, *Anal. Chim. Acta* 434 (2001) 43–50.
125. G.K. Nikolaos and M.R. Subrayal, *Analyst* 127 (2002) 368–372.

15

Lab-on-a-Chip and Microfluidic Technology

Nikolaos Tzamtzis, Georgia-Paraskevi Nikoleli,
Theodoros Varzakas, Dimitrios P. Nikolelis, Vassilios N. Psychoyios,
Nikolas Psaroudakis, and Stelios Liodakis

CONTENTS

15.1 Introduction

The ability to perform laboratory operations on a small scale using miniaturized lab-on-a-chip (LOC) devices is a promising biosensing technique. The advantages of LOC are the small time of analysis, the low reagent costs, and the reduced amount of chemical wastes. The application of portable, easy-to-use, and highly sensitive LOC biosensors for real-time detection could offer significant advantages over current analytical methods. Integrated optics-based biosensors have become the most suitable technology for LOC integration due to their ability for miniaturization; their extreme sensitivity, robustness, and reliability; and their potential for multiplexing and mass production at low cost.

An LOC is a device that integrates one or several laboratory functions on a single chip of only millimeters to a few square centimeters in size. LOCs deal with the handling of extremely small fluid volumes down to less than picoliters. LOC devices are a subset of microelectromechanical systems (MEMS) and are often indicated by "Micro Total Analysis Systems" (µTAS) as well. Microfluidics is a broader term that also describes mechanical flow control devices like pumps and valves or sensors like flowmeters and viscometers. However, strictly regarded "LOC" indicates generally the scaling of single or multiple lab processes down to chip format, whereas "µTAS" is dedicated to the integration of the total sequence of lab processes to perform chemical analysis. The term "lab-on-a-chip" was introduced later on, when it turned out that µTAS technologies were more widely applicable than only for analysis purposes.

After the invention of microtechnology (~1954) for realizing integrated semiconductor structures for microelectronic chips, these lithography-based technologies were soon applied in pressure sensor manufacturing (1966) as well. Due to further development of these usually complementary metal–oxide–semiconductor (CMOS)-compatibility limited processes, a tool box became available to create micrometer- or submicrometer-sized mechanical structures in silicon wafers as well: the MEMS era (also indicated with microsystem technology—MST) had started.

A big boost in research and commercial interest came in the mid 1990s, when µTAS technologies turned out to provide interesting tooling for genomics applications, like capillary electrophoresis and DNA microarrays. A big boost in research support also came from the military, especially from DARPA (Defense Advanced Research Projects Agency), for their interest in portable bio/chemical warfare agent detection systems. The added value was limited not only to the integration of lab processes for analysis but also the characteristic possibilities of individual components, and the application to other nonanalysis lab processes. Hence, the term "lab-on-a-chip" was introduced.

Although the application of LOCs is still novel and modest, a growing interest of companies and applied research groups is observed in different fields such as analysis (e.g., chemical analysis, environmental monitoring, medical diagnostics, and cellomics) and synthetic chemistry (e.g., rapid screening and microreactors for pharmaceutics). Besides further application developments, research in LOC systems is expected to extend towards downscaling of fluid handling structures as well, by using nanotechnology. Submicrometer and nanosized channels, DNA labyrinths, single-cell detection and analysis, and nanosensors might become feasible, allowing new ways of interaction with biological species and large molecules. Many books have been written that cover various aspects of these devices, including the fluid transport [1–3], system properties [4], and bioanalytical applications [5,6].

15.2 Chip Materials and Fabrication Technologies

The basis for most LOC fabrication processes is photolithography. Initially most processes were in silicon, as these well-developed technologies were directly derived from semiconductor fabrication. Because of demands, for example, specific optical characteristics, bio or chemical compatibility, lower production costs, and faster prototyping, new processes have been developed such as glass, ceramics, and metal etching; deposition and bonding; polydimethylsiloxane (PDMS) processing (e.g., soft lithography); and thick-film and stereolithography as well as fast replication methods via electroplating, injection molding, and embossing. Furthermore, the LOC field more and more exceeds the borders between lithography-based microsystem technology, nanotechnology, and precision engineering.

15.3 Advantages of LOCs

LOCs may provide advantages, which are specific to their application. Typical advantages are

- Low fluid volume consumption (less waste, lower reagents costs, and less required sample volumes for diagnostics)
- Faster analysis and response time due to short diffusion distances, fast heating, high surface-to-volume ratios, and small heat capacities
- Better process control because of a faster response system (e.g., thermal control for exothermic chemical reactions)
- Compactness of systems due to the integration of much functionality and small volumes
- Massive parallelization due to compactness, which allows high-throughput analysis

- Lower fabrication costs, allowing cost-effective disposable chips, fabricated in mass production
- Safer platform for chemical, radioactive, or biological studies because of integration of functionality, smaller fluid volumes, and stored energies

15.4 Disadvantages of LOCs

Some of the disadvantages of LOCs are

- Novel technology and therefore not yet fully developed.
- Physical and chemical effects—like capillary forces, surface roughness, and chemical interactions of construction materials on reaction processes—become more dominant on a small scale. This can sometimes make processes in LOCs more complex than in conventional lab equipment.
- Detection principles may not always scale down in a positive way, leading to low signal-to-noise ratios.
- Although the absolute geometric accuracies and precision in microfabrication are high, they are often rather poor in a relative way, compared to precision engineering, for instance.

15.5 LOCs and Global Health

LOC technology may soon become an important part of efforts to improve global health [7], particularly through the development of point-of-care testing devices. In countries with few health-care resources, infectious diseases that would be treatable in a developed nation are often deadly. In some cases, poor health-care clinics have the drugs to treat a certain illness but lack the diagnostic tools to identify patients who should receive the drugs. Many researchers believe that LOC technology may be the key to powerful new diagnostic instruments. The goal of these researchers is to create microfluidic chips that will allow health-care providers in poorly equipped clinics to perform diagnostic tests such as immunoassays and nucleic acid assays with no laboratory support.

15.5.1 Global Challenges

For the chips to be used in areas with limited resources, many challenges must be overcome. In developed nations, the most highly valued traits for diagnostic tools include speed, sensitivity, and specificity, but in countries where the health-care infrastructure is less well developed, attributes such as ease of use and shelf-life must also be considered. The reagents that come with the chip, for example, must be designed so that they remain effective for months even if the chip is not kept in a climate-controlled environment. Chip designers must also keep cost, scalability, and recyclability in mind as they choose what materials and fabrication techniques to use.

15.5.2 Examples of Global LOC Application

One active area of LOC research involves ways to diagnose and manage HIC infections. Around 40 million people are infected with HIV in the world today, yet only 1.3 million of these people receive antiretroviral treatment. Around 90% of people with HIV have never been tested for the disease. Measuring the number of CD4+ T lymphocytes in a person's blood is an accurate way to determine if a person has HIV and to track the progress of an HIV infection. At the moment, flow cytometry is the gold standard for obtaining CD4 counts, but flow cytometry is a complicated technique that is not available in most developing areas because it requires trained technicians and expensive equipment. Recently, such a cytometer was developed for just $5 [8].

15.6 LOCs and Plant Sciences

LOC devices could be used to characterize pollen tube guidance in *Arabidopsis thaliana*. Specifically, plant on a chip is a miniaturized device in which pollen tissues and ovules could be incubated for plant science studies [9].

15.7 LOC Made of Paper

While larger paper tests (like those for pregnancy) are common, shrinking the paper and minimizing the quantity of the required chemical reagents reduce manufacturing costs. The ability to direct the sample to a particular region of the paper enables the simultaneous performance of several tests, to look for multiple symptoms of a condition like kidney failure or infectious disease. And reducing the sample size is a particular advantage in developing countries, where noninvasively gathering small amounts of fluids avoids the need for syringes, which can be hard to clean and dispose of. Paper-based microfluidic devices could yield cheap and disposable diagnostic tests.

A new paper diagnostic test from Harvard University analyzes the glucose (left well) and protein (right well) content of urine; the top well is a control for the glucose assay. The beige part of the test paper has been treated with a hydrophobic polymer that channels the liquid into wells. In this test, the paper was dipped in an artificial urine solution that contained glucose and protein extracted from cow blood.

A pinprick of blood or drop of urine soaked up at the edge of the Whitesides device moves naturally through the paper, in much the way that wine will spread through a paper napkin. But the paper is treated with a hydrophobic polymer, which directs the liquid along prescribed channels. Once the liquid reaches the wells at the ends of the channels, it interacts with reagents, turning the paper into different colors. The colors can be matched to those on a color key, much as they are in a pH test. One test design that looks like a miniature three-branched geometric tree might have wells at the end of two branches for a glucose assay, and one at the end of the third for a protein assay, for example.

In recent years, scientists have been getting closer to developing LOCs, devices designed to manipulate pico- or even femtoliters of fluid. The chips could improve drug discovery, medical diagnostics, and even genomic testing; experiments that now require pipette-handling robots to manipulate large quantities of chemicals, for example, could be done on a tiny device. But the technology isn't quite there yet. The chips are small, but moving chemicals around inside them requires external pumps and valves. The state of the art, then, is closer to a lab around a chip than on one.

But now Neil Gershenfeld, the director of MIT's Center for Bits and Atoms, and graduate student Manu Prakash say they have figured out a way to get rid of those pumps and valves. They have designed a device, described in a recent issue of *Science*, that manipulates its liquid contents without external controls. And the technique is based entirely on the flow of bubbles.

The researchers discovered the technique by accident. Part of the center's mission is to help bring computing to remote areas by designing processors that use simple materials and don't require complicated power systems or electronics. Prakash and Gershenfeld were trying to develop a microfluidic-based microprocessor, but "the bubbles kept interfering," Gershenfeld says. "It eventually occurred to us that we should use them."

They did and found that the flow of the bubbles—where they go, how quickly they get there, and whether they merge or remain separate—can be controlled through the design of the channels in the chip. For example, the chip can synchronize the movement of two bubbles flowing along parallel paths, guaranteeing a meeting when the two avenues converge. If one bubble is far ahead of the other, the surrounding liquid, obeying the laws of fluid dynamics, floods through channels between the parallel paths. The altered flow drags on the leading bubble and accelerates the other one, helping it catch up.

"There are many potential applications of this work," Prakash says. The chip could serve as a microprocessor: intersections between the channels would act as logic gates, and the bubbles would stand in for electrons as the basic units of information. For drug discovery purposes, the bubbles would be replaced by drops of chemicals and reagents. Instead of being mixed on a lab bench, these materials would react within the chip, making the whole endeavor simpler. Prakash even envisions a bubble-based diagnostic device that could be used in remote communities to test patients' blood or urine for signs of disease. In this case, the doctor wouldn't have to send the samples to a lab. The lab would be right there, inside the chip.

15.8 Fluidic Connect 4515

The LOC platform Fluidic Connect 4515 offers a user-friendly way of creating your own LOC setup within minutes of time. The microfluidic chips within the platform enable several research subjects such as microreaction, cell analysis, and droplet generation. The products are affordable and meet the high-quality standards of Micronit. They can easily be used with standard laboratory equipment such as syringe pumps and microscopes. All the Fluidic Connect products are available off-the-shelf:

- User-friendly and leak-free connections
- Easy chip reconnection thanks to polymer cartridge
- Self-alignment of microfluidic chips
- Chemically inert materials
- Large chip area to detect with microscope objective
- Building block in a modular system
- Availability of many different off-the-shelf chip designs
- Possibility to have your own design manufactured

15.9 Microfluidics

Microfluidics deals with the behavior, precise control, and manipulation of fluids that are geometrically constrained to a small, typically submillimeter, scale. Typically, micro means one of the following features:

- Small volumes (μL, nL, pL, fL)
- Small size
- Low energy consumption
- Effects of the microdomain

Typically, fluids are moved, mixed, separated, or otherwise processed. Numerous applications employ passive fluid control techniques like capillary forces. In some applications, external actuation means are additionally used for a directed transport of the media. Examples are rotary drives applying centrifugal forces for the fluid transport on the passive chips. Active microfluidics refers to the defined manipulation of the working fluid by active (micro) components as micropumps or microvalves. Micropumps supply fluids in a continuous manner or are used for dosing. Microvalves determine the flow direction or the mode of movement of pumped liquids. Often processes that are normally carried out in a lab are miniaturized on a single chip in order to enhance efficiency and mobility as well as reducing sample and reagent volumes.

It is a multidisciplinary field intersecting engineering, physics, chemistry, microtechnology, and biotechnology, with practical applications to the design of systems in which such small volumes of fluids will be used. Microfluidics emerged in the beginning of the 1980s and is used in the development of ink-jet printheads, DNA chips, LOC technology, micropropulsion, and microthermal technologies.

Misiakos et al. [10] describes an optical affinity sensor based on a monolithic optoelectronic transducer, which integrates on a silicon die thin optical fibers (silicon nitride) along with self-aligned light-emitting diodes (LEDs), and photodetectors (silicon p/n junction). The LEDs are optically coupled to the corresponding photodetectors through silicon nitride fibers. Specially designed spacers provide smooth bending of the fiber at its end points and towards the light source, and the detector ensuring high coupling efficiency. The transducer surface is hydrophilized by oxygen plasma treatment, silanized with (3-aminopropyl) triethoxysilane, and bioactivated through the adsorption of the biomolecular probes. The use of a microfluidic module allows real-time monitoring of the binding reaction of the gold nanoparticle-labeled analytes with the immobilized probes. Their binding within the evanescent field at the surface of the optical fiber causes attenuated total reflection of the waveguided modes and reduction of the detector photocurrent. The biotin–streptavidin model assay was used for the evaluation of the analytical potentials of the device developed. Detection limits of 3.8 and 13 pM in terms of gold nanoparticle-labeled streptavidin were achieved for continuous and stopped-flow assay modes, respectively.

15.9.1 Microscale Behavior of Fluids

The behavior of fluids at the microscale can differ from "macrofluidic" behavior, in that factors such as surface tension, energy dissipation, and fluidic resistance start to dominate the system. Microfluidics studies how these behaviors change and how they can be worked around or exploited for new uses [1–3,11,12].

At small scales (channel diameters of around 100 nm to several hundred μm), some interesting and sometimes unintuitive properties appear. In particular, the Reynolds number (which compares the effect of momentum of a fluid to the effect of viscosity) can become very low. A key consequence of this is that fluids, when side by side, do not necessarily mix in the traditional sense; molecular transport between them must often be through diffusion [13].

High specificity of chemical and physical properties (concentration, pH, temperature, shear force, etc.) can also be ensured resulting in more uniform reaction conditions and higher-grade products in single- and multistep reactions [13–16].

15.9.2 Effects of Microdomain

- Laminar flow
- Surface tension
- Electrowetting
- Fast thermal relaxation
- Electric surface charges
- Diffusion

15.9.3 Key Application Areas

Microfluidic structures include micropneumatic systems, that is, microsystems for the handling of off-chip fluids (liquid pumps, gas valves, etc.), and microfluidic structures for the on-chip handling of nano- and picoliter volumes [17]. To date, the most successful commercial application of microfluidics is the ink-jet printhead. Significant research has been applied to the application of microfluidics for the production of industrially relevant quantities of material [18].

Advances in microfluidic technology are revolutionizing molecular biology procedures for enzymatic analysis (e.g., glucose and lactate assays), DNA analysis (e.g., polymerase chain reaction (PCR) and high-throughput sequencing), and proteomics. The basic idea of microfluidic biochips is to integrate assay operations such as detection, as well as sample pretreatment and sample preparation on one chip [19,20].

An emerging application area for biochips is clinical pathology, especially the immediate point-of-care diagnosis of diseases. In addition, microfluidic-based devices, capable of continuous sampling and real-time testing of air/water samples for biochemical toxins and other dangerous pathogens, can serve as an always-on "biosmoke alarm" for early warning.

15.10 Continuous-Flow Microfluidics

These technologies are based on the manipulation of continuous liquid flow through microfabricated channels. The actuation of liquid flow is implemented either by external pressure sources, external mechanical pumps, and integrated mechanical micropumps, or by combinations of capillary forces and electrokinetic mechanisms [21,22]. Continuous-flow microfluidic operation is the mainstream approach because it is easy to implement and less sensitive to protein fouling problems. Continuous-flow devices are adequate for many well-defined and simple biochemical applications and for certain tasks such as chemical separation, but they are less suitable for tasks requiring a high degree of flexibility or ineffective fluid manipulations. These closed-channel systems are inherently difficult to integrate and scale because the parameters that govern flow-field vary along the flow-path making the fluid flow at any one location, dependent on the properties of the entire system. Permanently etched microstructures also lead to limited reconfigurability and poor fault-tolerance capability.

Process monitoring capabilities in continuous-flow systems can be achieved with highly sensitive microfluidic flow sensors based on MEMS technology, which offers resolutions down to the nanoliter range.

15.11 Digital (Droplet-Based) Microfluidics

Alternatives to the earlier closed-channel continuous-flow systems include novel open structures, where discrete, independently controllable droplets are manipulated on a substrate using electrowetting. Following the analogy of digital microelectronics, this approach is referred to as digital microfluidics. Le Pesant et al. pioneered the use of electrocapillary forces to move droplets on a digital track [23]. The "fluid transistor" pioneered by Cytonix [22] also played a role. The technology was subsequently commercialized by the Duke University. By using discrete unit-volume droplets [24], a microfluidic function can be reduced to a set of repeated basic operations, that is, moving one unit of fluid over one unit of distance. This "digitization" method facilitates the use of a hierarchical and cell-based approach for microfluidic biochip design. Therefore, digital microfluidics offer a flexible and scalable system architecture as well as high fault-tolerance capability. Moreover, because each droplet can be controlled independently, these systems also have dynamic reconfigurability, whereby groups of unit cells in a microfluidic array can be reconfigured to change their functionality during the concurrent execution of a set of bioassays. Although droplets are manipulated in confined microfluidic channels, since the control on droplets is not independent, it should not be confused as "digital microfluidics." One common actuation method for digital microfluidics is electrowetting on dielectric (EWOD). Many LOC applications have been demonstrated within the digital microfluidic paradigm using electrowetting. However, recently, other techniques for droplet manipulation have also been demonstrated using surface acoustic waves, optoelectrowetting, mechanical actuation [25], etc.

15.12 Microfluidic Technology, Techniques, and Application

15.12.1 DNA Chips (Microarrays)

Early biochips were based on the idea of a DNA microarray, for example, the GeneChip DNA array from Affymetrix, which is a piece of glass, plastic, or silicon substrate on which pieces of DNA (probes) are affixed in a microscopic array. Similar to a DNA microarray, a protein array is a miniature array where a multitude of different capture agents, most frequently monoclonal antibodies, are deposited on a chip surface; they are used to determine the presence and/or amount of proteins in biological samples, for example, blood. A drawback of DNA and protein arrays is that they are neither reconfigurable nor scalable after manufacture. Digital microfluidics has been described as a means for carrying out digital PCR.

15.12.2 Molecular Biology

In addition to microarrays, biochips have been designed for 2-D electrophoresis [26], transcriptome analysis [27], and PCR amplification [28]. Other applications include various electrophoresis and liquid chromatography applications for proteins and DNA cell separation, in particular blood cell separation, protein analysis, cell manipulation, and analysis including cell viability analysis and microorganism capturing [19].

15.12.3 Evolutionary Biology

By combining microfluidics with landscape ecology and nanofluidics, a nano-/microfabricated fluidic landscape can be constructed by building local patches of bacterial habitat and connecting them by dispersal corridors. The resulting landscapes can be used as physical implementations of an adaptive landscape [29], by generating a spatial mosaic of patches of opportunity distributed in space and time. The patchy nature of these fluidic landscapes allows for the study of adapting bacterial cells in a metapopulation system. The evolutionary ecology of these bacterial systems in these synthetic ecosystems allows for using biophysics to address questions in evolutionary biology.

15.12.4 Microbial Behavior

The ability to create precise and carefully controlled chemoattractant gradients makes microfluidics the ideal tool to study motility, chemotaxis, and the ability to evolve/develop resistance to antibiotics in small populations of microorganisms and in a short period of time. These microorganisms include bacteria [30] and the broad range of organisms that form the marine microbial loop [31], responsible for regulating much of the oceans' biogeochemistry.

15.12.5 Cellular Biophysics

By rectifying the motion of individual swimming bacteria [32], microfluidic structures can be used to extract mechanical motion from a population of motile bacterial cells [33]. This way, bacteria-powered rotors can be built [34,35].

15.12.6 Acoustic Droplet Ejection

Acoustic droplet ejection (ADE) uses a pulse of ultrasound to move low volumes of fluids (typically nanoliters or picoliters) without any physical contact. This technology focuses acoustic energy into a fluid sample in order to eject droplets as small as a millionth of a millionth of a liter (picoliter = 10^{-12} L). ADE technology is a very gentle process, and it can be used to transfer proteins, high molecular weight DNA, and live cells without damage or loss of viability. This feature makes this technology suitable for a wide variety of applications including proteomics and cell-based assays.

15.12.7 Fuel Cells

Microfluidic fuel cells can use laminar flow to separate the fuel and its oxidant to control the interaction of the two fluids without a physical barrier as would be required in conventional fuel cells [36,37].

15.12.8 Tool for Cell Biological Research

Microfluidic technology is creating powerful tools for cell biologists to control the complete cellular environment, leading to new questions and new discoveries [38]. Following are the many diverse advantages of this technology for microbiology:

- Microenvironmental control: ranging from mechanical environment [39] to chemical environment [40].
- Precise spatiotemporal concentration gradients [41].
- Mechanical deformation.

- Force measurements of adherent cells.
- Confining cells [42].
- Exerting a controlled force [43,44].
- Fast and precise temperature control [44,45].
- Electric field integration [43].
- Cell culture.
- Plant on a chip and plant tissue culture [46].
- Antibiotic resistance: microfluidic devices can be used as hetero-geneous environments for microorganisms. In a heterogeneous environment, it is easier for a microorganism to evolve. This can be useful for testing the acceleration of evolution of a microorganism for testing the development of antibiotic resistance.

15.12.9 Optofluidics

The merger of microfluidics and optics is typical known as optofluidics. An example of an optofluidic device is a tunable microlens array [47,48].

Acknowledgments

The authors express their gratitude for the financial help of the Greek Ministry of Development, General Secretariat of Research and Technology and the European Commission (in particular the European Regional Development fund and National Resources) (Contract: 12SLO_ET30_1036) that co-funded the present research project in the framework of Greece–Slovakia bilateral projects.

References

1. B.J. Kirby, *Micro- and Nanoscale Fluid Mechanics: Transport in Microfluidic Devices*, Cambridge University Press, Cambridge, U.K., 2010.
2. H. Bruus, *Theoretical Microfluidics*, Oxford University Press, Oxford, U.K., 2007.
3. G.M. Karniadakis, A. Beskok, and N. Aluru, *Microflows and Nanoflows*, Springer Verlag, Dordrecht, the Netherlands, 2005.
4. P. Tabeling, *Introduction to Microfluidic*, Oxford University Press, Oxford, U.K., 2005.
5. J. Berthier, P. Silberzan, *Microfluidics for Biotechnology*, Artech House, Norwood, MA, 2006.

6. F.A. Gomez, *Biological Applications of Microfluidics*, Wiley-Interscience, New York, 2008.
7. P. Yager, T. Edwards, E. Fu, K. Helton, K. Nelson, M.R. Tam, and B.H. Weigl, *Nature*, 442 (2006) 412–418.
8. A. Ozcan, http://www.dailybruin.com/index.php/multimedia/43312, Retrieved on October 5, 2011.
9. A.K. Yetisen, L. Jiang, J.R. Cooper, Y. Qin, R. Palanivelu, and Y. Zohar, *J. Micromech. Microeng.*, 21 (2011) 054018.
10. K. Misiakos, S.E. Kakabakos, P.S. Petrou, and H.H. Ruf, *Anal. Chem.*, 76 (2004) 1366–1373.
11. S.C. Terry, J.H. Jerman, and J.B. Angell, *IEEE Trans. Electron. Devices*, 26 (1979) 1880–1886.
12. A. Manz, N. Graber, and H.M. Widmer, *Sensor Actuat. B-Chem.*, 1 (1990) 244–248.
13. V. Chokkalingam, B. Weidenhof, M. Kraemer, W.F. Maier, S. Herminghaus, and R. Seemann, *Lab Chip*, 10 (2010) 1700–1705.
14. I. Shestopalov, J.D. Tice, and R.F. Ismagilov, *Lab Chip*, 4 (2004) 316–321.
15. J. Greener, E. Tumarkin, M. Debono, C.-H. Kwan, M. Abolhasani, A. Guenther, and E. Kumacheva, *Analyst*, 137 (2012) 444–450.
16. J. Greener, E. Tumarkin, M. Debono, A.P. Dicks, and E. Kumacheva, *Lab Chip*, 12 (2012) 696–701.
17. N.T. Nguyen and S.T. Wereley, *Fundamentals and Applications of Microfluidics*, 2nd edn., Artech House, Boston, MA, 2006.
18. W. Li, J. Greener, D. Voicu, and E. Kumacheva, *Lab Chip*, 9 (2009) 2715–2721.
19. K.E. Herold and A. Rasooly (Eds.), *Lab-on-a-Chip Technology (Vol. 1): Fabrication and Microfluidics*, Caister Academic Press, Norfolk, U.K., 2009.
20. K.E. Herold and A. Rasooly (Eds.), *Lab-on-a-Chip Technology (Vol. 2): Biomolecular Separation and Analysis*, Caister Academic Press, Norfolk, U.K., 2009.
21. H.-C. Chang and L.Y. Yeo, *Electrokinetically Driven Microfluidics and Nanofluidics*, Cambridge University Press, Cambridge, U.K., 2009.
22. http://www.cytonix.com/fluid%20transistor.html
23. J.P. Le Pesant, M. Hareng, B. Mourey, and J.N. Perbet, U.S. Pat. No. 4569575, February 11, 1986.
24. V. Chokkalingam, S. Herminghaus, and R. Seemann, *Appl. Phys. Lett.*, 93 (2008) 254101–254103.
25. J. Shemesh, A. Bransky, M. Khoury, and S. Levenberg, *Biomed. Microdevices*, 12 (2010) 907–914.
26. E. Keith, K. Herold, and A. Rasooly (Eds.), *Lab-on-a-Chip Technology: Biomolecular Separation and Analysis*, Caister Academic Press, Norfolk, U.K., 2009, pp. 122–138.
27. N. Bontoux, L. Dauphinot, and M.-C. Potier, in: K.E. Herold and A. Rasooly (Eds.), *Lab-on-a-Chip Technology (Volume 2): Biomolecular Separation and Analysis*, Caister Academic Press, Norfolk, U.K., 2009, pp. 167–186.
28. N.C. Cady, in: K.E. Herold and A. Rasooly (Eds.), *Lab-on-a-Chip Technology (Volume 2): Biomolecular Separation and Analysis*, Caister Academic Press, Norfolk, U.K., 2009, Chapter 12.
29. J.E. Keymer, P. Galajda, C. Muldoon, S. Park, and R.H. Austin, *Proc. Natl. Acad. Sci. USA*, 103 (2006) 17290–17295.
30. T. Ahmed, T.S. Shimizu, and R. Stocker, *Integr. Biol.*, 2 (2010) 604–629.
31. J.R. Seymour, R. Simó, T. Ahmed, and R. Stocker, *Science*, 329 (2010) 342–345.
32. P. Galajda, J.E. Keymer, P. Chaikin, and R. Austin, *J. Bacteriol.*, 189 (2007) 8704–8707.

33. L. Angelani, R. Di Leonardo, and G. Ruocco, *Phys. Rev. Lett.*, 102 (2009) 048104.
34. R. Di Leonardo, L. Angelani, G. Ruocco, V. Iebba, M.P. Conte, S. Schippa, F. De Angelis, F. Mecarini, and E. Di Fabrizio, *Proc. Natl. Acad. Sci. USA*, 107 (2010) 9541–9545.
35. A. Sokolov, M.M. Apodacaca, B.A. Grzybowski, and I.S. Aranson, *Proc. Natl. Acad. Sci. USA*, 107 (2010) 969–974.
36. P. Berg, K. Promislow, J. St. Pierre, J. Stumper, and B. Wettonc, *J. Electrochem. Soc.*, 151 (2004) A341–A353.
37. E. Tretkoff, *Am. Phys. Soc. Sites*, 14 (2005) http://www.aps.org/publications/apsnews/200505/fuel.cfm
38. http://www.elvesys.com
39. A. Manbachi, S. Shrivastava, M. Cioffi, B.G. Chung, M. Moretti, U. Demirci, M. Yliperttula, and A. Khademhosseini, *Lab Chip*, 8 (2008) 747–754.
40. M. Yliperttula, B.G. Chung, A. Navaladi, A. Manbachi, and A. Urtti, *Eur. J. Pharm. Sci.*, 35 (2008) 151–160.
41. B.G. Chung, A. Manbachi, W. Saadi, F. Lin, N.L. Jeon, and A. Khademhosseini, *J. Vis. Exp.*, 7 (2007) 271.
42. J.-W. Choi, S. Rosset, M. Niklaus, J.R. Adleman, H. Shea, and D. Psaltis, *Lab Chip*, 10 (2010) 738–788.
43. G. Velve-Casquillas, M. Le Berre, M. Piel, and P.T. Tran, *Nano Today* 5 (2010) 28–47.
44. G. Velve-Casquillas, C. Fu, M. Le Berre, J. Cramer, S. Meance, A. Plecis, D. Baigl, J.-J. Greffet, Y. Chen, M. Piel, and P.T. Tran, *Lab Chip*, 11 (2011) 484–489.
45. http://www.elvesys.com/tempocell
46. A.K. Yetisen, L. Jiang, J.R. Cooper, Y. Qin, R. Palanivelu, and Y. Zohar, *J. Micromech. Microeng.*, 25 (2011) 054018 (9pp).
47. S. Grilli, L. Miccio, V. Vespini, A. Finizio, S. De Nicola, and P. Ferraro, *Opt. Exp.*, 16 (2008) 8084–8093.
48. P. Ferraro, L. Miccio, S. Grilli, A. Finizio, S. De Nicola, and V. Vespini, *Opt. Photon. News*, 19 (2008) 34.

16

Biosensors in Quality Assurance of Dairy Products

María Isabel Pividori and Salvador Alegret

CONTENTS

16.1 Introduction

16.1.1 Agents Affecting Safety in Dairy Products

Dairy products include not only milk, fermented milks, evaporated milks, and sweetened condensed milks but also other milk products, such as butter, dairy fat spreads, milk fat products (including butter oil, anhydrous milk fat, and ghee), cream and prepared creams, cheese and individual cheese varieties (mozzarella, cheddar, Danbo, Edam, Gouda, Havarti, Samsø, Emmental, Tilsiter, Saint-Paulin, provolone, cottage cheese, etc.), milk powders and cream powders, edible casein products, whey powders, and lactose.[1] As milk-based products contribute significantly to the overall human diet in many regions of the world, their contamination may cause concern. In recent years, a number of high-profile food-safety emergencies have shaken consumer confidence in the production of food and have focused attention on the way food is produced, processed, and marketed. The Codex Alimentarius refers today to the international food code established under the United Nations.[1] It is a collection of internationally adopted food standards that constitute a global reference point for national food legislators and control agencies, the international food trade, and food handlers and consumers. The code has a great impact on the approach to food quality management throughout the world. The code is being developed by the Codex Alimentarius Commission (CAC), which is an international organization run jointly by the Food and Agriculture Organization (FAO) and the World Health Organization (WHO). The CAC's objective is to establish standards, codes of practices, guidelines, and recommendations concerning foods aimed at protecting consumer's health, ensure fair practices in trade, and facilitate international trade. As contaminated food is one of the most widespread public health problems of the contemporary world causing considerable mortality,[2] the European Commission has identified food safety as one of its top priorities and has established plans for a proactive new food policy—modernizing legislation into a coherent and transparent set of rules, reinforcing controls from the farm to the table, and increasing the capability of the scientific advisory system—so as to guarantee a high level of human health and consumer protection. An effective food-safety policy requires assessment and monitoring of the risks to consumer health associated with contaminants in raw materials, farming practices, and processing activities.

Contaminants in dairy products can be grouped according to their origin and nature.[3] These potentially harmful compounds enter milk and dairy products through veterinary applications (antibiotics, hormones, agrochemicals) or accidentally from environment sources (air, feed, soil, and water) during growth, preparation (by milking utensils and teat treatment), storage, processing, and packaging.[3] For several reasons, their occurrence in milk is difficult to avoid and control, as is the case with the persistent environmental pollutants. Others result from agricultural, veterinary, and hygienic practices, which improve milk yield and quality but may also leave trace levels of residues in the finished products. While the term "contaminant" covers harmful substances or microorganisms that are not intentionally added, chemicals are also added during processing in the form of "additives."[4]

Additives were at one time a major concern, but today, microbiological issues are the greatest, followed by pesticide and animal drug residues and antimicrobial drug resistance.[5]

Many consumers are worried about the long-term impact of mixtures of chemical additives (such as pesticides, toxic metals, veterinary drug residues, flavorings, and colors) and chronic as well as acute effects on vulnerable groups.[6] Concerning the veterinary drug residues, antibiotics are added to reduce disease and improve the growth of farm animals; these are of less concern for their chemical effect and more for their ability to increase antimicrobial resistance in strains that might subsequently infect humans. There is evidence that antimicrobial resistance is increasing worldwide but particularly in developing countries. The human effect of consuming foods with these drugs is still being debated, but many countries refuse to accept products derived from animals given these drugs. The withdrawal times of such drugs are critical to keep the residues in food as low as possible.

Milk and dairy product quality and safety can only be ensured through the enforcement of quality-control systems throughout the entire food chain. They must be implemented at the farm level through (i) the use of good veterinary practices at the production level, (ii) good manufacturing practices at the processing level, and (iii) good hygienic practices at the retail and catering levels. One of the most effective ways for the food sector to protect public health is to base their food management programs on hazard analysis critical control point (HACCP). This systematic approach to the control of potential hazards within a food operation aims to identify problems before they occur.[6]

16.1.2 Detection Methods for Food Residues

As pesticides and many drugs produce undesirable residues, their use has been regulated by various national and international authorities, which have set maximum residue levels (MRLs) for food.[2] The MRL is the maximum concentration of a food residue and/or its toxic metabolites that is legally permitted in food commodities and animal feed. Authorities have also introduced a definition for residues of veterinary medicinal drugs, which, according to EU Council Regulation 2377/90, states: "All pharmacologically active substances whether active principles, excipients or degradation products, and their metabolites which remain in food stuffs obtained from animals to which the veterinary medicinal product in question has been administered." Monitoring of such residues in foods is often at the microgram per kilogram level.

Multiresidue analysis has been carried out using conventional chromatographic methods, such as gas chromatography (GC), high-performance liquid chromatography (HPLC), and capillary electrophoresis (CE). GC has proved to be an excellent technique for the detection of volatile pesticides and drug residues. Previously, thermal conductivity, flame ionization, and, in certain applications, electron capture and nitrogen phosphorus detectors (NPDs) were commonly used in GC analysis. In current GC-based residue detection methods, mass spectrometry (MS) in combination with electron-impact ionization (EI) is by far preferred due to the universality, selectivity, and specificity of this technique. HPLC is increasingly being employed in the determination of pesticide and drug residues, as it is especially suited to the analysis of nonvolatile, polar, and thermally labile residues that are difficult to analyze using GC. Besides physicochemical methods, microbiological growth-inhibition assays have long been used to test meat and milk for the presence of antibiotic residues.[7] These tests use antibiotic-sensitive bacterial reporter strains, such as *Bacillus subtilis* and *Bacillus stearothermophilus var. calidolactis.* These bacteria are inoculated under optimal conditions with and without sample. After culturing, results are read from visible inhibition zones or from the color change of the bacterial suspension in agar gels.

Immunochemical methods are used for rapid screening of an individual or a group of closely related residues. These methods require little or no sample cleanup, require no expensive instrumentation, and are suitable for field use. The development of immunoassays (IAs) for the detection of food components and contaminants has progressed rapidly in the last few years.[8] Antibodies against almost all the important food residue compounds are currently available. Classical immunochemical methods such as immunodiffusion and agglutination methods for food analyses generally involve no labeled antigen or antibody. The concentration of the antigen–antibody complex is estimated from the secondary reaction that leads to precipitation or agglutination. These methods are not sensitive, are subject to nonspecific interference, and can only be used for the analysis of high molecular weight proteins. However, the development of radioimmunoassay (RIA) has widened the scope of IAs. This method combines the unique properties of specific antibody–antigen interaction and the use of a radioactive labeled marker to monitor complex formation. Thus, RIA provides specificity, sensitivity, and simplicity and can be used for the analysis of both antigen and haptens. RIA involves the use of a radioactive marker that competes with an analyte in the sample for binding to an antibody. For RIA of a high molecular weight antigen, either the antigen or antibody molecules can be labeled. It is also common to use a radiolabeled second antibody, that is, antibody against the primary antibody. In contrast, labeled hapten is typically used in RIA for low molecular weight substances. Using different nonradioactive labeled markers, a variety of IAs, including fluorescence immunoassay (FIA), time-resolved FIA, FIA polarization IA, enzyme immunoassay (EIA), luminescent immunoassay (LIA), metalloimmunoassay (MIA), and viroimmunoassay (VIA), have been developed. Since no radioactive substances are used, the assays avoid the problems encountered in handling radioactivity. EIA is a general term for immunoassays involving the use of an enzyme as a marker for the detection of immunocomplex formation. Enzyme labeling can be achieved by conjugation of the enzyme to an antigen or antibody via periodate oxidation with subsequent reductive alkylation method, cross-linking using glutaraldehyde, or others. Some of the methods used in the conjugation of hapten to proteins can also be used. However, to avoid nonspecific interaction, methods for coupling of protein/hapten to enzyme should be different from the one that had been used for conjugating the hapten to protein for immunization purpose. Although horseradish peroxidase (HRP) and alkaline phosphatase are the two enzymes most commonly used, others, such as glucose-6-phosphate dehydrogenase, coupled with oxidoreductase and luciferase, glucose oxidase, beta-galactosidase, and urease, have also been employed. Depending on whether or not the immunocomplex is separated from the free antigen, EIAs are further divided into two types. One type is a homogeneous system, which is called enzyme multiplied immunoassay (EMIT), and is based on the modification of enzyme activity occurring when an antibody binds with the enzyme-labeled antigen/hapten in solution. Because the modification of enzymatic activity is generally not significant, this system is not very sensitive and has not been widely used in food analysis. The other is a heterogeneous system involving separation of free and bound antigen–antibody. In this system, either antigen or antibody is covalently or noncovalently bound to the solid matrix. Nonreacted antibody or antigen is simply removed by washing or centrifugation. The term "enzyme-linked immunosorbent assay" (ELISA) is used for this type of assay. Solid phases such as microtiter plates, cellulose, nylon beads/tubes, nitrocellulose membrane, polystyrene tubes/balls, and modified magnetic beads have been used. In some cases, staphylococcal protein A (ProtA) or protein G (ProtG) is coated on the solid surface, entrapping the antibody for subsequent analysis.

This method is further divided into two major types. One type is competitive ELISA, which can be used for the analysis of both hapten and macromolecule; the other is noncompetitive sandwich-type ELISA, which is only used for divalent and multivalent antigens. Two major types, that is, direct competitive ELISA (dC-ELISA) and indirect competitive ELISA (inC-ELISA), are used most commonly in food analysis.

In the dC-ELISA, specific antibodies are first coated to a solid phase. The sample or standard solution of analyte is generally incubated simultaneously with enzyme–conjugate or incubated separately in two steps. The amount of enzyme bound to the plate is then determined by incubation with a chromogenic substrate solution. The resulting color/fluorescence, which is inversely proportional to the analyte concentration present in the sample, is then measured instrumentally or by visual comparison with the standards.

There are a number of quick screening tests based on the ELISA principle described earlier. For example, microtiter plate ELISA assays can be completed in less than 20 min. Other approaches involve immobilizing the antibody on a paper disk or other membrane that is mounted on a plastic card, in a plastic cup, or on the top of a plastic tube. In the "dipstick" assay, antibody or antigen is coated on a stick that is then dipped in various reagents and the reactions monitored. Most of these screening tests are very simple, are easily performed, and are designed to provide semiquantitative information at defined cutoff concentrations for the substance of interest. Immunoscreening tests have gained wide application for monitoring residues in foods, with versatile assay kits commercially available. Immunoassays are available for almost all the important antibiotic residues that might be present in foods. For example, β-lactone antibiotics such as ampicillin, cloxacillin, and penicillin G could easily be measured in milk. Likewise, immunoassays for many pesticides are also available, including 2,4-D, aldicarb, carbendazim, thiabendazole, chlorpyrifos, diazinon, endosulfan, and metalaxyl. The triazine immunoassay is now available commercially, and like most IAs, it is specific, sensitive, rapid, and cost-effective.[9]

16.1.3 Detection Methods for Food Pathogen

Conventional bacterial identification methods usually include morphological evaluation of the microorganism as well as tests of its ability to grow in various media under a variety of conditions. This is accomplished by (i) preenrichment, (ii) selective enrichment, (iii) biochemical screening, and (iv) serological confirmation.[10] However, these growth and enrichment steps are relatively time-consuming, with a total assay time of up to 1 week for certain food pathogens.[11] Although classical culture methods can be sensitive, they are greatly restricted by the assay time. As long as 96 h may be needed to obtain a negative result if the entire culture process has to be applied, while presumptive positive results may be seen within 48–96 h. Accordingly, many alternative methods have been introduced in recent years to reduce analytical and staff time as well as media requirements. These rapid methods are designed to avoid the need for selective culturing and serological/biochemical identification.

Among the new instrumental methods developed using various principles of detection are chromatography, infrared or fluorescence spectroscopy, bioluminescence, flow cytometry, and other techniques based on electrical conductance and impedance. Currently, however, these methods are centralized in large stationary laboratories as they require complex instrumentation and highly qualified technical staff.[12] Furthermore, the capital cost involved in these instrumental analyses is high, which restricts its use.[11]

During the past decade, immunological detection of bacteria has become more sensitive, specific, reproducible, and reliable with many commercial IAs available for the detection of a wide variety of microbes and their products in food. Advances in antibody production have stimulated this technology, since polyclonal antibodies can be now quickly and cheaply obtained, and do not require the time or expertise associated with the production of monoclonal antibodies. Many commercial kits are available, including immunodiffusion, ELISA, and the use of specific antibodies to "capture and concentrate" the organism.[13] Moreover, immunoassays have shown the ability to detect not only contaminating organisms but also their biotoxins. However, immunoassays give total bacterial load rather than just the number of viable cells.

Nucleic-acid-based detection coupled with polymerase chain reaction (PCR) has distinct advantages over culture and other standard methods for the detection of microbial pathogens, such as specificity, sensitivity, rapidity, accuracy, and the capacity to detect small amounts of target nucleic acid in a sample.[14] In addition, multiple primers can be used to detect different pathogens in a single multiplex reaction.[15] This kind of methods requires an enrichment period that delays results, but sensitivity can be of the order of 3 CFU/25 g of food.[12] This approach cannot be applied directly to samples, because of the fact that DNA is heat stable. Thus, intact DNA will be present in processed foods. One possible way to overcome this problem is to include a culture step as a means of detecting viable cells, being this step not sufficient to allow the growth of more severely damaged cells.

The use of nucleic acid recognition layers represents a new and exciting area in analytical chemistry, which requires an extensive study. Besides classical methodologies to detect DNA, novel approaches have been designed, such as the DNA chips[16] and lab on a chip based on microfluidic techniques.[17] However, these technologies are still out of the scope of food industry, since it requires simple, cheap, and user-friendly analytical devices.

16.2 Biosensing of Dairy Products

The development of biosensors is a growing area, in response to the demand for rapid real-time, simple, selective, and low-cost techniques for food residues.[15] Biosensors are compact analytical devices, incorporating a biological sensing element, either closely connected to or integrated within a transducer system. The combination of the biological receptor compounds (antibody, enzyme, nucleic acid) and the physical or physicochemical transducer produces, in most cases, "real-time" observation of a specific biological event (e.g., antibody–antigen interaction).[18] Depending on the method of signal transduction, biosensors can also be divided into different groups: electrochemical, optical, thermometric, piezoelectric, or magnetic.[19] They allow the detection of a broad spectrum of analytes in complex sample matrices and have shown a great promise in areas such as clinical diagnostics, food analysis, bioprocess, and environmental monitoring.[20,21] Hazard analysis and critical control point (HACCP) systems, that is, generally accepted as the most effective system to ensure food safety, can utilize biosensor to verify that the process is under control.[22]

The sensitivity of each of the sensor systems may vary depending on the transducer's properties and the biological recognizing elements. An ideal biosensing device for the rapid detection of food contaminants should be fully automated, inexpensive, and routinely used in both the field and the laboratory. Optical transducers are particularly attractive as they can allow direct "label-free" and "real-time" detection, but they lack sensitivity. The phenomena of surface plasmon resonance (SPR) have shown good biosensing potential, and many commercial SPR systems are now available. The Pharmacia BIAcore™ (a commercial SPR system) is by far the most reported method for biosensing of food residues in food, and it is based on optical transducing. For example, the SPR biosensor was compared with existing methods (microbial inhibitor assays, microbial receptor assays, ELISA, HPLC) for the detection of sulfamethazine residues in milk by an immunological reaction.[23,24] BIAcore has indicated the occurrence of sulfamethazine at a concentration below the detection limit of HPLC and offered sufficient advantages (no sample preparation, high sensitivity, rapid and full analysis in real time) to be an alternative for the control of residues and contaminants in food. A similar commercial transducer system was also reported for the determination of β-lactam antibiotics,[25–27] multi-sulfonamide residues,[28] and chloramphenicol residues.[29] In the case of pesticides in food, they are mainly based on both the biosensing of the enzyme inhibition by the pesticide[30,31] and the immunosensing of the pesticide performed with the specific antibody with SPR transducer.[32–34] The detection of food pathogens by SPR, however, does not reach the required LOD to allow food safety without performing a preenrichment step.[35]

However, electrochemically based transduction devices are more robust, easy-to-use, portable, and inexpensive analytical systems.[36,37] Furthermore, electrochemical biosensors can operate in turbid media and offer comparable instrumental sensitivity.

Regarding the molecular recognition of food residues, immunological reagents are mainly used as a molecular receptor in order to obtain a useful signal,[38,39] while in the case of food pathogens, not only immunological reagents[40,41] but also DNA probes[42] can be used as molecular receptor in order to obtain a useful signal after a biological reaction on the transducer element (antigen/antibody and hybridization, respectively).[34]

Electrochemical immunosensors and genosensors can meet the demands of food control, offering considerable promise for obtaining information in a faster, simpler, and cheaper manner compared to traditional methods. Such devices possess great potential for numerous applications, ranging from decentralized clinical testing to environmental monitoring, food safety, and forensic investigations.

16.2.1 Electrochemical Biosensing Based on Graphite–Epoxy Composites

Rigid conducting graphite–epoxy composites (GECs) based on graphite microparticles have been extensively used in our laboratories and shown to be suitable for electrochemical (bio)sensing due to their unique physical and electrochemical properties.[4,43] Carbon composites result from the combination of carbon with one or more dissimilar materials. Each component maintains its original characteristics while conferring upon the composite distinctive chemical, mechanical, and physical properties. The user's ability to integrate various materials is one of their main advantages.

An ideal material for electrochemical biosensing should allow the effective immobilization of bioreceptor on its surface, a robust biological reaction between the target and the bioreceptor, a negligible nonspecific adsorption of the label, and a sensitive detection of the biological event. GECs fulfill all these requirements. Other advantages of GEC-based biosensing devices over more traditional carbon-based materials are higher sensitivity, robustness, and rigidity in addition to greater simplicity of preparation. Additionally, the GEC surface can be regenerated by a simple polishing procedure. Unlike carbon paste and glassy carbon, the malleability of the GEC material before the curing step permits different configurations with respect to shape and size, which are then fixed after the curing step. Moreover, the surface of the composite can be easily modified by dry and wet adsorption of the bioreceptor (DNAs, oligonucleotides, proteins, antibodies), yielding a reproducible and stable layer of bioreceptor on the transducer surface[44] that can be used in electrochemical detection (Figure 16.1A).

The direct adsorption of DNA and proteins on solid supports has the main advantages of simplicity as well as applicability to almost any type of macromolecule or solid support, since no reagents or reactive functions are required. However, the main disadvantage is that the biomolecule is not oriented because it is bound to the surface at multiple sites.[44] Based on its dual nature, with hydrophobic and hydrophilic domains, the GEC platform has demonstrated excellent food-safety features with respect to the stable adsorption of different biomolecules (mainly DNA and proteins) and without a loss in biorecognition capacity, at least for DNA biosensors and immunosensors.

16.2.2 Electrochemical Biosensing Based on Graphite–Epoxy Biocomposites

An additional interesting property of GECs is their biocompatibility. This feature allows not only adsorption but also integration of the bioreceptor into the bulk of the GEC without subsequent loss of the receptor's biological properties, thus generating a rigid and renewable transducing material for biosensing, namely, a graphite–epoxy biocomposite (GEB). With the bioreceptor integrated within its bulk, the biocomposite acts as a reservoir for the biomolecule while retaining all the interesting electrochemical and physical features previously described for GECs. The main advantage of GEBs is that they can be easily prepared by adding the

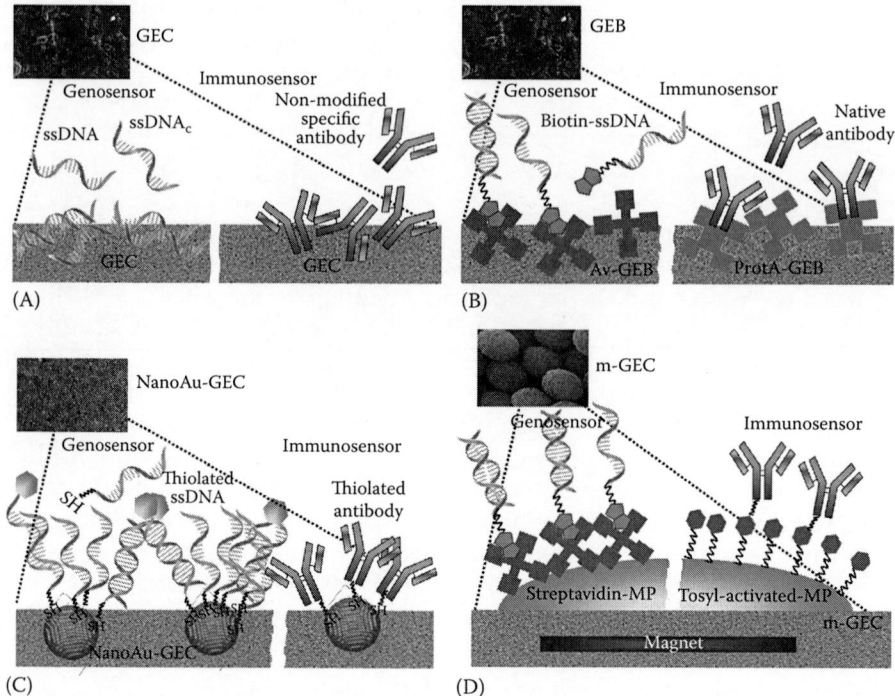

FIGURE 16.1
Schematic representation of the different strategies for the immobilization in genosensor and immunosensors in biosensors based on GEC platforms. (A) Physisorption of oligonucleotides (left) and antibodies (right) on GEC transducer. (B) Left: Single-point immobilization of biotinylated DNA on Av–GEB. Right: Oriented immobilization of the nonmodified antibody based on protein A biocomposite (ProtA–GEB). (C) Single-point immobilization of thiolated oligonucleotides (left) and antibodies (right) on gold nanocomposite (nanoAu–GEC). (D) Left: Single-point immobilization of biotinylated DNA on streptavidin magnetic particles. Right: Single-point attachment based on covalent immobilization of the nonmodified antibody on tosyl-activated magnetic particles, in both cases, to be captured on a magneto electrode (m-GEC)

bioreceptor to the composite formulation using dry-chemistry techniques, thereby avoiding tedious, expensive, and time-consuming surface immobilization procedures. Moreover, the surface of GEB electrodes can be easily modified with DNA, oligonucleotides, proteins, and antibodies for electrochemical detection.

The use of affinity proteins such as avidin (Av), protein A (ProtA), or protein G (ProtG), in the biocomposite, provides a robust platform for the oriented immobilization of DNA or immunospecies that improves the performance of the electrochemical biosensing devices by ensuring exposure of the bioreceptor to the complementary sites of the target molecule (Figure 16.1B).

With advancements in our knowledge about Av–biotin interactions, this system has become an extremely versatile tool. Strept(avidin) can be considered as a universal affinity biomolecule based on its ability to link not only biotinylated DNA and oligonucleotides but also biotinylated enzymes or antibodies. The extremely specific and high-affinity reaction between biotin and the glycoprotein avidin (association constant Ka = 10^{15} M) leads to strong associations, similar to the formation of a covalent bond.[45] Protein A, produced by *Staphylococcus aureus*, is a highly stable receptor capable of binding to the Fc region of immunoglobulins, such as IgG, from a large number of species.[46] When the antibodies are immobilized through their Fc fragment to protein A (or G), their Fab binding sites are mostly oriented away from the solid phase. As protein A is able to link the Fc region of different antibodies, there is no need to previously modify the antibody.[47,48] A rigid and renewable transducing material for electrochemical immunosensing or genosensing, based on bulk-modified GEBs, can be easily prepared by adding a small amount (2%) of an affinity protein, such as avidin, protein A, or protein G, to the formulation of the composite to obtain, in this case, Av–GEB, ProtA–GEB, or ProtG–GEB, respectively (as outlined in Figure 16.1B). After its use, the electrode surface can be renewed by a simple polishing procedure, thus allowing multiple uses—a further advantage of these materials with respect to surface-modified approaches such as classical biosensors and other common biological assays. These transducers therefore represent excellent alternatives for biosensing in dairy products.

16.2.3 Electrochemical Biosensing Based on Graphite–Epoxy Nanocomposites

Due to their unique size effects and enhanced chemical and physical properties, nanosized gold particles have been the focus of recent interest with respect to their potential applications in physics, chemistry, biology, medicine, and material science.[49] Gold nanoparticles (AuNPs) are nanostructured materials that bridge the gap between "bottom-up" synthetic methods and "top-down" fabrication.[50] The ability of AuNP to provide a suitable environment for the immobilization of biomolecules while preserving the biological activity of the latter has led to their intensive use in the construction of biosensors with improved performances.[51]

Chemisorption based on self-assembled monolayers (SAMs) is a single-point immobilization strategy that allows the oriented attachment of a wide range of biomolecules on gold-based transducer surfaces. One of the main drawbacks of SAMs is their highly ordered compact layer, which dramatically reduces the diffusion of electroactive species toward the transducer's surface. Moreover, the tightly packed layer may also produce steric hindrance and, as a consequence, a lower rate of reaction between the probe and the DNA target. The use of AuNPs in a GEC (nanoAu–GEC) has been proposed as an alternative to continuous gold surface films as this strategy avoids the need for stringent control of surface coverage parameters during the immobilization of thiolated oligos or antibodies (Figure 16.1C).

In this novel transducer, islands of chemisorbing material (AuNPs) surrounded by a rigid, non-chemisorbing, conducting GEC are obtained.[52] The spatial resolution of the immobilized thiolated DNA can be easily controlled by varying the percentage of AuNPs in the composition of the composite. The electrode surface is easily renewed, as described earlier. This favors the use of these transducers over continuous gold transducers, as the latter require complex surface pretreatment. Moreover, as with GEBs, the surface of nanoAu–GEC electrodes can be easily modified with DNAs, oligonucleotides, proteins, antibodies, etc., for electrochemical detection. These transducers can thus be easily adapted for electrochemical genosensing of food pathogens.

16.2.4 Electrochemical Biosensing Based on Magnetic Particles Coupled with a Magneto Electrode

One of the most promising materials in bioanalysis is biologically modified magnetic particles, the use of which is based on the concept of magnetic bioseparation. Magnetic particles offer several novel attractive possibilities in biomedicine and bioanalysis since they can be coated with biological molecules and manipulated by an external magnetic field gradient. As such, the biomaterial, that is, specific cells, proteins, or DNA, can be selectively bound to the magnetic particles and then separated from its biological matrix by applying an external magnetic field. Moreover, magnetic beads of a variety of materials and sizes—even nanosized magnetic particles—and modified with a wide variety of surface functional groups are now commercially available (Figure 16.1D). The integration of magnetic beads and electrochemical biosensing strategies improves analytical performance. Instead of direct modification of the electrode surface, both the biological reactions (immobilization, hybridization, enzymatic labeling, or affinity reactions) and the washing steps can be successfully performed on magnetic beads. After the modifications, the particles are easily captured by applying a magnetic field onto the surface of the GEC electrodes, which contain a small magnet (m-GEC) designed in our laboratories. This procedure constitutes a versatile platform for electrochemical biosensing (both DNA biosensors and immunosensors) in milk and dairy products (Figure 16.1D).

16.2.5 Electrochemical Detection Strategies for Milk and Dairy Products

Electrochemical detection of the biorecognition event should be considered, involving the transduction of the biological reaction into a useful and easy-to-amplify electrical signal.

The direct electrochemical detection of DNA was initially proposed by Paleček[53,54] who recognized the capability of both DNA and RNA to yield reduction and oxidation signals after being adsorbed. The oxidation of DNA was shown to be strongly dependent on both the DNA adsorption conditions and the substrate on which DNA is being absorbed, thus requiring a meticulous control of the DNA-adsorbed layer. Although it is very simple, this strategy requires multisite attachment—such as adsorption—as immobilization technique. Figure 16.2, A depicts the electrochemical determination of DNA based on its intrinsic oxidation signal. Here, electrochemical determination is performed by differential pulse voltammetry (DPV), in which the oxidation signal of guanine (or adenine) is measured by scanning from +0.30 to +1.20 V at a pulse amplitude of 100 mV and a scan rate of 15 mV/s. The DNA recognition event for electrochemical transducing can be also detected by means of external electrochemical markers such as electroactive indicators[55,56] or enzymes. Figure 16.2 also shows the different strategies used in electrochemical genosensing (Figure 16.2B and C) and immunosensing (Figure 16.2D) based on an enzymatic tag (usually HRP). Enzyme labeling has been transferred from nonisotopic DNA classical methods to electrochemical genosensing. In electrochemical

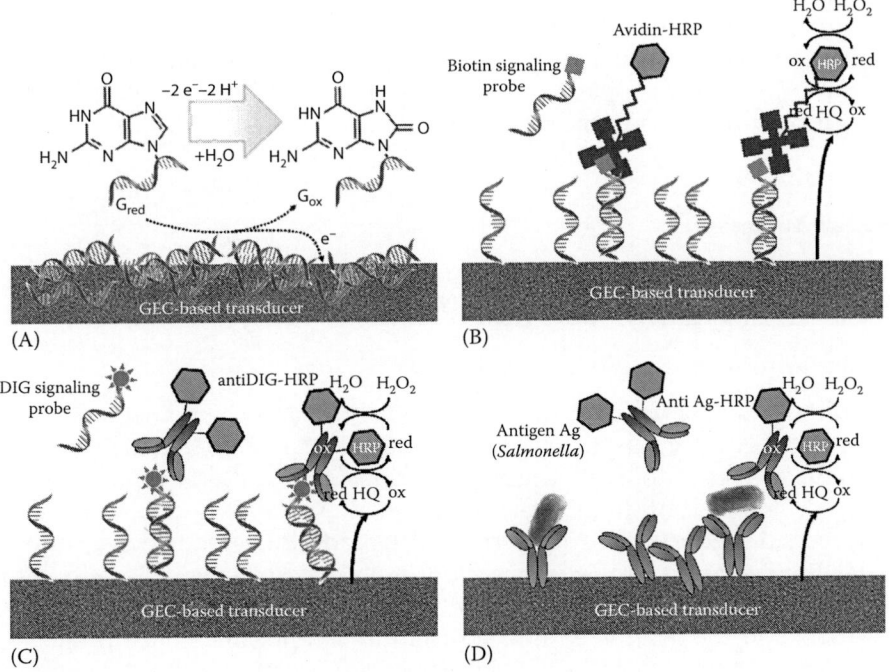

FIGURE 16.2
Schematic representation of the different electrochemical detection strategies for biosensing in dairy products. (A) Electrochemical genosensing based on the intrinsic oxidation signal of DNA. (B and C) Electrochemical genosensing based on enzymatic tags (Av–HRP and antiDIG-HRP, respectively). (B and C) Electrochemical genosensing based on enzymatic tags (Av–HRP and antiDIG-HRP, respectively). (D) Electrochemical immunosensor based on enzymatic tags, in this case the antibody–HRP conjugate. The strategies (B) to (D) require further addition of a mediator (HQ) and the enzyme substrate H_2O_2.

genosensing, the DNA duplex can be labeled with either strept(avidin)–HRP (Figure 16.2B) or antiDIG-HRP (Figure 16.2C) conjugates, depending on the tag of the DNA signaling probe (biotin or digoxigenin, respectively). Although a second incubation step is usually required for labeling, higher sensitivity and specificity have been reported for the enzyme labeling method compared with the other reported methods.[57,58] In electrochemical immunosensing, the enzymatic tag depends on the format of the immunoassay. In competitive immunoassays for small haptenic molecules, it is usually a conjugate obtained with HRP and the hapten. In other immunological formats, such as in sandwich assays or indirect approaches, the enzymatic label is typically a conjugate obtained with HRP covalently linked to the Fc part of the specific antibody (Figure 16.2D).

In all cases, amperometric determination is finally based on HRP activity following the addition of H_2O_2 and using hydroquinone as mediator. The modified electrode is immersed in the electrochemical cell containing hydroquinone, and under continuous magnetic stirring, a potential of -0.100 V vs. Ag/AgCl is applied. When a stable baseline is reached, H_2O_2 is added into the electrochemical cell (to a concentration able to saturate the total amount of enzyme employed in the labeling procedure), and the current is measured until steady state is reached (normally after 1 min of H_2O_2 addition). Finally, the use of metal nanoparticles—especially AuNPs[59–62]—as labels for biosensing devices is also gaining importance.

16.3 Electrochemical Immunosensing of Agents Affecting Milk and Dairy Product Safety

16.3.1 Antibiotic Residues in Milk

Over the past few decades, the use of antibiotics and chemotherapeutics in animal husbandry has increased considerably.[63] Antimicrobial drugs are administered both to treat bacterial infections or employed prophylactically to prevent spread of disease and to augment growth and yield in animals and animal products. All antimicrobial drugs administered to dairy animals enter the milk to a certain degree, and each drug is given a certain withdrawal (waiting) period, during which time the concentration in the tissues declines and the drug is eliminated from the animal. The most frequently and commonly used antimicrobial drugs are antibiotics, used to combat mastitis-causing pathogens.[3] The occurrence of residues of antimicrobials in milk besides being of interest in the context of consumer health and the development of antibiotic resistance has both economical and technological impact on the dairy industry. Ultimately, residues of antimicrobials may lead to a deterioration of quality and to monetary losses in the dairy industry by inhibiting starter cultures in dairy technological processes. Moreover, there is evidence that antimicrobial resistance is increasing worldwide but particularly in developing countries. The human effect of consuming foods containing antibiotics is still being debated, but many countries refuse to accept products derived from animals given these drugs. The withdrawal times of antibiotics are critical to ensure that the residues in food are as low as possible. The presence of certain antimicrobial agent residuals in milk constitutes a potential hazard for the consumer and may cause allergic reactions, interference in the intestinal microbiota, and resistance in populations of bacteria, thereby rendering subsequent treatment with the antibiotic ineffective.

The electrochemical magneto immunosensing of sulfonamides can be successfully performed in raw full cream as well as in all varieties of ultrahigh-temperature (UHT) milk, including full cream (~3.25% fat), semi-skimmed (~1.5%–1.8% fat), and skimmed (0.1% fat), by using class-specific antisulfonamide (anti-SFM) antibodies immobilized on magnetic beads and an SFM–HRP tracer for electrochemical detection.[64] Specificity studies have shown that at least 11 sulfonamide antibiotics can be sensitively detected using a class-specific anti-SFM antibody. This strategy is schematically outlined in Figure 16.3A. Although the anti-SFM antibodies can be successfully attached on different types of magnetic beads (–COOH modified magnetic nanoparticles, protein A magnetic beads), the best attachment performance is again achieved with tosylactivated magnetic beads. Raw full-cream samples spiked with sulfapyridine are diluted five times with PBST, while all UHT samples are processed without treatment.[64] As the European legislation has set the MRL of sulfonamide at 100 µg/kg, the detection limit of the electrochemical magneto immunosensing strategy (0.26 µg/L) is sufficient to assay any type of milk according to the requirements of the EC legislation.

16.3.2 Folic Acid in Vitamin-Fortified Food

Folates, including various forms of tetrahydrofolate (THF), are part of the vitamin B group. In order to prevent folate deficiency in individuals, folic acid is added to many food products.[65] Supplementation is particularly important in pregnant women, as folic acid insufficiency can cause neural-tube defects in the developing fetus. Moreover, folate deficiency is the most common cause of anemia after iron deficiency. Although folic acid is not thought to be toxic, it may contribute to the potential masking of pernicious anemia in the elderly, as well as interfering with anticonvulsive therapy or cancer treatment with antifolates. In addition, a rapid assay for folic acid may be mandatory to control the practice of "overage" during food preparation.[66]

The electrochemical magneto immunosensing of folic acid in vitamin-fortified milk samples can be successfully performed, as shown in Figure 16.3.[65] In the direct assay, the enzyme tracer conjugated with folic acid (folic-HRP) competes with folic acid for the binding sites of a specific antibody immobilized on the solid support (Figure 16.3B). In the indirect assay, the protein conjugate (folic-BSA) is immobilized on the solid support and competition for the specific antibody is established between the folic acid in the sample and immobilized folic-BSA (Figure 16.3C). In this case, the amount of specific antibody specifically bound on the solid support was determined using a secondary antibody conjugated with HRP as enzyme label (anti-IgG-HRP). As seen in Figure 16.3, the indirect assay achieved a better performance in terms of LOD, which for skimmed milk was 5.8 μg/L. Commercial vitamin-fortified milk samples were also evaluated, obtaining good accuracy in the results.

16.3.3 Allergens in Milk

Gliadin is the toxic protein fraction of gluten responsible for the intolerance underlying celiac disease. Its detection is of great interest for the

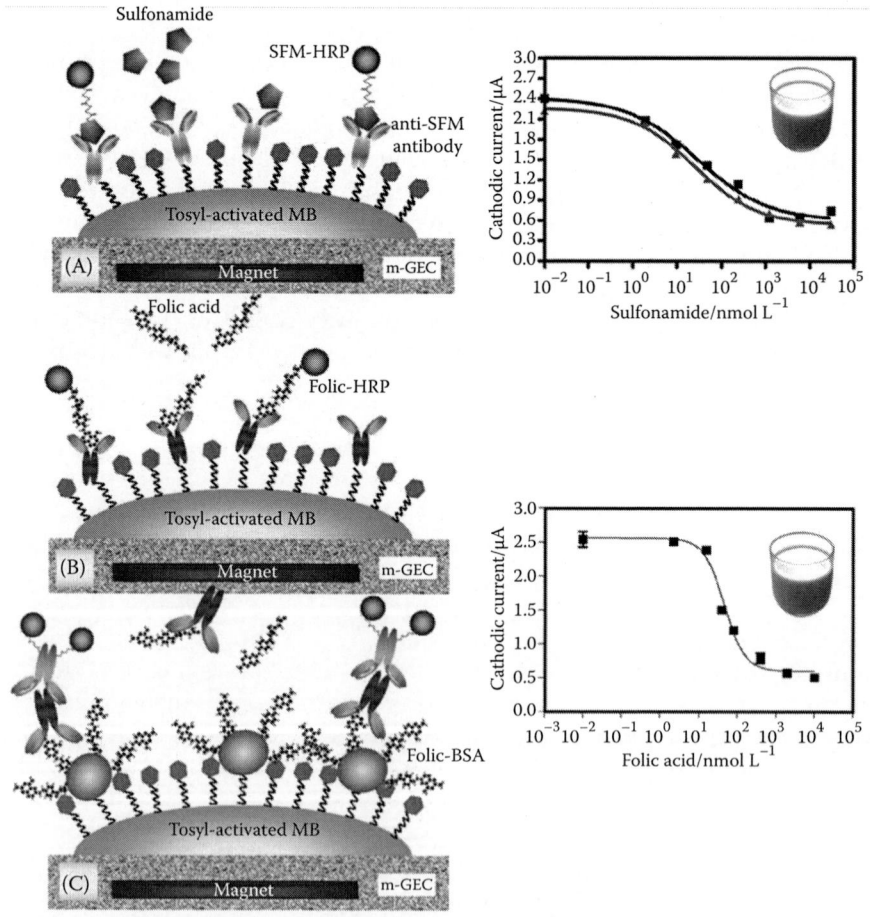

FIGURE 16.3
Electrochemical magneto immunosensing of different residues in milk using, in all cases, tosyl-activated magnetic particles. (A) Sulfonamide detection based on a direct competitive immunoassay. Folic acid determination based on (B) a direct and (C) indirect competitive immunoassay.

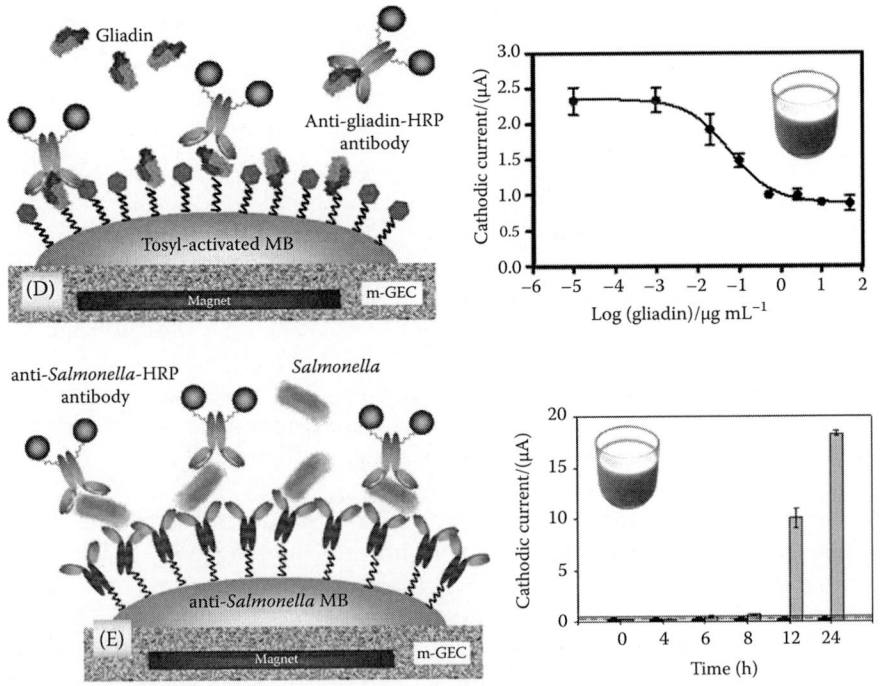

FIGURE 16.3 (continued)
Electrochemical magneto immunosensing of different residues in milk using, in all cases, tosyl-activated magnetic particles. (D) Gliadin detection based on direct competitive immunoassay. (E) *Salmonella* detection based on a sandwich immunoassay. Other experimental conditions are detailed in Refs. [64,66,68,70], respectively.

food safety of celiac patients, for whom the only treatment currently known is a lifelong dietary avoidance of this cereal protein.[67] As a result, gluten content has been included in food regulations, as mandated by the recent statement that foodstuffs labeled "gluten-free" may not exceed 20 ppm of gluten content. Therefore, an easy, rapid, and reliable method of analysis is essential to control gliadin content in gluten-free milk and dairy products and to permit the on-site monitoring during industrial processing. Gliadin detection in safe, gluten-free food is also successfully performed with an electrochemical magneto immunosensing strategy (Figure 16.3D) on micro- and nanostructured magnetic beads as solid supports, with the antigen, in this example gliadin, covalently immobilized on the activated surfaces. In all cases, the biorecognition strategy is based on a direct competitive assay using an antigliadin antibody–peroxidase (HRP) conjugate as the enzymatic label. Subsequent detection is achieved through an appropriate substrate and mediator for the HRP enzyme. Excellent LODs (on the order of 5 µg/L) have been achieved, consistent with the requirements for gluten-free products. The matrix effect on different samples (milk) as well as the assay's performance has been successfully evaluated using spiked-food samples, with good recovery values obtained in the results.[68]

16.3.4 Food Pathogens in Milk

Salmonella is one of the most frequently occurring foodborne pathogens affecting the microbial safety of foods.[69] Official agencies charged with ensuring food safety, such as the U.S. Food and Drug Administration (FDA), the U.S. Department of Agriculture (USDA), the Association of Official Analytical Chemist International (AOACI), and the International Organization of Standardization (ISO), recommend classical culture methods for recovering *Salmonella* spp. from food. However, the development of new methodologies, with advantages of rapid response, sensitivity, and ease of multiplexing, is a challenge for food hygiene inspection aimed at screening out negative samples.

A very simple and rapid method for the detection of *Salmonella* in milk is performed by electrochemical magneto immunosensing with m-GEC electrodes (Figure 16.3E).[70] In this approach, the bacteria are captured and preconcentrated from milk samples with magnetic beads by immunological reaction with a specific antibody against *Salmonella*. A second polyclonal antibody labeled with peroxidase is used as serological confirmation, with electrochemical detection based on a magneto electrode. Among the different procedures, better performances have been obtained using one-step immunological reactions. The "immunomagnetic separation step (IMS)/m-GEC electrochemical immunosensing" approach was employed, for the first time, in the detection of *Salmonella* artificially inoculated into skim milk samples. A limit of detection of $7.5 \cdot 10^3$ CFU mL^{-1} in milk was obtained in 50 min without any pretreatment. If the skim milk is preenriched for 6 h, the method can detect as low as 1.4 CFU mL^{-1}, while following preenrichment for 8 h as few as 0.108 CFU mL^{-1} (2.7 CFU in 25 g of milk) is detected, thus complying with legislative criteria (Figure 16.3E). The immunomagnetic separation and detection with a second specific antibody can effectively replace "selective enrichment/differential plating" and "biochemical/serological testing" assays, respectively. Moreover, the assay time is considerably reduced, from 4–5 days to 50 min.

16.4 Electrochemical Genosensing of Agents Affecting Milk and Dairy Product Safety

16.4.1 *Salmonella*

The multisite physical adsorption of DNA extracted from pathogenic bacteria on GEC has the main advantage of simplicity, since no reagent or reactive function is required. Once immobilized on the GEC, genomic

DNA preserves its unique hybridization properties, which can be revealed using different strategies: (i) enzymatic labeling based on HRP conjugates (Figure 16.2BC)[71,72] and (ii) intrinsic signal coming from DNA oxidation (Figure 16.2C).[73,74]

The detection of the *Salmonella* IS200 element by a DNA biosensing strategy in capture format based on a GEC is shown in Figure 16.4A. Briefly, the protocol consists of the following steps: (i) capturing probe immobilization by dry adsorption, (ii) hybridization of the DNA target in one step with the complementary digoxigenin signaling probe, (iii) enzyme labeling of the DNA duplex with antiDIG-HRP, and (iv) amperometric determination based on enzyme activity following the addition of H_2O_2 and using hydroquinone as mediator. As low as 22 pmol of DNA target—with a signal-to-nonspecific adsorption of 3.2—can be easily and cost-effectively detected with this strategy.

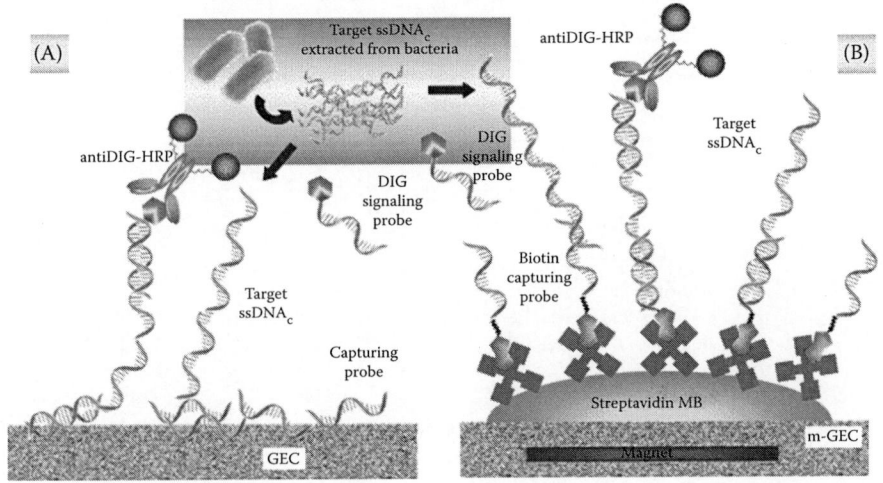

IS200 sequence specific of *Salmonella*
5 CAC ACC CGATGG AAC TGT AAA TAT CAC ATA GTT TTC GCG CCC AAA TAC CGAAGA CAA GCG TT 3;
Capturing probe: 5 GTG ATA TTT ACAGTT CCA TCG GG-biotin 3
Signaling probe: 5 DIG-CTT GTC TTC GGT ATT TGG GCGCG 3

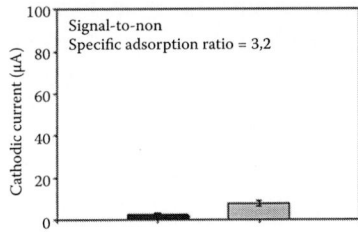
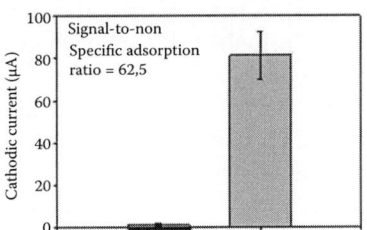

FIGURE 16.4
Comparative results for electrochemical genosensing of *Salmonella* enterica serovar Typhimurium ATCC 14028 using different strategies and transducers. (A and B) Electrochemical DNA sequence-specific detection in capture format for the *Salmonella* IS200 element.

(*continued*)

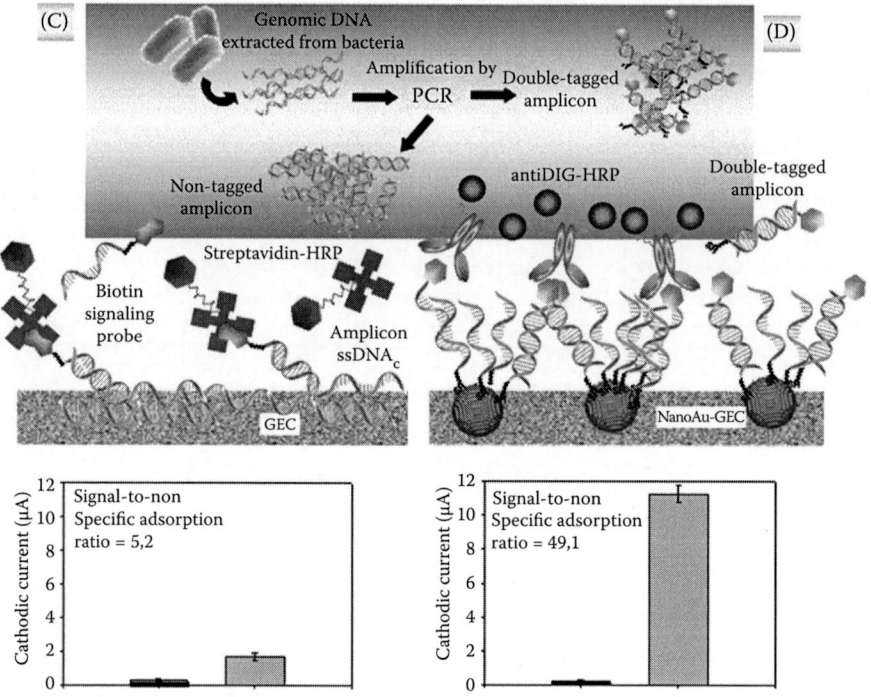

FIGURE 16.4 (continued)
Comparative results for electrochemical genosensing of *Salmonella* enterica serovar Typhimurium ATCC 14028 using different strategies and transducers. (C and D) PCR-amplified IS200 element of *Salmonella*. In all cases, gray bars show the specific signal, and black bars the corresponding nonspecific adsorption omitting either the DNA target (A and B) or the DNA template (C and D) during PCR. Error bars indicate the standard deviation (n = 3). Other experimental conditions are detailed in Refs. [52,74,75,72], respectively.

Moreover, to increase the assay's sensitivity, the amplification of the bacterial genome by PCR can be coupled with a DNA electrochemical approach. In this case, the amplicon is directly adsorbed on the GEC transducer. As the amplicon is double stranded, a denaturing alkaline procedure is mandatory to break the hydrogen bonds for further hybridization with the complementary signaling probe.[72] Briefly, the protocol consists of the following steps, as schematically outlined in Figure 16.4C: (i) amplicon immobilization by dry adsorption and alkaline treatment, (ii) hybridization with the complementary biotin signaling probe, (iii) enzyme labeling with streptavidin–HRP conjugate, and (iv) amperometric determination based on the enzyme. As low as 2.9 pmol of amplicon—with a signal-to-nonspecific adsorption of 5.2—can be easily and cost-effectively detected with this strategy. As shown in Figure 16.4A and C, and although tightly adsorbed on the GEC electrode, DNA preserves its unique hybridization properties in GEC platforms, suggesting that DNA bases are not fully committed in the adsorption mechanism but mostly available for hybridization.[71] The results comparatively presented in Figure 16.4 suggest that the oriented single-point immobilization of DNA achieved by different strategies, such as the integration of magnetic beads in m-GEC electrodes (Figure 16.4B) and AuNPs in graphite–epoxy nanocomposites (AuNP–GEC) (Figure 16.4D), provides, in all cases, improved results in terms of signal-to-nonspecific adsorption, as discussed in the succeeding text.

Streptavidin-modified magnetic beads are useful platforms for DNA biosensing, when combined with a biotinylated capturing probe complementary to the DNA target (Figure 16.4B). The *Salmonella* IS200 element can be easily detected in a one-step capture format, as schematically outlined in Figure 16.4B. The procedure consists of the following steps: (i) one-step immobilization/hybridization procedure, in which the biotin-labeled capturing probe is immobilized on streptavidin magnetic beads, while hybridization with both the target and a second complementary probe—in this case, labeled with digoxigenin—occurs simultaneously; (ii) enzymatic labeling using the antibody antiDIG-HRP as enzyme label; (iii) magnetic capture of the modified magnetic particles on the m-GEC electrode; and (iv) amperometric determination based on enzyme activity following the addition of H_2O_2 and using hydroquinone as mediator.[75] This approach (streptavidin magnetic beads integrated within m-GEC electrodes) clearly provides better analytical performances in terms of signal-to-nonspecific adsorption than achieved with other strategies involving adsorption on GEC (Figure 16.4A). The strategy exploits the advantages of magnetic beads, such as improved and more effective biological reactions, washing procedures, and magnetic separation after each step. This assay also benefits from the increased size of the active area due to the integration of magnetic beads within the m-GEC transducer. In addition to electrochemical detection based on the enzyme label, the DNA target immobilized on the magnetic beads can be successfully detected by the intrinsic DNA oxidation signal coming from the guanine moieties (Figure 16.2A).[76] To further increase the sensitivity of detection of foodborne pathogens, a double-tagged PCR strategy is coupled to an electrochemical magneto genosensing approach, based on streptavidin magnetic beads integrated within m-GEC electrodes.[75] Rapid electrochemical verification of the amplicon derived from the *Salmonella* IS200 element is performed by double labeling the amplicon during PCR with a set of two labeled PCR primers—one with biotin and the other with digoxigenin. During PCR, not only amplification of the bacterial genome is achieved but also double labeling of the amplicon ends with (i) the biotinylated capture primer, to achieve immobilization on the streptavidin-modified magnetic bead, and (ii) the digoxigenin signaling primer, to achieve electrochemical detection. The procedure briefly consists of the following steps:[75] (i) DNA amplification of the bacterial genome and double labeling, (ii) immobilization of the doubly labeled amplicon in which the biotin end of the double-stranded DNA (dsDNA) amplicon is immobilized on the streptavidin magnetic beads, (iii) enzymatic labeling using the antibody antiDIG-HRP as an enzyme label capable of binding the other labeled end of the dsDNA amplicon, (iv) magnetic capture of the modified magnetic particles, and (v) amperometric determination. Rapid and sensitive verification of the *Salmonella*-related PCR amplicon can be achieved with 2.8 fmol of amplified product.[75] Interestingly, the PCR also can be performed directly on the magnetic beads by using a magnetic primer, thus allowing real-time electrochemical detection of the bacteria.[75]

The detection of the *Salmonella IS200* element is also possible using other strategies based on graphite–epoxy nanocomposites (AuNP–GEC). In these materials, isolated AuNPs generate bioactive chemisorbing islands for the immobilization of thiolated DNA probes (Figure 16.1C). Less compact layers are thus achieved, favoring the biological reaction on biosensing devices. Hybridization efficiency is expected to be higher on the edge of the AuNPs surrounded by nonreactive GEC, as shown in Figure 16.1C. Briefly, the procedure consists of the following steps:[52] (i) thiolated probe immobilization by chemisorption, (ii) hybridization with the complementary probe modified with digoxigenin, (iii) enzyme labeling of the DNA duplex using antiDIG-HRP, and (iv) amperometric determination based on enzyme activity following the addition of H_2O_2 and using hydroquinone as mediator. The chemisorbing ability of AuNPs in the AuNP–GEC was shown to have an excellent LOD (9 fmol/60 pM of ssDNA) in hybridization studies aimed at the detection of DNA from *Salmonella*.[52] Moreover, and for the first time, a double-tagging PCR strategy was performed using a thiolated primer for the detection of *Salmonella* sp.[52] The results are shown in Figure 16.4D. Rapid electrochemical verification of the amplicon coming from the genome of pathogenic *Salmonella*, as performed by PCR using a set of two labeled primers, readily allows thiolation of the PCR product.[52] The thiolated end facilitates the immobilization of the amplicon on the AuNP–GEC electrode (Figure 16.4D). The procedure consists of the following steps[52]: (i) DNA amplification and double labeling of the *Salmonella* IS200 insertion sequence; (ii) immobilization of the doubly labeled amplicon, with the –SH end of the dsDNA amplicon immobilized on the AuNP–GEC nanocomposites by chemisorption; (iii) enzymatic labeling using as enzyme label the antibody antiDIG-HRP, capable of binding the other labeled end of the dsDNA amplicon; and (iv) amperometric determination. The detection using this strategy is as low as 200 fmol, with an electrochemical signal of almost 3 µA.[52]

Figure 16.4 compared the results obtained for the detection of 2.7 pmol of amplicon using AuNP–GEC sensors (Figure 16.4D) and GEC electrodes (Figure 16.4C). The oriented single-point immobilization of DNA achieved by chemisorption on AuNP–GEC clearly provides improved results in terms of signal-to-nonspecific adsorption. This double-tagging PCR strategy opens new routes not only for immobilization purposes but also as an easy strategy for labeling with gold or quantum dots during PCR. Moreover, the AuNP–GEC material shows interesting properties for electrochemical genosensing in hybridization experiments and very promising features for electrochemical biosensing of a wide range of biomolecules, such as dsDNA, PCR products, affinity proteins, antibodies, and enzymes.

16.4.2 *E. coli* O157:H7

Like *Salmonella, Escherichia coli* is one of the most frequent pathogens implicated in human bacterial infections. *Enterohemorrhagic E. coli* (EHEC) O157:H7 is one of the most dangerous serotypes of the bacterium, causing hemorrhagic colitis and severe hemolytic uremic syndrome, either of which may result in death due to acute or chronic renal failure. Outbreaks of EHEC O157:H7 infections have been associated with contaminated food products, such as ground beef and raw milk.[77]

Rapid electrochemical verification of the amplicon coming from the *eaeA* gene of *E. coli* O157:H7 is performed by double tagging the amplicon during PCR with a set of labeled PCR primers—one with biotin and the other with digoxigenin[78] (Figure 16.5A). During PCR, not only is amplification of *E. coli* achieved but also double labeling of the amplicon ends with (i) the biotinylated capturing primer, resulting in the immobilization on a biosensor based on a bulk-modified avidin biocomposite (Av–GEB), and (ii) the digoxigenin

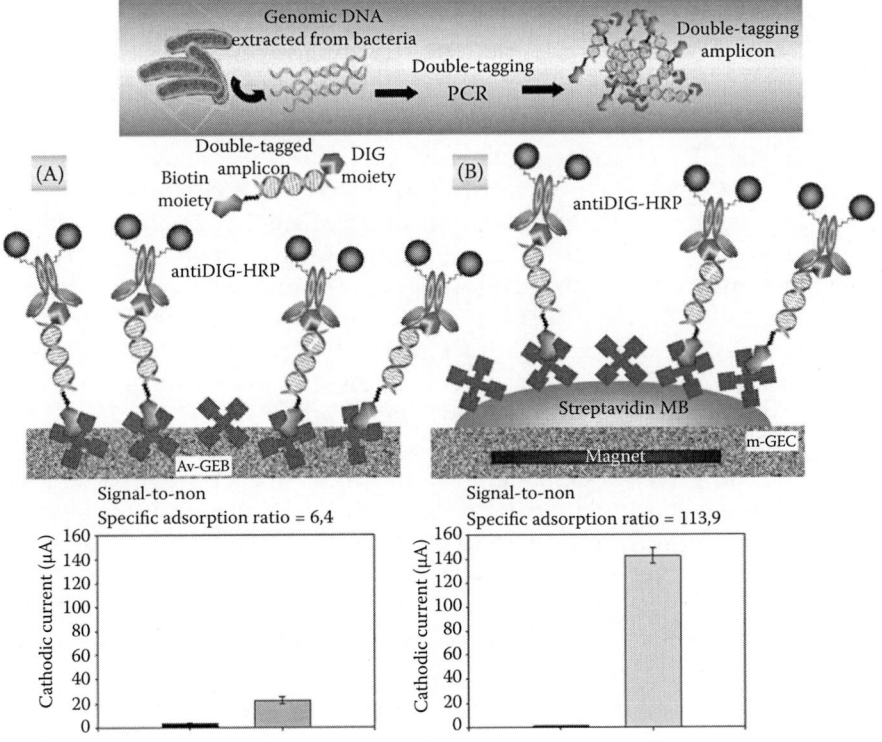

FIGURE 16.5
Electrochemical detection of 2 ng/μL DNA template of *E. coli* by double-tagging PCR followed by electrochemical genosensing based on Av–GEB (A) and electrochemical magneto genosensing based on m-GEC (B). In all cases, 60 μg antiDIG-HRP was used. Other experimental conditions are detailed in Ref. [78].

signaling primer, enabling electrochemical detection. The procedure consists of the following steps: (i) DNA amplification and double tagging of eaeA, the gene associated with the pathogenic activity of *E. coli* O157:H7; (ii) immobilization of the double-tagged amplicon, in which the biotin end of the dsDNA amplicon is immobilized on the Av–GEB biosensor; (iii) enzymatic labeling with antiDIG-HRP, which is capable of binding the other labeled end of the dsDNA amplicon; and (iv) amperometric determination.[78]

As shown in Figure 16.5A, the assay is very sensitive, detecting as little as 2 ng DNA template of *E. coli* μL^{-1} in 13 cycles—with a signal-to-nonspecific adsorption of 6.4. Moreover, 4.5 ng of the original bacterial genome μL^{-1} can be feasibility detected after only 10 cycles of PCR amplification.[78] DNA biosensors based on Av–GEB for amplicon detection are more sensitive than Q-PCR strategies based on fluorescent labels such as TaqMan probes. In addition, this strategy can be used for the electrochemical real-time quantification of amplicon since a linear relationship with the amount of amplified product is obtained.[78]

To increase the sensitivity of *E. coli* detection, a double-tagged PCR strategy is coupled to an electrochemical magneto genosensing approach, based on streptavidin magnetic beads integrated within m-GEC electrodes. Rapid electrochemical verification of the amplicon coming from the eaeA gene is performed by double labeling the amplicon during PCR with a set of two labeled PCR primers—one with biotin and the other with digoxigenin (Figure 16.5B)—as previously explained. The procedure consists of the following steps:[78] (i) DNA amplification of the bacterial genome and double labeling; (ii) immobilization of the doubly labeled amplicon, in which the biotin end of the dsDNA amplicon is immobilized on the streptavidin magnetic beads; (iii) enzymatic labeling using as enzyme label the antibody antiDIG-HRP, capable of binding the other labeled end of the dsDNA amplicon; (iv) magnetic capture of the modified magnetic particles; and (v) amperometric determination. The rapid and sensitive verification of the PCR amplicon derived from *E. coli* allows the detection of 0.45 ng of the original bacterial genome μL^{-1} after only 10 cycles of PCR amplification.[78] Moreover, as previously mentioned, electrochemical strategies for amplicon detection have proven to be more sensitive than Q-PCR strategies based on fluorescent labels such as TaqMan probes. Electrochemical magneto genosensing of the double-tagged amplicon clearly provides better analytical performance in terms of signal-to-nonspecific adsorption than obtained with electrochemical genosensing based on Av–GEB (Figure 16.5), when both are carried out with a similar single-point-oriented DNA attachment through Av–biotin linkage. This strategy can be used for the electrochemical real-time quantification of amplicon since the relationship with the amount of amplified product is linear.[78] However, this strategy is useful only when a unique and specific band is observed by gel electrophoresis, because of the high specificity of the set of primers used in PCR amplification of the bacterial genome. If the primer set amplifies not only the sequence of interest but also other nonspecific fragments, it is necessary to confirm the internal sequence of the amplicon by a second hybridization using a digoxigenin signaling probe.[78]

16.4.3 *Mycobacterium bovis* in Raw Contaminated Milk on Dairy Farms

Tuberculosis (TB) in humans and other mammals is usually caused by *Mycobacterium tuberculosis or Mycobacterium bovis*. While, worldwide, *M. tuberculosis* is the single greatest cause of infectious disease in humans, *M. bovis* affects the largest number of animals. In humans, the global prevalence of TB infection involves about one-third of the world's population, a number that is expected to grow steadily.[79] *M. bovis* causes bovine TB, which is easily transmitted between farm animals. This disease is also an important zoonosis, targeting not only workers on dairy farms but also the general public following the consumption of contaminated dairy products.

A very sensitive assay for the rapid screening out of TB based on electrochemical genosensing can be performed by specific amplification and double tagging of the *IS6110* fragment, highly related to *M. bovis*, followed by the electrochemical detection of the amplified product. PCR amplification is performed using a labeled set of primers, yielding a double-tagged amplicon with biotin and digoxigenin at the respective ends, as explained earlier for other pathogens. Two different electrochemical platforms for the detection of double-tagged amplicon can be used:[80] (i) an Av–GEB and (ii) a magneto electode (m-GEC) combined with streptavidin magnetic beads. In both cases, the immobilization of the double-tagged amplicon is achieved through the biotinylated end of the amplicon, and the electrochemical detection through the digoxigenin end by using an antiDIG-HRP conjugate. The assay has proven to be very sensitive, as it is able to detect 620 and 10 fmol of PCR amplicon for Av–GEB and m-GEC strategies, respectively. Compared with interlaboratory PCR assays and the "gold standard" tuberculin skin test, the m-GEC assay has shown promising features for the detection of TB on dairy farms, through the determination of *M. bovis* DNA in milk samples.

16.5 Electrochemical Biosensing Approaches Combining both Immunological and Genetic Information for Milk and Dairy Product Safety

The double-tagged PCR strategy with electrochemical genosensing can be also combined with an immunoseparation step of the bacteria to improve the LOD for detecting pathogenic bacteria.[81] The procedure consisted briefly

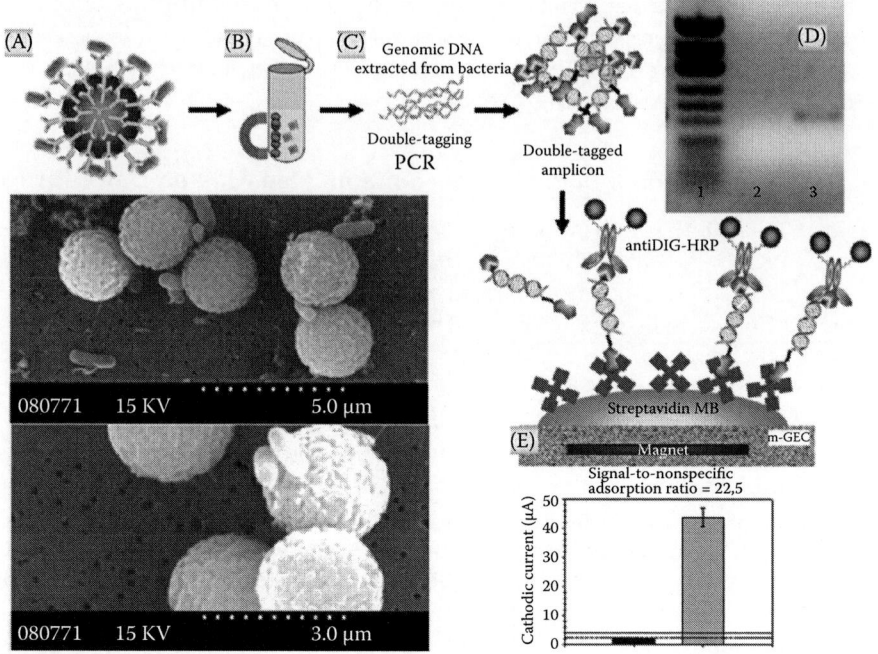

FIGURE 16.6

"IMS/double-tagging PCR/electrophoresis" approach showing the (A) IMS step of the bacteria from food samples, (B) the lysis of the bacteria and DNA release, (C and D) the DNA amplification and double labeling of the *Salmonella* IS200 insertion sequence, and (E) electrochemical signals for the IMS/double-tagging PCR/m-GEC electrochemical genosensing approach, with a preenrichment step of 6 h for artificially inoculated skim milk (0.04 CFU/mL or 1 CFU in 25 g of milk) (gray bar) and for negative controls (black bar). In all cases, n = 4. The agarose gel electrophoresis of double-tagged PCR amplicon was obtained with a preenrichment step of 6 h (0.04 CFU/mL or 1 CFU in 25 g of milk, lane 3). A negative control (0 CFU/mL, lane 2) and molecular mass markers (ΦX174-Hinf I genome, lane 1) are also shown. The scanning electron microscopy images of the IMS of 10^4 CFU/mL of *Salmonella* are also shown.

of the following steps, as depicted in Figure 16.6: (i) IMS of the bacteria from food samples (Figure 16.6A), (ii) lysis of the bacteria and DNA separation (Figure 16.6B), (iii) DNA amplification and double labeling of *Salmonella* IS200 insertion sequence (Figure 16.6C), (iv) immobilization of the doubly labeled amplicon in which the biotin extreme of the dsDNA amplicon is immobilized on the streptavidin magnetic beads, (v) enzymatic labeling using as enzyme label the antibody antiDIG-HRP capable of bonding the other labeled extreme of the dsDNA amplicon, (vi) magnetic capture of the modified magnetic particles, and (vii) amperometric determination.[81]

In this approach, the bacteria are captured and preconcentrated from food samples with magnetic beads by immunological reaction with the specific antibody against *Salmonella*, as shown in the SEM images in Figure 16.6. After the lysis of the captured bacteria, further amplification of the genetic material by PCR with a double-tagging set of primers is performed to confirm the identity of the bacteria. Both steps are rapid alternatives to the time-consuming classical selective enrichment and biochemical/serological tests. The double-tagged amplicon is then detected by electrochemical magneto genosensing using m-GEC electrodes. The "IMS/double-tagging PCR/m-GEC electrochemical genosensing" approach can be used for the sensitive detection of *Salmonella* artificially inoculated into skim milk samples. A limit of detection of 1 CFU mL^{-1} is obtained in 3.5 h without any pretreatment, in LB broth and in milk diluted 1/10 in LB. When the skim milk is preenriched for 6 h, the method is able to feasibly detect as low as 0.04 CFU mL^{-1} (1 CFU in 25 g of milk) with a signal-to-background ratio of 20.[81]

Interestedly, the specificity of this approach is conferred by both the antibody in the IMS and the set of primer during the double-tagging PCR, in this case for detecting *Salmonella* spp. The same approach could be also designed for detecting different *Salmonella* or *E. coli* serotypes by selecting a specific pair of primers or antibody.

16.6 Final Remarks

As food regulatory agencies have established strict control programs aimed at preventing contaminants from entering the food supply, official laboratories must be able to efficiently process a high number of samples. This has led to a demand for routine, rapid, and efficient food control procedures. Consequently, there is an urgent need to develop rapid, cost-effective, sensitive, high sample throughput, and on-site analytical strategies that can be used as an "alarm" to rapidly detect the risk of contamination by food pathogens in a wide variety of food matrices, especially milk and dairy products, since standard methods do not meet these detection requirements.

As demonstrated herein for many agents affecting safety in milk, the converging of technologies such as nanotechnology and biotechnology is opening new horizons in electrochemical biosensors. The integration of micro- and nanostructured materials in biosensing devices (such as graphite microparticles, bioreceptors, AuNPs, magnetic micro- or nanoparticles) provides excellent analytical performances for the detection of contaminants affecting the safety in dairy products. One of the key contributions of these technologies in the electrochemical biosensing field relies on the design of novel transducers, not only with enhanced transducing features, but also with improved immobilization of biomolecules while retaining their biological activity. Future trends are focused on the integration of materials and procedures for the design of powerful nano–bio–magneto–electroanalytical systems.

References

1. Heggum, C. (2011) Codex alimentarius. In: Fuquay, J.W., Fox, P.F., Mcsweeney, P.L.H. (eds.). *Encyclopedia of Dairy Sciences* (2nd edn.). Academic Press, London, U.K., pp. 312–321.
2. Nawaz, S. (2003) Pesticides and herbicides. Residue determination. In: Caballero, B., Trugo, L., Finglas, P. (eds). *Encyclopedia of Food Sciences and Nutrition* (2nd edn.). Academic Press, Boston, MA, pp. 4487–4493.
3. Heggum, C. (2011) Contaminants of milk and dairy products. In: Fuquay, J.W., Fox, P.F., Mcsweeney, P.L.H. (eds). *Encyclopedia of Dairy Sciences* (2nd edn.). Academic Press, London, U.K., pp. 887–897.
4. Roberts, D. (2003) Food poisoning. In: Caballero, B., Trugo, L., Finglas, P. (eds). *Encyclopedia of Food Sciences and Nutrition* (2nd edn.). Academic Press, Boston, MA, pp. 5654–2658.
5. Todd, E. (2003) Contamination of food. In: Caballero, B., Trugo, L., Finglas, P. (eds). *Encyclopedia of Food Sciences and Nutrition* (2nd edn.). Academic Press, Boston, MA, pp. 1593–1600.
6. Rooney, R. and Wall, P.G. (2003) Food safety. In: Caballero, B., Trugo, L., Finglas, P. (eds). *Encyclopedia of Food Sciences and Nutrition* (2nd edn.). Academic Press, Boston, MA, pp. 2682–2688.
7. Bergwerff, A.A. and Schloesser, J. (2003) Antibiotics and drugs. Residue determination. In: Caballero, B., Trugo, L., Finglas, P. (eds). *Encyclopedia of Food Sciences and Nutrition* (2nd edn.). Academic Press, Boston, MA, pp. 254–261.
8. Chu, F.S. (2003) Immunoassays. Radioimmunoassay and enzyme immunoassay. In: Caballero, B., Trugo, L., Finglas, P. (eds). *Encyclopedia of Food Sciences and Nutrition* (2nd edn.). Academic Press, Boston, MA, pp. 3248–3255.
9. Au, A.M. (2003) Pesticides and herbicides. Types, uses, and determination of herbicides. In: Caballero, B., Trugo, L., Finglas, P. (eds). *Encyclopedia of Food Sciences and Nutrition* (2nd edn.). Academic Press, Boston, MA, pp. 4483–4487.
10. Tietjen, M. and Fung, D.Y.C. (1995) Salmonella and food safety. *Crit. Rev. Microbiol.* 21, 53–83.
11. Ivnitski, D., Abdel-Hamid, I., Atanasov, P., and Wilkins, E. (1999) Biosensors for detection of pathogenic bacteria. *Biosens. Bioelectron.* 14, 599–624.
12. Humphrey, T. and Stephens, P. (2003) Salmonella detection. In: Caballero, B., Trugo, L., Finglas, P. (eds). *Encyclopedia of Food Sciences and Nutrition* (2nd edn.). Academic Press, Boston, MA, pp. 5079–5084.
13. Luk, J.M., Kongmuang, U., Tsang, R.S.W., and Lindberg, A.A. (1997) An enzyme-linked immunoadsorbent assay to detect PCR products of the rfbS gene from serogroup D salmonellae: A rapid screening prototype. *J. Clin. Microbiol.* 35, 714–718.
14. Wan, J., King, K., Craven, H., Mcauley, C., Tan, S.E., and Coventry, M.J. (2000) Probelia™ PCR system for rapid detection of Salmonella in milk powder and ricotta cheese. *Lett. Appl. Microbiol.* 30, 267–271.
15. Leonard, P., Hearty, P., Brennan, J., Dunne, L., Quinn, J., Chakraborty, T., and O'Kennedy, R. (2003) Advances in biosensors for detection of pathogens in food and water. *Enzyme Microb. Technol.* 32, 3–13.
16. Bowtell, D.D.L. (1999) Options available—From start to finish—For obtaining expression data by microarray. *Nat. Gen. Suppl.* 21, 25–32.

17. Sanders, G.H.W. and Manz, A. (2000) Chip-based microsystems for genomic and proteomic analysis. *Trends Anal. Chem.* 19, 364–378.
18. Deisingh, A.K. and Thompson, M. (2004) Biosensors for the detection of bacteria. *Can. J. Microbiol.* 50, 69–77.
19. Terry, L.A., White, S.F., and Tigwell, L.J. (2005) The application of biosensors to fresh produce and the wider food industry. *J. Agric. Food Chem.* 53, 1309–1316.
20. Velasco-Garcia, M.N. and Mottram, T. (2003) Biosensors for detection of pathogenic bacteria. *Biosyst. Eng.* 84, 1–12.
21. Patel, P.D. (2002) (Bio)sensors for measurements of analytes implicated in food safety: A review. *TRAC* 21, 96–115.
22. Mello, L.D. and Kubota, L.T. (2002) Review of the use of biosensors as analytical tools in the food and drink industries. *Food Chem.* 77, 237–256.
23. Mellgren, C., Sternesjo, A., Hammer, P., Suhren, G., Bjorck, L., and Heeschen, W. (1996) Comparison of biosensor, microbiological, immunochemical and physical methods for detection of sulfamethazine residues in raw milk. *J. Food Protect.* 59(11), 1223–1226.
24. Sternesjo, A., Mellgren, C., and Bjorck, L. (1995) Determination of sulphametazine residues in milk by a surface resonance based-biosensors assay. *Anal. Biochem.* 226, 175–181.
25. Gustavsson, E., Bjurling, P., and Sternesjö, Å. (2002) Biosensor analysis of penicillin G in milk based on the inhibition of carboxypeptidase activity. *Anal. Chim. Acta* 468, 153–159.
26. Cacciatore, G., Petz, M., Rachid, S., Hakenbeck, R., and Bergwerff, A. (2004) Development of an optical biosensor assay for detection of β-lactam antibiotics in milk using the penicillin-binding protein 2x. *Anal. Chim. Acta* 520, 105–115.
27. Gaudin, V., Fontaine, J., and Maris, P. (2001) Screening of penicillin residues in milk by a surface plasmon resonance-based biosensor assay: Comparison of chemical and enzymatic sample pre-treatment. *Anal. Chim. Acta* 436, 191–198.
28. Haasnoot, W., Bienenmann-Ploum, M., Lamminmäki, U., Swanenburg, M., and van Rhijn, H. (2005) Application of a multi-sulfonamide biosensor immunoassay for the detection of sulfadiazine and sulfamethoxazole residues in broiler serum and its use as a predictor of the levels in edible tissue. *Anal. Chim. Acta* 552, 87–95.
29. Ferguson, J., Baxter, A., Young, P., Kennedy, G., Elliott, C., Weigel, S., Gatermann, R., Ashwin, H., Stead, S., and Sharman, M. (2005) Detection of chloramphenicol and chloramphenicol glucuronide residues in poultry muscle, honey, prawn and milk using a surface plasmon resonance biosensor and Qflex® kit chloramphenicol. *Anal. Chim. Acta* 529, 109–113.
30. Vakurov, A., Simpson, C.E., Daly, C.L., Gibson, T.D., and Millner, P.A. (2005) Acetylcholinesterase-based biosensor electrodes for organophosphate pesticide detection: II. Immobilization and stabilization of acetylcholinesterase. *Biosens. Bioelectron.* 20, 2324–2329.
31. Andreescu, S. and Marty, J.-L. (2006) Twenty years research in cholinesterase biosensors: From basic research to practical applications. *Biomol. Eng.* 23, 1–15.
32. Shimomura, M., Nomura, Y., Zhang, W., Sakino, M., Lee, K.-H., Ikebukuro, K., and Karube, I. (2001) Simple and rapid detection method using surface plasmon resonance for dioxins, polychlorinated biphenylx and atrazine. *Anal. Chim. Acta* 434, 223–230.
33. Mullett, W.M., Lai, E.P.C., and Yeung, J.M. (2000) Surface plasmon resonance-based immunoassays. *Methods* 22, 77–91.

34. Mauriz, E., Calle, A., Lechuga, L.M., Quintana, J., Montoya, A., and Manclús, J.J. (2006) Real-time detection of chlorpyrifos at part per trillion levels in ground, surface and drinking water samples by a portable surface plasmon resonance immunosensor. *Anal. Chim. Acta* 561, 40–47.
35. Barlen, B., Mazumdar, S.D., Lezrich, O., Kämpfer, P., and Keusgen, M. (2007) Detection of *Salmonella* by surface plasmon resonance. *Sensors* 7, 1427–1446.
36. Ivnitski, D., Abdel-Hamid, I., Atanasov, P., Wilkins, E., and Stricker, S. (2000) Application of electrochemical biosensors for detection of food pathogenic bacteria. *Electroanalysis* 12, 317–325.
37. Mehervar, M. and Abdi, M. (2003) Recent development, characteristics, and potential applications of electrochemical biosensors. *Anal. Sci.* 20, 1113–1126.
38. Baeumner, A.J. (2003) Biosensors for environmental pollutants and food contaminants. *Anal. Bioanal. Chem.* 377, 434–445.
39. Nakamura, H. and Karube, I. (2003) Current research activity in biosensors. *Anal. Bioanal. Chem.* 377, 146–168.
40. Gehring, A.G., Crawford, C.G., Mazenko, R.S., Van Houten, L.J., and Brewster, J.D. (1996) Enzyme-linked immunomagnetic electrochemical detection of Salmonella typhimurium. *J. Immunol. Methods* 195, 15–25.
41. Croci, L., Delibato, E., Volpe, G., and Palleschi, G. (2001) A rapid electrochemical ELISA for the detection of Salmonella in meat samples. *Anal. Lett.* 34, 2597–2607.
42. Croci, L., Delibato, E., Volpe, G., De Medici, D., and Palleschi, G. Comparison of PCR, electrochemical enzyme-linked immunosorbent assays, and culture method for detecting Salmonella in meat products. *Appl. Environ. Microbiol.* 70, 1393–1396.
43. Alegret, S. (1996) Rigid carbon–polymer biocomposites for electrochemical sensing: A review. *Analyst* 121, 1751–1758.
44. Pividori, M.I. and Alegret, S. (2005) DNA adsorption on carbonaceous materials. In: Wittman, C. (ed.). *Topics in Current Chemistry*, Vol. 260. Springer, Berlin, Germany, pp. 1–36.
45. Jones, M.L. and Kurzban, G.P. (1995) Noncooperativity of biotin binding to tetrameric streptavidin. *Biochemistry* 34, 11750–11756.
46. Sjoquist, J., Meloun, B., and Hjelm, H. (1972) Protein A isolated from Staphylococcus aureus after digestion with lysostaphin. *Eur. J. Biochem.* 29, 572–578.
47. Pividori, M.I., Lermo, A., Zacco, E., Hernández, S., Fabiano, S., and Alegret, S. (2007) Bioaffinity platforms based on carbon-polymer biocomposites for electrochemical biosensing. *Thin Solid Films* 516, 284–292.
48. Zacco, E., Pividori, M.I., Llopis, X., del Valle, M., and Alegret, S. (2004) Renewable protein A modified graphite-epoxy composite for electrochemical immunosensing. *J. Immunol. Methods* 286, 35–46.
49. Guo, S. and Wang, E. (2007) Synthesis and electrochemical applications of gold nanoparticles. *Anal. Chim. Acta* 598, 181–192.
50. Shenhar, R. and Rotello, V.M. (2003) Nanoparticles: Scaffolds and building blocks. *Acc. Chem. Res.* 36, 549–561.
51. Pingarrón, J.M., Yáñez-Sedeño, P., and González-Cortés, A. (2008) Gold nanoparticle-based electrochemical biosensors, *Electrochim. Acta* 53, 5848–5866.
52. Oliveira Marques, P.R.B., Lermo, A., Campoy, S., Yamanaka, H., Barbé, J., Alegret, S., and Pividori, M.I. (2009) Double-tagging polymerase chain reaction with a thiolated primer and electrochemical genosensing based on gold nanocomposite sensor for food safety. *Anal. Chem.* 81, 1332–1339.

53. Paleček, E. (1958) Oscillographic polarography of nucleic acids and their build-ingblocks. *Naturwiss* 45, 186–187.
54. Paleček, E. (1960) Oscillographic polarography of highly polymerized deoxyri-bonucleic acid. *Nature* 188, 656–657.
55. Carter, M.T., Rodriguez, M., and Bard, A.L. (1989) Voltammetric studies of the interac-tion of metal chelates with DNA. 2. Tris-chelated complexes of cobalt(III) and iron(II) with 1,10-phenanthroline and 2,2%-bipyridine. *J. Am. Chem. Soc.* 111, 8901–8911.
56. Erdem, A., Kerman, K., Meric, B., and Ozsoz, M. (2001) Methylene blue as a novel electrochemical hybridization indicator. *Electroanalysis* 13, 219–223.
57. Alfonta, L., Singh, A.K., and Willner, I. (2001) Liposomes labelled with biotin and horseradish peroxidase: A probe for the enhanced amplification of antigen–antibody or oligonucleotide–DNA sensing processes by the precipitation of an insoluble product on electrodes. *Anal. Chem.* 73, 91–102.
58. Paleček, E., Kizek, R., Havran, L., Billova, S., and Fotja, M. (2002) Electrochemical enzyme-linked immunoassay in a DNA hybridization sensor. *Anal. Chim. Acta.* 469, 73–83.
59. González-García, M.B., Fernández-Sánchez, C., and Costa-García, A. (2000) Colloidal gold as an electrochemical label in streptavidin–biotin interaction. *Biosens. Bioelectron.* 15, 315–321.
60. Dequaire, M., Degrand, C., and Limoges, B. (2000) *Anal. Chem.* 72, 5521–5528.
61. Wang, J., Polsky, R., and Xu, D. (2001) *Langmuir* 17, 5739–5741.
62. Wang, J., Xu, D., Kawde, A.-N., and Polsky, R. (2001) Metal nanoparticle-based electrochemical stripping potentiometric detection of DNA hybridization. *Anal. Chem.* 73, 5576–5581.
63. Woodward, K.N. (2003) Antibiotics and drugs. Uses in food production. In: Caballero, B., Trugo, L., Finglas, P. (eds). *Encyclopedia of Food Sciences and Nutrition* (2nd edn.). Academic Press, Boston, MA, pp. 249–254.
64. Zacco, E., Adrian, J., Galve, R., Marco, M.-P., Alegret, S., and Pividori, M.I. (2007) Electrochemical magneto immunosensing of antibiotic residues in milk. *Biosens. Bioelectron.* 22, 2184–2191.
65. Bates, C.J. (2003) Folic acid. Properties and determination. In: Caballero, B., Trugo, L., Finglas, P. (eds). *Encyclopedia of Food Sciences and Nutrition* (2nd edn.). Academic Press, Boston, MA, pp. 2559–2564.
66. Lermo, A., Fabiano, S., Hernández, S., Galve, R., Marco, M.-P., Alegret, S., and Pividori, M.I. (2009) Rapid electrochemical magneto immunosensing of folic acid in vitamin-fortified food products. *Biosens. Bioelectron.* 24, 2057–2063.
67. Howdle, P.D. (2003) Celiac (coeliac) disease. In: Caballero, B., Trugo, L., Finglas, P. (eds). *Encyclopedia of Food Sciences and Nutrition* (2nd edn.). Academic Press, Boston, MA, pp. 987–994.
68. Laube, T., Kergaravat, S.V., Fabiano, S.N., Hernández, S.R., Alegret, S., and Pividori, M.I. (2011) Magneto immunosensor for gliadin detection in gluten-free foodstuff: Towards food safety for celiac patients. *Biosens. Bioelectron.* 27, 46–52.
69. D'Aoust, J.Y. (1994) Salmonella and the international food trade. *Int. J. Food Microbiol.* 24, 11–31.
70. Liébana, S., Lermo, A., Campoy, S., Cortés, M.-P., Alegret, S., and Pividori, M.I. (2009) Rapid detection of Salmonella in milk by electrochemical magneto-immunosensing. *Biosens. Bioelectron.* 25, 510–513.
71. Pividori, M.I. and Alegret, S. Graphite-epoxy platforms for electrochemical genosensing. *Anal. Lett.* 36, 1669–1695.

72. Pividori, M.I., Merkoçi, A., Barbé, J., and Alegret, S. (2003) PCR-genosensor rapid test for detecting *Salmonella*. *Electroanalysis* 15, 1815–1823.
73. Erdem, A., Pividori, M.I., del Valle, M., and Alegret, S. (2004) Rigid carbon composites: A new transducing material for label-free electrochemical genosensing. *J. Electroanal. Chem.* 567, 29.
74. Pividori, M.I. and Alegret, S. (2005) Electrochemical genosensing based on rigid carbon composites: A review. *Anal. Lett.* 38, 2541–2565.
75. Lermo, A., Campoy, S., Barbé, J., Hernández, S., Alegret, S., and Pividori, M.I. (2007) In situ DNA amplification with magnetic primers for the electrochemical detection of food pathogens. *Biosens. Bioelectron.* 22, 2010–2017.
76. Erdem, A., Pividori, M.I., Lermo, A., Bonanni, A., del Valle, M., and Alegret, S. (2006) Genomagnetic assay based on label-free electrochemical detection using magneto-composite electrodes. *Sens. Actuat. B* 114, 591–598.
77. De Buyser, M.L., Dufour, B., Maire, M., and Lafarge, V. (2001) Implication of milk and milk products in food-borne diseases in France and in different industrialised countries. *Int. J. Food Microbiol.* 67, 1–17.
78. Lermo, A., Zacco, E., Barak, J., Delwiche, M., Campoy, S., Barbé, J., Alegret, S., and Pividori, M.I. (2008) Towards Q-PCR of pathogenic bacteria with improved electrochemical double-tagged genosensing detection. *Biosens. Bioelectron.* 23, 1805–1811.
79. Bloom, B.R. and Murray, C.J.L. (1992) Tuberculosis: Commentary on an reemergent killer. *Science* 257, 1055–1064.
80. Lermo, A., Liébana, S., Campoy, S., Fabiano, S., García, M.I., Soutullo, A., Zumárraga, M.J., Alegret, S., and Pividori, M.I. (2010) A novel strategy for screening-out raw contaminated milk with mycobacterium bovis in dairy farms by double-tagging PCR and electrochemical genosensing. *Int. Microbiol.* 13, 91–97.
81. Liébana, S., Lermo, A., Campoy, S., Barbé, J., Alegret, S., and Pividori, M.I. (2009) Magneto immunoseparation of pathogenic bacteria and electrochemical magneto genosensing of the double-tagged amplicon. *Anal. Chem.* 81, 5812–5820.

17

Determination of Water-Soluble Vitamins and Drug Residues

Andreea Olaru and Camelia Bala

CONTENTS

LIST OF ABBREVIATIONS

AGCE	aminated glassy carbon electrode
AO	ascorbate oxidase
AOAC	Association of Official Analytical Chemists
AuE	gold electrode
AuNP	gold nanoparticle
BSA	bovine serum albumin
CE	capillary electrophoresis
CNT	carbon nanotube(s)
CV	cyclic voltammetry
DNA	deoxyribonucleic acid
EDC	endocrine disrupting compound(s)
EIS	electrochemical impedance spectroscopy

ELISA	enzyme-linked immunosorbent assay
EPA	Environmental Protection Agency
EU	European Union
FBP	folate binding protein
FDA	Food and Drug Administration
GC	gas chromatography
HorRatr	Horwitz ratio
HPLC	high-performance liquid chromatography
HRP	horseradish peroxidase
LC	liquid chromatography
LOD	limit of detection
MS	mass spectrometry
MTHF	methyltetrahydrofolate
MWCNT	multiwalled carbon nanotube(s)
NC	nitrocellulose
PABA	para-amino benzoic acid
PANI	polyaniline
PEDOT	poly(3,4-ethylenedioxythiophene)
PGA	pteroylglutamic acid
PGE	pencil graphite electrode
PMMA	poly(methyl methacrylate)
PMS	polymaleimidostyrene
PS	polystyrene
$PteGlu_n$	folate polyglutamates
RNA	ribonucleic acid
RSD	relative standard deviation
SPR	surface plasmon resonance
SWCNT	single-walled carbon nanotubes
SWV	square wave voltammetry
UCR	Unregulated Contaminant Regulation
UPLC	ultra-performance liquid chromatography
WWTP	waste water treatment plant
WIOS	wavelength-interrogated optical system

17.1 Introduction

Food and water safety are without a doubt high concerns of our present times, not only from the perspective of the consumers but also for the public health institutes and governmental agencies. Thus, the different categories of contaminants are classified and continuously monitored, contributing to the safety of the products.

 Among several categories of chemical compounds possessing high risks for animal and human health (classified as priority substances, e.g., pesticides, heavy metals, benzene derivatives, etc. [1]), pharmaceuticals are a special category. These organic compounds, very active at low concentrations, are a new concern for public health mainly due to their uncontrolled use and adverse effects [2]. Moreover, their presence in food is strictly under control [3–5] or even prohibited in food producing animals [6], while in environmental samples they are referred with an emphasis merely on endocrine disrupting compounds [7].

On the other hand, vitamin detection and accurate quantification are required for establishing vitamin concentrations in food products, quality control, and labeling, as well as for compliance monitoring, especially in fortified foods [8]. As organic components, vitamins are essential constituents of food, playing a key role in normal organism growth and functioning of animal bodies, with direct functions in the health of the nervous and circulatory systems, tissues, bones and skin, eyes, and liver [9].

Water-soluble vitamins detailed in Table 17.1 comprise the group of B vitamins and vitamin C (ascorbic acid). They have a major role in the body, and their deficiencies could cause perturbations of normal metabolism or even severe diseases. Regular consumption of fruits, vegetables, and meat products as natural rich sources of these vitamins prevents the occurrence of deficiencies, even though some precautions must be taken during processing and storage due to their relatively high instability. Some representatives are used to fortify different types of foods: addition of folic acid to cereal products or in baby milk formulae and fruit drinks is a widespread procedure, preventing the neural tube defects in newborn babies [10]. At the same time, folate enrichment will correct the anemia symptoms induced by the vitamin B_{12} deficiency, but it may hasten the development of the nerve damage found in this case, putting some people at risk [11].

Other examples include L-ascorbic acid, frequently added to processed foods as an antioxidant [12], and biotin (known also as Vitamin B_8, B_7, or H) addition to baby-milk formula to satisfy the nutritional requirements of infants during the first months of life [13], since this vitamin cannot be synthesized in the organism and must be obtained from dietary intake [14]. The European Union has published regulations relating to supplementation using vitamins, minerals, and other substances and also specifies the purposes of fortification (EU Regulation No. 1925/2006, [15]).

Currently, the methodologies used for water-soluble-vitamin detection are as follows:

a. Microbiological assays, based on the growth response of various vitamin-dependent lactobacilli (*Lactobacillus plantarum, Saccharomyces cerevisiae, Lactobacillus rhamnosus,* or *Streptococcus faecalis* [10]).

b. Traditional instrumental analysis: high-performance liquid chromatography (HPLC), in-capillary enzyme reaction methods, electrochemical methods, and enzyme-linked immunosorbent assay (ELISA) [9]. An excellent review of current assays is authored by Blake [8], which presents in detail the official international methods for water-soluble vitamins in foods.

TABLE 17.1

Water-Soluble Vitamins

Vitamin	Structure	Major Functions	Vitamin Deficiency	Food Sources
Vitamin C (*ascorbic acid*)	L-Ascorbic acid formula	Formation of collagen, wound healing, production of brain hormones, immune factors; antioxidant Increases the activity of enzymes in vitro, though this is a nonspecific reducing action; oxygen radical quencher	Signs of vitamin C deficiency in scurvy include skin changes, fragility of blood capillaries, gums decay; tooth loss, and bone fractures Increased infections	Citrus fruits, broccoli, strawberries, melon, green pepper, tomatoes, dark green vegetables, potatoes
Vitamin B$_1$ (*Thiamin*)	Thiamin formula	Central role in energy-yielding metabolism, especially in the metabolism of carbohydrates Normal functioning of the nervous system	Chronic peripheral neuritis, (beriberi) which may or may not be associated with heart failure and edema Acute pernicious (fulminating) beriberi and Wernicke's encephalopathy with Korsakoff's psychosis, which is associated especially with alcohol and drug abuse Impaired growth	Pork, liver, whole grains, enriched grain products, peas, meat, legumes

Vitamin	Formula	Function	Deficiency	Sources
Vitamin B$_2$ (riboflavin)	Riboflavin formula	Central role in energy-yielding metabolism, as the coenzymes flavin mononucleotide (FMN) and flavin adeninedinucleotide (FAD) Promotes good vision, healthy skin	Cheilosis, lingual desquamation and seborrheic dermatitis	Liver, milk and dairy products, dark green vegetables, whole and enriched grain products, eggs
Vitamin B$_3$ (niacin/nicotinic acid/ nicotinamide)	Niacin formula	Energy production from foods. Promotes healthy skin	Pellagra (photosensitive dermatitis), dementia, and diarrhea concomitant with the disease progress	Liver, fish, poultry, meat, peanuts, whole and enriched grain products
Vitamin B$_5$ (pantothenic acid)	Pantothenic acid formula	Role in the acyl group metabolism and fatty acid synthesis	Uncommon due to widely distribution in most foods	Liver, kidney, meats, egg yolk, whole grains, legumes; also made by intestinal bacteria

(continued)

TABLE 17.1 (continued)

Water-Soluble Vitamins

Vitamin	Structure	Major Functions	Vitamin Deficiency	Food Sources
Vitamin B$_6$ (pyridoxine/ pyridoxal/ pyridox-amine and their 5'-phosphates)	Pyridoxal-phosphate formula	Important role in protein metabolism and red blood cell formation, also in the steroid hormone action	Skin disorders, dermatitis, anemia; kidney stones; nausea; smooth tongue Moderate deficiency results in abnormalities of tryptophan and methionine metabolism	Pork, meats, whole grains and cereals, legumes, green, leafy vegetables
Vitamin B$_7$ (biotin)	Biotin formula	Regulation of the cell cycle and fat synthesis	Unknown or very rare (people with parenteral nutrition)	Liver, kidney, egg yolk, milk, most fresh vegetables, also made by intestinal bacteria
Vitamin B$_9$ (folate/folic acid/ tetrahydrofolate)	Folic acid formula	Role in protein metabolism. Supplements taken before and during pregnancy reduce the risks of neural tube defects (spina bifida)	Megaloblastic anemia	Liver, kidney, dark green leafy vegetables, meats, fish, whole grains, fortified grains and cereals, legumes, citrus fruits

Vitamin B₁₂ (cobalamin)	Role in DNA synthesis and regulation, fatty acid synthesis, and energy production as well in the development of normal red blood cells	Pernicious anemia, (megaloblastic) anemia; neurological disorders. Degeneration of peripheral nerves	Found only in foods of animal origin: meats, liver, kidney, fish, eggs, milk and milk products, oysters, shellfish

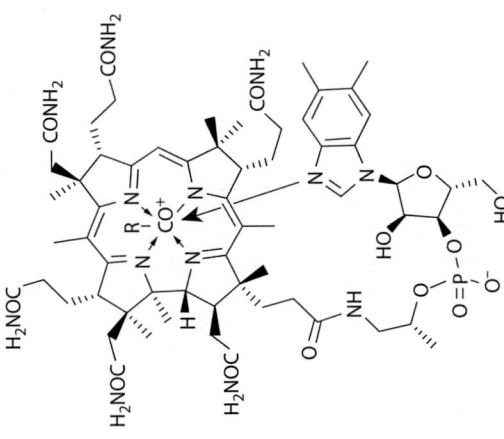

R = 5´-deoxyadenosyl, Me, OH, CN

Sources: Based on Murray, R. et al., *Harper's Illustrated Biochemistry,* Lange Medical Books-McGraw Hill, New York, 2003; Rucker, R.B. et al., *Handbook of Vitamins,* CRC Press, Boca Raton, FL, 2007.

Many of the traditional analysis methods are outdated or present several disadvantages: microbiological techniques require long analysis times and experienced personnel, provide relatively low precisions, while a relative measurement uncertainty of ±20% is fairly common.

Vitamin analysis performed in accredited laboratories often possesses a high failure rate due to the poor growth or contamination. Chromatographic techniques based on colorimetric or fluorometric detection are revealing more accurate results [16], but specificity remains their problem in trace analysis [17]. Fast LC-MS or UPLC-MS/MS seems to be promising technique, especially for rapid multivitamin determinations but requires expensive equipment and spare parts. Moreover, these methods must be validated and collaboratively tested before using within regulatory compliance scopes.

Synthetic drugs are a particular category of environmental pollutants, mainly due to their biological activity and lesser degree of biodegradability [19]. Their incidence in the domestic, municipal, industrial, or agricultural sources is not negligible, but the monitoring of these "emerging contaminants" is not frequently employed [20] and less regulated by the specific legislation: actual legislation did not specify any limits or test for specific drug residues in the potable or bottled water, with the exception of endocrine disrupting compounds [7]. By contrast, the drug residue detection in feed and food sources is strictly monitored and controlled by legal authorities [6].

Several water quality surveys and monitoring studies completed in the past years in different countries and geographic areas have identified ~160 types of pharmaceutical residues of various drug origins [19,20]. The main drug categories with high incidence in environment are presented in Table 17.2.

Pharmaceutical residues and their metabolites are in general present at low concentrations (ng/L–μg/L range) and are considered as micropollutants [21,22], but even at these trace levels they could induce toxic effects. In particular, this is the case of antibiotics and steroids that cause resistance in bacterial populations or endocrine disruption effects [23,24].

Identification and subsequent quantification of pharmaceutical drug residues from different environmental samples are traditionally completed by chromatographic methods coupled to mass spectrometry (gas or liquid chromatography: GC-MS, LC-MS/MS, or high-performance liquid chromatography HPLC-MS) as methods of choice [2,23]. Capillary electrophoresis (CE) has also been employed, but more rarely than HPLC due to its low sensitivity.

State of the art in this field, together with some upgrades of traditional assays, has been reported [25–28]. Anyhow, many disadvantages are present in chromatographic methods as well: use of expensive equipment, trained personnel, and higher periods of time required for regular screening, also the complex preparation step are among them [2].

TABLE 17.2

Categories of Drug Residues and Environmental Presence

Category /Examples	Environmental Presence
Antibiotics	Sewage sludge from WWTP
Sulfonamides (sulfamethoxazole)	Surface waters
Tetracyclines	
Ciprofloxacin	
Erythromycin	
Analgesics and anti-inflammatory drugs	Domestic and medical sewage
Acetaminophen/paracetamol	Wastewater effluents of municipal
Diclofenac	WWTP
Ibuprofen, ketoprofen	Surface waters
Naproxen	
Antiepileptics and blood lipid regulators	Waste water
Bezafibrate	Ground and river waters
Clofibric acid	
Gemfibrozil	
Carbamazepine	
Antitumoral drugs	Medical effluents
Cyclophosphamide and ifosfamide	WWTP effluents and influents Surface waters
Cardiovascular drugs	Domestic and medical sewage
β-Blockers and β2-Sympathomimetics	
Drugs of abuse	Domestic and medical sewage
Cocaine	Wastewater effluents of municipal
Heroin	WWTP
LSD	Surface waters
Natural and synthetic hormones	Ground waters
Sexual Hormones, Phytoestrogens	Domestic sewage
Contraceptives	

Sources: Based on Sanvicens, N. et al., *Trends Anal. Chem.*, 30, 541, 2011; Aga, D.S., *Fate of Pharmaceuticals in the Environment and in Water Treatment Systems*, CRC Press, Boca Raton, FL, 2007.

Thus, there is an increasing need to develop specific, rapid, and sensitive quantitative assays for the determination of water-soluble vitamins and drug residue content in both food matrices and environmental water samples. Moreover, a high degree of portability would be a plus, facilitating the direct analysis of target analytes for on-site applications. Biosensor technology could overcome these issues and offer a reliable, sensitive, and low-cost solution, even for in-field operation.

17.2 Biosensors for Water-Soluble Vitamins

17.2.1 Vitamin C

Vitamin C (ascorbic acid) content in foodstuffs is an indicator of its freshness and nutritive value. Its natural availability is as L-Ascorbic acid (since the D-ascorbic acid does not occur in nature), possessing a significant activity in the processes of oxidation and reduction in living organisms. Due to its extensive use in the food and beverages industries, a high number of articles are mentioning its quantification using biosensors, comparing with other vitamins. Nevertheless, vitamin detection as interfering compound in different matrices is widely analyzed with biosensors, and the future development of biosensing devices for ascorbate determination in food must carefully consider these approaches.

Enzymatic biosensors based on the specific enzyme, ascorbate oxidase (AO), are one of the most used devices for the detection of vitamin C, due to their multiple advantages such as high sensitivity, fast responses, and simple construction as well as easy miniaturization. Different materials have been used as immobilization matrices for the enzyme, and a special category is represented by the conducting polymers such as polyaniline, polypyrrole, and poly (3,4-ethylenedioxythiophene) due to their dual ability to act as both redox mediators and immobilization matrices [29], either as single incorporation elements [30–32] or merely in conjunction with nanomaterials [33].

Some examples are represented by the successful construction and characterization of an amperometric vitamin C biosensor, based on AO immobilization in polypyrrole-MWCNT composites [29] or into a sandwich-type composite film [31]. In the first case, the developed biosensor showed a linear range of 5×10^{-5} to 2×10^{-2} M with a detection limit of 0.3 µM and a fast response time, exhibiting also a high bioaffinity and good enzyme stability. The possibility of using the biosensor for vitamin C quantification in real food and agricultural samples is also being investigated.

In the second example, the biocompatible conducting poly(3,4-ethylenedioxythiophene)/PEDOT composite film and highly stable and selective MWCNTs–Nafion membrane is prepared as inner and outer films of the biosensor, and the enzyme molecules are immobilized between these two composite films. Thus, the content of ascorbic acid in commercial juices was determined, the constructed biosensor showing very good bioelectrocatalytic performance toward the oxidation of vitamin C in solution, besides other good characteristics such as fast current response, low working potential, and high sensitivity (187 mA $\times$ M^{-1} cm^{-2}) and also a wide linear range (4.0×10^{-7} to 1×10^{-3} M) and low detection limit (0.087 µM), better performances than other similar biosensors.

Electrochemical biosensor based on PEDOT-ethyl sulfate matrix (the preparation procedure and mechanism are indicated in Figure 17.1) involves

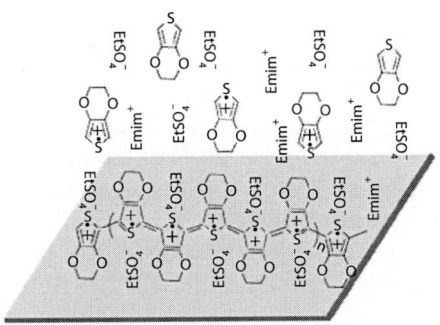

Electrosynthesis and mechanism

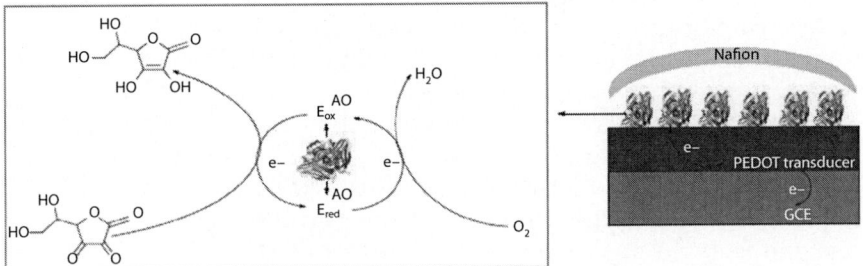

Working mechanism of fabricated biosensor

FIGURE 17.1
Electrochemical biosensor based on PEDOT-ethyl sulfate matrix fabricated for Vitamin C determination in commercial fruit juices: electrochemical preparation procedure and associated working mechanism. (Reproduced from Wen, Y.P. et al., *J. Solid State Electrochem.*, 16, 3725, 2012. With permission.)

a facile one-step synthesis and distinct characteristics: rapid response less than 2 s, at a low potential of 0.2 V over a wide range of concentrations from 8.0×10^{-7} M to 1×10^{-3} M and a limit of detection of 0.147 μM are claimed by the authors [34]. Another possible polymeric template reported by the same group, biocompatible conducting poly(3,4-ethylenedioxythiophene)-sodium N lauroyl-sarcosinate film, [32] represents a promising alternative electrode material for the construction of an electrochemical sensor and direct, rapid, and specific determination of vitamin C content in real samples without any sample pretreatment.

AO enzyme (E.C.1.10.3.3) was covalently immobilized onto a carboxylated multiwalled carbon nanotube and polyaniline (c-MWCNT/PANI) layer, electrochemically deposited on the surface of a gold electrode [35]. Thus, an ascorbate biosensor was fabricated using AO/c-MWCNT/PANI/Au electrode as a working electrode, Ag/AgCl (3 M/saturated KCl) as standard electrode, and Pt wire as an auxiliary electrode connected through a potentiostat. The detection method is characterized by a wide linear range of 2–206 mM, very short response time (2 s), and 0.9 mM detection limit, concomitant with good storage stability and no observed enzyme leakage. Ascorbate content in fruit juices and other matrices (sera and tablets) was determined, in good correlation with the results obtained with standard titrimetric method.

Other enzyme immobilization materials are mentioned in, where the ascorbate oxidase from *Lagenaria siceraria* is immobilized onto an egg shell membrane (through glutaraldehyde coupling) and then mounted over a gold electrode as the working electrode for ascorbate [36]. A linear relationship between L-ascorbic acid concentration in the range of 1×10^{-5} M and 4×10^{-4} M and current was obtained and further applied for amperometric quantification of L-ascorbic acid in various samples, including fruit juices. The authors mention distinct benefits of the developed biosensor, as the fast response time (10 s), good repeatability (200 assays), and long-term stability (a decrease of 50% of its initial sensitivity was observed after 4 months of storage). Enzyme micelle membrane and its application in the construction of an amperometric L-ascorbic acid biosensor is mentioned by Wang [37] and has been successfully tested for the determination of the Vitamin C content in fruit juices. An inexpensive polystyrene (PS) membrane served as the support of enzyme micelles, previously bounded to polymaleimidostyrene (PMS) in a stable structure. Oxygen penetrated the permeable enzyme micelle membrane and reduced at aminated glassy carbon electrode (AGCE) and gold electrode (AuE). The detection of ascorbic acid at both modified electrodes exhibits a good response of catalytic reduction current of oxygen with high sensitivity and short response time (within 1 min.). A good linear relationship was observed in the concentration range of 5 μM–0.4 mM, when AGCE was used at an applied potential of −0.5 V versus Ag/AgCl. Determinations at cathodic potential are used, to avoid interferences from the reducing agents.

Good results of vitamin detection in fruit juices are indicated by Vermeir [38], who developed a microplate differential microcalorimetric biosensor. Despite a relatively high limit of detection of 0.8 mM compared with other enzymatic biosensors [31], the main advantages are the low analysis cost in respect to the reagents and enzyme consumptions and the possibility of further integration into a fully automated device.

A new graphite/PMMA composite electrode was fabricated and applied to study the electrochemiluminescence behavior of luminol, which is inhibited in the presence of vitamin C. The developed electrochemiluminescent biosensor [39] also provides rapid and accurate results and a wide linearity range for detection of vitamin C.

Another promising approach, relying on the excellent properties of semiconducting materials, offers a rapid, ultrasensitive, and highly selective detection of vitamin C based on a "turn-on" fluorescent method [40]. The fluorescent sensing system is established on a colloidal $CePO_4$:Tb nanocrystalline solution, whose fluorescence is first efficiently quenched in the presence of $KMnO_4$ and turns on rapidly upon addition of vitamin C. Future improvements and/or additional tests of these latest methods on real samples could enlarge their applications in the area of food safety.

17.2.2 Group of B Vitamins

Despite the increased importance of accurate, sensitive, and robust determinations of each representative of "B" vitamins in food products, our literature survey of biosensor applications for vitamin detection shows unbalanced situations: while folic acid (Vitamin B_9), cobalamin (Vitamin B_{12}), and riboflavin (Vitamin B_2) are largely quantified in foods with biosensing assays (merely optical biosensors), the other representatives (Vitamins B_1, B_3, B_5, B_6, B_7) are less or only poorly detected with biosensors. This aspect could not be neglected and should be retained for future developments of biosensor devices for the categories of interest.

17.2.2.1 Vitamin B_9, Vitamin B_{12}, and Vitamin B_2

Folate enrichment of certain foods is required by most health agencies worldwide [6,41], thus the addition of commercial compounds such as PGA (pteroylglutamic acid) and 5-MTHF (5-methyltetrahydrofolate) is a common practice in the food industry. But the naturally present content of several folate forms in food, at trace concentrations, brings to attention the importance of analysis methods without neglecting the issues of low stability and complex matrices [10,42]. Thus, the biosensor-based immunoassay and SPR optical biosensors are considered the most reliable and sensitive assays for folate content in breakfast, cereals, drinks, milk, and soy-based infant formulas [43,44]. Low detection limit of 1 ng/mL and a reduced overall analysis time (20 samples in 6 h) are characteristics of the Qflex® kit for folic acid based on Biacore technology. Evaluations of optical biosensor for vitamin B_9 detection in milk-based infant formulae and cereal samples [9,10,45] indicate recoveries of 95.5%–109%, RSD in the range of 3.14%–4.06%, and some cross reactivity with related compounds (5-methyl-tetrahydrofolic acid 100%, dihydrofolic acid 17%, and tetrahydrofolic acid 8%) [9,10].

Milk folate binding protein (FBP), highly specific for folate, was recently confirmed as an alternative binder for the quantitation of folic acid in nutritional dairy products utilizing an SPR-based biosensor platform [46] and subsequently used in single laboratory validation studies [47]. Automated analysis was performed with the Biacore Q equipment; briefly, the measuring procedure supposes the injection of FBP (1 µg/mL in HEPES buffer, mixed either with calibrant or with sample extract) over a functionalized sensor surface (functionalization agent folate polyglutamates, PteGlu$_n$) and surface regeneration with NaOH in the end. Quantification of the folate content in foods (typically expressed as µg/100 g) was accomplished following interpolation of the relative binding response (SPR response units, RU) from a four-parameter logistic calibration curve. Based on the excellent results obtained for determination of folate in infant formula and adult/pediatric nutritional formula by the developed optical biosensor assay (LOD of 0.1 ng/mL, repeatability of 3.48% and intermediate reproducibility of 4.63% RSD), the method was submitted for adoption by the AOAC [48]. Moreover, an extent of the developed biosensing assay to supplemented and non-supplemented foods (milk, cereal, flour, broccoli, egg, fish meal, liver) and the method's performance are latter reported [49].

A highly selective disposable DNA electrochemical biosensor was proposed as a screening device for the rapid analysis of folic acid using a pencil graphite electrode modified with salmon sperm ds-DNA [50].

Changes in the electrochemical signal of adenine in the salmon sperm ds-DNA are used as an analytical signal for folic acid determination in fortified wheat flour and spinach samples (Figure 17.2). The effect of folic acid concentration on the adenine and guanine peak currents at the DNA modified electrodes.

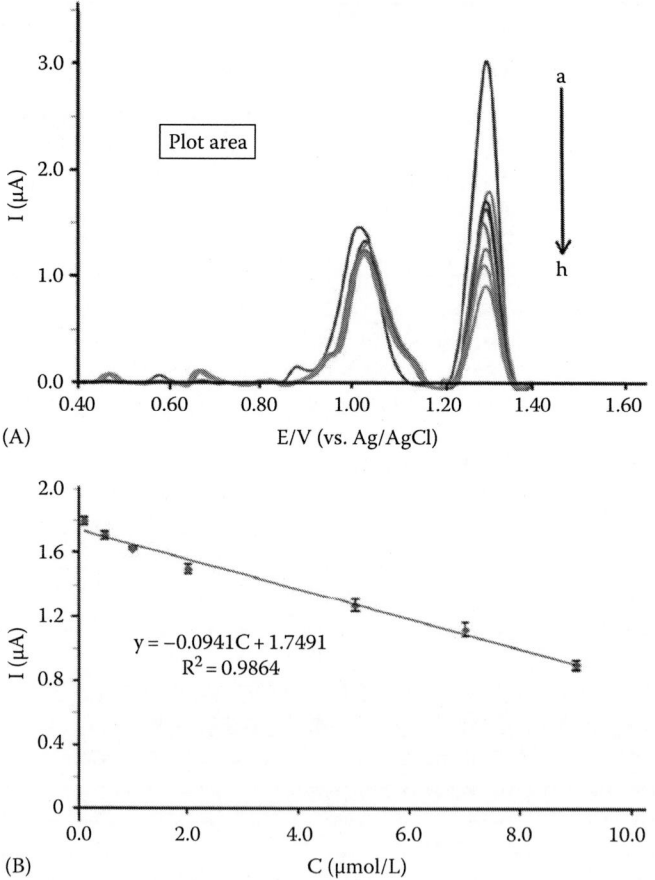

FIGURE 17.2

(A) Effect of folic acid concentration on the adenine and guanine peak current at DNA modified-PGE: (a) in acetate buffer (pH 4.8), (b) 0.10 μmol/L folic acid; (c) 0.50 μmol/L folic acid; (d) 1.0 μmol/L; (e) 2.0 μmol/L folic acid; (f) 5.0 μmol/L folic acid; (g) 7.0 μmol/L folic acid; and (h) 9.0 μmol/L folic acid; all (b–g) in acetate buffer. (B) Calibration curve for folic acid at the surface of the ds-DNA modified-PGE (error bars show the standard deviation for five replicate measurements). (Reproduced from Mirmoghtadaie, L. et al., *Mater. Sci. Eng. C*, 33, 1753, 2013. With permission.)

A folic acid sensor, based on single-walled carbon nanotube paste coated glassy electrode and the ionic liquid $OMIMPF_6$ (1-octyl-3-methylimidazolium hexafluorophosphate) as binding agent, was developed and optimized [51] various food samples analyzed (wheat flour, fruit juice, and milk samples) for the determination of vitamin content. High recovery rates (93%–108%) and low LOD (1.0×10^{-9} mol/L) mentioned by authors indicate the better performance of the proposed method compared with other electrochemical methods. Single-walled CNT film electrode has also been reported by Wang [52] for trace determination of folic acid but in some food supplements.

For vitamin B_{12}, the specifications of the commercial SPR Qflex kit indicate the smallest limit of detection, as low as 0.06 ng/mL, and a throughput of 20 samples in 12 h. Perhaps this performance is yet to be achieved with other biosensing devices, and even better LODs have obtained more recently [53,54]. Evaluation of vitamin B_{12} in a range of foods (milk, infant formula, meat, and liver) was reported with very good performance results by Indyk et al. in 2002 and more recently by Gao [9,55]. SPR biosensor analysis was also used to evaluate the thermodynamics and binding kinetics of naturally occurring and synthetic cobalamins interacting with vitamin B_{12} binding proteins. A direct binding assay, where recombinant human transcobalamin is conjugated to a biosensor chip, allows kinetic analysis of cobalamin binding [56].

A more novelistic approach for detection of vitamin B_{12} in energy drinks [54] is a dipstick-based immunochemiluminescence biosensor (Figure 17.3). Analytical methods based on chemiluminescence are rapid, specific, and cost-effective, providing sensitive and reliable results in food and environmental applications.

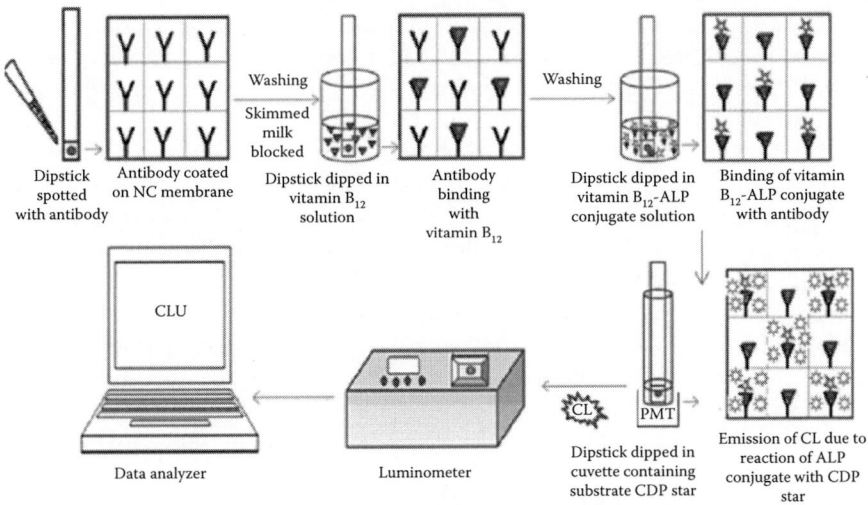

FIGURE 17.3
Schematic representation of immunochemiluminescence-based dipstick technique for detection of Vitamin B_{12}. (Reproduced from Selvakumar, L.S. and Thakur, M.S., *Anal. Chim. Acta*, 722, 107, 2012. With permission.)

The immunoassay method (direct competitive format) involves immobilization of vitamin B_{12} antibody onto a nitrocellulose membrane, followed by a specific treatment with vitamin B_{12} and vitamin B_{12}–alkaline phosphatase conjugate to facilitate competitive binding. The dipstick is further treated with a chemiluminescent commercial substrate (CDP-Star) to generate chemiluminescence. The method relies on the dependence between the numbers of generated photons and the vitamin B_{12} concentration in the analyzed beverage samples. The low detection limit obtained after method optimization (1 ng/mL) and a comparison study between the dipstick method and ELISA indicate that the dipstick-based detection can be a better alternative to conventional immunological methods in terms of cost, time duration, robustness, and ease of handling.

A rapid electrochemical micro bioassay (the response time for a rate measurement of 2 h) for vitamin B_{12} using *E.coli* and oxygen electrode was reported by Karube [57]. When the sample solution is injected into the microbial electrode system, the increased consumption of oxygen by the microorganisms causes a decrease in the dissolved oxygen around the porous membrane of the oxygen electrode, and the current decreases gradually with time until a steady state is reached.

Promising nanostructured materials such as carbon nanotubes have been used as electrode material for a disposable pencil graphite electrode (PGE) used in electrochemical detection of Vitamin B_{12} [58]. A single-walled carbon nanotube–chitosan (SWCNT–chitosan) modified PGE characterized by various electrochemical methods (SWV, EIS, CV) was used for the quantification of vitamins in pharmaceutical products, showing good recovery and a high selectivity and sensitivity in vitamin detection. Although it has not yet been tested for other real samples, the prepared electrodes could be promising candidates for other biosensing applications of interest, such as food monitoring or clinical diagnosis.

For vitamin B_2, the SPR biosensing assay Qflex kit specifies a limit of detection of 17.1 ng/mL, short analysis time (20 samples in 6 h), and a wide variety of foodstuffs (cereals, milk powders, fortified beverages) that could be analyzed. Quantification of Vitamin B_2 [9] in milk-based infant formula was evaluated in terms of the method's performance parameters (accuracy, recovery, precision); the RSD was below 2% for all the vitamins considered in the study and recoveries ranged from 94.7% to 109.1%.

The detection method based on optical biosensor technology is sensitive, reliable, and selective but not very specific: some cross-reactivity occurs with other vitamins or analogues (e.g., for riboflavin, cross-reactivity of 18.6% in the presence of Riboflavin 5′-monophosphate, 26.5% for lumichrome, and 51% for lumiflavin). Even so, this method is a practical alternative to other techniques for food quality control.

Different types of biosensors were developed for trace analysis of vitamin B_2: potentiometric biosensor based on binding protein [59,60], SPR biosensor with on-chip measurement [61], chemiluminescence sensor, or cyclodextrin-based optosensor [62], but only for vitamin quantification in pharmaceutical or biological samples. Electrodes modified with macrocyclic compounds (several crown ethers, with the best response obtained for 1,4,8,11-tetraazacycloocta-decane), exhibiting a relatively high linear working range and low detection limit (0.2 ng/cm^3), were used for riboflavin detection in milk and fruit juice samples by square wave technique [63]. Functionalized biosensors based on DNA hybrid film titanium-tungsten oxide (WO$_3$–TiO$_2$) modified electrodes have been discussed more recently [64,65] and to our knowledge, not yet tested on foods. Other examples are DNA biosensor based on self-assembly of carbon nanotubes and DNA, which exhibits a high sensitivity and low detection limit for vitamin B$_2$ in solution [66]; DNA immobilized CNT mixed paste electrode applied for vitamin assay in pharmaceutical and biological samples [67] and some synthetized water-soluble ZnS nanoparticles [68] as highly sensitive chemosensors for riboflavin detection based on the nanoparticle luminescence quenching in the presence of vitamin.

17.2.2.2 Vitamin B$_1$, Vitamin B$_3$, Vitamin B$_5$, Vitamin B$_6$, and Vitamin B$_7$

A limited number of studies report vitamin B$_1$ (thiamine) quantification with chemosensors or biosensors and apparently only in pharmaceutical forms or clinical probes [69–72]. Anyhow, these methods must be considered as promising methods for food analysis, due to their cumulative characteristics in terms of sensitivity, reproducibility, and ease of implementation. Among these, we must mention the methodology based on a novel DNA-functionalized electrochemical biosensor for thiamin quantification in clinical samples [72]. Vitamin B$_1$ was sensitive determined on a pretreated multiwalled carbon nanotube paste electrode by means of adsorptive stripping differential pulse voltammetry, and the decrease in the intensity of the guanine and adenine oxidation signals after interaction with vitamin B$_1$ was used as indicator signals.

Akilmaz [73] reports the results obtained with a novel amperometric biosensor based on the activation effect of thiamine on the pyruvate oxidase enzyme (which increases proportionally with vitamin concentrations), tested for thiamine determination in vitamin tablets. Whole cell biosensors based on *Saccharomyces cerevisiae* were employed for selective vitamin B1 quantification in some pharmaceutical forms [70]. Yeasts used in the construction of biosensors are offering multiple advantages, such as easy handling, robustness, and high speed of growth, but even so their application for monitoring the vitamin content in food remains limited.

Determination of vitamin B$_3$ (niacin/nicotinamide) in food products is also less investigated with biosensor devices, while the incorporation of nicotinamide as coenzyme in the development of several types of biosensors is a widespread procedure in the literature [74–78]. The detection of vitamin, in the form of 3-pyridinecarboxamide, is mentioned by Verma using an SPR optical fiber sensor by combining colloidal crystal and molecular-imprinted hydrogel [79]. Several advantages of this sensor, such as small probe size, low cost, reusable probe, remote sensing, and fast response, indicate its possible application for sensing low concentration of vitamin in food or pharmaceutical samples.

Other examples of sensors worth mentioning only in respect of the principle of construction/detection are as follows: a whole-cell amperometric biosensor for nicotinic acid that was early developed [80] for its determination in biological samples and a carbon paste electrode modified with macrocyclic compounds for the voltammetric detection of PABA, where vitamin B_3 is studied as interferent [81].

Optical biosensors based on SPR technology are currently employed for vitamin B_5 (pantothenic acid) in different matrices. Detection of pantothenic acid in foodstuffs using the commercial kit from Biacore® is characterized by a limit of detection of 4.4 ng/mL and is applicable to a large range of matrices (cereals, milk powder, milk, or soya-based infant formula, fortified beverages, and vitamin premixes), with a total analysis time of 12 h for 20 samples. The SPR immunoassay method, based on polyclonal antibodies for pantothenic acid, was recently compared with a novel monoclonal antibodies biosensing assay [82]. Analysis of a large number of food products (cereals, drinks, baby food) using both SPR-based assays showed good correlation between the results. Moreover, improved sample preparation steps, based on enzymatic digestion, enhanced the final results by eliminating the effects occurring due to the nonspecific binding. The good results indicate produced monoclonal antibodies as potential candidates for commercial application, based on the advantage of a constant and long-term supply of identical antibody.

In respect to vitamin B_6 (pyridoxine) detection in food, a rapid determination in marine products was achieved based on a microbial biosensor, one of the pioneer works in the field of vitamin detection in food [83]. The biosensing system was based on an immobilized microorganism (*Saccharomyces uvarum*), a Clark-type oxygen electrode, and an oxygen permeable Teflon membrane. The assay is based on the respiratory activity of the microorganism in the presence of vitamin B_6, providing fast and simple determinations. More recent studies [84–86] investigate only the interaction between small molecules such as pyridoxine and DNA due to the increased importance of rational design and construction of new and more efficient drugs targeted to DNA, or the substrate (pyridoxamine)-enzyme (pyridoxal kinase) interaction.

For vitamin B_7 (biotin) detection, the Biacore Qflex kit specifies a limit of detection of 1 ng/mL and throughput of 20 samples in 6 h. The evaluation of this SPR biosensing assay with very good results in terms of RSD%, recovery rate, and HorRat$_r$ values is reported in the works of Indyk [45] and Gao [9]; for infant milk fortified products, even some cross reactivity with analogous compounds was identified (10% for biocytin, 37% for biotinyl 4-amidobenzoic acid, and 38% for lumichrom [9]. The affinity biosensor system for determination of biotin and its derivatives in fruit juices is also a reliable approach for vitamin quantification [87]. Streptavidin-biotin bridge is often used for monitoring biochemical reactions and is employed for the development of a competitive assay setup (streptavidin is immobilized onto a controlled pore glass, providing an active support for competitive reactions between biotin derivatives and biotin labeled with peroxidase). The strong and specific noncovalent interaction between vitamin B$_7$ and streptavidin protein generates a high affinity complex, which could be easily visualized using enzymatic derivatization. Low concentrations of vitamin (0.09 and 0.46 µg/L) and recoveries of 97.3%–101.2% are indicated for peach and tomato juices analyzed. Some of the advantages of this biosensing system are an easy immobilization of the streptavidin-reactor via BSA/glutaraldehyde procedure, high selectivity and ease of implementation setup, and low sample and reagent consumption.

Biotin was also successfully detected using an electrochemical magneto biosensor in biotin-fortified commercial dietary supplements and infant formula samples [13]. The detection procedure is based on a competition scheme where biotin and the biotin-horseradish peroxidase (HRP) compete for the binding sites of streptavidin, previously immobilized on the surface of some magnetic beads. Subsequently, the beads are captured by a magnetoelectrode and put in the reaction cell along with the enzymatic substrates. Enzymatic product reduction currents, measured using SWV, were inversely related to biotin concentrations from the samples. Low limits of detection (20 µg/L) and a total analysis time of 40 min for 20 assays are mentioned in the study.

17.3 Accredited Methods Based on Biosensors for Detection of Vitamins

SPR biosensor technology using a ligand-binding protein interaction with the BIAcore® system (GE Healthcare Lifesciences) is already a golden standard for the analysis of several water-soluble vitamins [8]. These organic compounds are widely detected with optical sensors based on SPR method, due to its multiple benefits such as robustness, sensitivity, and versatility, delivering reliable, rapid, and relatively low-cost testing [88]. From the currently available instruments for SPR applications [89], the Biacore Q equipment is dedicated to the qualitative and/or quantitative determination of analytes in food-related products and can be used in combination with specially developed Qflex kits [88], delivering rapid, reliable, and automated quantification of vitamin content. Applications have been developed for vitamin B$_7$ (biotin), vitamin B$_9$ (folic acid), vitamin B$_5$ (pantothenic acid) and vitamin B$_{12}$ (cobalamin) and have been certified as Performance Tested Methods by the AOAC Research Institute.

SPR analysis with Qflex kits is based on an inhibition immunoassay principle, since it was demonstrated as an accurate, rapid, and sensitive technique for the detection of small analytes (<1000 Da) as vitamins [45,55]. In the inhibition assay format, the target antigen is immobilized on the sensor surface. Sample solution containing the antigen is mixed with specific antibodies in excess and incubated with the sensor surface. Antibodies will bind both to antigen in solution and to immobilized antigen from the sensor surface. The difference in signal between a blank sample (without antigen) and the sample solution indicates the amount of antigen in the sample. In this assay, high antigen concentrations in the sample result in low signals (less antibodies remain to bind to the surface) [89].

Sample preparation and analysis times are significantly reduced compared to traditional techniques [90], and the vitamin content could be quantified in almost any food matrix, including colored or opaque samples. Several studies report the application of Qflex kits for quantification of these four vitamins in supplemented infant formulas [9,45] and within a collaborative study for determination of Vitamin B_{12} content in fortified bovine infant milk, fortified soya-based infant formula powder, vitamin premix, and dietary supplements [91,92].

17.4 Biosensors for Drug Residue Detection

17.4.1 Environmental Samples

As mentioned in the previous section, among the drug residues with high incidence in environmental samples (surface and ground water, wastewater, or sewage sludge from WWTP), antibiotics and hormones are particular examples due to their irreversible toxic effects in aquatic population and terrestrial animals [93]. Exposure to both natural and synthetic chemicals at low levels has been demonstrated to be associated with diseases such as breast and uterine cancer in humans and hermaphroditism in wildlife [94,95]. Also, the regular monitoring of endocrine disrupting compounds (EDCs) is one of the requirements of EPA (EPA Unregulated Contaminant Regulation, UCR3 [7]). Endocrine disruptors are basically chemicals with the potential to interfere with the normal function of endocrine systems; the large group of EDCs have been defined as exogenous agents that interfere with the production, release, transport, metabolism, binding, action, or elimination of the natural hormones in the body, responsible for homeostasis and regulation of developmental processes; this includes besides the pesticides, plasticizers, and phytoestrogens, the pharmaceuticals or hormones that are excreted in animal or human waste.

Much progress was achieved within the framework of the European research projects River Analyser (RIANA) and Automated Water Analyser Computer Supported System (AWACSS) [96]. Quick, intelligent, and costs effective immunosensors were developed and tested for monitoring of different contaminants such as pharmaceuticals and endocrine disrupting compounds in environmental samples.

Single or multi-analyte determinations using the automated biosensing platforms have been reported for EDCs: estradiol-based hormones [97–102], progesterone [103] and testosterone [99], determined with limit of detections of 0.2–0.37 ng/L in drinking water, surface water, and river water and antibiotics (sulfonamides—[104] with LODs of 2.7–6.6 ng/L in drinking, ground, and surface water [99,100,103,105]). The main achievements and major draw-backs of these immunosensors are detailed in Ref. [96]. A reusable, portable evanescent-wave aptamer-based biosensor detects rapidly and with high selectivity the content of 17 β-estradiol in water samples, with a low limit of 2.1 nM (0.6 ng/mL) [106].

Many of the EDCs could bind to the natural estrogen receptor as agonists or antagonists [93], and this interaction could be exploited in biosensing: an SPR biosensor based on BIA core technology was applied in the determination of estrogens and xenoestrogens [107,108]. Other optical biosensors incorporating recombinant cells for co-expression of the human estrogen receptor have been developed for determination of estrogenic activity in water samples [102,109]. Electrochemical [110,111] and piezoelectric [112] biosensors based on estrogen receptors have also been developed.

A relatively limited number of studies report the construction of biosensing devices for detection of antibiotics only in water, since most of the developed biosensors for their detection in food could be potentially applied for environmental samples. Tetracycline was determined with a high specificity, at a low level of 10 nM, with an electrochemical aptasensor [113]. Biotinylated ssDNA aptamer was immobilized onto a streptavidin-modified screen-printed gold electrode, and further bindings of tetracycline to the aptamer were analyzed by cyclic voltammetry and square wave voltammetry. Whole-cell biosensors for tetracycline detection using three different recombinant cells modified with a tetracycline-inducible promoter were reported as tested in milk products [114] and are amenable to water samples.

17.4.2 Food Samples

Synthetic hormone-like substances are illegally used as growth promoters in cattle and calves to increase the weight gain of animals. Therefore, these substances have been banned in the European Union since 1988 and also in China in foods of animal origin. Because of the low levels (pg/kg to ng/kg) in the tissues of animal origin, the stringently regulated LODs, and the complexity of bio-sample matrices, the analysis of hormones and hormone-like substances is a challenging task. Similarly, the use of antibiotics (other than coccidiostats and histomonostats, based on the EC Regulation 1831/2003[115]), has been prohibited by the EU since 2005, thus reliable analytical methods are required to monitor trace concentrations of antibiotics and to assure product safety.

BIAcore Qflex kits® are also available for screening of several veterinary drugs of potential high incidence in food products, all with detection limits in the ng/mL domain: sulfonamides, sulfadiazine, and sulfamethazine (16.9 ppb in tissue to 0.6 and 0.3 ng/mL in bile), β agonists (0.02–1.46 ng/g), tylosin (5.7 μg/kg), chloramphenicol (0.02–0.073 ppb), and streptomycin (28 μg/L in milk) [116]. The net advantage of SPR bioassays is minimal sample preparation (simple centrifugation or dilution, depending on the sample nature), as well as ease of use and an overall reduced analysis time compared to the traditional analytical or microbiological assays. Moreover, the SPR method is environmental friendly because potentially hazardous organic solvents are not used during the analysis by comparison with the instrumental methods. Exemplifications of successful application of SPR commercial bioassays for quantification of drug residues in food are given in the extensive work of Haasnoot's group [117–120] as well as in Ref. [121]. Some improvements of the SPR assays are also mentioned, either in respect to the sample preparation procedure [122] or to the surface regeneration [123], both developed assays being tested and validated for rapid antibiotics screening (benzimidazole carbamate and chloramphenicol) in food samples.

Interestingly, very promising results have also been reported with other types of biosensors: a wavelength interrogated optical multiplexed biosensor comprising an array of waveguide gratings, in which different bioreceptors are immobilized for the detection of a wide range of antibiotic residues in food samples [124]. The sensing system and its structure are indicated in Figure 17.4, as a label free biosensor for fast screening of antibiotic residues

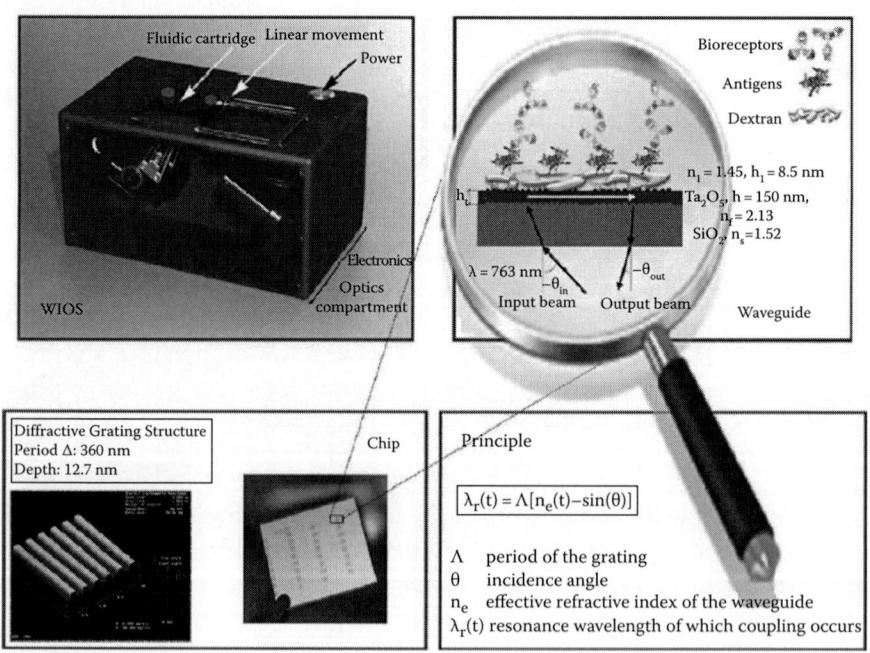

FIGURE 17.4

Wavelength-interrogated optical system (WIOS). The chip is a customized platform that contains 24 sensing sites. Each site is waveguide grating. Light from a VCSEL laser emitting at around 763 nm is incident on the first grating. The waveguide mode is excited and propagates into the wave-guiding layer. The second grating sends out the guided light (at a different angle), which is collected by large plastic optical fibers. Electronics controls the laser-wavelength modulation and amplifies the signals detected. A computer acquires and processes the data. (Reproduced from Adrian, J. et al., *Trends Anal. Chem.*, 28, 769, 2009. With permission.)

(sulfonamide, fluoroquinolone, β-lactam, and tetracycline) in contaminated milk samples. Other examples are a novel portable six channel SPR biosensor based on the plasmon of gold diffraction grating surface for simultaneous multianalyte antibiotic detection in milk sample [125] and an immunostrip biosensor system, based on a combination of immunochromatography assay and sandwich enzyme linked immunosorbent assay (ELISA) techniques, developed to detect enrofloxacin residues in the range of 100–10,000 ppb [126].

17.5 Conclusions and Future Trends

Miniaturization of the biosensor systems is one of the most important requirements for on-site applications, such as environmental monitoring or quality control of foods and beverages. Advances in microelectronics and microfluidics permitted the miniaturization of analytical systems, offering at the same time multiple advantages such as low volume samples, reduction in consumption of reagents and waste, decrease in analysis time concomitant with an increased reliability and sensitivity [93].

Since the development of Spreeta® as a portable version of commercial SPR biosensor technology from Texas Instruments [127], successfully used for various sensing applications [128,129], not much progress has been made: two-channel SPR refractometers, Biosuplar-6 [130], and the portable Sensia b-SPR [131] were used for the analysis of enrofloxacin in milk [132]. An upcoming instrumental alternative for SPR concentration analysis is the modular Reichert SR7000 system [88,133]. A novel palm-sized SPR biosensor device was recently reported [134] and tested by monitoring the binding of kallikrein 3 (synonym with PSA, the prostate specific antigen) to its antibody in real time to verify its potential applicability in analyzing biomolecular interactions. The results obtained with the developed prototype confirmed the practicality of the sensor for the on-site detection of a variety of substances in biology, diagnosis, environment, and defense.

Recent trends show that four types of biosensors are going to be extensively used in the field of food analysis: (1) antibody- or antigen-based biosensors, (2) aptamer-based sensors, (3) whole cell-based biosensors, and (4) photosynthetic protein-based biosensors [42]. For exemplification, a simple and stable RNA aptamer-based colorimetric sensor for the detection of vitamin B_{12} using gold nanoparticles (AuNPs) has been proposed in the work of Selvakumar [53]. Colorimetric aptasensors based on AuNPs have been considered as a method with high specificity and sensitivity in on-site detection, because of their easy preparation, simple operation, and detection. Aggregation of AuNPs was specifically induced by desorption of the vitamin B_{12} binding RNA aptamer from the surface of AuNPs as a result of the aptamer–target interaction, leading to the color change from red to purple. Vitamin B_{12} was analyzed in pharmaceutical samples available on the market (injections, tablets, and capsules) up to a minimum concentration of 0.1 μg/mL, and it is amenable to other vitamin types or matrices.

Several directions must be exploited and further considered in the design of novel biosensors: lack of biosensing assays for a certain group of B vitamins; the increased demands for validated and/or accredited bioanalysis methods as valuable alternatives for the replacement of the traditional assays, as well the potential of improvement for the existing (accredited) methods, without omitting the possibilities of miniaturization and/or portability for the developed biosensors.

Acknowledgment

This work was supported by a grant from the Romanian National authority for scientific research, project PN-II-ID-PCE-2011-3-0286.

References

1. Directive 2000/60/EC of the European Parliament and of the Council of October 23, 2000 establishing a framework for Community action in the field of water policy, *Official Journal of the European Communities*, L 327 (2000) 1.
2. N. Sanvicens, I. Mannelli, J. Pablo Salvador, E. Valera, M. Pilar Marco, *Trends in Analytical Chemistry*, 30 (2011) 541.
3. Council Directive 96/23/EC of April 29, 1996 on measures to monitor certain substances and residues thereof in live animals and animal products and repealing Directives 85/358/EEC and 86/469/EEC and Decisions 89/187/EEC and 91/664/EEC, *Official Journal of the European Communities*, L 125 (1996) 10.
4. 2002/657/EC: Commission Decision of August 12, 2002 implementing Council Directive 96/23/EC concerning the performance of analytical methods and the interpretation of results (Text with EEA relevance) (notified under document number C(2002) 3044), *Official Journal of the European Communities*, L 221 (2002) 8.
5. Regulation (EC) No 470/2009 of the European Parliament and of the Council of May 6, 2009 laying down Community procedures for the establishment of residue limits of pharmacologically active substances in foodstuffs of animal origin, repealing Council Regulation (EEC) No 2377/90 and amending Directive 2001/82/EC of the European Parliament and of the Council and Regulation (EC) No 726/2004 of the European Parliament and of the Council (Text with EEA relevance), *Official Journal of the European Communities*, L 152 (2009) 11.
6. FDA (Ed.), Post-approval monitoring of animal drugs, feeds and devices, in: *Food and Drug Administration Compliance Programe Guidance Manual*, 2006, p. 1.

7. *EPA (Ed.), Unregulated Contaminant Monitoring Rule 3 (UCMR3),* 77 FR 26071, 2012, p. 26071.
8. C.J. Blake, *Analytical and Bioanalytical Chemistry,* 389 (2007) 63.
9. Y.L. Gao, F. Guo, S. Gokavi, A. Chow, Q.H. Sheng, M.R. Guo, *Food Chemistry,* 110 (2008) 769.
10. M.B. Caselunghe, J. Lindeberg, *Food Chemistry,* 70 (2000) 523.
11. R. Murray, V. Rodwell, D. Bender, K.M. Botham, P. Anthony Weil, P.J. Kennelly, *Harper's Illustrated Biochemistry,* Lange Medical Books-McGraw Hill, New York (2003).
12. K. Rekha, B.N. Murthy, *Food and Agricultural Immunology,* 21 (2010) 103.
13. S.V. Kergaravat, G.A. Gomez, S.N. Fabiano, T.I.L. Chavez, M.I. Pividori, S.R. Hernandez, *Talanta,* 97 (2012) 484.
14. K.D. Bhalerao, S.C. Lee, W.O. Soboyejo, A.B. Soboyejo, *Journal of Material Science: Materials in Medicine,* 18 (2007) 3.
15. Regulation (EC) No 1925/2006 of the European Parliament and of the Council of December 20, 2006 on the addition of vitamins and minerals and of certain other substances to foods, *Official Journal of the European Communities,* L 404 (2006) 26.
16. L.F. Russel, *Handbook of Food Analysis,* Dekker, New York (2004).
17. S.S. Kumar, R.S. Chouhan, M.S. Thakur, *Analytical Biochemistry,* 398 (2010) 139.
18. R.B. Rucker, J. Zempleni, J.W. Suttie, D.B. McCormick (Eds.), *Handbook of Vitamins,* CRC Press, Boca Raton, FL (2007).
19. K. Kummerer (Ed.), *Pharmaceuticals in the Environment: Sources, Fate, Effects and Risks,* Springer Verlag, Berlin, Germany (2008).
20. D.S. Aga (Ed.), *Fate of Pharmaceuticals in the Environment and in Water Treatment Systems,* CRC Press, Boca Raton, FL (2007).
21. F. Sacher, F.T. Lange, H.J. Brauch, I. Blankenhorn, *Journal of Chromatography A,* 938 (2001) 199.
22. F. Van Hoof, P. Van Wiele, A. Bruchet, I. Schmitz, I. Bobeldiji, F. Sacher, F. Ventura et al., *Journal of AOAC International,* 84 (2001) 1420.
23. M.D. Hernando, E. Heath, M. Petrovic, D. Barcelo, *Analytical and Bioanalytical Chemistry,* 385 (2006) 985.
24. M.D. Hernando, M. Mezcua, A.R. Fernandez-Alba, D. Barcelo, *Talanta,* 69 (2006) 334.
25. E. O'Brien, D.R. Dietrich, *Trends in Biotechnology,* 22 (2004) 326.
26. T. Heberer, K. Reddersen, A. Mechlinski, *Water Science Technology,* 46 (2002) 81.
27. S. Thiele-Bruhn, T. Seibicke, H.R. Schulten, P. Leinweber, *Journal of Environmental Quality,* 33 (2004) 1331.
28. A.K. Sarmah, M.T. Meyer, A.B. Boxall, *Chemosphere,* 65 (2006) 725.
29. D. Li, Y.P. Wen, H.H. He, J.K. Xu, M. Liu, R.R. Yue, *Journal of Applied Polymer Science,* 126 (2012) 882.
30. Y.P. Wen, J.K. Xu, D. Li, M. Liu, F.F. Kong, H.H. He, *Synthetic Metals,* 162 (2012) 1308.
31. Y.P. Wen, J.K. Xu, M. Liu, D. Li, H.H. He, *Applied Biochemistry and Biotechnology,* 167 (2012) 2023.
32. Y.P. Wen, J.K. Xu, M. Liu, D. Li, L.M. Lu, R.R. Yue, H.H. He, *Journal of Electroanalytical Chemistry,* 674 (2012) 71.
33. Y.P. Wen, D. Li, Y. Lu, H.H. He, J.K. Xu, X.M. Duan, M. Liu, *Chinese Journal of Polymer Science,* 30 (2012) 460.
34. Y.P. Wen, X.M. Duan, J.K. Xu, R.R. Yue, D. Li, M. Liu, L.M. Lu, H.H. He, *Journal of Solid State Electrochemistry,* 16 (2012) 3725.
35. N. Chauhan, J. Narang, C.S. Pundir, *Analyst,* 136 (2011) 1938.

36. N. Chauhan, T. Dahiya, Priyanka, C.S. Pundir, *Journal of Molecular Catalysis B-Enzymatic*, 67 (2010) 66.
37. X.Y. Wang, H. Watanabe, S. Uchiyama, *Talanta*, 74 (2008) 1681.
38. S. Vermeir, B.M. Nicolai, P. Verboven, P. Van Gerwen, B. Baeten, L. Hoflack, V. Vulsteke, J. Lammertyn, *Analytical Chemistry*, 79 (2007) 6119.
39. H. Dai, X.P. Wu, Y.M. Wang, W.C. Zhou, G.N. Chen, *Electrochimica Acta*, 53 (2008) 5113.
40. W.H. Di, N. Shirahata, H.B. Zeng, Y. Sakka, *Nanotechnology*, 21 (2010).
41. A.J. Wright, J.R. Dainty, P.M. Finglas, *British Journal of Nutrition*, 98 (2007) 667.
42. T. Lavecchia, A. Tibuzzi, M.T. Giardi, in: M.T. Giardi, G. Rea, B. Berra (Eds.), *Bio-Farms for Nutraceuticals: Functional Food and Safety Control by Biosensors*, Landes Bioscience and Springer Science+Business Media, New York (2010), p. 267.
43. A. Ferancova, L. Heilerova, E. Korgova, S. Silhar, I. Stepanek, J. Labuda, *European Food Research and Technology*, 219 (2004) 416.
44. P. Puwastien, N. Pinprapai, K. Judprasong, T. Tamura, *Journal of Food Composition and Analysis*, 18 (2005) 387.
45. H.E. Indyk, E.L. Filonzi, *Australian Journal of Dairy Technology*, 55 (2000) 99.
46. H.E. Indyk, *International Dairy Journal*, 20 (2010) 106.
47. H.E. Indyk, *International Dairy Journal*, 21 (2011) 783.
48. H.E. Indyk, D. Dowell, *Journal of AOAC International*, 95 (2012) 298.
49. H.E. Indyk, D.C. Woollard, *Journal of Food Composition and Analysis*, 29 (2013) 87.
50. L. Mirmoghtadaie, A.A. Ensafi, M. Kadivar, P. Norouzi, *Materials Science and Engineering C*, 33 (2013) 1753.
51. F. Xiao, C. Ruan, L. Liu, R. Yan, F. Zhao, B. Zeng, *Sensors and Actuators B*, 134 (2008) 895.
52. C. Wang, C. Li, L. Ting, X. Xu, C. Wang, *Microchim Acta*, 152 (2006) 233.
53. L.S. Selvakumar, M.S. Thakur, *Analytical Biochemistry*, 427 (2012) 151.
54. L.S. Selvakumar, M.S. Thakur, *Analytica Chimica Acta*, 722 (2012) 107.
55. H.E. Indyk, B.S. Persson, M.C. Caselunghe, A. Moberg, E.L. Filonzi, D.C. Woollard, *Journal of AOAC International*, 85 (2002) 72.
56. M.J. Cannon, D.G. Myszka, J.D. Bagnato, D.H. Alpers, F.G. West, C.B. Grissom, *Analytical Biochemistry*, 305 (2002) 1.
57. I. Karube, Y. Wang, E. Tamiya, M. Kawarai, *Analytica Chimica Acta*, 199 (1987) 93.
58. F. Kuralay, T. Vural, C. Bayram, E.B. Denkbas, S. Abaci, *Colloids and Surfaces B-Biointerfaces*, 87 (2011) 18.
59. B.A. Morris, A. Sadana, *Sensors and Actuators B: Chemical*, 106 (2005) 498.
60. T. Yao, G.A. Rechnitz, *Analytical Chemistry*, 59 (1987) 2115.
61. I. Caelen, A. Kalman, L. Wahlstrom, *Analytical Chemistry*, 76 (2004) 137.
62. Z. Gong, Z. Zhang, *Analyst*, 121 (1996) 1119.
63. R.M. Kotkar, P.B. Desai, A.K. Srivastava, *Sensors and Actuators B-Chemical*, 124 (2007) 90.
64. A.A. Ensafi, E. Heydari-Bafrooei, M. Amini, *Biosensors and Bioelectronics*, 31 (2012) 376.
65. Y. Li, P.C. Hsu, S.M. Chen, *Sensors and Actuators B-Chemical*, 174 (2012) 427.
66. J. Li, Y. Zhang, T. Yang, H. Zhang, Y. Yang, P. Xiao, *Materials Science and Engineering C*, 29 (2009) 2360.
67. S.Y. Ly, H.S. Yoo, J.Y. Ahn, K.H. Nama, *Food Chemistry* 127 (2011) 270.
68. A. Chatterjee, A. Priyam, D. Ghosh, S. Mondal, S.C. Bhattacharya, A. Saha, *Journal of Luminescence*, 132 (2012) 545.

69. E. Aboul-Kasim, *Journal of Pharmaceutical and Biomedical Analysis*, 22 (2000) 1047.
70. E. Akyilmaz, I. Yasa, E. Dinckaya, *Analytical Biochemistry*, 354 (2006) 78.
71. J. Du, Y. Li, J. Lu, *Talanta*, 57 (2002) 661.
72. P.K. Brahman, R.A. Dar, K.S. Pitre, *Sensors and Actuators B*, 177 (2013) 807.
73. E. Akyilmaz, E. Yorganci, *Biosensors and Bioelectronics*, 23 (2008) 1874.
74. A. Arvinte, L. Rotariu, C. Bala, A.M. Gurban, *Bioelectrochemistry*, 76 (2009) 107.
75. A.-M. Gurban, T. Noguer, C. Bala, L. Rotariu, *Sensors and Actuators B: Chemical*, 128 (2008) 536.
76. X. Ren, L. Yang, F. Tang, C. Yan, J. Ren, *Biosensors and Bioelectronics*, 26 (2010) 271.
77. A. Vasilescu, S. Andreescu, C. Bala, S.C. Litescu, T. Noguer, J.-L. Marty, *Biosensors and Bioelectronics*, 18 (2003) 781.
78. J.L. Zhou, P.P. Nie, H.-T. Zheng, J.-M. Zhang, *Chinese Journal of Analytical Chemistry*, 37 (2009) 617.
79. R. Verma, B.D. Gupta, *Sensors and Actuators B*, 177 (2013) 279.
80. K. Takayama, T. Ikeda, T. Nagasawa, *Electroanalysis*, 8 (1996) 765.
81. R.M. Kotkar, K.S. Ashwini, *Sensors and Actuators B*, 119 (2006) 524.
82. S.A. Haughey, C.T. Elliott, M. Oplatowska, L.D. Stewart, C. Frizzell, L. Connolly, *Food Chemistry*, 134 (2012) 540.
83. H. Endo, A. Kamata, M. Hoshi, T. Hayashi, E. Watanabe, *Journal of Food Science*, 60 (1995) 554.
84. S.Q. Liu, M.L. Cao, S.L. Dong, *Bioelectrochemistry*, 74 (2008) 164.
85. C.C. Fong, W.P. Lai, Y.C. Leung, S.C.L. Lo, M.S. Wong, M.S. Yang, *Biochimica Et Biophysica Acta-Protein Structure and Molecular Enzymology*, 1596 (2002) 95.
86. H.C.M. Yau, S.Y. Wub, H.P. Hob, M. Yang, *Sensors and Actuators B: Chemical*, 85 (2002) 227.
87. F. Delgado Reyes, J.M. Fernández Romero, M.D. Luque de Castro, *Analytica Chimica Acta*, 436 (2001) 109.
88. M. Petz, *Monatshefte Fur Chemie*, 140 (2009) 953.
89. B.M.R. Schasfoort, A. McWhirter, in: B.M.R. Schasfoort, J.A. Tudos (Eds.), *Handbook of Surface Plasmon Resonance*, RSC Publishing, Cambridge, U.K, (2008), p. 35.
90. G.E. Healthcare, Biacore Food Analysis home page, Vitamin Content, Qflex kits, http://www.biacore.com/food/food/vitamin_content/qflex_kits/index.html (2013).
91. P. Vyas, A.A. O'Kane, *Journal of AOAC International*, 94 (2011) 1217.
92. P. Vyas, A.A. O'Kane, D. Dowell, *Journal of AOAC International*, 95 (2012) 329.
93. S. Rodriguez-Mozaz, M.J. Lopez de Alda, D. Barcelo, *Analytical and Bioanalytical Chemistry*, 386 (2006) 1025.
94. B.E. Henderson, R. Ross, L. Bernstein, *Cancer Research*, 48 (1988) 246.
95. A.L. Herbst, P. Cole, T. Colton, S.J. Robboy, R.E. Scully, *American Journal of Obstetrics and Gynecology*, 128 (1977) 43.
96. S. Rodriguez-Mozaz, M.J. de Alda, D. Barceló, *Handbook of Environmental Chemistry*, 5J (2009) 33.
97. S. Rodriguez-Mozaz, M.J. de Alda, D. Barcelo, *Journal of Chromatography A*, 1045 (2004) 85.
98. S. Rodriguez-Mozaz, S. Reder, M.L. de Alda, G. Gauglitz, D. Barcelo, *Biosensors and Bioelectronics*, 19 (2004) 633.
99. J. Tschmelak, M. Kumpf, N. Kappel, G. Proll, G. Gauglitz, *Talanta*, 69 (2006) 343.
100. J. Tschmelak, G. Proll, G. Gauglitz, *Analytical and Bioanalytical Chemistry*, 378 (2004) 744.

101. J. Tschmelak, G. Proll, G. Gauglitz, *Talanta*, 65 (2005) 313.
102. E. Wozei, S.W. Hermanowicz, H.Y. Holman, *Biosensors and Bioelectronics*, 21 (2006) 1654.
103. P. Hua, J.P. Hole, J.S. Wilkinson, G. Proll, J. Tschmelak, G. Gauglitz, M. Jackson, R. Nudd, J. Kaiser, P. Kraemmer, *Optics Express*, 13 (2005) 1124.
104. J. Tschmelak, M. Kumpf, G. Proll, G. Gauglitz, *Analytical Letters*, 37 (2004) 1701.
105. J. Tschmelak, G. Proll, G. Gauglitz, *Analytica Chimica Acta*, 519 (2004) 143.
106. N. Yildirim, F. Long, C. Gao, M. He, H.C. Shi, A.Z. Gu, *Environmental Science and Technology*, 46 (2012) 3288.
107. M. Usami, K. Mitsunaga, Y. Ohno, *Journal of Steroid Biochemistry and Molecular Biology*, 81 (2002) 47.
108. B. Hock, M. Seifert, K. Kramer, *Biosensors and Bioelectronics*, 17 (2002) 239.
109. T. Hahn, K. Tag, K. Riedel, S. Uhlig, K. Baronian, G. Gellissen, G. Kunze, *Biosensors and Bioelectronics*, 21 (2006) 2078.
110. M. Murata, M. Nakayama, H. Irie, K. Yakabe, K. Fukuma, Y. Katayama, M. Maeda, *Analytical Science*, 17 (2001) 387.
111. E. Dempsey, D. Diamond, A. Collier, *Biosensors and Bioelectronics*, 20 (2004) 367.
112. M. Zhihong, L. Xiaohui, F. Weiling, *Analytical Communications*, 36 (1999) 281.
113. Y.J. Kim, Y.S. Kim, J.H. Niazi, M.B. Gu, *Bioprocess and Biosystems Engineering*, 33 (2009) 31.
114. L.H. Hansen, S.J. Sorensen, *FEMS Microbiology Letters*, 190 (2000) 273.
115. Regulation (EC) No 1831/2003 of the European Parliament and of the Council of September 22, 2003 on additives for use in animal nutrition (Text with EEA relevance), *Official Journal of the European Communities*, L 268 (2003) 29.
116. A. McWhirter, L. Wahlstrom, in: B.M.R. Schasfoort, J.A. Tudos (Eds.), *Handbook of Surface Plasmon Resonance*, RSC Publishing, Cambridge, U.K. (2008), p. 333.
117. W. Haasnoot, H. Gercek, G. Cazemier, M.W. Nielen, *Analytica Chimica Acta*, 586 (2007) 312.
118. W. Haasnoot, E.E.M.G. Loomans, G. Cazemier, R. Dietrich, R. Verheijen, A.A. Bergwerff, R.W. Stephany, *Food and Agricultural Immunology*, 14 (2002) 15.
119. W. Haasnoot, G.R. Marchesini, K. Koopal, *Journal of AOAC International*, 89 (2006) 849.
120. W. Haasnoot, P. Stouten, G. Cazemier, A. Lommen, J.F. Nouws, H.J. Keukens, *Analyst*, 124 (1999) 301.
121. M. Bienenmann-Ploum, T. Korpimaki, W. Haasnoot, F. Kohen, *Analytica Chimica Acta*, 529 (2005) 115.
122. J. Keegan, M. Whelan, M. Danaher, S. Crooks, R. Sayers, A. Anastasio, C. Elliott et al., *Analytica Chimica Acta*, 654 (2009) 111.
123. J. Yuan, J. Addo, M.I. Aguilar, Y. Wu, *Analytical Biochemistry*, 390 (2009) 97.
124. J. Adrian, S. Pasche, D.G. Pinacho, H. Font, J.-M. Diserens, F. Sanchez-Baeza, B. Granier, G. Voirin, M. Pilar Marco, *Trends in Analytical Chemistry*, 28 (2009) 769.
125. F. Fernandez, K. Hegnerova, M. Piliarik, F. Sanchez-Baeza, J. Homola, M.P. Marco, *Biosensors and Bioelectronics*, 26 (2010) 1231.
126. Y.-K. Kim, H. Kim, *Journal of Industrial and Engineering Chemistry*, 15 (2009) 229.
127. J. Melendez, R. Carr, D.U. Bartholomew, K. Kukanskis, J. Elkind, S. Yee, C. Furlong, R. Woodbury, *Sensors and Actuators B*, 35 (1996) 212.
128. J. Hu, F. Luo, W. Li, G. Jiang, Z. Li, R. Zhang, *Biosensors and Bioelectronics*, 24 (2009) 1974.

129. A.N. Naimushin, S.D. Soelberg, D.K. Nguyen, L. Dunlap, D. Bartholomew, J. Elkind, J. Melendez, C.E. Furlong, *Biosensors and Bioelectronics*, 17 (2002) 573.
130. Analytical microsystems (Department of Mivitec GmbH), Biosuplar Specifications, http://www.biosuplar.com (2013).
131. Fagor Electrónica, S. Coop. home page, Sensia Products Specifications, http://www.sensia.es (2013).
132. L. Cao, H. Lin, V.M. Mirsky, *Analytica Chimica Acta*, 589 (2007) 1.
133. Reichert Technologies home page, SPR Systems, http://reichertspr.com (2013).
134. Y.-B. Shin, H.M. Kim, Y. Jung, B.H. Chung, *Sensors and Actuators B*, 150 (2010) 1.

18

Optical Biosensors in Food Safety and Control

Theodoros Varzakas, Georgia-Paraskevi Nikoleli,
Nikolaos Tzamtzis, and Dimitrios P. Nikolelis

CONTENTS

18.1 Introduction

These types of sensors are based on measuring responses to light emission or to illumination. Optical biosensors are based on well-founded methods including fluorescence, phosphorescence, light absorbance, photothermal techniques, chemiluminescence, surface plasmon resonance (SPR), total internal reflectance, light rotation, and polarization and could employ a number of techniques to detect the presence of a target analyte. As an example, these technical usages have been demonstrated to detect the presence of allergens, particularly peanuts, during food production [1,2].

18.2 Optical Biosensors

The development of optical-fiber sensors during recent years is related to two of the most important scientific advances: the laser and modern low-cost optical fibers. Recently, optical fibers have become an important part of sensor technology. Their use as a probe or as a sensing element is increasing in clinical, pharmaceutical, industrial, and military applications. Excellent light delivery, long interaction length, low cost, and ability not only to excite the target molecules but also to capture the emitted light from the targets are the main points in favor of the use of optical fibers in biosensors. Optical fibers transmit light on the basis of the principle of total internal reflection (TIR). Fiber-optic biosensors are analytical devices in which a fiber-optic device serves as a transduction element. The usual aim is to produce a signal that is proportional to the concentration of a chemical or biochemical to which the biological element reacts. Fiber-optic biosensors are based on the transmission of light along silica glass fiber or plastic optical fiber (POF) to the site of analysis. Optical-fiber biosensors can be used in combination with different types of spectroscopic technique, for example, absorption, fluorescence, phosphorescence, and SPR. Optical biosensors based on the use of fiber optics can be classified into two different categories: intrinsic sensors, where interaction with the analyte occurs within an element of the optical fiber, and extrinsic sensors, in which the optical fiber is used to couple light, usually to and from the region where the light beam is influenced by the measurand [3].

Organophosphorus (OP) neurotoxins comprise a unique class of contaminants and chemical warfare (CW) agents, which have high acute toxicity. These neurotoxins are powerful inhibitors of esterase enzymes, such as acetyl- and butyrylcholinesterases or neurotoxic esterase. Simple methods for OP neurotoxin detection using fluorescence assays based on specific recognition of OP by organophosphate hydrolase (OPH) enzyme have been developed.

A biosensor for direct detection of OP neurotoxins such as paraoxon has been reported by Simoniana et al. [4].

The biosensing method has been based on the change in fluorescence of a competitive inhibitor of the OPH enzyme when the inhibitor is displaced by the OP substrate. The change in fluorescence intensity was correlated with the concentration of paraoxon presented in the solution. The sensitivity to paraoxon was obtained when enzyme inhibitor and OPH–gold nanoparticle (NP) conjugates were present at near equimolar levels.

An optical glutamate biosensor test strip based on stacked membranes of Nafion/solgel (bottom layer) and chitosan (uppermost layer) was fabricated on a piece of paper as support to form a test strip by Muslim et al. [5]. The use of a stacked membrane system allows multiple immobilizations of sensing components directly without any covalent attachment via straightforward procedures. The uppermost membrane consisted of immobilized enzymes L-glutamate oxidase (GLOD) and horseradish peroxidase (HRP), which sensed the presence of L-glutamate and the bottom membrane contained the indicator dye, 3,3',5,5'-tetramethylbenzidine (TMB). The test strip can be used to measure L-glutamate quantitatively by observing a color change from light green to dark green with increasing L-glutamate concentrations. Quantitative analysis could be performed by measuring the reflectance intensity of the color change at 550 nm. The glutamate biosensor test strip gave a linear response range of 0.01–0.30 mM to L-glutamate with a limit of detection (LOD) of 5 µM. The strips were successfully applied for the estimation of L-glutamate in common food items such as sauces, soups, processed food, and flavor enhancers. The results of the analysis of L-glutamate in various food samples using the glutamate test strip were comparable to a standard procedure employing HPLC method.

The uniqueness of this optical biosensor design is a simultaneous immobilization of several sensing agents (one indicator dye, TMB, and two enzymes, i.e., GLOD and HRP) via a stacked membrane system without any covalent attachment of the sensing components. Avoiding covalent attachment of sensing materials on the test strip support enables a simple and straightforward fabrication procedure for the device apart from preventing deactivation of the enzymes during chemical attachment [6]. In addition to that, the application of reflectance spectrophotometry in the detection of color change of the test strip in the presence of L-glutamate has the advantage of not only allowing opaque support material to be used for the test strip construction but also enabling nontransparent sample matrices, especially food samples to be analyzed directly without the need of any sample pretreatments.

POF biosensors consist in a viable alternative for rapid and inexpensive scheme for detection. In order to study the sensitivity of tapers for microbiological detection, geometric parameters are studied, such as the taper waist diameter, since the formation of taper regions is the key sensing element in this particular type of sensors. Beres et al. [7] prepared a series of POF taper sensors using a specially developed tapering machine, and the dispersion of geometric dimensions is evaluated, aiming to achieve the best tapering characteristics that will provide a better sensitivity on the sensor response. The fiber tapers that presented the finest results were those constructed in U-shaped (bended) configurations, with taper waist diameters ranging from 0.40 up to 0.50 mm. These fiber tapers were used as the main section of the monitoring device, when chemically treated as immunosensors for the detection of bacteria, yeast, and erythrocytes [7]. A variety of studies approach the application of straight silica optical fibers in the manufacture of tapered sensors; nevertheless, the employment of U-shaped tapers in POF can afford several advantages, such as increased sensitivity, smaller taper length, economic use of reagents, an improved handling and fabrication, and a greater mechanical resistance [8,9].

Fiber-optic sensors can be combined with antibodies that are able to recognize and bind to a defined antigen, which induces immediate environmental changes, such as the refraction index (RI), around the probe containing the antibody. Also, the large diameter of POFs facilitates installation and alignment, unlike their glass counterparts in which a few micron misalignment results in heavy losses. Other well-known advantages are the efficient light coupling owing to the large numerical aperture, high ductility, low cost of production, and easy handling.

18.3 Surface Plasmon Resonance Biosensors

SPR is a modern analysis technique based on the changes in the refractive index of material on the metal surface. In 1982, Liedberg first realized the biosensing potential of a prism SPR sensor with an immunoglobulin G (IgG) antibody adsorbed overlayer on the gold-sensing film, which allowed selective binding detection of IgG [10]. Originally, the SPR technique was applied in the analysis of gases, liquids, and solids. In recent years, SPR technique has an increasing application in biochemistry [11,12], clinical diagnosis [13], food analysis [14], environmental monitoring [15], and so on. The popularization of SPR is due to its properties of label-free, real-time detection and high sensitivity. SPR can not only be used in analyte detection but also provide rich information on the specificity, affinity, and kinetics of biomolecular interactions. In the last 10 years, much attention has focused on the detection of low molecular weight analyses in food and environmental fields using SPR biosensor. In present review, we address the basic principle of SPR, the existing detection methods, and the progress on mycotoxin detection using SPR biosensor.

SPR is a physical optics phenomenon based on the change in the refractive index on the metal surface. Several reviews have described the basic principle and operation of SPR [16,17] plane-polarized light that shines directly through the prism to the metal/solution dielectric interface over a wide range of incident angles; the evanescent wave will be generated under TIR condition. At a selected incident light wavelength or angle, the evanescent waves can resonate with surface plasmons (SPs) produced by free electrons on the metal film of the sensor surface, and the energy of incident light will be absorbed by SP, which results in a narrow dip in the spectrum of reflected light. The angle at which the drop is maximum (minimum of reflectivity) is denoted as the "SPR angle." This "SPR angle" is extremely sensitive to the refractive index of the sample contacting with the metal surface, so that it is also highly influenced by the species and amount of biomolecules immobilized on the gold layer. Furthermore, the kinetics information of the interaction between molecules can also be obtained [18].

As one of the relatively new analytical techniques, SPR has been proven particularly advantageous for rapid, label-free, sensitive analyte detection. Using SPR, qualitative and quantitative analysis can be performed in real time. Mycotoxins are a group of small, toxic products formed as secondary metabolites by a few fungal species. They can contaminate foodstuffs on a large scale and consequently threaten human health through food chain. Thus, rapid, sensitive, and selective determination of mycotoxin is of great significance for the food safety. This contribution addresses the basic principle of SPR, the existing detection methods, and the progress on mycotoxin detection using SPR biosensor [19].

SPR biosensing has matured into a valuable analytical technique for measurements related to biomolecules, environmental contaminants, and the food industry. Contemporary SPR instruments are mainly suitable for laboratory-based measurements. However, several point-of-measurement applications would benefit from simple, small, portable, and inexpensive sensors to assess the health condition of a patient, potential environmental contamination, or food safety issues. This Trend article by Breault-Turcot and Masson [20] explores nanostructured substrates for improving the sensitivity of classical SPR instruments and NP-based colorimetric substrates that may provide a solution to the development of point-of-measurement SPR techniques. Novel nanomaterials and methodology capable of enhancing the sensitivity of classical SPR sensors are destined to improve the limits of detection of miniature SPR instruments to the level required for most applications. In a different approach, paper- or substrate-based SPR assays based on NPs are a highly promising topic of research that may facilitate the widespread use of a novel class of miniature and portable SPR instruments [20].

Nanotechnology involves the characterization, fabrication, and/or manipulation of structures, devices, or materials that are between 1 and 100 nm in size. One of the major advantages of using nanomaterials for biosensing is due to their large surface area, allowing a greater number of biomolecules to be immobilized, and this consequently increases the number of reaction sites available for interaction with a target species. This property, coupled with excellent electronic and optical properties, facilitates the use of nanomaterials in "label-free" detection and in the development of biosensors with enhanced sensitivities and improved response times [21].

The advances in the manipulation of nanomaterials have permitted the development of nanobiotechnology with enhanced sensitivities and improved response times. Low levels of infection of the major pathogens require the need for sensitive detection platforms, and the properties of nanomaterials make them suitable for the development of assays with enhanced sensitivity, improved response time, and increased portability.

Nanobiotechnologies focusing on the key requirements of signal amplification and preconcentration for the development of sensitive assays for foodborne pathogen detection in food matrices have been described and evaluated by Gilmartin and O'Kennedy [22]. The potential that exists for the use of nanomaterials as antimicrobial agents has also been examined by the same authors.

Cantilevers function by detecting differences in the stress, forces, or vibration frequency, occurring when molecules bind to the surface. For example, bacteria can be detected due to the additional mass resulting from the binding of specific antibodies immobilized on the cantilever surface and the bacterial cell. There are several reports of sensitive detection of bacteria using cantilever technology such as a single *Listeria innocua* cell [23] and *Escherichia coli* O157:H7 in spinach, spring lettuce mix, and ground beef (LOD 100 CFU/mL) [24] and an antibody-functionalized piezoelectric-excited millimeter-sized cantilever sensor capable of identification of 1 *E. coli* O157:H7 cell/mL [25].

However, differences in pathogen adherence to food matrices, which affect target binding to the sensor surface, can affect sensitivity [26]. The scaling down of this technology to the nanoscale has increased the capacity of nano-cantilevers for ultrasensitive, faster detection due to higher frequencies and better mass resolution.

SPR technology has been used for the detection of foodborne pathogens using gold surfaces. Gold nanorods (NRs) are elongated NPs with distinct optical properties that depend on their shape. Compared to the single plasmon adsorption band of NPs, the excitation of SPs by light in NRs can be seen as two plasmon absorption bands, one corresponding to light absorption and scattering along the width of the particle and the other along the length of the particle. Changes in the aspect ratio (ratio of the width of the object to its length) of an NR give rise to plasmon adsorption bands at different positions; hence, different-sized NRs can be used as labels in a multiplex assay. The position and intensity of these bands can be affected by changes, for example, binding events, in the dielectric constant around the vicinity of these NRs, known as localized SPR (LSPR) or nanoSPR [27].

The sensing capacity of the detection systems is being improved lastly by using nanomaterials such as magnetic nanoparticles (MNPs), carbon nanotubes (CNTs), NRs, quantum dots (QDs), nanowires (NWs), and nanochannels (NCs). These nanomaterials, used in electrical biosensors, have a very high capacity for charge transfer, which makes them suitable to reach lower detection limits and higher sensitivity values. Nanomaterials can contribute as labels or transducer modifiers so as to improve the performance of the biosensor. Some of the reported nanomaterials are QDs. QDs are crystalline clusters with a nanosize [28] that can be synthesized from semiconductor materials (e.g., cadmium sulfide [29], cadmium selenide [30], cadmium telluride [31], indium phosphide [32], or gallium arsenide [33]).

"Nanosized" and nanomaterial-based biosensors, called also nanobiosensors, are a modern and efficient class of detection systems [34–36]. The application of nanobiosensors in food industry could lead to immense improvements in quality control, food safety, and traceability.

Nanobiosensors can achieve very low detection limits (even single molecule or cell). In addition, they offer multidetection possibilities and may ensure a high stability (i.e., NPs such as QDs are more stable than enzymes or fluorescence dyes). The main advantage besides the reduction of reagent volumes, detection time, and keeping the same sensitivity is the user-friendly applicability: there is no need for professional users. The idea is to develop one-push button like devices that can give a fast "yes–no" response or ensure a similar simple communication with the end user.

Although some interesting nanobiosensors based on the use of NPs and techniques such as optical microscopy (i.e., based in light absorption, scattering, and fluorescence of NPs) and electrochemistry (i.e., stripping analysis, potentiometry) have been developed and reported in several publications

[29,35,37–42], Perez-Lopez and Mercoci [43] gave a general overview of some of the most important nanomaterial-based biosensing systems based on various detection technologies and applied in food field. In addition, they revised and gave opinions on the current status of detection systems and the obstacles and some suggestions for the future development of this technology.

Previous studies [44,45] employed Biacore Q SPR biosensor and CM5 sensor chips from Biacore AB (Uppsala, Sweden). Although they produced good results, their instrument and chips were still expensive.

An SPR biosensor inhibition immunoassay for determination of ractopamine (Rac) residue in pork was constructed by immobilizing Rac derivative on the SPR-2004 biosensor chip. After extraction with perchloric acid, pork sample was cleaned by ethyl acetate and analyzed by SPR-2004 biosensor. The LOD was 0.6 μg/kg for pork sample. Recoveries of Rac were higher than 80% with relative standard deviations below 10%. Although the same pretreatment was applied for both the UPLC–MS/MS and the biosensor, the biosensor showed little matrix interference by constructing pure solution and matrix-match calibration curves. Compared with Biacore Q made by Biacore AB, SPR-2004 biosensor made by Chinese Academy of Sciences is of low cost and showed minimal matrix interferences. Accordingly, this biosensor was a promising screening instrument to be used for the detection of Rac in supermarkets, food factories, and food regulatory organizations.

One of the most successful optical-based biosensor systems introduced has been the range of instruments supplied by Biacore (Uppsala, Sweden). This instrument can be employed to study a wide range of biological interactions, both automatically and in real time. The instrument is based on SPR, whereby biomolecular binding events cause changes at a metal/liquid interface, usually involving a complex that includes a specific antibody against a target analyte. On binding, these changes (in the refractive index) are recognized by a shift in the SPR signal, indicating a presence of the target analyte in a sample solution. One particular advantage for sensors based on SPR is that the system does not require the presence of a labeled ligand (e.g., enzyme conjugated antibody) to function [46]. SPR sensor systems have been used extensively to investigate the presence of harmful contaminating microorganisms in food and to determine food quality. For example, an optically based biosensor was recently used to screen poultry liver and eggs for the presence of the drug nicarbazin, a feed additive used to prevent outbreaks of coccidiosis in boiler chickens [47]. The limits of detection for the sensor system were 17 and 19 ng/g for liver and eggs, respectively.

Panagopoulou et al. [48] reported for the first time the development of kappa-casein (κ-CN)-based electrochemical and SPR biosensors for the assessment of the clotting activity of rennet. κ-CN is the "active" component of milk micelles. Using κ-CN as a sensing layer, they managed to overcome problems associated with the short storage stability of the artificial casein micelle (ACM) stock suspension (1 week) and the ACM size-dependent response of the resulting biosensors.

The electrochemical biosensors were developed over gold electrodes modified with a self-assembled monolayer of dithiobis-N-succinimidyl propionate, while SPR measurements were performed on regenerated carboxymethylated dextran gold surfaces. In both types of biosensor, κ-CN molecules were immobilized onto modified gold surfaces through covalent bonding. In electrochemical biosensors, interactions between the immobilized κ-CN molecules and chymosin (the active component of rennet) were studied by performing cyclic voltammetry, differential pulsed voltammetry, and electrochemical impedance spectroscopy (EIS) measurements, using hexacyanoferrate(II)/(III) couple as a redox probe. κ-CN is cleaved by rennet at the Phe105–Met106 bond, producing a soluble glycomacropeptide, which is released to the electrolyte, and the positively charged insoluble para-κ-casein molecule, which remains attached to the surface of the electrode. This induced reduction of the net negative charge of the sensing surface, along with the partial degradation of the sensing layer, results in an increase of the flux of the redox probe, which exists in the solution, and, consequently, to signal variations, which are associated with the increased electrocatalysis of the hexacyanoferrate(II)/(III) couple on the gold surface. SPR experiments were performed in the absence of the redox probe, and the observed SPR angle alterations were solely attributed to the cleavage of the immobilized κ-CN molecules. Various experimental variables were investigated, and under the selected conditions, the proposed biosensors were successfully tried to real samples. The ratios of the clotting power units in various commercial solid or liquid samples, as they are calculated by the EIS-based data, were almost identical to those obtained with a reference method. In addition, EIS measurements showed an excellent reproducibility, lower than 5%.

Pantothenic acid (PA), vitamin B5, is an essential B vitamin that may be fortified in food and as such requires robust and accurate methods of detection to meet compliance legislation. Haughey et al. [49] reported the production and characterization of the first monoclonal antibody (MAb) specific for PA and the subsequent development of an SPR biosensor assay for the quantification of PA. The developed assay was compared with an SPR-based commercial kit, which utilized a polyclonal antibody (PAb). Foodstuffs, including cereals (n = 43), infant formulas, and baby food (n = 10) and fruit juices (n = 48), were analyzed by both the MAb and PAb biosensor assays, and comparison plots showed good correlation (R^2 0.77–0.99). The results indicate that the MAb-based biosensor assay is suitable for the measurement of PA in foodstuffs and has the added advantage of facilitating a constant, long-term supply of identical antibody. Preliminary matrix studies suggest that the MAb-based assay is an excellent candidate for further validation studies and routine quality assurance-based analysis.

18.4 Optical Biosensor Test Strips

An optical glutamate biosensor test strip based on stacked membranes of Nafion/solgel (bottom layer) and chitosan (uppermost layer) was fabricated on a piece of paper as support to form a test strip as reported by Muslim et al. [5]. The use of a stacked membrane system allows multiple immobilizations of sensing components directly without any covalent attachment via straightforward procedures. The uppermost membrane consisted of immobilized enzymes GLOD and HRP, which sensed the presence of L-glutamate, and the bottom membrane contained the indicator dye, 3,3',5,5'-TMB. The test strip can be used to measure L-glutamate quantitatively by observing a color change from light green to dark green with increasing L-glutamate concentrations. Quantitative analysis could be performed by measuring the reflectance intensity of the color change at 550 nm. The glutamate biosensor test strip gave a linear response range of 0.01–0.30 mM to L-glutamate with a LOD of 5 µM. The strips were successfully applied for the estimation of L-glutamate in common food items such as sauces, soups, processed food, and flavor enhancers. The results of the analysis of L-glutamate in various food samples using the glutamate test strip were comparable to a standard procedure employing HPLC method.

The underlined concept for the design of biosensor for L-glutamate analysis is based on the catalysis of L-glutamate to form hydrogen peroxide by the enzyme GLOD, which occurs in the uppermost layer of the stacked membrane of the test strip. The H_2O_2 generated in the presence of HRP will then catalyze the oxidation of TMB in the second membrane via electron transfer from the reduced TMB (light green) to the H_2O_2 to yield the oxidized TMB (dark green) product.

18.5 Advantages and Disadvantages of Labeled and Unlabeled Binding Proteins in Optical Biosensors

There are advantages and disadvantages in the use of fluorescently labeled binding proteins (antibodies and receptors) and aptamers over unlabeled equivalents:

- Advantages include the ability to improve the sensitivity of the test by choosing a dye with an intense signal and/or changing the ratio of dye to binding protein. A larger ratio should create an improvement in the relative sensitivity since the signal from the binding of a smaller amount of antibody will be increased.

- Disadvantages include an additional labeling step that needs to be well characterized to provide a reproducible result using different batches of labeled binding protein.

Contamination may also become a problem. In certain samples, such as bile, it may not be possible to use labeled binding proteins since bile itself contains compounds that will fluoresce [50].

References

1. Rana, J.S., J. Jindal, V. Beniwal, and V. Chhokar. 2010. Utility biosensors for applications in agriculture—A Review. *Journal American Science* 6(9): 353–375.
2. Otles, S. and B. Yalcin. 2012. Review on the application of nanobiosensors in food analysis. *Acta Scientiarum Polonorum Technologia Alimentaria* 11(1): 7–18.
3. Bosch, M.E., A.J. Ruiz Sánchez, F. Sánchez Rojas, and C. Bosch Ojeda. 2007. Recent development in optical fiber biosensors. *Sensors* 7: 797–859.
4. Simoniana, A.L., T.A. Good, S.-S. Wang, and J.R. Wild. 2005. Nanoparticle-based optical biosensors for the direct detection of organophosphate chemical warfare agents and pesticides. *Analytica Chimica Acta* 534: 69–77.
5. Muslim, N.Z.M., M. Ahmad, L.Y. Heng, and B. Saad. 2012. Optical biosensor test strip for the screening and direct determination of L-glutamate in food samples. *Sensors and Actuators B* 161: 493–497.
6. Wong, F.C.M., M. Ahmad, L.Y. Heng, and L.B. Peng. 2006. An optical biosensor for dichlovos using stacked sol–gel films containing acetylcholinesterase and a lipophilic chromoionophore. *Talanta* 69: 888–893.
7. Beres, C., F.V.B. de Nazare, N.C.C. de Souza, M.A.L. Miguel, and M.M. Werneck. 2011. Tapered plastic optical fiber-based biosensor—Tests and application. *Biosensors and Bioelectronics* 30: 328–332.
8. Frazao, O., J.M. Baptista, J.L. Santos, and P. Roy. 2008. *Applied Optics* 47(13): 2520–2523.
9. de Nazare, F.V.B., C. Beres, N.C.C. Souza, M.M. de Werneck, M.A.L. Miguel, 2011. *2011 IEEE International Instrumentation and Measurement Technology Conference Proceedings*, Hangzhou, China, pp. 959–963.
10. Liedberg, B., C. Nylander, and I. Lundstrom. 1983. A new carrier-domain magnetometer. *Sensors and Actuators* 4: 229–236.
11. Green, R.J., R.A. Frazier, K.M. Shakesheff, M.C. Davies, C.J. Roberts, and S.J.B. Tendler. 2000. Surface plasmon resonance analysis of dynamic biological interactions with biomaterials. *Biomaterial* 21: 1823–1835.
12. Homola, J., S.S. Yee, and G. Gauglitz. 1999. Surface plasmon resonance sensors: Review. *Sensors and Actuators B: Chemical* 54: 3–15.
13. Kanoh, N., M. Kyo, K. Inamori, A. Ando, A. Asami, A. Nakao et al. 2006. SPR imaging of photo-cross-linked small-molecule arrays on gold. *Analytical Chemistry* 78: 2226–2230.

14. Spadavecchia, J., M.G. Manera, F. Quaranta, P. Siciliano, and R. Rella. 2005. Surface plasmon resonance imaging of DNA based biosensors for potential applications in food analysis. *Biosensors and Bioelectronics* 21: 894–900.

15. Inamori, K., M. Kyo, Y. Nishiya, Y. Inoue, T. Sonoda, E. Kinoshita, T. Koike, and Y. Katayama. 2005. Detection and quantification of on-chip phosphorylated peptides by surface plasmon resonance imaging techniques using a phosphate capture molecule. *Analytical Chemistry* 77:3979–3985.

16. Hodnik, V. and G. Anderluh. 2009. Review: Toxin detection by surface plasmon resonance. *Sensors* 9: 1339–1354.

17. Shankaran, R., K.V. Gobi, and N. Miura. 2007. Recent advancements in surface plasmon resonance immunosensors for detection of small molecules of biomedical, food, and environmental interest. *Sensors and Actuators B: Chemical* 121: 158–177.

18. Huang, Y., M.C. Bell, and I.I. Suni. 2008. *Analytical Chemistry* 80: 9157–9162.

19. Li, Y., X. Liu, and Z. Lin. 2012. Recent developments and applications of surface plasmon resonance biosensors for the detection of mycotoxins in foodstuffs. *Food Chemistry* 132(3): 1549–1554.

20. Breault-Turcot, J. and J.-F. Masson. 2012. Nanostructured substrates for portable and miniature SPR biosensors. *Analytical and Bioanalytical Chemistry* 403(6): 1477–1484.

21. Xu, K., J. Huang, Z. Ye, Y. Ying, and Y. Li. 2009. Recent development of nanomaterials used in DNA biosensors. *Sensors* 9(7): 5534–5557.

22. Gilmartin, N. and R. O'Kennedy. 2012. Nanobiotechnologies for the detection and reduction of pathogens. *Enzyme and Microbial Technology* 50: 87–95.

23. Li, B., A.D. Ellington, and X. Chen. 2011. Rational, modular adaptation of enzyme-free DNA circuits to multiple detection methods. *Nucleic Acids Research* 16: e110–e114.

24. Gupta, A., D. Akin, and R. Bashir. 2004. Detection of bacterial cells and antibodies using surface micromachined thin silicon cantilever resonators. *Journal of Vacuum Science Technology B* 6: 2785–2791.

25. Maraldo, D. and R. Mutharasan. 2007. Preparation-free method for detecting *E. coli* O157:H7 in the presence of spinach, spring lettuce mix, and ground beef particulates. *Journal of Food Protection* 11: 2651–2655.

26. Campbell, G.A. and R. Mutharasan. 2007. A method of measuring *E. coli* O157:H7 at 1 cell/mL in 1 liter sample using antibody functionalized piezoelectric-excited millimeter-sized cantilever sensor. *Environmental Science Technology* 5: 1668–1674.

27. Perez-Juste, J., I. Pastoriza-Santos, L. Liz-Marzan, and P. Mulvaney. 2005. Gold nanorods: Synthesis, characterization and applications. *Coordination Chemistry Reviews* 1(7–18): 1870–1901.

28. Murphy, C.J. 2002. Optical sensing with quantum dots. *Analytical Chemistry* 74: 520–526.

29. Merkoci, A., S. Marın, M.T. Castaneda, M. Pumera, J. Ros, and S. Alegret. 2006. Crystal and electrochemical properties of water dispersed CdS nanocrystals obtained via reverse micelles and arrested precipitation. *Nanotechnology* 17: 2553–2559.

30. Steigerwald, M.L. and L.E. Brus. 1990. Semiconductor crystallites: A class of large molecules. *Accounts of Chemical Research* 23: 183–188.

31. Eychmuller, A. and A.L. Rogach. 2000. Chemistry and photophysics of thiol-stabilized II-VI semiconductor nanocrystals. *Pure and Applied Chemistry* 72: 179–188.

32. Guzelian, A.A., Katari, J.E.B., Kadavanich, A.V., Banin, U., Hamad, K., Juban, E., et al. 1996. Synthesis of size-selected, surface-passivated InP nanocrystals. *Journal of Physical Chemistry* 100: 7212–7219.

33. Olshavsky, M.A., A.N. Goldstein, and A.P. Alivisatos. 1990. Organometallic synthesis of gallium-arsenide crystallites, exhibiting quantum confinement. *Journal of the American Chemical Society* 112: 9438–9439.

34. Sanvicens, N., C. Pastells, N. Pascual, and M.-P. Marco. 2009. Nanoparticle-based biosensors for detection of pathogenic bacteria. *Trends in Analytical Chemistry*, 28(11): 1243–1252.

35. Lin, Y.-H., S.-H. Chen, Y.-C. Chuang, Y.-C. Lu, T. Y. Shen, C. A. Chang et al. 2008. Disposable amperometric immunosensing strips fabricated by Au nanoparticles-modified screenprinted carbon electrodes for the detection of foodborne pathogen *Escherichia coli* O157:H7. *Biosensors and Bioelectronics* 23: 1832–1837.

36. Delmulle, B.S., S.M. De Saeger, L. Sibanda, I. Barna-Vetro, and C.H. Van-Peteghem. 2005. Development of an immunoassay-based lateral flow dipstick for the rapid detection of aflatoxin B1 in pig feed. *Journal of Agricultural and Food Chemistry* 53: 3364–3368.

37. Ambrosi, A., de la Escosura-Muñiz, A., Castañeda, M.T., and Merkoçi. A. 2009. Gold Nanoparticles: A Versatile Label for Affinity Electrochemical Biosensors. Chapter 6. In *Biosensing Using Nanomaterials* (ed A. Merkoçi), John Wiley & Sons, Inc., Hoboken, NJ. doi:10.1002/9780470447734.ch6.

38. Antiochia, R., I. Lavagnini, and F. Magno. 2004. Amperometric mediated carbon nanotube paste biosensor for fructose determination. *Analytical Letters* 37(8): 1657–1669.

39. De la Escosura-Muniz, A., A. Ambrosi, and A. Merkoci. 2008. Electrochemical analysis with nanoparticle-based biosystems. *Trends in Analytical Chemistry* 27(7): 568–584.

40. Dungchaia, W., W. Siangprohb, W. Chaicumpac, P. Tongtawed, and O. Chailapakula. 2008. Salmonella typhi determination using voltammetric amplification of nanoparticles: A highly sensitive strategy for metalloimmuno-assay based on a copper-enhanced gold label. *Talanta* 77: 727–732.

41. Ozdemir, C., F. Yeni, D. Odaci, and S. Timur. 2010. Electrochemical glucose biosensing by pyranose oxidase immobilized in gold nanoparticle-polyaniline/AgCl/gelatin nanocomposite matrix. *Food Chemistry* 119: 380–385.

42. Yang, L.J. and Y.B. Li. 2005. Quantum dots as fluorescent labels for quantitative detection of Salmonella typhimurium in chicken carcass wash water. *Journal of Food Protection* 68(6): 1241–1245.

43. Perez-Lopez, B. and A. Merkoci. 2011. Nanomaterials based biosensors for food analysis applications. *Trends in Food Science & Technology* 22: 625–639.

44. Shelver, W.-L. and D.-J. Smith. 2003. Determination of ractopamine in cattle and sheep urine samples using an optical biosensor analysis: Comparative study with HPLC and ELISA. *Journal of Agricultural and Food Chemistry* 51(13): 3715–3721.

45. Thompson, C.-S., S.-A. Haughey, I.-M. Traynor, T.-L. Fodey, C.-T. Elliott, J.-P. Antignac et al. 2008. Effective monitoring for ractopamine residues in samples of animal origin by SPR biosensor and mass spectrometry. *Analytic Chimica Acta* 608(2): 217–225.

46. Terry, L.A., S.F. White, and L.J. Tigwell. 2005. The application of biosensors to fresh produce and the wider food industry. *Journal of Agricultural and Food Chemistry* 53: 1309–1316.
47. McCarney, B., I.M. Traynor, T.L. Fodey, S.R.H. Crooks, and C.T. Elliot. 2003. Surface plasmon resonance biosensor screening of poultry liver and eggs for nicarbazin residues. *Analytica Chimica Acta* 483: 165–169.
48. Panagopoulou, M.A., D.V. Stergiou, I.G. Roussis, G. Panayotou, and M.I. Prodromidis. 2012. Kappa-casein based electrochemical and surface plasmon resonance biosensors for the assessment of the clotting activity of rennet. *Analytica Chimica Acta* 712: 132–137.
49. Haughey, S.A., C.T. Elliott, M. Oplatowska, L.D. Stewart, C. Frizzell, and L. Connolly. 2012. Production of a monoclonal antibody and its application in an optical biosensor based assay for the quantitative measurement of pantothenic acid (vitamin B5) in foodstuffs. *Food Chemistry* 134: 540–545.
50. McGrath, T.F., C.T. Elliott, and T.L. Fodey. 2012. Biosensors for the analysis of microbiological and chemical contaminants in food. *Analytical and Bioanalytical Chemistry* 403: 75–92.

19

Biosensors in Express Control of Quality Assurance of Products

Nickolaj F. Starodub, Julia O. Ogorodnijchuk,
Nelja F. Slishek, and Mykola Mel'nichenko

CONTENTS

19.1 Introduction

Currently, the quality and safety characteristics of food products are becoming a priority due to deterioration of the environment and as a result of reducing conditions of sanitary.

Questions of methodology and characteristics of analysis methods used to control the quality are also associated with the solution of the complex problems of the material—both technical and economic, as well as the duration of the pilot studies. Ways to test the quality of food systems are associated with many indicators; they are time-consuming and, above all, do not always take into account the risk of hazards, changes in quality, and biological activity of the product occurring during processing and storage. At the same time, the integrated safety feature of raw materials, their products, and monitoring the formation of the quality of food systems are very important.

The massive impact of anthropogenic origin for many decades has led to the appearance of about 50,000–60,000 species of xenobiotics in the biosphere, which are involved in the natural cycle of substances. As they undergo various changes—oxidization, hydroxylation, hydrolysis, isomerization, and transformations—they form a huge number of substances that act on the body. To predict and identify each of this set of compounds, which are often more toxic than the original substance, is almost impossible. In addition, the level of contamination of food and feed products by various xenobiotics depends not only on the content of the external environment but also on the technological processes. Among the different chemical substances, numerous problems arise in the determination of food quality and results in bacterial contamination. This is connected with the food production technology as well as with preservation of products in the markets and in home conditions. At any time, it is an actual problem that requires immediate and effective solutions.

Therefore, to prevent the fully integrated effect of food on the body there is a necessity for improving the methods of quality control and safety. The application of methods based on the principles of biosensorics is most important in modern practice demands. This chapter is devoted to considering the efficiency of different types of biosensors for the control of such microorganisms as *Salmonella* as well as for revealing strong and widespread toxins that are produced by big groups of fungi affecting many plants. It is important to mention that plant products are the dominant components of human diet.

19.2 Bacterial Contamination of Foods and Their Express Control

Bacteria, viruses, and other microorganisms are a big concern in different branches of industry such as food production, veterinary medicine, and health care. Some of them are pathogenic for people and animals and become extremely dangerous because of a few more reasons: first of all, it is of course their small size, which makes them invisible and difficult to identify, second their wide prevalence in the environment, hardness to the live conditions, a high rate of reproduction, survival, and fast adaptation. Considering these characteristics, microorganisms also have all chances to be used as biological weapons to threaten the security of the country or regions.

Conventional methods remain the most common for microorganism detection despite their long turnover times because of their high selectivity and sensitivity. Recently, other modern laboratory techniques and a wide range of their modifications have found frequent use because of shorter time of fulfillment. Still these methods require previous sample preparation, they are multistep, and demand expensive lab equipment, reagents, and qualified staff.

Biosensors are very promising analytical devices for microorganism detection and identification, which have the potential to shorten the time between sample uptake and results and allow diagnosis of pathogens in real-time, highly sensitive, and specific format and serve as easy to use and low-cost devices. Several branches of industry sharply require equipment of this kind and can provide a good market of their application. Highly sensitive and reliable biosensor systems have great prospects of application in medical diagnostics, food quality control, environmental monitoring, defense, and other industries especially in the case of multichannel devises that could provide multiple analyte analysis simultaneously.

19.2.1 Conventional Methods for Bacteria Detection

Conventional methods despite their disadvantages are the most popular methods for bacteria detection because of their high sensitivity, selectivity, and reliability. Traditional methods for the detection of bacteria include the following basic steps: preenrichment, selective enrichment, biochemical screening, and serological confirmation.

The most common of conventional methods are culture and colony counting (plate) techniques, which remain the standard detection methods.

These methods are based on morphological evaluation (microscopy) of the microorganisms and include tests for the detection of microorganisms' ability to grow in selective media under a variety of conditions. Different selective media can contain inhibitors that suppress the growth of nontargeted species or include particular substances that only the targeted species can degrade or change the color under the influence of bacteria vital activity. This method provides a basis for the approximate diagnosis and usually requires further confirmation with the help of the bacteriological and serological methods. The last ones include the following reactions: agglutination, precipitation, bacteriolysis, hemolysis, complement fixation, and others. Bacteriological method is particularly important in the diagnosis of most infectious diseases, and it demands obtaining a pure culture of the causative agent of disease and further investigations. This method is key in the diagnosis of bacterial diseases (Zarickiy, 1988; Laboratory diagnostics of salmonellosis, 2010; Odumeru and León-Velarde, 2012).

Although standard microbiological techniques allow the detection of single bacteria, this process is relatively time-consuming because of the growth of a single cell into a colony. For example, in the case of Salmonella microbiological method includes 16–20 h preenrichment, 24 h Salmonella-specific enrichment, 24–28 h identification step, and 3–6 days for serological confirmation in case of positive outcome (Schneider et al., 2002). For Campylobacter it takes 4–9 days to obtain a negative result and between 14 and 16 days for positive result confirmation (Lazcka et al., 2007). Such insuperable disadvantages make these approaches very inconvenient for industrial application especially in food and medicine sphere and inapplicable for long-term monitoring.

19.2.2 Modern Methods of Laboratory Diagnosis of Microorganisms

Among modern methods of infection detection, we focus our attention on the most widely used approach based on polymerase chain reaction (PCR), immunochemical ways of identification of the infectious agents, and on application in this case of some instrumentation devices.

19.2.2.1 PCR

The plate method is the oldest bacterial detection technique. However, in the last few decades, there has been progress in the sphere of laboratory diagnosis and some methods have been successfully modified for identification of pathogenic microorganisms. As a result, a wide range of molecular biology as well as immunology-based methods are the most common tools used for pathogen detection nowadays. These methods involve DNA analysis and antigen–antibody interactions.

PCR is a nucleic acid amplification technology that is widely used in bacterial detection. The PCR technique is based on the isolation, amplification, and determination of a short DNA sequence including the targeted bacteria's genetic material. Different PCR methods are used for bacteria detection such as real-time PCR, multiplex PCR, and reverse transcriptase PCR.

The PCR method consists of different cycles of extracted and purified DNA denaturation by heat, followed by an extension phase using specific primers and a thermostable polymerization enzyme. Then each new double-stranded DNA amplifies and acts as target for a new cycle. The detection of the targeted bacteria's sequence is provided by gel electrophoresis.

Real-time PCR technique is based on the specific dye detection in fluorescent emission and requires spending less time for obtaining results. The advantage of this method is eliminating of postamplification processing steps such as gel electrophoresis due to the possibility of following the amplification using fluorescence intensity, which is proportional to the amount of amplified product (Poltronieri et al., 2009; *Salmonella*—A Dangerous Foodborne Pathogen, 2012). Real-time PCR technique is very popular and has been developed by commercial organizations in a wide range of ready-to-use kits for microorganism detection, which provides high sensitivity, minimal hands-on, and takes about hours to fulfill. This technique's sensitivity lies in a limit 10^2–10^5 CFU and depends on preliminary sample preparation and can noticeably vary (Lazcka et al., 2007; Haakensen et al., 2008; Poltronieri et al., 2009; *Salmonella*—A Dangerous Foodborne Pathogen, 2012). Patel et al. used molecular beacon PCR procedure for Salmonella detection, and they were able to detect 2 CFU in 25 g chicken samples following 18 h preenrichment (Patel et al., 2006). Chen et al. were able to detect 3 CFU in 25 g food samples using a fluorogenic TaqMan™ probe (Chen et al., 1997). Whyte et al. detected 10 cells without a DNA extraction step (Whyte et al., 2002). Lin et al. reduced the enrichment time to 1 day, and they were able to detect 1 CFU/g of Salmonella in inoculated chicken meat after 8 h of enrichment (Lin et al., 2004).

Multiplex PCR is a variant of the PCR technique in which two or more loci are simultaneously amplified in the same reaction. In this method, several specific primer sets are combined into a single PCR assay. Multiplex PCR is very useful as it allows the simultaneous detection of several organisms to save time and effort in the laboratory. Recently, a real-time PCR technique has been described, which allows detecting more than two gene sequences in a single reaction by using spectrally distinct dye-labeled probes (*Salmonella*—A Dangerous Foodborne Pathogen, 2012).

DNA is always present whether the cell is dead or alive, and one of the limitations of PCR techniques lies in that the user cannot discriminate between viable and nonviable cells. Reverse transcriptase PCR (RT-PCR) is different from other PCR techniques by its ability to detect viable cells. Furthermore, this technique gives sensitive results without any time-consuming preenrichment step (Lazcka et al., 2007; *Salmonella*—A Dangerous Foodborne Pathogen, 2012).

In some cases, the PCR method can be more sensitive than the culture method, but application of PCR for the detection of pathogens in food samples is often limited by the presence of substances that inhibit the PCR reaction, poor quality of target DNA, or insufficient enrichment of target DNA. Furthermore, it requires pure samples and hours of processing (from 5 to 24 h, and this does not include any previous enrichment steps).

19.2.2.2 Immunology-Based Methods

Many virulent enteric pathogens can be identified via specific lipopolysaccharides (LPS), flagellar antigens, or secreted toxins due to the potential to cause an antigenic reaction and be recognized by specific antibodies.

Immunological methods for pathogenic bacteria detection are based on specific antigen–antibody reaction. This approach is widely used as an independent technique but can also be applied as a sample preparation step or combined with other techniques.

One of the most accepted and laboratory applied immunological methods is the enzyme-linked immunosorbent assay (ELISA) using microtitration plates. ELISAs combine the specificity of antibodies and the sensitivity of simple enzyme assays by using antibodies or antigens coupled to an easily assayed enzyme. The most common kind of ELISA is "sandwich method." ELISA provides highly specific quantitative and qualitative results but still has a wide list of disadvantages for direct pathogen detection: the small sample volume (200 mL) that the microtitration plate holds and the long incubation time required for each ELISA step. In addition, the sensitivity of ELISA methods is insufficient for direct measurement of bacteria and other microorganisms in the original samples. Because low numbers of pathogenic bacteria are often present in a biological sample, an analytical standard often used for pathogenic bacteria is to detect cells in 25 g of food. It is not possible to put a 25 g sample in a microtitration plate; therefore, it is necessary to concentrate on the bacteria by growing a single cell into a colony or using additional methods. As a pretreatment step for target cells concentration from the bacterial suspension, immunomagnetic separation (IMS) can be used by introducing antibody-coated magnetic beads in it (Ivnitski et al., 1999; Lazcka et al., 2007; *Salmonella*—A Dangerous Foodborne Pathogen, 2012).

Recently, chemiluminescent detection has been coupled with the selectivity of antibodies, magnetic microparticle separation or isolation, and enzymatic signal amplification in order to develop a rapid method, termed "enzyme-linked immunomagnetic chemiluminescence (ELIMCL)." Gehring et al. reported that they applied ELIMCL to the detection of *E. coli* O157:H7 in pristine buffered saline with a detection limit of 7.6×10^3 for live cells in approx. 75 min assay time. After enrichment, ELIMCL was demonstrated to detect *E. coli* O157:H7 inoculated in ground beef at 10 CFU/g in a total assay time of about 7 h (Gehringa et al., 2004). Hao and Bruno combined IMS with electrochemiluminescence (ECL) detection and evaluated this technique for detection of *E. coli* O157 and *Salmonella typhimurium* in foods. The authors obtained the detection limits in the range of 10^2–10^3 bacteria/mL in pristine buffer for *E. coli* O157 and *S. typhimurium*, respectively, or 10^3–2×10^3 bacteria/mL in food samples (depending on the sample) and the total processing and assay time was rapid (<1 h) even in food samples (Hao et al., 1996).

19.2.2.3 Instrumental Methods for Bacteria Identification Based on Fluorescence and Bioluminescence

Fluorescence in situ hybridization (FISH), the assay of specific nucleic acids sequence localization in native context, is a technology that appeared a few decades ago and has developed continuously. This method is widely used because of its variety of applications and the relative ease of implementation and performance of *in situ* studies. As a technique allowing simultaneous visualization, identification, enumeration, and localization of individual microbial cells, FISH is useful for many applications in all fields of microbiology (Levsky and Singer, 2003; Wendeberg, 2010; *Salmonella*—A Dangerous Foodborne Pathogen, 2012). FISH is a molecular technique that is often used to identify and enumerate specific microbial groups. It consists in hybridization of the rRNA sequence of immobilized cells by a fluorescently labeled 16S rRNA oligonucleotide probe. FISH with rRNA-targeted oligonucleotide probes facilitates the rapid and specific identification of individual microbial cells in their natural environments. This method has the advantage that bacteria do not need to be cultured before detection, and this would lead to a reduced time in identification of the infecting organism (Waar et al., 2005). The sensitivity of the technique is such that threshold levels of detection are in the region of 10–20 copies of mRNA per cell. Therefore, it allows detecting specific nucleotide sequences within even a single cell. FISH has found an application in clinical practice and is used in situations in which quick identification of the infection organism has an advantage in the treatment of the patient (Hogardt et al., 2000; Waar et al., 2005). Fluorescent *in situ* hybridization was also used in food research to indicate the possibility of its application of Salmonella spp. detection and was recommended as very sensitive, fast, and cheap (Fang et al., 2003; *Salmonella*—A Dangerous Foodborne Pathogen, 2012).

Flow cytometry is a powerful technique for analyzing large populations of single cells. In this technique, a sample is injected into a fluid that passes through a sensing region of a flow cell. The system includes a pressurized hydrodynamic system, laser beam(s), flow cell/laser intercept, a series of light detectors, and a data analysis station. Cells are carried by flow through a focus of light. Each cell emits a pulse of fluorescence, and the scattered light is collected and then analyzed. The light scattering of the cells gives information about their size, shape and structure, cell mass, and bacterial growth. Some cells have natural autofluorescence abilities because of their pigments, but in most cases different dyes or fluorescent probes specific for cellular substances of target cells are used. Autofluorescence can cause inconveniences if it is emitted at a similar wavelength as the dyed cells since it is often impossible to separate them. Typical rates for flow cytometry measurements are about 1000 cells/s, and it provides an ability to make rapid, quantitative measurements of multiple parameters of each cell within a large number of cells; however, flow cytometry has been frequently applied for measurements of mammalian or eukaryotic cells. The method's application in microbiology is limited because of several reasons: small bacteria size and the low number of DNA molecules to be stained require high sensitivity of instruments, and biological characteristics of many bacteria cells vary depending upon the growth conditions used, or the source from which the organisms were obtained (Davey et al., 1993; Davey, 1994; Ivnitski et al., 1999).

Adenosine triphosphate (ATP) bioluminescence is another technique that can be used for bacteria detection. ATP is present in all living cells. Enzyme Luciferase utilizes the energy from ANP to oxidize D-luciferin and produce light, which can be measured by sensitive luminometers and is proportional to the amount of ATP present in the sample. The majority of ATP-based systems are qualitative. This technique is used to detect the presence or absence of microorganisms after preenrichment of the sample and reduces the test time by approximately one third of that taken by traditional methods (Moldenhauer, 2008; Leon and Albrecht, 2007).

19.2.3 Biosensors for Microorganisms

Biosensors are analytical devices that combine a biological sensing element associated within a physicochemical transducer or transducing microsystem to produce a signal proportional to the analyte concentration (Ivnitski et al., 1999; Lei et al., 2006; Lazcka et al., 2007). They are postulated as an alternative technology for fast and rapid detection of pathogen microorganisms. As biological sensing elements, different biomolecules such as enzymes, antibodies, nucleic acids, organelles, engineered proteins, aptamers as well as animal and plant cells or tissues can be used. Usually, biosensors are classified according to the principle of transduction methods and may be divided into four basic groups such as optical, mass, electrochemical, and thermal sensors or according to the biological recognition element used. There are techniques that allow direct measurement of physical phenomena occurring during the biochemical reactions on a transducer surface. Depending on the biochemical reactions on a transducer surface and measuring parameter, biosensors can also be divided into direct and indirect detection systems. Direct detection biosensors are designed in such a way that the biospecific reaction is directly determined in real time via measuring the physical changes caused by reagents interaction. Indirect detection biosensors are those in which a preliminary biochemical reaction takes place and the products of that reaction are then detected by a sensor.

Areas for which biosensors show particular promise are clinical diagnostics, food analysis, bioprocess, and environmental monitoring. These devices have to combine next critical characteristics such as specificity and high sensitivity. In addition, biosensors have to provide short time analysis without any preenrichment steps, be a chip, portable and easy to use. It is not an easy task to combine all the mentioned characteristics in a single device; therefore, the majority of new developments do not find practical application. Because of lack of sensitivity and results reliability, new devices cannot compete with conventional methods and never leave the laboratories. However, this branch develops very fast all over the world, gives successful techniques and devises that are widely used in practice, and eventually replaces the standard methods.

19.2.3.1 Biological Sensing Elements and Immobilization Strategies

Biosensor specificity and sensitivity directly and strongly depends on the kind of biological material that will be used as recognition element. There are several critical factors that influence biosensors detection, especially when low limits are required such as affinity of the target for the molecular recognition elements, nonspecific binding of analyte at the binding sites, and the sensitivity of the detection method (Sensor systems for biological agent attacks, 2005; Liébana et al., 2009). There are several groups of biological substances that are the most frequently used as biological sensing elements in biosensor engineering such as enzymes, antibodies, and nucleic acids. For this purpose, enzymes are supposed to function as labels rather than actual bacterial recognition elements. In addition, they can be used to label either antibodies or DNA probes functioning as in an ELISA assay. More advanced techniques such as SPR, piezoelectric, or impedimetric biosensors operate without labeling the recognition element. Because of high specificity and relative cheapness, antibodies are the most frequently applied for biosensor designing than DNA probes. Antibodies may be polyclonal, monoclonal, or recombinant, depending on their selective properties and the way they are synthesized. Polyclonal antibodies can be raised quickly and cheaply. With the development of hybridoma and recombinant antibody phage display techniques, immunological detection of microbial contamination has become more sensitive, specific, reproducible, and reliable with many commercial immunoassays available for the detection of a wide variety of microbes and their products (Leonard et al., 2003). Nucleic acid-based detection can provide more specific and sensitive results than an immunological one, but the latter is faster and more robust and detects not only contaminating organisms but also their biotoxins that may not be expressed in the organism's genome (Iqbal et al., 2000; Leonard et al., 2003). In addition, immunosensors' sensitivity directly depends on the method of antibody immobilization on transducer surface (Starodub et al., 2005). The reason is that antibodies have specific sites that recognize and bind antigens. The main task is to find the immobilization technique that provides maximal antibodies–antigen interaction. As an alternative to direct adsorption of antibodies on polymer or metal surfaces, the site-directed approach is more preferable. For this purpose, dextran matrices have been used for chip creation. Also, avidin-biotin systems are used in ELISA sandwich method, but the high cost of reagents makes this technique less suitable for biosensor creation (Lazcka et al., 2007; Albers and Vikholm-Lundin, 2011). In most of the studies for obtaining bimolecular layers with a higher degree of orientation, site-oriented or site-directed immobilization is more preferable. In that case, intermediate linking molecules have been used such as the Fc-receptors protein A (Starodub et al., 2011; Starodub and Ogorodnijchuk 2012a,b), protein G (Oh et al., 2004b), and new hybrid proteins with highly specific immunoglobulin binding properties such as protein LG and LA (Kihlberg et al., 1992; Svensson et al., 1998). When antibodies are covalently coupled to self-assembled layers (SALs) or Langmuir-Blodgett films via thiol groups, antibody activity can increase up to >70% (Albers and Vikholm-Lundin, 2011).

19.2.3.2 Optical Biosensors

Optical biosensors are probably the most popular in bioanalysis, due to their selectivity, sensitivity, and direct (label-free) detection of bacteria. These sensors are able to detect changes of refractive index or thickness of surface layers, which occur when cells bind to receptors immobilized on the transducer surface providing the real-time analysis.

SPR is one of the optical techniques for microorganism detection widely applied for biosensor development. SPR-based instruments use an optical method to measure the refractive index (within ~300 nm) near a sensor surface that is presented as thin (about 50 nm) metal, usually gold film (Oh et al., 2004b; Son et al., 2007; Syam et al., 2012). SPR-based devises operate as flow systems and have a flowing cell where liquids are pumped permanently. The angle position of SPR minimum is determined by the properties of the gold-solution interface. In addition, adsorption phenomena and kinetics can be monitored using the SPR technique. Antigen–antibody reaction is the most widely used biological sensing element of transducer preparation for SPR biosensors. SPR has successfully been applied for pathogen bacteria detection by means of immunoreactions. Table 19.1 demonstrates detection level of SPR biosensors for Salmonella identification.

TIRE is another powerful technique for monitoring and analyzing adsorption and desorption on thin semitransparent metal films as well as for analyzing protein adsorption and adsorption from opaque liquids on metal surfaces (Arwin et al., 2004).

In our research, we made an attempt to work out highly specific biosensors for *Salmonella typhimurium* detection. As a registering part for our research, the devices based on the surface plasmon resonance and total internal reflection ellipsometry were used. First, experiments were carried out using Spreeta biosensor (TI, USA) (Starodub et al., 2010, 2011) and the SPR device with flowing cell named as "Plasmonotest" (created in V.M. Glushkov

TABLE 19.1

Comparison of Results of *S. typhimurium* Determination in Solutions Obtained by Optical Immune Biosensors

Type of Immune Biosensor	Detection Limit (Cells/mL)	Working Range (Cells/mL)
"Spreta" module, intermediate layers—polyelectrolyte, protein A (Starodub et al., 2011, 2010)	10^3–10^4	10^3–10^7
"Spreta" module, intermediate layers—polyelectrolyte, protein G (Oh et al., 2004a)	2×10^2–10^3	2×10^2–10^7
SPR, intermediate layer— mercapto-undecanoic acid, protein G (Oh et al., 2004b)	10^2	10^2–10^9
"Biacore," intermediate layer—developed dextran (Bokken et al., 2003)	1.7×10^3	—
SPR, intermediate layer— neutravidin (Son et al., 2007)	10^5	—
SPR, direct physical adsorption (Kubova et al., 2001)	1×10^6	—

Institute of Cybernetics of National Academy of Sciences, Ukraine). The principle of "Plasmonotest" is very close to Spreeta module organization. It is an optical device of angular type based on the SPR principle that is equipped with a CCD array of 2048 pixels (Figure 19.1).

The problems of information processing and data acquisition have been solved by developing software in the Institute of Cybernetics (Budnyk et al., 2011). Angular resolution of the devise is 0.001°. Additionally, the flowing cell for liquor pumping and sample temperature stabilization could be mounted. The accuracy of temperature stabilization in the flow cell is at least 0.1°C.

One more difference is that the detecting layer is formed on the glass plate surface covered by 1–2 nm niobium adhesive film and 50 nm gold plasmon

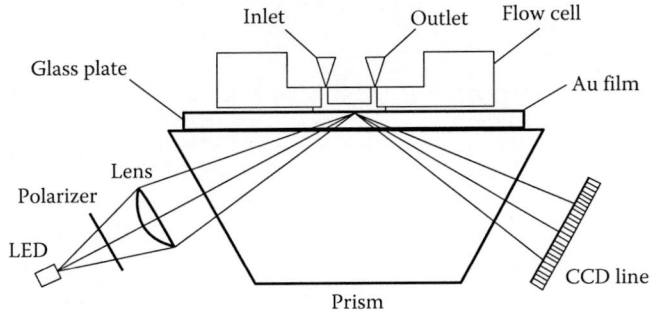

FIGURE 19.1
Scheme of "Plasmonotest" device with flow cell.

supporting film connected with prism using immersion liquid. This is more favorable because glass plates can be easily changed and/or renewed. Furthermore, it allows providing of all surface preparations in advance.

As a basis for TIRE biosensor, a commercial spectroscopic rotating analyzer instrument was used operating in the 350–1000 nm wavelength range. The detecting layer was formed on the chromium/gold-coated glass plate attached to a 68 trapezoidal glass prism (BK7, n = 1.515). This prism provides a total internal reflection effect between glass and aqueous solutions (n = 1.33). The sample analysis was provided using a specially designed 1.5 cm³ cell attached through a rubber ring. The injection of different solutions into the cell was carried with inlet and outlet tubes. Using the prism, polarized white light was coupled to the sample. After passing through another polarizing element (analyzer), the reflex beam was collected with a photodetector array.

According to antigen–antibody reaction, the transducer surface is covered by a specific antibody layer that provides selective binding of relative antigen from added liquid. However, it was observed that previous surface preparation leads to an increase in biosensor sensitivity level (Starodub et al., 2001a; Starodub and Starodub 2001). The procedure of transducer preparation included several sequential steps: (1) covering of surface by poly alyllamine hydrochloride and (2) immobilization of protein A from *Staphylococcus aureus*, and, finally, the oriented binding of the specific antibodies. Such a scheme of transducer preparation provides site-directed immobilization of antibodies toward the antigen solution.

In the case of Spreeta-based biosensor, it has been defined that device sensitivity was on the level 10^3–10^7 cells/mL. The diagram of the obtained results is presented in Figure 19.2 (the changes in microorganism concentrations are indicated by the pointers mentioned previously and the start of washing—by ones mentioned later).

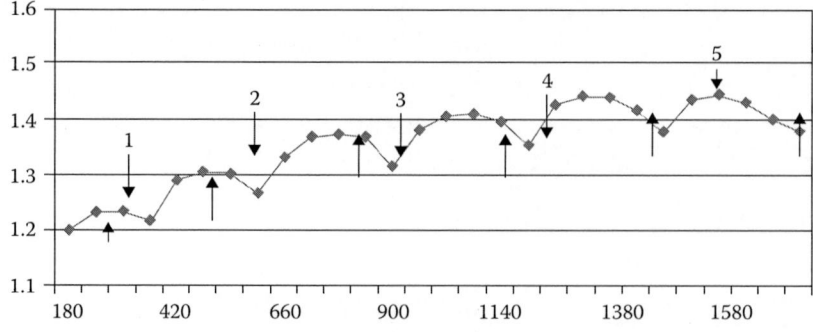

FIGURE 19.2
Sensor diagram of *S. typhimurium* solutions analysis. 1–5 × 10^3 × 10^7 cells/mL. Abscissa—time (s) and ordinate—change of resonant angle.

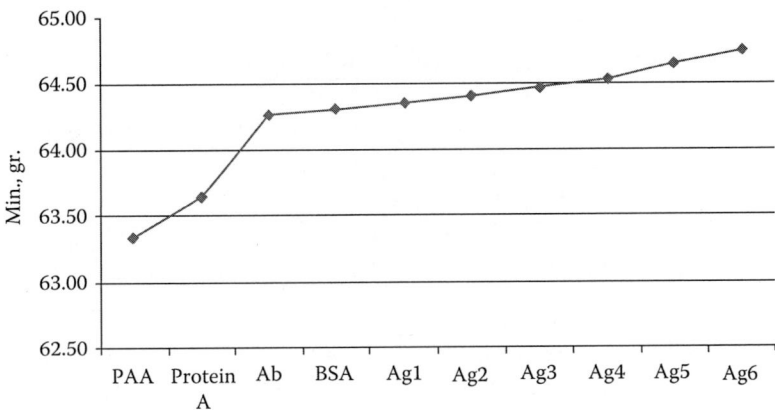

FIGURE 19.3
Diagram of *S. typhimurium* solutions analysis using "Plasmonotest" (Ag—row of *S. typhimurium* solutions from $10^{-1} \times 10^6$ cells/mL).

"Plasmonotest" sensitivity was within 10^1–10^6 cells/mL. It is higher than in Spreeta case but it is not sufficient for all practice situations. The obtained results are shown in Figure 19.3.

A biosensor based on TIRE has shown higher sensitivity than the one based on SPR. The maximal level of sensitivity was on the level of several cells in 10 mL (up to less than 5 cells). TIRE-based biosensor sensitivity is the closest for practice cases of bacteria detection without previous sample preparation.

One promising type of detector is the quartz crystal microbalance (QCM). The QCM is a piezoelectric mass-sensing device. It works by sending an electrical signal through a gold-plated quartz crystal, which causes a vibration at some resonant frequency. The QCM then measures the frequency of oscillation in the crystal. In the case of biosensor application, the QCM can detect changes in the frequency of the crystal due to changes in mass on the surface of the crystal. This technique has advantages such as the sensor can be used for the direct, marker-free measurement of specific interactions between immobilized molecules and analytes in solution. Binding of a soluble analyte to the immobilized ligand causes a shift in the resonance frequency, and this signal can be recorded using a frequency counter with high resolution. Despite a long period of existence, this method has only recently been developed for immunological measurements in a flow-through system. As real-time measurements are performed, the sensorgrams are capable of deducing equilibrium binding constants as well as the affinity rate constants (Zhihong et al., 2007). In microbiology, QCM can be applied or divided for direct detection of a microbe or spore, as well as for detection of an associated antigen or toxin and biofilms formed by microorganism characterization. QCM can be used to provide a rapid, specific assay using whole bacteria or associated

TABLE 19.2

Comparison of QCM Assays for Pathogenic Bacteria Detection

Microorganism and Matrix	Sensitivity (Cells or CFU/mL)	Assay Time (min)
S. typhimurium (PBS) (Cho and Li, 2004)	10^3	30
S. typhimurium (BHY broth/chicken meat solution) (Su and Li, 2005)	10^2	60
E. coli (Zhihong et al., 2007)	7.5×10^2	—
E. coli (BHY broth) (Su and Li, 2004)	10^3	30–50
E. coli (Food samples) (Kim and Park, 2003)	1.7×10^5	20–30
S. paratyphi (PBS) (Fung and Wong, 2001)	170	50
M. tuberculosis (Sputum) (He et al., 2002)	200	30

spores, most often using polyclonal or monoclonal antibodies specific for the pathogen of interest. There is a huge number of studies that provide a review of different microorganism detection. Analyzing then, we can conclude that the type of biosensor provides sensitivity about 10^2 CFU/mL with assay time from 20 to 60 min. In Table 19.2, results of different authors are presented in an attempt to detect microorganisms using the QCM technique.

19.2.3.3 Electrochemical Biosensors

Electrochemical biosensors are mainly based on the observation of current or potential changes due to interactions occurring at the sensor–sample matrix interface. Techniques are generally classified according to the observed parameter: current (amperometric), potential (potentiometric), or impedance (impedimetric). In fact, electrochemical immunosensors are an extension of conventional antibody-based enzyme immunoassays (ELISA), where catalysis of substrates by an enzyme conjugated to an antibody produces in the form of pH change, ions, and oxygen consumption that generate electrical signals on a transducer (Warsinke et al., 2000). In amperometric detection, for example, alkaline phosphatase (AP) conjugated to an antibody hydrolyzes *p*-nitrophenyl phosphate to phenol, which is detected by voltammetry. In light-addressable potentiometric sensors (LAPS), urease-conjugated antibody hydrolyzes urea, resulting in the production of carbon dioxide and ammonia that change the pH of the solution. Compared to optical methods, electrochemistry allows working with turbid samples, and the capital cost of equipment is much lower. On the other hand, electrochemical methods demonstrate more limited selectivity and sensitivity than optical approaches.

Amperometry is perhaps the most common electrochemical detection method used in biosensors. It works on the principle of an existing linear relationship between analyte concentration and current. In the case of biosensors, where direct electron exchange between the electrode and either the analyte or the biomolecule is not permitted, redox mediators are required (Eggins, 2002).

Redox mediators are small size compounds able to reversibly exchange electrons between both the sensor and the enzyme of choice (e.g., ferricyanide, osmium or ruthenium complexes, dyes). Amperometric biosensors have found wide application in clinical monitoring (about 92%) providing glucose, lactate, urea, etc. control. About 2% of the devices are used for environment monitoring and penetrate the food industry because of considerable opportunities in food analysis (Prodromidis and Karayannis, 2002). In addition, amperometry shows promising results for bacterial biosensor creation. For instance, Abdel-Hamid et al. reported in their work with *E. coli* determination during 30 min the detection range between 100 and 600 cells/mL using flow-through immunofiltration method coupled to amperometry. Gau et al. presented their work in which they highlight a system for amperometric detection of *E. coli* based on the integration of microelectromechanical systems (MEMS), self-assembled monolayers (SAMS), DNA hybridization, and enzyme amplification. The detection system was capable of detecting 10^3 *E. coli* cells with high specificity for *E. coli*, and analysis was conducted with solution volumes on the order of a few microliters and completed in 40 min (Gau et al., 2001).

In potentiometric type of sensors, the measured parameter is oxidation or reduction potential occurring during electrochemical reaction. Such a biosensor consists of a permselective outer layer and a bioactive material, usually an enzyme. The working principle is based on the fact that when a voltage is applied to an electrode in solution, a current flow occurs because of electrochemical reaction. The voltage at which enzyme-catalyzed reactions occur indicates a particular reaction and particular species. Potentiometry provides a logarithmic concentration dependence (Leonard et al., 2003; Syam et al., 2012). A field effect transistor (FET) is a device where the transistor amplifier is adapted to be a miniature transducer for the detection and measurement of the potentiometric signal produced by a sensor process at the gate of the FET. This allows miniaturization and increased sensitivity due to the minimal amount of circuit wiring. This method makes use of an ion-sensitive field effect transistor (ISFET) built on standard technology that produces source, drain, and gate regions. The gate uses an ion-sensitive membrane that renders ISFET capable of biochemical recognition in the presence of the analyte with the increase in local ion concentration (Chauhan et al., 2004). Next, we present the results obtained with the application of the ISFET based on cerium oxide for the development of the immune biosensor.

A new type of immune biosensor was proposed based on ISFET with CeOx instead of Si_3N_4 gate surface, which promotes the sensitivity and stability of the analysis. This biosensor was used for the determination of Salmonella typhimurium in the model solution.

First, the ISFET surface was cleaned by sequential immersion into the mixture of potassium dichromate and sulfuric acid (10 s), water (vigorous washing), acetone, ethanol, and phosphate 10 mM Na-phosphate buffer (PB), pH 7.3, containing 100 mM NaCl. By drop coating, an aqueous solution of cystamine (10 mg/mL) was placed on the transducer surface. After

30 min at room temperature (~23.8°C), cystamine was placed again using the same procedure. Then, the surface was twice activated by 2.5% water solution of glutaraldehyde (GA) from Sigma for 20 min. After this, protein A from *Staphylococcus aureus* (Sigma) (20 mg/mL) or human serum at 1:2 dilutions was placed on the gate surface of the ISFET and kept for 30 min at humid conditions. In the next step, the chip was kept 1 h at 4°C, and then, after vigorous washing with the PB was immersed into the glycine solution (10 mg/mL in the PB at the room temperature) to block nonreacted groups of the GA. After 30 min at room temperature, the ISFET was washed with the PB. When the protein A on the working chip and nonspecific antiserum on the reference one were immobilized, the specific antibodies were added. The anti-salmonella serum (St. Petersburg Research Institute of Vaccines and Serums, Russia) at 1:2 dilutions in 0, 85% of NaCl solution was placed on the working chip, and it was incubated for 30 min in humid conditions and then washed with the PB. The prepared chips were stored in dry state at 4°C.

The analysis was performed by the "sandwich" method when the immobilized specific antibodies interact with *S. typhimurium* dissolved in 0, 85% of NaCl solution for 20 min. Then, bound cells on the transducer surface were treated by the specific antibodies labeled by the horseradish peroxidase (HRP) for 10 min. The activity of HRP was registered in the presence of the special working buffer containing 5 mM *tris*-HCl (pH 7.5), 100 mM NaCl, 15 mM ascorbic acid, and 5 or 10 mM H_2O_2. The substrate conversion causes a local basic pH shift, because the dehydroascorbic acid formed is a more neutral compound compared to ascorbic acid. The signals (dV/dt) of the ISFETs were registered by an electronic device providing signal amplification and its processing on the basis of a custom-made computer program. After every assay, the chip was treated 5 min by 0.1 M HCl and then it was carefully washed with the previously mentioned *tris*-HCl buffer.

Change of the ISFET potential of the immunobiosensor as a function of the salmonella concentration as presented in Figure 19.4 (mean values

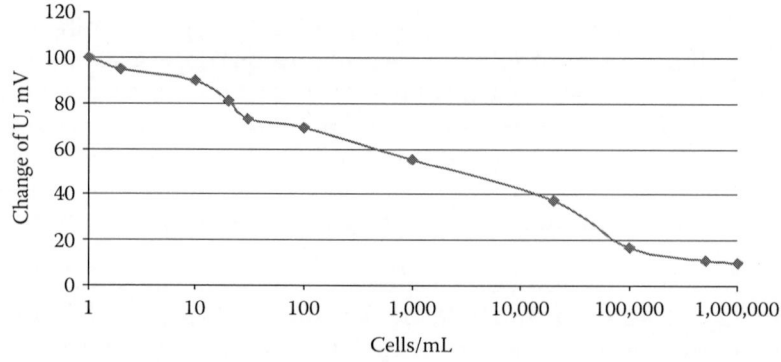

FIGURE 19.4
Immune biosensor response at different concentrations of *Salmonella* in the model solution.

from 6 to 8 measurements). A reliable decrease in the sensor signal was observed down to 2.0–3.0 cells/mL of *S. typhimurium* in the analyzed solution. It should be noted that the biosensor output depends on the quantity of antigen-binding sites on the ISFET surface. Therefore, oriented immobilization of antibodies via Staphylococcal protein A is an effective way to reach a high signal. The signal of the immune biosensor gradually decreased with increasing salmonella concentrations. It was observed in the range of the concentrations from 2 to 5×10^5 cells/mL. In this range, the potential of the ISFET gate varied from 95 to 5 mV. The standard deviation was on average about 5% (Starodub and Ogorodnijchuk 2012a,b).

The possibility to reuse the developed immune biosensor up to five times without signal decreasing was demonstrated. The comparison of the characteristics of the immune biosensors based on SPR, TIRE, quartz crystal acoustic wave, amperometry, chemiluminescence, and on ISFET with CeOx gate surface testifies that they have similar sensitivity, but the last approach may help in achieving a low cost of analysis. To have the sensitivity of analysis in respect of Salmonella and other bacteria on the level of infection dose (Ivnitski et al., 1999), there is a necessity to have a special system of the analyzed sample and in particular, accumulation of cells through application of bioaffinic columns, as was early demonstrated by some authors (Croci et al., 2001).

The recently developed light addressable potentiometric sensor (LAPS), based on the FET, has been used for the detection of microbial contamination. It consists of an *n*-type silicon semiconductor-based sensor and an insulating layer that is in contact with an aqueous solution where an immunoreaction takes place. Changes in potential at the silicon interface are detected by the difference in charge distribution between the surface of the insulator and the FET.

Gehring et al. used LAPS for detection of *E. coli* O157:H7 capturing target cells previously using membrane or magnetic beads with specific antibodies (Gehring et al., 1998; Tu et al., 1999). The reaction was performed in a sandwich method, so a fluorescent-labeled antibody was bound to the target cells. A urease-labeled anti-fluorescent antibody was then added. In the presence of urea, NH3 was produced, which changed the pH of the solution on the *n*-type sensor coated with a pH sensitive insulator that recorded the voltage change. The detection limit for such a sensor was 7.1×10^2 cells/mL for heat-killed cells and 2.5×10^4 cells/mL for live cells. In food samples, Tu et al. showed that *E. coli* O157:H7 could be detected 1 CFU/g after 6 h of enrichment (Tu et al., 1999, 2002).

Dill et al. have used LAPS technique for the detection of *Y. pestis* and *Bacillus globigii* spores with a limit of detection of 10 cells or spores per sample (Dill et al., 1997). Later, they successfully used the same technique to detect *S. typhimurium* and obtained the detection level as low as 119 CFUs (Dill et al., 1999).

Based on LAPS technology, a commercially available device called the Threshold Immunoassay System® is available from molecular devices.

This system is capable of detecting eight agents simultaneously within 15 min. The limit of detection for *B. subtilis* is 3×10^3 CFU/mL (Uithoven et al., 2000).

Electrical impedance is one of the powerful techniques for biosensor creation. Its main advantage compared with other biosensors is its ability to detect viable cells in a sample matrix via their metabolism processes. Furthermore, it can provide real-time detection and monitoring of concentration, growth, and the physiological state of cells. Therefore, this method is well suited for the detection of bacteria in clinical specimens, to monitor quality and detect specific food pathogens, also for industrial microbial process control, and for sanitation microbiology (Swaminathan and Feng, 1994; Silley and Forsythe, 1996). Impedance is based on the changes in conductance in a medium due to the microbial breakdown of inert substrates into electrically charged ionic compounds and acidic by-products. Microbial metabolism usually results in an increase in conductance and capacitance, causing a decrease in impedance. The main effect of cells on the sensor signal is due to the insulating property of the cell membrane. Current instruments detection range is 10^6–10^7 active metabolizing bacteria per milliliter (Ivnitski et al., 1999). Deng et al. developed a reusable Bulk Acoustic Wave (BAW)–Impedance Sensor for continuous detection of growth and numbers of Proteus vulgaris. Using the proposed method, bacteria can be detected in the range of 3.4×10^2–6.7×10^6 cells/mL. Ruam et al. described an impedance immunosensor method for detection of *E. coli* by immobilizing an anti-*E. coli* antibody on the electrode surface and applying an impedance measurement. The sensor demonstrated identification of the *E. coli* O157:H7 cells with the detection limit of 6×10^3 cells/mL (Ruan et al. 2002).

19.3 Express Control of Some Mycotoxins in Food and Feeds

The presence of mycotoxins in agricultural products necessitates large-scale testing of a wide range of sample material to ensure the safety of food and feed. Form feed and food spoilage leads to great economic losses around the world. It is believed that 5%–10% of world food production is wasted because of contamination of fungal degradation (Pitt and Hocking, 1997).

Among the many environmental toxic substances, mycotoxins formed by microscopic fungi have recently attracted increasing attention. Different generations of micromycetes are able to produce aflatoxins, ochratoxin A, fumonisin, and trihotetseteny. Mycotoxins are a group of low-molecular weight, non-immunogenic compounds, many of which are relatively heat resistant. Today, mycotoxins are given special attention due to many reasons, namely, their wide distribution (in the desert, salt marshes, and high mountains, although, in general, they are most common in temperate latitudes), their high toxicity to living organisms (embryo toxic, mutagenic, carcinogenic effects), and they recently created a danger of their use by bioterrorists (such as their representative, as the T-2, showing several orders of magnitude greater lethality than mustard or lewisite). Different varieties of mycotoxins specifically affect organs and tissues: liver, kidney, esophagus and intestinal mucosa, and brain tissue, and genitals. Mycotoxins are included in the list of substances as subjects for regulation of their content in food, feed, and raw materials. Today's practical requirements for methods of food control have increased dramatically. They include a number of provisions, namely, a high level of selectivity and sensitivity analysis, its rapidity and low cost, and the ability to conduct research in the field, in the modes on line and off line. Also, the ability to simultaneously test for multiple similar samples or samples of different origin is very important, and, finally, providing electronic processing of the results, as well as their automatic transfer to specialized laboratories and supervisory authorities. Fulfillment of these requirements is possible with the development and introduction of new tools of analysis. Such a perspective means that one has to include those that are based on the principles of biosensors.

Since the discovery of mycotoxins, several methodologies for their determination have been developed. Methods routinely used are mainly based on thin-layer chromatography, gas chromatography, or high-performance liquid chromatography (Gilbert, 2002). Traditionally employed analytical methods to detect mycotoxins involve lengthy extraction procedures, expensive chemical cleanup, and use of hazardous materials (Donnelly et al., 2003).

19.3.1 Traditional Instrumental Approaches for the Express Registration of Mycotoxins

Among these analytical methods, we will pay attention to two that are most dispersed in practice, namely thin-layer chromatography (TLCh) and gas chromatography (GCh).

19.3.1.1 Thin-Layer Chromatography

It is a low-cost, rapid analytical technique, yielding qualitative or semiquantitative estimations by visual inspection, but with densitometry measurements and also reliable quantitative results (Lin et al., 1998; Krska et al., 2007).

TLCh is the most commonly used physicochemical test because more than one mycotoxins can be detected for each test sample. TLCh is based on the separation of compounds by how far they migrate on a specific matrix with a specific solvent. The distance that a compound will travel is a unique identifier for specific compounds, and a retention factor (Rf) has been determined for most mycotoxins.

As with any detection system, a positive control containing purified mycotoxins must be run in parallel to ensure accuracy, since different chemicals can have a similar Rf.

19.3.1.2 Gas Chromatography

Gas chromatography combines superior separation on the capillary columns with a variety of general and specific detectors. GCh is often used in more technical laboratories for some of the mycotoxins and in particular for the analysis of type-A trichothecenes (T-2 toxin, HT-2 toxin, neosolaniol, and diacetoxyscirpenol) that do not render themselves readily amenable to HPLC analysis. Components are separated using the relative affinity of the compounds for a stationary column and a mobile, inert gas.

Analytes separated on the column and eluted with the inert gas are detected by chemical or physical means. Various detection systems may be used as coupled to GC, but in most cases electron-capture detection (ECD) and mass spectrometry (MS) have been employed. GC-ECD and GC-MS are highly sensitive methods that enable the simultaneous determination of several trichothecenes even in complex food matrices in the lower μg/kg range (Krska et al., 2001).

19.3.2 Biosensors in Control of Mycotoxins

In this chapter we limit our discussion to our experience in the determination of patulin, aflatoxin, and T-2 mycotoxin. (All reagents were produced by Sigma, USA). It should be noted that most of the proposed biosensor devices to control and mycotoxins are based on using the phenomenon of SPR or TIRE. In addition, they also offer an immune electrochemical biosensor based on ISFET with the cerium oxide gate surface.

Several algorithms were used for the analysis, namely: (1) the "direct" method when specific antibodies are immobilized on a preprepared surface and then measured for cell contributed analyte; (2) the "competitive" method in which the specific antibodies were immobilized on the surface of the corresponding protein conjugate, or specific antibodies and the free analyte compete with free antibody or corresponding protein conjugate for binding sites with other immune components, which had been immobilized on the surface transducer, and (3) "the saturate" method in which specific antibodies reacted with the analyte, and then the remaining binding sites on the surface of the corresponding protein "saturate" by some conjugate. Thus, to determine aflatoxin by the developed biosensor based on SPR, the analysis was conducted in the two previously mentioned algorithms: the "competitive" and "saturate" methods. In both cases, polyclonal antibodies were used.

To ensure optimal operation of the immune biosensor, different approaches were considered to immobilize antibodies on the surface of the transducer. To meet this objective, antibodies were immobilized on the surface of gold and through intermediate layers of dodecanethiol and dextran. It was found that the use of the dodecanethiol monolayer promotes the stabilization of the conjugated antigen. In the case of surface modification of dextran sulfate, the immune biosensor was less stable and reproducible than when it was surface prepared by dodecanethiol and in the case of the application of gold transducer. The differences between the results from the plate to the plate were in the range of 15%–20%.

There was also a determination of some mycotoxins using an immune biosensor based on TIRE. A typical set of $\Delta(\lambda)$ experimental spectra show the spectral shift caused by successive adsorption layers of polyamine hydrochloride protein (both as intermediate layers for the selective immobilization of structures) and specific Ab, as well as by binding to different concentrations.

19.3.2.1 Determination of Patulin and T-2 Mycotoxin in Real Samples by Immune Biosensor Based on SPR

The results of patulin determination in real samples (tomato juice) by "competitive" methods are presented in Figure 19.5. It can be seen that there is a possibility to reveal patulin at the concentration starting from 0.05 mg/L. The linear dispersion is in the range of 0.05–10 mg/L. We think that the difference observed between standard solution and real sample is stipulated by some nonreversible sorption of patulin in the organic phase.

In subsistence farming at the National university of Life and Environmental Sciences of Ukraine, samples of maize and bran affected by fungi were chosen. With the help of the ELISA method, it was determined that the concentration of T-2 mycotoxin in these samples was in the range of 220–180 ng/g, respectively. At the same time by using the "competitive" analysis with the immune biosensor, a T-2 mycotoxin concentration of about 230 and 188 ng/g was determined. There is necessity to mention that the maximal permitted concentration of this mycotoxin in food is 100 ng/g (Kononenko et al., 1999).

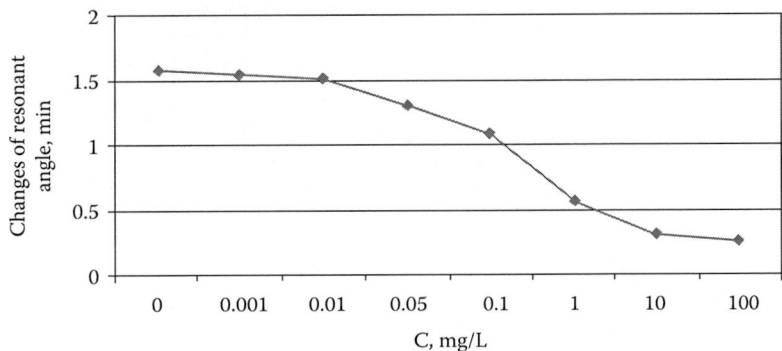

FIGURE 19.5
Determination of patulin in tomato juice by SPR immune biosensor at "competitive" analysis.

19.3.2.2 Determination of Some Mycotoxins in the Model
Solution by the Immune Biosensor Based on TIRE

A typical set of experimental spectra shown in Figure 19.6 demonstrates the spectral shift caused by consecutive adsorption of layers of polyamine hydrochloride, protein A (both as intermediate layers at the immobilization of selective structures), and specific Ab as well as by binding different concentrations of T-2 mycotoxin to Ab (from 0.15 up to 300 ng/mL).

Ellipsometry data fitting allows the evaluation of thickness values of the adsorbed layer. Since the refractivity increments caused by adsorption of different biomolecules represent only 0.1%–0.14% of the refractive index, the spectral changes were associated mainly with the thickness. The resulted calibration curve for T-2 mycotoxin showed a possibility of detection of T-2 mycotoxin in concentrations down to 0.15 ng/mL (or 0.15 ppb).

The calibration curves (i.e., thickness changes vs. mycotoxin concentration) obtained from the TIRE experiments for the other two mycotoxins, zearalanone and aflatoxin are shown in Figure 19.7. The response to aflatoxin appeared to be about three times less than that for zearalanone; and both are smaller than that for T-2 (compare with Figure 19.6). This could be due to the limited concentration (or activity) of antibodies. Another explanation may be related to the hydrophobicity of these mycotoxins and thus their ability to form aggregates in aqueous solutions (Starodub et al., 2008).

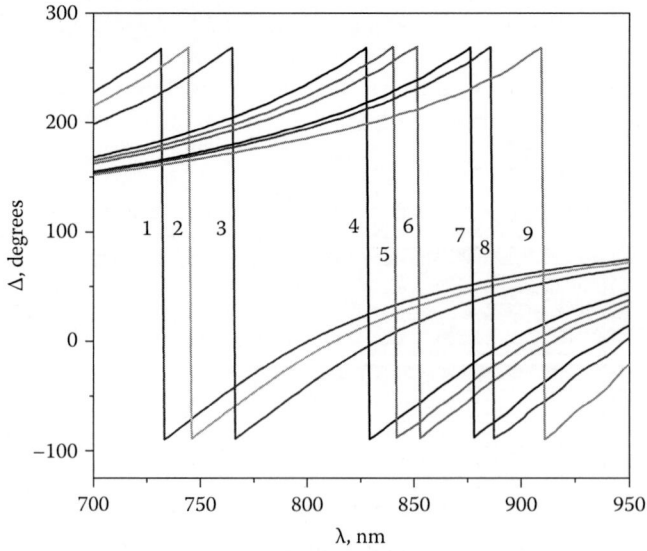

FIGURE 19.6

Typical spectra of Δ for the bare Au surface (1), after consecutive adsorption of PAH (2), Protein A (3), antibodies to T-2 (4), and after binding T-2 mycotoxin in different concentrations of 0.15 mg/mL (5), 1.5 mg/mL (6), 7.5 mg/mL (7), 75 mg/mL (8), 300 mg/mL (9).

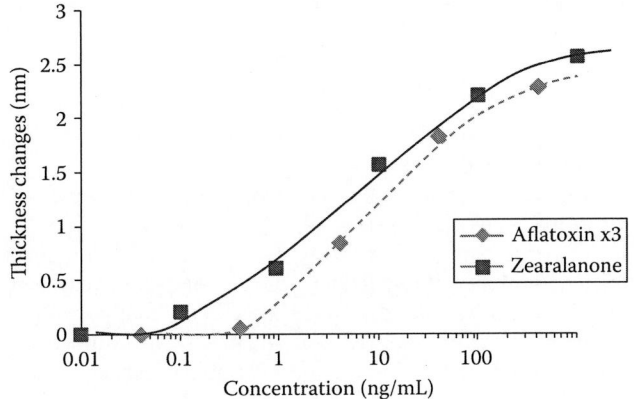

FIGURE 19.7
TIRE calibration curves for direct immunosensors for zearalanone and aflatoxin. The thickness values for aflatoxin were threefold increased.

Similarly to T-2, the minimal detected concentrations for both zearalanone and aflatoxin are about 0.1 and 0.4 ng/mL, respectively, which is quite a remarkable achievement (no other optical direct immunoassays provide such high sensitivity) (Nabok et al., 2011).

19.4 Conclusion

Conventional methods for the detection and identification of microbial and mycotoxin contaminants remain the gold standard among analytical approaches used in practice. Although they can be very sensitive, inexpensive, and present both qualitative and quantitative information, they often require several days to yield results. As an alternative to the more traditional methods, biosensors offer exciting results, allowing rapid "real-time" and multiple analyses that are essential for the detection of both types of contaminations in food, environment, and clinical specimens. As shown previously, the sensitivity of each of the sensor systems discussed in this review may vary depending on the transducer's properties, previous surface treatment, sample preparation, etc. Nevertheless, the biosensor system must have the specificity to distinguish the target bacteria in a multiorganism matrix, the adaptability to detect different analytes, the sensitivity to detect bacteria directly, on-line without preenrichment, and the rapidity to give real-time results. Obviously, enhancing the specificity of biosensor systems and incorporation of all these features within one biosensor device is a very complicated task. Although there is no biosensor system that has a bacterial specificity and reliability as that of the plate culture method, some biosensors were successfully applied in practice and today are produced on an industrial scale.

References

Albers W.M., Vikholm-Lundin I. Surface plasmon resonance on nanoscale organic films. In: *Nano-Bio-Sensing*. Ed. S. Carraro., Springer, New York, 2011, pp. 83–125. ISBN 978-1-4419-6168-6.

Arwin H., Poksinski M., Johansen K. Total internal reflection ellipsometry: Principles and applications. *Appl. Opt.*, 2004, 43, N15, 3028–3036.

Board on Manufacturing and Engineering Design Division on Engineering and Physical Sciences National Research Council. *Sensor Systems for Biological Agent Attacks. Protecting Buildings and Military Bases.* 2005, 208pp, ISBN 0-309-09576-X.

Bokken G., Corbee R.J., Knapen van F., Bergwerff A.A. Immunochemical detection of Salmonella group B, D and E using an optical surface plasmon resonance biosensor. *J FEMS Microbiol. Lett.*, 2003, 222, 75–82.

Budnyk M., Frolov Yu., Kurlov S., Lebyedyeva T., Minov Yu., Sutkovyy P., Shpylovyy P. Modeling and data processing for thin-film optical sensors. In: *Proceedings of the 6th IEEE International Conference on Intelligent Data Acquisition and Advanced Computing Systems: Technology and Applications IDAACS'2011*, September 15–17, Prague, Czech Republic, 2011, Vol. 1, pp. 119–124.

Chauhan S., Rai V., Singh H.B. Biosensors *Resonance*, 2004, 9(12), 33–45.

Chen S., Yee A., Griffiths M., Larkin C., Yamashiro C.T., Behari R., Paszko-Kolva C., Rahn K., De Grandis S.A. The evaluation of fluorogenic polymerase chain reaction assay for the detection of Salmonella species in food commodities. *Int. J. Food Microbiol.*, 1997, 35, 239–250.

Cho Y.K., Kim S., Kim Y.A., Lim H.K., Lee K., Yoon D.S., Lim G., Pak Y.E., Ha T.H., Kim K. Characterization of DNA immobilization and subsequent hybridization using in situ quartz crystal microbalance, fluorescence spectroscopy, and surface plasmon resonance. *J. Colloid Interface Sci.*, 2004, 278, 1, 44–52.

Croci L., Delibato E., Volpe G., Palleschi G. A rapid electrochemical ELISA for the detection of salmonella in meat sample. *Anal. Lett.*, 2001, 34, 2597–2607.

Darby, I. A. (Ed.). *In Situ Hybridization Protocols (Methods in Molecular Biology (Cloth), Methods in Mol. Biol.*, vol. 123. Humana Press, New York, 2012. http://www.genedetect.com/insitu.htm

Davey H.M. Flow cytometry of microorganisms. Thesis, Institute of Biological Sciences, University of Wales, Aberystwyth, Wales, U.K., February, 1994.

Davey H.M., Davey C.L., Kell D.B. SKATGRAF: A stand-alone program for the calibration and plotting of flow cytometric data. *Binary*, 1993, 5, 165–170.

Dill K., Song J.H., Blomdahl J.A., Olson J.D. Rapid, sensitive and specific detection of whole cells and spores using the light-addressable potentiometric sensor. *J. Biochem. Biophys. Methods*, 1997, 34, 161–166.

Dill K., Stanker L.H., Young C.R. Detection of salmonella in poultry using a silicon chip-based biosensor. *J. Biochem. Biophys. Methods*, 1999, 41, 61–67.

Donnelly C., Marley E., Dunnigan P., Gallagher M. Diagnostic test systems for mycotoxins. In: *Mycotoxins in Food Production Systems Aspects of Applied Biology.* Vol. 68. Association of Applied Biologists, Warwick, U.K., 2003.

Eggins B.R. *Chemical Sensors and Bisensors.* John Wiley & Sons Ltd., New York, 2002; Emde K.M.E., Mao H.Z., Finch G.R. *Water Environ. Res.*, 1992, 64, 4, 641–647.

Fang Q., Brockmann S., Botzenhart K., Wiedenmann A. Improved detection of Salmonella sp. in foods by fluorescent in situ hybridization with 23S rRNA probes: A comparison with conventional cultural methods. *J. Food Prot.*, 2003, 66, 723–731.

Fung Y.S., Wong Y.Y. Self-assembled monolayers as the coating in a quartz piezo-electric crystal immunosensor to detect Salmonella in aqueous solution. *Anal. Chem.*, 2001, 73, N21, 5302–5309.

Gau J.-J., Lan E.H., Dunn B., Ho C.M., Woo J.C.S. A MEMS based amperometric detector for *E. coli* bacteria using self-assembled monolayers. *Biosens. Bioelectron.*, 2001, 16, 745–755.

Gehring A.G., Patterson D.L., Tu S.I. Use of a light-addressable potentiometric sensor for the detection of *Escherichia coli* O157:H7. *Anal. Biochem.*, 1998, 258, 293–298.

Gehringa A.G., Irwina P.I., Reeda S.A., Tua S., Andreotti P.E., Akhavan-Tafti H., Handley R.S. Enzyme-linked immunomagnetic chemiluminescent detection of *Escherichia coli* O157:H7. *J. Immunol. Methods*, 2004, 293, 97–106.

Gilbert J. Validation of analytical methods for determining mycotoxins in foodstuff. *Trends Anal. Chem.*, 2002, 21, 468–486.

Haakensen M., Dobson1 C.M., Deneer H., Ziola B. Real-time PCR detection of bacteria belonging to the Firmicutes Phylum. *Int. J. Food Microbiol.*, July 31, 2008, 125, 3, 236–241.

Hao Yu., Bruno J.G. Immunomagnetic-electrochemiluminescent detection of *Escherichia coli* O157 and *Salmonella typhimurium* in foods and environmental water samples. *Appl. Environ. Microbiol.*, 1996, 62, 587–592.

He F.J., Zhang L.D., Zhao J.W., Hu B.L., Lei J.T. A TSM immunosensor for detection of M-tuberculosis with a new membrane material. *Sens. Actuators*, 2002, B85, 3, 284–290.

Hogardt M., Trebesius K., Geiger A.M., Hornef M., Rosenecker J., Heesemann J. Specific and rapid detection by fluorescent in situ hybridization of bacteria in clinical samples obtained from cystic fibrosis patients. *J. Clin. Microbiol.*, 2000, 38, 2, 818–825.

Iqbal S.S., Mayo M.W., Bruno J.G., Bronk B.V., Batt C.A., Chambers J.P. A review of molecular recognition technologies for detection of biological threat agents. *Biosens. Bioelectron.*, 2000, 15, 11–12, 549–578.

Ivnitski D., Abdel-Hamid I., Atanasov P., Wilkins E. Biosensors for detection of pathogenic bacteria. *Biosens. Bioelectron.*, 1999, 14, 599–624.

Kihlberg B.M., Sjobring U., Kasternl W., Bjorck L., Protein L.G. A hybrid molecule with unique immunoglobulin binding properties. *J. Biol. Chem.*, December 15, 1992, 267, 35, 25583–25588.

Kim N., Park I.S. Application of a flow-type antibody sensor to the detection of *Escherichia coli* in various foods. *Biosens. Bioelectron.*, 2003, 18, 9, 1101–1107.

Kononenko G.P., Burkin A.A., Soboleva N.A., Zotova E.V. Enzyme immunoassy for determination of T-2 toxin in contaminated grain. *Appl. Biochem. Microbiol.*, 1999, 35, 457–462.

Krska R., Baumgartner S., Josephs R. The state-of-the-art in the analysis of type-A and type-B trichothecene mycotoxins in cereals. *Fresenius J. Anal. Chem.*, 2001, 371, 285–299.

Krska R., Welzig E., Boudra H. Analysis of Fusarium toxins in feed. *Anim. Feed Sci. Technol.* 2007, 137, 241–264.

Kubova V., Karasova L. et al. Detection of foodborne pathogens using surface plasmon resonance diosensors. *Sens. Actuators B Chem.*, 2001, 74, 100–105.

Laboratory Diagnostics of Salmonellosis. The detection of Salmonella in food and environmental samples. Guidelines MU 4.2.2723-10 (approved by the Chief State Sanitary Doctor of Russia, M., August 13, 2010). http://www.garant.ru/products/ipo/prime/doc/4091056/#review

Lazcka O., Del Campo F.J., Munoz F.X. Pathogen detection: A perspective of traditional methods and biosensors. *Biosens. Bioelectron.*, 2007, 22, 1205–1217.

Lei Yu., Chen W., Mulchandani A. Microbial biosensors. *Anal. Chim. Acta*, 2006, 568, 200–210.

Leon M.B., Albrecht J.A. Comparison of adenosine triphosphate (ATP) bioluminescence and aerobic plate counts (APC) on plastic cutting boards. *J. Foodservice*, 2007, 18, 145–152.

Leonard P., Hearty S., Brennan J., Dunne L., Quinn J., Chakraborty T., O'Kennedy R. Advances in biosensors for detection of pathogens in food and water. *Enzyme Microbial Technol.*, 2003, 32, 3–13.

Levsky J.M., Singer R.H. Fluorescence in situ hybridization: past, present and future. *J. Cell Sci.*, 2003, 116, 2833–2838.

Liébana S., Lermo A., Campoy S., Cortès M.P., Alegret S., Pividori M.I., Rapid detection of Salmonella in milk by electrochemical genosensing. *Biosens. Bioelectron.*, 2009, 25, 510–513.

Lin C.-K., Hung C.-L., Hsu S.-C., Tsai C.-C., Tsen H.-Y. An improved PCR primer pair based on 16S rDNA for the specific detection of Salmonella serovars in food samples. *J. Food Prot.* 2004, 67, 1335–1343.

Lin L., Zhang J., Wang P., Wang Y., Chen J. Thin-layer chromatography of mycotoxins and comparison with other chromatographic methods. *J. Chromatogr. A*, 1998, 815, 3–20.

Mahmoud B.S.M. *Salmonella—A Dangerous Foodborne Pathogen*. InTech, January 20, 2012 under CC BY 3.0 license 450pp. DOI: 10.5772/1308, ISBN 978-953-307-782-6.

Moldenhauer J. Overview of rapid microbiological methods in book. In: *Principles of Bacterial Detection*, eds. M. Zourob, S. Elwary, A. Turner. Springer, New York, 2008.

Nabok A., Tsargorodskaya A., Mustafa M.K., Szekacs I., Starodub N.F., Szekacs A. Detection of low molecular weight toxins using an optical phase method of ellipsometry. *Sens. Actuat. B Chem.*, 2011, 154(2), 232–237.

Odumeru J.A., León-Velarde C.G. Salmonella detection methods for food and food ingredients, in *Salmonella—A Dangerous Foodborne Pathogen*, Barakat S.M. Mahmoud (Ed.). InTech, Rijeka, Croatia, 2012, pp. 373–392. http://www.intechopen.com/books/salmonella-a-dangerous-foodbornepathogen/salmonella-detection-methods-for-food-and-food-ingredients

Oh B.K., Kim Y.K., Park K.W., Lee W.H., Choi J.W. Surface plasmon resonance immunosensor for the detection of *Salmonella typhimurium*. *Biosens. Bioelectron.*, 2004b, 19, 1497–1504.

Oh B.K., Lee W., Kim Y.K., Lee W.H., Choi J.W. Surface plasmon resonance immunosensor using self-assembled protein G for the detection of Salmonella paratyphi. *J. Biotechnol.*, 2004a, 111, 1–8.

Patel J.R., Bhagwat A.A., Sanglaya G.C., Solomon M.B. Rapid detection of Salmonella from hydrodynamic pressure-treated poultry using molecular beacon real-time PCR. *Food Microbiol.*, 2006, 23, 39–46.

Pitt J.I., Hocking A.D. *Fungi and Food Spoilage*. Blackie, London, U.K., 2nd edn., 1997.

Poltronieri P., de Blasi M.D., D'Urso O.F. Detection of Listeria monocytogenes through real-time PCR and biosensor methods. *Plant Soil Environ.*, 2009, 55, 9, 363–369.

Prodromidis M.I., Karayannis M.I. Enzyme based amperometric biosensors for food analysis. *Electroanalysis*, 2002, 14, 4, 241–261.

Ruan C., Wang H., Li Y. A bienzyme electrochemical biosensor coupled with immunomagnetic separation for rapid detection of Escherichia coli O157:H7 in food samples. *Trans. ASAE*, 2002, 45, 249–255.

Schneider A., Gronevald C. et al., Real-time detection of the genus Salmonella with the light cycler system. *Biochemica*, 2002, 4, 19–21.

Silley P., Forsythe S. Impedance microbiology: A rapid change for microbiologists. *J. Appl. Bacteriol.*, 1996, 80, 233–243.

Son J.R., Kim G., Kothapalli A., Morgan M.T., Ess D. Detection of Salmonella enteritidis using a miniature optical surface Plasmon resonance biosensor. *J. Phys.*, 2007, 61, 1086–1090.

Starodub N.F., Kanjuk M.I., Ivashkevich S.P., Pilipenko L.N., Egorova A.V., Pilipenko I.V. Patulin toxicity and determination of this toxin in environmental objects by optical biosensor systems. In: *Proceedings of 3th International Schentific-Technical Conference Sensor Electronics and Microsystem Technologies (SMEST-3)*, Odessa, Ukraine, June 2–6, 2008, pp. 237–238.

Starodub N.F., Nabok A.V. et al. Immobilization of biocomponens for immune optical sensor. *Ukr. Biochem. Zhurn.*, 2001a, 73, 4, 55–64.

Starodub N.F., Ogorodnijchuk Yu.A. et al., Optical immune biosensor based on the surface Plasmon resonance for the control o Salmonella tiphimurium level in solutions. *Scientifc Herald*, 151, *Ser. Vet. Med. Biosafety Foods*, 2010, 2, 183–189.

Starodub N., Ogorodnijchuk Ju. Efficiency of immune biosensor based on total internal reflection ellipsometry at the determination of Salmonella. In: *Proceedings of the 14th International Meeting on Chemical Sensors*, Nurnberg, Germany, May 20–23, 2012a, pp. 170–179.

Starodub N.F., Ogorodnijchuk J.O. Immune biosensor based on the ISFETs for express determination of Salmonella typhimurium. *Electroanalysis*, 2012b, 24, 3, 600–606.

Starodub N.F., Ogorodnijchuk Ju.A., Romanov V.O. Optical immune biosensor based on SPR for the detection of Salmonella typhimurium. In: *Abstract book: The SENSOR + TEST 2011 Conference*. Nurenberg, Germany, 2011, p. 7.

Starodub N.F., Pirogova L.V., Demchenko A., Nabok A.V. Antibody immobilisation on the metal and silicon surfaces. The use of self-assembled layers and specific receptors. *Bioelectrochemistry*, 2005, 66, 111–115.

Starodub V.M. and Starodub N.F. Electrochemical and optical biosensors: origin of development, achievements and perspectives of practical application, in *Novel Processes and Control Technologies in the Food Industry*, F. Bozoglu et al. (Eds.), NATO series. Kluwer Academic Publishing, Amsterdam, the Netherlands, 2001, pp. 63–94.

Su X.L., Li Y.B. A self-assembled monolayer-based piezoelectric immunosensor for rapid detection of Escherichia coli O157: H7. *Biosenss. Bioelectron.*, 2004, 19, 6, 563–574.

Su X.L., Li Y.B. A QCM immunosensor for Salmonella detection with simultaneous measurements of resonant frequency andmotional resistance. *Biosens. Bioelectron.*, 2005, 21, 6, 840–848.

Svensson H.G., Hoogenboom H.R., Sjobring U. Protein LA, a novel hybrid protein with unique single-chain Fv antibody- and Fab-binding properties. *Eur. J. Biochem.*, 1998, 258, 8902896.

Swaminathan B., Feng P. Rapid detection of food-borne pathogenic bacteria. *Annu. Rev. Microbiol.*, 1994, 48, 401–426.

Syam R., Davis K.J., Pratheesh M.D., Anoopraj R., Joseph B.S. Biosensors: A novel approach for pathogen detection. *VETSCAN*, 2012, 7, 1, 14–18.

Tu S.I., Uknalis J., Gehring A. Detection of immunomagnetic bead captured *Escherichia coli* O157:H7 by light addressable potentiometric sensor. *J. Rapid Methods Automat. Microbiol.*, 1999, 7, 69–79.

Tu S.I., Uknalis J., Gore M., Irwin P. The capture of Escherichia coli O157:H7 for light addressable potentiometric sensor (LAPS) using two different types of magnetic beads. *J. Rapid Methods Automat. Microbiol.*, 2002, 10, 185–195.

Uithoven K.A., Schmidt J.C., Ballman M.E. Rapid identification of biological warfare agents using an instrument employing a light addressable-potentiometric sensor and a flow-through immunofiltration-enzyme assay system. *Biosens. Bioelectron.*, 2000, 14, 761–770.

Waar K., Degener J.E., vanLuyn M.J., Harmsen H.J.M. Fluorescent in situ hybridization with specific DNA probes offers adequate detection of Enterococcus faecalis and Enterococcus faecium in clinical samples. *J. Med. Microbiol.*, 2005, 54, 10, 937–944.

Warsinke A., Benkert A., Scheller F.W. Electrochemical immunoassays. *Fresenius J. Anal. Chem.*, 2000, 366, 622–634.

Wendeberg A. Fluorescence in situ hybridization for the identification of environmental microbes. *Cold Spring Harb Protoc.*, 2010, 1–8. DOI: 10.1101/pdb.prot5366.

Whyte P., McGill K.C., Collins J.D., Gormley E., The prevalence and PCR detection of Salmonella contamination in raw poultry. *Vet. Microbiol.*, 2002, 89, 53–60.

Zarickiy A.M. Salmonellosis. K.: Zdorovja, 1988, 160pp.

Zhihong S., Mingchuan H., Caide X., Yun Z., Xiangqun Z., and Peng G.W. Nonlabeled quartz crystal microbalance biosensor for bacterial detection using carbohydrate and lectin recognitions. *Anal. Chem.*, 2007, 79, 6, 2312–2319.

20

Efficiencies of Biosensors in Environmental Monitoring

Nickolaj F. Starodub

CONTENTS

RESUME   The efficiency of the application of electrochemical and optical biosensors in controlling some heavy metal ions (HMI), phosphororganic pesticides (PhOrPe), and herbicides in various environmental objects is discussed in this chapter. Among the different types of biosensors, the main attention is given for those based on ion sensitive field effect transistors (ISFETs), electrolyte-insulator-semiconductor structures (EIS), surface plasmon resonance (SPR), porous silicon (PS), and optical fibers, as well as biosensors that work on the effects of fluorescence, evanescent waves, nonemitting energy transfer, and enhanced chemical luminescence. The principles of development and application of multiparametrical (multienzyme and multiimmune) and combined (multi-immune-enzyme) sensors are analyzed as well. For all types of biosensors, the integration of biologically sensitive materials with transducer surfaces is presented. All biosensor systems are analyzed from the point of view of their sensitivity, reproducibility, application for mass screening, and individual analytical analyses. The possibilities of improvement in biosensors and their prospective application in mass screening and individual analytical analyses as well as the possibilities of commercialization of biosensors are also discussed.

KEY WORDS   Biosensors, Electrochemical, Optical, Application, Control toxic substances, Environmental monitoring

20.1 Introduction

A necessity in the quality control process is the elaboration of analytical methods, their development taking into account diagnostics problems, quality control of food products, and environmental monitoring, which is categorically imperative at present. These predetermine success in response to the challenges thrown by the present to humanity. Increasing globalization of economy requires their international unification. At the end of 50-th of last century the radioimmune analysis (RIA) for the determination of insulin in blood was developed (Yalow and Benson, 1959). That stimulated the development of immunochemical analysis (IChA) with the use of different labels for immunological components, such as enzymes, fluorescent and luminescent labels (Edvall and Perlmann, 1971; Soini and Hemmila, 1979; Van Weemen and Schuurs, 1971). The development of the IChA methods widened the researcher's ability to solve many problems connected with medical, biological, and ecological analyses. Unfortunately, the difficulty, the duration of the analysis, and requirements of the qualification of the investigators limit the perspectives of wide practical introduction of the IChA methods. Biosensors are an alternative for them.

First, an enzymatic sensor for determination of blood glucose was developed by Clark in 1957, which predetermined progress in biosensor technology (Turner et al., 1989). Today, many types of biosensors exist and they are divided into separate groups according to the species of biological materials, transducers, principles of measurements, and the kind of substances they may provide control. In this chapter, we focus our main attention on biosensors that work on the electrochemical and optical principles and are intended for the control of different toxic substances among environmental objects. Among these substances, we mainly analyze HMI and pesticides.

20.2 Immobilization of Biological Materials on Transducer Surface

In the creation of biosensors, the main problem is in the integration of biological material with the transducer on some physical surface. Among the existing methods in this respect we can distinguish the following: (1) direct immobilization without using bifunctional reagents (by physical sorption), (2) direct immobilization as indicated in point 1 but after preliminary treatment of transducer surface by chemical agents, and (3) use of special membranes with or without inclusion of bifunctional reagents. In our work, we have used all these methods depending on the development of concrete biosensors. In any case, the transducer surface (silicon nitride, or some metallic, or others types) at the biological material immobilization should be cleaned by its subsequent immersion into a mixture of potassium bichromate and sulfuric acid, water (vigorous washing), acetone, ethanol, and phosphate-buffer saline (PhBS) solution (10 mM phosphate buffer, pH 7.3, containing 100 mM NaCl).

20.2.1 Immobilization of Enzymes

For the purpose of immobilization, we used different methods (Starodub et al., 1996a,b, 1997a–c, 1998a,b, 1999a–d). According to the first method, enzyme was polymerized by glutaraldehyde (GA) on the gate surface. The mixture of enzyme–water solution and GA was deposited on the gate surface, and the sensor was left in the refrigerator. The second procedure was direct covalent immobilization of enzyme on the silicon nitride surface. In this case, the surface was preliminary treated (30 min) by GA (25%). Next, the surface was washed with distilled water and a drop of enzyme solution was deposited on the sensitive layer. Nonattached enzyme molecules were washed from the surface by vigorously stirring 10 mM *tris*-HCl buffer solution (pH 7.3). The third procedure was accomplished by enzyme embedding in alginate gel. The protocol of its fulfillment was as follows: sodium alginate (1%) was mixed with the enzyme solution in a ratio of 1:1. The mixture (about 1 mL) was deposited on the surface, and a solution of calcium chloride (approximately 0.5 mL, 1%) was applied on top of it. The surface was left in the refrigerator at humid atmosphere for 30 min. The formed membrane was washed with calcium chloride (10 mmol/L) buffer at pH 7.3 to remove free enzyme molecules out of the membrane. Besides, the special enzymatic nitrocellulose strip (NC-strip) biosensor was prepared by the following procedure. NC sheet (10 × 10 cm) was soaked in the enzymatic solution (5 mL at the concentration of 10 mg/mL in the 50 mmol/L, pH 7.3 *tris*-HCl buffer containing 140 mmol/L sodium chloride). The sheet was washed with the indicated buffer for 30 min and dried at room temperature. Then it was slit into single strips. Their dimensions were chosen in an experimental way.

The use of the biomolecule polymerization by GA and their entrapping in alginate gel are very effective ways of enzyme immobilization on the sensor surface. The maximal residual activities for β-glucose oxidase (GOD), urease, and acetyl-cholineesterase (AChE) in the case of the examined enzymes immobilization with the use of GA are obtained for 2.5% GA concentration for enzyme/BSA ratio about 1 and for polymerization time of 1 h. Alginate gel and NC strips are suitable materials for creation of replaceable enzymatic membranes for the sensors and the systems for repeated use. They are very perfect for analysis of substances that are irreversible inhibitors such as PhOrPe and HMI, since it is not possible to achieve the initial enzyme activity by chemical reactivation of membranes. For example, it is proposed to use pyridine-2-aldoximide methiodide (PAM-2) for reactivation of AChE after its interaction with PhOrPe. Unfortunately, we cannot achieve the initial enzyme activity and enzymatic sensor response after such reactivation (Figure 20.1).

The response time of an enzymatic sensors based on the alginate gel membrane is of the same order to that of the biosensors that contained the membrane with polymerized GA. The main part of the output signal (about 90% of its maximal value) was reached in 2 min. This time for the measuring system based on the NC strip is much longer (about 10–20 min depending on the strip dimensions). The standard deviation of the output signals of the enzymatic sensors based on alginate gel and NC-strip membranes for series of measurements and for different membrane castings did not exceed 10%. All these membranes were examined about preservation of the enzymatic activity of GOD, urease, and AChE. It was stated that the best way was immersing membranes in the solution containing 20 mM phosphate buffer (PhB), 0.1% sodium azide, and 1 mM EDTA, pH 7.3. EDTA prevents the formation of sulfides that may appear as a result of the presence of small concentrations of metal ions even

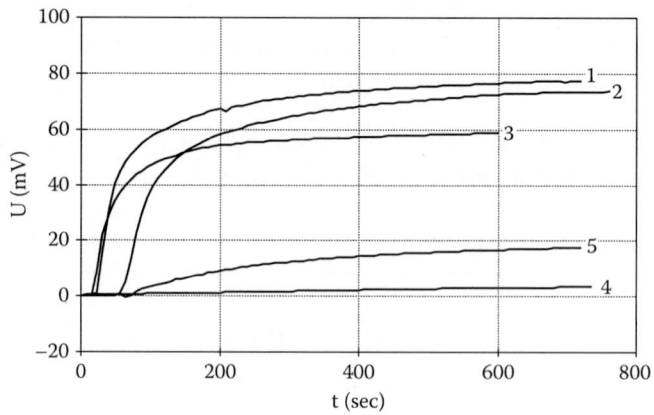

FIGURE 20.1
Enzymatic sensor output in the presence of the different concentrations of PhOrPe (chlorpyrifos) and after reactivation by 2-PAM iodide. 1-substrate 1 mmoL/L AChI alone, 2–4-in the presence of 0.5, 5 and 50 μM, pesticide respectively, 5-after reactivation.

in distilled water. It is important to mention that the addition of low-weight thiol-containing substances to the urease membrane does not promote enzyme stabilization. Maybe, it results in the formation of disulfide bounds with the thiol group of active center.

We have developed special approaches based on the incorporation of biomaterial (enzymes, antibodies, antigens, and cells) into polymeric membrane, which is among the simple and effective methods of immobilization. The first attempts with the application of polymers for the biological material inclusions were made several decades ago and shortly after discovering such type of analytical devices. Generally, it was polymers produced chemically on the basis of acrylic acids. Attempts were made for inclusions in such gel as enzymes and cells. Simultaneous with the validation of the advantages of this approach, their defects that should be eliminated by the application of some initial components and realization of the specific ways for their polymerization to achieve desirable effects were determined. We have proposed the application of liquid polymerizable compositions (LPC) on the basis of monomer-olygomeric substances at the biological membrane creation, which may be considered as a prospective approach directed at providing the previously mentioned practice demands (Rebrijev, 2000, 2002; Rebrijev and Starodub, 2001; Rebrijev et al., 2001, 2002 a,b). These compositions give the possibility of forming sensitive membranes with adjustable physical-chemical and mechanical abilities without strong temperature and chemical destructive effects on biological molecules. Among the most widely dispersed LPC, it is necessary to mention a number of monomeric and olygomeric acrylate compounds (acrylic, metacrylic acids, their ethers, and derivatives) as well as urethane olygomers and vinyl copolymers (sterol, vinyl acetate, vinylidenchloride, vinylpyrrolidone, and others). At the varying of chemical origin and concentration of some components, there is a possibility of regulating many parameters of biological membranes obtained on the basis of these components.

The use of the LPC in biosensors supposes that they should be characterized by the number of indexes, namely: they should be nonactive concerning biological substances, permeable in respect of determined analytes, as well as have defined hydrophobic–hydrophilic balance and sufficient level of adhesion to the transducer surface. The liquid photopolymerizable composition (LPhPC) causes special interest in biosensorics. However, its wide application is restricted by the practice demands mentioned previously. As a rule, at the biosensor creation the influence of supported substances on the biological materials is not especially observed. Usually, the excess of biological material is taken, and for the estimation of its state nondirect approaches are used, namely: the determination of biosensor response, the rate of product formation, and others. At the same time, the change in the structure of biological molecules at the creation of biochips or during their preservation reflects disproportionately on the intensity of response and lifetime of biosensor work. Moreover, at the multilayer immobilization of biological material the inner layers may work with the small productivity in comparison with the external ones due to the diffusive restrictions.

Today many investigations have been conducted in this direction, and the analysis of the formation of biological membranes by photochemicals is the main purpose of this chapter. Attention is paid to the general approaches of fulfillment of polymerization procedures, in particular, on the specifics of biomaterial inclusion and its photo-cross linking with the transducer surface. Special consideration is given to urethane and acrylate derivatives as well as the use of photopolymers in biosensors and their practical application. A detailed analysis of the application of LPhPC at the creation of enzymatic biosensors has been presented in Starodub (2011).

20.2.2 Immobilization of Immune Components

In this case a number of methods were used (Starodub et al., 1992a–d, 1994a,b, 1996a,b, 1997a–d; Wang, 1999). At the direct immobilization of one of the components of immune reaction on the silicon nitride surface, the following procedure was fulfilled. By drop coating, 2.5% water solution of the GA was placed on the prepared surface. After 90 min at room temperature (25°C), the GA was placed again according to the same procedure. Then, after 90 min from the second GA deposition the chip was washed with distilled water. After this, bovine serum albumin (BSA) at a concentration of 20 g/mL in 10 mM PhB, or nondiluted antiserum, or protein A (20 µg/mL in PhB) was placed on the gate surface of the ISFET and kept for 1 h in humid conditions. Treatment of the chip with the GA was carried out at room temperature. At last, the chip was kept overnight at 4°C, and then, after vigorous washing with PhB, was immersed into the glycine solution (10 mg/mL in PhB and room temperature) in order to block nonreacted groups of the GA. After 30 min at room temperature the surface of the chip was washed with PhB. Another approach with the direct immobilization of one of the components of immune reaction at the surface of the optrodes or on the inside wall of the special capillary was fulfilled in two ways. In one case before immobilization the optrode surface was activated with cyanogen bromide at—150°C in the presence of triethylamine. Then, the optrodes were immersed into the antigen or the antibody solution for about 30 min and in 0.1 M glycine solution for 40 min to block the remaining free functional groups. In the other case, the optrodes were previously silanized and then activated with GA. The process of silanization was accomplished as follows. Aminopropyltriethoxysilane (APTES) was frozen and placed in a vacuum, where the optrodes were placed. The vacuum container was filled with APTES vapor in which the optrodes were held for no less than 12 h. Then it was placed for 3 h in vapor of GA. At last, the optrode with the activated surface was immersed into the solution of one of the immune components (for 20–30 min) and after washing by PhB the remaining free functional groups were blocked by 0.1 M glycine solution. The biological material bound 1.8 times more effectively if BrCN was used instead of APTES and GA.

NC strips were used too for immobilization of immune components. They were immersed in the biomaterial solution at a concentration of 0.1 µg/mL.

Sorption lasted for 1 h at room temperature. After that the strips were dessicated and the remaining free functional groups were blocked.

In the case of the use of the PS as transducer, its surface was washed several times by distilled water and then with alcohol, and then it was dried at room temperature. The antibodies or antigens were deposited by spontaneous sorption. For that purpose PS samples were immersed into the buffer solution (20 mmol/L PhBS, pH 7.3) with a biological material of a certain concentration and retained there for 15–60 min. After this time PS surface was washed with PhBS buffer (pH 7.3). According to the data obtained, 60 min was chosen as adequate time for long experiments. Such a conclusion was made, since the data showed that periods of 15 and 30 min of biological material immobilization are rather short to have on the surface such a quantity of antibodies that is sufficient for the sensitive detection of analyte (Mb). At the same time, periods of 60 and 90 min are sufficient for efficient antibody immobilization, and they showed similar results.

A number of other approaches of immobilization of some immune components we have used in the development of immune biosensors have been described in other chapters of this book.

20.2.3 Immobilization of Enzymes and Immune Components on Siliconorganic Polymers

The sol-gel technology is widely used in the fabrication of biosensors (Samodumova et al., 1976; Samodumova, 1983; Samodumova and Starodub, 1995). The sol–gel process is a method of material preparation by room temperature reaction of organic precursors. Typically, a low-molecular weight metal alkoxide precursor molecule (usually tetramethoxysilane: TMOS; $Si(OCH_3)_4$ or tetraethoxysilane (TEOS; $Si(OC_2H_5)_4$) is hydrolyzed first in the presence of water, acid catalyst, and mutual solvent, generally ethanol. Hydrolysis causes Si-OH group formation. In the next stage Si-O-Si polymers form in a condensation reaction. As result of it, a colloidal suspension (sol) and eventually a gel are in solution. Next, the solvents are removed from the interconnected porous network (Figure 20.2.).

Hydrolysis

$$M(OR)_4 + xH_2O \longrightarrow M(OH)_4 + xROH$$

$M + Si, Ti, Al, B, Zr, Ce$

Condensation

$$- M - OH + RO - M - \longrightarrow - M - O - M - + ROH$$
$$- M - OH + OH - M - \longrightarrow - M - O - M - + H_2O$$

Polycondensation

$$x(- M - O - M -) \longrightarrow (- M - O - M -)x$$

FIGURE 20.2
Steps of sol–gel process.

The selectivity of siliconorganic polymers with respect to biomolecule sorption may be achieved in two ways (Samodumova et al., 1976; Samodumova, 1983; Samodumova and Starodub, 1995). One of them is connected with geometric modification by alternation of the parameters of adsorbent structure under appropriate conditions of synthesis when the chemical structure of the polymers remains unchanged. The second way is based on chemical modification of siliconorganic polymers so that the following could take place: (a) alternation of the character of the group linked with silicon atoms (-Si-CH_3, -Si-OH, -Si-CH = CH_2, Si-(CH_2)-CH-NH_2 etc.; (b) including heteroatoms in the silicon chain (Al, Zn, Cu, Co, Mn, Fe). We have obtained the following modifications of siliconorganic polymers: (1) polymethylsiloxane (PMS), a hydrophobic adsorbent; (2) silicapolymethylsiloxane (S-PMS), a polymer with a given ratio of hydrophobic (CH_3) and hydrophilic (OH) groups on the surface; (3) polymetalmethylsiloxane (PolyMet-PMS), when metal is included in the structure of the siloxane chain; (4) polyheterometalmethylsilixane (polyheteroMet-PMS), when the structure of the polymer chain consists of two different atoms of metals instead of some silicon groups; (5) PMS with included aminogroups or modified by aminosilanes for covalent immobilization of biomolecules with the help of GA.

It was shown (Samodumova and Starodub, 1995; Starodub et al., 1995c) that PolyMet-PMS (containing Co-ions) is very effective as membrane for immobilization of urease on the ISFET gate surface. This membrane increases sensor signal in comparison with the protein polymerized by GA. Probably, NH_3 ions formed during enzymatic reaction selectively bound with the Co-contained organosilane and more intensively changed local pH. The same effect was observed in the case of development of immune sensor based on ISFET and with the use of urease-labeled conjugates.

20.3 Enzymatic Sensors Based on ISFETs

We have developed different types of such sensors, in particular, for the determination of glucose, urea, PhOrPe, and HMI (Starodub, 1990; Shul'ga et al., 1992; Starodub et al., 1997a, 1998b). These investigations were fulfilled together with Prof. W. Torbicz and collaborators from the Institute of Biocybernetics and Biomedical Engineering of Polish Academy of Sciences (Warsaw, Poland). The main principles of this sensor work are the same. Since the volume of this article is restricted, we will pay our attention to the sensors intended for PhOrPe and HMI determination only.

In the development of these biosensors, we examined all the previously mentioned types of membranes and their main characteristics indicated in 2.1. The output signal value of enzymatic sensors depends on the pH of the medium,

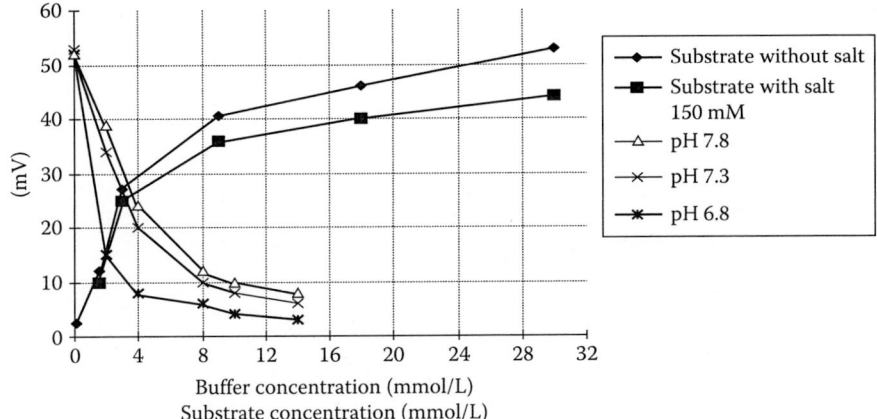

FIGURE 20.3
Dependence of sensor response on substrate concentration, capacity, pH, and ionic strength of buffer.

buffer capacity, as well as salt and substrate concentrations (Figure 20.3). Before measuring of some analytes, we have chosen the optimal conditions. To prevent nonspecific effects, the analyzed samples were diluted by standard buffer solution (3–5 mM, pH 7.3) containing 140 mmol/L NaCl. This dilution had to precede analysis and was determined by experimental way (from 20 up to 100 times of initial sample volume).

Nondiluted vegetable sap cannot be analyzed by ISFET-based enzymatic sensor due to strong influence of sap components on the enzyme activity and ISFET response as well (Figure 20.4). It was shown that the dilution of cabbage and potato saps by the previously mentioned standard buffer 40–60 times allows to sharply decrease nonspecific influence of sap

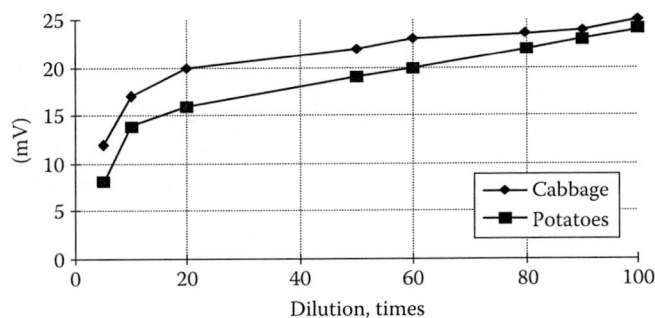

FIGURE 20.4
Dependence of the AChE-based biosensor output signal on dilution of cabbage and potato saps.

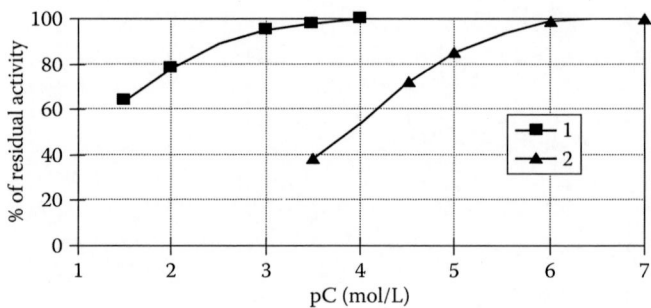

FIGURE 20.5
Dependence of the residual activity of BChE (1) and urease (2) on Co^{2+}-ions concentration.

components on the sensor response. Various types of cholinesterase (ChE) sensors have a different sensitivity to PhOrPe. This sensitivity decreases in sequence: AChE, butyryl-cholinesterase (BChE), and total ChE. Urease sensor is not sensitive to PhOrPe and more sensitive to HMI than sensors with membranes containing any ChE. The sensitivity of ChE sensors depends on HMI to be analyzed. The upper limit of BChE sensor sensitivity was about several mmol/L of metal ion concentration (Figure 20.5). The upper limits of urease sensitivity to these ions were at the concentration range of 10^{-4}–10^{-6} mol/L (depending on the kind of ion to be analyzed). Urease sensor was inhibited irreversibly by HMI. At the same time, inhibition of ChE sensor is reversible, and their activity may be restored by removing HMI from enzymatic membranes. It is very important to underline the following differences in the reaction of urease and ChE sensors on the presence of HMI in solution.

20.4 Multienzymatic Sensors

To diminish nonspecific signal in the case of PhOrPe analysis, which is generated by ChE sensors in the presence of HMI in solution, it is necessary to have information from urease sensor. The level of specificity of analysis may be sharply increased by multifactorial analysis with the help of multi-biosensor.

We, together with Prof. Yu. Shirshov and collaborators from the Institute of Physics of Semiconductors of Ukrainian National Academy of Sciences (Kyiv, Ukraine), have developed a multichannel enzymatic sensor based on EIS structure with silicon nitride ion-sensitive layers as transducers (Starodub et al., 1997a,c, 1998a,b, 1999a–c; Kukla et al., 1998, 1999). The overall scheme of the six-channel sensor array with a flow-through injection

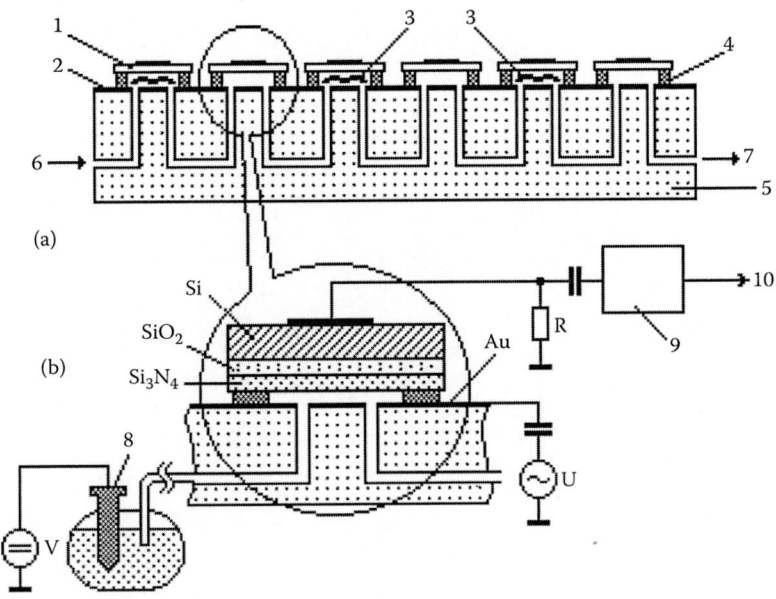

FIGURE 20.6
Schematic view of EIS sensor (a): 1-sensitive structures, 2-gold contrary electrode, 3-enzymatic membrane, 4-rubber sailings, 5-organic glass support, 6-flow in, 7-flow out and the measuring cell structure (b): 8-reference electrode, 9-C/V convertor, 10-connection to PC.

system is shown in Figure 20.6. Each measuring cell (microreactor) consists of a pH-sensitive multilayer structure of type $Si-SiO_2-Si_3N_4$-electrolyte and a metallic counter electrode.

The operational principle of measuring cell is based on the measurement of high-frequency C–V curves for multilayer structure and the determination of flat band potential U_{fb} for the silicon surface. C–V curves were measured for all channels simultaneously. The capacitances of measuring cells used usually were in the 3–10 nF range. The ac voltage (V_{ac}) of 20 mV with a frequency of 3 kHz and the dc bias voltage (V_b) swept in the –2.5 to +2.5 V range vs the Ag/AgCl reference electrode were applied to the measuring cells.

In this type of biosensor, we used NC-sheet membrane. The sensitivity of HMI and PhOrPe determination essentially depends on the incubation time of enzyme membranes in the environment of these analytes. We tested two different approaches: (1) registration of the sensor output signal in the mixture of a substrate and an analyte and (2) separation of the inhibition reaction from the following measurement of the residual enzyme activity. In the latter case, the threshold sensitivity of toxin analysis was about 10 times higher. The time of incubation was chosen experimentally and it was 15 min.

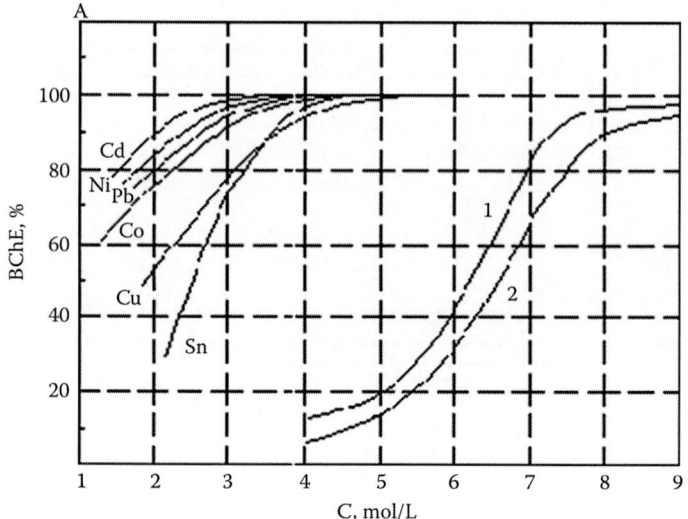

FIGURE 20.7
Residual activity (R) of BChE as a function of the concentrations of HMI and DVDP (1) and phasolone (2).

The results of the measurements of residual activity of enzymes were used as a function of concentration for some inhibition reagents. Actually, such plots can serve as the calibration plots.

The residual activity of BChE inhibited by the same HMI and PhOrPe (2,2-dichlorovinyl dimethyl phosphate–DVDP and phasalone) is shown in Figure 20.7. The effects of both pesticides are very similar. The limit of detection of pesticides indicated here was 10^{-7} M. The range of the linear response was from 10^{-5} to 10^{-7} M. At the same time, the sensitivity of BChE to HMI was substantially lower than that of urease.

Inhibition of the enzyme activity by all HMI investigated is typical for the urease. They strongly and irreversibly suppress this enzyme activity. There is, however, a peculiarity in the case of tin ions. When their concentration increases from 10^{-4} to 10^{-3} M, a substantial (about 125%) growth of urease activity and sensor response occurs. At a further increasing of tin ion concentration, the activity drops abruptly. The concentration of HMI that could be determined by the urease channel of the sensor array lay within the range from 10^{-4} to 10^{-7} M, depending on the type of the metal used. The range of linear detection covered 2–3 orders of the concentration change.

We have preliminary results (Kukla et al., 1998; Starodub et al., 1999a) that it is possibility to identify the presence of individual metal ions in solution by using the multienzyme analysis. It was demonstrated for Co, Ni, and Pb ions in mixture with high probability—all three of them or without the Pb ions. To make such identification (as well as the content determination) more accurate, one needs additional information from the sensor array. Such data may be obtained by employment of an increased number of enzymes as well as by variation of such measurement conditions as pH value and type of buffer (of course, additional calibration curves are to be used in these cases).

It is necessary to note that this analysis had essential limitations in respect of the analyzed samples, namely: (1) a set of components of a mixture should not exceed the number of preliminary calibrated metals for each of the enzyme channels and (2) the concentration range for the analyzed metal ions in general must not exceed 10^{-1}–10^{-6} M, which corresponds to the region of sensitivity for the enzymes used. (To widen this range toward lower concentrations, one needs previously to concentrate the analyzed substances). In spite of the above limitations, the multienzymatic analysis method, in conjunction with the advantages of the capacitance EIS-sensor array, is best suitable for express environmental analysis.

20.5 Immune Biosensors

We have developed different types of immune sensors, including electrochemical and optical ones. Among the electrochemical sensors, we will pay attention in this chapter to those based on ISFETs only.

20.5.1 ISFET-Based Immune Sensors

These sensors were created for determination of Mb and herbicides: simazine and 2,4-D (Starodub and Starodub, 1999a,b; Starodub et al., 2000 a,b). The development of immune sensor for the determination of the previously indicated herbicides was accomplished together with Prof. B. Dzantiev and collaborators from the Institute of Biochemistry, Russian Academy of Sciences (Moscow, Russia).

Biological material was directly immobilized at the gate surface of ISFET by GA. In order to conduct the analysis, the working buffer (WB) containing 10 mM *tris*-HCl buffer, 100 mM NaCl, and 15 mM ascorbic acid was prepared. The pH of the WB was adjusted to 7.5 using the solution of 0.1 N NaOH. The electrochemical activity of horseradish peroxidase (HRP) conjugated with specific antibodies was measured in the presence of hydrogen peroxide in WB according to the registered shift in pH value on the ISFET gate. The conversion of ascorbic acid into dehydroascorbic acid during the enzymatically catalyzed H_2O_2 reduction causes a local basic pH shift, because the dehydroascorbic acid formed is a neutral compound compared to ascorbic acid. The concentration of hydrogen peroxide added to the measuring cell (volume—1 mL) was chosen experimentally, and it was equal to 5 and 10 mM.

In the case of myoglobin (Mb) determination, it was immobilized on the gate surface. Then, in the presence of the specific monoclonal antibodies in

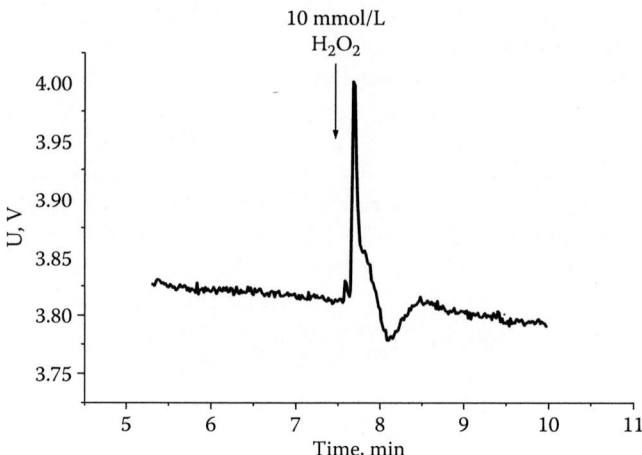

FIGURE 20.8
Typical response of the immune sensor based on ISFET.

solution, only the initial level of the signal was registered. In order to register free Mb, the competitive technique was applied. Concentrations of Mb taken for the analysis were in the range from 1 µg/mL to 1 ng/mL. During registration of the sensor signals, all solutions were stirred and measurements were made at the constant temperature equal to 23°C. After every cycle of measurements of the sensor output to the addition of hydrogen peroxide, the chip was treated with 0.1 N HCl (5 min) and then carefully washed with PhB. The typical response of such a type of immune sensor is given in Figure 20.8.

It was found that the lowest concentration of free Mb, which is possible to be registered by the biosensor, is equal to 1 ng/mL. The sensor output is linear in the range of free Mb concentrations from 10 ng/mL to 1 µg/mL. Such Mb concentrations gave changes in the potential of the ISFET gate in the range from 180 to 20 mV. Since ISFETs are very sensitive to the different factors of the working environment, we investigated the influence of both WB and hydrogen peroxide on the sensor output. For this purpose the chip with two ISFETs, containing on their surfaces Mb and BSA respectively, was immersed into WB. Then, the sensor output was registered at the addition of WB (50 µL) and H_2O_2 (in the concentration which was used during the Mb measurements). It was shown that WB does not influence the sensor output, while the sensor showed small sensitivity (less than 15 mV) to the addition of hydrogen peroxide. H_2O_2 caused acidic shift in pH on the gate surface of the working ISFET (as reminder, at the determination of the enzyme activity the pH moved to more alkaline range). These results testify that neither WB nor hydrogen peroxide contributes to the development of the sensor signal with basic pH shift during the registration of Mb concentration.

The overall time of one measurement was 40–50 min. The limiting stage of the analysis (30 min) is competition among immobilized and free Mb for the binding sites on the antibodies. According to our investigations, the time of the chip immersion in the solution containing antibodies and free Mb can be shortened for 20–25 min. In this case, the sensitivity of the analysis may decrease till 50 or even 100 ng/mL. At the same time, it is important to note that even with such a level of the sensitivity it will be possible to use in practice the immunosensor proposed for the Mb determination, since the concentration of this hemoprotein in the serum of patients with myocardial infarction varies in the range 100–1000 ng/mL (Apple et al., 1995; Woo et al., 1995). So, decreasing the sensitivity the shorter time of the analysis can be achieved.

In the case of herbicide determination, the specific polyclonal antibodies were attached to the ISFET gate via staphylococcal protein A. The analysis was fulfilled by two methods: (i) competitive, when native (detected) and peroxidase-labeled molecules competed for binding with antibodies on the ISFET surface and (ii) sequential saturation of antibodies, left unbound after their exposition to native molecules in the investigated sample, with appropriate labeled herbicides.

The special investigation of the optimal concentration of the hydrogen peroxide indicated that the sensor output was about 15%–20% higher when 10 mM of H_2O_2 was used compared to the signal value at the addition of 5 mM of H_2O_2. Moreover, the maximal level of sensor signal was obtained at this hydrogen concentration.

The concentration of the HRP-labeled simazine during the competitive detection of the simazine was equal to 0.25 μg/mL (by the HRP). A reliable decrease in the sensor signal was observed down to 1.25 ng/mL of simazine in the analyzed mixture. The linearity of signal decrease accorded to simazine concentrations from 5 to 175 ng/mL; in this range, the potential of the ISFET gate varied from 10 to 74 mV. The standard deviation was on average about 5%.

The developed method allowed to reveal a few lower concentrations of 2,4-D in comparison with that which was detected for simazine. In this case, the limit of the detected concentration was 0.5 ng/mL, and the linear plot was in the range of 0.5–150 ng/mL. According to our opinion, the differences in the sensitivity of analysis of simazine and 2.4-D by the proposed immune sensor are connected with the various affinities of specific antibodies.

The overall time of the assay, including the duration of all preparation stages, was 50 min. The limiting stage of the analysis is the competition between the native and the HRP-labeled herbicide for binding with the antibodies (30 min). We performed experiments in which the duration of this stage was reduced to 10 min, while the other assay conditions remained unchanged. In this case, the sensor showed a sensitivity of about 2–6.5 ng/mL for 2,4-D and simazine, respectively. It was more than five times lower compared to the sensitivity of the assay, procedures of which were described earlier. Therefore, reduction of the assay time may lead to a decrease in its sensitivity.

The previously indicated sensitivity of analysis by proposed immune sensor does not respond to practical demands. That is why, we tried to find an approach for its increase. At first, we changed the protocol of analysis and tried to use a so-called "saturation" immunoassay (Starodub et al., 2000a,b). The chip with immobilized antibodies was incubated with native herbicide in the investigated sample and then in order to saturate unbound antibodies—with the HRP-labeled one (0.25 µg/mL, by HRP). Both stages lasted 10 min. This approach allowed reaching a sensitivity of 0.65 ng/mL and linearity in the range 1.25–185 ng/mL at the simazine analysis. In the case of 2,4-D revealing, we obtained a sensitivity of about 0.05 ng/mL and linear plot in the range of 0.1–130 ng/mL.

Then, we examined the second approach for increasing analysis sensitivity. As mentioned here, the level of response of a sensor based on the ISFET depends on the capacity of the measuring buffer. According to our investigations, the sensitivity of the sensor can be increased by decreasing the WB capacity. Therefore, it was shown that the level of the sensor signal may be intensified two times in the case of decreasing of WB buffer concentration from 10 to 5 mM. All these analyzed approaches open the possibility to analyze with confidence both investigated herbicides on the level (0.1 µg/L) as demanded by the United States Environmental Protection Agency and the European regulations (US Environmental Protection Agency, 1992). According to our investigations (after taking into account all peculiarities of analysis by immune sensor based on ISFET), the real level of detection of simazine in the solution to be analyzed is 0.1 µg/L. As for concentration of 2,4-D, this level is 0.05 µg/L. These levels are comparable with that obtained by the ELISA method. It is very important to mention that the overall time of analysis by immune sensor is much less than that with standard immune chemical method.

Destruction of antigen–antibody bonds by the treatment of the chips with 0.1 M HCl for 4 min made it possible to use the chips for 2–3 measurements without signal decrease. Between the measurements, chips were stored in dry state at 4°C. Before every new assay, they were washed with PhB for 2 h.

The proposed analytical system may be compared with other approaches. Some summarized data are presented in Table 20.1. It can be seen that the sensitivity of most proposed immune sensors responds to practical demands. The ISFET-based immune sensor developed by us has the sensitivity to atrazine a little less (about 0.1 ng/mL) than some amperometrical, optical, and acoustic immune sensors, but it is more simple in application, provides more cheap analysis, and promises its prospective use in automatization of measurements. It is necessary to mention that the sensitivity of this method may be sharply increased with the use of specific monoclonal antibodies instead of polyclonal ones. It has been confirmed by our experiments with 2,4-D when we had the sensitivity higher (about 0.05 ng/mL) than in the case of simazine. This opinion conforms with the reports of many investigators in this field.

TABLE 20.1

Comparison of Sensitivity of Different Biosensors in the Analysis of Atrazine

Methods of Analysis	Sensitivity (μg/L)	References
Capillary assay with the amperometric detection and alkaline phosphatase conjugates	0.1	Jiang et al. (1995)
Immunosensor based on surface plasmon resonance (BIACore)	0.05	Minunni and Mascini (1993)
Potentiometric electrodes	0.02	Engel and Baumann (1993)
Surface transverse wave acoustic sensor	0.06	Tom-Moy et al. (1995)
Direct piezoelectric immunosensor	0.1–1	Skladal et al. (1997) and Steegborn and Skladal (1997)
Piezoelectric crystal immunosensor	0.03	Guilbault et al. (1992)
ImmunoFET with ametryn-glucose oxidase conjugates	1.0	Colapicchioni et al. (1991)
Reflectometric interference spectroscopy	0.5	Brecht et al. (1995)
Amperometrical immune sensor with screen printing electrodes and system of HRP with the GOD conjugates	0.01	Keay and McNeil (1998)
Developed ISFET immune sensor for simazine	0.1	Starodub et al. (2000a,b)
Developed ISFET immune sensor for 2,4-D	0.05	Starodub et al. (2000a,b)

20.5.2 Optical Immune Biosensors

Optical immune biosensors developed by us are based on such effects as "evanescent" wave, nonemitting energy transfer, enhanced chemiluminescence, and PS PhL.

20.5.2.1 Fiber-Optic Immune Biosensors

Modern fiber-optic immune biosensors may be divided into two types: the first one is based on the use of fluorescent labels either for single or for double components (antigen or antibodies) of immune reaction and the second one, where the immune chemical reaction is accompanied by fluorescence or chemiluminescence. The first type of devices is based on direct label registration, nonemitting energy transfer between two labels, and "evanescent" wave. As an example of the second type of devices is that based on the principle of enhanced chemiluminescence.

20.5.2.1.1 Immune Biosensors Based on the Evanescent Wave

Evanescent-wave biosensors were developed for simultaneous detection of such drugs as phenytoin and lidocaine (Starodub et al., 1993). Specific antibodies to both types of antigens were immobilized with the help of GA on the optrode surface, which was preliminary treated by monofunctional 4-aminobutyldimethylmethoxysilane. The analysis was carried out in a competitive way with the use of phenatoine labeled with fluorescein isothiocyanate (FITC) and lidocaine

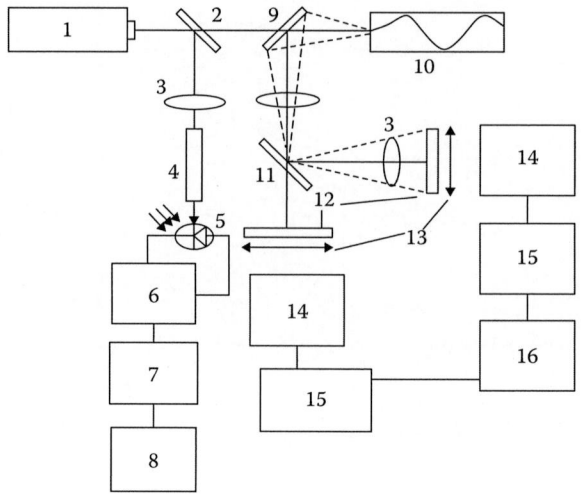

FIGURE 20.9
Scheme of the immune sensor based on the effect of "evanescent" wave (detailed explanation in the text).

labeled with B-phycoerythrin. The scheme of the developed immune sensor is given in Figure 20.9. The light from argon laser (1) spreads through the flat parallel plate (2) and semitransparent mirror (9) and then falls on the entrance butt-end of the optrode (10). The fluorescent signal from the lateral optrode caused by the "evanescent" wave is tunneled through the back of the optrode. Then, this signal is reflected from the semitransparent mirror (9) and falls on the higher effective beamsplitter. One part of the reflected light is directed to the photoelectric cell with a maximum band-pass at 575 nm. The other part of the rays (passed through 11) falls on the second photoelectric cell through an interference heliofilter with maximum band-pass at 520 nm (12). The signal from each photoelectric cell (15) comes into the two-channel photon counter and is then passed to the two-channel recorder (16) through the analogous transformer. To remove artifacts caused by instability of the light source, part of the laser rays reflected on the flat parallel plate is arranged by lenses (3) and with the help of fiber conductors (4) reaches the photodiode (5). The photodiode signal is amplified and registered by a voltmeter (7) and recorder (8). The value of the specific signal (B) is calculated as follows: $B = (I - I_0)/I$, where I = maximum value of the fluorescent signal, I_0 = the light background of photoelectric cell. An optrode is fixed at the distal end only. It permits maximum use of the energy of the "evanescent" wave and rapid replacement of the optrode. The inner diameter of the microcell is 1.6 mm, providing a useful capacity of up to 20 μL.

It was shown that minimal sensor response time is 50 and 75 s for lidocaine and phenatoine, respectively, when measuring their concentration in the 20 ng/mL to 20 mg/mL range. The statistically reliable response was achieved in 2 min at a concentration of 100 ng/mL for both drugs.

TABLE 20.2

Response Level (%) of Optical Immune Sensor
Based on "Evanescent" Wave in the Presence of
Different Substances in Solution (Concentration
of 200 ng/mL)

Substance	Response	Substance	Response
Lidocaine	100	Theophylline	6.4
Phenatoine	100	Gentamicin	5.1
Digoxin	5.4	Digitoxin	4.3
Ferritin	3.8	Methotrexate	2.5
Human IgG	4.2	Phenobarbital	6.7

This concentration is much less than that required for a therapeutic effect. The level of specificity of immune sensor response at the presence of different substances in solution is demonstrated in Table 20.2.

For reutilization of optrode, it was treated by 50 mmol/L glycine-HCl buffer at pH 3.5. After the 10-th measurements, the immune sensor sensibility decreased by almost an order of one. It may be connected with incomplete optrode regeneration and conformational alteration of the immobilized antibodies. Through 2 months of optrode storing in dry and sterile condition at 4°C, the response time increased to 15 s and the sensitivity decreased by 2%.

The created immune sensor offers the very attractive possibility for development of multisensors, which may work on the basis of the principle of homogenous immune assay that is method without division of initial components from ones of immune reaction. It allows to sharply shorten the time of analysis.

20.5.2.1.2 Immune Biosensors Based on the Nonemitting Energy Transfer

Theoretically, the idea of such a sensor is illustrated by the following scheme:

1. $[Ab^1] + [Ab^2] \longrightarrow [Ab^1 - Ag^2]$;
2. $[Ab^1 - Ag^2] + [Ag^2] \longrightarrow [Ab^1 - Ag] + [Ag^2]$,

where
Ab^1 is the immobilized antibody connected with one type of label (donor)
Ag^2 is the antigen with the other type of label (acceptor)
Ab^1-Ag^2 is the complex of labeled and immobilized antibody with the labeled antigen
Ab^1-Ag is the complex of labeled and immobilized antibody with the analyzed antigen

Two labels: FITC and tetramethylrhodamine isothiocyanate (TRIC) were used (Starodub et al., 1992a,b). Specific antibodies were immobilized on the optrode surface after it was treated by BrCN or APTES and GA. See Section 2.2 for information on the efficiency of the utilized approaches for antibodies immobilization.

The scheme of devices is very similar to that presented earlier for immune sensors based on the effect of "evanescent" wave. In this case, only one working optical channel is used (for registration of one type of label of fluorescence).

Specific IgG (anti-influence antibodies) may be detected by this immune sensor at a concentration of about 10 ng/mL. The time of analysis is 15 min, but it may be sharply (up to 2 min) increased with the addition of polyethyleneglycol (MW 6000) to the final concentration of 0.1% in the analyzed medium. If TRIC was replaced by B-phycoerythrin, the sensitivity of analysis increased approximately 10 times.

It is necessary to pay attention both the previously presented types of fiber optical immune sensors as they are able to provide fulfillment of very fast and sensitive analysis. The last index may be essentially increased by the use of effective labels (which have characteristics close to ideal labels). At the same time, these sensors have disadvantages since they demand highly efficient fluorescent labels and expensive lasers and special devices.

20.5.2.1.3 *Immune Biosensors Based on Enhanced Chemiluminescence*

This type of immune sensor lacks the previously mentioned disadvantages (Arenkov et al., 1994a,b; Starodub et al., 1994a,b). Moreover, it allows using commonplace immune reagents. The main elements of construction of such a sensor are given in Figure 20.10. The optrode with the immobilized biological material (see Section 2.2) was placed in a teflon cell of 8 μL capacity containing as a rule the specific antiserum and HRP-labeled antibodies or other appropriate components that are needed for immune analysis. The washed optrode (2) was immersed in the special nontransparent cell (1) containing the substrate for enhanced chemiluminescence. The chemiluminescence signal resulting from the agitated luminol passes through the opened shutter (3) to the photodiode (4), photon counter (5) then processed by the recorder (6) and the computer program.

The succession of events at the detection of enzymatic label is presented in Figure 20.11 HRP conjugated with an immune component (which takes part in

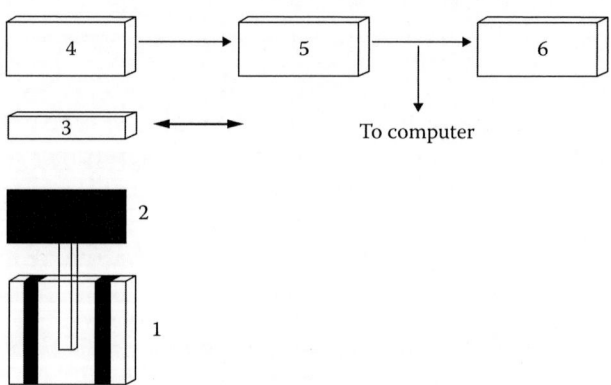

FIGURE 20.10
Block scheme of the chemiluminescent immune sensor (see text for explanation).

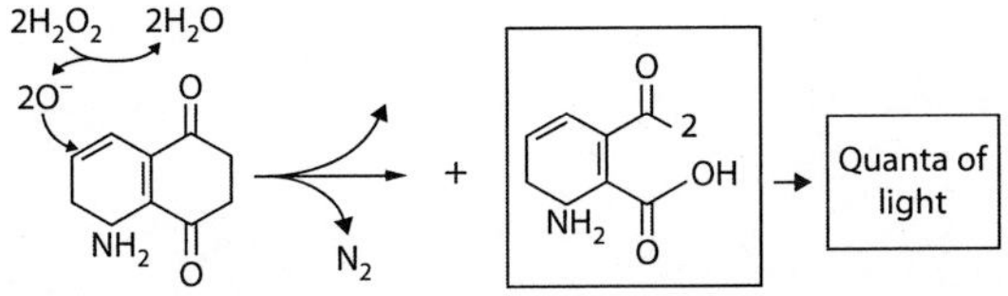

FIGURE 20.11
Chemical basis of the enhanced luminescence.

the immune reaction) catalyxes breaking up of hydrogen peroxide into oxygen radicals, which oxidize luminol to yield aminophtalic acid (1). In this case, the exited part of the oxidized molecules of luminol emits the quanta of light (3). The number of flashes may be drastically increased in the presence of some chemical compounds. The effect of such an enhancer is *p*-iodophenol (2). The level of chemiluminescence depends upon a number of factors: concentration of hydrogen peroxide, *p*-iodophenol, luminol, pH media, etc. According to our data, the optimal conditions are: 2 mM hydrogen peroxide e, 0.06 mM *p*-iodophenol, 0.06 mM luminol, and 50 mM *tris*-HCl buffer pH 8.1.

This type of immune sensor was used for quantitative determination of several antigens such as estradiol-17, α-2-interferon, chorionic gonadotropin, total IgG, components of lysine producing cells, *Salmonella typhimurium*, and antibodies to the influenza virus. In these cases, the sensitivity of this immune sensor was comparable with that of the standard ELISA method, but the rate of analysis using the immune sensor was much faster.

20.5.2.1.4 *Immune Biosensors Based on Photoluminescence (PhL) of Porous Silicon (PS)*

We used the effect of PS PhL for creation of immune biosensor intended for the control of external biological contamination of air around the plant produced (Starodub et al., 1995c; Starodub and Starodub, 1996, 1999b; Starodub et al., 1997a, 1998a, 1999a,b, 2000c). This work was fulfilled together with Dr. L. Fedorenko and collaborators from the Institute of Physics of Semiconductors of Ukrainian National Academy of Sciences (Kiev, Ukraine).

The samples of PS, the size of which was about 4×4 mm^2, were obtained from mono-crystalline silicon of *p*-type with specific resistance $\rho = 10\ \Omega$ cm. Samples of mono-crystalline silicon were treated by laser beam YAG:N-laser ($\lambda = 1.06\ \mu$m; impulse duration $t = 5 \times 10^{-4}$ c; *energy* E = 0.5 J) and then chemically etched in a solution of HF-HNO$_3$: H$_2$O (volume ratio 1:3:5) during 5–40 min at room temperature. Having been etched samples were then passivated.

To study the structure of PS, we use atom force microscopy (AFM). It was shown that the dip of pores was in the range of 10–200 nm, and the diameter and dimension of silicon crystallite varied from 50 to 400 nm. The immune components (antibodies and antigens) were immobilized by physical sorption and were legibly identified on the PS surface by AFM (Figure 20.12). The block scheme of the setup for PS PhL measurement is given in Figure 20.13

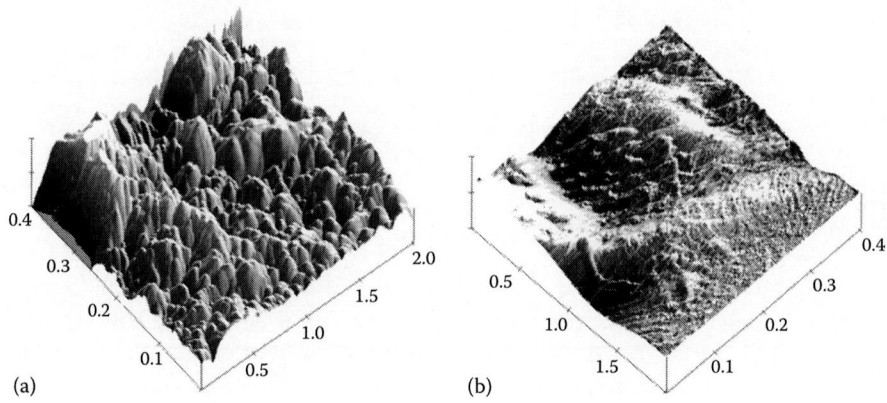

FIGURE 20.12
The view of PS before (a) and after (b) biological material immobilization.

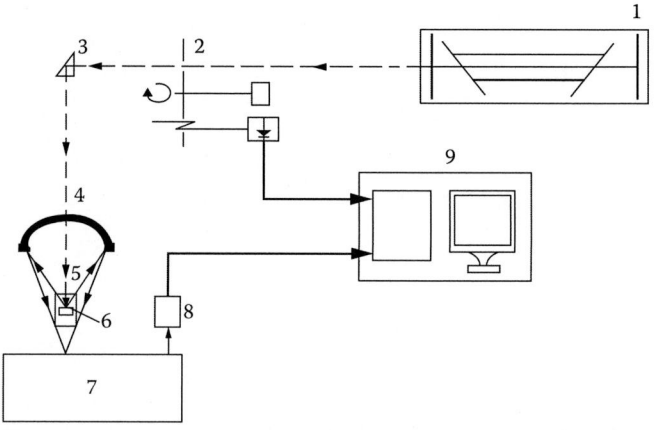

FIGURE 20.13
Scheme of immune sensor based on PS PhL. 1-He-Cd- laser, 2-modulator, 3-prism, 4-mirror, 5-quartz cell, 6-PS-sample, 7-spectrophotometer, 8-photomultiplier, 9-PC IBM.

PS photoluminescence was excited by He-Cd laser (λ = 440 nm, P = 0.001 W) and measured by standard monochromator. The visible photoluminescence observed had a wavelength equal to 650 nm, and its time decay was described by "stretch" exponent.

As demonstrated previously (Starodub et al., 1996a), the PhL did not change at the contact of the PS with distilled water, or 10 mmol/L *tris*-HCl buffer (pH 7.3), or 140 mmol/L buffered sodium chloride solution. The same situation was observed at the separate immersions of the PS in the solution of antigens or antibodies (over a wide range of concentrations, from 10 to 1000 μg/mL) prepared in the previously mentioned buffer containing sodium chloride. The intensity of arbitrary degradation of PL in time after the immersion of the PS surface in different solutions

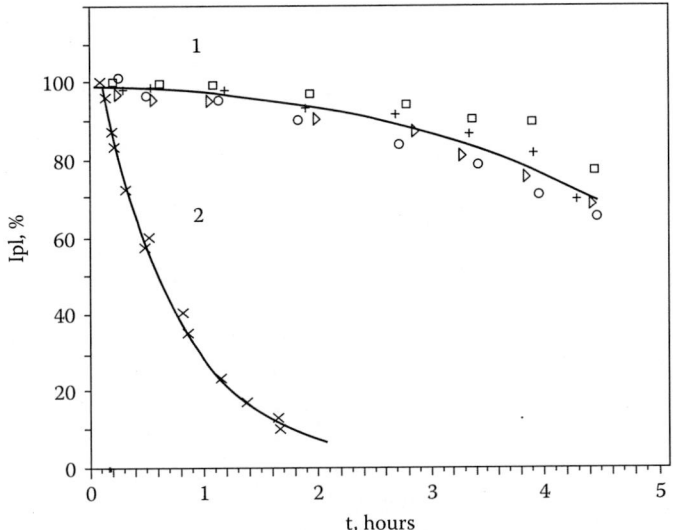

FIGURE 20.14
PhL time delay of PS (see explanations in the text).

is shown by curve 1 in Figure 20.14. The intensity and shape of the PhL spectra remain almost immutable after the exposition of PS to these solutions during no less than 2–2.5 h. At the same time, a large reduction of PhL intensity was observed during the immune complex formation on the PS surface (curve 2 in Figure 20.14). The degree of the PhL intensity reduction depends on the time of exposition and on the concentration of components in the solution analyzed. It was found that after completion of the antibody–antigen reaction, the intensity of PhL was reduced more than 3 times. No other deformations of PhL spectra shape were observed.

According to the existing ideas (Lin et al., 1997; Unal and Bayliss, 1998) about the nature of the PS visible photoluminescence, extinguishing of the photoluminescence can be explained by a process of dehydrogenization of the PS surface, which takes place after a specific immune complex has been formed. Hydrogen is released from silicon bonds and subsequently captured by the immune complex. Torn Si bonds are known to intensify nonradioactive channel of recombination, which leads to the decrease in photoluminescence intensity.

The discovered effect of the extinguishing of PhL of PS after the immune complex formation on the surface of PS was used for the creation of a new type of optical immune sensors. Such sensors were developed for the determination of Mb in solution and for the monitoring of specific biological components in air.

As the first model for the analysis of antigen–antibody complex formation, it was used for the determination of Mb. The process of monoclonal anti-Mb antibodies immobilization was carried out during 15, 30, 60, and 90 min. According to the data obtained, the time of 60 min was chosen as an optimal time for further experiments.

In order to test the work of the sensor in the conditions similar to real ones, human serum of three different dilutions (undiluted, diluted to 1:10, and to 1:100) was used. PS samples with antibodies on the surfaces were immersed into those three samples of serum for 15 min, and then the PhL intensity was measured. The results showed that the PhL intensity of PS samples retained in undiluted serum was about 50%, while PhL intensity of PS samples immersed into diluted serum samples approached 100%. Such changes in the PhL intensity in the case of undiluted serum can be explained either by the fact that the serum of healthy people can contain Mb in concentrations up to 100 ng/mL or by the influence of other proteins. The latter is less probable, since according to our previous results it was shown that sorption of albumin or IgG on the PS surface did not affect the PhL intensity. At the same time, the difference between data obtained for undiluted serum and Mb solution (100 ng/mL) was observed. In the case of Mb solution, the PhL intensity decreased to 40%, while for undiluted serum this value was 50%. Such a difference in the magnitude of the decrease in the PhL intensity can testify about the probable influence of other proteins present in the serum on the intensity of PhL when the immune complex is formed on the PS surface. In order to check such an influence, the following experiments were carried out. Different quantities of Mb were added to the samples of diluted serum (1:10) and then the obtained solutions were used to measure the changes in the PhL intensity. The obtained results are a little different form those that was obtained for the Mb solution. This difference can be explained by the presence of initial quantities of Mb in the serum.

To get an idea about the correlation of the data obtained with the help of the bioaffinic sensor with the results showed by standard ELISA method, the following measurements were undertaken. Samples of serum (dilution 1:10) from three patients, suffering from heart failure disease of different degree of seriousness, were taken for analysis by both methods. The results obtained are presented in Table 20.3. Since the usual concentration of Mb in the serum

TABLE 20.3

Detection of Mb Concentration in Human Serum from People Suffering from Heart Failure Disease of Different Degrees of Seriousness

Number of Experiments	Mb Concentration (ng/mL)	
	ELISA	Bioaffinic Sensor
1	210	200
2	420	400
3	580	600

of healthy people is about 0.1 µg/mL, and in heart failure disease it can increase up to 1 µg/mL, the range of concentrations between 0.1 and 1 µg/mL was used for the construction of a calibration curve. As can be seen from Table 20.3, the data obtained with the help of both methods correlate and their difference is no more than 5%.

Taking into account the fact that the durability of the analysis accomplished by the ELISA method (no less than 4–6 h) is significantly higher than overall time of measurements performed by the bioaffinic sensor (about 15 min), it is possible to assume that the sensor created has high potential for the further development of the area of biological sensors.

In order to study the operational stability of the immune sensor, the PS surface was treated with HCl (0.1 N) or acetate buffer (pH 2.2) after every cycle of measurements. It was found that the PhL intensity decreased dramatically after the first cycle of the PS sample use in the experiments. It may be explained by either possible destruction of the PS surface or incomplete removal of the immune complex from the surface. The first assumption is more real than the second one. Moreover, it has been shown (Demidovich et al., 1998; Sailor et al., 1998) that PS is highly sensitive to some types of vapors and acceptor molecules.

The creation of an immune sensor based on PS PhL for the control of the level of biological substances in air was the next step. Air pollution belongs to one of the most important environmental problems. There are many sources of the contamination of the environment by biologically active substances, such as particles of protein nature. Different types of protein particles exhausted into the air from brewery, dairy, biotechnological plants etc. cause numerous diseases, including different types of allergies, among people working at those industrial units and living in the surrounding areas (Artamonova and Cherednik, 1988; Binnie, 1991). The status of this problem is difficult to determine with existing time-consuming methods, which are incapable of separating specific proteins coming from a particular industrial unit from the total level of protein appearing in the air naturally.

When the sensitivity of the biological sensor was studied, solutions containing protein particles taken from the pipes for dust removal of the plant were prepared in different concentrations over a range from 0.001 to 0.1 mg/L. PS samples, containing specific antibodies on the surface, were immersed into these solutions for 60 min. The data of these measurements show that a biological sensor is capable of monitoring the content of specific proteins in the buffer solution starting from the concentration of 0.1 mg/L at the time of immersion equal to 60 min. This sensitivity allows the detection of 0.33 µg of the specific protein in 1 m^3 of the air. In order to monitor specific proteins in the air, calibration curves were constructed. For this purpose, a certain quantity of microorganisms was used to get a water-soluble fraction of specific proteins. In the previous experiments, it was shown that, when on the surface of PS sample, an interaction between antibodies and antigen takes place, the PhL intensity of PS decreases significantly already during the first 30–60 min. Thus, two groups of experiments were carried out. In the first group, changes of the PhL intensity of PS samples under the conditions of their immersion into solutions of specific proteins of concentrations from 0.1 to 10 mg/L during 60 min were investigated. In the second group, measurements of the PhL intensity were undertaken when PS samples were retained in the specific protein solutions with concentrations over the range 10–260 mg/L for 30 min. A linear plot of this calibration curve lies in the range of specific protein concentrations between 40 and 240 mg/L.

The results of measurements of the concentration of specific proteins in the air inside the biotechnological plant with the help of an immune sensor based on the PS PhL and standard ELISA method are presented in Table 20.4. It may be seen that the sensitivity of the sensors proposed is comparable with the sensitivity of the ELISA method, but the overall time analysis is much shorter in the case of sensor approaches.

Moreover, the data obtained by the developed immune sensor based on the PS PhL were compared with another approach, involving the use of chemical luminescence in combination with fiber optics. With the help of this technique, protein particles were determined in two ways. According to the first one, optrodes were immersed into the solution containing proteins from the air samples taken, while the second way utilized pulverization of dry particles into a special box, after which the concentration of specific proteins was determined. After the calibration curves were constructed, it was found that the first way of the use of optrodes allowed determination of specific biological pollutants at the lowest concentration equal to 5–10 ng/mL with a linear plot of the curve in the range of concentrations from 10–20 to 200 ng/mL. For the way when pulverization was utilized, the level of the sensitivity

TABLE 20.4

Content of Total and Specific Proteins in the Air of Different Workrooms of Biotechnological Plant Determined with the Help of the Sensors and ELISA Method

Place of Air Sampling	Content of Specific Proteins (mg/m³) Determined with the Help of		
	Sensor Based on the PS PhL	Fiber-Optic Sensor	ELISA Method
1. Workroom of fermentation	0.03 ± 0.005	0.035 ± 0.007	0.033 ± 0.004
2. Workroom of lysine concentration	0.045 ± 0.007	0.052 ± 0.009	0.053 ± 0.010
3. Workroom of drying of lysine	0.45 ± 0.006	0.56 ± 0.016	0.050 ± 0.015
4. Workroom of packing of finished product	0.75 ± 0.05	0.87 ± 0.06	0.82 ± 0.10

was a little lower, namely, it was about 15–20 ng/mL and the linear plot of concentration values was in the range from 20–30 to 300 ng/mL.

The results obtained testify that although the technique utilizing pulverization is closer to real conditions, it is a little less sensitive than the analogous method applied with the help of optrodes. The lower sensitivity can be explained by partial loss of protein particles due to the pulverization and their subsequent dissolution in buffer solution. Having compared the sensitivity of the sensor based on the PS Ph and chemical luminescence in combination with fiber optics, it is possible to note that the first sensor is less sensitive than the second one. At the same time, the sensitivity shown by the immune sensor utilizing PS Ph is sufficient for its frequent and fast use for the determination of the content of biological pollutants in the air.

It was shown that the workrooms where lysine is dried and packed are the most contaminated by both total and specific proteins. Moreover, coming from the beginning of the lysine production to the end of the process, the values of concentrations of specific and total proteins in the air increased more than 10 times. Such changes in values of protein concentrations can be connected with different production and preservation conditions.

Then, the developed immune sensor based on the PS PhL was used for determination of the biological pollutants in the air outside the plant. The obtained results (Table 20.5) suggest that on the first day of the experiment

TABLE 20.5

Mean Concentrations of the Specific Antigens in the Air of Different Districts Surrounding the Plant

Place of Air Sampling	First Day with Moderate Wind (μg/m^3)		Second Day with Almost No Wind (μg/m^3)	
	Sensor Based on the PS Ph	Fiber-Optic Sensor	Sensor Based on the PS Ph	Fiber-Optic Sensor
1. On the territory of the biotechnological plant in the vicinity of the chimney	0.50 ± 0.07	0.52 ± 0.04	3.33 ± 0.37	3.37 ± 0.31
2. At a distance of 500 m from the chimney	0.50 ± 0.08	0.48 ± 0.06	2.17 ± 0.18	2.19 ± 0.16
3. In a residential area at a distance of 2 km from the chimney	0.33 ± 0.05	0.36 ± 0.03	1.50 ± 0.12	1.47 ± 0.09
4. In a residential area at a distance of 4 km from the chimney	0.33 ± 0.06	0.35 ± 0.03	1.33 ± 0.16	1.34 ± 0.13
5. In a residential area at a distance of 6 km from the chimney	<0.33	0.29 ± 0.03	0.83 ± 0.09	0.81 ± 0.06
6. At a distance of 2 km from the chimney in the direction opposite to the center	0.33 ± 0.07	0.35 ± 0.04	1.21 ± 0.11	1.24 ± 0.10

the concentrations of the specific biological pollutants at a distance 2 and 4 km from the chimney were very close to the minimum antigen content detectable by the biological sensor based on the Ph PS.

The concentration of the specific antigens at a distance 6 km from the chimney was beyond the sensitivity of the sensor. It was found that the sensor based on fiber optics in combination with the enhanced chemical luminescence was capable of detecting the content of specific biological contaminants in the air of that place. This can be explained by the slightly higher sensitivity of this sensor. Since low permissible doses for the yeast particles and their components coming from a biotechnological plant producing lysine have not been established yet, there is no clear solution whether the sensitivity of the Ph PS sensor should be increased or not. Taking into account that with the help of this sensor it is possible to detect as less as 0.33 μm in 1 m^3 of the air, it seems less probable that this sensitivity should be increased.

It was determined that the concentration of the specific antigens in the air on a windy day was lower than that in the still air. This can be explained by the fact that volatile antigen particles coming into the air from the chimney can be either picked up by wind and carried to other places diluting the antigen concentration or just settle down in the region close to the chimney in the absence of wind. The data for the concentrations of the specific biological pollutants in the air inside and outside the plant (Tables 20.4 and 20.5, respectively) showed that the specific antigen content was much higher in the air of workrooms than in the ambient air. This is due to the fact that the air inside the plant is closed in the workrooms and cannot easily escape from there.

The obtained data testify that the difference in the results produced by both sensors was not statistically significant. This suggests that the proposed immune sensors can be efficiently used for the monitoring of specific biological pollutants in ambient air.

As can be seen from Tables 20.4 and 20.5, the sensitivities of the sensors and the ELISA method are comparable. Nevertheless, when more precise determination is necessary, then chemical luminescence in combination with fiber optics can be used.

20.6 Biosensor for the Control of Some Surfactants

In recent times, surface active substances (SAS) have taken a leading place among pollutants of the environment (Oosterkamp et al., 1997). Nonylphenol (NPh – $C_6.H_4(OH)C_9H_{19}$, M.m. = 250) being a stable final product of degradation of nonionic SAS is arrived together with the industrial waste water and may be found in water, air, soil, and sedimentary complex. NPh is a main component for the preparation of a number of materials. Its production at the end of last century in the United States was on the level 104,000 tons and increases by 2% each year. The concentration of NPh in rivers situated near plants produced several dozens of mg/kg of water, and this level is preserved for several years after stopping the work (Bennett and Metcalfe, 1998, 2000). The dissolving NPh in water oscillates depends on the pH and temperature. The concentration of NPh in seawater may achieve 3, 63 g/L in common conditions. It dissolves in organic solvents (Ahel and Giger, 1993; Coldham et al., 1997). NPh and nonylphenol ethoxylates (NPhE) are used in industrial production (preparation of paints, detergents, agrarian chemicals, antioxidants for resins, and some oils) and for different types of cleaning at home (Abelsohn et al., 2002; Matsuya et al., 2002a,b). NPhE is obtained by ethoxylation of NPh with ethylene, and the length of the ethoxylate chain may have varieties that determine the abilities of the synthesized product. Its biodegradation goes through the shortening ethoxylate chain with the formation of the lipophilic metabolites, which are more toxic and stable to microbial destruction. The final products of the NPhE degradation are NPh and nonylphenol-2-ethoxylates, which are much more toxic than initial substances. Acute toxicity of NPhE for fish was equal to 4–12 mg/L (LD_{50}); at the same for NPh, it was on the level of 1.3–2.3 mg/L (Abelsohn et al., 2002).

NPh has estrogenic activity since it has a structure similar to estrogens. That is why NPh belongs to the substances named as endocrine disrupters (Lee et al., 2002; Mantovani, 2002). The last (xenohormones) are denoted as the substances that influence directly or indirectly on the hormonal system destroying the interactions in its frame. This group of substances include phytohormons (for example, genistein), contraceptives (ethinylestradiol), pesticides (DDT), and industrial chemicals such as bisphenol, tributilyn, and others (Mantovani et al., 1999). Their effects depend on age and sex of organisms (Pryor at al., 2000). The main place of their activity is the nuclear receptors of the steroid hormones through which the regulation of the transcription takes place (Lee, 1998). In particular, NPh interacts with the estrogen receptors; in other words, it is the competitive inhibitor. The binding affinity of NPh to the estrogen receptors is much more than for the progesterone ones (Mantovani et al., 1999; Lee et al., 2002). For the estrogen activity of NPh is the optimal structure that contains alkyl group with 6 of 8 carbons in *p*-position on the nonreplaced phenol ridge. In case of association with human receptors, the optimal variant of alkyl group is presented by 9 carbons as it exists in NPh. NPhE with the short chains and *p*-NPh are not as strong as estrogens, but they simulate action of 17-β-estradiol (Bolt et al., 2001; Preuss et al., 2006). The negative effect of NPh on the process of reproduction and the ability to induce feminization and hermaphrodites in fish were demonstrated (Berryman et al., 2004). Estrogenic activity of NPh was studied *in vitro*. NPh is able to induce cell proliferation and associate with the receptors of progesterone in estrogen-sensitive cells of MCF7 of cancer. It can affect the mitogenic activity of the endothelium of uterus of rats (Soto et al., 1991; Watanabe et al., 2004). With the application of DNA analysis, it was demonstrated that most genes activated by NPh in high doses (50 mg/kg) may be activated by estradiol. In the case of doses of 0.5 and 5 mg/kg, NPh has a small effect on the genes, which are activated by estradiol. It was shown that NPh has effect on gene expression similar to estradiol in the tissues of uteri but not in the liver. This underlines the tissue specificity of the influence of alkylphenols (Servos, 1999). Acute and chronic toxicity of NPh for a wide set of water organisms and its potential effect on their endocrine functions were investigated. NPh has toxicity on fish (0.017–3 mg/L), nonvertebrates (0.02–2 mg/L), and aquatic (0.027–2.5 mg/L). In case chronically conditions it was not revealed effects of accumulation at the concentration low than 6 µg/L for fish and 3.7 µg/L for invertebrates (Seki et al., 2003). To determine acute NPh toxicity, 18 species and 2 semi species of organisms were investigated, and it was stated that the effect of acute toxicity was observed at the NPh concentration from 66.72 µg/L for amphipods (Hyalella azteca) to 774 µg/L for mollusca (*Physella virgata*). Freshwater fishes have a middle sensitivity (from 133.9 for *Pimephales promelas* to 289.3 µg/L for *Gila elegans*). Nonvertebrates have the smallest level of sensitivity (LD_{50} for *Lumbriculus variegates* was on the level of 324 and for *Physella virgata*—774 µg/L). In experiments with fish (*Oryzias latipes*) it was established that its abnormal sex differentiation induced by the alkylphenols including NPh is constant or has reversible character at the return organism in the clear water. In reality, the transformation of famine characteristics in male was observed in returning fish in the clear water. The histological investigations have shown that the hermaphrodite gonads were preserved for a long time (up to 2 months). This testifies that feminization of the sex characteristics of fish as result of NPhE does not have a constant character, but the changes in the gonads may be during a long period (Smith et al., 2001).

The semidegradation of NPh is equal from 28 to 104 days (Naylor et al., 1996). Aminophenol ethoxylates may be destructed by anaerobic and aerobic ways. Microorganisms that have degradable abilities to number of the aromatic substances are revealed. They belong to *Pseudomonas* and *Sphingomonas* species. The presence of the latter is needed for the NPh degradation. More than 95% of it may be destroyed in 10 days. NPh forms from NPhE, which are less stable than the first (Katsuhiko et al., 2002).

20.6.1 General Traditional Methods for NPh Determination

To prevent the nondesirable effects of NPhE, it is necessary to control their content in the water of rivers, lakes, in the tissues of fish, and water plants (Soto et al., 1991). For the last 20 years, many approaches for sample preparation, chromatographic separation, and very sensitive determination of NPhE were published. Liquid and gas chromatography in combination with mass spectroscopy are the most applicable (Roland et al., 2002; Loyo-Rosales et al., 2003; Esperanza et al., 2004). Testing with the use of water organisms is wide spread too. This approach may give information about the acute and chronic influence of NPhE on the line of organisms in the frame of taxonomic and ecological conditions. For the determination of chronic toxicity, such organisms as *C. dubia, Chironomus tentans, Mysidopsis bahia,* and *Rana catestebeiana* were used. At the same time for the control of the acute toxicity, *Chironomus tentans, Hyalella azteca, Selenastrum capricornutum, Mysidopsis bahia, Cyprinodon variegates,* and *Skeletonema costatum* are often used (Jobling and Sumpter, 1993, Jobling et al., 1996; Matsuya et al., 2002a,b).

For the determination of alkylphenols, a method based on the use of the highly effective liquid chromatography with the measurement of the fluorescence of europium chelate complex was described. The minimal NPh concentration that could be determined is 0.99 ng/L (Desbrow et al., 1998). Three types of analysis on the basis of the application of submitochondrial electron-transport components and intended for the determination of chemical toxicity were developed. The level of the toxicity was presented by the concentration value at which the intensity of the NAD^+ transformation is decreased by 50%. These values are 1.8 and 1.3 mg/L for NPh and NPhE, respectively. These differences between them are a result of specificity in the structure of the analyzed components. In general, these data have a good correlation with that obtained at the biological testing (Jobling et al., 1996). A highly sensitive and selective method based on the solid phase extraction for the simultaneous determination of 4-NPh and bisphenol in the human milk was created. Its sensitivity to NPh was 0.5 ng/g (Otaka et al., 2003).

To control endocrine disruptors, a number of new methods based on fish were proposed. Vitellogenin is synthesized at the influence of the estradiole and transported in the ovaries through blood. That is why the induction of vitellogenin is an important marker of estrogenicity. With the use of the cell lines of trout (*Oncorhynchus mykiss* and *Poeciliopsis lucida*), a system for the identification of estrogenic-like components in the water medium was developed. Two strategies were proposed: (1) determination of mRNA of the vitellogenin, the level of the induction of the protein synthesis and expression of the reporter genes that are controlled by the estrogen regulating elements; (2) estimation of estrogen-induced mRNA with the help of polymerase chain reaction and application of the competitive reverses transcription (Matsuya et al., 2002a,b). The influence of NPh on trout liver on the molecular level was studied with the use of infrared Fourier spectroscopy, and strong differences between the control tissues and those treated by NPh were determined. They appeared in increasing triglycerides and lipids as well as in decreasing proteins in the latter case. Moreover, the glycogen level decreases and the relative content of the nucleic acids increases. The same effects were observed in trout liver at the influence of 17β-estradiol (Odum et al., 1997; Matsuya et al., 2002a,b). These experiments confirm the efficiency of the previously mentioned fish species for biotesting. Two hybrid systems of yeast for revealing natural and synthetic estrogens were created. They were based on the interaction between the receptor of the human estrogen (ligand-binding domain) and co-activator of the nuclear receptor-binding domain (Estévez et al., 2006).

20.6.2 Traditional Immune Analysis and Immune Biosensors Used for NPh Determination

Immune-chemical analysis in the form of the ELISA method is very sensitive (Dzantiev et al., 1996; Jobling et al., 1996; Samsonova et al., 2003, 2004; Starodub et al., 2005a,b). Its sensitivity is in the range of 10–100 ng/mL. The application of enhanced chemiluminescence allows increasing this sensitivity up to 0.06 ng/mL (Mart'ianov et al., 2004; Lobanova, 2006). It was demonstrated (Starodub et al., 2005a,b) that the polarization-fluorescence immune analysis has the possibility of revealing NPh at a concentration of 4 ng/mL.

In spite of the high sensitivity of these analytical approaches, they have a number of disadvantages such as the long time of analysis (several hours), the necessity to fulfill several procedures, impossibility of their application in field conditions, and on line regime. Today, the practice demands to use express, inexpensive, simple, portable, and sensitive methods for the monitoring of microquantities of contaminating substances, in particular NPh and NPhE, in natural waters and in wastewaters. The creation of such devices may be achieved using the principles of biosensorics only (Bennie, 1999). As some step on the way, immune analysis with the application of "Biocore," based on the principle of surface plasmon resonance (SPR) was developed (Usami et al., 2002). Estradiol served as the selective element that was immobilized on the transducer surface preliminary treated by carboxymethyl dextran. Human recombinant receptor (hrERα) served as the test element. The binding constant in respect of NPh was 7.49×10^{-6} M. At the application of Biocore and monoclonal or polyclonal antibodies, a sensitivity of 2 and 5 ng/mL, respectively, was obtained. In this case, the competitive analysis was carried out when NPh and 9-(*p*-hydroxyphenyl)nonionic acid competed for immobilized antibodies (Samsonova et al., 2004). A similar competitive analysis (but with immobilization of some conjugate on the surface) was fulfilled by the ELISA method at the determination of NPh and 4-octylphenol with the application of the antibodies raised against 4-*n*-alkylphenol hapten. The sensitivity of the 4-*n*-NPh analysis was about 11.5 ng/mL and without any cross-reactivity for the linear alkylbenzene sulfonates and phenolic components (Zeravik et al., 2004). The last approach demands application of the additional antibodies labeled by the horseradish peroxidase (HRP) and special high effective days. The analogous situation appeared at the development of the amperometric immune biosensor, which had a sensitivity of 10 ng/mL with a linearity of 20 ng/mL–44 µg/mL (Evtugyn et al., 2006). Another type of the immune biosensor was created on the basis of piezo crystals when the conjugate of aminophenol with some protein was immobilized on the silanized surface of the transducer and it competed together with free NPh for the antibodies in the solution. The calibration curve was linear in the range of 1–20 ng/mL, and the minimal concentration achieved was 0.8 ng/mL (Ermolaeva et al., 2006).

All the previously mentioned approaches had a high sensitivity, which is in accordance with the practice demands, but the instrumental analytical device could provide express analysis only. To completely fulfill all practice demands, there is a necessity for investigations for searching the most optimal variants of the analysis and the application of the most effective physical structures for the registration of the antigen–antibody complex formation. That is why we tried to combine both these positions in the development of the different types of immune biosensors, the results of which have been discussed earlier in this chapter.

We have conjugated free NPh (from Sigma-Aldrich) and 7-(p-hydroxyphenil)-heptanic acid (HPhHA) with bovine serum albumin (BSA), or ovalbumin (OV), or soybean inhibitor of tripsin (SIT) with the help of Munich reaction in the presence of formaldehyde and 7-(p-hydroxyphenil)-heptanic acid was bound by carbodiimide/succinimide method (Hirose et al., 1998; Mikhura et al., 2000). Antiserums were obtained by rabbit immunization, and then IgG fraction was prepared by precipitation in 50% of sulfate ammonia (Starodub et al., 2004a). The specificity of antibodies was characterized according to their ability to bind different variants of the conjugates in which HPhHA or NPh was as hapten. Cross-reactivity of IgG was estimated with the application of the different classes of structural analogues and metabolites of NPh. Triton X-100 gives about 20% of cross-reactivity with the NPh and among other substances it was on a negligible level.

Two types of optical biosensors based on SPR and total internal reflection ellipsometry (TIRE) as well as a thermal one were used (Starodub et al. 2004a,b; Nabok et al., 2005, 2007; Demchenko, Starodub, 2007; Demchenko et al., 2007). At the optimization of the protocol of the analysis by the ELISA method, it was stated that the optimal concentration of the antigen in the form of NPh-OVA on the sorption stage was 5 μg/mL. The level of the dissolving antiserum was 1:7000, and the overall time of the immune chemical reaction −60 min. The most suitable variant for the NPh detection is the application of the antiserums against NPh-BSA and NPh-OVA. With its help the sensitivity in 20–50 ng/mL and working diapason in 50–1000 ng/mL were achieved. The overall time of the analysis was about 4 h. To determine much lower concentrations of NPh that may result in nondesirable effects, there is a necessity for the application of a special procedure of the sample preparation for providing preliminary concentration of the obtained extracts.

In biosensor analysis, the efficiency of the third algorithms of the analysis was estimated: (a) "direct" when the antibodies are immobilized on the transducer surface and the NPh in the solution is to be analyzed; (b) "competitive"-1 when the free NPh and some NPh-conjugate compete for the binding sites on the antibodies immobilized on the transducer surface; (c) "competitive"-2 when the conjugate is immobilized on the transducer surface and the specific antibodies and the free NPh are in the solution; and (d) "to saturated" when after the interaction of the immobilized antibodies with the free NPh they contact with the NPh-conjugate. It was demonstrated that the "direct" algorithm of the analysis has sensitivity on the level of 80–100 hg/mL. This algorithm of the analysis is very simple, but it is not sufficiently sensitive to fulfill the practice demands. To understand the reason for such a situation and to find the outcome in this case, we studied in what way the preliminarily prepared transducer surface may affect the sensitivity of the analysis. Therefore, it was stated that the effective thickness of the immobilized immune component layer and the formed specific immune complex is increased if the selective antibodies are situated on the transducer surface: (a) indirectly (by physical sorption); (b) through the intermediate layer from the polyelectrolyte's (polyalylamine hydrochloride—PAA and polysterylsulfate—PSS) and protein A; (c) through the dodecanethiol and protein A; and (d) through the dextran sulfate and glutaraldehyde (GA).

According to our investigations, the variants of the analysis with the immobilization of the specific antibodies on the surface were preliminary consequentially treated by the dodecanethiol and protein A was most effective since it allowed achieving a sensitivity up to 10 ng/mL. Moreover, the application of the procedure of the transducer surface treatment by dodecanethiol and protein A increased the stability of the immune biosensor work. Therefore, it was stated that the best way for the preliminary preparation of the transducer surface at the NPh determination by the SPR immune biosensor is the application of dodecanethiol and protein A. In case of the "competitive" analysis, a sensitivity of about 7–10 ng/mL was obtained with a working range of 7–1000 hg/mL. It is necessary to mention that the sensitivity of the analysis in this case is higher than that with the ELISA method. In addition, it is necessary to take into account that the overall time of the analysis is shortened in the case of the immune biosensor application in comparison with the traditional immune-chemical methods. Moreover, there is no need to use the additional immune chemical reagents as the "second" antibodies labeled by HRP. At the carried out analysis by saturating the sensitivity was on the level of 10 ng/mL. At the determination of the NPh by the SPR immune biosensor in the "competitive" or the "to saturate" regimes with the preliminary treatment of the transducer surface by the polyelectrolytes (PAA/PSS/PAA), the signal was stabilized, which testifies the formation of NPh-conjugate layers with the high density. This modification of the surface did not affect the sensitivity of the immune biosensor. The treatment of the surface by dodecanethiol promotes the same in obtaining stable results, but it decreases the level of the immune biosensor response. The creation of the transducer surface of the additional layers of PAA/PSS had a negative effect on the process of immobilization of the NPh-conjugate that connected with the presence a lot of negative charges. The same results were obtained in the fulfillment of both regimes of analysis: the "competitive" and the "to saturate" (Starodub et al., 2004a,b; Nabok et al., 2005, 2007; Demchenko and Starodub, 2007; Demchenko et al., 2007). The calibration curve for the determination of NPh by SPR immune biosensor in regimes of "to saturate" and "competitive" ways is presented in Figures 20.15 and 20.16, respectively. It can be seen that the sensitivity of analysis in "to saturate" regime is slightly higher that in "competitive" one. Nevertheless, in both cases the sensitivity sufficient for practice demands was achieved.

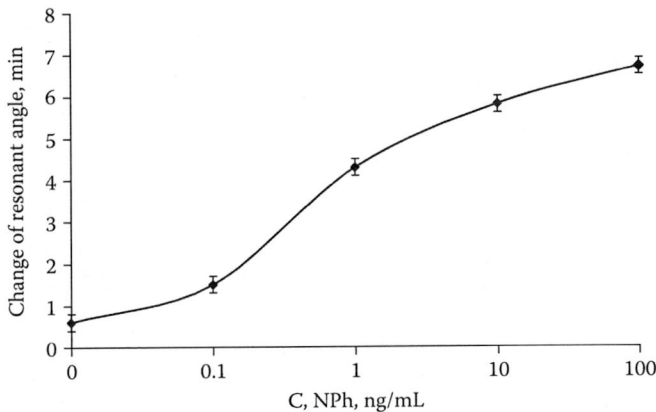

FIGURE 20.15
Determination of NPh by SPR-based immune biosensor in "to saturate" regime.

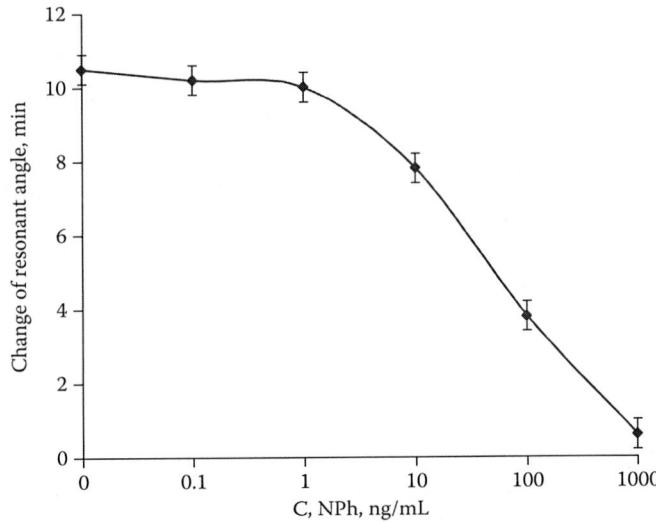

FIGURE 20.16
Determination of NPh by SPR-based immune biosensor in "competitive" regime.

We have developed immune biosensors based on photoluminescence (PhL) (Starodub et al., 1999a, 2000c; Starodub and Starodub, 2001, 2002, 2004; Starodub et al., 2010a) and photoresistance (Starodub et al., 2010a,b) of porous silicon intended for the determination of some biochemical quantities. The principle of the immune biosensor work on the basis of PhL of porous silicon was realized at the registration of NPh level in the solution. It was demonstrated that the sensitivity of this approach may be reached up to 10 ng/mL (Figure 20.17). It is necessary to mention that in this case we worked with the "direct" algorithm of the analysis, and moreover this approach is very simple

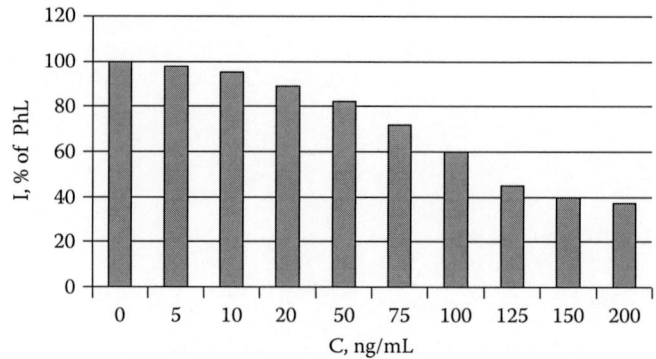

FIGURE 20.17
Dependence of immune sensor signal (intensity of PhL of nPS) on the concentration of NPh in the solution to be analyzed ("direct" way of analysis).

and may provide the express determination of low molecular substances at their screening in the environment.

It is well known that the interaction of antibodies with antigen is accomplished by an exothermic reaction. In particular, the value of ΔH for the reaction of anti-dinitrophenyl antibodies with the dinitrophenyl hapten is equal 73.5 kJ/mol and in the case of the interaction of the immunoglobulin with di-, trinitrophenyl groups it may range from 84 to 50.4 kJ/mol. It gives the possibility of directly registering the interaction of the antigen antibodies (Bevsa et al., 2002). We have created a calorimetric biosensor, the general scheme of which is given in Figure 20.18.

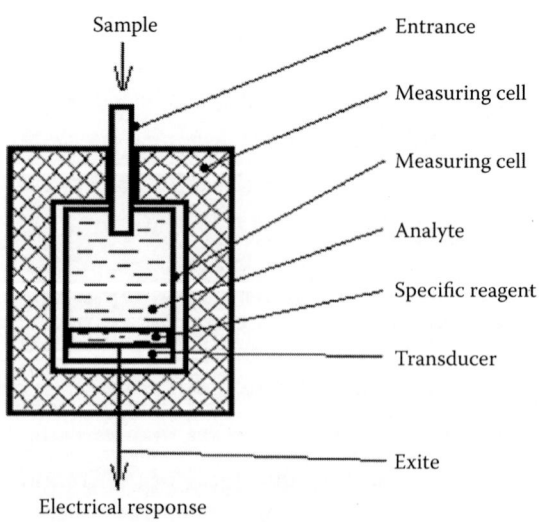

FIGURE 20.18
Scheme of calorimetric biosensors.

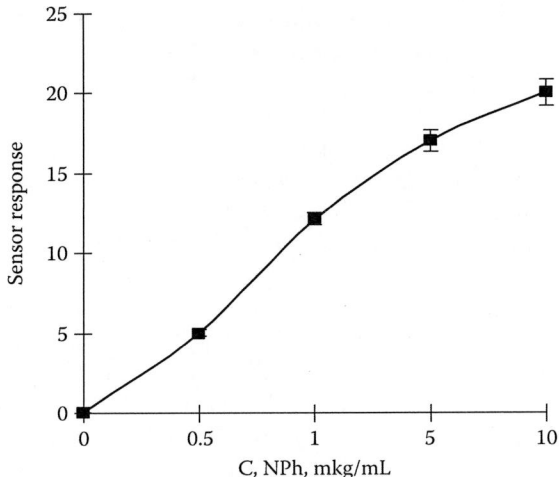

FIGURE 20.19
"Direct" detection of NPh by calorimetric biosensor.

At the fulfillment of the analysis, 150 µL antiserum at a concentration of 10 µg/mL was introduced in the measuring cell and it was kept for 15 min to establish base line. Then 50 µL of the solutions of OV or NPh at a concentration of 1 µg/mL was introduced in the measuring cell. It was demonstrated that the response of biosensor was observed in the case of the formation of the specific immune reaction only. It was stated that the optimal concentration of antiserum was 10 µg/mL. At those conditions, the developed immune biosensor was able to register NPh at a concentration of 1 µg/mL (Figure 20.19). Certainly, it is not enough sensitivity for the practice but it must be taken into account that this biosensor is very simple and can be recommended for screening control of environmental objects. Nevertheless, further investigations are necessary to increase the sensitivity of this device.

20.7 Conclusion

The presented experimental material illustrates huge possibilities of biosensors in providing sensitive and rapid biochemical analysis in the field of medical diagnostics, control of biotechnological processes, and environmental monitoring including food quality control in the presence of a number of toxic agents. These instrumental devices have sensitivity no less than the standard biochemical and immunochemical methods, but they are simpler in performance and provide fulfillment of analysis in a shorter time. Among the different types, the electrochemical and the number of optical biosensors are very widespread, since they are simpler and more developed in comparison with other devices of such type. Today it is very difficult to enumerate all substance for determination of which the biosensors are developed. Nevertheless, a small quantity of them finds a wide practical application. Above all, it concerns of the sensors for the determination of pesticides, HMI, nonylphenol, and other substances of environment. At the present time, other types of biosensors are at the stage of laboratory examination. To expedite the practical application of biosensors, expansion of investigations in following direction is necessary: (a) progress in improvement of transducers and creation of selective sites using biological or synthetic elements; (b) search of new principles of physical-chemical signals registration, especially, to directly reveal specific interaction of biological molecules; (c) development of multiparametrical analysis on the basis of use of sensor array; (d) development of special computer program for discrimination of nonspecific signal and amplification of specific one; and (e) working out of optimal algorithms of analyses to exclude its labor-intensiveness and to achieve a maximal selectivity and sensitivity of analyte determination. It is especially important to pay attention to solving complex problems with the development of multi-biosensors. It is necessary to have a portable mini-biochemical analyzer that could work with enzyme substrates and immune components (Starodub, 1996, 2011).

References

Abelsohn A., Gibson B.L., Sanborn M.D., and Weir E. Identifying and managing adverse environmental health effects: Persistent organic pollutants. *Can. Med. Assoc., J.* 2002, 166(12), 1549–1554.

Ahel M. and Giger W. Aqueous solubility of alkylphenols and alkylphenol polyethoxylates. *Chemosphere*, 1993, 26, 1461–1470.

Apple F.S., Henry T.D., Berger C.R., and Landt Y.A. Cardiac troponin, CK-MB and myoglobin for the early detection of acute myocardial infarction and monitoring of reperfusion following thrombolytic therapy. *Clin. Chim. Acta*, 1995, 237, 59–66.

Arenkov P.I., Starodub A.N., and Beresin V.A. Fiber optic immunosensors based on enhanced chemiluminescence and their application to determine different antigens. *Sens. Actuat.*, 1994a, 18B(1–3), 161–165.

Arenkov P.I., Starodub A.N., and Beresin V.A. Construction and biomedical application of immunosensors based on fiber optics and enhanced chemiluminescence. *Opt. Eng.*, 1994b, 33(9), 2958–2963.

Artamonova V.G. and Cherednik A.I. Hygienic aspects of the study of biological contamination of the environment. *Hyg. Sanit.*, 1988, 2, 3–4.

Bennett E.R. and Metcalfe C.D. Distribution of alkylphenol compounds in Great Lakes sediments. United States and Canada. *Environ. Toxicol. Chem.*, 1998, 17, 1230–1235.

Bennett E.R. and Metcalfe C.D. Distribution of degradation products of alkylphenol ethoxylates near sewage treatment plants in the Lower Great Lakes, North America. *Environ. Toxicol. Chem.*, 2000, 1, 784–792.

Bennie D.T. Review of the environmental occurrence of alkylphenols and alkylphenol ethoxylates. *Water Qual. Res. J.*, 1999, 34(1), 411–422.

Berryman D., Houde F., DeBlois C., and O'Shea M. Nonylphenolic compounds in drinking and surface waters downstream of treated textile and pulp and paper effluents: A survey and preliminary assessment of their potential effects on public health and aquatic life. *Chemosphere*, 2004, 56(3), 247–255.

Bevsa O.V., Shmyryeva O.M., and Starodub N.F. Thermo biosensors: Particularity of design, functioning and perspective practical application *Ukr. Biochem. J.*, 2002, 74(2), 1–20.

Binnie P.W.H. Biological pollutants in the indor environment. In: *Indoor Air Pollution*, eds. J.G. Kay et al., Chelsea, MI, Lewis Publishers, 1991.

Bolt H.M., Janning P., Michna H., and Degen G.H. Comparative assessment of endocrine modulators with oestrogenic activity: I. Definition of a hygiene-based margin of safety (HBMOS) for xeno-oestrogens against the background of European developments. *Arch. Toxicol.*, 2001, 74(11), 649–662.

Brecht A., Piehler J., Lang G., and Gauglitz G. A direct optical immunosensor for atrazine detection. *Anal. Chim. Acta*, 1995, 311, 289–299.

Colapicchioni C., Barbaro A., Porcelli F., and Giannini I. Immunoenzymatic assay using CHEMFET devices. *Sens. Actuat.*, 1991, B4, 245–250.

Coldham N.G., Sauer M.J., Sivapathasundaram S., Ashfield L., Pottinger T., and Goodall C. Tissue distribution, metabolism and excretion of 4-nonylphenol in rainbow trout. *J. Vet. Pharmacol. Therap.*, 1997, 20(1), 256–257.

Demchenko A.V., Mel'nik V.G., and Starodub N.F. Thermal biosensor for detecting nonylphenol in the environment. *Ukr. Biochem. J.*, 2007, 79(5), 173–176.

Demchenko A.V. and Starodub N.F. Modification of surface of immune biosensor on the basis of surface Plasmon resonance at the analysis of nonylphenol. *Biopymers Cell.*, 2007, 23(2), 143–147.

Demidovich V.M. et al. The influence of an adsorption on the charge transport in the porous silicon and the oxidized porous silicon. In: *Extended Abstracts of International Conference Porous Semiconductors—Science and Technology*, Malorca, Spain, March 16–20, 1998, pp. 177–178.

Desbrow C., Routledge E.J., Brighty G.C., Sumpter J.P., and Waldock M. Identification of estrogenic chemicals in STW effluent: 1. Chemical fractionation and in vitro biological screening. *Environ. Sci. Technol.*, 1998, 32, 1549–1558.

Dzantiev B.B., Zherdev A.V., Romanenko O.G., and Sapegova L.A. Development and comparative study of different immune enzyme techniques for pesticides detection. *Int. J. Environ. Anal. Chem.*, 1996, 65, 95–111.

Edvall E. and Perlmann P. Enzyme-linked immunosorbent assay (ELISA). Quantitative assay for immunoglobulin G. *Immunochemistry*, 1971, 8, 71–854.

Engel L. and Baumann W. Direct potentiometric immunoelectrodes. 3. A graphite based atrazine immunoelectrode. *Fresenius J. Anal. Chem.*, 1993, 346, 745–751.

Ermolaeva T.N., Dergunova E.S., Kalmykova E.N., and Eremin S.A. Flow-injection determination of nonylphenol in liquid media using a piezoelectric immunosensor. *J. Anal. Chem.*, 2006, 61(6), 609–613.

Esperanza M., Suidan M.T., Nishimura F., Wang Z.M., and Sorial G.A. Determination of sex hormones and nonylphenol ethoxylates in the aqueous matrixes of two pilot-scale municipal wastewater treatment plants. *Environ. Sci. Technol.*, 2004, 38(11), 3028–3035.

Estévez M.C., Kreuzer M., Sánchez-Baeza F., and Marco M.P. Analysis of nonylphenol: Advances and improvements in the immunochemical determination using antibodies raised against the technical mixture and hydrophilic immunoreagents. *Environ. Sci. Technol.*, 2006, 40(2), 559–568.

Evtugyn A., Eremin S.A., Shaljamova R.P., Ismagilovaa A.R., and Budnikov H.C. Amperometric immunosensor for nonylphenol determination based on peroxidase indicating reaction. *Biosens. Bioelectron.*, 2006, 22(1), 56–62.

Guilbault G.G., Hock B., and Schmid R. A piezoelectric immunobiosensor for atrazine in drinking water. *Biosens. Bioelectron.*, 1992, 7, 411–419.

Hirose K., Akizawa T., and Yoshioka M. Preparation of monoclonal antibody against D-3-methoxy-4-hydroxyphenylglycol. *Anal. Chim. Acta.*, 1998, 36, 129–135.

Jiang T.B., Halsall H.B., Heineman W.R., Giersch T., and Hock B. Capillary enzyme immunoassay with the electrochemical detection for the determination of atrazine in water. *J. Agric. Food Chem.*, 1995, 43, 1098–1104.

Jobling S., Sheahan D., Osborne J.A., Mathiessen P., and Sumpter J.P. Inhibition of testicular growth in rainbow trout (Oncorhynchus mykiss) exposed to estrogenic alkylphenol chemicals. *Environ. Toxicol. Chem.*, 1996, 15, 194–202.

Jobling S. and Sumpter J.P. Detergent components in sewage effluent are weakly oestrogenic to fish: an in vitro study using rainbow trout (Oncorhynchus mykiss) hepatocytes. *Aquat. Toxicol.*, 1993, 27, 361–372.

Katsuhiko F., Naoto U., Hideki U., and Masataka S. Profile of a nonylphenol-degrading microflora. *J. Biochem.*, 2002, 131, 399–405.

Keay R.W. and McNeil C.J. Separation-free electrochemical immunosensor for rapid determination of atrazine. *Biosens. Bioelectron.*, 1998, 13, 963–970.

Kukla A.L., Kanjuk N.I., Starodub N.F., and Shirshov Yu.M. Use of multienzyme analysis for determination of heavy metal ions and phosphororganic pesticides in solutionsı-o. In: *Proceedings of the XII European Conference on Solid-State Transducers and the IX UK Conference on Sensors and their Applications*, September 13–16, 1998, Southampton, U.K., White, N.M. (Ed.), Bristol, U.K., Institute of Physics, 1998, Vol. 1, pp. 521–524.

Kukla A.L., Kanjuk N.I., Starodub N.F., and Shirshov Yu.M. Multienzyme electrochemical sensor array for determination of heavy metal ions. *Sens. Actuat.*, 1999, 57, 213–218.

Lee P.C. Disruption of male reproductive tract development by administration of the xenoestrogen, nonylphenol, to male newborn rats. *Endocrine*, 1998, 9, 105–111.

Lee H.S., Miyauchi K., and Nagata Y. Employment of the human estrogen receptor—Ligand-binding domain and co-activator SRC1 nuclear receptor-binding domain for the construction of a yeast two-hybrid detection system for endocrine disrupters. *J. Biochem.* 2002, 131, 399–405.

Lin V.S.-Y., Motesharei K., Dancil K.-P., Sailor M.J., and Ghadiri M.R., A porous silicon-based optical interferometric biosensor. *Science*, October 31, 1997, 278, 840–943.

Lobanova A.Yu. Studying effect of structure of antigens labeled by fluorochromes on their interaction with antibodies and the development of methods for immune analysis of pesticides, surfactants and polyaromatic carbohydrates for the control of environment and foods. Autoref. of PhD dissertation of Candian Chemistry Science 03.00.04., A.V. Bach Institute of Biochemistry, Moscow, Russia, 2006, 24pp.

Loyo-Rosales J., Schmitz-Alfonso I., Rice C., and Torrents A. Analysis of octyl- and nonylphenol and their ethoxylates in water and sediments by liquid chromatography-tandem mass spectrometry. *Anal. Chem.*, 2003, 75, 4811–4817.

Mantovani A. Hazard identification and risk assessment of endocrine disrupting chemicals with regard to developmental effects. *Toxicology*, 2002, 181, 367–370.

Mantovani A., Stazi A.V., Maranghi F., Macrì C., and Ricciardi C. Problems in testing and risk assessment of endocrine disrupting chemicals with regard to developmental toxicology. *Chemosphere*, 1999, 39, 1293–1300.

Mart'ianov A.A., Zherdev A.V., Eremin S.A., and Dzantiev B.B. Preparation of antibodies and development of enzyme-linked immunosorbent assay for nonylphenol. *Int. J. Environ. Anal. Chem.* 2004, 84(13), 965–978.

Matsuya T., Ohtake K., Hoshino N., Ogasawara M., Harita T., Arao S., and Matsumoto K. Highly sensitive time—Resolved fluorometric determination for alkylphenols by high perfomance liquid chromatography using β-diketonate Europium chelate. *Chromatography*, 2002a, 23(2), 432–445.

Matsuya T., Ohtake K., Hoshino N., Ogasawara M., Harita T., Arao S., and Matsumoto K. Highly sensitive time—Resolved fluorometric determination for alkylphenols by high perfomance liquid chromatography using β-diketonate Europium chelate. *Chromatography*, 2002b, 23(2), 453–462.

Mikhura I.V., Formanovsky A.A., Nikitin A.O., Yakovleva J.N., and Eremin S.A. Synthesis of w-(4-hydroxyphenyl)alkanecarboxylic acids. *Mendeleev Commun.*, 2000, 10, 1–2.

Minunni M. and Mascini M. Detection of pesticide in drinking water using real-time biospecific interaction analysis (BIA). *Anal. Lett.*, 1993, 26, 1441–1460.

Nabok A.V., Tsargorodskaya A., Hassan A.K., and Starodub N.F. Total internal reflection ellipsometry and SPR detection of low molecular weight environmental toxins. *Appl. Surf. Sci.*, 2005, 246, 381–386.

Nabok A., Tsargorodskaya A., Holloway A., Starodub N.F., and Demchenko A. Specific binding of large aggregates of amphiphilic molecules to the respective antibodies. *Langmuir*, 2007, 6, 17616154.

Naylor C.G., Williams J.B., Varineau P.T., and Webb D.A. Nonylphenol ethoxylates in an industrial river. In: *Proceedings of 4th World Surfactants Congress*, Barcelona, Spain, June 3–7, Cambridge Press: Cambridge, U.K., 1996, Vol. 4, pp. 378–391.

Odum J., Lefevre P.A., Tittensor S., Paton D., Routledge E.J., Beresford N.A., Sumpter J.P., and Ashby J. The rodent uterotrophic assay: critical protocol features, studies with nonyl phenols, and comparison with a yeast estrogenicity assay. *Regul. Toxicol. Pharmacol.*, 1997, 25, 176–188.

Oosterkamp A.J., Hock B., Seifert M., and Irth H. Novel monitoring strategies for xenoestrogens. *Trends Anal. Chem.*, 1997, 16, 544–553.

Otaka H., Yasuhara A., and Morita M. Determination of bisphenol A and 4-nonylphenol in human milk using alkaline digestion and cleanup by solid-phase extraction. *Anal. Sci.*, 2003, 19, 1663.

Preuss T.G., Gehrhardt J., Schirmer K., Coors A., Rubach M., Russ A., Jones P.D., Giesy J.P., and Ratte H.T. Nonylphenol isomers differ in estrogenic activity. *Environ. Sci. Technol.*, 2006, 40(16), 5147–5153.

Pryor J.L., Hughes C., Foster W., Hales B.F., and Robaire B. Critical windows of exposure for children's health: The reproductive system in animals and humans. *Environ. Health Perspect.*, 2000, 108(3), 491–503.

Rebrijev A.V. Enzyme sensor on the basis of photopolimeric membranes for the urea determination. *Ukr. Biochem. J.*, 2000, 72(6), 141–142.

Rebrijev A.V. Optimization of conditions of the immobilization of enzymes in photo polymerizable membrane. *Ukr. Biochem. J.*, 2002, 74(4b addition 2), 194–195.

Rebrijev A.V., Ivashkevich S.P., Starodub N.F., Kercha S.F., and Masljuk A.F. Electrochemical sensor on the basis of photo polymeric membrans for the determination of urea. *Ukr. Biochem. J.*, 2001, 73(1), 133–142.

Rebrijev A.V. and Starodub N.F. Photopolymers as immobilization matrix in biosensorics. *Ukr. Biochem. J.*, 2001, 73(6), 5–17.

Rebrijev A.V., Starodub N.F., and Masljuk A.F. Optimization of the conditions for the immobilization of enzymes in photopolymeric membrane. *Ukr. Biochem. J.*, 2002a, 74(3), 82–87.

Rebrijev A.V., Starodub N.F., and Masljuk A.F. Liquid photo polymerizable compositions as immobilization matrix in biosensorics. *Ukr. Biochem. J.*, 2002b, 74(4), 101–106.

Roland J., Meesters W., and Schröder H.F. Simultaneous determination of 4-nonylphenol and bisphenol A in sewage sludge. *Anal. Chem.*, 2002, 74(14), 3566–3574.

Sailor M.J. et al. The chemistry of nanocrystalline porous silicon surfaces and applications for chemical and biological sensing. In: *Extended Abstracts of International Conference, Porous Semiconductors—Science and Technology*, Malorca, Spain, March 16–20, 1998, pp. 114–115.

Samodumova I.M. Adsorptional properties of new group of elementorganic-silicon sorbents with the bivalent metals. *Theor. Exp. Chem. (USSR)*, 1983, 19, 748–751.

Samodumova I.M., Slinjkova I.B., and Kisileva L.I. Influence of some factors on the porous structure of xerogels of polyaluminomethylsiloxanes. *Colloid. J. (USSR)*, 1976, 37, 502–507.

Samodumova I.M. and Starodub V.N. Usage of organosilanes for integration of enzymes and immunocomponents with electrochemical and optical transducers. *Sens. Actuat.*, 1995, B176, 173–176.

Samsonova J.V., Rubtsova M.Y., and Franek M. Determination of 4-p-nonylphenol in water by enzyme immunoassay. *Anal. Bioanal. Chem.* 2003, 375(8), 1017–1019.

Samsonova J.V., Uskova N.A., Andresyuk A.N., Franek M., and Elliott C.T. Biocore biosensor immunoassay for 4-nonylphenols: Assay optimization and applicability for shellfish analysis. *Chemosphere*, 2004, 57, 975–985.

Seki M., Yokota H., Maeda M., Tadokoro H., and Kobayashi K. Effects of 4-nonylphenol and 4-tert-octylphenol on sex differentiation and vitellogenin induction in medaka (oryzias latipes). *J. Environ. Toxicol. Chem.*, 2003, 22(7), 1507–1516.

Servos M.R. Review of the aquatic toxicity, estrogenic responses and bioaccumulation of alkylphenols and alkylphenol polyethoxylates. *Water Qual. Res. J. Can.*, 1999, 34(1), 123–177.

Shul'ga A.A., Sandrovsky A.K., Strikha V.I., Soldatkin A.P., Starodub N.F., El'skaya A.V. Overall characterization of ISFET-based glucose biosensor. *Sens. Actuat.*, 1992, B10, 41–46.

Soini E. and Hemmila I., Fluoroimmunoassay: Present status and key problems. *Clin. Chem.*, 1979, 25, 253–361.

Skladal P., Horacek J., and Malina M. Direct Piezoelectric immunosensors for pesticides. In: *Biosensors for Direct Monitoring of Environmental Pollutants in Field*. Nikolelis D.P. et al. (Eds.), Dordrecht, the Netherlands, Kluver Academic Publishers, 1997, Vol. 38, pp. 145–153.

Smith E., Ridgwayb I., and Coffeya M. The determination of alkylphenols in aqueous samples from the Forth Estuary by SPE-HPLC-fluorescence. *J. Environ. Monit.*, 2001, 3, 616–620.

Soto A.M., Justicia H., Wray J.W., and Sonnenschein C. p-Nonyl-phenol: An estrogenic xenobiotic released from "modified" polystyrene. *Environ. Health Perspect.*, 1991, 92, 167–176.

Starodub N.F. *Lecture Notes of the International Sensors Center of Biocybernetics of the Academy of Science of the Socialist Countries*, Jablonna, Poland, 1990, Vol. 3, pp. 173–202.

Starodub N.F., Arenkov P.Ya., Rachkov A.E., and Berezin V.A. Optoimmunosensors for analysis of specific and non-specific classes of immunoglobulins. *Sens. Actuat.*, 1992a, B7, 371–375.

Starodub N.F. New variants of electrochemical and optical biosensors: Achievements and main modern directions of development. In: *Lecture Notes of the ICB Seminars. Biomeasurements. Electrochemical Measurements of Biochemical Quantities.* P. Bergveld and W. Torbicz. (Eds.), April 1995, Warsaw, Poland, 1996, pp. 159–181.

Starodub N.F. Photopolymers and formation of selective membranes in biosenssor technology. Chapter 18, In: *Environmental Monitoring*, Ema O. (Ed.), Ekundayo, ISBN 978-953-307-724-6, InTech, November 11, 2011.

Starodub N.F., Arenkov P.I., Rachcov A.E., and Beresin V.A. Fiber optic immunosensors for detection of some drugs. *Sens. Actuat.*, 1993, B13–14, 728–731.

Starodub N.F., Arenkov P.J., and Starodub A.N. Fiber optic immune sensors based on enhanced chemiluminescence and their application to determine different antigens. *Sens. Actuat.*, 1994b, B18–19, 161–165.

Starodub N.F., Arenkov P.I., Starodub A.N., and Beresin V.A. Construction and biomedical application of immunosensors based on fiber optics and enhanced chemiluminescence. *Opt. Eng.*, 1994a, 33, 2958–2963.

Starodub N.F., Demchenko A.V., Piven' N.V., Goncharik A.V., Orlova E.E., Burakovsky A.I., Martjanov A.A., Gherdev A.V., and Dzantijev B.B. The development of new methods for control of water quality. *Chem. Water Technol.* 2005a, 27(6), 591–599.

Starodub N.F., Demchenko A., Starodub V.M., Dmitrenko N.P., Zherdev A.V., Dzantiev B.B., Goncharik A.V., and Piven'N.V. Optical immune sensors for the determination of low weight toxic substances in environment. In: *Proceedings of International Scientific-Technical Conference on Sensors Electronics and Microsystems Technology*, SMEST-1, Odessa, Ukraine, June 1–5, 2004b, pp. 150–151.

Starodub N.F., Dibrova T.I., Shirshov Yu.M., and Kostiukevich K.V. Control of heavy metal ion level in some vegetables by enzymatic sensors based on the ISFETs. In: *Proceedings of Eurosensors XI. The 11th European Conference on Solid-State Transducers*, September 21–24, 1997, Warsaw, Poland, 1997a, Vol. 3, pp. 1233–1236.

Starodub N.F., Dzantiev B.B., Starodub V.M., and Zherdev A.V. Immunosensor for the determination of the herbicide simazine based on an ion-selective field effect transistor. *Anal. Chim. Acta*, 2000a, 424, 37–43.

Starodub V.M., Dzantiev B.B., Zherdev A.V., and Starodub N.F. ISFET based immune sensor for Simazine. In: *Proceedings of the 14th European Conference on Solid-State Transducers*, August 27–30, 2000, Copenhagen, Denmark, 2000b, pp. 795–798.

Starodub N.F., Fedorenco L.L., Starodub V.M., Dikiy S.P., and Svechnicov S.V. Extinguishing visible photoluminescence of porous silicon stimulated by antigen-antibody immunocomplex formation. In: *SPIE Proceedings Optical Organic and Semiconductor Inorganic Materials*, August 26–29, 1996, Riga, Latvia, 1997b, Vol. 2968, pp. 73–76.

Starodub V.M., Fedorenko L.L., Sisetskiy A.P., Starodub N.F. Use of the silicon crystals photoluminescence to control immune complex formation. *Sens. Actuat.*, 1996a, B35–36, 44–47.

Starodub V.M., Fedorenko L.L., and Starodub N.F. Control of a myoglobin level in solution by the bioaffinic sensors based on the photoluminescence of porous silicon. In: *Proceedings of the XII European Conference on Solid-State Transducers and the IX UK Conference on Sensors and Their Applications*, September 13–16, 1998, Southampton, U.K., White, N.M. Bristol, U.K., Institute of Physics, 1998a, Vol. 1, pp. 817–820.

Starodub V.M., Fedorenko L.L., Sisetskiy A.P., and Starodub N.F. Control of myoglobin level in a solution by an immune sensor based on the photoluminescence of porous silicon. *Sens. Actuat.*, 1999a, B58, 409–414.

Starodub V.M., Fedorenko L.L., and Starodub N.F. Optical immune sensor for the monitoring protein substances in the air. *Sens. Actuat.*, 2000c, B68(1–3), 40–47.

Starodub N.F., Kanjuk N.I., Kukla A.L., Kanjuk M.I., and Shirshov Y.M. Multienzymatic electrochemical sensor: Field measurements and their optimization. *Anal. Chim. Acta*, 1999b, 385, 461–466.

Starodub N.F., Kanjuk N., Starodub V., and Ternovoj K. Development of enzymatic sensors for monitoring of the phosphor-organic pesticide levels in the environment. In: *Third International Symposium and Ekzhibition on Environmental Contamination in Central and Estern Europe*, September 1996, Warsaw, Poland, 1996b, pp. 51–53.

Starodub N.F., Kukla A.L., and Shirshov Y.M. Simultaneous control of phosphororganic pesticides and heavy metal ions by multienzymatic sensor. In: *Proceedings of the 9th International Trade Fair and Conference for Sensors, Transducers & Systems*, May 18–20, 1999, Exhibition Centre Nuernberg, Germany, 1999c, Vol. 2, pp. 105–110.

Starodub N.F., Piven' N.V., Demchenko A.V., Goncharik A.V., Orlova E.E., Burakovsky A.I., Martjanov A.A., Gherdev A.V., and Dzantijev B.B. Immune enzymatic analysis of non-ionic surface active substances in water. *Ukr. Biochem. J.*, 2005b, 77(6), 116–121.

Starodub N.F., Rachkov A.E., and Beresin V.A. Optoimmunosensors for analysis of specific and non-specific classes of immunoglobulins. *Sens. Actuat.*, 1992b, B7, 371–375.

Starodub N.F., Samodumova I.M., and Starodub V.M. Usage of organosilanes for integration of enzymes and immunocomponents with electrochemical and optical transducers. *Sens. Actuat.*, 1995c, B24–25, 173–176.

Starodub N.F. and Starodub V.N. Silicium crystal photoluminescence as transducer for biosensors. In: *Proceedings of NATO ASI on Advanced Electronic Technologies and Systems Based on Low-Dimensional Quantum Devices*, Vol. 42 of High Technology, September 17–29, 1996, Sozopol, Bulgaria. Balkanski M. and Andreev N. (Eds.), Dordrecht, the Netherlands, Kluwer Academic Publishers, 1996, pp. 91–92.

Starodub V.M. and Starodub N.F. Electrochemical and optical biosensors: Origin of development, achievements and perspectives of practical application. In: *Book of NATO Series Novel Processes and Control Technologies in the Food Industry*, Bozoglu F. et al. (Eds.), Kluwer Acadamic Publishing, Amsterdam, the Netherland, 2001, pp. 63–94.

Starodub N.F. and Starodub V.N. Express control of toxic substances and pathogenic microorganisms. *Ukr. Biochem. J.*, 2002, 74(4), 5–23.

Starodub N.F., and Starodub V.M. Biosensors based on the photoluminescence of porous silicon. General characteristics and application for medical diagnostics. *Sens. Electron. Microsyst. Technol.*, 2004, 2, 63–83.

Starodub N.F., Shirshov Yu.M., Starodub V.M., Kukla A.L., and Kanjuk M.I. Multienzymatic sensor for simultaneous determination of phosphororganic pesticides and heavy metal ions. In: *Electrochemical Society Proceedings*, 1997c, Vol. 97–19, pp. 799–808.

Starodub N.F., Torbicz W., Starodub V.M., and Kanjuk M.I. Stabilization of activity and repeated usage of biomaterial during integration with transducers and analysis of irreversible inhibitors. In: *SPIE Proceedings Micromachined Devices and Components III*, 1997d, Vol. 3224, pp. 273–282.

Starodub V.M. and Starodub N.F. Electrochemical immune sensor based on the ion-selective field effect transistor for the determination of the level of myoglobin. In: *Book of abstracts of the 13th European Conference on Solid-State Transducers*, September 12–15, 1999, the Hague, the Netherlands, 1999a, pp. 180–184.

Starodub V.M. and Starodub N.F. Optical immune sensors for the monitoring protein substances in the air. In: *Proceedings of the 13th European Conference on Solid-State Transducers*, September 12–15, 1999, the Hague, the Netherlands, 1999b, pp. 181–184.

Starodub N.F., Shirshov Yu.M., Torbicz W., Kanjuk N.I., Starodub V.M., and Kukla A.L. Biosensors for in field mesurements: Optimization of parameters to control phosphororganic pesticides in water and vegetables. In: *Proceedings of the NATO Advanced Research Workshop on Biosensors for Direct Monitoring of Environmental Pollutants in Field*, May 4–8, 1998, Smolenice, Slovakia. Nikolelis D.P. et al. (Eds.), the Netherlands, Kluwer Academic Publishers, 1998b, pp. 209–219.

Starodub N.F., Starodub V.M., Demchenko A.V., Zherdev A.V., Dzantiev B.B., Goncharic A.V., and Piven N.V. Creation of optical immune sensors for the determination of low weight toxic substances in environment. In: *Proceedings of International Trade Fair for Optical and Microtechnology Products with Conferences*, May 25–27, 2004, Nuremberg, Exhibition Center, Germany, 2004a, pp. 335–340.

Starodub N.F., Sitnik J., and Mel'nichenko M.M. Efficiency of the nanostructured silicon based biosensors at the express biochemical diagnostics of the retroviral bovine leucosis. In: *Lecture Notes of the ICB*, 2010a, Vol. 86, pp. 63–74.

Starodub N.F., Shulyak A.M., Shmyryeva A.B., Pylypenko I.V., Pylypenko L.N., and Mel'nichenko M.M. Nanostructured silicon and its application as a transducer in immune biosensor. In: *NATO Book Biodefence*, S. Mikhalovsky and A. Khajibaev (Eds.), Sciences for Peace and Science, series. A, Chemistry and Biology, Springer Science + Business Media, 2010b, Vol. 9, pp. 87–98.

Starodub N.F., Torbicz W., Pijanovska D., Starodub V.M., Kanjuk M.I., and Dawgul M. Optimization methods of enzyme integration with transducers for analysis of irreversible inhibitors. *Sens. Actuat.*, 1999d, B58, 420–426.

Steegborn C. and Skladal P. Construction and characterization of the direct piezoelectric immunosensor for atrazine operating in solution. *Biosens. Bioelectron.*, 1997, 12, 19–27.

Tom-Moy M., Baer R.L., Spira-Solomon D., and Doherty T.P. Atrazine measurements using surface transverse wave device. *Anal. Chem.*, 1995, 67, 1510–1516.

Turner A.P.F. et al. *Biosensors: Fundamentals and Applications*, Oxford, U.K., Oxford University Press, 1989, pp. 3–12.

Unal B. and Bayliss S. Electrical characterization of photovoltaic porous Si. In: *Extended Abstracts of International Conference, Porous Semiconductors—Science and Technology*, Malorca, Spain, March 16–20, 1998, pp. 181–182.

US Environmental Protection Agency, National survey of pesticides in drinking water wells, Phase II report, EPA 570/9-91-020, National Technical Information Service, Springfield, VA, 1992.

Usami M., Mitsunaga K., and Ohno Y. Estrogen receptor binding assay of chemicals with a surface plasmon resonance biosensor. *J. Steroid Biochem. Mol. Biol.*, 2002, 81(1), 47–55.

Van Weemen B.K. and Schuurs A.N.W. Immunoassay using antigen enzyme conjugates. *FEBS Lett.*, 1971, 15, 232–236.

Wang J. Sol-gel materials for electrochemical biosensors. *Anal. Chim. Acta*, 1999, 399, 21–27.

Watanabe H., Suzuki A., Goto M., Lubahn D.B., Handa H., and Iguchi T. Tissue-specific estrogenic and non-estrogenic effects of a xenoestrogen, nonylphenol. *J. Mol. Endocrinol.*, 2004, 331, 243–252.

Woo J., Lacbawan F.L., Sunheimer R., LeFever D., and McCabe J.B. Is myoglobin useful in the diagnosis of acute myocardial infarction in the emergency department setting? *Am. J. Clin. Pathol.*, 1995, 103, 725–729.

Yalow R.S. and Benson S.A. Assay of plasma insulin in human subjects by immunological methods. *Nature*, 1959, 184, 1648–1649.

Zeravik J., Skryjová K., Nevoranková Z., and Fránek M. Development of direct ELISA for the determination of 4-Nonylphenol and octylphenol. *Anal. Chem.*, 2004, 76(4), 1021–1027.

21

Application of Biosensors on Air Pollution Monitoring

Evangelos Bakeas

CONTENTS

21.1 Introduction

The increasing number of potentially harmful pollutants in the environment calls for fast and cost-effective analytical techniques to be used in extensive monitoring programs. Additionally, over the last few years, a growing number of initiatives and legislative actions for environmental pollution control have been adopted in parallel with increasing scientific and social concern in this area [1–4]. The requirements for application of most traditional analytical methods to environmental pollutant analysis often constitute an important impediment for their application on a regular basis. The need for disposable systems or tools for environmental applications, in particular for environmental monitoring, has encouraged the development of new technologies and more suitable methodologies. In this context, biosensors appear as a suitable alternative or as a complementary analytical tool.

Biosensors can be considered as a subgroup of chemical sensors in which a biological mechanism is used for analyte detection [1,3,4]. A biosensor is defined by the International Union of Pure and Applied Chemistry (IUPAC) as a self-contained integrated device that is capable of providing specific quantitative or semiquantitative analytical information using a biological recognition element (biochemical receptor), which is retained in contact direct spatial with a transduction element [5]. Biosensors should be distinguished from bioassays where the transducer is not an integral part of the analytical system [4–7]. Biosensing systems and methods are being developed as suitable tools for different applications, including environmental applications.

Biosensors are usually classified according to the bioreceptor element involved in the biological recognition process (e.g., enzymes; immunoaffinity recognition elements; whole cells of microorganisms, plants, or animals; or DNA fragments) or according to the physicochemical transducer used (e.g., electrochemical, optical, piezoelectrical, or thermal). The main classes of bioreceptor elements that are applied in environmental analysis are whole cells of microorganisms, enzymes, antibodies, and DNA. Additionally, in most of the biosensors described in the literature for environmental applications, electrochemical transducers are used [5].

For environmental applications, the main advantages offered by biosensors over conventional analytical techniques are the possibility of portability, miniaturization, work on-site, and the ability to measure pollutants in complex matrices with minimal sample preparation. However, there are technical and commercial obstacles that must be addressed before biosensors can have a significant impact on environmental monitoring. The most important obstacle includes the large number of potential pollutants and the broad range of their chemical classes, because the majority of biosensors are specific for only one pollutant or a very limited number of pollutants. In addition, a biosensor should be rigorously validated before considering it a suitable alternative to laboratory methods. Although many of the developed systems cannot compete yet with conventional analytical methods in terms of accuracy and reproducibility, they can be used by regulatory authorities and by industry to provide enough information for routine testing and screening of samples [1,3].

Several reviews on biosensors or optical sensors for different analytical fields have been published in recent years [8–15]. These reviews described technological advances, the fundamental theory, and the main advantages of these bioassays (i.e., short times of analysis, low cost, portable equipment, real-time measurements, and the use as remote devices).

Biosensors can be used as environmental quality monitoring tools in the assessment of biological/ecological quality or for the chemical monitoring of both inorganic and organic priority pollutants. Pollutants are usually classified into groups according to their chemical structure but can also be divided into groups according to their mode of action, such as endocrine disruption, cytotoxicity, carcinogenicity, mutagenicity, or genotoxicity. The following sections describe the biosensors that have been developed for air pollution monitoring. It must be mentioned that the use of biosensors in air pollution monitoring is still limited comparing to their application in other environmental parts such as water and soil.

21.2 Organic Air Pollutants

A few studies have been performed on the determination of volatile organic compounds (VOCs) in air using biosensors. The compounds that have been studied are mainly formaldehyde (FA), benzene, and methanol.

A bacterial biosensor based on flow-injection analysis (FIA) has been developed for the determination of benzene in workplace air samples. Benzene can be used by the bacteria *Pseudomonas putida* ML2 as a sole carbon source, and its aerobic degradation can be measured using a dissolved oxygen electrode. The bacterial cells were immobilized between two cellulose acetate membranes and fixed onto a Clark dissolved oxygen probe, which was inserted into a custom-made flow cell. The applicability of the biosensor for the analysis of air samples containing benzene was investigated. Air samples were collected from a controlled exposure room using charcoal adsorption tubes and benzene extracted with solvent desorption using dimethylformamide (DMF). The biosensor displayed a linear detection range between 0.025 and 0.15 mM benzene based on standard solutions containing a maximum of 2% DMF, with a response time of 6 min. This linear detection range allows the analysis of air containing between 3 and 16 ppm benzene based on a 60 min sampling period. DMF proved to be compatible for the use with the biosensor, causing minimal interference with the sensor response and causing no toxic effects on the bacterial cells. The FIA system was easily transported to an *in situ* location, and a correlation was obtained between the biosensor and gas chromatography (GC) results for the preliminary air samples investigated. Moreover, the biosensor displayed no interference to other benzene-related compounds in the benzene, toluene, ethylbenzene, xylene (BTEX) range. It was shown that the biosensor has potential applications for the analysis of benzene in workplace air samples, with the added advantages over the conventional GC methods of low operation costs, ease of use, and portability for *in situ* measurements [16].

The development of two recombinant bacterial systems that can be used to monitor environmental benzene contamination based on *Escherichia coli*, which carry genes coding for benzene dioxygenase and benzene dihydrodiol dehydrogenase from P. putida MST, has been reported. *E. coli* strains express these two enzymes under the control of the Ptac promoter or without any induction. These activities can be detected electrochemically or colorimetrically and used to monitor benzene pollution in environmental air samples collected from an oil refinery assessing benzene by different laboratory experimental procedures. The procedures involving whole-cell bioassays determine the concentration of benzene through benzene dioxygenase activity, which allows for direct correlation of oxygen consumption, and through the benzene dihydrodiol dehydrogenase that causes catechol accumulation and restores NADH necessary for the activity of the first enzyme.

Oxygen consumption and catechol production deriving from both enzymatic activities are related to benzene concentration, and their measurements determined the sensitivity of the system. It was indicated that the sensitivity was enough to detect the benzene vapor at a lower concentration level of 0.01 mM in about 30 min. The possibility for online monitoring of benzene concentration by our new recombinant cells results from the fact that no particular treatment of environmental samples is required. This was a major advantage over other biosensors or assays. Moreover, the development of microbial cells that did not require any addition or effectors for the transcription of the specific enzymes allowed these systems to be more versatile in automated environmental benzene monitoring [17].

The simultaneous detection and identification of several targets by one biosensor are not possible in the majority of the biosensor systems. In a study was proved the concept of the detection and identification of two different volatile toxic compounds with a nonselective biochip-based algal biosensor. For that purpose, array plate biochips were produced to utilize three membrane-immobilized algal strains of genus *Klebsormidium* and *Chlorella* in one biosensor system. A novel IMAGING-PAM chlorophyll fluorometer was applied to measure the impact of VOCs on photosynthesis of chip-immobilized algae in terms of quantum efficiency of electron transport. FA vapor was detectable with statistical significance in concentrations relevant to human health from 10 ppb to 10 ppm. The biosensor response recorded within minutes was concentration dependent and reversible. Moreover, vapors of FA (0.05–1 ppm) and methanol (MeOH) (200–1000 ppm) were significantly identified by the compound-specific response rate as a quotient of the biosensor responses of the respective algal strains. Using the IMAGING-PAM chlorophyll fluorometer, data sampling proved to be highly efficient. It was concluded that the principle of the algal sensor chip (ASC) suggests further research on the detection and identification of VOCs and other toxic substances in gaseous environment with that biochip system [18].

The development of a novel detection method, based on the coupling of a biosensor measuring device and a flow-injection system, using the enzyme FA dehydrogenase and a Os(bpy)2-poly(vinylpyridine) (POs-EA) chemically modified screen-printed electrode has been reported for the determination of FA in air. The sensor can detect 30 ng mL^{-1} of FA in aqueous solution (corresponding to sub-ppb atmospheric concentrations of FA). The sensor is selective, inexpensive, stable over several days, and disposable, as well as simple to manufacture and operate. The system can easily be adapted to other substrates using their corresponding dehydrogenases [19].

An ion-sensitive field-effect transistor (ISFET) has been used in biosensor for FA. The ISFET monitors H+ produced when FA is oxidized by FADH with NAD+ as a cofactor. The FA was removed from the atmosphere by pumping air through a glass coil together with an aqueous solution. The solution dissolved the FA and acted as a carrier of the FA to the ISFET. A membrane containing FADH covered the ISFET and the solution containing FA was transferred directly to the surface. The enrichment factor of this sampling technique was 8000-fold, but there were some problems with the immobilization of the enzyme, which complicated the evaluation of the sensor [20].

Another biosensor for FA and alcohols has been described. This biosensor utilized enzymes and cofactors immobilized in a reversed micelle medium on screen-printed electrodes. The biosensor used FADH for the determination of FA, but a similar construction was also used to determine alcohols with ADH. The reoxidation of NADH to NAD+ was measured amperometrically at 0.8 V versus Ag/AgCl. The reversed micelle medium was used to prevent water loss as the silicone oil acted as a barrier against evaporation. The biosensor was found to be suitable for gas phase sensing when it was tested in controlled atmospheres [21].

A biosensor based on an electrochemical cell divided into two parts by a dialysis membrane to prevent migration of the enzyme has been reported. FADH was put on the working electrode, and prior to sampling, the cell was filled with electrolyte containing cofactor and mediator. Since the electrolyte was not added until the time of sampling, the device could be stored for a long time. The device was tested in a controlled atmosphere above an aqueous FA solution and the limit of detection was 0.3 ppm [22].

FADH was one of the enzymes used in the diffusion badges developed [23]. Various enzymes lyophilized onto sintered glass rods put into vessels containing buffer–reagent solutions and covered with gas-permeable membranes were used. The diffusion badges contained the buffer solutions to overcome the problem of drying. Since the device was constructed from two parts, the glass rod with enzyme and the vessel with buffer–reagent solution, they could be stored separately. The enzyme could be stored dry, which increased the storage time. In addition to FA, the compounds determined using this type of construction were hydrogen peroxide, acetaldehyde, and ethanol, and the enzymes used were diaphorase, aldehyde dehydrogenase, alcohol dehydrogenase (ADH), and horseradish peroxidase. The reaction between analytes and enzymes caused a dye to change a color. This color change was documented photographically and the color was stable for hours after exposure.

The use of biosensors has also been described for the determination of phenols in air. In one study, the device was constructed by immobilizing PPO on a gold microbiosensor using a glycerol-based gel [24]. The phenol vapor reacts with the enzyme, and the product (catechol) takes part in a redox recycling reaction at the electrode surface. Good sensitivity was achieved partly by this recycling of the catechol–quinone redox couple. The limit of detection was estimated to be 29 ppb and the response was linear up to 13 ppm, both at 40% relative humidity.

In another two studies, a biosensor for monitoring phenol in both the liquid and gas phases has been reported [25,26]. The first one mainly described the performance of the sensor in the liquid phase, but preliminary experiments in the gas phase were also described. The second one described the experiments performed in the gas phase, also investigating with a mixed p-cresol and 4-chlorophenol vapors. The enzyme used in the sensor was tyrosinase, which catalyzes both the reaction of phenol to catechol and the reaction of catechol to o-quinone. The electrochemical reduction of o-quinone back to catechol produced a measurable signal at the electrode.

The possibility of developing a simple, inexpensive, and specific personal passive real-time air sampler incorporating a biosensor for formic acid was investigated in a study [27]. The sensor is based on the enzymatic reaction between formic acid and formate dehydrogenase (FDH). With nicotinamide adenine dinucleotide NAD+ as a cofactor and Meldola's blue as mediator, an effective way to immobilize the enzyme, cofactor, and Meldola's blue on screen-printed, disposable, electrodes was found to be in a mixture of glycerol and phosphate buffer covered with a gas-permeable membrane. The steady-state current was reached after 4–15 min and the limit of detection was calculated to be below 1 mg m^{-3}. This study has demonstrated the potential for a simple, inexpensive, and specific personal passive real-time sampler based on biosensor technology, for measurement of formic acid in air. However, further development is required to create a device with sufficient storage stability for commercial application. To improve the storage properties, the enzyme may have to be stabilized in a more rigid structure, for example, a polymer, or stored dry and rehydrated shortly before use, but the use of glycerol does have the advantage of preventing loss of water during operation of the biosensor. In other respects, the system described shows promise in meeting the considerable demand for a simple, specific, and inexpensive method for personal exposure monitoring.

An enzyme-based biosensor monitoring system provided the basis for continuous sampling of organophosphate contamination in air [28]. The enzymes butyrylcholinesterase (BuChE) and organophosphate hydrolase (OPH) are stabilized by encapsulation in biomimetic silica nanoparticles, entrained within a packed bed column. The resulting immobilized enzyme reactors (IMERs) were integrated with an impinger-based aerosol sampling system for the collection of chemical contaminants in air. The sampling system was operated continuously, and organophosphate detection was performed in real time by single wavelength analysis of enzyme hydrolysis products. The resulting sensor system detects organophosphates based on either enzyme inhibition (of BuChE) or substrate hydrolysis (by OPH). The system proved suitable for the detection of a range of organophosphates including paraoxon, demeton-S, and malathion.

A cell-based gas biosensor presented was used for the detection and investigation of gaseous organic compounds in air [29]. The response of living human nasal cells (RPMI 2650) and human lung cells (A549) toward the direct exposure of gaseous substances for 10 min is monitored with a multiparametric sensor system. Changes in the cellular impedance, oxygen consumption rate, and acidification rate can be recorded after the exposure and represent the cytotoxicity of the present gas. The sensor was able to notify the presence of acetone in aqueous solution (2%) but in notably lower concentrations in the gas phase (100–333 ppm) within 30–60 min after the end of the gas exposure. Cell viability was not affected by a sequential exposure to humidified synthetic air (60% r.h.) with a flow rate of 300 mL min^{-1} and therefore offers the possibility for a continuous air monitoring. In addition, the exposure to synthetic air has no influence on the signals of consecutive acetone exposure. The system might be used in the future for the monitoring of ambient air in work spaces.

21.3 Inorganic Air Pollutants

An enzyme-based carbon monoxide sensor has been described [30]. The biosensor was based on the oxidation of carbon monoxide to carbon dioxide by the enzyme carbon monoxide oxidoreductase. Carbon monoxide oxidoreductase was placed on a conducting gel and covered with a membrane. The conducting gel consisted of graphite, mediator, and liquid paraffin and was in contact with a platinum electrode. The gas-permeable membrane was used to keep the enzyme at the surface and to make it possible for carbon monoxide to pass through to the enzyme. 1,1-Dimethylferrocene was used as a mediator and was oxidized at the electrode surface at 150 mV versus Ag/AgCl. The amperometric response reached a steady-state current in less than 15 s and the current in the device decreased by 12% h^{-1}. Most of the reported experiments, however, were performed in solutions.

A novel whole-cell-based biosensor for the detection of toxic substances in air was established [31,32]. The adaption of a commercial sensor chip (Bionas®) for the measurement of pollutants in liquids enables the direct exposure of cells with air. Cells of the respiration tract (A549, RPMI2650, V79), which tend to survive at a gas phase, are used as biological receptors. Three physiological cell parameters are monitored continuously in parallel (acidification, respiration, morphology). Water-insoluble gases (e.g., CO) as well as water-soluble gases (e.g., NH_3) were used as model gases to test the feasibility of the novel sensor system. MIR measurements proofed the reproducibility of the draining method. This sensor system provides a basis for many sensing applications such as environmental monitoring, building technology, and public security.

A biosensor for the determination of NO_2 in air has been developed [33]. Nitrite oxidizing bacteria were immobilized on an acetyl cellulose membrane. The membrane was then attached to an oxygen electrode and covered with permeable Teflon membrane. The sample was prepared in a gas bag, pumped into the system, and dissolved in a buffer that was pumped through the sample cell of the biosensor. The decrease in oxygen, caused by an increased activity of the microorganisms when NO_2 was present, was measured. The minimum concentration that could be determined is 0.51 ppm.

A biosensor for nitrogen monoxide was described in another study [34]. It consisted of a sol–gel containing cytochrome c spin coated onto a glass substrate. A gas flow through cell covered the sol–gel for the gaseous sample to come in contact with the enzyme. When the NO attached to cytochrome c, a shift of the absorption wavelength occurred, which was measured spectrophotometrically. Since the bond between NO and cytochrome c was reversible, the sensor could be used for repeated exposures of NO. The limit of detection was calculated to be 1 ppm and the range of detecting NO was 1–25 ppm.

Sensors for monitoring sulfur dioxide in air have also been described [35,36]. A mixture of agarose and carboxymethylcellulose was chosen from a range of matrices as the medium for immobilization of sulfite oxidase. The linear range was 0–13.5 ppm and the LOD was 73.9 ppb [36].

References

1. K.R. Rogers and C.L. Gerlach, *Environ. Sci. Technol.* 30 (1996) 486.
2. S. Rodriguez-Mozaz, M.-P. Marco, M.J.L. Alda, and D. Barcel, *Pure Appl. Chem.* 76 (2004) 723.
3. K.R. Rogers, *Anal. Chim. Acta* 568 (2006) 222.
4. S. Rodriguez-Mozaz, M.-P. Marco, M.J.L. Alda, and D. Barcel, *Talanta* 65 (2005) 291.
5. D.R. Thıvenot, K. Toth, R.A. Durst, and G.S. Wilson, *Pure Appl. Chem.* 71 (1999) 2333.
6. P. Leonard, S. Hearty, J. Brennan, L. Dunne, J. Quinn, T. Chakraborty, and R. O'Kennedy, *Enzyme Microb. Technol.* 32 (2003) 3.
7. C. Ziegler and W. Gopel, *Curr. Opin. Chem. Boil.* 2 (1998) 585.
8. M. Farre, R. Brix, D. Barcelo, *Trend. Anal. Chem.* 24 (2005) 532.
9. M.A. Gonzalez-Martinez, R. Puchades, and A. Maquieira, *Anal. Bioanal. Chem.* 387 (2007) 205.
10. S. Rodriguez-Mozaz, M.J.L. de Alda, M.P. Marco, and D. Barcelo, *Talanta* 65 (2005) 291.
11. M. Tudorache and C. Bala, *Anal. Bioanal. Chem.* 388 (2007) 565.
12. C.R. Suri, R. Boro, Y. Nangia, S. Gandhi, P. Sharma, N. Wangoo, K. Rajesh, and G.S. Shekhawat, *Trend. Anal. Chem.* 28 (2009) 29.
13. S. Rodriguez-Mozaz, M.J.L. de Alda, and D. Barcelo, *Anal. Bioanal. Chem.* 386 (2006) 1025.
14. A.K. Wanekaya, W. Chen, and A. Mulchandani, *J. Environ. Monit. Assess.* 10 (2008) 703.
15. M. Farre, L. Kantiani, S. Perez, and D. Barcelo, *Trend. Anal. Chem.* 28 (2009) 170.
16. Y.H. Lanyon, G. Marrazza, I.E. Tothill, and M. Mascini, *Biosens. Bioelectron.* 20 (2005) 2089.
17. P. DiGennaro, N. Bruzzese, D. Anderlini, M. Aiossa, M. Papacchini, L. Campanella, and G. Bestetti, *Ecotoxicol. Environ. Safe.* 74 (2011) 542.
18. B. Podola, E.C.M. Nowack, and M. Melkonian, *Biosens. Bioelectron.* 19 (2004) 1253.

19. Y. Herschkovitz, I. Eshkenazi, C.E. Campbell, and J. Rishpon, *J. Electroanal. Chem.* 491 (2000) 182.
20. F. Vianello, A. Stefani, M.L. Di Paolo, A. Rigo, A. Lui, B. Margesin, M. Zen, M. Scarpa, and G. Soncini, *Sens. Actuat.* 341 (1996) 37.
21. M.J. Dennison, J.M. Hall, and A.P.F. Turner, *Analyst* 121 (1996) 1769.
22. M. Hammerle, E.A.H. Hall, N. Cade, and D. Hodgins, *Biosens. Bioelectron.* 11 (1996) 239.
23. K.-P. Rindt and S. Scholtissek, in *Biosensors Applications in Medicine, Environmental Protection and Process Control*, eds. R.D. Schmid and F. Scheller, GBF Monographs, Vol. 13, VCH, New York, (1989) p. 405.
24. M.J. Dennison, J.M. Hall, and A.P.F. Turner, *Anal. Chem.* 67 (1995) 3922.
25. A. Kaisheva, I. Iliev, R. Kazareva, S. Christov, U. Wollenberger, and F. Scheller, *Sens. Actuat.* 33 (1996) 39.
26. A. Kaisheva, I. Iliev, S. Christov, and R. Kazareva, *Sens. Actuat.* 44 (1997) 571.
27. K.J.M. Sandstrom, J. Newman, A.L. Sunesson, J.O. Levin, and A.P.F. Turner, *Sens. Actuat.* 70 (2000) 182.
28. H.R. Luckarift, R. Greenwald, M.H. Bergin, J.C. Spain, and G.R. Johnson, *Biosen. Bioelectron.* 23 (2007) 400.
29. U. Bohrn, E. Stutz, M. Fleischer, M.J. Schoning, and P. Wagner, *Biosen. Bioelectron.* 40 (2013) 393.
30. A.P.F. Turner, W.J. Aston, I.J. Higgins, J.M. Bell, J. Colby, G. Davis, and H.A.O. Hill, *Anal. Chim. Acta* 163 (1984) 161.
31. U. Bohrn, E. Stütz, K. Fuchs, M. Fleischer, M.J. Schöning, and P. Wagner, *Procedia Eng.* 25 (2011) 1421.
32. U. Bohrn, E. Stütz, K. Fuchs, M. Fleischer, M.J. Schöning, and P. Wagner, *Sens. Actuat.* 175 (2012) 208.
33. T. Okada, I. Karube, and S. Suzuki, *Biotechnol. Bioeng.* 25 (1983) 1641.
34. J.W. Aylott, D.J. Richardson, and D.A. Russell, *Chem. Mater.* 9 (1997) 2261.
35. W. Matuszewski and M.E. Meyerhoff, *Anal. Chim. Acta* 248 (1991) 379.
36. C.K. O'Sullivan, PhD thesis, Cranfield University, Bedford, U.K., 1996.

22

Oligonucleotide and DNA Microarrays as Versatile Tools for Rapid Diagnostics

Susana Campuzano, María Pedrero, and José M. Pingarrón

CONTENTS

22.1 Introduction

One of the most widely recognized tools for improving the evolution of research in molecular biology is the development of adequate bioanalytical technologies such as DNA microarrays. While traditional methods do not allow simultaneous research on a large number of genes, DNA microarray technologies enable the investigation and addressing of issues that were once thought to be non-traceable. Through this approach, one can analyze the expression of many genes in a single reaction quickly and in an efficient manner. Initiated in the 1990s [1], this technology has empowered the scientific community to understand the fundamental aspects underlining the growth and development of life as well as to explore the genetic causes of anomalies occurring in the functioning of the human body.

A DNA array is an orderly arrangement of samples where matching of known and unknown DNA chains is done based on base-pairing rules. In fact, DNA microarrays work on the principle of base pairing that allows probes to hybridize targets on the microarray. Traditional array experiments make use of common assay systems such as microplates or standard blotting membranes. The sample spot sizes are typically lower than 200 µm in diameter and a microplate usually contains thousands of spots.

A typical DNA microarray experiment involves the hybridization of a messenger RNA (mRNA) molecule to the DNA template from which it is originated [2]. The amount of mRNA bound to each site on the array indicates the expression level of the various genes. As mentioned earlier, this number may run in thousands. All the data are collected and a profile is generated for gene expression in the cell.

Thus, DNA microarrays are a high-throughput technology used to measure the expression levels of thousands of genes, in some cases all of the genes in a genome, simultaneously. The fundamental idea behind most microarrays is to exploit complementary base pairing to measure the amount of the different types of mRNA molecules in a cell, which indirectly allows measuring the expression levels of the genes that are responsible for the synthesis of the particular mRNA molecules.

Thousands of spotted samples known as probes (with known identity) are immobilized on a solid support (microscope glass slides, silicon chips, or nylon membranes). The probes can be DNA, complementary DNA (cDNA), or oligonucleotides. These are used for complementary binding of the unknown sequences, thus allowing parallel analysis for gene expression and gene discovery. Given that an experiment with a single DNA chip can provide information on thousands of genes simultaneously, an orderly arrangement of the probes on the support is important as the location of each spot on the array is used for the identification of a gene.

Although a general approach to DNA microarrays is intended in this chapter, we will focus in the last sections especially on electrochemical oligonucleotides and DNA microarrays.

22.2 Fundamentals and Fabrication of DNA Microarrays

A DNA microarray (commonly called gene chip, DNA chip, DNA array, or biochip) can be thought of either as a miniaturized form of dot blot, but in a high-throughput format, or as a miniaturized gene-hybridization or gene-detection assay [3]. DNA microarrays represent DNA fragments attached to a surface in a predefined ordered fashion at high density. The principle of microarray experiments is that mRNA or total RNA from given cells or tissues is used to generate a labeled sample, which is hybridized in parallel with a large number of DNA sequences immobilized on a solid surface in an ordered array. According to the recommended nomenclature, the "probe" is the tethered nucleic acid with known sequence immobilized onto the surface, whereas the "target" is the free nucleic acid sample, complementary to the probe, whose identity or abundance is being detected through hybridization to the immobilized probe (Figure 22.1). The probe DNAs are oligonucleotides or polymerase chain reaction (PCR) products amplified from an individual clone using specific primers or universal primers if all genes were cloned in the same vector. Transduction of the hybridization event can be made optically, electrochemically, or using mass-sensitive devices [4].

Each microarray experiment typically follows several steps in a defined order: array fabrication, target preparation, hybridization, signal capture, and data analysis [3].

22.2.1 Array Fabrication: Probe Immobilization

DNA microarray supports can be made from different materials such as glass, plastic, or silicon [4], although the immobilization of DNA probes directly on electrochemical transducer surfaces is also possible [5,6]. Microarrays are constituted of tens to thousands 10–100 μm reaction zones onto which individual oligonucleotide sequences are immobilized. The exact number of DNA probes varies in accordance with the application. Contrary to other DNA biosensors that allow single-shot measurements, DNA microarrays allow multiple parallel detection and pattern analysis of expression of thousands of genes in a single experiment.

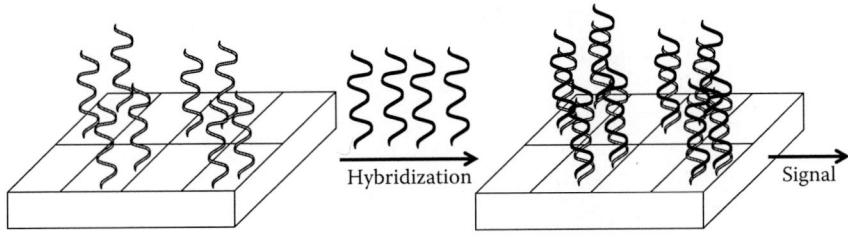

FIGURE 22.1
Steps involved in the detection of a DNA sequence.

A critical step is the DNA probe immobilization on the microarray surface. Depending on the application, the choice of the probe immobilization technique is extremely important. The achievement of high sensitivity and selectivity requires minimization of nonspecific adsorptions and stability of immobilized biomolecules. The control of this step is essential to ensure high reactivity, appropriate orientation and accessibility, and stability of the surface-confined probe and to avoid nonspecific binding. Immobilization techniques used in the development of DNA microarrays are based on two main strategies: the probes can be made base by base on the support or presynthesized and then spotted on the surface using different immobilization techniques such as adsorption, covalent immobilization, and avidin (or streptavidin)–biotin interaction.

22.2.2 Target Preparation/Labeling

Once prepared, the spots on a microarray contain single-stranded DNA (ssDNA) probes. Each of these spots will contain DNA that is of a complementary sequence to the specific mRNA molecule that corresponds to the targeting gene. The complementary mRNA molecule should hybridize to the corresponding probe forming a strong mRNA–DNA bond. If these mRNA molecules are previously labeled with, for example, a fluorescent dye, the amount of occurring hybridization can be measured by the level of fluorescence of the dye, which is examined with a scanner. This scanner then outputs a text file for each array, which contains the relevant data pertaining to that array, such as the level of fluorescence of each spot and the level of background noise. These text files are subsequently computationally analyzed. In theory, a spot with brighter fluorescence means that more mRNA was hybridized, which in turn infers that more mRNA was present in the sample extracted from the original cell and that the gene represented by this spot is experiencing a higher level of expression.

Thus, target DNA is defined as the labeled DNA that is applied to the microarray and hybridizes with the complementary probe DNA attached to the surface of the slide. In classical microarrays, target DNA is usually produced through reverse transcription of RNA into cDNA in such a way that a fluorescence nucleotide is incorporated. In fact, a cell or tissue RNA is extracted (targets), multiplied, converted to cDNA, and labeled [3]. Another traditional method of target labeling is radioactivity. Although this method is among the most sensitive ones, the use of radioisotopes, such as ^{32}P or ^{125}I, shows serious disadvantages. Other optical materials, apart from fluorescent dyes, such as quantum dots (QDs) involving different labels, are nowadays largely used for optical detection [7]. Moreover, electrochemistry has received considerable attention for the detection of DNA hybridization because it offers some relevant advantages, such as low cost, simple design, or small dimensions. The label can be an enzyme, an electroactive indicator, such as ferrocene (Fc), cationic metal complexes, and intercalating organic compounds (e.g., methylene blue (MB)), or nanoparticles.

A direct labeling strategy can be used where the immobilized DNA probe hybridizes with the labeled target DNA (Figure 22.2a). Also, a sandwich-type

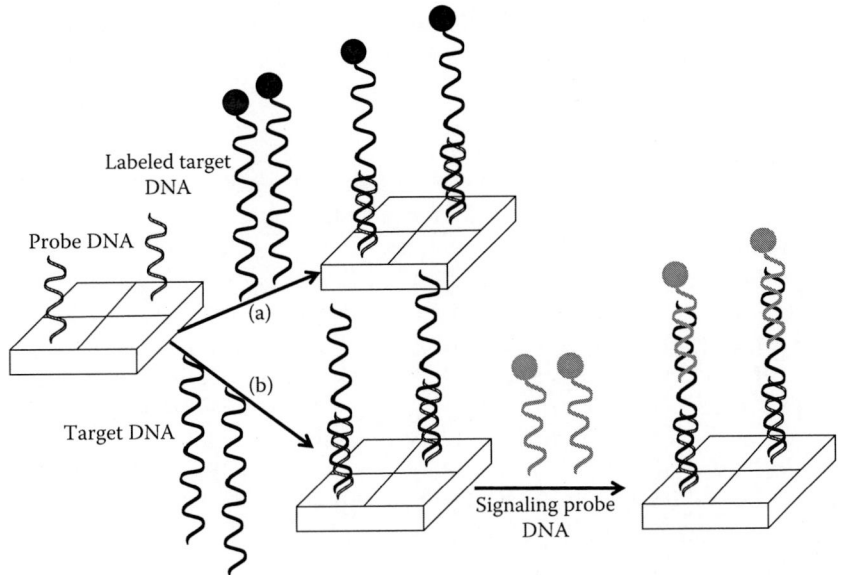

FIGURE 22.2
Main types of labeled DNA microarrays: (a) direct label between immobilized probe DNA and labeled target DNA and (b) sandwich-type system. A sandwich-type ternary complex is formed between immobilized probe DNA, target, and signaling DNA probe.

ternary complex can be formed in which the immobilized DNA probe hybridizes to a part of the target, whereas the other part of the target is complementary to a signaling DNA sequence that serves to label the target upon hybridization (Figure 22.2b). In a few cases, a competitive system can be used with a competition between the target and a labeled sequence both complementary to a DNA probe.

22.2.3 Nucleic Acid Hybridization Detection

Once the targets have been exposed to the microarray for a sufficient amount of time to allow for hybridization (typically 12–16 h), the array is washed. When fluorescent tags have been used for labeling, certain probes on the array will be fluorescent, because the fluorescent targets were hybridized to them, and a computer scanner will detect this fluorescence. The fluorescent probes correspond to the genes that were expressed in the cell. Advanced algorithms are used to provide data on which genes and how much each of those genes was expressed in the cells.

As mentioned earlier, nucleic acid hybridization can be detected according to different optical, electrochemical, or gravimetric (QCM) techniques and using tagging approaches. DNA hybridization can be optically detected using not only fluorescence but also surface plasmon resonance (SPR), chemiluminescence, colorimetry, interferometry, or surface-enhanced Raman scattering (SERS) spectroscopy [4]. However, label-free electrochemical detection of hybridization represents an attractive alternative approach for detecting DNA sequences. In this case, the detection is based on modifications of properties, such as capacitance, or an intrinsic electrochemical response due to DNA (e.g., oxidation of guanines).

Although optical biosensing based on fluorescence detection has arguably become the standard technique for quantifying extents of hybridization between surface-immobilized probes and fluorophore-labeled analyte targets in DNA microarrays, electrochemical detection techniques are emerging as attractive alternatives able to eliminate the need for physically bulky optical instrumentation and enabling the design of portable devices for *on-site* testing. Unlike fluorescence detection, which can function well using a passive substrate (one without integrated electronics), multiplexed electrochemical detection requires an electronically active substrate to analyze each array site and benefits from the addition of integrated electronic instrumentation to further reduce platform size and eliminate the electromagnetic interference that can result from bringing non-amplified signals off chip [8].

Multiple aspects of microarray designs contribute to their performance: characteristic of the oligonucleotide (provider, length, and purity), hybridization conditions, labeling method, etc. [9]. Variations often work in opposite directions regarding sensitivity and specificity, and therefore an appropriate compromise may need to be reached for each experimental setup and application. For example, 25-mer oligonucleotides can provide optimal discrimination at the cost of significant losses in sensitivity, which may be very detrimental for the detection of genes expressed at low levels. In contrast, 60-mers have a higher sensitivity but at the cost of more limited specificity.

It is important to be able to detect low DNA concentrations and to detect a point mutation. Thus, two types of systems can be developed: systems for DNA hybridization and systems for detection of DNA damage. A perfect match (PM) in the target sequence produces very stable double-stranded DNA (dsDNA), whereas one or more base mismatches (MMs) decrease the stability, causing a signal modification.

Despite its broad applicability in biology, classical microarray technology has some technical limitations, mainly imposed by the need for fluorescent labeling of the sample to be analyzed. This has triggered the development of alternative, nonoptical microarray-based detection techniques that avoid fluorescent labeling of the target DNA. Some of these rely on the use of nucleic acid analogs as probe molecules, such as peptide nucleic acid (PNA), locked nucleic acid (LNA), and dendritic nucleic acid structures, molecular beacons, and stem-and-loop structures [10].

Soon after DNA microarrays were available, the improved stability of PNA and its unique hybridization features encouraged the development of PNA-based microarrays. PNA can form Watson–Crick complementary duplexes with DNA. In comparison to DNA duplexes, PNA hybrids show higher thermal stability that is strongly affected by the presence of imperfect matches.

Moreover, PNA hybrids can be formed at low ionic strengths, and they exhibit a greater resistance to both nuclease and protease digestion [11]. The peptidomimetic nature of the PNA backbone also enables label-free monitoring of DNA hybridization, by use of analytical techniques that detect either physicochemical signatures of the phosphate and/or sugars present in DNA and RNA or the net increase in negative charge that occurs upon hybridization. This was soon evaluated for PNA microarray-based detection of unlabeled DNA molecules, thus circumventing one of the aforementioned limitations of DNA microarrays.

22.3 Classification of DNA Microarrays

DNA microarrays can be classified by the type of arrayed DNA fragments, the protocol used for attaching DNA to the chip, or the information fetched [3] (Scheme 22.1).

22.3.1 Immobilized Probes

According to the *immobilized probes*, the types of DNA microarrays most widely used today can be broadly divided into three main categories: cDNA, oligonucleotides, and microRNA (miRNA) arrays.

22.3.1.1 cDNA Arrays

In these microarrays each spot corresponds entirely to a specific gene. Sometimes duplicate spots will target the same gene, but these spots are exact copies of each other. The probes are of varying length but are generally hundreds of bases long. cDNA arrays are often used in RNA expression analysis, while oligonucleotide arrays are additionally used for sequence analysis. Instead of mRNA levels being directly measured, these arrays measure cDNA, because this is a more stable molecule than mRNA at these large sizes. mRNA from the

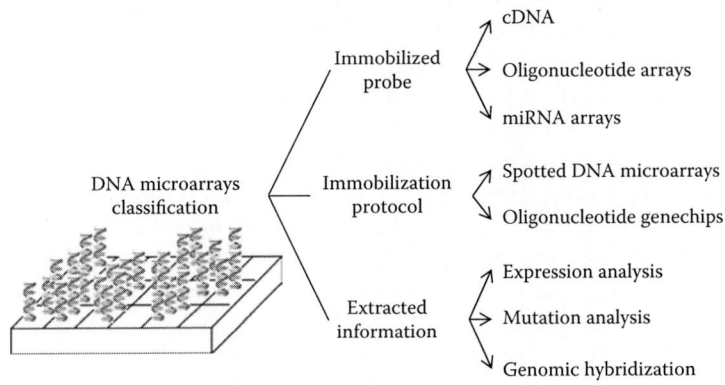

SCHEME 22.1
DNA microarrays classification.

original sample is reverse transcribed in a laboratory to create an equivalent number of the more stable cDNA molecules that are then hybridized to the microarray. These cDNA molecules are usually more than 500 bases long. One disadvantage of using these long DNA probes is the difficulty of generating reliable MM controls that will assess the specificity of hybridization [9]. Each of the probes contained on the spots in the microarray will be complementary to a cDNA molecule that represents a given gene. Thus, the measure of how much cDNA binds to its corresponding spot gives an accurate measure of the expression level of the gene in question, assuming that nothing has gone wrong.

Also instead of expression levels of an individual sample being measured directly, two separate samples are hybridized to the same array at the same time. One of these samples is generally a control sample, while the other one is a sample of interest such as tumor tissue. For example, using fluorescent detection, each of these samples is labeled with a particular dye, either a red-fluorescent dye, Cyanine 5 or Cy5, or a green-fluorescent dye, Cyanine 3 or Cy3. When the array is read by the scanner, the differential expression level of a given gene is measured by the difference in intensity between the red and green channel at the spot that corresponds to the gene in question.

cDNA microarrays are initially read by a scanner, which produces a Tagged Image File Format (TIFF) image of the array. These images are then interpreted by one of a number of image analysis software packages, all of which output data in slightly different formats. This system supports data analysis from the major platforms, including Spotfire, GenePix, BlueFuse, and Agilent.

22.3.1.2 Oligonucleotide Arrays

Oligonucleotide arrays are commonly used in biology laboratories and in clinical research. They differ slightly in operation from other kinds of arrays. Whereas the previously mentioned cDNA microarrays use long DNA strands as probes, the oligonucleotide arrays use typically around 25 base pairs in length probe oligonucleotide sequences. Each array contains hundreds of thousands of probe spots, and each of these spots contains millions of copies of an individual 25-base long DNA oligonucleotide. The Affymetrix GeneChip is the most widely used oligonucleotide array.

The whole genome of an organism can be placed on a single microarray as oligonucleotide probes. Each targeted gene is represented by typically (but not necessarily always) 11 pairs of these probes. This set of probes contains 11 PM probes, which are exactly complementary to the DNA sequence of a subset of 25 bases of the target gene. Each PM probe has a corresponding MM probe, which contains the same 25-base long sequence as the PM probe except for the middle base, or the 13th base in the chain, which is substituted for the complement of the 13th base of its corresponding PM probe; so, for example, a G in the 13th base of a PM probe will be replaced with a C in the MM probe. This is intended to give an estimate of nonspecific binding, which occurs when mRNA that is not targeted binds to a PM probe.

Oligonucleotide-based microarrays offer a number of advantages over cDNA microarrays including [9]

- More controlled hybridization specificity that makes them particularly useful for the analysis of single-nucleotide polymorphism (SNP) or mutational analysis
- Versatility to address subtle questions about transcriptome composition such as the presence and prevalence of alternatively spliced or polyadenylated transcripts
- Capacity to systematically screen whole genomic regions for gene discovery
- The fact that only sequence information (not biological samples or cDNA collections) is required to generate custom-made microarrays

22.3.1.3 miRNA Arrays

Microarrays can be also used for detection of miRNA expression levels. miRNAs are short RNA molecules, generally about 22 nucleotides (nt) in length, that regulate numerous critical cellular processes. They are encoded in genes but are not translated into proteins; instead, these molecules downregulate the expression of certain genes through binding to partially complementary sequences of specific mRNA molecules created in a cell. The miRNA molecules bind to the complementary sections of these mRNAs and stop them from being converted into proteins.

Exiqon manufactures microarrays for detection of miRNA expression. The spots on these microarrays consist of LNA probes. LNA is a modified RNA nucleotide that, because of the short length of miRNA molecules, forms a more stable bond with miRNA than standard DNA probes meaning that accuracy of measurements is increased. The miRNA molecules that are being targeted bind to its complementary LNA probe.

22.3.2 Immobilization Protocol

Regarding the *protocol used for the probe immobilization*, there are two types of DNA microarrays that have been developed so far:

22.3.2.1 Spotted DNA Microarrays

DNA probe is delivered onto the array by robotic machines. Probe DNA may be represented by ssDNA or dsDNA:

- Oligonucleotide microarrays (spotted oligoarrays)—presynthesized oligos
- cDNA microarrays—cDNA libraries, expressed sequence tags (ESTs), PCR products, etc.

22.3.2.2 Oligonucleotide GeneChips

DNA probe is *in situ* synthesized on the surface of the DNA chip.

22.3.3 Extracted Information

Taking into account the *information extracted*, microarray experiments can be categorized in

22.3.3.1 Microarray Expression Analysis

In this experimental setup, the cDNA derived from the mRNA of known genes is immobilized. The sample has genes from both the normal and the diseased tissues. Spots with more intensity are obtained for diseased tissue gene if the gene is overexpressed in the diseased condition. This expression pattern is then compared to the expression pattern of a gene responsible for a disease.

22.3.3.2 Microarray for Mutation Analysis

For this analysis, genomic DNA (gDNA) is used. The genes might differ from each other by as less as a single-nucleotide base (SNP).

22.3.3.3 Comparative Genomic Hybridization

It is used for the identification in the increase or decrease of the important chromosomal fragments harboring genes involved in a disease.

One great advantage of microarrays is their flexibility. Many different platforms exist, and many more can be created. The platforms previously outlined may be modified and easily implemented as needed for a specific application. While these variations are a great advantage in research, one of the potential holdups in implementing microarray technology in the clinical setting is the lack of standards for microarray platforms.

22.4 Uses of Microarrays

The use of DNA microarrays has extended beyond the boundaries of basic biology into diagnosis, environmental monitoring, pharmacology, toxicology, and biotechnology. Applications of these microarrays range from global analyses of transcriptional programs in yeast or mammals to establishment of novel criteria for the classification and evaluation of clinical course of tumors and to accelerated discovery of drug targets.

Oligonucleotide arrays have a wide range of possible applications, in a wide range of species. In many cases, they have been employed for the same purposes as other DNA microarrays. The main areas of application in which oligonucleotide arrays can speed up genomic studies are [3]

a. *Genome-wide transcriptional profiles and gene discovery*: DNA microarray technology helps in the identification of new genes, knows about their functioning and expression levels under different conditions, and allows us to produce a "gene-expression profile" for a particular organism grown at different developmental stages or under certain environmental conditions. Instead of comparing the mRNA levels of different samples, DNA microarrays can detect the presence of an mRNA transcript and can estimate its abundance relative to other mRNA species within the same sample. This type of study addresses the question of whether a gene or pathway is expressed under a given condition and at what level these genes are transcribed. It provides useful hints on the metabolic activity of the cell. Genome-wide transcriptional profiling generates patterns that are specific for a growth condition, developmental stage, or drug treatment.

b. *Comparative genomics and genotyping*: Genomic hybridization of a whole-genome array detects the presence or absence of similar DNA regions in other microorganisms, allowing genome-wide comparison of their genetic contents. Strain comparison by hybridizing gDNA to microarrays (genomotyping) is a more realistic approach than the whole-genome sequencing of dozens of strains of DNA microarrays and can facilitate a better understanding of the genetic differences between closely related organisms, providing useful information for the identification of virulence factors, exploration of molecular phylogeny, improvement of diagnostics, and development of vaccines. DNA microarray technology is also an excellent way to identify changes in the genetic contents of the same strain after long-term adaptation or strain optimization. In an array used for genotyping, oligonucleotides represent all the possible variations of a certain gene sequence that can have a number of applications, particularly also in the identification of polymorphisms. With this type of array, it is possible to rapidly screen a gene of unknown sequence, determining whether a large number of deleterious changes have occurred. This allows identification of polymorphisms in genes whose sequences have been determined for the first time. It is important to understand the impact of these polymorphisms on biological process, such as in disease. The GeneChips used in SNP mapping assays have a number of advantages, including speeding up the process of genetic analysis by decreasing labor, time to run an assay, and time to analyze the data. These types of assays allow a better understanding of the connections between these polymorphisms or mutations and disease. This will lead to improved knowledge of the mechanisms that induce disease and also the responses of patients to treatment, allowing a better medical care. Unfortunately, genotyping using an oligonucleotide array requires the complete sequencing of the oligonucleotides involved. Associated with this problem is that the longer the target DNA sequence, the longer the nucleotide sequence must be to eliminate any possible inaccuracies or ambiguities. The size of the fragment to be analyzed is limited by the fact that any alteration in the length of the fragment results in a change in the number of probe cells fourfold. It is also well known that this technique is weak for those sequences that contain direct repeats or inverted repeats, that is, nonrandom sequences. As an example, Chen et al. [12] have developed recently a hairpin-DNA probe-based microarray for the detection of a mutation (the methylenetetrahydrofolate reductase (MTHFR), gene C677T mutation) in subjects with coronary heart disease.

c. *Gene expression*: Oligonucleotide arrays can be effectively used to generate accurate data concerning the expression of certain gene sequences. This has implications for a number of biological assays and also for improving knowledge of cellular pathways. They can be used to study thousands of mRNA molecules, whether genes or ESTs, quantitatively and simultaneously. This greatly increases the ease with which large genomic analysis can be carried out, simplifying genomic research. Oligonucleotide arrays have the advantage of being very specific and very sensitive. This enables the detection of mRNA that is only present in a few copies per cell, as well as in several hundred thousand copies. Each probe cell is made to contain millions of copies of a particular oligonucleotide probe, and this also means that the detection of low levels of mRNA is sensitive and accurate. Oligonucleotide arrays used for gene expression have an advantage over those used for genotyping: they can be used to analyze longer fragments. A further benefit with gene-expression arrays is that although knowing what each array spot is can be advantageous, it is not a necessity. It is quite common to use probe that represents genes with unknown sequence or function.

Some commercial examples of these gene-expression arrays are Affymetrix's GeneChip; Atlas arrays (by Clontech) used in the detection of expression of specific genes such as regulators of the cell cycle, cytokines, and transcription factors; DisplayARRAY membranes (by Display Systems Biotech) designed to search for novel homologous genes or study the expression patterns of specific genes; and Panorama gene arrays (by Genosys Biotechnologies) specific for use with *Escherichia coli* (*E. coli*) and providing a method for quantifying the expression levels from all 4,290 *E. coli* genes and under any growth condition.

d. *Mapping genomic libraries*: GeneChips have been also used for mapping genomic libraries by determining the order of the overlapping clones.

e. *Predicting biochemical pathways*: Information obtained by the DNA microarrays can help pathway engineering and process optimization in several ways:

- Regulatory circuitry and coordination of gene expression among different pathways under different growth conditions
- Physiological state of the cells during fermentation
- Identification of genes involved in a production process
- Detection of differences in genetic contents
- Expression profiles among wild-type (WT) and improved strains

f. *Diagnostic applications and research into diseases*: DNA microarrays could potentially be an assay method to address multiple questions for species identification in both clinical and environmental settings. It identifies the phylogenetic status based on unique 16S ribosomal RNA (rRNA) sequences and provides information related to the presence of antibiotic markers and pathogenicity regions. Alternatively, as we mentioned before, mutant (MUT) alleles or SNPs can be detected with microarrays. Microarray technology is an efficient tool in the identification of gene-expression patterns that define disease states and that may represent prognostic indicators in different diseases such as heart, mental, and infectious diseases and especially in the study of many different types of cancer. Until recently, different types of cancer have been classified on the basis of the organs in which the tumors develop. Nowadays, with the evolution of microarray technology, it is possible to further classify the types of cancer on the basis of the patterns of gene activity in the tumor cells. This will tremendously help the pharmaceutical community to develop more effective drugs as the treatment strategies targeted directly to the specific type of cancer. One particular example of this is the use of the GeneChip with 96,000 oligonucleotide probes for the detection of all the possible heterozygous mutations in the BRCA1 breast and ovarian cancer gene. Microarray technology also has a vast number of possibilities for the detection of mutations with particular relevance to humans. These include the use of GeneChips to investigate the human immunodeficiency virus (HIV) and the genes that make up the human mitochondrial genome, to detect mutations in the cystic fibrosis transmembrane conductance regulator, and to perform studies into beta-thalassemia mutations in the beta globin gene in blood.

Novel microarrays based on the immobilization of carboxyalkylated oligonucleotides onto the epoxy-functionalized glass microslides have been developed and applied recently for the base MM study and detection of human infectious disease, meningitis [13].

A label-free detection strategy using signaling aptamer-/protein-binding complex for platelet-derived growth factor (PDGF-BB) oncoprotein detection involving amino-terminated 3D carbon micro-arrays fabricated by pyrolyzing patterned photoresist was developed very recently by Penmatsa et al. [14]. The detection mechanism is based on the release of fluorophore (TOTO intercalating dye) from the target-binding aptamer's stem structure when PDGF is captured. This simple detection technique offered high sensitivity with PDGF detection in the sub-nanomolar range (detection limit of 5 pmol) and good selectivity against different proteins, and it could be extended for the detection of other biomarkers and proteins for potential application in the preliminary diagnosis of cancer.

g. *Host responses to microbial infection*: Microarrays have been used to study the effect of viral or bacterial infection in host RNA expression. The basic model consists of an *ex vivo* measurement of gene expression of the host cells before and after they have been infected with a microorganism. By following the pattern of gene expression at different times, it is possible to elucidate which host genes are up- or downregulated over the course of the infection. Identification of genes that are differentially regulated and the characterization of their functions provide a promising tool for the understanding of pathogenicity. Several studies have used cDNA arrays to examine host cellular responses to bacterial pathogens.

h. *Analysis of microbial evolution and epidemiology*: DNA microarrays can be used to explore the variability in genetic contents and in gene-expression profiles within a natural population of the same or related species and between the ancestor and the descendents. As a result it provides very comprehensive information on the molecular basis of microbial diversity, evolution, and epidemiology. Microarray technology enables us to trace the appearance, disappearance, and reappearance of genes or their close homologs in closely related genomes.

i. *Detection of virulence factors of microbial pathogens, bioterrorism/biohazard agents, and food-safety monitoring*: The microarray technology has been used to identify the presence of specific genetic markers in bacterial genomes associated with pathogenesis and epidemiologic studies. DNA microarrays have also enormous potential for virus and bacteria detection and genotyping and are needed for rapid effective treatment, environmental monitoring, and the detection of bioterrorism agents [15,16]. Microarray technology offers also a powerful strategy for the analysis of chemical contaminants [17]. Their application to the analysis of mycotoxins, biotoxins, pesticide residues, and pharmaceutical residues has been described. Microarray technology is a promising tool for the monitoring of food safety and provides a greater understanding of food-borne chemical contaminants.

j. *Drugs and toxicological research*: Inhibition of a particular cellular process may result in a regulatory feedback mechanism leading to changes in gene-expression patterns. Simply on the basis of the transcriptional changes made by drugs, inhibitors, or other toxic compounds, the prediction of the mode of action becomes possible. Microarrays can be used to identify drug candidates that are effective in controlling bacterial growth and infectivity. In one approach, the microarrays are used to monitor bacterial gene expression in response to selected environmental conditions. This microarray analysis of gene expression enables understanding the mechanistic basis of many drugs and designs novel drugs based on newly identified regulatory pathways or networks. In addition to the alteration in gene-expression patterns related to the drug's mode of action, drugs can induce changes in genes related to stress responses that are linked to the toxic consequences of the drug. The secondary effects of a drug may reveal information of the potential resistance mechanism, which may help design drugs that have fewer side effects, but have higher efficacy by reducing the bacterium's ability to neutralize the drug. Microarray technology has extensive application in *pharmacogenomics* (the study of correlations between therapeutic responses to drugs and the genetic profiles of the patients). Comparative analysis of the genes from a diseased and a normal cell will help the identification of the biochemical constitution of the proteins synthesized by the diseased genes. This information can be used to synthesize drugs that combat with these proteins and reduce their effect. Microarray technology provides a robust platform for the research of the impact of toxins on the cells and their passing on to the progeny. *Toxicogenomics* establishes correlation between responses to toxicants and the changes in the genetic profiles of the cells exposed to such toxicants.

22.5 Electrochemical Oligonucleotides and DNA Microarrays

As mentioned earlier, most of the methodologies involving oligonucleotides and DNA microarrays use optical detection of fluorescent, luminescent, or gold (Au) nanoparticle-labeled probes. In some cases, multicolor labeling approaches are applied to allow for multiplex recognition. In general, optical detection-based approaches require large, delicate, and expensive scanners or charge-coupled-device-based imaging systems. Besides, fluorescence readout of these chips involves highly precise and sophisticated methods to analyze and interpret data. Recent development of DNA-based electrochemical sensing offers a lot of promise such as simple, accurate, and inexpensive platforms for patient diagnosis and is easier to miniaturize since electrochemical-based DNA microarrays are compatible with micro-manufacturing technology [18].

Furthermore, scanning of microarrays is typically labor intensive and a significant source of system noise complicates the use of microarrays significantly; consequently, it has hampered the transition of these technologies into high-throughput clinical scenarios by adding undue instrumentation cost and complexity. The electrochemical approach for deriving signal from a microarray is extremely simple and provides a clear path to making micro-array processing truly hands-off throughout the entire process [16].

To develop cost-effective instrumentation and miniaturized point-of-care DNA diagnostic systems, *in vitro* diagnostic (IVD) companies have taken a number of different approaches for the electronic detection of DNA hybridization. Microarrays for such systems can be produced inexpensively with silicon-, ceramic-, or printed circuit board (PCB)-based fabrication technologies.

Electrochemical test methods are being extensively applied in blood glu-cose monitoring for diabetes care. However, although numerous studies have described molecular diagnostic technologies based on electrochemical detection, none have achieved practical implementation in clinical chemistry until recently.

In these electrochemical microarrays, commonly known as eSensors, the detection of the hybridization event involves monitoring of an increased cur-rent signal from a redox indicator (which recognizes the DNA duplex) or from other hybridization-induced changes in electrochemical parameters such as capacitance or conductivity. Electrochemical detection is conducted mostly by three techniques, namely, direct electrochemistry of DNA, indirect electrochemistry of DNA, and usage of DNA-specific redox indicator [18]. Thus, in general, electrochemical signaling strategies are based on direct or catalyzed oxidation of DNA bases or on redox reactions of reporter mol-ecules or enzymes recruited to the electrode surface by specific DNA probe target interactions. The electrochemical tag can be an enzyme, ferrocene or ferrocene derivatives, an interactive electroactive substance (a groove binder or an intercalating compound), or nanoparticles [4]. The most common elec-trochemical strategies for detecting hybridization of DNA rely on interacting electroactive substances such as groove binders (e.g., $Co(phen)_3^{3+}$, Hoechst 33258, or $Co(bpy)_3^{3+}$) or intercalating organic compounds (e.g., acridine orange) that interact in different ways with ssDNA and dsDNA (Figure 22.3). The electrochemical detection of DNA via an interacting electroactive sub-stance is an attractive approach for oligonucleotide hybridization measure-ments because the target DNA does not need to be chemically modified.

An example of an eSensor is the one composed of a PCB consisting of an array of Au electrodes [19]. Each electrode is modified with a multicompo-nent, self-assembled monolayer that includes presynthesized oligonucleotide

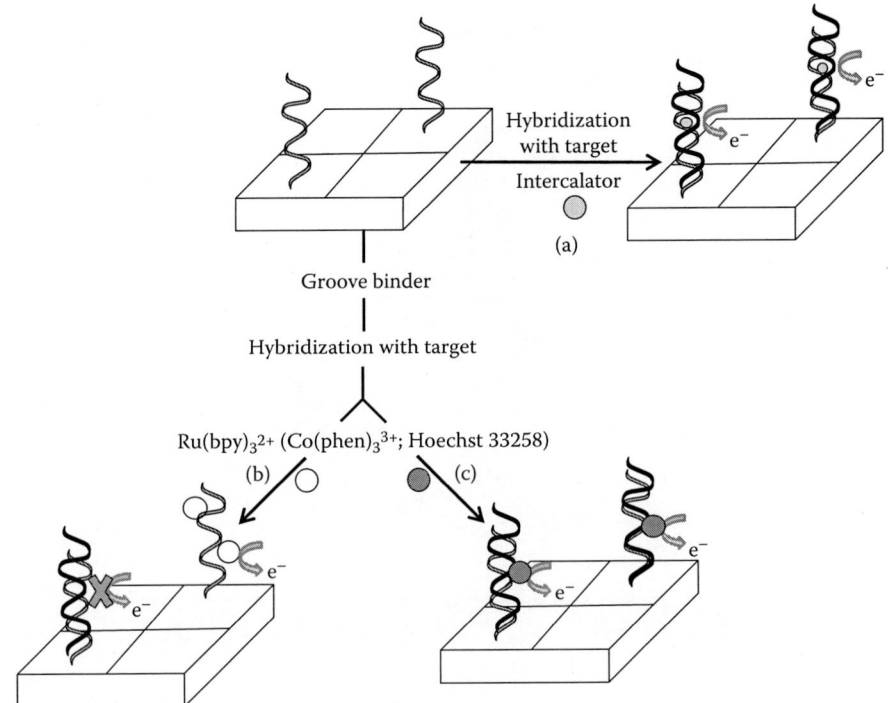

FIGURE 22.3
Different strategies to develop redox indicator-based electrochemical DNA biosensors.
(a) Intercalator interacts with the DNA duplex. (b) $Ru(bpy)_3^{2+}$ interacts with guanines of
ssDNA, whereas the formation of the double helix precludes the collision of $Ru(bpy)_3^{2+}$ with
guanine bases. (c) $Co(phen)_3^{3+}$ or Hoechst 33258 interacts with the DNA duplex, thus allowing
DNA detection.

capture probes (Figure 22.4). Nucleic acid detection is based on a sandwich
assay principle. Signal and capture probes are designed with sequences
complementary to immediately adjacent regions on the corresponding target
DNA sequence. A three-member complex is formed between capture probe,
target, and signal probe based on sequence-specific hybridization. This process brings the 5′ end of the signal probe containing electrochemically active
ferrocene labels into close proximity to the electrode surface.

The ferrous ion in each ferrocene group undergoes cyclic oxidation and
reduction that is measured as current at the electrode surface using alter-
nating-current voltammetry (ACV). Higher-order harmonic signal analysis
also facilitates discrimination of ferrocene-dependent faradaic current from
background capacitive current. Using this approach, the sensitivity of the
eSensor technology has been estimated at ~10 pM that is sufficient for genotyping applications when combined with PCR amplification.

The combination of electrochemical detection with ACV, harmonic signal
analysis, and self-assembled monolayers is highly resistant to interference

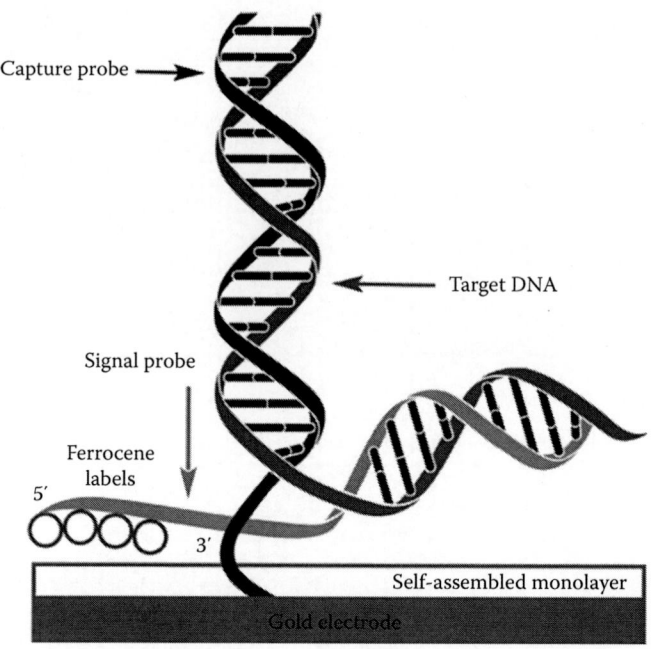

FIGURE 22.4
Electrochemical detection of DNA using eSensor technology. Target is bound to the electrode via sequence-specific hybridization to the capture probe. Signaling probe hybridizes to the target sequence adjacent to the base of the capture probe, and the associated ferrocene labels are detected at the electrode surface by ACV. (Reprinted from Liu, R.H. et al., *IVD Technol.*, in http://www.ivdtechnology.com/article/electrochemical-detection-based-dna-microarrays, 2008. With permission.)

from sample constituents. Blood constituents that would normally interfere with fluorescence detection (e.g., hemoglobin, bilirubin) have no effect on signal detection using the eSensor technology. In addition, oxidation of the ferrocene labels and transmission of current through the monolayer depend on proximity of the label to the monolayer surface. As a result, an unbound signal probe is not detected, and washing steps are not required to remove unbound reagents prior to the ACV measurement, even when a large number of signal probes representing multiple target sequences are present. The assay process is simplified, which allows hybridization and detection to be done in a small-footprint instrument without fluid handling or waste containers. In contrast, conventional microarrays require robotic instrumentation to automate multistage fluidic handling processes. Such instruments are bulky, complicated, and expensive and limit microarrays to high-cost applications.

In eSensors for genotyping of mutations or polymorphisms, capture probes covalently bound to the electrode hybridized equally to target DNA with both WT and MUT sequences. Allele-specific signal probes complementary

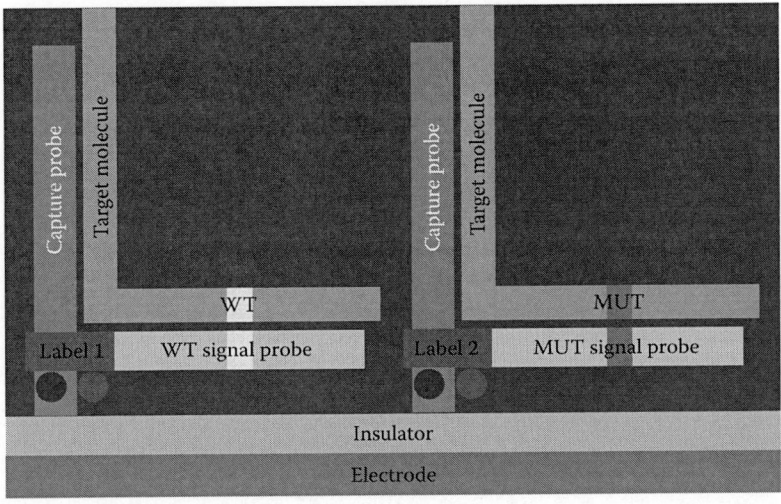

FIGURE 22.5
Genotyping assay principle in eSensors devoted to the detection of mutations or polymorphisms. (Reprinted from Liu, R.H. et al., *IVD Technol.*, in http://www.ivdtechnology.com/article/electrochemical-detection-based-dna-microarrays, 2008. With permission.)

to the WT and MUT target sequences are present in the hybridization buffer and contain ferrocene labels with different redox potentials. Both the WT and MUT targets bind to the capture probe at a site adjacent to the mutation (Figure 22.5). The signal probes compete for binding to the region of the target DNA containing the mutation site, and binding of the PM signal probe (i.e., WT signal probe to WT target) predominates. The genotype is determined by measuring the ratio of electrochemical signals from the WT and MUT signal probes. Genotyping boundaries are established based on statistical analysis of data from a large number of samples. Subsequent identification of unknown samples requires no further calibration of the instrument or cartridge lot. This approach can discriminate single- or multiple-base changes, insertions, and deletions. A mutation site with multiple alleles, or two adjacent mutation sites, can be genotyped using additional ferrocene labels.

A different approach has been considered by Chow et al. [20] by developing a wireless electrochemical DNA microarray sensor. The functioning of this sensor was controlled using two wires regardless of various individual sensing electrodes. This was enabled by confining bipolar sensing electrodes within a microfluidic channel and exerting potential control over the electrolyte solution rather than on individual electrodes. Two driving electrodes control the potential difference between the sensing electrodes and the solution, and the current at the sensing electrodes is indirectly measured by electrogenerated chemiluminescence (ECL) at the anode end of each bipolar electrode.

A real-time, multiplexed electrochemical DNA detection using an active complementary metal–oxide–semiconductor (CMOS) biosensor array with integrated sensor electronics has been developed by Levine et al. [8]. This CMOS-integrated electrochemical biosensor array allows performing quantitative DNA hybridization detection on chip using targets conjugated with ferrocene redox labels. Because only labeled target molecules localized immediately at the electrode surface allow facile electron transfer, no washing step is required and DNA hybridization can be monitored in real time. A 4 × 4 array of Au working electrodes and integrated potentiostat electronics, consisting of control amplifiers and current-input analog-to-digital converters, on a custom-designed 5 mm × 3 mm CMOS chip drive redox reactions using cyclic voltammetry, sense DNA binding, and transmit digital data off chip for analysis. The authors demonstrate multiplexed and specific detection of DNA targets as well as real-time monitoring of hybridization, a task that is difficult, if not impossible, with traditional fluorescence-based microarrays. Furthermore, these fully integrated CMOS electrochemical biosensors could reduce the cost of nucleic acid diagnostic platforms, leveraging the tremendous economies of scale associated with the semiconductor industry.

A digital CMOS-based 24 × 16 sensor array platform for fully automatic electrochemical DNA detection has been developed by Kruppa et al. [21]. The DNA sensor chip is implemented in a 0.35 µm standard CMOS technology and is extended by an additional back-end process dedicated to the Au electrodes. The whole chip with a total of 384 sensor positions captures an area of 15.8 mm^2 and dissipates less than 102 mW. This chip allows flexible, manual or fast, fully automatic working mode in which a complete electrochemical DNA detection can be done in multiple of milliseconds for the whole sensor array and the measurement can be accomplished in a tube without using microfluidics. Several electrochemical techniques, such as cyclic voltammetry and chronocoulometry, can be applied, making the chip multifunctional and flexible but still easy to handle. This chip is suitable as a platform for a wide range of applications in new fields and markets like medical diagnosis in doctors' office, food control, and environmental monitoring.

A novel microelectrode array was designed for the multipoint addressable detection of DNA hybridization [22]. Row and column electrode arrays were orthogonally arranged, and the microwells were assembled on the crossing points of the row/column electrodes to form a 4 × 4 microwell array. Amperometric signals at the individual microwells could be detected separately on the basis of redox cycling of localized electroactive species occurring between the electrodes. The detection principle was based on immobilization, and hybridization of DNA inhibited the redox cycling of $Fe(CN)_6^{4-}/Fe(CN)_6^{3-}$ at the designated microwells due to the negatively charged sugar phosphate backbone of DNA, resulting in the reduction of current response. This device was used to detect DNA hybridization with excellent sensitivity (0.03 µM) and selectivity. This methodology allows the implementation of inexpensive platforms for comprehensive, high-throughput assays in clinical diagnosis.

22.5.1 Applications of Electrochemical Oligonucleotides and DNA Microarrays for Rapid Diagnostics

Multiplexed DNA target detection is of great significance in many fields including clinical diagnostics, environmental monitoring, biothreat detection, and forensics [23].

22.5.1.1 Clinical/Bacterial Diagnosis and Pharmacogenetic Testing

Until recently, multiplex technology, such as real-time PCR, could analyze only a small number of polymorphisms in a single test. The multiplexing capabilities of DNA microarrays have been improved, thereby increasing the number of independent genotypes that can be determined simultaneously. Early microarrays focused on high-density applications that were well suited for gene discovery. Recent developments of new detection methods and simplified methodologies have facilitated the transition from expensive high-density arrays to cost-effective, low-to-medium-density systems for clinical diagnostic and pharmacogenetic applications.

Takahashi et al. [24] developed an electrochemical DNA array for simultaneous genotyping of four SNPs associated with the therapeutic effect of interferon in the treatment of hepatitis C virus (HCV). The electrochemical detection was based on the use of Hoechst 33258 that specifically binds to dsDNA. The electrochemical DNA array could genotype clinical samples as accurately as the direct sequencing method and could help in the development of individualized therapies for HCV patients.

A rapid and sensitive genotyping assay for identification of upper respiratory tract pathogens using an electrochemical semiconductor-based oligonucleotide microarray was developed by Lodes et al. [16]. The assay was applied to the detection of four bacterial pathogens (*Bordetella pertussis, Streptococcus pyogenes, Chlamydia pneumoniae,* and *Mycoplasma pneumoniae*) and nine viral pathogens (adenovirus 4; coronavirus OC43, 229E, and HK; influenza A and B; parainfluenza types 1, 2, and 3; and respiratory syncytial virus). The rapid identification of upper respiratory pathogens allowed a significant decrease in time and cost for the identification of potential lethal virus and bacterial strains, leading to better treatment and management of infections.

Henry et al. [25] developed a low-density electrochemical DNA sensor microarray for the diagnosis of predisposition to breast cancer via the BRCA1 gene. They used as model target the synthetic control gene amplicon sequence lymphotoxin-α (LTA), 107 nt long, and a corresponding surface of 20 nt long, immobilized probe. An array of 15 Au electrodes was used to detect the formation of the hybridized duplex following interaction of non-hybridized guanine bases with MB present in solution. Upon hybridization, the number of free guanines present at the electrode surface increases significantly from 8, in the immobilized probe, to a total of 25 after formation of the surface hybrid. These additional guanine bases present in the target sequence did not participate in hybridization and remained free to interact directly with MB, concentrating the electrochemical reporter directly at the electrode surface. Consequently, signal amplification resulted from the enhanced presence of guanines at the surface following hybridization, as depicted in Figure 22.6, directly measurable in a low-concentration, low-ionic-strength MB solution. The assay was quantitative and linear in the range of 6.25–50 nM target DNA exhibiting an LOD of 17.5 nM. The method was rapid and easy to perform, with no need for lengthy incubations with the MB label or requirement for washing steps.

Au electrode arrays including 16 sensors modified with ternary monolayers, composed of thiolated capture probes (specific to 16S bacterial rRNAs), mercaptohexanol and dithiothreitol, have been successfully applied to the ultra-sensitive (zeptomole) chronoamperometric detection of nucleic acid hybridization without signal amplification and to the identification of real uropathogenic clinical isolates [5].

Electrochemical interdigitated electrode (IDE) arrays functionalized by means of biomolecule friendly photolithography have been developed by Mir et al. [26]. In this work, the problem of fabricating closely spaced microelectrodes (20 μm sensor diameter and 20 μm spaced IDE array) that can be modified selectively in order to create multi-analyte sensor arrays is addressed by employing a biomolecule friendly photolithographic procedure for the sequential immobilization of different biomolecules onto separated electrodes of the same array. Different biorecognition elements were patterned to perform a DNA sensor for the selective detection of oligonucleotides connected with the BRCA1 gene mutation, related with predisposition to breast/ovarian cancer, an immunosensor for the hormone T4, and an enzymatic sensor for sarcosine and glucose. This work demonstrated that IDE arrays

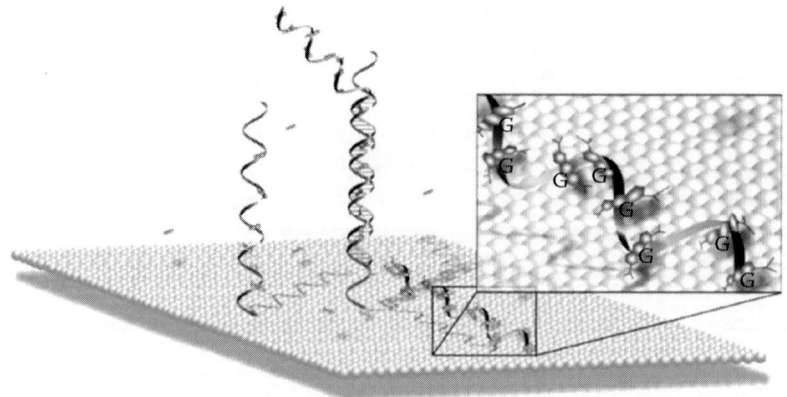

FIGURE 22.6
Electrochemical hybridization detection using surface hybrid-free guanine interaction with MB. (Reprinted from *Biosens. Bioelectron.*, 24, Henry, O.Y.F., Acero Sanchez, J.L., Latta, D., and O'Sullivan, C.K., 2064–2070, Copyright 2009, with permission from Elsevier.)

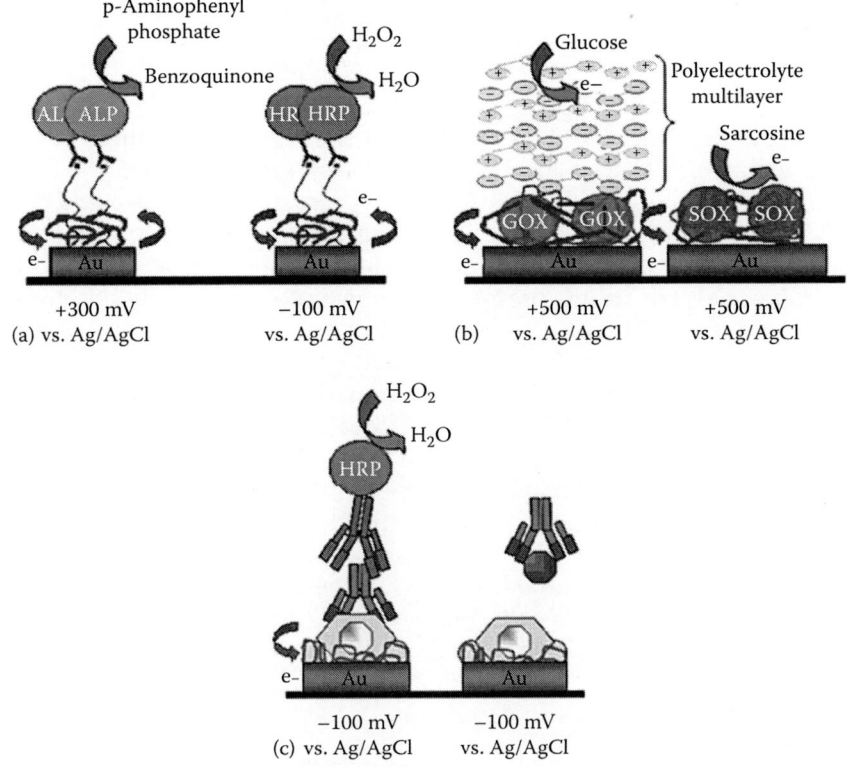

FIGURE 22.7
Schematic representation of the three configurations constructed using IDE arrays: DNA with different enzyme labels, ALP and HRP, immobilized on each electrode (a), sarcosine and glucose enzymatic sensors, using a polyelectrolyte multilayers blocking on the first set of electrodes in order to avoid the nonspecific adsorption from the second enzyme (b), and competitive immunoassay for T4 detection (c). (Reprinted from *Biosens. Bioelectron.*, 25, Mir, M., Dondapati, S.K., Duarte, M.V.. Chatzichristidi, M., Misiakos, K., Petrou, P., Kakabakos, S.E., Argitis, P., and Katakis, I., 2115–2121, Copyright 2010, with permission from Elsevier.)

are compatible with all the types of biorecognition molecules used in biosensors (Figure 22.7). The biosensor selective response and the cross talk of the modified IDEs were tested chronoamperometrically by injecting the specific substrate. The integrated device showed high specificity and sensitivity, possibility of multi-biorecognition, absence of cross talk, and complete suppression of electroactive interferences. Since current densities remain high along with the absence of cross talk, the possibility of creating a sensor that is competitive with current market technology is envisioned.

Electrochemical arrays performing gap hybridization assays have been developed for the detection of miRNAs, recently associated with cancer development and early diagnosis by acting as tumor suppressors or oncogenes [27]. The rapid, selective, and sensitive gap hybridization assay established for detection of mature miRNAs is based on four DNA/RNA hybridization components and electrochemical detection with enzymatic and redox-recycling amplification using esterase 2-oligodeoxynucleotide conjugates. Through complementary binding of miRNA to a gap built of capture and detector oligodeoxynucleotide, the reporter enzyme is brought to the vicinity of the electrode and produces an electrochemical signal (Figure 22.8). In the absence of miRNA, the gap between capture and detector oligodeoxynucleotide is not filled, and missing base stacking energy destabilizes the hybridization complex resulting in the absence of electrochemical signal. The results presented demonstrate the feasibility of this gap hybridization assay for sensitive (LOD of 2 pM or 2 amol of miR-16) and specific detection of synthetic miRNAs including the possibility of single-MM discrimination. In addition, the developed method was successfully applied for parallel detection of miRNAs (miR-21 and miR-16) in total RNA isolated from a human breast adenocarcinoma cell line (MCF-7 cells). Including RNA isolation, the gap hybridization assay was developed in a total assay time of 60 min and without the need for reverse transcription PCR amplification of the sample. The characteristics of the assay could satisfy the need for rapid, inexpensive, and easy methods to monitor miRNA expression levels of cancer patients near bedside, offering a valuable tool for early cancer marker detection in clinical diagnostics and drug discovery.

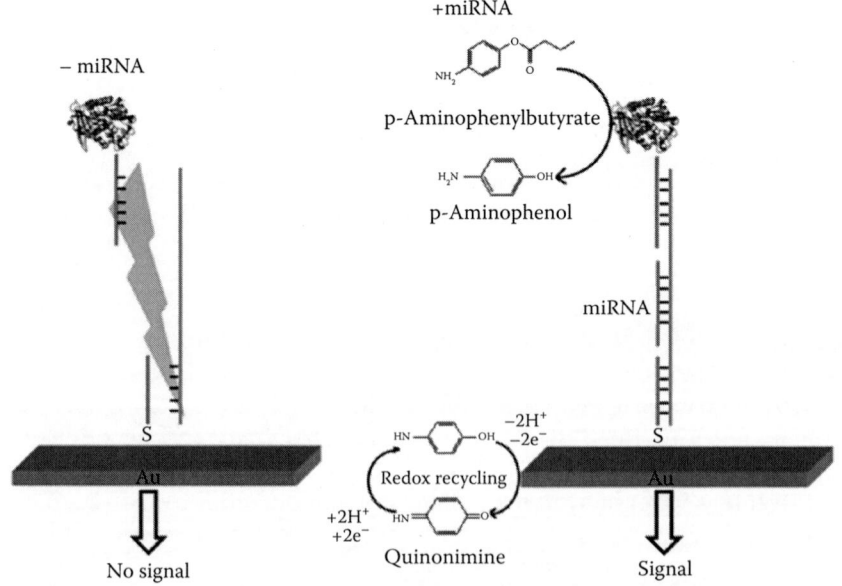

FIGURE 22.8

Electrochemical detection of gap hybridization among immobilized capture probe, miRNA, EST2-miR conjugate, and complementary RNA probe. (Reprinted with permission from Pöhlmann, C. and Sprinzl, M., *Anal. Chem.*, 82, 4434–4440, 2010. Copyright 2010 American Chemical Society.)

A simple, inexpensive, and stable label-free electrochemical DNA biosensor array was developed as a model system for simultaneous detection of multiplexed DNAs using very low sample volumes (µL) [28]. This multielectrode array comprised of six Au working electrodes and a Au auxiliary electrode fabricated by Au-sputtering technology and a printed silver/silver chloride (Ag/AgCl) reference electrode fabricated by screen-printing technology. The DNA biosensor array was used for the simultaneous detection of the HIV oligonucleotide sequences: HIV-1 and HIV-2. It was prepared in sequence by self-assembling each of two kinds of thiolated hairpin-DNA probes onto the surfaces of the corresponding three working electrodes, respectively. The hybridization events were monitored by square wave voltammetry using MB as a hybridization redox indicator (Figure 22.9). The oxidation currents of MB accumulated on the array decreased with increasing the concentration of HIVs due to the higher affinity of MB for ssDNA rather than dsDNA. The peak currents were linear over the range from 20 to 100 nM, for HIV-1 and HIV-2, and LODs of 0.1 nM were achieved for both targets. The biosensor array showed a good specificity without the obvious cross-interference. Furthermore, single-base mutation oligonucleotides and random oligonucleotides could be discriminated from complementary target DNAs. This work demonstrates that different hairpin-DNA probes can be used to design label-free electrochemical biosensor arrays for simultaneous detection of multiplexed DNA sequences for clinical applications, furthering the application in proteomics research.

Ultrasensitive and direct electrochemical DNA arrays were implemented by using Ag nanoparticle aggregates and differential pulse voltammetry (DPV) [23] (Figure 22.10a). The multiprobes containing oligo(d)A and the

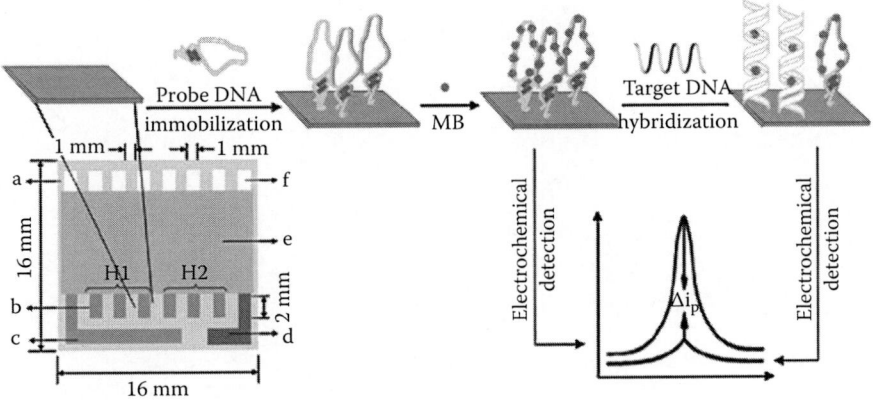

FIGURE 22.9
Schematic diagrams of multielectrode array and representation of biosensor array with fabrication steps and performance. a—Epoxy film. b—Au working electrode. c—Au auxiliary electrode. d—Ag/AgCl reference electrode. e—Insulating dielectric. f—Electric contact (Ag layer). (Reprinted from *Biosens. Bioelectron.*, 25, Zhang, D., Peng, Y., Qi, H., Gao, Q., and Zhang, C., 1088–1094, Copyright 2010, with permission from Elsevier.)

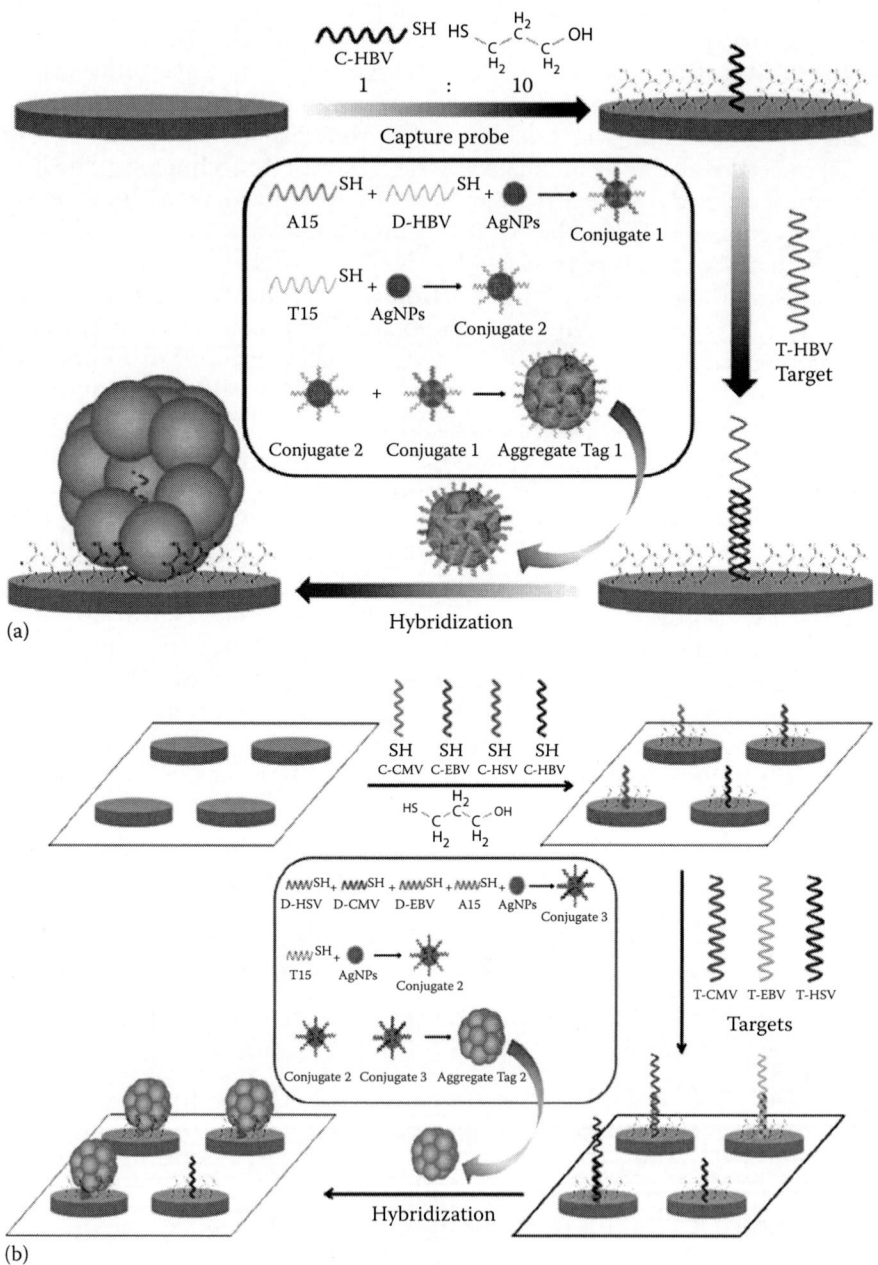

FIGURE 22.10
Schematic illustration of the electrochemical assay (a) and multiplexed assay (b) with Ag nanoparticle conjugates; preparation of the aggregates is shown as well. (Reprinted with permission from Li, H., Sun, Z., Zhong, W., Hao, N., Xu, D., and Chen, H.Y., *Anal. Chem.*, 82, 5477–5483, 2010. Copyright 2010 American Chemical Society.)

reporting probes were anchored onto Ag nanoparticles, followed by hybridizing with Ag nanoparticle conjugates modified with oligo(d)T (Figure 22.10b). The hybridization-induced tags were successfully applied to bind with the target DNA via a sandwich hybridization mode and offered direct and amplified readout by DPV. The sensitivity using the aggregate tags was improved by three orders of magnitude as compared to the single Ag nanoparticle labels, and a detection limit of 5 amol L^{-1} was obtained. The multiplexed DNA target detection was demonstrated using array chips functionalized with herpes simplex virus (HSV), Epstein–Barr virus (EBV), and cytomegalovirus (CMV) sequences, showing effective recognition of the relative sequences individually or simultaneously. This strategy based on the use of multiprobe-modified Ag nanoparticles and hybridization-induced Ag nanoparticle aggregates may open new avenues for ultrasensitive, PCR-less detection of low concentrations of genomics materials.

Mach et al. [29] and Halford et al. [6] have recently developed eSensors for rapid antimicrobial susceptibility testing (AST) directly from clinical samples. The electrochemical biosensor platform developed by Mach et al. is based on the quantization of bacterial 16S rRNA levels to monitor bacterial growth after short-term pathogen culture in the presence and absence of antibiotics using bacterial-specific 16S rRNA probes. This platform can provide culture and susceptibility information directly from a urine sample within 3.5 h in contrast to conventional culture and susceptibility tests that can take up to 72 h. Moreover, Haake's group has reported another eSensor for AST based on the sensitive detection of precursor rRNA (pre-rRNA) levels [6].

22.5.1.2 Biodefense-Related Diagnostics

Elsholz et al. [30] developed a multiplex PCR microarray using interdigitated array Au electrodes embedded into a silicon chip and enzyme-coupled electrochemical detection for the recognition of PCR-amplified gene segments corresponding to four biowarfare agents of the highest threat potential (*Bacillus anthracis, Yersinia pestis, Francisella tularensis,* and orthopoxviruses). The fully automated analysis could be carried out in 27 min.

A multispecific oligonucleotide array based on electrochemical impedance spectroscopy (EIS) was designed for label-free genetic sensing of 2.7 kb long target *Yersinia pestis* DNA and for protein sensing of ricin toxin A (RTA) chain [31]. This multispecific electrochemical array, involving immobilization of DNA and aptamer probes, has eight individually addressable 2 mm diameter Au working electrodes to incorporate multiple negative controls in the course of a single binding experiment, as well as to perform parallel identical experiments to improve the reliability of the detection, and allows rapid biosensing data accumulation by EIS in the presence of a redox agent. The array is fit with an attached electrochemical cell including an Ag/AgCl mini-reference electrode and can be used to process macro-samples of 0.5–1 mL or micro-samples of 5 μL in a dropwise fashion. Eight individual EIS measurements are completed in 15 min. The reported array is disposable, economical, and easy to use.

22.5.2 Commercial eSensors

The Osmetech Molecular Diagnostics Company (Pasadena, CA) has developed a couple of DNA microarray products based on electrochemical detection. One of them is the cystic fibrosis carrier detection system eSensor, which has received the Food and Drug Administration (FDA) clearance and commercial application in the clinical laboratory. This system consists of the eSensor 4800 instrument and the reagents and cartridges for genotyping 23 mutations and one polymorphism in the cystic fibrosis transmembrane regulator gene. This process is recommended for carrier screening by the American College of Obstetricians and Gynecologists and the American College of Medical Genetics.

The same company has developed recently another eSensor system that consists of the eSensor XT-8 instrument and microfluidic cartridges. The eSensor XT-8 cartridge device consists of a PCB chip, a cover, and a microfluidic component. The PCB chip contains 72 Au-plated working electrodes (as compared with 36 electrodes for the cystic fibrosis carrier detection chip), a Ag/AgCl reference electrode, and two Au-plated auxiliary electrodes. Each working electrode has a contact pad on the opposite side of the chip for electrical connection to the eSensor XT-8 instrument. The cartridge also contains an electrically erasable programmable read-only memory component, a memory device that stores information related to the cartridge (e.g., assay identifier, cartridge lot number, and expiration date). The microfluidic component is composed of a plate and a multilayer laminate, which includes a diaphragm pump and check valves in line with a serpentine channel that forms the hybridization chamber above the array of electrodes. The PCB chip is prepared for an eSensor assay by depositing DNA capture probes and insulator molecules on the working electrodes. Each specific deposition solution is dispensed on the appropriate electrode using a robotic pipetting system. The capture probe and insulator react with the Au surface to form an insulating self-assembled monolayer. After capture-probe dispensing, the PCB chips are washed, dried, and assembled with the laminate, plate, and plastic cover into a cartridge to form a microfluidic circulating system that can hold ~140 μL. The diaphragm pump in the cartridge is connected to a pneumatic source from the eSensor XT-8 instrument and provides unidirectional pumping of the hybridization mixture through the microfluidic channel during hybridization. Using microfluidic technology to circulate the hybridization solution minimizes the unstirred boundary layer at the electrode surface and continuously replenishes the volume above the electrode that has been depleted of complementary targets and signal probes. This process reduces hybridization time from 2 h, for the cystic fibrosis carrier detection test, to 30 min or less.

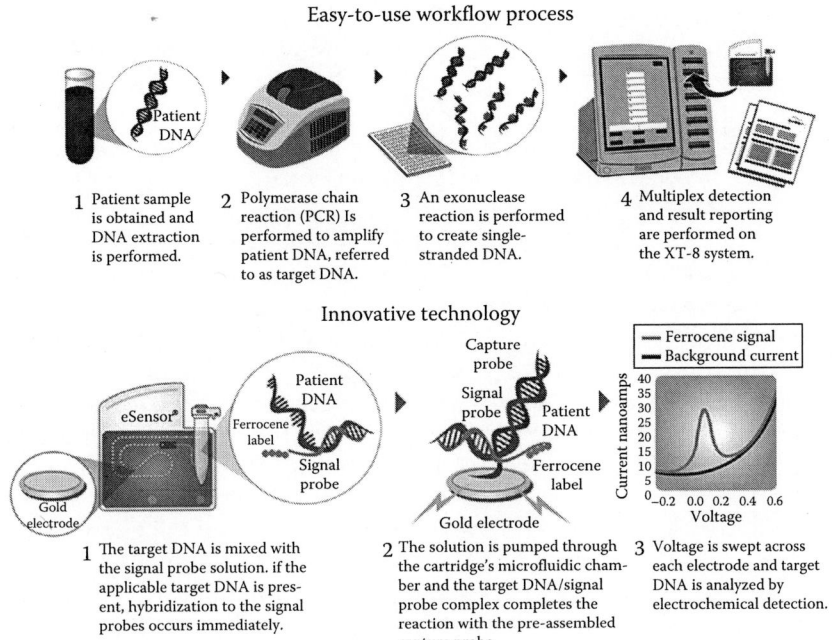

FIGURE 22.11
Schemes showing the experimental procedure used by the eSensor XT-8 systems. (Reproduced from GenMark Diagnostics Inc., in http://www.genmarkdx.com/technology/esensor.php, 2012. With permission.)

A video showing the technology of this product can be found in http://www.genmarkdx.com/technology/video.php, Figure 22.11, showing schemes for the experimental procedure it follows.

The first clinical application of this eSensor XT-8 system was the eSensor warfarin sensitivity test [19]. Warfarin is the most commonly prescribed anticoagulant in the United States. However, it exhibits a narrow therapeutic range, a wide interindividual variation in dosage required to reach optimal therapeutic effect, and severe adverse effects from overdosage, primarily due to excessive bleeding. The test genotypes three polymorphisms in two genes that correlate with warfarin dose and allows individualization of therapy based on genotype. The warfarin sensitivity assay kit consists of a PCR buffer containing primers and high-quality deoxyribonucleotide triphosphates (dNTPs), thermostable DNA polymerase, exonuclease, genotyping reagent containing signal probes, cartridges, and ancillary buffer ingredients. The test is initiated by adding the PCR master mix to the gDNA sample. This mixture is subjected to thermal cycling in a standard instrument. After PCR, the entire product is treated briefly with bacteriophage 1 exonuclease that specifically recognizes and digests the nontarget amplicon strand. The entire product of PCR and exonuclease digestion is used for genotyping with no further purification required. Genotyping is performed by adding the reagents containing the allele-specific, ferrocene-labeled signal probes to the digested PCR product, then placing the mixture into the sample reservoir of the cartridge. The entire warfarin sensitivity test procedure can be completed in 4 h, starting with isolated DNA. The performance of the warfarin sensitivity test on the eSensor XT-8 instrument has been evaluated for accuracy, reproducibility, assay range, and the effects of interfering substances. Testing 40 replicates of a gDNA sample gave 100% call rates at 10, 100, and 1000 ng DNA/PCR, indicating a broad tolerance for input sample. This range is sufficient to span the lower and upper limits of expected DNA yield for most commercial whole-blood gDNA isolation kits. Accuracy was evaluated by comparing genotyping of 101 gDNA samples by the eSensor method and bidirectional DNA sequencing. Agreement was 100%, with 97% of the samples giving results after the first round of testing, and a 100% calling rate after retesting of no-call results. Reproducibility was evaluated by testing a panel of 20 gDNA samples and three plasmid controls representing all possible panel genotypes. The panel was tested on 5 separate days with three lots of warfarin sensitivity test kits. This protocol was performed using three instruments for a total of 345 tests. Agreement of the eSensor results with DNA sequencing was 100%, with 100% of the samples giving results after the first round of testing. The effects of potential interfering substances on the warfarin test were evaluated by adding to whole blood the following: human serum albumin (3 g dL^{-1}), human IgG (3 g dL^{-1}), bilirubin (0.3 mg mL^{-1}), triglycerides (500 mg dL^{-1}), hemoglobin (20 g dL^{-1}), and ethylenediaminetetraacetic acid (a concentration four times higher than what is used to prevent coagulation in order to simulate a low-volume blood draw). None of these substances affected PCR efficiency, genotyping call rate, or accuracy. Although the performance of this eSensor warfarin sensitivity test has been evaluated in internal and clinical studies for FDA submission, its performance characteristics have not yet been fully established. Pending FDA clearance, this test can identify patients at increased risk of adverse effects of warfarin and aid in determining the optimal initial warfarin dosage based on published algorithms.

Other tests at varying development stages include genotyping tests for additional cytochrome P450 markers related to general drug metabolism, tests for infectious diseases, and combinations of markers related to the metabolism and efficacy of specific drug therapies.

The recent efforts by FDA to encourage identification and validation of new pharmacogenetic biomarkers are leading to new diagnostic opportunities in genetics, cancer, infectious diseases, and pharmacogenetics. Such new markers can be adapted to a platform suitable for the clinical laboratory and submitted for FDA clearance, thus providing easy-to-use, cost-effective tests that can be run in hospitals of all sizes.

22.6 Conclusions, Emerging Applications, and Future Prospects

Microarray analysis has demonstrated to be a foundational technology with broad and extremely important applications in areas including genetic screening, proteomics, safety assessment, and diagnostics [33].

DNA microarrays have enabled biology researchers to conduct large-scale quantitative experiments. This capacity has produced qualitative changes in the breadth of hypotheses that can be explored. In what has become the dominant mode of use, changes in the transcription rate of nearly all the genes in a genome, taking place in a particular tissue or cell type, can be measured in disease states, during development, and in response to intentional experimental perturbations, such as gene disruptions and drug treatments. The response patterns have illuminated mechanisms of disease and identified disease sub-phenotypes, predicted disease progression, assigned function to previously unannotated genes, grouped genes into functional pathways, and predicted activities of new compounds. Directed at the genome sequence itself, microarrays have been used to identify novel genes, binding sites of transcription factors, changes in DNA copy number, and variations from a baseline sequence, such as in emerging strains of pathogens or complex mutations in disease-causing human genes [34].

Recent successes in benchmarking microarrays by the U.S. FDA suggest also broad applications for assessing the safety of food, drugs, vaccines, medical devices, and other products of consumer interest. Identification of food-borne bacterial pathogens by microarray would reduce the incidence of food poisoning, illness, and death associated with bacterial contamination of meat, seafood, dairy products, and other foods. A related use of "bacto-chips" would be in clinical settings, to establish the identity of organisms in patients admitted to hospitals with systemic bacterial infections. The capacity to type unambiguously all the common bacteria on a single chip within a few hours of sampling would allow high-speed testing in agricultural, manufacturing, and clinical settings.

The efficacy and safety of pharmaceuticals is also amenable to microarray analysis, and both genotyping and gene-expression assays can be envisioned in clinical trials. Genotyping microarrays could be used to parse the clinical trial population into responders and nonresponders, enhancing the accuracy of drug-testing results and allowing drugs to be tailored to specific subsets of the population according to clearly identifiable markers in the patient population. Gene-expression microarrays could be used to examine the physiological effects of drug administration, allowing the analysis of pathways and the identification of side reactions in which drugs bind promiscuously to cellular proteins, producing toxic side effects.

Population-based screening also holds great promise for identifying disease carriers, thereby reducing the incidence of genetic disease. Particularly in the case of common genetic disorders that exhibit carrier frequencies of $\geq 1\%$, affordable microarray-based genotyping assays could be used to identify disease-causing alleles in the population and reduce the chances of passing along inherited disorders. Both oligonucleotide and multipatient microarrays can be used for patient screening, and the availability of this technology may promote the availability of home testing kits for breast cancer, cystic fibrosis, sickle cell anemia, and other common genetic diseases.

On the other hand, additional advantages provided by electrochemical detection will allow further miniaturization of electronic components and integration of upstream sample extraction and amplification processes. Previous studies have demonstrated the feasibility of a single-use, sample-to-answer eSensor device, and recent developments in microfluidics [35] provide additional tools to perform the required functions. As an example, developing a point-of-care microarray system using electrochemical detection of nucleic acids will meet critical healthcare needs, including rapid genotyping to support accurate warfarin dosage in the cardiac care unit and detection of infectious diseases in the hospital and near-patient settings.

Despite the predictable impact and demonstrated widespread use of oligonucleotide-based microarrays, there is a paucity of publicly available information regarding the design and use of this technology. Although a diversity of microarrays for diagnostic and therapeutic applications has been described in research laboratories worldwide, only some of these chips have entered the clinical market, and more chips are awaiting commercialization. As with every emerging technology, standards need to be established to avoid the doubts about the lack of reproducibility, repeatability, and compatibility across platforms and laboratories. There is a need for standardization that will facilitate comparison of microarray data. Basic questions such as the number of oligonucleotides required for reliable detection of RNA can have profound practical consequences for the quality and financial feasibility of specific experiments or projects [9]. For that, the microarray community and regulatory agencies have developed a consortium to establish a set of quality assurance and quality control criteria to ensure data quality, to identify critical factors affecting data quality, and to optimize and standardize microarray procedures [4].

Acknowledgments

The financial support of the Spanish Ministerio de Economía y Competitividad Research Projects, CTQ2012-34238, 120215-Cajal4EU Project, and the AVANSENS Program from the Comunidad de Madrid (S2009PPQ-1642) is gratefully acknowledged.

References

1. J. Sobek, K. Bartscherer, A. Jacob, J. D. Hoheisel, P. Angenendt, *Com. Chem. High T. Scr.* 9 (2006) 365–380.
2. C. S. Vasamsetty, S. R. Peri, A. A. Rao, K. Srinivas, C. Someswararao, *J. Theor. App. Inf. Technol.* 27 (2011) 43–53.
3. T. Matján, G. Bukovská, *J. Timko. Folia Microbiol.* 49 (2004) 635–664.
4. A. Sassolas, B. D. Leca-Bouvier, L. J. Blum, *Chem. Rev.* 108 (2008) 109–139.
5. J. Wu, S. Campuzano, C. Halford, D. A. Haake, J. Wang, *Anal. Chem.* 82 (2010) 8830–8837.
6. C. Halford, R. Gonzalez, S. Campuzano, B. Hu, J. T. Babbitt, J. Liu, J. Wang, B. M. Churchill, D. A. Haake, *Antimicrob. Agents Chemother.* 57 (2013) 936–943.
7. G. Giraud, H. Schulze, T. T. Bachmann, C. J. Campbell, A. R. Mount, P. Ghazal, M. R. Khondoker et al., *Int. J. Mol. Sci.* 10 (2009) 1930–1941.
8. P. M. Levine, P. Gong, R. Levicky, K. L. Shepard, *Biosens. Bioelectron.* 24 (2009) 1995–2001.
9. A. Relógio, C. Schwager, A. Richter, W. Ansorge, J. Valcárcel, *Nucleic Acids Res.* 30 (2002) e51, 1–10.
10. C. Briones, M. Moreno, *Anal. Bioanal. Chem.* 402 (2012) 3071–3089.
11. R. P. Singh, B.-K. Oh, J.-W. Choi, *Bioelectrochemistry.* 79 (2010) 153–161.
12. Q. Chen, Y. Sun, L. Zhang, K. Deng, H. Xia, H. Xing, Y. Xiang, B. Ran, M. Zhang, X. Xu, W. Fu, *Biosens. Bioelectron.* 28 (2011) 84–90.
13. S. Patnaik, S. K. Dash, D. Sethi, A. Kumar, K. C. Gupta, P. Kumar, *Bioconjugate Chem.* 23 (2012) 664–670.
14. V. Penmatsa, A. R. Ruslinda, M. Beidaghi, H. Kawarada, C. Wang, *Biosens. Bioelectron.* 39 (2013) 118–123.
15. S. A. Hashshama, L. M. Wickc, J. M. Rouillarde, E. Gularie, J. M. Tiedjeb, *Biosens. Bioelectron.* 20 (2004) 668–683.
16. M. J. Lodes, D. Suciu, J. L. Wilmoth, M. Ross, S. Munro, K. Dix, K. Bernards et al., *Plos One* 9 (2007) e924, 1–13.
17. Z. Zhang, P. Li, X. Hu, Q. Zhang, X. Ding, W. Zhang, *Sensors* 12 (2012) 9234–9252.
18. T. A. Joshi, *Int. J. Res. Pharm. Chem.* 1 (2011) 1015–1027.
19. R. H. Liu, W. A. Coty, M. Reed, G. Gust, *IVD Technol.* in http://www.ivdtechnology.com/article/electrochemical-detection-based-dna-microarrays (2008), accessed on June 2013.
20. K. F. Chow, F. Mavré, R. M. Crooks, *J. Am. Chem. Soc.* 130 (2008) 7544–7545.
21. P. Kruppa, A. Frey, I. Kuehne, M. Schienle, N. Persike, T. Kratzmueller, G. Hartwich, D. Schmitt-Landsiedel, *Biosens. Bioelectron.* 26 (2010) 1414–1419.
22. X. Zhu, K. Ino, Z. Lin, H. Shiku, G. Chen, T. Matsue, *Sensor Actuat. B-Chem.* 160 (2011) 923–928.
23. H. Li, Z. Sun, W. Zhong, N. Hao, D. Xu, H. Y. Chen, *Anal. Chem.* 82 (2010) 5477–5483.
24. M. Takahashi, J. Okada, K. Ito, M. Hashimoto, K. Hashimoto, Y. Yoshida, Y. Furuichi, Y. Ohta, S. Mishiro, N. Gemma, *Clin. Chem.* 50 (2004) 658–661.
25. O. Y. F. Henry, J. L. Acero Sanchez, D. Latta, C. K. O'Sullivan, *Biosens. Bioelectron.* 24 (2009) 2064–2070.

26. M. Mir, S. K. Dondapati, M. V. Duarte, M. Chatzichristidi, K. Misiakos, P. Petrou, S. E. Kakabakos, P. Argitis, I. Katakis, *Biosens. Bioelectron.* 25 (2010) 2115–2121.
27. C. Pöhlmann, M. Sprinzl, *Anal. Chem.* 82 (2010) 4434–4440.
28. D. Zhang, Y. Peng, H. Qi, Q. Gao, C. Zhang, *Biosens. Bioelectron.* 25 (2010) 1088–1094.
29. K. E. Mach, R. Mohan, E. J. Baron, M.-C. Shih, V. Gau, P. K. Wong, J. C. Liao, *J. Urology* 185 (2011) 148–153.
30. B. Elsholz, A. Nitsche, J. Achenbach, H. Ellerbrok, L. Blohm, J. Albers, G. Pauli, R. Hintsche, R. Wörl, *Biosens. Bioelectron.* 24 (2009) 1737–1743.
31. E. Komarova, K. Reber, M. Aldissi, A. Bogomolova, *Biosens. Bioelectron.* 25 (2010) 1389–1394.
32. GenMark Diagnostics Inc. (2012) in http://www.genmarkdx.com/technology/esensor.php, accessed on June 2013.
33. R. L. Stears, T. Martinsky, M. Schena, *Nat. Med.* 9 (2003) 140–145.
34. R. B. Stoughton. *Annu. Rev. Biochem.* 74 (2005) 53–82.
35. L. Wang, P. C. H. Li, *Anal. Chim. Acta* 687 (2011) 12–27.

23

Biosensors for Pesticides and Foodborne Pathogens

Gennady A. Evtugyn

CONTENTS

23.1 Introduction

An improvement of the preservation conditions and increase of the yield in agriculture area are commonly achieved by application of herbicides, insecticides, antibiotics, and hormones. However, the wide use of these compounds like other xenobiotics represents a potential risk for consumers and the environment related to unexpected consequences of their spreading in food chain and direct effect on the human health. To some extent, this can be also related to microbial contamination, which is responsible for numerous cases of acute poisoning, crop losses, and livestock mortality.

Although modern tools of analytical chemistry and microbiology offer broad opportunities for the sensitive detection of food contaminants and estimation of potential hazards related to their bioaccumulation and prolonged effect, the accessibility of analytical methods for food quality assessment and the timeliness of the data obtained remain a subject of special concern. The progressing food contamination and fundamental risks of such processes call for the development and application of new approaches and portable devices based on biorecognition assay. In comparison with conventional analytical techniques, they are intended for the detection of target biochemical reactions and hence provide information closely related to the possible effects (poisoning, immunomodulation, cancer and mutagenesis risks, etc.).

Biosensors are portable analytical devices with biochemical elements (enzymes, antibodies (*Abs*), DNA) integrated with the transduction system intended for direct detection of chemical compounds [1]. The biosensors offer unique opportunities related to their application in field and high sensitivity provided by biochemical elements in their assembly. Even though biosensors do not always show high stability and reliability of the response and are mainly recommended for preliminary testing, they can significantly extend the area of the monitoring of potential hazards.

In this chapter, the assembly and signal readout will be considered for the most common biosensors devoted for the detection of pesticides and foodborne pathogens with particular emphasis to those adapted for the food testing. The examples of particular biosensor-based approaches and techniques mostly cover the literature within the last decade (2002–2012).

23.2 Pesticide Determination

In accordance with EPA definition [2], a *pesticide* is any substance or mixture of substances intended for preventing, destroying, repelling, or mitigating any pest. Insects, mice and other animals, unwanted plants (weeds), fungi, or microorganisms like bacteria and viruses are considered as target for pesticides, which are classified in accordance with this criterion as insecticides, herbicides, fungicides, acaricides, algaecides, and others (biopesticides, antimicrobial, and pest control devices). Various chemical compounds exert pesticide properties. Among others, organophosphates, carbamates, organochlorine, and pyrethroid derivatives are most common and commercially important. The overview of the pesticides classified in accordance with their chemical structure is given in Table 23.1 [3].

Biopesticides are derived from natural materials, for example, animals, plants, bacteria, and certain minerals. They involve also plant-incorporated

TABLE 23.1

Major Classes of Pesticides in Accordance with Their Chemical Structure

Chemical Class	Code	Pesticide Group	Chemical Class	Code	Pesticide Group
Arsenic compounds	AS	Fungicides, insecticides, herbicides	Organotin compounds	OT	Fungicides, herbicides
Bipyridine derivatives	BP	Herbicides	Phenoxyacetic acid derivatives	PZ	Insecticides
Carbamates	C	Acaricides, fungicides	Pyrazole derivatives	PZ	Insecticides
Coumarins	CO	Rodenticides	Pyrethroids	PY	Acaricides, insecticides
Copper compounds	CU	Algaecides, fungicides, insecticides	Triazine derivatives	T	Herbicides
Inorganic and organic mercury compounds	HG	Fungicides, rodenticides	Thiocarbamates	TC	Herbicides
Organochlorine compounds	OC	Fungicides, insecticides			

protectants that are produced by plants from introduced genetic material (genetically modified organisms [GMOs]) and biochemical, which control pests by nontoxic mechanisms (pheromones, behavior modification, etc.). The importance of biopesticides grows, and about 780 products were registered in United States at the end of 2001 [4]. Nevertheless, their contribution to the world market is still rather low, and their relation to the pesticide family is a subject of further discussions.

Many pesticides are highly toxic for a human being and can exert various hazards especially on nervous system [5]. Some pesticides were found to act as endocrine disruptors [6] and potentially carcinogens [7]. The toxic effect of common pesticides like other pollutants can increase due to the accumulation of persistent pesticides during rather long period of their application in agriculture. The transfer of the pesticide residues in the food chain not only leads to the increased levels of pesticides in food but also enhances the potential area of their effect among the population and hence directly affects human health all over the world [8].

For this reason, the residuals of pesticides in fresh and drinking water, food, and soil are strictly regulated by national environmental agencies and some international organizations, that is, WHO and FDA. The maximal threshold concentrations, permissible content, and recommended daily intakes become gradually tougher with understanding negative influence of pesticides on the human health. Presently, sub-ppb levels of the pesticide concentrations are permitted in the water. The limits of the pesticide residuals in food depend not only on their toxicity but also on the bioavailability of the pesticides and daily consumption of appropriate food [9]. Briefly, the variation of the permissible levels depends on the lipophilicity of a contaminant and content of lipids and fat in the sample. The same factors are most important for the extraction of the pesticides by organic solvent required for most traditional instrumental techniques recommended for the pesticide detection.

Pesticides have been traditionally determined by chromatography techniques, starting from paper thin layer chromatography 40 years ago and finishing with the most sophisticated HPLS/MS and GC/MS today [10,11]. Modern analytical instrumentation provides high reliability and accuracy of the determination of the pesticide residues on ppt–ppb levels. However, these techniques require preliminary extraction and careful purification and concentration of the extracts, often followed by derivatizing an analyte. Such procedures assume the use of large volumes of water and highly purified organic solvents and the assistance of qualified personnel as well as the use of expensive laboratory equipment.

In parallel to such stationary equipment, portable sensing devices have been developed for preliminary testing and semiquantitative estimation of the pesticide contamination in various samples [12]. Many of them utilize the biorecognition principles when the pesticides are detected by their influence on cells and enzymes or due to affinity interactions with specific *Ab*s or specially designed aptamers. Although such biosensors are less stable and less accurate than conventional analytical instruments, they have some advantages, for example, direct relation between a pesticide content and its biological effect or the field application prospects. Pesticide biosensors are promising in screening of the new preparations with reduced toxicity for warm-blooded organisms and with higher efficiency against pests.

The number of articles devoted to the biosensors for pesticide detection is enormously high. Such biosensors were first fabricated for the detection of the so-called nerve agents, primarily chemical weapons suppressing the nerve impulse transmission (sarin, soman, VX) [13]. The military devices were then adapted for the detection of organophosphates and carbamates, which exert similar effect on insects and meanwhile show relatively high acute toxicity on warm-blooded organisms including humans. Although the application of such pesticides is continuously decreasing, the appropriate biosensors retain their importance because of the numerous accidences among the agriculture workers, especially in developing countries, and potential hazards related to the chemical terrorism threats.

The *immunosensors* are able to recognize pesticides due to their specific interaction with *Ab*s. They are considered as another promising approach to detecting the food and water contamination in the field [14]. Being as much sensitive as conventional immunoassay protocols, immunosensors offer prospects related to the automation of measurements and to their more reliable response especially in complex media. To some extent, this can be also referred to *DNA sensors* employing aptamers specifically designed for the recognition of particular analytes [15]. Being more sensitive toward specific antigens (*Ags*) than enzyme sensors, immuno- and aptasensors are less effective in detecting metabolites of the pesticides or in assessing certain groups of pesticides present in the environment.

23.2.1 Enzyme-Based Biosensors

The detection of pesticides with enzyme sensors is commonly based on two techniques, that is, (1) the determination of the inhibition exerted by a pesticide and (2) the conversion of a pesticide as a substrate of appropriate enzyme [16].

In the first case, the reaction is performed in standard conditions favorable for the substrate detection. The substrate is added prior to or together with an inhibitor, and the decrease in the signal is estimated as a measure of the inhibitor concentration. This protocol assumes at least two stages of the measurement and the necessity to carefully control the conditions of the substrate conversion. The selectivity of the inhibitor detection is not as high as that of the substrate. Usually a group of structurally relative compounds is detected simultaneously. However, this drawback is compensated for by extremely high sensitivity of inhibition measurement, which is at least comparable to that established by immunoassay technique. Thus, the use of cholinesterases makes it possible to detect down to 0.1 nM of organophosphates and carbamates allowing direct sample monitoring without any preconcentration [17].

The measurement of pesticides involved in biochemical conversion as enzyme substrates is limited by milli- and micromolar range of their concentrations. This is insufficient for most purposes of food safety control and environmental monitoring. To some extent, the limitation is leveled by the multiuse format of such enzyme sensors. Contrary to that, the inhibitory biosensors once inhibited need to be replaced or reactivated by auxiliary agents.

The enzymes used for the detection of pesticides are summarized in Table 23.2. The number of appropriate references exceeds that given in the table, which contains mainly reviews; some others will be considered later regarding specific aspects of the biosensor applications.

It should be mentioned that most part of the enzyme sensors developed for the detection of inhibitory effect caused by pesticides were tested in standard solutions and spiked samples of natural waters. Their application for food contamination assessment is still a subject of future consideration. The influence of organic solvents used for pesticide extraction and that of matrix components complicates the interaction of an enzyme with a pesticide so that the estimates made for standard solutions and even spiked samples can significantly differ from that of real samples in their inhibitory influence.

TABLE 23.2

Application of Enzyme Biosensors for Pesticide Determination

Enzyme	Pesticide Class	Application Area	References
AChE	Insecticides (organophosphates and carbamates)	Air, fresh and drinking waters, food, and soil contamination	[18–20]
OPH	Insecticides (organophosphates)	Air, water	[21]
Tyrosinase	Herbicides (atrazines)	Organic extracts from corn, barley, and lentils	[22]
	Insecticides (organophosphates and carbamates)	Natural water	[23,24]
Aldehyde dehydrogenase	Fungicides (dithiocarbamates)	Water (standard solutions of an inhibitor)	[25]
Peroxidase	Herbicides (glyphosate), insecticides (thiodicarb)	Spiked drinking water, organic extracts from apple, potato, and strawberry	[26,27]
ALP	Herbicides (2,4-D), insecticides (organophosphates)	Natural water	[28,29]
Acid phosphatase	Insecticides (organophosphates)	Water (standard solutions of an inhibitor)	[30]

23.2.1.1 Cholinesterase Sensors

As was mentioned earlier, cholinesterase sensors are most intensively investigated for pesticide detection. Indeed, their progress was first affected by military purposes requiring compact robust devices for chemical weapon detection. Besides direct applications for the detection of individual organophosphate and carbamate pesticides, cholinesterase sensors have been successfully applied as models for the consideration of the factors affecting the inhibition in heterogeneous conditions (e.g., the influence of enzyme immobilization on the sensitivity of inhibition, the effect of organic solvents, the optimization of the signal detection, and the signal shifts during the storage).

Cholinesterases belong to the family of hydrolases that promote the hydrolysis of organic esters, namely, acetylcholine, an important neurotransmitter. Similar enzymes called carboxylesterases are present in plants, liver, and muscle tissues of many animals, but their sensitivity and selectivity to both the substrates and inhibitors are far from the characteristics of the cholinesterases [31,32]. In accordance with the specific rate of the substrate hydrolysis, the cholinesterases are divided into several groups. Two of them, that is, acetylcholinesterase (AChE) and butyrylcholinesterase (BChE), are mainly used in the biosensor assembly. The AChE, also called true cholinesterase, shows the highest efficiency in the hydrolysis of acetylcholine, while the BChE, a serum cholinesterase, is more specific toward butyrylcholine (Equation 23.1).

Several mixed forms with predominant affinity toward benzoylcholine and propionylcholine have been also described but have not found any application in the biosensor design:

$$
(CH_3)_3N^+CH_2CH_2OCR + H_2O \xrightarrow{\text{Cholinesterase}} (CH_3)_3N^+CH_2CH_2OH + \text{Acylated}
$$

$$
\underset{O}{\|}
$$

R=CH$_3$: Acetylcholine
R=C$_3$H$_7$: Butyrylcholine

Choline   cholinesterase

$$
\Big\downarrow \begin{array}{l} H_2O \\ RCOOH \end{array} \qquad (23.1)
$$

Cholinesterase

Only acetylcholine is converted by cholinesterases in living beings. Other substrates were synthesized for the estimation of the enzyme activity and kinetic characterization of enzyme–substrate and enzyme–inhibitor specificity. Thiocholine esters [33], indoxyl acetate, resorufin butyrate [34], and other artificial substrates are employed in the colorimetric, fluorescent, and electrochemical sensors and assay protocols.

The reaction between cholinesterase E and a substrate S starts with the reversible formation of the molecular complex E–S converted to the so-called acylated cholinesterase E–A followed by the choline (XH) release (Equation 23.2). The acylated enzyme E–A is easily hydrolyzed by water to recover the initial enzyme molecule able to the reaction with the next substrate molecule:

$$
E + S \underset{k_{-1}}{\overset{k_1}{\rightleftharpoons}} E\text{-}S \xrightarrow[HX]{k_2} E\text{-}A \xrightarrow[HA]{k_3} E \qquad (23.2)
$$

The catalytic constant k_{cat} and the Michaelis constant K_M of the reaction (23.2) can be expressed by the rate constants of partial stages of the reaction:

$$
k_{cat} = k_{-1} + \frac{k_2}{k_3}; \quad K_M = \frac{k_3(k_{-1} + k_2)}{k_1(k_2 + k_3)} \qquad (23.3)
$$

For AChE from human erythrocytes, the K_M value for acetylcholine and acetylthiocholine is equal to 0.09 ($k_{cat} = 7 \cdot 10^5$ min^{-1}) and 0.057 mM, respectively [35].

All the cholinesterase types have similar molecular structure that does not significantly alter with their biological origin so that the results obtained with electric eel or horse cholinesterases do not differ dramatically from those obtained with the human enzymes. This allows preliminary estimation of relative toxicity of the newly synthesized compounds prior to their laboratory testing on animals. Nevertheless, the sensitivity of pesticide detection varies with the enzyme source, and this is one of the routes to the pesticide discrimination described for the multienzyme sensor arrays [36,37].

The active site of the AChE globule involves anionic and esteratic centers, which differ in their pH-dependent affinity toward the substrate, choline, and H⁺ ions. The distance between the aforementioned centers does not change, while free enzyme is acylated by acetylcholine. This follows rather small conformational changes of the enzyme and hence the high efficiency of the substrate conversion [38]. The anionic center of AChE lies at the bottom of a sterically restricted, hydrophobic cleft and remains open after acylation. The interaction of the anionic center with acetylcholine maximizes the rate of its hydrolysis, whereas the interaction with the second substrate molecule suppresses the cleavage of the acetylated AChE and the enzyme recovery. For this reason, AChE exerts allosteric inhibition by a substrate excess [39,40]. The BChE enzyme differs from the AChE by the relative affinity of the active site toward the substrate/product and the pH dependence of the rates of the particular reaction stages. Butyrylcholine is not involved in the allosteric regulation of the BChE activity, either. The biochemical functioning of the BChE remains unclear. Probably this enzyme can compensate for the lack of AChE activity or fulfill scavenging functions in organism.

Both AChE and BChE are used in the biosensor assembly because of different sensitivities toward certain groups of organophosphates and carbamates [41,42]. From the very beginning, BChE from horse and human serum was mainly applied due to its high stability and sufficient sensitivity toward the organophosphates [43]. Later on, AChE from *electric eel* and *Drosophila melanogaster* became preferable due to higher specific activity [44]. Nowadays, many types of the cholinesterases commercially available are produced by genetically engineered microorganisms, for example, *Escherichia coli*. In addition to natural products, genetically modified cholinesterases with increased activity to certain insecticides have been introduced in biosensor applications [45,46]. The behavior of various groups of cholinesterases and the kinetic and thermodynamic aspects of their interaction with substrates and inhibitors have been reviewed [47,48].

The carbamate and organophosphate insecticides irreversibly inhibit cholinesterase by replacing the substrate in the esterification of the active site. Contrary to the acylated enzyme, phosphorylated and carbamoylated cholinesterases are rather stable toward hydrolysis and hence are excluded from the catalytic cycle of the enzyme conversion (23.4):

$$\begin{array}{ccc}
\underset{R'}{\overset{R'}{\diagdown}}P\text{—OR} & \xrightarrow[\text{R'OH}]{\text{Cholinesterase}} & \underset{R'}{\overset{R'}{\diagdown}}P\text{—O—ChE} \xrightarrow[\text{Slowly}]{\text{H}_2\text{O}} \underset{R'}{\overset{R'}{\diagdown}}P\text{—OH}
\end{array} \quad (23.4)$$

Phosphorylated cholinesterase

The kinetic scheme of the reaction (23.4) is very similar to that of substrate conversion (23.2), but the final product slowly reacts with water (23.5):

$$E + I \underset{k_{-1}}{\overset{k_1}{\rightleftharpoons}} E\text{–}I \overset{k_2}{\underset{I'}{\searrow}} E\text{–}I' \overset{k_3}{\underset{I''}{\searrow}} E \qquad (23.5)$$

Here, $E\text{–}I$ is an enzyme–inhibitor complex, and I' and I'' are the parts of an inhibitor molecule released from the complex in its partial decomposition. If $k_2 \gg k_{-1}$, the reaction describes fully irreversible inhibition, whereas the opposite relation ($k_2 \ll k_{-1}$) is typical for reversible competitive inhibition. The last stage of the enzyme recovery is called reactivation. The formation of phosphorylated cholinesterase is commonly fully irreversible in the time scale of signal measurement, whereas carbamoylated enzyme can spontaneously recover while left to stay in working media. The efficiency of reactivation can be quantified by the k_3 value. Thus, the reaction of the hydrolysis of acetylated AChE from electric eel is characterized with $k_3 = 6 \cdot 10^5$ min^{-1}. Similar reaction of the carbamoylated enzyme $(CH_3)_2NC(O)\text{–}AChE$ shows $k_3 = 0.012$ min^{-1} [48]. Phosphorylated AChE yields the k_3 values of about and lower than $n \cdot 10^{-5}$ min^{-1}.

The spontaneous reactivation can be accelerated by the treatment of the cholinesterase sensor with reactivators or antidotes [49]. Two oximes, that is, 1,1'-trimethylene-bis-4-formylpyridinium bromide dioxime (TMB-4) and pyridine 2-aldoxime methiodide (2-PAM), are mostly used for immobilized cholinesterases [50]. The efficiency of reactivation depends on the inhibition degree and the "age" of phosphorylated cholinesterase. The lower the influence of the pesticide on the enzyme and the faster the cholinesterase sensor is treated with a reactivator after its contact with the sample, the higher the recovery of the initial biosensor signal. The phosphorylated cholinesterase left after the inhibition for a certain period of time tends to lose its activity irreversibly. This process called "aging" is related to covalent modification of enzyme active site by irreversible transfer of the methyl group in the enzyme–inhibitor complex [51]. The reactivation can affect the further measurement of the pesticide residues. First, the reactivators themselves slightly inhibit the activity of cholinesterases. Then, the sensitivity of the immobilized enzyme toward irreversible inhibitor commonly decreases with each reactivation cycle. This might be due to preferential inactivation of enzyme molecules most sensitive to inhibitors or due to increasing influence of the mass transfer stages, which diminish the access of the enzyme molecules to the analytes. Previously, the reactivation was considered as one of the promising ways directed to the diminishing cost of the pesticide detection. Now, the amounts of the enzymes required for a single measurement and development of cheap transducers (screen-printed electrodes [52,53] or field-effect transistors [FETs] [54,55]) make the requirements of reactivation of the phosphorylated enzyme not obligatory.

The irreversible inhibition is quantified using the Aldridge equation [56] (23.6) assuming the inhibitor concentration exceeds the concentration of enzyme active sites:

$$\ln \frac{v_0}{v_t} = k_{II} c_I \tau \tag{23.6}$$

where
v_0 and v_t are the rates of enzymatic reaction prior to and after the incubation
c_I is an inhibitor concentration
τ is the incubation period

The bimolecular inhibition constant k_{II} is a measure of the inhibition efficiency, which depends on the nature of an inhibitor, the enzyme source, and the incubation conditions but not on the inhibitor concentration. The Aldridge equation (Equation 23.6) gives an estimated of the limit of detection (LOD) for irreversible inhibitor. If the quantification limit corresponds to the shift of the enzymatic rate of reaction by 15%–25%, the LOD is equal to $(0.13 \div 0.22)/(k_{II} \cdot \tau)$. In some cases, the Aldridge equation is represented by (Equation 23.7), which is more convenient for the application. It expresses the dependence of the remained enzyme activity (a) on an inhibitor concentration:

$$\log a = \log 100 - \frac{k_{II} \tau}{2.303} \tag{23.7}$$

Some typical k_{II} values of are given in Table 23.3.

TABLE 23.3

Bimolecular Inhibition Constants k_{II} (M^{-1} s^{-1}) for Some Typical Organophosphate and Carbamate Pesticides and Chemical Weapons

Enzyme Biological Origin	AChE		BChE	
	Electric Eel	Bovine Erythrocytes	Human Serum	Horse Serum
Paraoxon	$2.2 \cdot 10^4$	$6.0 \cdot 10^5$	$1.5 \cdot 10^6$	$8.0 \cdot 10^5$
Dichlorvos	$4.2 \cdot 10^4$	$2.3 \cdot 10^4$	$2.3 \cdot 10^5$	$8.0 \cdot 10^4$
Aldicarb	$5.0 \cdot 10^4$	$1.3 \cdot 10^4$	$2.4 \cdot 10^4$	$1.0 \cdot 10^4$
Carbaryl	$3.3 \cdot 10^4$	$1.8 \cdot 10^4$	$1.9 \cdot 10^3$	$7.0 \cdot 10^3$
Carbofuran	$1.7 \cdot 10^6$	$8.0 \cdot 10^5$	$6.0 \cdot 10^3$	$3.1 \cdot 10^4$
Sarin	$2.5 \cdot 10^7$	$6.3 \cdot 10^6$	$6.7 \cdot 10^6$	$1.2 \cdot 10^6$
Soman	$1.0 \cdot 10^8$	$3.5 \cdot 10^7$	$1.9 \cdot 10^7$	$7.3 \cdot 10^6$

Source: From Alfthan, K. et al., *Anal. Chim. Acta*, 217, 43, 1989; Herzsprung, P. et al., *Int. J. Environ. Anal. Chem.*, 47, 181, 1992.

The use of genetically modified cholinesterases makes it possible to selectively increase the inhibition constant to selected pesticides and hence increase the sensitivity and selectivity of detection. This was shown for three carbamates, that is, carbaryl, aldicarb, and pirimicarb, using wild-type enzyme from *D. melanogaster* and three mutant enzymes with extended selectivity. The inhibition constants varied by more than one order of magnitude [59]. In a similar manner, carbofuran and paraoxon were classified with multisensors involving mutant and wild AChE from *D. melanogaster* and natural AChE from electric eel, bovine erythrocytes, and rat brain [45]. The design of cholinesterases by gene expression and protein engineering with particular emphasis to the appropriate changes in the inhibition constants toward paraoxon and some other insecticides is summarized in Refs. 60 and 61.

The irreversible inhibition can be recognized by the dependence of the relative decay of the enzyme activity on the incubation period. The inhibition reaches 100% with increasing inhibitor concentration. In some reports, lower limit (60%–90%) is reported. This might be due to nonenzymatic paths of the substrate conversion or due to the decomposition of the inhibitor on the enzyme support. The impurities that reversibly affect the activity of cholinesterase suppress irreversible inhibition due to the so-called protecting effect [62].

The protecting effect is quantified by the coefficient γ calculated from the shift of the bimolecular inhibition constants k_{II} after the addition of a reversible inhibitor to the solution of irreversible inhibitor added to the enzyme:

$$\gamma = \frac{(k_{II})_{I=0}}{(k_{II})_I} - 1 = \frac{\{\ln(v_0/v_i)\}_{I=0}}{\{\ln(v_0/v_i)\}_I} - 1 = \frac{c_I}{K_i} \tag{23.8}$$

Here, "$I = 0$" corresponds to the measurement series performed in the absence of reversible inhibitor and that marked with "I" to the same parameters obtained in the presence of a constant amount of reversible inhibitor c_I. K_i is the constant of competitive inhibition that corresponds to the ratio of enzyme involved in enzyme–substrate and enzyme–inhibitor complexes. Ideally, K_i corresponds to the inhibitor concentration resulting in 50% decrease of the enzyme activity.

For the quantification of irreversible inhibition, the biosensor is first incubated in the solution of an inhibitor. Then it is transferred into the standard solution of a substrate. Alternatively, the substrate can be directly added to the sample tested after the incubation. In this case, the reaction of the enzyme with an inhibitor is stopped by its involvement in the enzyme–substrate complex, which is insensitive to the presence of an inhibitor. The characteristics of the inhibitor detection do not depend on the format of the signal detection and the substrate concentration. This means that the researcher can choose the signal transduction system in accordance with the accuracy and specificity of the response, which are more important for the achievement of the lowest detection limits. This measurement protocol assumes two separate stages of the signal measurement, that is, prior to and after the contact with an inhibitor, and hence increased influence of the reproducibility of the signal on the inhibition calculation. The simplified protocol assumes simultaneous addition of the substrate and inhibitor. In this case, the inhibition kinetics corresponds to competitive inhibition. Instead of bimolecular inhibition constant k_{II}, the sensitivity of inhibitor detection is expressed by inhibition constant K_i. The competition of the substrate and inhibitor for the same active site of the cholinesterase results in the dependence of the inhibition measured on the substrate concentration. The lower the substrate concentration, the higher the sensitivity of inhibitor detection. For this protocol, the measurement techniques most sensitive toward the substrate concentration have an advantage [63]. This should be taken into account when different biosensors are compared or the biosensor performance is discussed especially for field applications or semiautomatic systems.

All the equations presented earlier correspond to ideal homogeneous conditions with no effect of mass transfer of the reactants. In biosensor assembly, the enzyme is incorporated in the support whose permeability toward the substrate/product of the reaction can significantly vary. The immobilization of cholinesterase influences the characteristics of inhibitor determination. Most valuable results were obtained for rather thin membranes with enzyme covalently attached to the support by covalent bonds or implemented in self-assembled multilayered films retaining the natural structure of an enzyme. The choice of enzyme support depends also on the target analytes and samples to be tested. Thus, the use of hydrophilic supports compensates for the influence of organic solvents, which can inactivate the cholinesterase even in the absence of the pesticides.

Among other possible effects, the relative saturation of the enzymatic layer is of special importance. If the specific enzyme activity is rather high and the permeability of the surface layer is limited, some of the enzymes incorporated in the surface layer remain free in the conditions of the signal measurement. After the contact with an inhibitor, the number of active sites in the membrane reduces, but this is compensated for by the involvement in the reaction of the centers remained unoccupied with the substrate molecule prior to incubation. This results in apparent decrease of the inhibition measured by the shift of the signal prior to and after the inhibition. The effect of partial saturation of the enzymatic layer obviously follows from the heterogeneous condition on the transducer interface. The pH shift due to buffer properties of the enzyme support and steric limitation of the substrate/inhibitor access in the membrane are mostly mentioned in the discussion of the influence of immobilization on the cholinesterase sensor performance.

For these reasons, the inhibition of cholinesterases immobilized in the biosensor assembly is quantified by the semilogarithmic dependence of the inhibition on an inhibitor concentration. The absolute values of the sensor signal proportional to the enzyme activity can be used for such plotting. In electrochemical sensors, the current related to the substrate/product oxidation/reduction is used in such dependencies.

The cholinesterase activity can be quantified by various approaches. The reaction (Equation 23.1) of the acetylcholine hydrolysis is monitored by the amounts of the acid released [64]. In the simplest way, pH shift is visualized by appropriate pH indicator immobilized together with an enzyme on the paper [65]. Similarly, fiber-optic sensor with enzyme immobilized by sol–gel technique and bromothymol blue pH indicator in the sensing layer has been employed for the detection of chlorpyrifos with the LOD of 0.04 mg/L [66].

Potentiometric biosensors based on pH glass electrodes [67,68] or choline selective sensors and *conductometric* devices with interdigitated electrodes [69] are described for the same purposes. Although the sensitivity of the substrate detection of potentiometric biosensors is lower than that of voltammetric devices, the potentiometric and conductometric cholinesterase sensors showed the sensitivity of the detection of common pesticides comparable to those of other electrochemical techniques based on similar enzymatic membranes. Recording pH shifts of cholinesterase hydrolysis is sensitive toward the sample acidity and its buffer properties.

Choline formed from a natural substrate, that is, acetylcholine, is recorded using specific potentiometric sensors [70], by fluorescent detectors [71] or as a substrate of choline oxidase, a second enzyme implemented in the biosensor assembly. Such bi-enzyme (cholinesterase–choline oxidase) sensors with *amperometric* detection of the hydrogen peroxide, a final product of acetylcholine conversion (23.9), showed remarkable sensitivity toward both the substrate and inhibitors.

$$(CH_3)_3N^+CH_2CH_2OCCH_3 + H_2O \xrightarrow{\text{Cholinesterase}} (CH_3)_3N^+CH_2CH_2OH + CH_3COOH$$

(23.9)

with the choline oxidase step consuming O_2 and producing H_2O_2 to give $(CH_3)_3N^+CH_2C\begin{smallmatrix}O\\OH\end{smallmatrix}$

The oxidation of the hydrogen peroxide released in the reaction (Equation 23.8) can be promoted by mediators of electron transfer, for example, Prussian blue [72], Pt nanoparticles [73], MnO_2 [74], and phthalocyanine complexes [75]. Horseradish peroxidase (HRP) is also described for the same purpose [76,77]. A principal scheme of the reactions conducted in such tri-enzyme layer is presented in Figure 23.1.

The use of additional enzymes complicates the immobilization procedure because of a significant difference in the specific activity of AChE and choline oxidase and in their pH optima of the activity. To overcome this limitation,

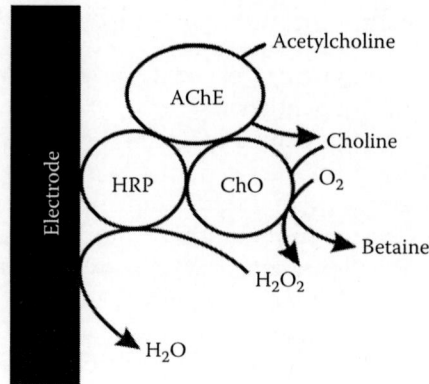

FIGURE 23.1
Detection of the AChE activity with tri-enzyme sensor involving AChE, choline oxidase (ChO), and HRP.

choline oxidase can be immobilized on appropriate transducer (oxygen or peroxidase sensors), while AChE is added in soluble form to the sample containing an inhibitor. Although such a protocol assumes a single-use application of the enzyme, the consumption of the AChE preparation is very small (down to 0.001 U per measurement) and the accuracy and sensitivity of the response toward acetylcholine and inhibitors are very high.

Artificial substrates of cholinesterases were synthesized especially to meet the requirements of particular detection systems. Thus, chromogenic substrates produce colored products of enzymatic reactions, fluorogenic substrates enhance the fluorescence, and electrochemically active substrates produce the current or potential shift without any additional enzymes described for choline detection (see Equation 23.9).

In *optical detection systems*, indoxyl acetate forms in the presence of cholinesterases leucoindigo, which undergoes oxidation by dissolved oxygen to indigo, a blue dye (23.10) [78,79].

$$\underset{\text{Indoxylacetate}}{\overset{\text{OC(O)CH}_3}{\boxed{}}} \xrightarrow[\text{−CH}_3\text{COOH}]{\text{Cholinesterase}} \underset{\text{}}{\overset{\text{OH}}{\boxed{}}} \xrightarrow{\text{O}_2}$$

$$\longrightarrow \underset{\text{Indigo}}{\boxed{}}$$

(23.10)

The reaction can be performed in the paper dipstick format with visual detection of the color intensity or time required for the color change as a measure of enzyme activity and hence inhibition of cholinesterase. Indoxyl acetate and 2-naphtyl acetate were also used for fluorescent detection based on the detection of the products of their enzymatic hydrolysis. The approach was successfully applied for the determination of fenitrotion [80].

Standard method for cholinesterase activity assay proposed by G.L. Ellman [81] includes the reaction of thiocholine ester with 5,5′-dithiobis-(2-nitrobenzoic acid) (Ellman reagent) followed by monitoring of a yellow product at 405 nm (23.11).

$$(23.11)$$

The Ellman method is used for the standardization of the enzyme preparations including the estimation of residual enzyme activity in the immobilization protocols and expression of enzyme activity units. Similar approach has been proposed for the chemiluminescent detection of the AChE activity with dioxetane derivative (12, R = thiocholine) [82].

$$(23.12)$$

Acetylthiocholine is employed in the electrochemical biosensors that represent a majority of the cholinesterase sensors. The oxidation results in the formation of dimeric disulfide with the consumption of one electron per thiocholine molecule:

$$(CH_3)_3N^+CH_2CH_2SH \xrightarrow{-e^-,-H^+} \tfrac{1}{2}(CH_3)_3N^+CH_2CH_2S\text{-}SCH_2CH_2N^+(CH_3)_3$$

$$(23.12a)$$

The potential of oxidation on unmodified electrodes varies from 680 mV (Pt) to 450–500 mV (carbonaceous materials). The electrode reaction is well characterized and commonly requires additional mediators of the electron transfer to decrease the overvoltage and prevent electrode poisoning with sulfur containing by-products. Metallocyanate [83,84] and phthalocyanine [85,86] complexes, carbon nanotubes [87,88], nanoparticles of noble metals and oxides of transient metals [89,90], and 7,7,8,8-tetracyanoquinodimethane (TCNQ) [91] can be used for the amplification of the signal related to thiocholine oxidation.

The use of thiocholine esters simplifies the biosensor assembly in comparison with bi-enzyme (cholinesterase–choline oxidase) biosensors. However, the affinity of the artificial substrate toward the enzyme active site can differ from that of acetylcholine. This may have an effect if the substrate and inhibitor contact with the enzyme simultaneously.

The characteristics of the pesticide detection with cholinesterase sensors are summarized in Table 23.4. They mostly illustrate the performance of the biosensors in testing food and drink samples, in some cases in relation to those obtained in standard aqueous solutions. The number of articles in which the cholinesterase sensors are characterized by standard solutions of pesticides or spiked waters is much higher.

In most cases, the validation of the biosensor response toward pesticides is performed for a limited number of the samples spiked with known amounts of pesticides. Commonly rather high amounts of the analytes in comparison with their quantification level are taken for this purpose. This is insufficient for the estimation of the reliability of the biosensor results in the whole range of the concentrations determined. In such experiments, the recovery calculated from the ratio of the determined and added quantities of pesticides is given as the only measure of their applicability for real sample testing. The use of the reference methods (HPLC or GC–MS) confirms the conclusions about the robustness of the pesticide detection in biological matrices to a much higher extent.

Sample treatment required usually involves the extraction of the pesticides by polar organic solvents (ethanol, methanol, acetonitrile, cyclohexanone, etc.) followed by the extract treatment and partial or full evaporation of the organic solvent. In many cases, the extraction follows the protocols elaborated for HPLC analysis. The solid-phase extraction on columns or cartridges containing the extracting systems on solid supports is also described.

The direct detection of the pesticides in juices or in aqueous extracts without any pretreatment often is limited by the high inhibition of AChE by the so-called natural enzyme inhibitors. Thus, in grape juice, vine, and must, the inhibition is related to biogenic phenolic compounds. A strong reversible inhibition of the AChE by tomato [115] and potato [116,117] glycoalkaloids was established by direct measurements with free and immobilized enzyme.

Some of interferences can be eliminated by special sample treatment. Thus, phenols can be removed by electrolysis with Al anode. The Al^{3+} ions formed in the anodic dissolution of the anode precipitate the phenols in insoluble complexes removed from the solution by filtration [95]. In fruit juices, the inhibition is

TABLE 23.4

Cholinesterase Biosensors for the Detection of Organophosphate and Carbamate Pesticides in Food

Pesticide	Detection Mode	Sample	Detection Characteristics	References
Paraoxon	Screen-printed carbon electrode modified with TCNQ	Orange juice, peach, and apple baby food	LOD 1 µg/kg (3.6 nM) in standard solution and 10 µg/kg in isooctanol extract from orange juice, conc. range 1–60 µg/kg	[92]
Paraoxon	Screen-printed carbon electrode modified with Pd/Pt nanoparticles	Grapes	LOD 0.1 µg/L, conc. range 10–75 µg/L	[93]
Paraoxon	Screen-printed carbon electrode modified with TCNQ	Milk	LOD 1–20 µg/L (various types of engineered AChE from *Nippostrongylus brasiliensis*), conc. range 1–100 µg/L	[94]
Paraoxon	Glassy carbon electrode modified with Prussian blue and Nafion	Grape juice, must	LOD 10 (paraoxon) and 5 (parathion-methyl) µg/L, conc. range 14–173 (paraoxon) and 7–26 (parathion-methyl) µg/L	[95]
Carbofuran, methiocarb, pirimicarb, carbaryl, paraoxon, chlorpyrifos-ethyl, chlorpyrifos-methyl, dichlorvos, malathion, methidathion	Screen-printed carbon electrode modified with TCNQ	26 fruit and vegetable samples and 23 samples of processed infant food	3–50 µg/kg of pesticides confirmed by GC-MS, samples spiked with paraoxon: conc. range 2–20 µg/kg	[96]
Methyl parathion, monocrotophos	Carbon paste electrode	Orange, tomato, banana	LOD 0.04 and 47 ppb, conc. range 0.1–1.0 and 90–590 ppb for methyl parathion and monocrotophos, respectively (aqueous solutions), spiked fruits (orange, tomato, banana, 0.2 ppm and 5110 0.4 ppb): recovery 73%–94%	[97]

(continued)

TABLE 23.4 (continued)

Cholinesterase Biosensors for the Detection of Organophosphate and Carbamate Pesticides in Food

Pesticide	Detection Mode	Sample	Detection Characteristics	References
Paraoxon, chlorpyrifos oxon, malaoxon	Screen-printed carbon electrode modified with Co phtalocyanine, FIA mode	Milk	LOD 5 pM, 5 nM, and 0.5 nM for chlorpyrifos oxon, paraoxon, and malaoxon, respectively (milk). Conc. range 5 pM–5 µM (10 min incubation)	[98]
Coumaphos, chlorpyrifos-methyl, carbofuran	Screen-printed carbon electrode modified with TCNQ	Grape juice	LOD 0.02, 0.05, and 0.008 µM, conc. range 0.03–6.0, 0.1–5.5, and 0.01–0.9 µM for chlorpyrifos-methyl, coumaphos, and carbofuran, respectively	[99]
Paraoxon, carbofuran	Thick-film planar carbon electrodes	Orange juice	LOD $10^{-9.5}$–$10^{-8.9}$ M (paraoxon in orange juice, AChE and BCHE from various sources) recovery 93%–110%	[100]
Parathion-methyl, carbaryl	—	Bovine, egg, milk, honey	LOD 2 (carbaryl) and 1 (parathion-methyl) ng/mL, conc. range 2–90 and 1–100 ng/mL, respectively; recovery in acetone–hexane extracts against HPLC 86%–101% (10 and 30 ng/g)	[101]
Dimethoate	Screen-printed carbon electrodes modified with carbon nanotubes, Prussian blue, and Nafion in ZrO$_2$ matrix	Chinese cabbage	LOD 5.6·10^{-4} ng/mL, conc. range 1.0·10^{-3}–10 ng/mL (water solutions), recovery in acetone extract against HPLC 88%–105%	[102]
Chlorfenvinfos	Screen-printed carbon electrode modified with Co phtalocyanine	Wheat, cabbage, apple, orange, and cherry	Spiked methanol extracts containing 10^{-5} and 10^{-7} M chlorfenvinfos, recovery 78% and 93%, respectively	[103]

Pesticide	Biosensor	Sample	Results	Ref.
Dichlorvos	Screen-printed carbon electrode modified with Co phtalocyanine	Apple skin	LOD $7 \cdot 10^{-11}$ M and $1 \cdot 10^{-8}$ M for mutant AChE and $6 \cdot 10^{-7}$ M for wild AChE from *D. melanogaster*, 99% recovery in 5% acetonitrile extract from apple skin	[104]
Malathion	Au electrode modified with carbon nanotubes and Fe_3O_4	Nonfat dry milk, whole milk	Recovery 106%–109% against malathion calibration curve in water solution, spiked samples contained 20, 40, and 80 nM of pesticide	[105]
Coumaphos	Screen-printed carbon electrode modified with Prussian blue and choline oxidase	Honey	LOD 10 ng/mL. conc. range 10–100 ng/mL in methanol extract, recovery against LC–MS 86% (samples spiked with 10, 50, 100, 150, 300, 600, 1000 ng/mL of coumaphos)	[106]
Chlorpyrifos-methyl	Screen-printed carbon electrode modified with TCNQ	Grapes, vine leafs	Conc. range 5–22 µg/g (extract from vine leafs), 180–850 ng/mL (grape), variation coefficient against HPLC 10%–20%	[107]
Chlorpyrifos	Glassy carbon modified with exfoliated graphite nanoplatelets	Broccoli	LOD $1.58 \cdot 10^{-10}$ M (water solution), conc. range 10^{-8} to 10^{-7} M (spiked ethyl acetate extract from broccoli), recovery against HPLC 95%	[108]
Chlorpyrifos-oxon	Screen-printed carbon electrode	Tea, orange juice, milk, pepper	LOD 0.5 µg/L for 1 mU of AChE from *D. melanogaster*, samples spiked with 10 µg/mL, recovery against HPLC–MS up to 13%	[109]

(continued)

TABLE 23.4 (continued)

Cholinesterase Biosensors for the Detection of Organophosphate and Carbamate Pesticides in Food

Pesticide	Detection Mode	Sample	Detection Characteristics	References
Carbaryl	Carbon paste carbon electrode containing Co phtalocyanine	Tomato	LOD $4.0 \cdot 10^{-4}$ g/L, conc. range 0.5–750 µM, spiked samples of tomato pulp: recovery against HPLC 83.4%	[110]
Aldicarb, carbaryl, carbofuran, methomyl, propoxur	Screen-printed carbon electrodes modified with Co phtalocyanine	Potato, carrot, sweet pepper Apple, peach, orange, carrot, green bean, sweet pepper	Conc. range from $1 \cdot 10^{-4}$ to 35 mg/kg (AChE from electric eel, bovine erythrocytes, human erythrocytes, BChE from horse serum and human serum, pesticide recovery from extracts from 79% to 96% except propoxur) Spiked samples extracted with methanol, 0.5–3.0 mg/kg. Recovery against HPLC results 60%–125%	[111] [112]
Paraoxon, chlorpyrifos, diazinon, carbaryl, carbofuran	Photothermal detection with Ellman reagent, FIA mode	Spiked apple and orange juices	LOD 0.2 ng/mL (paraoxon, water solution), 2.8 (paraoxon, orange juice), 4 ng/mL (paraoxon, apple juice). Chlorpyrifos, diazinon, carbaryl, carbofuran: LOD 1 ng/mL–4 mg/mL (water solutions). AChE from bovine erythrocyte enzyme immobilized in polyurethane foam	[113]
Carbofuran, propamocarb, oxydemeton-methyl, parathion-ethyl	Photothermal detection with Ellman reagent, FIA mode	Salad, iceberg lettuce, onion	Estimation of the total amount of pesticides by carbofuran equivalent, satisfactory agreement with GC–MS except parathion-ethyl. BChE from horse serum, AChE from electric eel	[114]

related to acidic media affecting both the stability of the substrate toward non-enzymatic hydrolysis and enzyme activity. The addition of the buffers can be recommended to shift the pH to more convenient neutral and basic media, but this dilutes the sample and diminishes real concentration of an inhibitor.

The availability of the pesticides toward enzyme active site in complex matrices can be limited by its distribution between the aqueous phase and microheterogeneous components (tissue particles, parts of biological membranes, lipid fraction, etc.). For milk, the hydrophobic pesticides are mainly concentrated in fatty fraction.

The extraction of the pesticide residues in accordance with the protocols elaborated for chromatographic detection seems to be a universal but time- and labor-consuming approach to sample treatment. Methanol, isooctanol, cyclohexane, acetone, hexane, and acetonitrile have been applied for this purpose. This is quite understandable for the investigations with the chromatography used as a reference method. But, from the point of view of biosensor applications, especially in field, this contradicts the main advantage of biosensor, that is, easy and fast application. Indeed, the biosensors, as compact portable devices, are mainly intended for field application where there is no place for column cleaning, solvent evaporation, and residue resuspension or reextraction. These stages are commonly mentioned in typical extraction protocols.

The determination of inhibitory activity directly in organic solvents is also described. Previously it was established for free AChE that most of the solvents inhibit its activity irreversibly [118]. Water-miscible solvents exert higher effect on free enzyme than immiscible. The immobilization of the enzyme in hydrophilic matrix protects them from the solvent inhibition. Thus, the implementation of BChE in κ-carrageenan gel made it possible to directly detect paraoxon and aldicarb in water-saturated chloroform/hexane (50% v/v) mixture [119,120]. The linear range of concentration determined was found to be 9.4–56.4 and 10–60 µg/L, respectively.

The description of the influence of organic solvents on the determination of pesticides by cholinesterase sensors is contradictory. On the one hand, interference of the hexane with paraoxon detection was reported [121]. On the other hand, small amounts of cyclohexane (up to 10%), acetonitrile (5%–10%), and ethanol (10%) enhance the concentration range of the organophosphates determined and increase their concentration [122,123]. Moreover, acetonitrile increases the signal of cholinesterase biosensor toward the substrate prior to its contact with an inhibitor [123,124]. Assuming additive summation of the solvent and pesticide inhibition, simplified protocols have been proposed for the determination of the pesticide residues in diluted extracts not treated or cleaned by simplified scheme. Such an approach can be used if the contribution of the solvent does not exceed 15%–20% of the initial enzyme activity. Meanwhile, the influence of the organic solvent on the biosensor lifetime and efficiency of its recovery after the measurement is not discussed. For this reason, such approaches can be mainly recommended for disposable sensors, which do not assume repeated application for inhibition quantification.

Most of the organophosphates mentioned in Table 23.4 are not used in agricultural purposes. They are too toxic and exert acute toxicity on the workers. Instead of that, thionic analogs are commercially available, which are oxidized to real anticholinesterase agents directly in insects by mixed function oxidases. In a human being, this biochemical path is not so efficient, and hence, such pesticides are less dangerous in use. The oxidation of parathion to paraoxon is given in Equation 23.13 as an example of such reaction, which belongs to metabolic activation of xenobiotics.

$$H_3C-O \atop H_3C-O \Big\rangle \!\! P-O-\!\!\!\bigcirc\!\!\!-NO_2 \longrightarrow {H_3C-O \atop H_3C-O} \Big\rangle \!\! P-O-\!\!\!\bigcirc\!\!\!-NO_2 \qquad (23.13)$$

| Parathion | Paraoxon |

In the measurement conditions, thionic pesticides should be preliminary oxidized to phosphoryl analogs. For this purpose, bromine [125], cytochrome P_{450} [126], chloroperoxidase [127], and N-bromosuccinimide [128] have been applied. The oxidation of thionic pesticide can be combined with its column preconcentration by solid-phase extraction [128]. In chemical oxidation, the excess of the oxidant should be removed prior to inhibition measurement by appropriate reagent, for example, formic acid or sodium thiosulfate. Then the pH should be corrected, again. All of these stages complicate the analysis but are necessary to establish the sensitivity of the pesticide measurement required. Indeed, thionic organophosphates are themselves weak reversible inhibitors of cholinesterases. Their inhibition reported in some articles can be referred to the impurities of oxon forms present in the samples or spontaneous oxidation of thiophosphates by dissolved oxygen during the sample treatment. Thus, the efficiency of the reaction of thiophosphate/phosphate conversion is important for the final success of the assay.

As could be seen from the description of cholinesterase biosensors, they cannot be employed for the determination of selected pesticides. The total amount of the anticholinesterase agents will be detected, and there are no simple ways to distinguish the signal. Indeed, the irreversible inhibition in the mixture of the pesticide depends on both their partial concentration and individual inhibition constants. The Aldridge equation does not allow presenting the total decay of the biosensor signal by a sum of partial inhibition degrees related to individual compounds.

This might be useful if the biosensor is considered as an analog of the toxicity test performed on daphnia or infusoria (mortality testing for the detection of acute toxicity in waters). Some attempts have been performed on real samples of wastewaters in industrial area to correlate the biosensor signals with other bioassay techniques [129,130]. The inhibition can be expressed in concentration units of a standard inhibitor, for example, paraoxon or carbaryl. Such an approach was adopted from toxicology where relative toxicity of various compounds is compared. For the same purpose, the kinetic parameters of inhibition, that is, k_{II}, K_i, and derived parameters (IC_{50}, a concentration exerting 50% inhibition), are applied.

The use of several enzyme sources in the assembly of cholinesterase sensors has been described for the discrimination of selected pesticides and their semiquantitative quantification [45,131,132]. The discrimination is achieved by the use of chemometric approaches (artificial neuron nets, partial least-squares analysis), which make it possible to classify the samples in accordance with the similarity of the biosensor approaches to the sets of standard solutions characterized in advance. Such an approach called "electronic tongue" was developed for the arrays of potentiometric sensors intended for the determination of inorganic ions [133] and then extended to voltammetric devices and biosensors. The biosensor arrays show remarkable results on the mixtures of two pesticides taken in the amounts exerting similar inhibition, but their application for real samples needs additional investigations. Anyway, the extension of the number of pesticides used in particular area will be a problem for such multisensor systems.

One can see from Table 23.4 that the analytical characteristics of pesticide detection depend on the signal transduction system even though the Aldridge equation does not assume such an influence. This seems rather unusual especially for the model aqueous solutions of the pesticides when the matrix effect can be excluded from the consideration. As an example, Table 23.5 summarizes the characteristics of the determination of paraoxon,

TABLE 23.5

Comparison of the LOD Values for the Paraoxon Determination with Different AChE Biosensors

Enzyme Source	Detection Mode	LOD/Incubation Time	References
Electric eel	Screen-printed electrode with carbon nanotubes as mediator	0.5 nM/30 min	[87]
Electric eel	Screen-printed electrode modified with TCNQ	3.6 nM/10 min	[92]
Electric eel	Screen-printed electrodes covered with Nafion, measurements in the presence of 5% acetonitrile	19 nM/10 min	[134]
Electric eel	FET, bovine serum albumin as matrix	0.5 μM/20 min	[135]
Electric eel	Conductometric, Pt planar interdigitated electrodes	0.5 μM/15 min	[136]
Electric eel	Voltammetric detection, Pt electrodes, flow-through regime	10 nM/30 min	[137]
D. melanogaster	Screen-printed electrode modified with TCNQ	2.0 nM/10 min	[138]
D. melanogaster	Screen-printed electrode modified with Ni/NiO nanoparticles	1.0 pM/20 min	[139]

one of the most investigated anticholinesterase pesticides. The following reasons of the diversity of the biosensor performance can be mentioned:

- Nonenzymatic paths. The hydrophobicity of some enzyme supports could improve sensitivity due to adsorption of pesticides near the enzyme active sites within the surface layer. The hydrolysis of the substrate on the components of the surface player can also take place.

The difference in the LOD values by about five orders of magnitude cannot be related to the immobilization protocol only. The following aspects should be also taken into consideration:

- Substrate/inhibitor distribution. A nonuniform structure of the surface layer can promote the access of the pesticide and substrate to the enzyme active due to electrostatic or lipophilic interactions within the layer.
- Metrological characteristics. The measurement assumes the calculation of the inhibition degree from the signals obtained prior to and after the incubation step. This means the deviation is doubled, and the minimal detectable decay of the signal is at least twice higher than that of the substrate determination. This is especially important for nonlinear and semilogarithmic calibration plots.
- Enzyme loading. The relative decay of enzyme activity directly depends on the initial enzyme activity. Modern transduction system decreases the amounts of the enzyme required in comparison with the results published about 10 years ago.

Summarizing the efforts made in the development of cholinesterase-based biosensors, it should be noted that they solve both problems of direct detection of the dangerous concentrations of organophosphate pesticides and total assessment of the toxicity of the sample for nervous system. The LOD value achieved is comparable and even below the limited threshold levels. Meanwhile, low selectivity of the response and moderate stability of the biosensor signal within the storage period remain a weak point of cholinesterase biosensors. This can be referred to many other enzyme sensors, though.

23.2.1.2 Other Enzymes Employed for Pesticide Detection

Cholinesterase biosensors for organophosphates and carbamates cover more than 90% of all the publications devoted to enzyme sensors for pesticide determination. This is especially true if the application of biosensors for food safety is considered. Meanwhile, there are some interesting examples of other approaches utilizing different enzymes able to convert pesticides as specific substrates or change their activity in their presence.

23.2.1.2.1 Organophosphate Hydrolase

This enzyme was selected for the detection of organophosphate nerve agents as an alternative to cholinesterases. At the first stages of the investigations, bacterial enzymes were used in biosensor assembly. Their activity was rather low, and most efforts have been devoted to the establishment of immobilization protocols. At present, the organophosphate hydrolases (OPHs) from recombinant microorganisms, for example, *E. coli*, are mainly used. Mutant OPHs with preferential hydrolysis of P–O, P–S, P–CN, and P–F bonds were isolated [140]. The main idea of OPH application was to avoid a principal drawback of cholinesterase biosensors referred to the necessity of the replacement of the sensing element after its irreversible inhibition. The OPH hydrolyzes the esters of organophosphorus acids as shown for paraoxon in the following:

$$(23.14)$$

Nitrophenol released from this process can be detected spectrophotometrically by the absorption of phenolate ion [141] and amperometrically [142,143]. The second product, that is, organophosphoric acid, can be detected potentiometrically with pH sensor [144], ion-selective FET [145], and conductometric devices [70]. The paraoxon LOD values obtained with OHP biosensor vary from 10^{-6} to 10^{-9} M. The fluorogenic label (7-isothiocyanato-4-methylcoumarin) increases the sensitivity of optical detection of paraoxon to 0.5 nM [146]. The sensitivity of detection can be improved by electrocatalytic cycling of the reaction product [147]. Thus, the cathodic reduction of nitrophenol results in the formation of *p*-aminophenol, which is reversibly oxidized to quinone imine at rather low potential. The reaction (23.15) takes place within the surface layer onto Au nanoparticles, which increase the specific surface of the electrode and exert synergic effect on the oxidation–reduction of the reactants:

$$(23.15)$$

The OPH can be used for direct hydrolysis of thionic pesticides without their preoxidation to phosphoryl analogs as required for cholinesterases. In the earlier example, methyl parathion was used as model toxicant, and the LOD of 0.3 ng/mL was obtained for standard solutions. The biosensor was tested on the aqueous ethanol extracts from garlic. The recovery of 95%–102% for 1–1000 ng/mL of the pesticide was reported.

Dehydrohalogenase is a bacterial enzyme catalyzing the reaction of dehydrochlorination of some pesticides. As OHP, this is a rare case of detection of pesticide involved in enzymatic conversion. The dehydrohalogenase from *Pseudomonas putida* is specific to DDT and that from *Burkholderia pseudomallei* to hexachlorocyclohexane. The enzyme activity and hence the concentration of the pesticides were determined by chloride selective ion-selective electrode [148]. The glutathione-S-transferase from mosquito *Aedes aegypti* also exerts dehydrochlorinase activity. The Cl^- ions released from the DDT decomposition were detected using pH sensor or colorimetrically with pH indicator [149]. The LOD of the assay was 3.8 μg/mL, and the linear range of quantification from 12 to 250 μg/mL. Similar detection with bromocresol green as pH indicator was realized in fiber-optic biosensor based on glutathione-S-transferase from maize immobilized by sol–gel technology. The enzyme catalyzes dehydrochlorination of the atrazine. The signal is linear to the analyte concentration in the range from 2.52 to 125 μM (LOD 0.84 μM) [150].

Tyrosinase (polyphenoloxidase) catalyzes the regioselective aerobic oxidation of monophenols to *o*-diphenols and their eventual dehydrogenation to *o*-quinones. The interest to this enzyme is stimulated by its high activity in predominantly organic media, for example, chloroform saturated with buffer solution. The activity of tyrosinase (Tyr) is measured by the current of quinone reduction (Equation 23.16).

Catechol and phenol are mainly used for inhibitor measurements. Direct mediated oxidation of the enzyme active site by a mediator, pyrroloquinoline quinone (PQQ), covalently attached to the electrode surface is also described. The same products can be detected spectrophotometrically or by fluorescence intensity.

The inhibition of tyrosinase is recorded by simultaneous addition of the substrate and an inhibitor to the enzyme either free or immobilized on appropriate support. The decay of the enzyme activity is reversible, and the signal of the biosensor can be recovered by washing without any reactivators. The consideration of the dependence of the kinetic parameters on the concentration of the reactant confirmed competitive and in some cases mixed inhibition when both the substrate and inhibitor compete for the same binding site of the enzyme. This requires the use of small substrate concentrations to reach high sensitivity toward the inhibitor:

$$(23.16)$$

Indeed, most methods for the direct substrate detection are not sensitive, and usually the concentrations of the pesticides detected are higher than those achieved with cholinesterase-based biosensors. The performance of tyrosinase biosensors can be improved by the modification of the sensors with mediators of electron transfer as was shown for PQQ and Co phthalocyanine modified transducers. The characteristics of pesticide detection with tyrosinase-based biosensors are presented in Table 23.6.

Alkaline phosphatase (ALP) like cholinesterases belongs to the hydrolase family and catalyzes the hydrolysis of organic esters of phosphoric acid. The maximum of enzyme activity corresponds to pH 9–11. The detection can be performed by the substrates converted to colored or electrochemically active products. 3-Indoxylphosphate, ascorbate-2-phosphate, and 1-naphtylphosphate are mostly mentioned in the description (Equations 23.17 and 23.18):

$$(23.17)$$

TABLE 23.6

Analytical Characteristics of the Pesticide Determination with Enzyme Biosensors

Pesticide	Detection Mode	Sample	Detection Characteristics	References
Tyrosinase (polyphenoloxidase)				
Atrazine	Spectrophotometric detection in FIA mode, substrate: catechol	Standard solutions in chloroform saturated with 0.05 M phosphate buffer, chloroform extracts from spiked corn	LOD 0.5 mg/L, conc. range 1.0–7.0 mg/L, recovery for atrazine added to corn 91%–107% for 9–27 ng	[151]
2,4-Dichlorophenoxyacetic acid (2,4-D)	Screen-printed electrode, substrate: catechol	Standard aqueous solutions	LOD $9 \cdot 10^{-6}$ M, conc. range 0–0.8 mM	[152]
2,4-D	Glassy carbon electrode covered with Au nanoparticles modified with SAMs with covalently attached PQQ, direct electron transfer measurement	Standard aqueous solutions	Conc. range 0.5–100 ppt in batch and 0–10 ppt in continuous flow conditions (LOD 0.66 ppt)	[153]
Simazine, propazine, terbuthylazine, azinphos-methyl	Clark-type oxygen electrode covered with the enzyme implemented in κ-carrageenan gel, substrate: phenol	Standard solutions in water and chloroform, corn, barley, lentils	LOD $0.5 \cdot 10^{-9}$ M, recovery 83%–94% for real chloroform extracts from samples containing 0.05%–0.10% (w/w) of the pesticides	[22]
Paraoxon, malathion, parathion-ethyl, aldicarb, carbaryl, carbofuran, pirimicarb, dimethoate	Standard solutions in chloroform		LOD $0.5 \cdot 10^{-5}$ (paraoxon, malathion, parathion-ethyl, aldicarb, carbaryl, carbofuran), $1.0 \cdot 10^{-5}$ (pirimicarb), $1.0 \cdot 10^{-6}$ (dimethoate) mM; conc. range $1.0 \cdot 10$–5–10 (paraoxon, malathion, parathion-ethyl, aldicarb, carbaryl, carbofuran), $2.0 \cdot 10^{-6}$– $2.0 \cdot 10^{-1}$ (dimethoate), $2.0 \cdot 10^{-5}$–5 (pirimicarb) mM	[154]

Atrazine, dichlorvos	Glassy carbon electrode, substrates: 1,2-naphthoquinone-4-sulfonic acid, 1,2-naphthoquinone, 3,5-di-tert-butyl-1,2-benzoquinone	Standard aqueous solutions	LOD 0.106–0.17 μM, conc. range 0.8–10 μM for various substrates	[155]
Methyl parathion, diazinon, carbaryl, carbofuran	Carbon paste electrodes with graphite–cellulose composite modified with Co phtalocyanine, substrate: catechol	Standard solutions in chloroform, spiked natural waters	Conc. range 6–100 (methyl parathion), 19–50 (diazinon), 5–90 (carbofuran) and 10–50 (carbaryl) ppb, recovery for 30 ppb in natural water sample 92%–98%	[24]
Thiodicarb	Carbon paste electrode with polyphenoloxidase *Caryocar brasiliense* included in the electrode material; substrate: hydroquinone	Standard solutions, homogenates of peach, grape, and lettuce	LOD $1.58 \cdot 10^{-7}$ M, conc. range $3.75 \cdot 10^{-7}$ to $2.23 \cdot 10^{-6}$ M, recovery for 0.2–0.6 mg/L of spiked homogenates 94%–111% against HPLC	[23]
Ziram, diram, zinc diethyldithiocarbamate	Graphite disk electrode with adsorbed tyrosinase, substrate: phenol	Chloroform/water (reversed micelles)	LOD 0.074 (ziram), 1.3 (diram), 1.7 (and zinc diethyldithiocarbamate) μM	[156]
ALP				
Malathion, 2,4-D	H_2O_2 sensor (Pt electrode), substrate: indoxylphosphate, ascorbate-2-phosphate, and phenylphosphate	Standard aqueous solutions	LOD 0.25–0.5 (2,4-D) and 0.1–0.25 (malathion) μg/L, conc. range 2–75 mg/L for various substrates	[28]
2,4-D, 2,4,5-trichlorophenoxyacetic acid (2,4,5-T), carbofuran, endosulfan	Screen-printed carbon electrode, substrate: ascorbate-2-phosphate	Standard aqueous solutions	Conc. range 1–60 μg/L (2,4-D and 2,4,5-T), <200 μg/L (carbofuran)	[157]

(continued)

TABLE 23.6 (continued)

Analytical Characteristics of the Pesticide Determination with Enzyme Biosensors

Pesticide	Detection Mode	Sample	Detection Characteristics	References
Paraoxon	Enzyme immobilized in film of polythiophene derivative on glass, chemiluminescent detection, substrate: chloro-3-(4-methoxyspiro[1,2 dioxetane-3-2'-tricyclo-3.3.1.1-decan]-4-yl) phenyl phosphate	Standard aqueous solutions	LOD 50 ppb	[29]
Metham sodium, tetradifon	Enzyme immobilized by sol-gel technology, chemiluminescent detection, substrate: 1-naphtylphosphate	Standard aqueous solutions	LOD 4.9 (metham sodium), 29.3 (tetradifon), conc. range 194–774 and 3.5–28 μM	[158]
Peroxidase				
Thiodicarb	Au electrode modified with SAM of cysteine; substrate: hydroquinone	Potato, apple, strawberry	LOD $5.75 \cdot 10^{-7}$ M, conc. range $2.27 \cdot 10^{-6}$–$4.40 \cdot 10^{-5}$ M (standard solutions), vegetable extracts, recovery 99%–101% against HPLC	[27]
Glyphosate	Carbon paste electrode with peroxidase from atemoya implemented in the paste	Standard aqueous solutions	LOD 30 μg/L, conc. range 0.10–4.55 mg/L	[26]
Laccase				
Methomyl	Ceramic carbon electrode, enzyme immobilized by sol-gel technology; substrate: esculetin	Carrot, cucumber, lettuce, pepper, potato, and tomato (ethanol extracts)	LOD 02 μM, conc. range 0.5–12.2 μM (standard solutions), vegetable extracts, recovery 95% against HPLC	[159]
Methomyl	Screen-printed carbon electrode modified with Pt nanoparticles and the 1-butyl-3-methylimidazolium tetrafluoroborate, substrate: dopamine	Carrot, tomato (ethanol extracts)	LOD $2.35 \cdot 10^{-7}$ M, conc. range $9.8 \cdot 10^{-7}$–$9.0 \cdot 10^{-6}$ M (standard solutions), vegetable extracts, recovery 85%–105% against HPLC	[160]

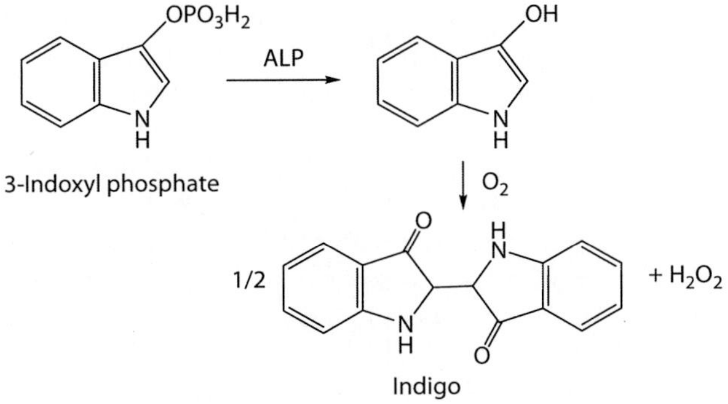

3-Indoxyl phosphate

Indigo

(23.18)

The inhibition is performed in the presence of a substrate followed by equalization and recording signal. Competitive and uncompetitive inhibition was established from the kinetics consideration. The concentrations of the substrates are mainly determined from the sensitivity of the readout systems. The reversible inhibition makes it possible to recover more than 98% of the initial enzyme activity. Unfortunately, there are no cases of the application of the ALP biosensors for real sample testing. The characteristics of the pesticide detection are summarized in Table 23.6.

23.2.1.2.2 Other Enzymes

A limited number of other enzyme sensors have been developed for pesticide determination. Most of them cannot compete with cholinesterase sensors and to some extent tyrosinase sensors in the sensitivity and operational characteristics. Meanwhile, they extend the list of contaminants determined. As in the case of ALP, most of these biosensors were tested on standard solutions, and their application for real samples has not been confirmed by independent analysis results. Peroxidase sensor [27] utilizing hydroquinone oxidation (23.20) is a rare exception. Trace amounts of thiodicarb, a carbamate pesticide exerting also anticholinesterase activity, were detected in the vegetable ethanol extracts, and the recovery was determined against HPLC results. Acid phosphatase was applied in bi-enzyme sensor together with glucose oxidase for the detection of malathion, methyl parathion, and paraoxon [30]. Acid phosphatase catalyzes the hydrolysis of glucose-6-phosphate; the concentration of glucose is measured due to its oxidation to gluconic acid. The hydrogen peroxide released in the latter reaction is recorded with commercial amperometric sensor. Instead of enzyme preparation, a thin slice of potato was used as a source of acid phosphatase activity. The LOD values down to 1 μg/L were achieved. NAD-dependent aldehyde dehydrogenase was inhibited with dithiocarbamate fungicides. The activity of the enzyme was monitored by the ferricyanide ion as a mediator of electron transfer. Diaphorase promoted the reversed oxidation of the NADH formed in the oxidation of propionaldehyde. Both diaphorase and aldehyde dehydrogenase were immobilized in

PVA–SbQ polymeric gel on the surface of Pt electrode. The biosensor makes it possible to detect zineb and maneb on ppb levels [25,161].

Laccase catalyzes the reduction of dissolved oxygen to water without the formation of the hydrogen peroxide. This reaction can be monitored by direct electron transfer from the electrode or be meditated with catechol derivatives (Equation 23.19), for example, dopamine or esculetin.

Methomyl was found to be the only able to inhibit laccase among pesticides although inorganic inhibitors are also known [162]. Like peroxidase, laccase is rather stable in organic solvents, and this offers good opportunities for its application for food samples. A was shown for some vegetables (see Table 23.6), direct extraction of pesticide residues from chopped vegetables by ethanol provides high recovery measured against HPLC as reference method:

Dopamine

Esculetin

(23.19)

One could see that the detection limits of most pesticides determined with enzymes different from cholinesterases are higher than the limited threshold values and hence the application of such biosensors assumes the preconcentration steps. Two advantages are mainly mentioned for this group of biosensors, that is, the extended list of the pesticides determined and the compatibility of the measurement protocol with organic solvents used for extraction. The latter is true for peroxidase, tyrosinase, and laccase. However, the advantages of direct operation in predominantly organic media (solvents saturated with buffer solutions) have not yet found an adequate attention of researchers. The OPH biosensors are the only exception. The attention to this system was first initiated by the necessity to replace cholinesterase in warning devices directed to the detection of the chemical weapons. Such systems are not required to be very sensitive because of very high acute toxicity of potential targets. Meanwhile, the genetic engineering and sophisticated detection systems decreased the LOD values of pesticides to those typical for cholinesterase biosensors. This offers very good opportunities for the development of multiuse enzyme sensors for the detection of organophosphates in a field where the replacement of biochemical components of biosensors is questionable.

23.2.2 Immunosensors for Pesticide Detection

The immunosensors involve biorecognition element *Ab* or *Ab* fragments able to form with analytes called also *Ag*-specific complexes. Contrary to enzyme sensors especially those based on cholinesterase inhibition, the immunosensors show very high specificity of the response, for example, they can distinguish the thiophosphates and their phosphoryl analogs. This is an important advantage and meanwhile a weak point of immunosensors because each analyte requires a specific set of immunoreagents for its quantification. Recently, the immunosensors specific to a group of chemically relative compounds containing a certain functional fragment (*Ag* determinant) have been developed [163] but preferably to pharmaceuticals. In many immunoassay protocols, the components of the reaction are first attached to a bulky polymeric support, for example, protein and synthetic polymer. The products of such interactions are called conjugates and can bear labels, the *Ag*s (pesticide or its derivative) and auxiliary agents. The optimization of the *Ag–Ab* pair and their immobilization and implementation in appropriate conjugates are the main goals of the immunosensor development together with the choice of signal transduction and suppression of interferences especially related to nonspecific adsorption of the matrix components on the transducer surface.

Considering the detection of pesticide residues, most of the developed immunosensors are based on a *competitive assay* [14]. Two approaches can be followed for the development of competitive immunosensors. In the first one, *Ab*s react with a mixture of free and labeled *Ag*s, and the signal recorded is related to their ratio in solution. In the second approach, the free *Ag* first reacts with an excessive amount of *Ab*s to form *Ag–Ab* complex in solution. Then, the *Ab*s remained free react with the *Ag* analogs (high molecular compounds containing the derivatives of the pesticides responsible for immune signal). They are attached to the insoluble support or the transducer surface. After that, the formation of the complex on the sensor interface is registered by the so-called secondary *Ab*s bearing labels, for example, enzymes, redox active compounds, colored particles, and fluorogenes are applied as labels. The protocols of direct and indirect competitive immunoassay are schematically represented in Figure 23.2.

The competitive immunosensors are commonly adopted from similar ELISA formats of conventional immunoassay. Contrary to the enzyme sensors, the advantages of immunosensors are not so obvious because the incubation steps and washing require a long time incomparable to that required for the signal detection itself. This depreciates the main advantage of the biosensors related to their fast and simple operation. On the other hand, the automation of the stages required for *Ab* immobilization and specific interactions in flow regime of operation improve the accuracy of the measurement and hence decrease the LODs against related conventional techniques.

In *label-free* approaches, the *Ag–Ab* interaction is directly referred to the appropriate changes of the properties of the transducer surface. Label-free techniques are one of the modern trends in the development of the biosensors. Their application is not limited with the conditions of the label detection and makes it possible to employ universal approaches to the biosensor design compatible with various targets and *Ab*s. The formation

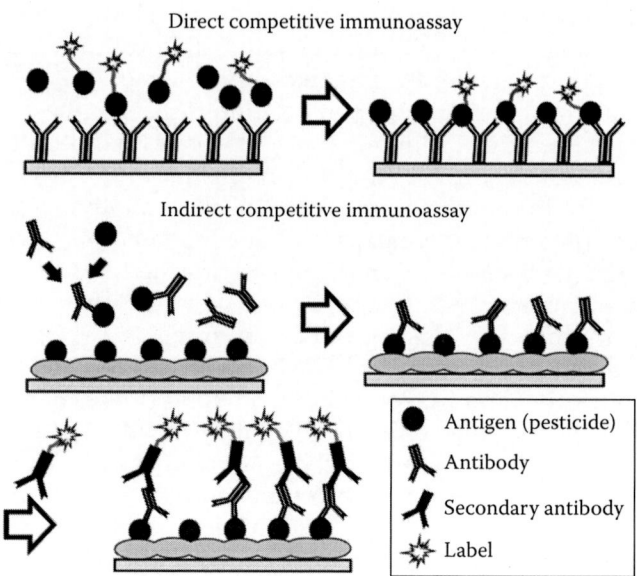

FIGURE 23.2
Principal scheme of direct and indirect competitive immunoassay of low molecular compounds.

of the immune complex changes the permeability of the surface layer for the charge carriers measured by electrochemical techniques, especially impedance spectroscopy [164,165]. Direct mass changes are recorded by piezosensors, for example, quartz crystal microbalances (QCMs) [166] or surface acoustic wave (SAW) sensors [167,168]. From modern optical transduction, surface plasmon resonance (SPR) [169] and total internal reflection fluorescence (TIRF) [170] provide both detection of the analytes and kinetic evaluation of the surface interactions. It should be mentioned that regarding the pesticide detection, label-free approaches are often less sensitive and specific than label-based protocols because of rather small size of the analyte molecules. The classification of the immunosensors for pesticide detection depending on the transduction system is given in review [171].

The sensitivity of the label-free detection can be improved by the use of auxiliary reagents co-immobilized together with *Ab* (*Ag*) molecules on the sensor surface. Thus, the use of polyaniline and other redox active polymers makes it possible to detect changes in their conductivity and redox activity resulted from uncoupling the electron transfer chain followed by the *Ag–Ab* interaction. For enzymes added to the surface layer, the increased sensitivity toward the *Ag–Ab* interactions is due to the suppression of the substrate access. Although such approaches are mostly applied for the detection of bulky *Ag* molecules, for example, proteins and DNAs, they can be useful for the determination of the pesticides, either.

HRP is mainly used as a label for both competitive and label-free formats of the immunosensors. Its activity can be monitored by luminescence [172], photometrically and amperometrically [173], using various mediators of the electron transfer (Equation 23.20). Other aspects of electrochemical detection of enzyme labels in electrochemical immunosensors are summarized in review [174]:

$$H_2N-\langle\rangle-OH \quad \xrightarrow{\substack{H_2O_2, HRP \\ oxidation}} \quad HN=\langle\rangle=O$$

$$HO-\langle\rangle-OH \quad \xleftarrow{Cathodic\ reduction} \quad O=\langle\rangle=O$$

(23.20)

The sensitivity of the pesticide detection based on immunosensor approach depends mainly on the way of the *Ab* production and their implementation in the biosensor assembly. Polyclonal *Ab*s and antisera containing such *Ab*s provide satisfactory results in many cases. They can be modified by linkers to avoid steric limitations in the immobilization stages and interactions with analyte and secondary *Ab* molecules. The same approach is used for covalent attachment of specific labels detected by signal transducers [175].

Polyclonal *Ab*s are produced by traditional immunization, commonly in rabbits, goats, sheep, and pigs. This method does not provide identical set of *Ab*s even in two animals of the same species [176].

In some cases, the recognition specificity of polyclonal *Ab*s is insufficient for specifying individual analytes. Thus, triazinic pesticides show considerable cross selectivity and hence are determined with polyclonal *Ab* simultaneously (see [177] as an example). In these cases, monoclonal *Ab*s are preferred due to much higher selectivity of immune reactions. The hybridoma technology guarantees the unlimited production of monoclonal *Ab* with constant characteristics [178]. The Fab fragments consisting of variable parts of the protein chains of the *Ab* molecules are also tested in the biosensor assembly. Although their application is complicated with the necessity of the additional stages of their production, the results obtained with the Fab fragments are less sensitive to their coordination on the sensor interface and more stable than those related to whole *Ab* molecules. It should be also mentioned that the detection of pesticides is much less exacting to the steric factors than conventional biopolymers (proteins, cells) determined with immunoassay techniques.

Contrary to other biosensors, the calibration curve obtained with immunosensors as well as other heterogeneous immunoassay formats cannot be linearized in the whole range of the concentrations to be determined. Instead of that, the following nonlinear four-parameter logistic approximation is used:

$$f(x) = \frac{a-d}{1+(x/c)^b} + d \qquad (23.21)$$

where
   a and d are the asymptotic maximum and minimum values
   c is the value of x at the inflection point (IC_{50})
   b is the slope

The sensitivity of the detection is characterized by IC_{50} value and the LOD estimated from $s/n = 3$ ratio. The cross selectivity can be expressed as the ratio of analyte concentrations exerting the same signal of the immunosensor or the ratio of the signals measured at the same concentration of the analyte and interferences in separate solutions.

The examples of the immunosensors for the detection of pesticide residues are given in Table 23.7 covering the last 10 years of investigations. One can see that there are only few examples of the application of immunosensors for the detection of real food samples.

In these cases, the influence of the matrix was found to be much smaller than that for enzyme sensors. High specificity of the immune reaction and intermediate washing stages that remove potential interferences are possible reasons. The influence of nonspecific sorption is successfully suppressed by coating the transducer with inert protein (bovine serum albumin, ovalbumin). For gold (Au) electrodes, self-assembled monolayers (SAMs) with charged components (mercaptopropionic or mercaptooctanoic acids) play the same role. The prevention of adsorption can be performed on the stages of the *Ab* immobilization or after the incubation of the immunosensor in the sample tested. In the latter case, the proteins or surfactants can be directly added to the diluted sample.

Organic solvents used for the extraction of the pesticide residues exert insignificant influence on the affinity of *Ab* molecules implemented in the surface layers. To some extent, this could be related to the washing steps removing excessive amounts of unbounded immune reagents, which limit the contact of the immunosensor with organic phase. Meanwhile, the use of HRP resistant to polar organic solvents positively affects the application of immunosensors to organic extracts of pesticide residues.

The optimization of the specific concentration of the *Ab* performed on the base of their titer in the antiserum or dilution curves obtained in ELISA or relative immunoassay formats offers broad opportunities for tuning the sensitivity of the *Ag* detection, which take into account the interferences of the solvents and other additives required for sample pretreatment. The use of enzyme labels gives rise to concern, but HRP, a common label, can function

TABLE 23.7
Examples of Immunosensors Developed for the Pesticide Determination

Pesticide	Immunoassay Mode	Transduction System	Abs, Immobilization Protocol	Detection Characteristics	References
Atrazine	Direct competitive	Screen-printed electrode covered with polyaniline–poly(vinylsulfonic acid) layer, direct electron transfer from HRP label, amperometric detection in batch and flow conditions	Recombinant single-chain fragments of *Ab*, implementation in the multilayers onto the surface	LOD 0.1 ppb (0.5 nM), conc. range 10 μM–5 nM	[179]
Atrazine	Label-free	Au thin layer on silicon waver covered with polypyrrole with covalently attached nitrilotriacetic groups, impedimetric signal	Monoclonal Fab fragments of IgG immobilized via histidine tags via chelate immobilization protocol	LOD 10 pg/mL, conc. range 10 pg/mL 1 μg/L	[180]
Atrazine	Label-free	Glassy carbon covered with electropolymerized N-(6-(4-hydroxy-6-isopropylamino-1,3,5-triazin-2-ylamino)hexyl)5-hydroxy-1,4-naphthoquinone-3-propionamide, square wave voltammetry of the own redox activity of the modifier	Monoclonal anti-atrazine *Ab*, interaction with atrazine fragment of the surface coating	LOD 1 pM, conc. range up to 10 nM	[181]
Atrazine	Label-free	Au electrode modified with biotinylated SAM, impedimetric detection amplified by neutravidin treatment	Biotinylated Fab fragment, avidin–biotin binding	LOD 20 ng/mL, conc. range up to 300 ng/mL	[182]

(continued)

TABLE 23.7 (continued)

Examples of Immunosensors Developed for the Pesticide Determination

Pesticide	Immunoassay Mode	Transduction System	Abs, Immobilization Protocol	Detection Characteristics	References
Atrazine	Direct and indirect competitive	H_2O_2 sensor based on Clark oxygen electrode, amperometric detection of the HRP label	Monoclonal anti-atrazine Ab conjugated with albumin, electrostatic immobilization onto Immobilon® nylon membrane	LOD $5 \cdot 10^{-11}$ M, conc. range $1 \cdot 10^{-10}$–$3 \cdot 10^{-5}$ M, measurements in buffalo milk, extracts from field grass and olives	[183]
Triazinic pesticides	Direct competitive	HRP and O_2 diffusional sensor based on Clark electrode, amperometric detection of the HRP label with *tert*-butyl hydroperoxide as substrate in hexane/chloroform (50% v/v)	Monoclonal anti-atrazine Ab conjugated with albumin, electrostatic immobilization onto Immobilon nylon membrane	LOD 7–12 nM, conc. range 50 nM–5.0 μM in the presence of olive oil, cross selectivity toward simazine (60%), atrazine–desethyl (42%), *tert*-buthylazine (46%), aldicarb (27%), carbaryl (28%), azinphos-ethyl (7%), paraoxon (37%)	[177]
Atrazine	Indirect competitive	Interdigitated Au microelectrodes with Ab immobilized in the gaps between the electrodes onto SiO_2 coating, conductometric and impedimetric detection after deposition of Au nanoparticles as labels	Antisera partially purified, covalent attachment to the lysine or arginine side chains	LOD 2–3 μg/L, conc. range up to 100 μg/L (impedimetric sensor), red wine assay: LOD 0.05–0.5 μg/kg, up to 50 μg/kg (conductometric sensor)	[184–186]

Analyte	Format	Description	Immobilization	Performance	Ref.
Atrazine	Indirect competitive	Florescent and luminometric detection of HRP activity in flow regime, measurements in 50% methanol	Antisera, affine immobilization by protein G	LOD 0.15 µg/L, measurements in extracts from broccoli, green bean, tomato, celery, watermelon, and lettuce	[187]
Atrazine	Direct competitive	Graphite–epoxy composite electrode, magnetic separation of the Fe_3O_4 particles modified with *Ab*, amperometric detection of the HRP activity as label in conjugate with *Ab*, substrate hydroquinone	Polyclonal *Abs* immobilized onto magnetic beads via carbodiimide binding or affine binding to protein A	LOD 0.027 nM, conc. range 0.04–2.87 and 0.38–5.43 nM for carbodiimide binding and affine immobilization, respectively	[188]
Trifluralin	Label-free	Optical sensor based on optical waveguide light mode spectroscopy (OWLS)	Polyclonal *Abs*, glutaraldehyde cross-binding on glass substrate	Conc. range $2 \cdot 10^{-7}$–$3 \cdot 10^{-5}$ ng/mL, IC_{50} $1.05 \cdot 10^{-6}$ ng/mL	[189]
Simazine	Direct and indirect competitive	Ion-selective FET, detection of HRP activity (direct format)	Polyclonal *Abs* immobilized via staphylococcal protein A	LOD 1.25 ng/mL (direct competitive) and 0.65 ng/mL (indirect competitive format), conc. range 5–175 ng/mL	[190]
Estrone, isoproturon, atrazine	Indirect competitive	Solid-phase fluoroimmunoassay at an optical transducer chip with total internal reflection (River Analyzer [191], RIANA®), signal of HRP as label of secondary *Ab*	Polyclonal *Ab* toward selected pesticides, covalent binding to glass via 3-aminopropyltriethoxysilane	LOD 0.155, 0.046, and 0.084 µg/L (atrazine, isoproturon, and estrone)	[192]

(continued)

TABLE 23.7 (continued)

Examples of Immunosensors Developed for the Pesticide Determination

Pesticide	Immunoassay Mode	Transduction System	Abs, Immobilization Protocol	Detection Characteristics	References
Atrazine, diuron	Direct competitive	Solid-phase fluoroimmunoassay in flow-injection regime with single-use exchangeable column	Monoclonal *Ab*, physical adsorption on polystyrene	IC_{50} 0.4 (diuron), 0.7 μg/L (atrazine)	[193]
Diuron	Indirect competitive	Au electrode on polystyrene substrate covered with Prussian blue and Au nanoparticles, amperometric detection of the ALP activity	Physical adsorption of diuron conjugate	LOD 1 ppt, conc. range 1 ppt and 10 ppm	[194]
Carbofuran	Label-free	Au electrode modified with a composite consisting of Au nanoparticles modified with Prussian blue, multiwalled carbon nanotubes, chitosan, and protein A, detection by the changes in the voltammetric signal of Prussian blue	Monoclonal *Abs*, affine binding to protein A	LOD 0.021 ng/mL, conc. range 0.1–1 μg/mL	[195]
Carbofuran	Label-free	Au electrode modified with multilayers of L-cysteine and Au colloidal nanoparticles and covered with membrane with immobilized HRP, H_2O_2 voltammetric signal	Monoclonal *Abs*	LOD 0.01 ng/mL, conc. range up to 50 ng/mL/spiked samples of lettuce and cabbage: recovery 93%–104% (1 and 5 ng/mL of carbofuran)	[196]

Carbofuran	Label-free	Glassy carbon electrode covered with composite material containing Au and Fe_3O_4 nanoparticles, carbon nanotubes, and chitosan in bovine serum albumin matrix	Monoclonal *Abs* physically adsorbed on the modified electrode	LOD 0.032 ng/mL, conc. range 1.0–100.0 ng/mL, 85% methanol extracts from cabbage: recovery 90%–110% (5 µg/mL of carbofuran)	[197]
Carbofuran	Label-free	Glassy carbon electrode, impedimetric measurements	Monoclonal *Abs* immobilized by sol–gel technology	LOD 0.33 ng/mL, conc. range 1 ng/mL–100 µg/mL, cabbage and lettuce spiked samples, recovery 89%–17%	[198]
2,4-D and 2,4,5-T	Indirect competitive	Peroxidase sensor based on Clark electrode, amperometric detection of the HRP label with H_2O_2 as substrate	Noncommercial *Abs* conjugated with albumin, electrostatic immobilization onto Ny+ Immobilon affinity membrane	LOD $8 \cdot 10^{-11}$ (2,4-D) and $2.8 \cdot 10^{-9}$ (2,4,5-T), conc. range $1.5 \cdot 10^{-10}$–$1.4 \cdot 10^{-5}$ (2,4-D), $3.5 \cdot 10^{-9}$–$1.0 \cdot 10^{-5}$ M (2,4,5-T). Cross selectivity (2,4-D 100%) toward dichlorprop (82%), 2,4,5-T (47%), 2,4,5-TP (59%), chlorfenac (40%)	[199]
2,4-D	Label-free	Interdigitated array and screen-printed disk and fingerlike Au electrodes, impedimetric measurements	Monoclonal *Abs* immobilized by carbodiimide binding onto various self-assembled layers	Conc. range 45 nM–0.45 mM	[200]

(continued)

TABLE 23.7 (continued)

Examples of Immunosensors Developed for the Pesticide Determination

Pesticide	Immunoassay Mode	Transduction System	Abs, Immobilization Protocol	Detection Characteristics	References
2,4-D	Direct competitive	Eight-channel screen-printed electrode array in the standard immunoplate, amperometric detection of the ALP activity as label of the conjugate, substrate p-aminophenyl phosphate	Monoclonal Abs, physical adsorption on the wells of the titer microplate	LOD 0.072 ng/mL, conc. range 0.1–330 ng/mL, recovery (0.1, 1.0, 10, and 100 ng/mL) 89%–116%	[201]
DDT, DDT metabolites	Indirect competitive	SPR sensor SENSIA®. Au thin layer modified with SAM of mercaptoundecanoic acid	Monoclonal Abs immobilized by carbodiimide binding to SAM layer	LOD 15 ng/L (DDT) 31 ng/L (DDT group) IC50 1 µg/L	[202]
Acetochlor	Indirect competitive	Piezosensor with Au electrodes modified with aminothiophenol ordithio bis(succinimidyl propionate)	Polyclonal Abs immobilized by glutaraldehyde cross-binding	LOD 0.2 µg/L, conc. range 0.2–100 µg/L	[203]
Chlorpyrifos	Indirect competitive	Portable SPR immunosensor, Au thin layer modified with the SAM of mercaptoundecanoic acid	Monoclonal Abs immobilized by carbodiimide binding to SAM layer	LOD 55 ng/L (standard solutions) and 45–64 ng/L in river and drinking waters, recovery 80%–120% against GC–MS	[204]

Analyte	Format	Description	Immobilization	Performance	Ref.
Chlorpyrifos	Label-free	Glassy carbon modified with multiwalled carbon nanotubes, thionine, and chitosan. detection of the changes in the voltammetric signal of thionine	Monoclonal *Abs* immobilized by cross-linking with glutaraldehyde	LOD 0.046 ng/mL, conc. range $0.1–10^5$ ng/mL	[205]
Chlorsulfuron	Direct competitive	Screen-printed carbon electrode with HRP implemented in the carbon ink, amperometric signal related to glucose oxidase as label in the chlorsulfuron conjugate	*Abs* adsorbed onto Biodyne A membrane	Conc. range 0.01–1.0 ng/mL	[206]
Endosulfan	Indirect competitive	Au electrode modified with vertically aligned carbon nanotubes with ferrocene derivatives, square wave voltammetric detection of ferrocene signal modulation	Monoclonal *Abs*, carbodiimide binding	LOD 0.01 ppb, conc. range 0.01–20 ppb	[207]
Azinphos-methyl	Indirect competitive	Glassy carbon electrode, voltammetric detection of HRP activity as label in anti-mouse IgG conjugate, hydroquinone as substrate	Monoclonal *Abs* cross-linked with ovalbumin	IC_{50} 1.2 nM, conc. range 06–500 nM, measurements in extracts from honeybees, recovery 98%–102%	[173]
Metsulfuron methyl	Indirect competitive	This fiber-optic sensor, absorbance measurement related to HRP activity as label in conjugate with goat anti-rabbit IgG, substrate: *o*-phenylene diamine	Microscopic slide, cross-linking with glutaraldehyde in ovalbumin matrix	Conc. range 0.3–100 ng/mL	[208]

after the contact with polar solvents. The immunosensors can be regenerated after use by breakage of the *Ag–Ab* complex with specific reagents. In some cases, it is easier and more reliable to remove all the immunoreagents from the transducer surface with following reestablishment of the recognition layer by pumping reagents in flow-through regime. The regeneration reagents should be carefully selected to avoid possible influence on the analyte detection. Organic solvents, surfactants, buffer systems, and denaturation reagents can be used for this purpose. Thus, the consideration of various treatment procedures for the regeneration of the QCM biosensor for ethyl-parathion detection made it possible to conclude that best results were obtained with glycine–HCl buffer containing 1% of dimethyl sulfoxide [209]. The complete removal of the proteins from the surface is achieved by extreme pH and high concentrations of electrolytes [202,210]. The use of affine immobilization offers additional possibilities. Thus, reversible interaction of concanavalin A with proteins [211] or avidin–biotin binding [212] can be mentioned.

A significant amplification of the immunosensor signal can be achieved by the application of hybrid biorecognition technologies. Thus, attomoles of coumaphos have been detected with the biosensor based on the ITO electrodes, which combine hybridization and *Ag–Ab* detection [213]. The principal scheme of the assay is presented in Figure 23.3.

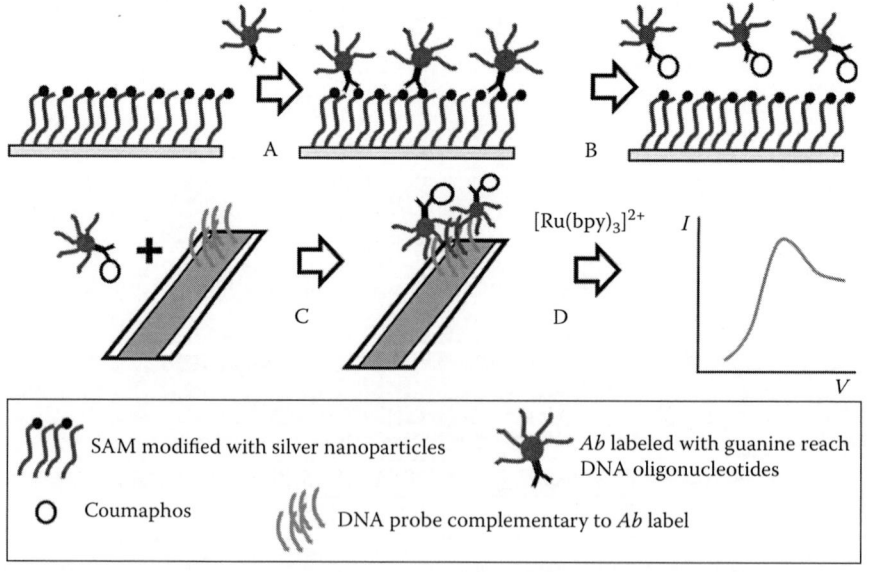

FIGURE 23.3
Hybrid immunosensor for coumaphos based on DNA hybridization detection. (A) SAM on Au electrode modified with silver nanoparticles is treated with specific *Ab* labeled with DNA; (B) the complex of *Ab* with coumaphos is removed from the surface; (C) the *Ab*–coumaphos complex is attached to the screen-printed electrode modified with DNA probe complementary to the label by hybridization; (D) the surface amount of the hybridization products is detected by the signal of Ru(II) bipyridine complex.

The Au electrode was first modified with SAM and silver nanoparticles. Then, the *Ab*s toward coumaphos were added. Guanine-rich DNA probe was covalently attached to the *Ab* molecule as a label. Due to interactions of guanine and silver nanoparticles, the labeled *Ab* molecules are closely attached to the sensor surface. After that, the sample is added, and the reaction of the *Ab* with coumaphos results in the formation of more stable *Ag–Ab* complex leaching to the solution. Then, the reaction is transferred to the ITO electrode modified with DNA sequence, which is complementary to the DNA label. As a result, hybridized double-stranded DNA sequence is format, and its presence is registered by Ru bipyridine complex by the oxidation current recorded in direct current mode. The hybrid DNA immunosensor makes it possible to determine 0.5–80 ng/L of coumaphos. The LOD reported (0.18 ng/L) is equal to 50 attomole of the analyte in 100 µL sample volume. The recovery of the analysis of spiked milk samples against GC–MS/MS was 96%–98%. The displacement protocol described can be extended to other pesticides, as well.

23.2.3 DNA Sensors and Aptasensors

Aptasensors are single-stranded DNA or RNA, which are selected from the library of randomly synthesized oligonucleotide sequences to specifically bind to various targets, including small molecules [214]. As an alternative to *Ab*s, aptamers are more stable toward various agents and can be easily modified with functional groups required for their immobilization and signal readout. In addition, they do not require the biological materials for their manufacture and hence are more convenient from the point of view of bioethics. Although many aptasensors have been described for the detection of small molecules in the past decade, the number of aptamers specific to pesticides is rather small due to complications of their selection [215,216]. For this reason, most works are performed for the isolation and characterization of aptamers toward individual pesticides. Thus, molecular beacons have been designed to detect organophosphate pesticides (phorate, profenofos, isocarbophos, omethoate) [217]. The aptamers were labeled with carboxyfluorescein (FAM) and 4-([4-(dimethylamino)phenyl]azo)benzoic acid (DABCYL) units. In a loop form, the fluorescence is suppressed by electron transfer between a fluorophore and quencher placed close to each other at terminal fragments of a loop (Figure 23.4). The reaction with complementary sequence increases the fluorescence intensity, whereas the presence of pesticides stabilizes the loop structure and hence inhibits the fluorescence. The dissociation constants that characterize the affinity of aptamers toward organophosphates varied between 0.9 and 2.5 µM.

The DNA aptamer to acetamiprid was selected using SELEX protocol and its affinity quantified in a similar manner by fluorescence quenching [218]. The results show preferably binding the acetamiprid against other common pesticides (imidacloprid, nitenpyram, and chlorpyrifos) with dissociation constant of about 4.98 µM.

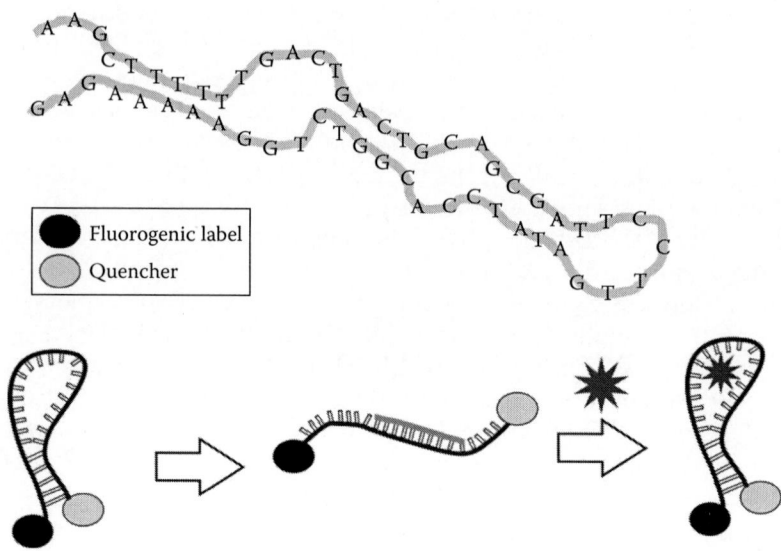

FIGURE 23.4
Structure of molecular beacon aptamers selected for pesticide detection by fluorescence quenching and principal scheme of competitive assay with complementary DNA strand.

The biosensors employing DNA sequences can be used for nonspecific detection of selected pesticides based on their interference with natural reaction of DNA, that is, its hybridization or accessibility for oxidation of guanine fragments. Thus, the organophosphate pesticides, chlorpyrifos and malathion increase the signal of guanine oxidation recorded on the electrode covered with polyelectrolyte complex of native double-stranded DNA, polyaniline, and poly(vinyl sulfonate) [219]. The increase of the peak current on square wave voltammogram recorded at 400–800 mV was proportional to the pesticide concentration within 0.05 ppb–0.01 ppm. The reaction can be monitored within 30 s after the pesticide injection.

Voltammetric and QCM sensor based on DNA hybridization control was used for the detection of several pesticides (atrazine, glufosinate, 2,4-D, diflubenzuron, carbofuran, paraoxon ethyl) [220,221]. The DNA sensor involved Au electrode of the QCM sensor covered with the SAM layer with DNA probe attached to the support by streptavidin–biotin binding. After hybridization, a ferrocene derivative was attached to the hybridization product. An increase in the mass of the surface layer and alteration of the ferrocene signal were recorded in QCM and voltammetric mode, respectively. The DNA probe used is derived from the *ssrA* gene of *Listeria monocytogenes*, a common food pathogen. For pesticide detection, the biosensor was first immersed in the 1 mM solution of the pesticide for the predetermined time, and then the ferrocene signal was measured. In accordance with the changes of the current, atrazine and paraoxon ethyl were found the most dangerous.

The DNA unwinding was considered as the main reason for the signal changes observed. The mechanism of interaction as well as the coordination of the pesticide has been confirmed by *ab initio* calculations for atrazine as an example [221].

The selection of aptamers can be considered as a preliminary stage of the investigations directed to the development of new aptasensors for the pesticide detection. However, the sensitivity of the detection reported is far from the immune and enzyme biosensors already applied for the same analytes.

23.2.4 Microbial Biosensors

There are two strategies in the application of microbial sensors for the detection of the contaminants. In the first one, specific interactions of the pesticides with enzymes or receptors are monitored as a part of signal transduction system. The second approach is intended to the general assessment of the bactericidal activity irrespective of the quantities and nature of particular hazards. This is similar to the biotesting principles with acute toxicity estimated on the basis of average mortality of the living cells of microorganisms or biological tissues. Conventional sensors developed for the measurement of the biochemical oxygen demand (BOD) are employed for the pesticide detection [222]. Active sludge, soil microorganisms, yeasts, and selected bacteria strains are used as degraders in such devices based mainly on Clark-type oxygen electrodes [223,224]. It should be noted that most of the respiratory toxicity testing is directed to the monitoring of wastewaters and is hardly adapted to food contamination control.

The microbial sensor for organophosphate detection employing microorganisms producing OPH can be mentioned as an example of the first strategy. A bacterium *Sphingomonas* sp. from field soil hydrolyzes the methyl parathion to *p*-nitrophenol detected by electrochemical and colorimetric methods [225]. Whole cells of *Sphingomonas* bacteria were immobilized directly onto the surface of the wells of polystyrene microplates by glutaraldehyde cross-linking. The microplate reader recorded changes in the optical density as a measure of the product formation. The optic sensor makes it possible to detect 4–80 μM methyl parathion. Similar results were obtained with *Flavobacterium* sp. catalyzing hydrolysis of methyl parathion with LOD of 0.3 μM and concentration range of 4–80 μM [226].

Recombinant cells of *E. coli* with high periplasmic expression of organophosphorus hydrolase were immobilized on the screen-printed carbon electrode by glutaraldehyde [227]. *p*-Nitrophenol formed in the enzymatic hydrolysis of methyl parathion was determined voltammetrically at 0.1 V by the current of its oxidation. The biosensor makes it possible to detect 2–80 μM of the pesticide in only 20 μL of the sample. The biosensor retains 80% of the response after 32 measurements.

Genetically engineered *Moraxella* sp. displaying OPH on the cell surface was used for the detection of nitroaromatic derivatives of organophosphates [228].

The activity of the pesticide degradation was monitored by oxygen electrode immersed in the suspension of the microorganisms. The LOD of parathion was 0.1 mM with no interference with some other pesticides (atrazine, diazinon, sutan, coumaphos). In a similar microbial sensor, genetically modified *P. putida* makes it possible to detect 0.28 ppb of paraoxon, 0.26 ppb of methyl parathion, and 0.29 ppb parathion [229]. These LOD values are comparable to those achieved with the biosensors based on cholinesterase inhibition. Unlike the inhibition measurements, microbial biosensors described selectively respond to organophosphate pesticides with a *p*-nitrophenyl substituent. This was related to the degradation of the *p*-nitrophenol released in primary enzymatic reaction (see Equation 23.14) in a sequence of the reactions (Equation 23.22) with electrochemically active intermediates:

$$(23.22)$$

Respiratory biosensor with oxygen electrode and *E. coli* suspension was applied for the quantification of toxic materials suppressing metabolic activity of microorganisms [230]. Relative toxicity was estimated by IC_{50} values against a standard contaminant, 2.4-dochlorophenol (8.0 mg/L). Ametryn inhibits the oxygen consumption by 50% at 6.5 mg/L, fenamiphos at 22, and endosulfan at 5.7 mg/L.

Wild-type *Flavobacterium* sp. was immobilized in the poly(carbamoyl sulfonate) hydrogel onto the surface of a glass pH electrode [231]. The biosensor can be utilized to monitor the microbial degradation of paraoxon and chlorpyrifos in millimolar range of their concentrations. The reaction is completed for about two days. The pH shift of the reaction media was suggested as a measure of pesticide content. The sensitivity of the signal can be increased if cytoplasmic membrane fractions were used instead of the whole cells. The characteristics of paraoxon detection remain quite stable within 3 weeks.

The respiratory test is not the only test intended for the general toxicity assessment. The bioelectric recognition assay (BERA [232]) sensors have been developed to record changes in the cell potential of the cells implemented in the gels onto the electrode surface. As in the case of the respiration, the response toward toxic species is complex and can be affected by ion permeability of the cellular channels, changes in the metabolism, and some other reasons. The selectivity of the response can be improved by the application of various microorganisms and consideration of the time resolution of the potential shifts. In general, the response of the BERA sensors is faster than that of BOD sensors.

Neuroblastoma cells are often applied for the detection of neurotoxic compounds. The cells can be easily immobilized in alginate gel. The measurement assumes the incubation of the gel beads in the sample tested followed by their attachment to the electrode surface or direct modification of the electrode and its incubation in the toxin solution. The potential shift is recorded against the reference electrode similarly to potentiometric ion-selective electrodes. The avermectin herbicide affects the permeability of γ-amino butyric acid- and glutamate-gated chloride channels and nicotinic acetylcholine receptors of the cells. The response makes it possible to detect 0.01–100 ppm of the herbicide within 100–180 s. The analysis of the kinetics of the response distinguishes possible influence of other chemicals, for example, pyrethroid pesticides [233].

BERA biosensors were successfully applied for the detection of the pesticide residues in tomatoes [234]. The organophosphate diazinon and propineb, a dithiocarbamate pesticide, were detected with the LOD of 3 nM with immobilized neuroblastoma cells and fibroblasts. The linear response in the range from 0.03 to 0.33 µM was confirmed by fluorescence estimation of the intracellular Ca^{2+} concentration. The latter one changed synchronously with the pesticide in the tomato extracts.

Summarizing the consideration of microbial sensors for pesticide detection, their low selectivity and sensitivity in comparison to the enzyme and immunosensors should be mentioned. Even though the use of genetically engineered microorganisms and those adapted to particular contaminants increases the signal toward certain groups of pesticides (organophosphates are described as the only example), the detection level is by two orders of magnitude lower than those of other biosensors considered earlier. The engineering of biorecognition elements, for example, the use of cell fragments with bonded enzymes involved in the metabolism of the xenobiotics, is one of the perspective paths to reach the detectable levels comparable to minimal threshold concentrations of the food contaminants.

23.2.5 Phytotoxicity Biosensors

Many of the herbicides presently applied in agriculture and pond fishery inhibit the photosynthetic activity due to direct interactions with the photosystems I and II or disjunction of the electron transfer chain [235]. The biosensors utilizing the fragments of photosynthetic systems provide highly sensitive detection of such toxic species. In general, using whole photosynthetic cells simplifies the preparation of a biosensor, whereas the application of the isolated chloroplasts or thylakoids offers a higher sensitivity toward herbicides due to the direct contact between the functional sites and the sample.

The photosystem II isolated from thermophilic cyanobacteria *Synechococcus elongatus* was immobilized on the electrode surface by cross-linking with glutaraldehyde, by implementation in albumin gel [236,237], or by affine immobilization via histidine tags or protein A binding [238]. In some experiments, an artificial electron acceptor, 1,5-diphenylcarbazide, increased the photocurrent. The inhibition of the signal by atrazine and 3-(3,4-dichlorophenyl)-1,1-dimethylurea (DCMU) was estimated for 10^{-7} M of herbicide. The same sensor was applied for monitoring of the degradation of isopruton in soil. Spiked samples contained 10^{-7} M of the herbicide [239].

The Au electrode modified with polymeric form of sulfobenzoquinone made it possible to detect direct electron transfer between the electrode and photosystem II isolated from thermophilic cyanobacterium *Synechococcus bigranulatus*. The biosensor showed high sensitivity toward DCMU (LOD 0.7 nM and IC_{50} 9 nM) [240].

Thylakoids from *Spinacia oleracea* were extracted from leafs and immobilized by physical adsorption on silicon septum or nitrocellulose membrane or implemented in the bovine serum albumin matrix [241]. The photocurrent related to the photosynthesis was recorded in screen-printed carbon electrode under illumination with the LED. The biosensor makes it possible to detect 10^{-8}–10^{-5} M of atrazine and linuron within 10 min. The same thylakoids were also immobilized in microtiter plates by entrapment in poly(vinyl alcohol) bearing styryl pyridinium groups. The inhibition of photosynthesis was monitored spectrophotometrically using 2,6-dichlorophenolindophenol as redox indicator. The herbicide residues were preconcentrated by solid-phase extraction. The biosensing device allows detecting $1 \cdot 10^{-10}$–$2 \cdot 10^{-5}$ M of atrazine, simazine, cyanazine, metribuzin, and diuron [242].

The photosystem I complex was immobilized on the gate of the FET together with naphthoquinone derivative attached to Au nanoparticles [243]. The biosensors showed remarkable response toward illumination, but its application for herbicide detection was not reported.

For online applications in wastewater monitoring, programmable electrochemical analyzer equipped with a flow cell and the LED sensor was designed [244]. *Synechococcus* shows the LOD levels below 200 ppb for the herbicide residues. The inhibition was recorded using hydroquinone–benzoquinone or ferricyanide ion as mediator.

The simple and cost-effective detection of herbicide has been proposed in Ref. 245. *Chlorella* cells were covered with magnetic nanoparticles. The suspension of the modified cells was diluted with ferricyanide solution, and one drop of the mixture was placed on the working surface of the screen-printed carbon electrode (Figure 23.5). The magnetized microalgae were concentrated on the electrode by external magnetic field. The current of the ferricyanide oxidation was recorded in the dark and light periods of illumination by natural light. The difference in the aforementioned values was proportional to the concentration of the pesticide in the range of 0.9–74 µM atrazine (LOD 0.7 µM) and 0.6–120 µM propazine (LOD 0.4 µM).

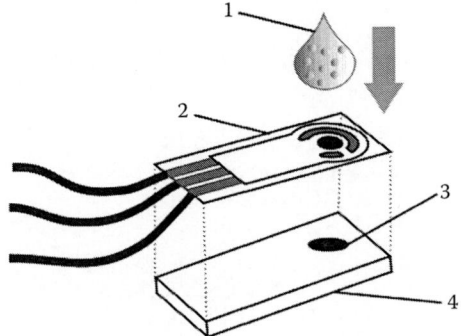

FIGURE 23.5
Determination of herbicides by magnetized *Chlorella vulgaris* cells. (1) Suspension of microalgae covered with Fe_3O_4 nanoparticles, (2) screen-printed carbon electrode, (3) constant magnet, and (4) polytetrafluoroethylene support.

23.3 Foodborne Pathogen Detection

Foodborne pathogens are disease-causing agents in which the influence on human health is related to the food. Most pathogens are infectious microbes (bacteria or viruses) as well as parasites (fungi and protozoans) [246] (Table 23.8).

As in the case of pesticides, the problem of foodborne pathogen detection is related to mass scale hazards for population, especially for children. Portable simple devices directed for the detection of potentially dangerous levels of food contamination are demanded to prevent the use of contaminated food and spreading infections. Despite strict regulations of food quality, the incidence of foodborne diseases continues to rise. Thus, *Campylobacter* was the highest reported foodborne-related disease in 2009 in Europe (198,252 confirmed cases), followed by salmonellosis (108,614 confirmed cases), and yersiniosis (7,595 confirmed cases) [248]. The number of cases of listeriosis and verotoxigenic *E. coli* (VTEC) infection increases by 15%–20% each year. Recent outbreak of *E. coli* serotype O104:H4 in Germany resulted in 3602 cases reported and 47 people died.

The biosensors for foodborne pathogen detection compete with traditional microbiological and immunoassay approaches, which offer higher reliability of the results but commonly need more time and specific equipment [249]. Some of them involve the same recognition elements as conventional assay tools (bacteria strains, specific *Abs*). This simplifies their practical use due to psychological reasons. Additionally, the aptasensors are intensively investigated to detect not only the bacteria cells but also specific toxins typical for food contamination. Both approaches assume rather short and simple sample treatment and integration of auxiliary stages and appropriate devices on the same substrate with the sensor itself.

TABLE 23.8

Selection of Pathogenic Bacterial, Fungal, and Viral Strains

Pathogen	Causative Agent
Bacteria	
Bacillus anthracis	Anthrax, toxin producer
Bacillus subtilis	Food poisoning
B. abortus	Brucellosis
Campylobacter spp. and *C. jejuni*	Campylobacteriosis
Clostridium botulinum	Botulism, producer of neurotoxins
E. coli O157:H7	Foodborne illness, producer of toxins, for example, verocytoxin or "shiga-like" toxin
Legionella pneumophila	Legionnaires' disease (legionellosis)
L. monocytogenes	Listeriosis
Mycobacterium tuberculosis	Tuberculosis
Neisseria meningitidis	Bacterial meningitis
S. typhimurium	Salmonellosis
S. aureus	Hospital-acquired infection, toxin producer
Yersinia enterocolitica	Yersiniosis
Fungi	
Candida albicans	Vaginal thrush
Trichophyton rubrum	Athlete's foot and ringworm

Source: Byrne, B. et al., *Sensors*, 9, 4407, 2009.

In conventional bacteriological analysis, pathogenic strains are transferred from soil or food sample to the growth medium. The amplification of microbial cell numbers permits their quantitative determination [250]. The application of antibiotics to suppress the growth of other strains and the use of different media generate the colonies distinguished by their morphologies by ocular inspection. The identification is confirmed by biochemical or DNA-based assays [251]. Colony counting provides inexpensive protocol of the bacterial pathogen detection. However, it requires up to 2 weeks. The analysis of pathogen-specific DNA by polymerase chain reaction (PCR) is an alternative to culturing, which can be realized in real-time scale with minimal biomass requirements and multiple pathogen assay formats [252,253].

However, the application of such methodologies for pathogen detection is complicated by external factors. Thus, strains may originate from complex matrices, for example, food with high level of fats, carbohydrates, etc., requiring a sample cleanup prior to assay. Then the DNA assay is indicative only for the presence of microorganisms but not for the toxins. This calls for the development of new alternative methods of pathogen analysis, which can be more useful. Biosensors are mainly intended for preliminary screening of the food contamination and often assume the semiquantitative detection of certain groups of microorganisms or toxins expressed in cell number or colony forming units (CFUs), typical characteristic of the microbiological assay [254].

23.3.1 Immunosensors for Foodborne Pathogen Detection

The majority of the immunosensors developed for the detection of pathogens utilize polyclonal and monoclonal *Ab*s selected against target analyte, isolated, purified, and immobilized on appropriate support or a signal transducer. Contrary to the hapten detection (see Figure 23.2 as examples), ELISA formats are mostly realized in appropriate immunosensors. They involve the detection of a label signal by the formation of the *Ag–Ab* complex followed by the binding of the secondary *Ab* molecules labeled with fluorogenic, optical, or electrochemical labels (sandwich assay format, see Figure 23.6). The examples of the immunosensors for foodborne pathogen detection are presented in Table 23.9.

Sandwich assay shows excellent sensitivity toward target analytes but is very lengthy and requires several intermediate washing stages. This complicates the automation of appropriate measurements and affects the productivity of the assay. As an alternative, label-free techniques have been also elaborated. In them, the permeability of the surface layer toward indicators, for example, ferricyanide ions, is measured after the incubation of the immunosensor in the sample tested. To some extent, direct QCM measurement of the pathogen interaction with the specific *Ab* molecules attached to the surface layer can be also referred to label-free techniques. The QCM response can be amplified by consecutive addition of secondary *Ab* labeled with metal nanoparticles to additionally increase the surface mass [257]. All the determinations can be performed with living cells, thermally killed cells. In selected cases, the surface-bound proteins are specified as targets for the *Ab* binding.

In food testing, the biosensor-based assay of liquids (milk [263–265], fruit juices [264]) is described. Also, chopped plant tissues can be used for extraction with buffer systems followed by bacteria detection.

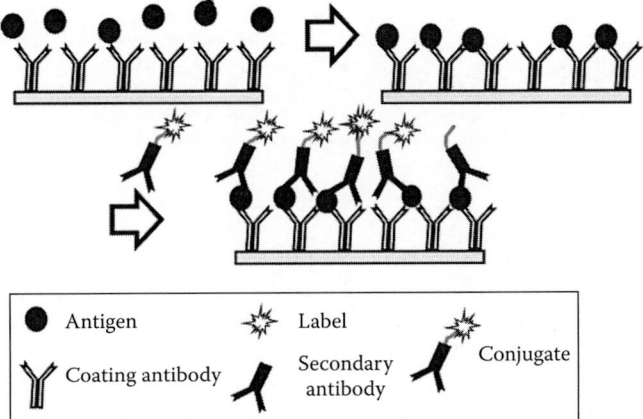

FIGURE 23.6
Principal scheme of a sandwich immunoassay.

TABLE 23.9

Immunosensors for the Detection of Foodborne Pathogens

Analyte	Immunoassay Mode	Transduction System	Immobilization	Detection Characteristics	References
Campylobacter sp.	Label-free	Lipid bilayer system, ion conductivity measurement under constant potential polarization	*Ab* implemented in the lipid bilayer system as ion channels	Ionic conductivity up to 10^{10} ions/s	[255]
E. coli O157:H7	Sandwich format	Screen-printed electrode modified with Au nanoparticles, voltammetric detection of HRP activity with ferrocene dicarboxylic acid as mediator, voltammetric signal	Monoclonal *Ab*, immobilization	LOD 6 CFU/mL, conc. range m 10^2–10^7 CFU/mL	[256]
E. coli O157:H7	Sandwich format	QCM chip covered with BSA film with immobilized *Ab*, amplification of the system via binding the Au nanoparticles bearing *Ab*	Polyclonal *Ab*, cross-linking with glutaraldehyde	Conc. range of 0–1 log CFU/mL	[257]
E. coli K-12	Sandwich format	Fluorescent microscopy, CCD measurements of fluorescence intensity with 25 µm microwire coated with *Ab*	Polyclonal *Ab* labeled with fluorescein isothiocyanate (FITC)	Conc. range 10–10^8 CFU/mL, baby spinach leafs	[258]
E. coli	Label-free	Glassy carbon electrode covered with double-layered coating consisting of polypyrrole–*Ab* and polypyrrole–alginate conjugates, amperometric signal related to the oxidation of *p*-aminophenol formed in the reaction of a bacterial enzyme galactosidase	Polyclonal *Ab* attached to polypyrrole by carbodiimide binding	LOD 10 CFU/mL, conc. range 10–10^6 CFU/mL	[259]

L. monocytogenes	Label-free	Screen-printed carbon electrodes covered with polyaniline, impedimetric measurements, detection of surface protein internalin B	Polyclonal *Ab*, avidin–biotin binding	Conc. range 1–100 pg/mL of internalin B	[260]
L. monocytogenes	Label-free	Au microelectrode modified with TiO_2 nanobundle, impedimetric measurements	Monoclonal *Ab* adsorbed on TiO_2 wire	LOD 470 CFU/mL	[261]
Salmonella spp.	Label-free	Glassy carbon electrode grafted with ethylene diamine and covered with the SAM of the Au nanoparticles, impedimetric measurements	Monoclonal *Ab* adsorbed onto Au nanoparticles	LOD 100 CFU/mL, conc. range $1.0 \cdot 10^2$–$1.0 \cdot 10^5$ CFU/mL	[262]
S. typhimurium LT2	Sandwich format	Screen-printed carbon electrode, magnetic separation with conjugates labeled with Au nanoparticles, differential pulse voltammetry	Monoclonal capture *Ab*, polyclonal second *Ab* immobilized by physical adsorption	LOD 143 cells/mL, conc. range 10^3–10^6 cells/mL; recovery in skimmed milk samples for $1.5 \cdot 10^3$ and $1.5 \cdot 10^5$ cells/mL is equal to 83% and 94%	[263]
S. typhimurium	Sandwich format	Carbon nanotubes/Au nanoparticles as labels, magnetic separation followed by signal enhancement by silver deposition, photometric scanning	Polyclonal *Ab*, immobilization by avidin–biotin binding	LOD 42 CFU/mL, conc. range 10^3–10^7 CFU/mL, spiked whole and low-fat milk, freshly squeezed orange juice, chicken broth, distinguishing thermally killed and living cells	[264]
S. typhimurium	Sandwich format	Screen-printed carbon electrode modified with carboxydextran, amperometric detection of HRP activity in conjugate by tetramethylbenzidine/H_2O_2 substrate system	Monoclonal capture *Ab* and polyclonal *Ab* in conjugate with HRP, adsorption and covalent immobilization via amino groups to carboxydextran	LOD 5000 and 20 cells/mL for physical and covalent immobilization, respectively, conc. range in optimized conditions 10^2–10^7 CFU/mL	[265]

(continued)

TABLE 23.9 (continued)

Immunosensors for the Detection of Foodborne Pathogens

Analyte	Immunoassay Mode	Transduction System	Immobilization	Detection Characteristics	References
S. enteritidis	Label-free	Conductance measurement with interdigitated planar electrodes of various geometries	Polyclonal *Ab* immobilized in the gap of interdigitated electrodes by avidin–biotin binding	LOD 1000 CFU/mL in pork meat extract	[266]
E. coli O157:H7 and *Salmonella*	Sandwich format	Lateral flow system with conductometric sensor consisting of two planar electrodes, signal (conductance increase) of polyaniline conjugates with secondary *Ab*	Polyclonal *Ab*, immobilization by glutaraldehyde binding	LOD 79 CFU/mL in 10 min, conc. range 10^1–10^5 CFU/mL	[267]
E. coli O157:H7, *Campylobacter* and *Salmonella*	Sandwich format	Graphite electrode modified with carboxylated carbon nanotubes with a mixture of *Ab*. Secondary *Ab* modified with nanocrystals of CdS, CuS, and PbS. The detection of metal ions by square wave anodic stripping voltammetry	Polyclonal capture *Ab* immobilized by glutaraldehyde cross-linking, secondary *Ab* modified with nanocrystals by carbodiimide binding	LOD 400, 400, and 800 cells/mL (*Salmonella*, *Campylobacter*, and *E. coli*), conc. range $1 \cdot 10^3$–$5 \cdot 10^5$ cells/mL, recovery in spiked bovine milk 86%–103%	[268]
L. pneumophila	Sandwich format	Screen-printed Au electrode modified with 6-mercaptohexanoic acid, impedimetric measurements	Polyclonal *Ab* covalently attached by carbodiimide binding	LOD 200 cells/mL, spiked tap water	[269]
Vibrio parahaemolyticus	Sandwich format	Screen-printed electrode modified with agarose–Au nanoparticle composite, amperometric detection of HRP activity in conjugate by thionine/H_2O_2 substrate system	Polyclonal *Ab* entrapped in agarose gel	LOD 7.4 10^4 CFU/mL, conc. range 10^5–10^9 CFU/mL	[270]

Source: Byrne, B. et al., *Sensors*, 9, 4407, 2009.

The multiplex detection of several pathogens remains one of the most attractions for the further improvement of the immunosensor approaches. For this reason, several approaches have been recently described. Thus, the *Ab* molecules were attached to the solid support, and the signal of fluorescent labels coupled with optical readout systems can improve the operational characteristics. Thus, the detection of $3 \cdot 10^6$–$9 \cdot 10^7$ cells/mL of *E. coli* O157:H7 was reported [271]. The biotinylated *Abs* were immobilized onto streptavidin-coated glass slide. The labeling products were visualized through the use of fluorescent microscopy. Parallel analysis of somatic and flagellar *Ags* on 117 *Salmonella* strains was achieved by polyclonal *Ab* from rabbit antiserum spotted on commercially obtained microarray slides eosin Y as fluorescent label [272]. Printing technique elaborated for array manufacture was also applied for simultaneous detection of *E. coli* and *Salmonella typhimurium* in aqueous suspension and ground beef extract with the LODs of $1 \cdot 10^6$ and $1 \cdot 10^7$ cell/mL, respectively [273].

Structurally diverse pathogens posing risks of bioterrorism were detected with multiplex formats of immunoarrays (*Burkholderia mallei, Francisella tularensis*, West Nile virus, and mycotoxins [274,275]). Multiplex electrochemical assay has been proposed in Ref. 268. Contrary to many other devices, all the *Ab* molecules toward various pathogens (*E. coli* O157:H7, *Campylobacter*, and *Salmonella*) were co-immobilized together on the signal transducer. To distinguish the signal in the mixtures, secondary *Abs* were labeled with different nanocrystals synthesized *in situ*. The sulfides of transient metals were dissolved after immunoseparation on the sensor surface, and the Cd(II), Pb(II), and Cu(II) ions were determined by anodic stripping voltammetry. The selectivity of the response is achieved due to different potentials required for the detection of metal ions on the last stage of the assay. The principal scheme of the analysis is presented in Figure 23.7.

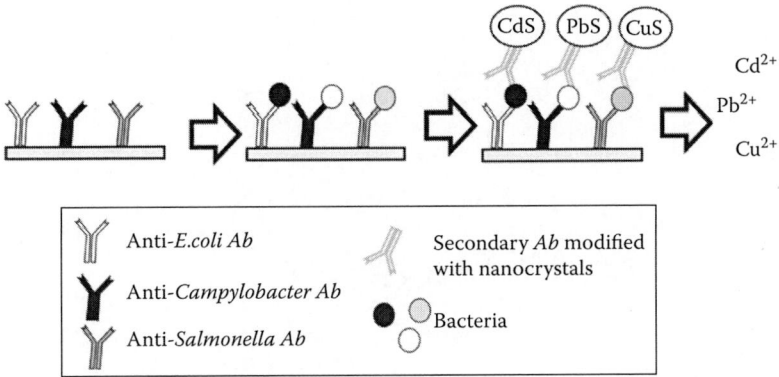

FIGURE 23.7
Electrochemical multiplex immunosensor based on nanocrystals of transient metal sulfides as labels.

Although most of the labels and immunoassay formats used for foodborne pathogen detection are similar to those previously elaborated for other *Ag* molecules, new approaches utilizing specific enzymes present in microorganisms were proposed. Thus, the activity of intracellular β-D-galactosidase was recorded as a measure of *E. coli* cells attached to the specific *Ab* in the biosensor assembly [259]. In this case, the formation of the surface layer is performed in two stages. First, the bacteria cells are separated from the solution on the *Ab* molecules covalently attached to the aminated polypyrrole on the surface of the glassy carbon electrode. After that, the cells are covered with the second polypyrrole layer and alginate gel. To simplify the access of the substrate to the enzyme, the biosensor is additionally treated with lysozyme. The signal of the immunosensor is detected after the addition of *p*-aminophenylgalactose, which is hydrolyzed to *p*-aminophenol, which is easily oxidized on the polypyrrole layer (Figure 23.8). The use of polypyrrole–alginate matrix suppresses the release of the product in the bulk solution and hence increases the sensitivity of the signal.

Colorimetric test system based on vesicles synthesized from 10,12-pentacosadyinoic acid and *N*-[(2-tetradecanamide)-ethyl]-ribonamide has been proposed for the detection of some pathogenic bacteria (*Staphylococcus aureus* and *E. coli*) [276]. The interaction of bacterial components with vesicles results in the color change from blue to red. The reaction was performed on cellulose strips within 30 min. The test strip was applied for preliminary screening of apple juices mixed with supernatants of the bacteria, and the color transition was confirmed though the selectivity of the response in complex matrix required further investigation. In this colorimetric device, no *Ab* molecules are used, and color changes were related to the direct interactions of the cells with the vesicle components.

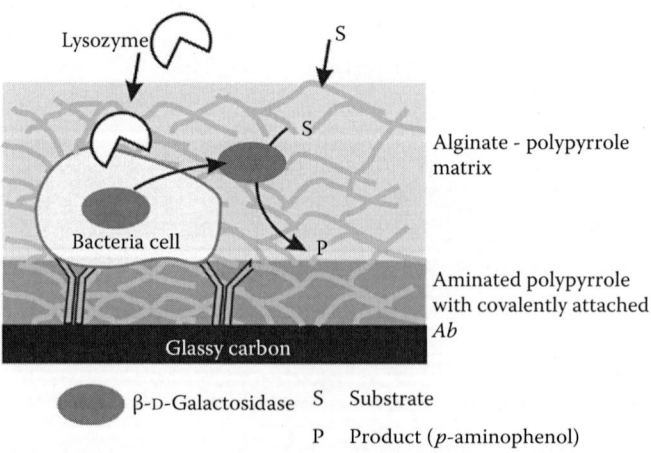

FIGURE 23.8
Determination of *E. coli* by β-D-galactosidase activity measured in polypyrrole matrix after immunoseparation of the cells.

Although the immunosensors for pathogen detection are based on well-developed platform of conventional immunoassay, its adaptation to the demands of on-site analysis remains a problem of significant efforts in the future. The detection of living cells or toxins is required for early warnings, which may help the farmer to perform early treatments or to apply other control strategies to prevent the disease from becoming epidemic. Similar reasons can be mentioned for food testing. The detection of pathogens in field enables site-specific treatments and assumes the application of portable devices, preferably with remote control. The design of transduction system and the stability of the biorecognition layer need special attention in the development of portable immunosensors.

The validation of immunosensors is a typical limitation of newly described assemblies. In most cases, the standard suspension and a limited number of spiked samples are used for estimating the recovery of their detection and the possible drawbacks of the matrix effects. The development of the immunosensors has not still taken into account rational design of the assemblies required for further mass production and the end-user preferences. This is especially important for complex matrices, for example, dirt samples, soil, and biological tissues.

Many current technologies, for example, SPR, QCM, and fluorescence detectors, will likely play major roles in future on-site detection, though some current limitations inherent in these technologies will be resolved. Thus, modern laboratory equipment required for the sensors mentioned takes up space comparable with small table. Sensing elements are rather small, but auxiliary equipment limits their portability. Nevertheless, there are some examples of miniaturization success. Thus, Spreeta™ SPR analyzers have a coin size and provide independent reference analysis in a real-time scale [277]. The standard version has three parallel channels, pump, source of optical interrogation, a diode array, and laptop software equipment for data processing. The Spreeta SPR analyzers were used for the detection of *Campylobacter*, *E. coli*, and *L. monocytogenes* [278–280].

The Spreeta SPR analyzer was applied for the detection of *S. typhimurium* in chicken carcass [281]. The *Abs* were immobilized by avidin–biotin binding; the response was sensitive toward the *Salmonella* cells in the range from 1.0 to 10^9 CFU/mL. The LOD established for chicken carcass testing was equal to $1 \cdot 10^6$ CFU/mL.

A handheld SPR sensor powered by 9 V battery has the dimensions of 15 cm × 8 cm and a weight of 600 g. The performance of the biosensor was characterized by the detection of 200 ng/mL of ricin within 10 min [282].

23.3.2 DNA Sensors and Aptasensors

The use of the DNA sensors for the detection of pathogenic microorganisms is mainly related to the hybridization reactions between the DNA probes and the target oligonucleotides specific for bacterial genes. Such reactions can be recorded by various techniques considered in previous chapters. Briefly, the following strategies for the hybridization detection are most important:

- Direct or mediated electrochemical oxidation of guanine in DNA probes and hybridization products
- Application of the electrochemically active mediators that suppress their activity due to involvement in the DNA moiety
- Sandwich assay with two DNA probes bearing the solid support and the label detected by appropriate detectors

In addition, a family of the SPR sensors has been developed for the detection of particular targets and multiplex assay of pathogens. The assembly of the sensing layer is optimized in order to reach the specificity and selectivity of the response required for the food quality assessment. However, all of the methods of DNA assay assume rather complex and time-consuming stages of the isolation and amplification of the DNA material. As was mentioned earlier for immunosensors, the necessity of the cell lysis and DNA materials enrichment complicates the biosensor operation and counterbalances the advantages of biosensors, for example, the real-time assay and fast and expensive design and operation. Even though the standard procedures of DNA assay based on PCR are compatible with the DNA-sensor protocols, this complicates the prospects of the DNA sensors for autonomous detection of bacterial pathogens. Some examples of the DNA sensors developed are given in the succeeding text to illustrate the approaches to the biorecognition layer assembling and the sensitivities of the target analyte detections reached with different detection systems.

23.3.2.1 Direct Guanine Oxidation

L. monocytogenes amplicons were detected by square wave voltammetry by the peaks of guanine oxidation [283]. PCR products were analyzed with electrochemical sensor obtained by carbodiimide binding of the aminated capture DNA 5′ GAT GCG TTG AAA CGC GCT TCC GC 3′ on graphite electrode. To amplify the signal, guanine residues were substituted with inosine analogs. In this case, the guanine signal can be referred to the hybridization product only. The formation of hybridization products was monitored by the charge transfer resistance measured with electrochemical impedance spectroscopy. The LOD of 267 pM was estimated from the calibration curve obtained in the range from 1 to 90 µg/mL. The biosensor was applied for the detection of pathogen bacteria in a wide range of foods (mushrooms, chicken, cheese, turkey, mayonnaise, doner kebab, calamari, cream cake, and ice cream).

23.3.2.2 Application of Electrochemically Active Intercalators

Methylene blue was applied for sensitive detection of *Aeromonas hydrophila*, a foodborne pathogen associated with production of cytotoxic enterotoxin aerolysin [284]. The thiolated DNA probe 5'-GTGGTGGGCTGGGCGATCAA-p-$(CH_2)_3$-SH specific for aerolysin gene was immobilized by self-assembling on Au electrode. The peak current of an intercalator was recorded by square wave voltammetry. The assembling of the surface layer and quantification of the DNA coverage were performed by chronocoulometry. The DNA sensor was tested on the PCR product obtained from isolated bacteria. For selectivity control, other pathogenic bacteria, *Propionibacterium jensenii*, were tested in the same conditions.

Single- and multielectrode system manufactured by photolithography and vacuum deposition of thin Au layers on the alumina wafer was developed for the signal transduction in the assembly of DNA sensors [285]. The advantages of the approach were demonstrated on the detection of selected bacteria pathogens (*Salmonella* spp., *L. monocytogenes*, and *E. coli*). The appropriate DNA probes were covalently attached to the electrode surface by terminal thiol groups. The hybridization was recorded by the signal of $[Ru(NH_3)_5L]^{2+}$, where L is [3-(2-phenanthren-9-yl-vinyl)-pyridine] in differential pulse voltammetry mode. The selectivity of the signal to *Salmonella* in the presence of other bacterial DNA and simultaneous detection of all the tree analytes was demonstrated with 10 µM samples of DNA.

A simplified test on *Salmonella* based on portable potentiostat has been proposed for simultaneous detection of respiratory activity and PCR amplification [286]. Screen-printed electrodes were used in both types of the analysis. For PCR product testing, the Hoechst dye H33258 was added to the PCR mixture, and the decay of its oxidation peak current was monitored in DPV mode. The semiquantitative assay of about 10^6 cells of *Salmonella* was confirmed by electrophoretic measurements in agarose gel.

23.3.2.3 Sandwich Format

The detection of *Salmonella* spp. was carried out in the DNA sensor with double-tagged oligonucleotides [287]. The probe was immobilized via –SH groups onto Au nanoparticles physically implemented in epoxy–carbon composite as signal transducer. Secondary oligonucleotide was labeled with digoxigenin. After hybridization, the biosensor was treated with a conjugate of antidigoxigenin *Ab* with HRP. The activity of the enzyme was recorded by the hydroquinone/benzoquinone redox pair (Figure 23.9).

The density of the surface coverage with DNA probe was controlled by fluorescence microscopy and the characteristics of the charge transfer with constant current voltammetry with ferricyanide ion as redox probe. The efficiency of the electrochemical response can be easily changed by variation of the composite content (increasing or decreasing portion of Au nanoparticles). The 201 bp fragment of the target DNA corresponds to the *Salmonella enterica* sv. Typhimurium-specific IS200 mobile element was amplified using PCR and used then as a biological target in the DNA-sensor testing. The LOD of 60 pM (9 amol in the sample tested) was achieved.

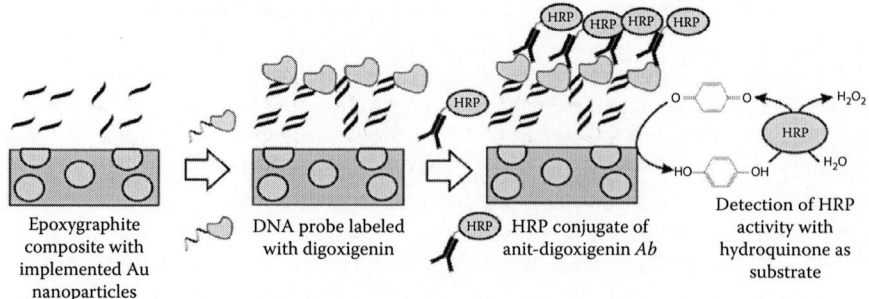

FIGURE 23.9
Principal scheme of the detection of *Salmonella* spp.-related gene sequence by DNA sensor based on nanoAu–epoxy–carbon composite and double-tagged oligonucleotides.

In a similar manner, the DNA fragment associated with the insertion element (*Iel*) gene of *S. enterica* sv. enteritidis was detected by bio-barcode techniques [288]. In this approach, the target DNA sequence reacts with two DNA oligonucleotides used for the attachment of the complex to the solid support and its labeling. To amplify the signal, Au nanoparticles were chosen for DNA probe immobilization, and magnetic separation of the complex formed has been achieved on Fe_3O_4 particles. The LOD of $2.15 \cdot 10^{-16}$ M (1 ng/mL) of target DNA was reported.

Sandwich assay of *E. coli* based on the ALP as DNA probe label is described [289]. The specific region of *E. coli* 16S rRNA gene was specified as a target. It reacts with thiolated DNA probe immobilized on the Au electrode. The product of the partial hybridization reacted with secondary DNA probe bearing biotin group. After the formation of the complex, streptavidin conjugated with enzyme label is added. The product of the enzymatic hydrolysis of *p*-aminophenyl phosphate, that is, *p*-aminophenol, is oxidized on the electrode. The product formed is chemically reduced back to *p*-aminophenol. This process called substrate recycling is aided by the NADH added to the solution. The LOD of 1 pM DNA corresponds to 4 amole in the 4 µL sample. This is 100-fold times better than the detection without NADH addition.

23.3.2.4 Optic DNA Sensors

The detection system based on blue LED for fluorescence excitation, pumping system, and flow-through reactor for real-time PCR has been developed for fast and inexpensive detection of pathogenic bacteria [290]. The detection of bacterial DNA is performed with SYBR Green fluorogenic dye. On example of *S. enterica* Newport, the sensitivity of the optic system was found to be sufficient for the detection of 0.1–1.0 µg/mL of DNA. This corresponds to approx. 10^6–10^7 Salmonella cells. Although these values are significantly higher than those of immunosensors, the combination of the real-time PCR microchamber with flow-through detector seems very promising from the point of view of automation and miniaturization of the biosensors for pathogenic bacteria detection.

Four pathogens, that is, *E. coli* O157:H7, *Salmonella choleraesuis* sv. *typhimurium*, *L. monocytogenes*, and *Campylobacter jejuni*, were detected using an eight-channel SPR sensor based on wavelength division multiplexing [291]. The calibration curves show the SPR response in the range of 10^3–10^7 CFU/mL with LODs of $3.4 \cdot 10^3$–$1.2 \cdot 10^5$ CFU/mL depending on the media and measurement conditions. Besides buffer solutions, apple juices with native and adjusted pH were tested. The DNA probes were immobilized in SAM by avidin–biotin binding.

High-throughput SPR biosensor for rapid and parallelized detection of nucleic acids identifying *Brucella abortus*, *E. coli*, and *S. aureus* with the LOD at 100 pM levels was developed [292]. The biosensor consists of a high-performance SPR imaging sensor with polarization contrast and internal referencing (refractive index resolution $2 \cdot 10^{-7}$ RIU) and an array of DNA probes microspotted on the surface of the SPR sensor. Short thiolated DNA sequences (20–23 bases) specific of the 16 S ribosomal RNAs were immobilized on Au thin film by self-assembling via thiol groups. In the flow-through regime with about 230 microspots poisoned on the sensors placed in the flow chamber, the whole measurement cycle takes 15 min.

23.4 Conclusion

The risk associated with contaminants, that is, pesticide residues and foodborne pathogens, is one of the great challenges in food safety. Besides some other contaminants, for example, those of industrial origin (perfluorinated compounds, polychlorinated dioxins, and furans), they are of most significance due to large scale of application, sever damage caused by poisoning, and biomagnification in the food chains [293]. Although the conventional instrumentation makes it possible to reliably detect all the contaminants mentioned, the technique is mostly time-consuming and requires sophisticated equipment and qualified personnel. This limits their on-site application and prompt response to dangerous contamination. The biosensors developed for preliminary testing of food and agriculture samples (soil, groundwater, etc.) are intended to fill the gap between the basic analytical instrumentation directed to quantitative analysis and biological toxicity assay devoted to the quick detection of potential hazards.

Three main directions of the further progress of the biosensors in the specified area can be named. They involve (1) the validation of the measurements, especially in complex media; (2) the miniaturization and automation of the biosensing devices; and (3) the methodology of the measurements directed on the selectivity of the response and significance of the information available.

The validation assumes the calibration of the biosensor signals against standard instrumentation, for example, HPLC in the pesticide analysis or microbiological assay in pathogen detection. In most cases, the biosensors reported satisfactory results in standard solutions of the analytes and spiked samples. However, the number of the works describing the application of the biosensors for real samples is very limited and covers either preferably aqueous mixtures (juices, milk, aqueous extracts from biological tissues, surface waters) or organic extracts. The efficiency of extraction is normally sufficient because the protocols are mainly elaborated for chromatographic assay and seem quite reliable. Meanwhile, the specific requirements of biosensor operation are not always taken into account. In some cases, their consideration complicates the sample treatment and levels the advantages of biosensors as portable and fast devices. This is particularly true for enzyme sensors with the measurement time incomparable with the duration of extraction, extract cleaning, and solvent evaporation. Immunosensors and DNA sensors are less sensitive to the sample treatment conditions because their application assumes long stages of incubation and washing.

The automation and miniaturization of the biosensors solve different problems depending on the biorecognition system. In enzyme sensors, the use of flow-through systems is mainly directed to the replacement of immobilized enzymes implemented in the replaceable membranes, columns, and cartridges after inhibition event. The recovery of cholinesterase activity by reactivators suppresses the repeatability of the signal and increases the probability of a wrong response about the safety of the sample. This is inappropriate for warning devices. The sensitivity of the response toward an inhibitor depends on the conditions of the mass transfer and is maximal for the enzyme closely attached to the transducer surface. Thus, the use of small chambers or thin-layered cells improves the performance of such sensors. The application of magnetic separation for the delivery of immobilized enzymes attached to magnetic particles is an elegant solution of the enzyme replacement quite compatible with microfluidic systems. In other cases, the flow-through regime with controlled injection of reactants, solution replacement, and cell washing improves the metrological characteristics of the response and decreases the total measurement duration.

In immuno- and DNA sensors, the automation and miniaturization are mostly intended to enhance the number of detection techniques potentially attractive for field applications but requiring bulky equipment sensitive to the operational conditions. Some of the recent advantages achieved in SPR biosensors and fluorescence techniques were considered earlier in the chapter. Microcantilevers, QCM, and most optical detections are very sensitive and reliable in laboratory but lose most of the advantages in portable devices. The use of ultra-small detectors together with new principles of biorecognition layer assembling can change the situation. Thus, the use of nanoparticles as carriers of DNA and proteins together with MEMS or microelectronic transducers (FETs, interdigitated electronic arrays, logic gates) can significantly increase the number of measurement techniques available for field application.

The value of information available with biosensor devices is a methodological problem actively discussed. Besides chemical and biological weapons, the impact of pesticides and foodborne pathogens in common situations is far from those assessed as harmful. For this reason, the detection of hazardous levels of contamination is demanded in rather narrow situations covering technological accidents, natural catastrophes, etc. In other cases, the identification of a risk should be provided together with its nature and possible source of intake. In other words, the selectivity of the signal becomes as much important as the sensitivity of the response. Enzymatic sensors cannot give a reliable response except very rare cases. The anticholinesterase pesticides are detected in total with rather limited information about the source of inhibition. In case of immune- and DNA sensors, the selectivity of the response is very high, but even minor modification of the analyte structure would suppress the signal. In many cases, like pesticide transformation, this is not always an advantage. In case of bacterial pathogens, a unique specificity of hybridization-based techniques is counterbalanced with a high genetic variety of the pathogenic bacteria and necessity in sterile conditions of sample treatment and measurement. Indeed, the drawbacks mentioned are typical for all the bioanalytical methods utilizing the proteins and DNA, and the biosensors are not an exception in this line.

Great promises are related to the new biorecognition elements [294]. The application of genetically engineered cholinesterases showed the advantages of such approaches in increasing sensitivity and selectivity of inhibitor determination. Aptamers called "synthetic *Abs*" demonstrate much higher stability and modification prospects in comparison with proteins. Their wider use in biosensor assembly is limited by rather complicated and labor-consuming selection and hence a short list of potential analytes detected. The use of other DNA-based components, for example, ribozymes and DNAzymes, with catalytic activity will extend the potentialities of appropriate bioanalytical instrumentation.

Acknowledgment

Financial support of Russian Ministry for Science and Education (contract no. 6.740.11.0496) is gratefully acknowledged.

References

1. B. Nagel, H. Dellweg, L.M. Gierasch, *Pure Appl. Chem.* 64 (1992) 143–168.
2. http://www.epa.gov/pesticides/.
3. G. Aragay, F. Pino, A. Merkoçj, *Chem. Rev.* 112 (2012) 5317–5338.
4. D.L. Sudakin, *Toxicol. Rev.* 22 (2003) 83–90.
5. T.R. Fukuto, *Environ. Health Perspect.* 87 (1990) 245–254.
6. W. Mnif, A.I.H. Hassine, A. Bouaziz, A. Bartegi, O. Thomas, B. Roig, *Int. J. Environ. Res. Public Health* 8 (2011) 2265–2303.
7. U. Saffiotti, *IARC Sci. Publ.* 25 (1979) 151–166.
8. S.K. Yadav, *J. Hum. Ecol.* 32 (2010) 37–45.
9. Regulation (EC) No 396/2005 of the European Parliament and of the council of February 23, 2005 on maximum residue levels of pesticides in or on food and feed of plant and animal origin and amending Council Directive 91/414/EC (2003).
10. V. Di Stefano, G. Avellone, D. Bongiorno, V. Cunsolo, V. Muccilli, S. Sforza, A. Dossena, L. Drahos, K. Vékey, *J. Chromatogr. A* 1259 (2012) 74–85.
11. L. Zhang, S. Liu, X. Cui, C. Pan, A. Zhang, F. Chen, *Cent. Eur. J. Chem.* 10 (2012) 900–925.
12. A. Sassolas, B. Prieto-Simón, J.-L. Marty, *Am. J. Anal. Chem.* 3 (2012) 210–232.
13. Y. Miao, N. He, J.-J. Zhu, *Chem. Rev.* 110 (2010) 5216–5234.
14. F. Ricci, G. Volpe, L. Micheli, G. Palleschi, *Anal. Chim. Acta* 605 (2007) 111–129.
15. E.J. Cho, J.-W. Lee, A.D. Ellington, *Annu. Rev. Anal. Chem.* 2 (2009) 241–264.
16. J.S. Van Dyk, B. Pletschke, *Chemosphere* 82 (2011) 291–307.
17. A. Amine, H. Mohammadi, I. Bourais, G. Palleschi, *Biosens. Bioelectron.* 21 (2006) 1405–1423.
18. G.A. Evtugyn, H.C. Budnikov, E.B. Nikolskaya, *Russ. Chem. Rev.* 68 (1999) 1041–1064.
19. F. Arduini, A. Amine, D. Moscone, G. Palleschi, *Microchim. Acta* 170 (2010) 193–214.
20. T.J.-L. Marty, B. Leca, T. Noguer, *Analusis* 26 (1998) M144–M149.
21. A. Mulchandani, W. Chen, P. Mulchandani, J. Wang, K.R. Rogers, *Biosens. Bioelectron.* 16 (2001) 225–230.
22. L. Campanella, R. Dragone, D. Lelo, E. Martini, M. Tomassetti, *Anal. Bioanal. Chem.* 384 (2006) 915–921.
23. F. De Lima, B. Lucca, A.M.J. Barbosa, V.S. Fereira, S.K. Moccelini, A.C. Franzoi, I.C. Vieira, *Enzym. Microb. Technol.* 47 (2010) 153–158.
24. Y.D. de Albuquerque, L.F. Ferreira, *Anal. Chim. Acta* 596 (2007) 210–221.
25. T. Noguer, B. Leca, G. Jeanty, J.-L. Marty, *Field Anal. Chem. Technol.* 3 (1999) 171–178.
26. G.C. Oliveira, S.K. Moccelini, M. Castilho, A.J. Terezo, J. Possavatz, M.R.L. Magalhães, E.F.G.C. Dores, *Talanta* 98 (2012) 130–136.
27. S.K. Moccelini, I.C. Vieira, F. De Lima, B. Lucca, A.M.J. Barbosa, V.S. Ferreira, *Talanta* 82 (2010) 164–170.
28. F. Mazzei, F. Botrè, S. Montilla, R. Pilloton, E. Podesta, C. Botrè, *J. Electroanal. Chem.* 574 (2004) 95–100.
29. M.S. Ayyagari, S. Kamtekar, R. Pande, K.A. Marx, J. Kumar, S.K. Tripathy, D.L. Kaplan, *Mater. Sci. Eng. C2* (1995) 191–196.
30. F. Mazzei, F. Botrè, C. Botrè, *Anal. Chim. Acta* 336 (1996) 67–75.

31. K.A. Lord, *Ann. Appl. Biol.* 43 (1055) 192–202.
32. C. Hou, K. He, L. Yang, D. Huo, M. Yang, S. Huang, L. Zhang, C. Shen, *World J. Microbiol. Biotechnol.* 28 (2012) 541–548.
33. G.L. Ellman, K.D. Courtney, V. Andres Jr., R.M. Featherstone, *Biochem. Pharm.* 7 (1961) 91–95.
34. G.G. Guilbault, D.N. Kramer, *Anal. Chem.* 37 (1965) 120–123.
35. L.P.A. De Jong, C.C. Groos, C. Van Dijk, *Biochim. Biophys. Acta* 227 (1071) 475–478.
36. N.F. Starodub, N.I. Kanjuk, A.L. Kukla, Yu.M. Shirshov, *Anal. Chim. Acta* 385 (1999) 461–466.
37. B.J. White, J.A. Legako, H.J. Harmon, *Sens. Actuators B* 89 (2003) 107–111.
38. J.L. Taylor, R.T. Mayer, C.M. Himel, *Mol. Pharmacol.* 45 (1994) 74–83.
39. E. Reiner, V. Simeon-Rudolf, *Pflugers Arch.* 440 (2000) R118–R120.
40. J.-P. Colletier, D. Fournier, H.M. Greenblatt, J. Stojan, J.L. Sussman, G. Zaccai, I. Silman, M. Weik, *EMBO J.* 25 (2006) 2746–2756.
41. L.W. Hazleton, *J. Agric. Food Chem.* 3 (1955) 312–319.
42. B. Nunes, in: D.M. Whitacre (ed.), The use of cholinesterases in ecotoxicology, in: *Reviews of Environmental Contamination and Toxicology*, Springer, New York, 2011, pp. 29–59.
43. J.L. Marty, D. Garcia, R. Rouillon, *Trends Analyt. Chem.* 14 (1995) 329–333.
44. P.D.B. O'Marques, G.S. Nunes, T.C.R. dos Santos, S. Andreescu, J.L. Marty, *Biosens. Bioelectron.* 20 (2004) 825–832.
45. T.T. Bachmann, B. Leca, F. Vilatte, J.-L. Marty, D. Fournier, R.D. Schmid, *Biosens. Bioelectron.* 15 (2000) 193–201.
46. Y. Boublik, P. Saint-Aguet, A. Lougarre, M. Arnaud, F. Villatte, S. Estrada-Mondaca, D. Fournier, *Protein Eng.* 15 (2002) 43–50.
47. D.M. Quinn, *Chem. Rev.* 87 (1987) 955–979.
48. P. Skládal, *Food Technol. Biotechnol.* 34 (1996) 43–49.
49. K. Musilek, O. Holas, A. Horova, M. Pohanka, J. Zdarova-Karasova1, D. Jun, K. Kuca, in: M. Stocheva (ed.), *Pesticides in the Modern World—Effects of Pesticides Exposure*, Intech, Manhattan, NY, 2011, pp. 341–358.
50. K.C. Gulla, M.D. Gouda, M.S. Thakur, N.G. Karanth, *Biochim. Biophys. Acta* 1597 (2002) 133–139.
51. H.J. Mason, C. Sams, A.J. Stevenson, R. Rawbone, *Hum. Exp. Toxicol.* 19 (2000) 511–516.
52. O.D. Renedo, M.A. Alonso-Lomillo, M.J.A. Martínez, *Talanta* 73 (2007) 202–219.
53. J.P. Hart, A. Crew, E. Crouch, K.C. Honeychurch, R.M. Pemberton, *Anal. Lett.* 37 (2004) 789–830.
54. Y. Ishige, S. Takeda, M. Kamahori, *Biosens. Bioelectron.* 26 (2010) 1366–1372.
55. A. Hai, D. Ben-Haim, N. Korbakov, A. Cohen, J. Shappir, R. Oren, M.E. Spira, S. Yitzchaik, *Biosens. Bioelectron.* 22 (2006) 605–612.
56. W.N. Aldridge, *Biochem. J.* 46 (1950) 451–460.
57. K. Alfthan, H. Kenttämaa, T. Zukale, *Anal. Chim. Acta* 217 (1989) 43–51.
58. P. Herzsprung, L. Weil, R. Niessner, *Int. J. Environ. Anal. Chem.* 47 (1992) 181–200.
59. B. Bucur, D. Fournier, A. Danet, J.-L. Marty, *Anal. Chim. Acta* 562 (2006) 115–121.
60. H. Schulze, S. Vorlová, F. Villatte, T.T. Bachmann, R.D. Schmid, *Biosens. Bioelectron.* 18 (2003) 201–209.
61. F. Vilatte, V. Marcel, S. Estrada-Mondaca, D. Fournier, *Biosens. Bioelectron.* 13 (1998) 157–164.

62. G.A. Petroianu, M.Y. Hasan, K. Arafat, S.M. Nurulain, A. Schmitt, *J. Appl. Toxicol.* 25 (2005) 562–567.
63. G.A. Evtugyn, H.C. Budnikov, E.B. Nikolskaya, *Talanta* 46 (1998) 465–484.
64. A.N. Ivanov, G.A. Evtugyn, R. Gyurcsanyi, K. Toth, H.C. Budnikov, *Anal. Chim. Acta* 404 (2000) 55–65.
65. M. Pohanka, J.Z. Karasova, K. Kuca, J. Pikula, O. Holas, J. Korabecny, J. Cabal, *Talanta* 81 (2010) 621–624.
66. B. Kuswandi, C.I. Fikriyah, A.A. Gani, *Talanta* 74 (2008) 613–618.
67. H.C. Budnikov, G.A. Evtugyn, *Electroanalysis* 8 (1996) 817–820.
68. C. Tran-Minh, *Ion-Sel. Electrode Rev.* 7 (1985) 41–75.
69. S.V. Dzyadevych, A.P. Soldatkin, J.-M. Chovelon, *Anal. Chim. Acta* 459 (2002) 33–41.
70. J. Dingabc, W. Qin, *Chem. Commun.* (2009) 971–973.
71. G.-P. Nikoleli, D.P. Nikolelis, N. Psaroudakis, T. Hianik, *Anal. Lett.* 44 (2011) 1265–1276.
72. F. Ricci, A. Amine, G. Palleschi, D. Moscone, *Biosens. Bioelectron.* 18 (2003) 165–174.
73. S. Upadhyay, G.R. Rao, M.K. Sharma, B.K. Bhattacharya, V.K. Rao, R. Vijayaraghavan, *Biosens. Bioelectron.* 25 (2009) 832–838.
74. E.A. Dontsova, Y.S. Zeifman, I.A. Budashov, A.V. Eremenko, S.L. Kalnov, I.N. Kurochkin, *Sens. Actuators B* 159 (2011) 261–270.
75. A.A. Ciucu, C. Negulescu, R.P. Baldwin, *Biosens. Bioelectron.* 18 (2003) 303–310.
76. M. Espinosa, P. Atanasov, E. Wilkins, *Electroanalysis* 11 (1999) 1055–1062.
77. A.L. Ghindilis, T.G. Morzunova, A.V. Barmin, I.N. Kurochkin, *Biosens. Bioelectron.* 11 (1996) 873–880.
78. S.M.Z. Hossain, R.E. Luckham, M.J. McFadden, J.D. Brennan, *Anal. Chem.* 81 (2009) 9055–9064.
79. M. Pohanka, *Anal. Lett.* 45 (2012) 367–374.
80. A.N. Díaz, F.G. Sánchez, V. Bracho, J. Lovillo, A. Aguilar, *Fresenius J. Anal. Chem.* 357 (1997) 958–961.
81. G.L. Ellman, K.D. Courtney, V. Andres, Jr., R.M. Featherstone, *Biochem. Pharm.* 7 (1961) 88–95.
82. S. Sabelle, P.-Y. Renard, K. Pecorella, S. de Suzzoni-Dézard, C. Créminon, J. Grassi, C. Mioskowski, *J. Am. Chem. Soc.* 124 (2002) 4874–4880.
83. F. Arduini, A. Cassisi, A. Amine, F. Ricci, D. Moscone, G. Palleschi, *J. Electroanal. Chem.* 626 (2009) 66–74.
84. Y. Song, M. Zhang, L. Wang, L. Wan, X. Xiao, S. Ye, J. Wang, *Electrochim. Acta* 56 (2011) 7267–7271.
85. L.G. Shaidarova, A.Yu. Fomin, S.A. Ziganshina, E.P. Medyantseva, G.K. Budnikov, *J. Anal. Chem.* 57 (2002) 150–156.
86. J.P. Hart, I.C. Hartley, *Analyst* 119 (1994) 259–263.
87. K.A. Joshi, J. Tang, R. Haddon, J. Wang, W. Chen, A. Mulchandani, *Electroanalysis* 17 (2005) 54–58.
88. L.-G. Zamfir, L. Rotariu, C. Bala, *Biosens. Bioelectron.* 26 (2011) 3692–3695.
89. J. Gong, L. Wang, L. Zhang, *Biosens. Bioelectron.* 24 (2009) 2285–2288.
90. A. Cagnini, I. Palchetti, I. Lioni, M. Mascini, A.P.F. Turner, *Sens. Actuators B* 24–25 (1995) 85–89.
91. L. Rotariu, L.G. Zamfir, C. Bala, *Anal. Chim. Acta* 748 (2012) 81–88.
92. H. Schulze, R.D. Schmid, T. Bachmann, *Anal. Bioanal. Chem.* 372 (2002) 268–272.

93. A. Boni, C. Cremisini, E. Magaro, M. Tosi, W. Vastarella, R. Pilloton, *Anal. Lett.* 37 (2003) 1683–1699.

94. Y. Zhang, S.B. Muench, H. Schulze, R. Perz, B. Yang, R.D. Schmid, T.T. Bachmann, *J. Agric. Food Chem.* 53 (2005) 5110–5115.

95. E. Suprun, G. Evtugyn, H. Budnikov, F. Ricci, D. Moscone, G. Palleschi, *Anal. Bioanal. Chem.* 383 (2005) 597–604.

96. H. Schulze, E. Scherbaum, M. Anastassiades, S. Vorlova, R.D. Schmid, T.T. Bachmann, *Biosens. Bioelectron.* 17 (2002) 1095–1105.

97. P. Raghu, T.M. Reddy, B.E.K. Swamy, B.N. Chandrashekar, K. Reddaiah, M. Sreedhar, *J. Electroanal. Chem.* 665 (2012) 76–82.

98. R.K. Mishra, R.B. Dominguez, S. Bhand, R. Muñoz, J.-L. Marty, *Biosens. Bioelectron.* 32 (2012) 56–61.

99. A. Ivanov, G. Evtugyn, H. Budnikov, F. Ricci, D. Moscone, G. Palleschi, *Anal. Bioanal. Chem.* 377 (2003) 624–631.

100. M. Albareda-Sirvent, A. Merçoci, S. Alegret, *Anal. Chim. Acta* 442 (2001) 35–44.

101. M. Del Carlo, M. Mascini, A. Pepe, G. Diletti, D. Compagnone, *Food Chem.* 84 (2004) 651–656.

102. N. Gan, X. Yang, D. Xie, Y. Wu, W. Wen, *Sensors* 10 (2010) 625–638.

103. A. Crew, D. Lonsdale, N. Byrd, R. Pittson, J.P. Hart, *Biosens. Bioelectron.* 26 (2011) 2847–2851.

104. G. Valdés-Ramírez, D. Fournier, M.T. Ramírez-Silva, J.-L. Marty, *Talanta* 74 (2008) 741–746.

105. N. Chauhan, C.S. Pundir, *Anal. Chim. Acta* 701 (2011) 66–74.

106. M. del Carlo, A. Pepe, M. Sergi, M. Mascini, A. Tarentini, D. Compagnone, *Talanta* 81 (2010) 76–81.

107. M. del Carlo, M. Mascini, A. Pepe, D. Compagnone, M. Mascini, *J. Agric. Food Chem.* 50 (2002) 7206–7210.

108. I. Ion, A.C. Ion, *Mater. Sci. Eng.* C 32 (2012) 1001–1004.

109. A. Hildebrandt, R. Bragós, S. Lacorte, J.L. Marty, *Sens. Actuators B* 133 (2008) 195–201.

110. J. Caetano, S.A.S. Machado, *Sens. Actuators B* 129 (2008) 40–46.

111. G.S. Nunes, P. Skladal, H. Yamanaka, D. Barcelo, *Anal. Chim. Acta* 362 (1998) 59–68.

112. G.S. Nunes, D. Barceló, B.S. Grabaric, J.M. Dýaz-Cruz, M.L. Ribeiro, *Anal. Chim. Acta* 399 (1999) 37–49.

113. L. Pogacnik, M. Franko, *Biosens. Bioelectron.* 14 (1999) 569–578.

114. L. Pogacnik, M. Franko, *Biosens. Bioelectron.* 18 (2003) 1–9.

115. S.V. Dzyadevych, V.N. Arkhypova, A.P. Soldatkin, A.V. El'skaya, C. Martelet, N. Jaffrezic-Renault, *Anal. Lett.* 7 (2004) 1611–1624.

116. I.V. Benilova, A.P. Soldatkin, C. Martelet, N. Jaffrezic-Renault, *Electroanalysis* 18 (2006) 1950–1956.

117. I.V. Benilova, V.N. Arkhypova, S.V. Dzyadevych, N. Jaffrezic-Renault, C. Martelet, A.P. Soldatkin, *Pestic. Biochem. Physiol.* 86 (2006) 203–210.

118. N. Mionetto, J.-L. Marty, I. Karube, *Biosens. Bioelectron.* 9 (1994) 463–470.

119. L. Campanella, G. Favero, M.P. Sammartino, M. Tomassetti, *Anal. Chim. Acta* 393 (1999) 109–120.

120. L. Campanella, S. de Luca, M.P. Sammartino, M. Tomassetti, *Anal. Chim. Acta* 385 (1999) 59–91.

121. G.L. Turdean, M.S. Turdean, *Pestic. Biochem. Physiol.* 90 (2008) 73–81.

122. S. Fennouh, V. Casimiri, C. Burstein, *Biosens. Bioelectron.* 12 (1997) 97–104.
123. T. Montesinos, S. Pérez-Monguia, F. Valdes, J.-L. Marty, *Anal. Chim. Acta* 431 (2001) 231–237.
124. G.A. Evtugyn, A.N. Ivanov, E.V. Gogol, J.-L. Marty, H.C. Budnikov, *Anal. Chim. Acta* 385 (1999) 13–21.
125. H.-S. Lee, Y.A. Kim, Y.A. Cho, Y.T. Lee, *Chemosphere* 46 (2002) 571–576.
126. M. Waibel, H. Schulze, N. Huber, T.T. Bachmann, *Biosens. Bioelectron.* 21 (2006) 1132–1140.
127. C.B.S. Roepcke, S.B. Muench, H. Schulze, T. Bachmann, R.D. Schmid, B. Hauer, *J. Agric. Food Chem.* 58 (2010) 8748–8756.
128. M.P. Dondoi, B. Bucur, A.F. Danet, C.N. Toader, L. Barthelmebs, J.-L. Marty, *Anal. Chim. Acta* 578 (2006) 162–169.
129. A.H. Lockwood, *Curr. Opin. Neurol.* 10 (1997) 507–511.
130. G.A. Evtugyn, E.P. Rizaeva, E.E. Stoikova, H.C. Budnikov, *Electroanalysis* 9 (1997) 1124–1128.
131. G. Istamboulie, M. Cortina-Puig, J.-L. Marty, T. Noguer, *Talanta* 79 (2009) 507–511.
132. G.A. Alonso, R.B. Dominguez, J.-L. Marty, R. Muñoz, *Sensors* 11 (2011) 3791–3802.
133. Y. Vlasov, A. Legin, A. Rudnitskaya, *Anal. Bioanal. Chem.* 373 (2002) 136–146.
134. S. Andreescu, T. Noguer, V. Magearu, J.-L. Marty, *Talanta* 57 (2002) 169–176.
135. S.V. Dzyadevych, V.N. Arkhypova, C. Martelet, N. Jaffrezic-Renault, J.-M. Chovelon, A.V. El'skaya, A.P. Soldatkin, *Electroanalysis* 16 (2004) 1873–1882.
136. S. Dzyadevych, A.P. Soldatkin, V.N. Arkhypova, A.V. El'skaya, *Sens. Actuators B* 105 (2005) 81–87.
137. G. Jeanty, J.-L. Marty, *Biosens. Bioelectron.* 13 (1998) 213–218.
138. B. Bucur, S. Andreescu, J.-L. Marty, *Anal. Lett.* 37 (2004) 1571–1588.
139. M. Ganesana, G. Istarnboulie, J.-L. Marty, T. Noguer, S. Andreescu, *Biosens. Bioelectron.* 30 (2011) 43–48.
140. D.P. Dumas, J.R. Wild, F.M. Raushel, *Biotech. Appl. Biochem.* 11 (1989) 235–243.
141. A. Mulchandani, S. Pan, W. Chen, *Biotechnol. Prog.* 15 (1999) 130–134.
142. A. Mulchandani, P. Mulchandani, W. Chen, J. Wang, L. Chen, *Anal. Chem.* 71 (1999) 2246–2249.
143. T. Laothanachareon, V. Champreda, P. Sritongkham, M. Somasundrum, W. Surareungchai, *World J. Microbiol. Biotechnol.* 24 (2008) 3049–3055.
144. P. Mulchandani, A. Mulchandani, I. Kaneva, W. Chen, *Biosens. Bioelectron.* 14 (1999) 77–85.
145. A.L. Simonian, A.W. Flounders, J.R. Wild, *Electroanalysis* 16 (2004) 1896–1906.
146. J. Orbulescu, C.A. Constantine, V.K. Rastogi, S.S. Shah, J.J. DeFrank, R.M. Leblanc, *Anal. Chem.* 78 (2006) 7016–7021.
147. S. Chen, J. Huang, D. Du, J. Li, H. Tu, D. Liu, A. Zhang, *Biosens. Bioelectron.* 26 (2011) 4320–4325.
148. M.R. Murthy, I.M. Mandappa, R. Latha, A.C. Vinayaka, M.S. Thakur, H.K. Manonmani, *Anal. Methods* 2 (2010) 1355–1359.
149. E. Morou, H.M. Ismail, A.J. Dowd, J. Hemingway, N. Labrou, *Anal. Biochem.* 378 (2008) 60–64.
150. V.G. Andreou, Y.D. Clonis, *Anal. Chim. Acta* 460 (2002) 151–161.
151. A. Hipolito-Moreno, M.E. Leo Gonzalez, L.V. Perez-Arribas, L.M. Polo-Dõez, *Anal. Chim. Acta* 362 (1998) 187–192.
152. J. Wang, V.B. Nascimento, S.A. Kane, K. Rogers, M.R. Smyth, L. Angnes, *Talanta* 43 (1996) 1903–1907.

153. G.-Y. Kim, M.-S. Kang, J. Shim, S.-H. Moon, *Sens. Actuators B* 133 (2008) 1–4.
154. L. Campanella, D. Lelo, E. Martini, M. Tomassetti, *Anal. Chim. Acta* 587 (2007) 22–32.
155. J.C. Vidal, L. Bonel, J.R. Castillo, *Electroanalysis* 20 (2008) 865–873.
156. M.T.P. Pita, A.J. Reviejo, F.J.M. de Villena, J.M. Pingarrón, *Anal. Chim. Acta* 340 (1997) 89–97.
157. L.K. Shyan, L.Y. Heng, M. Ahmad, S.A. Aziz, Z. Ishak, *Asian J. Biochem.* 3 (2008) 359–365.
158. F.G. Sánchez, A.N. Díaz, M.C.R. Peinado, C. Belledone, *Anal. Chim. Acta* 484 (2003) 45–51.
159. S.C. Fernandes, I.C. Vieira, A.M.J. Barbosa, V.S. Ferreira, *Electroanalysis* 23 (2011) 1623–1630.
160. E. Zapp, D. Brondani, I.C. Vieira, C.W. Scheeren, J. Dupont, A.M.J. Barbosa, V.S. Ferreira, *Sens. Actuators B* 155 (2011) 331–339.
161. T. Noguer, J.-L. Marty, *Anal. Chim. Acta* 347 (1997) 63–70.
162. D. Leech, K.O. Feerick, *Electroanalysis* 12 (2000) 1339–1342.
163. B.M. Kaufman, M. Clower Jr., *J. AOAC Int.* 78 (1995) 1079–1090.
164. G. Lillie, P. Payne, P. Vadgama, *Sens. Actuators B* 78 (2001) 249–256.
165. M.I. Prodromidis, *Electrochim. Acta* 55 (2010) 4227–4233.
166. X. Su, F.T. Chew, S.F.Y. Li, *Anal. Sci.* 16 (2000) 107–114.
167. W. Welsch, C. Klein, M. von Schickfus, S. Hunklinger, *Anal. Chem.* 68 (1996) 2000–2004.
168. K. Mitsakakis, A. Tserepi, E. Gizeli, *Microelectron. Eng.* 86 (2009) 1416–1418.
169. A. Abbas, M.J. Linman, Q. Cheng, *Biosens. Bioelectron.* 26 (2011) 1815–1824.
170. H.A. Engström, P.O. Andersson, S. Ohlson, *Anal. Biochem.* 11 (2006) 159–166.
171. X. Jiang, D. Li, X. Xu, Y. Ying, Y. Li, Z. Ye, J. Wang, *Biosens. Bioelectron.* 23 (2008) 1577–1587.
172. M.A. González-Martínez, J. Penalva, J.C. Rodríguez-Urbis, E. Brunet, A. Maquieira, R. Puchades, *Anal. Bioanal. Chem.* 384 (2006) 1540–1547.
173. A. Ivanov, G. Evtugyn, H. Budnikov, S. Girotti, S. Ghini, E. Ferri, A. Montoya, J.V. Mercader, *Anal. Lett.* 41 (2008) 392–405.
174. F. Ricci, G. Adornetto, G. Palleschi, *Electrochim. Acta* 84 (2012) 74–83.
175. T.R.J. Holford, F. Davis, S.P.J. Higson, *Biosens. Bioelectron.* 34 (2012) 12–24.
176. M. Franek, K. Hruska, *Vet. Med.* 50 (2005) 1–10.
177. M. Tomassetti, E. Martini, L. Campanella, *Electroanalysis* 24 (2012) 842–856.
178. B. Hock, *Anal. Chim. Acta* 347 (1997) 177–186.
179. K. Grennan, G. Strachan, A.J. Porter, A.J. Killard, M.R. Smyth, *Anal. Chim. Acta* 500 (2003) 287–298.
180. R.E. Ionescu, C. Gondran, L. Bouffier, N. Jaffrezic-Renault, C. Martelet, S. Cosnier, *Electrochim. Acta* 55 (2010) 6228–6232.
181. H.V. Tran, R. Yougnia, S. Reisberg, B. Piro, N. Serradji, T.D. Nguyen, L.D. Tran, C.Z. Dong, M.C. Pham, *Biosens. Bioelectron.* 31 (2012) 62–68.
182. S. Hleli, C. Martelet, A. Abdelghani, N. Burais, N. Jaffrezic-Renault, *Sens. Actuators B* 113 (2006) 711–717.
183. L. Campanella, S. Eremin, D. Lelo, E. Martini, M. Tomassetti, *Sens. Actuators B* 156 (2011) 50–62.
184. E. Valera, D. Muñiz, Á. Rodríguez, *Microelectron. Eng.* 87 (2010) 167–173.
185. E. Valera, J. Ramón-Azcón, A. Barranco, B. Alfaro, F. Sánchez-Baeza, M.-P. Marco, Á. Rodríguez, *Food Chem.* 122 (2010) 888–894.

186. E. Valera, J. Ramón-Azcón, F.-J. Sanchez, M.-P. Marco, Á. Rodríguez, *Sens. Actuators* 134 (2008) 95–103.
187. M.Á. González-Martínez, Á. Maquieira, R. Puchades, *Int. J. Environ. Anal. Chem.* 83 (2003) 633–642.
188. E. Zacco, M.I. Pividori, S. Alegret, R. Galve, M.-P. Marco, *Anal. Chem.* 78 (2006) 1780–1788.
189. A. Székács, N. Trummer, N. Adányi, M. Váradi, I. Szendro, *Anal. Chim. Acta* 487 (2003) 31–42.
190. N.F. Starodub, B.B. Dzantiev, V.M. Starodub, A.V. Zherdev, *Anal. Chim. Acta* 424 (2000) 37–43.
191. A. Klotz, C. Brecht, C. Barzen, G. Gauglitz, R.D. Harris, G.R. Quigley, J.S. Wilkinson, R.A. Abuknesha, *Sens. Actuators B* 51 (1998) 181–187.
192. S. Rodriguez-Mozaz, S. Reder, M.L. de Alda, G. Gauglitz, D. Barceló, *Biosens. Bioelectron.* 19 (2004) 633–640.
193. I.M. Ciumasu, P.M. Krämer, C.M. Weber, G. Kolb, D. Tiemann, S. Windisch, I. Frese, A.A. Kettrup, *Biosens. Bioelectron.* 21 (2005) 354–364.
194. P. Sharma, K. Sablok, V. Bhalla, C.R. Suri, *Biosens. Bioelectron.* 26 (2011) 4209–4212.
195. X. Sun, S. Du, X. Wang, *Eur. Food Res. Technol.* 235 (2012) 469–477.
196. S. Du, X. Wang, X. Sun, Q. Li, *Anal. Lett.* 45 (2012) 1230–1241.
197. X. Sun, Q. Li, X. Wang, S. Du, *Anal. Lett.* 45 (2012) 1604–1616.
198. X. Sun, S. Du, X. Wang, W. Zhao, Q. Li, *Sensors* 11 (2011) 9520–9531.
199. M. Tomassetti, E. Martini, L. Campanella, *Int. J. Environ. Anal. Chem.* 92 (2012) 417–431.
200. I. Navrátilová, P. Skládal, *Bioelectrochemistry* 62 (2004) 11–18.
201. A.-P. Deng, H. Yang, *Sens. Actuators B* 124 (2007) 202–208.
202. E. Mauriz, A. Calle, J.J. Manclús, A. Montoya, A. Hildebrandt, D. Barceló, L.M. Lechuga, *Biosens. Bioelectron.* 22 (2007) 1410–1418.
203. M.Yu. Lebedev, S.A. Eremin, P. Skládal, *Anal. Lett.* 36 (2003) 2443–2457.
204. E. Mauriz, A. Calle, L.M. Lechuga, J. Quintana, A. Montoya, J.J. Mancl¹us, *Anal. Chim. Acta* 561 (2006) 40–47.
205. X. Sun, Y. Cao, Z. Gong, X. Wang, Y. Zhang, J. Gao, *Sensors* 12 (2012) 17247–17261.
206. B.B. Dzantiev, E.V. Yazynina, A.V. Zherdev, Yu.V. Plekhanova, A.N. Reshetilov, S.-C. Chang, C.J. McNeil, *Sens. Actuators B* 98 (2004) 254–261.
207. G. Liu, S. Wang, J. Liu, D. Song, *Anal. Chem.* 84 (2012) 3921–3928.
208. W.-L. Xing, L.-R. Ma, Z.-H. Jiang, F.-H. Cao, M.-H. Ji, *Talanta* 52 (2000) 879–883.
209. V.B. Kandimalla, N.S. Neeta, N.G. Karanth, M.S. Thakur, K.R. Roshini, B.E.A. Rani, A. Pasha, N.G.K. Karanth, *Biosens. Bioelectron.* 20 (2004) 903–906.
210. J.P.M. Sardinha, M.H. Gil, J.V. Mercader, A. Montoya, *J. Immunol. Methods* 260 (2002) 173–182.
211. J. Švitel, A. Dzgoev, K. Ramanathan, B. Danielsson, *Biosens. Bioelectron.* 15 (2000) 411–415.
212. O. Ouerghi, A. Touhami, N. Jaffrezic-Renault, C. Martelet, H. Ben Ouada, S. Cosnier, *Bioelectrochemistry* 56 (2002) 131–133.
213. Z. Dai, H. Liu, Y. Shen, X. Su, Z. Xu, Y. Sun, X. Zou, *Anal. Chem.* 84 (2012) 8157–8163.
214. M. Mascini, *Anal. Bioanal. Chem.* 390 (2008) 987–988.
215. T. Hianik, J. Wang, *Electroanalysis* 21 (2009) 1223–1235.
216. M. Citartan, S.C.B. Gopinath, J. Tominaga, S.-C. Tan, T.-H. Tang, *Biosens. Bioelectron.* 34 (2012) 1–11.

217. L. Wang, X. Liu, Q. Zhang, C. Zhang, Y. Liu, K. Tu, J. Tu, *Biotechnol. Lett.* 34 (2012) 869–874.
218. J. He, Y. Liu, M. Fan, X. Liu, *J. Agric. Food Chem.* 59 (2011) 1582–1586.
219. N. Prabhakar, G. Sumana, K. Arora, H. Singh, B.D. Malhotra, *Electrochim. Acta* 53 (2008) 4344–4350.
220. A.M. Nowicka, A. Kowalczyk, Z. Stojek, M. Hepel, *Biophys. Chem.* 146 (2010) 42–53.
221. M. Stobiecka, K. Coopersmith, S. Cutler, M. Hepel, *ECS Trans.* 28 (2010) 1–12.
222. P. Melidis, E. Vaiopoulou, A. Aivasidis, *Bioprocess Biosyst. Eng.* 3 (2008) 277–282.
223. D.A. Schofield, C. Westwater, J.L. Barth, A.A. DiNovo, *Appl. Microbiol. Biotechnol.* 76 (2007) 1383–1394.
224. O.N. Ponomareva, V.A. Arlyapov, V.A. Alferov, A.N. Reshetilov, *Appl. Biochem. Microbiol.* 47 (2011) 1–11.
225. J. Kumar, S.F. D'Souza, *Biosens. Bioelectron.* 26 (2010) 1292–1296.
226. J. Kumar, S. Kumar Jha, S.F. D'Souza, *Biosens. Bioelectron.* 21 (2006) 2100–2105.
227. J. Kumar, S.F. D'Souza, *Biosens. Bioelectron.* 26 (2011) 4289–4293.
228. P. Mulchandani, W. Chen, A. Mulchandani, *Anal. Chim. Acta* 568 (2006) 217–221.
229. Y. Lei, P. Mulchandani, J. Wang, W. Chen, A. Mulchandani, *Environ. Sci. Technol.* 39 (2005) 8853–8857.
230. D. Yong, C. Liu, D. Yu, S. Dong, *Talanta* 84 (2011) 7–12.
231. S. Gaberlein, F. Spener, C. Zaborosch, *Appl. Microb. Biotechnol.* 54 (2000) 652–658.
232. S. Kintzios, E. Pistola, P. Panagiotopoulos, M. Bomsel, N. Alexandropoulos, F. Bem, G. Ekonomou, J. Biselis, R. Levin, *Biosens. Bioelectron.* 16 (2001) 325–336.
233. E. Voumvouraki, S. Kintzios, *Procedia Eng.* 25 (2011) 964–967.
234. K. Flampouri, S. Mavrikou, S. Kintzios, G. Miliadis, P. Aplada-Sarlis, *Talanta* 80 (2010) 1799–1804.
235. M.T. Giardi, M. Koblízek, J. Masojídek, *Biosens. Bioelectron.* 16 (2001) 1027–1033.
236. M. Koblízek, J. Maly, J. Masojídek, J. Komenda, T. Kucera, M.T. Giardi, F.K. Mattoo, R. Pilloton, *Biotechnol. Bioeng.* 78 (2002) 110–116.
237. M. Koblízek, J. Komenda, T. Kucera, J. Masojidek, R. Pilloton, A.K. Mattoo, M.T. Giardi, *Biotechnol. Bioeng.* 60 (1998) 664–669.
238. J. Maly, A. Masci, J. Masojidek, M. Sugiura, R. Pilloton, *Anal. Lett.* 37 (2004) 1645–1656.
239. J. Maly, K. Klem, A. Lukavská, J. Masojádek, *J. Environ. Qual.* 34 (2005) 1780–1788.
240. J. Maly, J. Masojidek, A. Masci, M. Ilie, E. Cianci, V. Foglietti, W. Vastarella, R. Pilloton, *Biosens. Bioelectron.* 21 (2005) 923–932.
241. K. Buonasera, G. Pezzotti, V. Scognamiglio, A. Tibuzzi, M.T. Giardi, *J. Agric. Food Chem.* 58 (2010) 5982–5990.
242. E.V. Piletskaya, S.A. Piletsky, T.A. Sergeyeva, A.V. El'skaya, A.A. Sozinov, J.-L. Marty, R. Rouillon, *Anal. Chim. Acta* 391 (1999) 1–7.
243. N. Terasaki, N. Yamamoto, K. Tamada, M. Hattori, T. Hiraga, A. Tohri, I. Sato et al., *Biochim. Biophys. Acta* 1767 (2007) 653–659.
244. D.M. Rawson, A.J. Willmer, M.F. Cardosi, *Toxic. Assess.* 2 (1987) 325–340.
245. A.I. Zamaleeva, I.R. Sharipova, R.V. Shamagsumova, A.N. Ivanov, G.A. Evtugyn, D.G. Ishmuchametova, R.F. Fakhrullin, *Anal. Methods* 3 (2011) 509–513.
246. X. Xu, Y. Ying, *Food Rev. Int.* 27 (2011) 300–329.
247. B. Byrne, E. Stack, N. Gilmartin, R. O'Kennedy, *Sensors* 9 (2009) 4407–4445.
248. A. Lahuerta, T. Westrell, J. Takkinen, F. Boelaert, V. Rizzi, B. Helwigh, B. Borck, H. Korsgaard, A. Ammon, P. Mäkelä, *Eurosurveillance* 13 (2011) 5–8.

249. N. Gilmartin, R. O'Kennedy, *Enzym. Microb. Technol.* 50 (2012) 87–95.
250. K.S. Gracias, J.L. McKillip, *Can. J. Microbiol.* 50 (2004) 883–890.
251. A.K. Bhunia, *Adv. Food Nutr. Res.* 54 (2008) 1–44.
252. M. Uyttendaele, R. Schukkink, B. van Gemen, J. Debevere, *Int. J. Food Microbiol.* 27 (1995) 77–89.
253. A. Nadal, A. Coll, N. Cook, M. Pla, *J. Microbiol. Methods* 68 (2007) 623–632.
254. R. Pochampally, in: D.J. Prockop, D.G. Phinney, B.A. Bunnell (eds.), Colony forming unit assays for MSCs, in: *Methods in Molecular Biology. Mesenchymal Stem Cells: Methods and Protocols*, Humana Press, Totowa, NJ, 2008, pp. 83–91.
255. D. Ivnitski, E. Wilkins, H.T. Tien, A. Ottova, *Electrochem. Commun.* 2 (2000) 457–460.
256. Y.-H. Lin, S.-H. Chen, Y.-C. Chuang, Y.-C. Lu, T.Y. Shen, C.A. Chang, C.-S. Lin, *Biosens. Bioelectron.* 23 (2008) 1832–1837.
257. X. Guo, C.-S. Lin, S.-H. Chen, R. Ye, V.C.H. Wu, *Biosens. Bioelectron.* 38 (2012) 177–183.
258. S. Kim, L. Lu, J.-H. Chung, K. Lee, Y. Li, S. Jun, *Innov. Food Sci. Emerg. Technol.* 12 (2011) 617–622.
259. K. Abu-Rabeah, A. Ashkenazi, D. Atias, L. Amir, R.S. Marks, *Biosens. Bioelectron.* 24 (2009) 3461–3466.
260. E. Tully, S.P. Higson, R. O'Kennedy, *Biosens. Bioelectron.* 23 (2008) 906–912.
261. R. Wang, C. Ruan, D. Kanayeva, K. Lassiter, Y. Li, *Nano Lett.* 8 (2008) 2625–2631.
262. G.-J. Yang, J.-L. Huang, W.-J. Meng, M. Shen, X.-A. Jiao, *Anal. Chim. Acta* 647 (2009) 159–166.
263. A.S. Afonso, B. Pérez-López, R.C. Faria, L.H.C. Mattoso, M. Hernández-Herrero, A.X. Roig-Sagués, M. Maltez-da Costa, A. Merkoçi, *Biosens. Bioelectron.* 40 (2013) 121–126.
264. M. Amaro, S. Oaew, W. Surareungchai, *Biosens. Bioelectron.* 38 (2012) 157–162.
265. F. Salam, I.E. Tothill, *Biosens. Bioelectron.* 24 (2009) 2630–2636.
266. G. Kim, M. Morgan, B.K. Hahm, A. Bhunia, J.H. Mun, A.S. Om, *J. Phys. Conf. Ser.* 100 (2008) 052044.
267. Z. Muhammad-Tahir, E.C. Alocilja, *Biosens. Bioelectron.* 18 (2003) 813–819.
268. S. Viswanathan, C. Rani, J.A. Ho, *Talanta* 94 (2012) 315–319.
269. N. Li, A. Brahmendra, A.J. Veloso, A. Prashar, X.R. Cheng, V.W.S. Hung, C. Guyard, M. Terebiznik, K. Kerman, *Anal. Chem.* 84 (2012) 3485–3488.
270. G. Zhao, F. Xing, S. Deng, *Electrochem. Commun.* 9 (2007) 1263–1268.
271. G. MacBeath, S.L. Schreiber, *Science* 289 (2000) 1760–1763.
272. H.Y. Cai, L. Lu, C.A. Muckle, J.F. Prescott, S. Chen, *J. Clin. Microbiol.* 43 (2005) 3427–3430.
273. A.G. Gehring, D.M. Albin, S.A. Reed, S.I. Tu, J.D. Brewster, *Anal. Bioanal. Chem.* 391 (2008) 497–506.
274. J.B. Delehanty, F.S. Ligler, *Anal. Chem.* 74 (2002) 5681–5687.
275. K.E. Sapsford, M.M. Ngundi, M.H. Moore, M.E. Lassman, L.C. Shriver-Lake, C.R. Taitt, F.S. Ligler, *Sens. Actuators B* 113 (2006) 599–607.
276. A.C. dos Santos Pires, N.F.F. Soares, L.H.M. da Silva, M.C.H. da Silva, M.V. De Almeida, M.L. Hyaric, N.J. Andrade, R.F. Soares, A.B. Mageste, S.G. Reis, *Sens. Actuators B* 153 (2011) 17–23.
277. S.D. Soelberg, T. Chinowsky, G. Geiss, C.B. Spinelli, R. Stevens, S. Near, P. Kauffman, S. Yee, C.E. Furlong, *J. Industr. Microb. Biotechnol.* 32 (2005) 669–674.

278. V. Nanduri, A.K. Bhunia, S.I. Tu, G.C. Paoli, J.D. Brewster, *Biosens. Bioelectron.* 23 (2007) 248–252.

279. J. Waswa, J. Irudayaraj, C. DebRoy, *Lwt-Food Sci. Technol.* 40 (2007) 187–192.

280. D. Wei, O.A. Oyarzabal, T.S. Huang, S. Balasubramanian, S. Sista, A.L. Simoman, *J. Microbiol. Methods* 69 (2007) 78–85.

281. Y. Lan, S. Wang, Y. Yin, W.C. Hoffmann, X. Zheng, *J. Bionic Eng.* 5 (2008) 239–246.

282. B.N. Feltis, B.A. Sexton, F.L. Glenn, M.J. Best, M. Wilkins, T.J. Davis, *Biosens. Bioelectron.* 23 (2008) 1131–1136.

283. Z. Urkut, P. Kara, Y. Goksungur, M. Ozsoz, *Electroanalysis* 23 (2011) 2668–2676.

284. M. Tichoniuk, D. Gwiazdowska, M. Ligaj, M. Filipiak, *Biosens. Bioelectron.* 26 (2010) 1618–1623.

285. T. García, M. Revenga-Parra, L. Añorga, S. Arana, F. Pariente, E. Lorenzo, *Sens. Actuators B* 161 (2012) 1030–1037.

286. K. Yamanaka, T. Ikeuchi, M. Saito, N. Nagatani, E. Tamiya, *Electrochim. Acta* 82 (2012) 132–136.

287. P.R.B. de Oliveira Marques, A. Lermo, S. Campoy, H. Yamanaka, J. Barbé, S. Alegret, M.I. Pividori, *Anal. Chem.* 81 (2009) 1332–1339.

288. D. Zhang, D.J. Carr, E.C. Alocilja, *Biosens. Bioelectron.* 24 (2009) 1377–1381.

289. A. Walter, J. Wu, G.-U. Flechsig, D.A. Haake, J. Wang, *Anal. Chim. Acta* 689 (2011) 29–33.

290. A.T. Csordas, M.J. Delwiche, J.D. Barak, *Sens. Actuators B* 134 (2008) 1–8.

291. A.D. Taylor, J. Ladd, Q. Yu, S. Chen, J. Homola, S. Jiang, *Biosens. Bioelectron.* 22 (2006) 752–758.

292. M. Piliarik, L. Párová, J. Homola, *Biosens. Bioelectron.* 24 (2009) 1399–1404.

293. K. Arora, S. Chand, B.D. Malhotra, *Anal. Chim. Acta* 568 (2006) 259–274.

294. B. Van Dorst, J. Mehta, K. Bekaert, E. Rouah-Martin, W. De Coen, P. Dubruel, R. Blust, J. Robbens, *Biosens. Bioelectron.* 26 (2010) 1178–1194.

24

Micro- and Nanopatterning for Bacteria- and Virus-Based Biosensing Applications

Gulsah Congur and Arzum Erdem

CONTENTS

24.1 Introduction

The diseases caused by bacteria or viruses are major public health concerns. A lot of fatal infections are associated with several pathogens or viruses, such as *Salmonella*, *Escherichia coli*, *Brucella*, and *Legionella*, and noroviruses (NoV), rotaviruses, (RoV), hepatitis A (HAV) E (HEV) viruses—well-known members of bacteria and viruses. Due to the living environment of viruses, they need specific host cells to grow and replicate [1]. Thus, many different recognition and screening platforms of viruses have been progressed including biosensors (shown in Scheme 24.1) for monitoring of bacteria and viruses to maintain public health safety.

Although there are some differences among these applications, biosensors have offered some advantages for the detection of biomolecules, such as sensitivity and selectivity detection, rapid response, easy to use, and time-saving. In recent years, the researchers have intensively endeavored to develop multiple microarray systems to monitor several bacteria and viruses at the same time [2–6].

Nanomaterial-based sensing platforms have recently received a growing interest since the applications of numerous nanomaterials such as

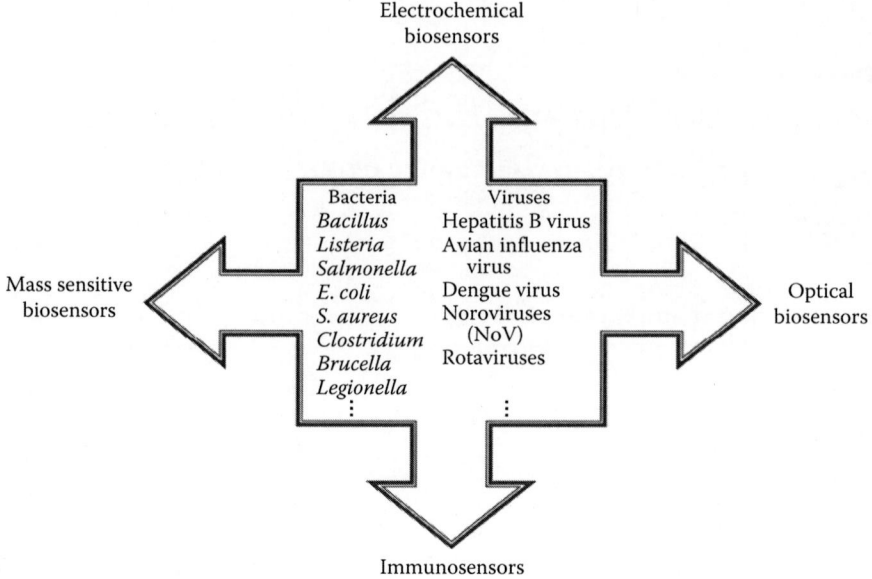

SCHEME 24.1
Schematic representation of detection of bacteria and viruses by using different biosensing strategies; electrochemical biosensors, optical biosensors, immunosensors, and mass sensitive biosensors.

nanoparticles [7–11], nanofibers [12–14], nanotubes [15–21], or nanorods [22–24] have been introduced to biosensor area. The nanostructured surfaces have enhanced the electrical, mechanical, and optical properties of sensing systems.

An overview about detection of bacteria and viruses onto micro- and nanopatterning surfaces combined with biosensing system is briefly summarized in Table 24.1 [3–6,25–54] and reported in the following four main groups: electrochemical biosensors, immunosensors, optical biosensors, and mass sensitive biosensors.

24.2 Biosensing Applications for Monitoring and Detection of Bacteria and Viruses

24.2.1 Electrochemical Biosensors

Electrochemical biosensors have received increasingly a considerable attention in recent years for a variety of practical applications in the fields of medical diagnostics, clinical genetic analysis, forensic identification, and environmental monitoring due to the fact that they provide accurate and sensitive detection platforms that have many advantages such as being cost-effective and time-saving.

TABLE 24.1

Biosensing of Bacteria and Viruses at Different Detection Platforms

Organisms	Detection Platform	Detection Method	LOD	References
Bacillus	Gold interdigitated microelectrodes	LSV	1 CFU/mL	[25]
	Immunosensor	DCT	12×10^1, 34×10^1 CFU/mL polyaniline protonated with hydrochloric acid and polyaniline protonated with perchloric acid, respectively	[26]
	AuNP–coated quartz crystals	QCM	3.5×10^2 CFU/mL	[27]
L. monocytogenes	CCD sensor	BARDOT	1 CFU/25 mg sample	[4]
	Fiber-optic sensor	Immunofluorescence assay	10^3 CFU/mL	[3]
	SPR-active film-coated glass chip	SPR	3.5×10^3 CFU/mL	[5]
Salmonella	CCD sensor	BARDOT	1 CFU/25 mg sample	[4]
	Fiber-optic sensor	Immunofluorescence assay	10^3 CFU/mL	[3]
	Glass interdigitated microelectrode	EIS	3.45×10^6 CFU/mL	[28]
	CNT–enhanced ELISA	Immunoassay	10^6 and 10^3 CFU/mL with different Ab/HRP ratio	[29]
	SPR-active film-coated glass chip	SPR	4.4×10^4 CFU/mL	[5]
	Gold layer–coated glass	SPR	1.25×10^5 cells/mL	[30]

(continued)

TABLE 24.1 (continued)

Biosensing of Bacteria and Viruses at Different Detection Platforms

Organisms	Detection Platform	Detection Method	LOD	References
E. coli	CCD sensor	BARDOT	1 CFU/25 mg sample	[4]
	Microfluidic chip	Impedance	9×10^5 CFU/mL	[31]
	PDMS microfluidic module	Fluorescence	Not reported	[2]
	Photomultiplier tube	Bioluminescence	Not reported	[32]
	Fiber-optic sensor	Immunofluorescence assay	10^3 CFU/mL	[3]
	SUT	Impedance	Not reported	[6]
	IDAM	Impedance	For pure culture, 7.4×10^4; for beef samples, 8.0×10^5 CFU/mL	[33]
	Antibody-coated paramagnetic beads	Fluorescence	Not reported	[34]
	Gold electrode	Impedance	Not reported	[35]
	Quartz crystals	QCM	23 CFU/mL and 53 CFU/mL in PBS and milk	[36]
	Quartz crystals	QCM-D	10^6 cells/mL	[37]
	AuNP–modified quartz crystals	QCM	1.2×10^4 CFU/mL and 1.2×10^2 CFU/mL, with and without AuNP modification, respectively	[38]
	CNT–modified gold microelectrode	LSV	10^5 and 10^3 CFU/mL for whole cells or lysates, respectively	[39]

TABLE 24.1 (continued)

Biosensing of Bacteria and Viruses at Different Detection Platforms

Organisms	Detection Platform	Detection Method	LOD	References
E. coli	Streptavidin-functionalized CdSe/ZnS core/shell quantum dots	Fluorescence	2.08×10^7 CFU/mL	[40]
	Graphene-modified quartz	FET	10 CFU/mL	[41]
	SPR-active film-coated glass chip	SPR	1.4×10^4 CFU/mL	[5]
	SPR microarray	SPR	<60 pM	[42]
S. aureus	Photomultiplier tube	Bioluminescence	Not reported	[32]
	SUT	Impedance	Not reported	[6]
	CdS quantum dots	SELS	Not reported	[43]
	SPR microarray	SPR	<60 pM	[42]
E. faecalis	SUT	Impedance	Not reported	[6]
P. aeruginosa	SUT	Impedance	Not reported	[6]
	Quartz crystals	QCM	1.5×10^2 CFU/mL	[44]
Clostridium leptum	PDMS microfluidic module	Fluorescence	Not reported	[2]
Ruminococcus obeum	PDMS microfluidic module	Fluorescence	Not reported	[2]
Eubacterium biforme	PDMS microfluidic module	Fluorescence	Not reported	[2]
Lactobacillus acidophilus	PDMS microfluidic module	Fluorescence	Not reported	[2]
Enterococcus faecium	PDMS microfluidic module	Fluorescence	Not reported	[2]
E. aerofaciens	PDMS microfluidic module	Fluorescence	Not reported	[2]
Bifidobacterium longum	PDMS microfluidic module	Fluorescence	Not reported	[2]

(*continued*)

TABLE 24.1 (continued)

Biosensing of Bacteria and Viruses at Different Detection Platforms

Organisms	Detection Platform	Detection Method	LOD	References
Bacteroides vulgatus	PDMS microfluidic module	Fluorescence	Not reported	[2]
Fusobacterium prausnitzii	PDMS microfluidic module	Fluorescence	Not reported	[2]
Sulfate reducing bacteria	ITO electrode	EIS	0.7×10^4 CFU/mL	[45]
A. hydrophila	Gold electrode	CV, CC, SWV	Not reported	[46]
	Gold evaporated quartz crystal	QCM	Not reported	[47]
Avian metapneumovirus	SWNT-modified Cr/Au-deposited silicon wafer	Fluorescence	100 $TCID_{50}$/mL	[48]
Dengue virus	AuNPs combined with magnetic separation	ICP/MS	80 zmol	[49]
Avian influenza virus	MWCNT/PPNWs/AuNPs–modified gold electrode	DPV, CV, EIS	4.3×10^{-13} M	[50]
Campylobacter	Quantum dot sandwich assay	ECL	2.5 CFU/mL in buffer, 10–250 CFU/mL in various food matrices	[51]
	SPR-active film-coated glass chip	SPR	1.1×10^5 CFU/mL	[5]
SEB	CNT–modified immunoassay	DPV	10 pg/mL	[52]
Legionella pneumophila	Optical fiber–modified gold layer	SPR	10 CFU/mL	[53]
Brucella	SPR microarray	SPR	<60 pM	[42]
P. ostreatus	Gold thin film chip	SPR	Not reported	[54]
HBV	PGE	DPV	74.8 fmole/mL in the 50 μL samples	[55]

TABLE 24.1 (continued)

Biosensing of Bacteria and Viruses at Different Detection Platforms

Organisms	Detection Platform	Detection Method	LOD	References
HBV	SWCNT/PGE	DPV, EIS	2.10×10^{-7} mol/L (210 nM)	[56]
	SWCNT–PVF$^+$/ PGE	DPV	32.46 µg/mL (5.14 µM)	[57]
	SWCNT– CHIT/PGE	DPV	13.25 µg/mL (2.09 µM)	[58]
	GO/PGE	DPV, EIS	9.06 µg/mL	[59]
	ZnP/PGE	DPV	11.7 µg/mL (1.85 µM)	[60]
	STR–SWCNT/ PGE	DPV, EIS	1.45×10^{-7} mol/L	[61]
	SnO$_2$–NP– PVF$^+$/PGE	DPV, EIS	1.82 µg/mL (279 nM)	[11]
	MWCNT/SPE	DPV	5.25 µg/mL (1.02 µM)	[62]
Microcystis spp.	MWCNT/SPE	DPV	96.33×10^{-9} mol/L	[63]

Abbreviations

PDMS, polydimethylsiloxane; SUT, sample under test; IDAM, interdigitated array microelectrode; ITO, indium tin oxide; PGE, pencil graphite electrode; AuNP, gold nanoparticle; SWCNT, single-walled carbon nanotube; MWCNT, multiwalled carbon nanotube; PPNW, polypyrrole nanowires; PVF$^+$, poly(vinylferrocenium); CHIT, chitosan; GO, graphene oxide; ZnP, zinc oxide nanoparticles; STR, streptavidin; SnO$_2$-NP, tin oxide nanoparticles; TCID$_{50}$/mL, 50% tissue culture infective dose; SEB, staphylococcal enterotoxin B; Ab/HRP ratio, antibody/horseradish peroxidase ratio.

Detection Methods

LSV, linear sweep voltammetry; DCT, direct-charge transfer; QCM, quartz crystal microbalance; QCM-D, quartz crystal microbalance with dissipation monitoring; BARDOT, bacteria rapid detection using optical scattering technology; EIS, electrochemical impedance spectroscopy; SPR, surface plasmon resonance; FET, field-effect transistors; ICP/MS, inductively coupled plasma/mass spectrometry; ECL, electrochemiluminescence; SELS, surface-enhanced light scattering.

There are several reports on electrochemical monitoring of bacteria and viruses [39–63]. The gold interdigitated microelectrodes were fabricated by García-Aljaro and collaborators for the detection of *Bacillus globigii* and they were used as the simulant of the threatening bioterrorism agent *B. anthracis*. [39]. In the study of García-Aljaro, the polypyrrole (PPy) nanowires were assembled onto the surface of microelectrodes to construct this chemiresistive biosensor. Linear sweep voltammetry measurements were then performed, and a linear correlation was obtained for spore concentrations ranging from 1 to 100 CFU (colony forming units)/mL by using a 30 min detection time [39]. Also, carbon nanotube (CNT)-based immunosensors were introduced by the same research group for detection of a pathogen; *E. coli* O157:H7 and a virus bacteriophage T7. The limit of detection (LOD) was determined as 10^5 CFU/mL, which equates a total of 10^3 CFU/chip in buffer and 10^3 CFU/mL, corresponding to 101 CFU/ chip in cell lysate for *E. coli* [39].

For the recognition and monitoring of marine pathogen sulfate-reducing bacteria (SRB), bacteria-mediated bioimprinted films were developed [45]. Indium tin oxide (ITO) electrode surface was modified with the chitosan doped with reduced graphene sheets. The surface characterization of modified electrode was done by using cyclic voltammetry (CV) and electrochemical impedance spectroscopy (EIS) techniques. LOD of this assay was reported as 0.7×10^4 CFU/mL.

An emerging foodborne hazard *Aeromonas hydrophila* was associated with a variety of virulence factors including production of cytotoxic enterotoxin aerolysin. Tichoniuk et al. reported a study about detection of *A. hydrophila* by using an electrochemical DNA biosensor [46]. The thiolated single-stranded DNA probe (ssDNA) was immobilized onto a gold electrode surface. The chronocoulometric quantitation of DNA was performed, and electroactive indicator, methylene blue (MB), signals were measured by using square wave voltammetry (SWV). The selectivity of the genosensor was also tested in the presence of noncomplementary sequence.

An electrochemical aptasensor has been developed to detect of avian influenza virus (AIV) H5N1 gene sequence. Gold working electrode surface was assembled with multiwalled carbon nanotubes (MWNTs), polypyrole nanowires (PPNWs), and gold nanoparticles (AuNPs). The voltammetric and impedimetric characterization were performed and LOD was reported as 4.3×10^{-13} M [50].

Erdem and her research group fabricated a label-free genomagnetic assay combined with disposable sensor technology for electrochemical monitoring of wild-type hepatitis B virus (HBV) DNA in polymerase chain reaction (PCR) amplicons [55]. LOD was found as 74.8 fmole/mL in the 50 μL samples by using DPV technique. There are several studies for detection of HBV by using disposable biosensor technology combined with modification of different nanomaterials by the group of Erdem, such as single-walled carbon nanotubes (SWCNTs) [56], SWCNT–poly(vinylferrocenium) (PVF$^+$) [57], SnO$_2$–NP–PVF$^+$ [11], SWCNT–chitosan (SWCNT–CHIT) [58], graphene oxide (GO) [59], zinc oxide nanoparticles (ZnO) [60], streptavidin-modified CNT (STR-CNT) [61], and MWNTs [62,63].

24.2.2 Immunosensors

Among the methods used to detect pathogenic bacteria and viruses, enzyme-linked immunosorbent assay (ELISA) is one of the most widely used technique. Chunglok and coworkers reported direct and sandwich assay ELISA immunosensor for detection of *Salmonella enterica* serovar and *Typhimurium* detection [29]. Antibody and horseradish peroxidase (HRP) were immobilized onto SWCNT surface. A typical ELISA yields a sensitivity of 10^6–10^7 CFU/mL. LOD was found to be 10^3 and 10^4 CFU/mL, respectively, for direct and sandwich ELISA, when Ab/HRP at 1:400 ratio was used. In terms of LOD, the results showed that 100 times sensitive detection platform was successfully fabricated by using SWCNTs due to their enhanced surface area.

B. cereus is defined as a foodborne pathogen. An immunosensor was employed for determination of *B. cereus* by using direct-charge transfer (DCT) technique [26]. Polyaniline nanowires were used as the molecular electrical transducer. They were conjugated with polyclonal antibodies and secondary antibodies. Experimental conditions such as polyaniline types and concentrations were optimized. The biosensor sensitivity in pure cultures of *B. cereus* was found to be 10^1–10^2 CFU/mL.

24.2.3 Optical Biosensors

There have been several applications of optical biosensors developed for bacteria and virus detection in the literature [3,4,30,54].

A fiber-optic sensor for detection of the most common foodborne bacterial pathogens, *Listeria monocytogenes*, *E. coli* O157:H7, and *S. enterica*, was introduced by Ohk et al. [3]. Biotinylated polyclonal antibodies were modified with the streptavidin-coated optical waveguides and exposed to the bacterial suspensions or enriched food samples for 2 h. Pathogens were detected after the interaction between pathogens and Alexa Fluor 647–labeled monoclonal antibodies. Each pathogen (w100 CFU/25 g)-inoculated ready-to-eat beef, chicken, and turkey meats were enriched in SEL (*Salmonella*, *E. coli*, *Listeria*). The multipathogen selective detection was also performed with a LOD as 10^3 CFU/mL.

Banada et al. described a light-scattering label-free sensor that was capable of real-time detection and identification of colonies of multiple pathogens. *Listeria*, *Staphylococcus*, *Salmonella*, *Vibrio*, and *Escherichia* were distinguished with an accuracy of 90%–99% for samples derived from food or experimentally infected animal by using this method [4].

A surface plasmon resonance (SPR) biosensor chip was developed for the rapid detection of the oyster mushroom spherical virus (OMSV) in edible mushroom, *Pleurotus ostreatus*. An anti-OMSV monoclonal antibody (mAb) was generated and immobilized onto chip surface, then the biosensor chip detected OMSV in the extract a concentration-dependent manner [54].

Salmonella is a well-known contaminant in all foods including pasteurized milk. For the rapid detection of *S. typhimurium* in spiked milk, an SPR-based immunoassay was developed [30] with a LOD as 1.25×10^5 cells/mL.

24.2.4 Mass-Sensitive Biosensors

The quartz crystal microbalance (QCM) biosensor systems have become a promising tool for *in situ* measurements that have especially required on-line detection into biofluids [64]. Some representative studies including QCM-based biosensing systems for recognition of bacteria and viruses have been explained in this section [27,36,38,47,65–67].

A QCM immunosensor was described for detecting *E. coli* O157:H7 by Shen and collaborators [36]. Streptavidin–gold was combined with beacon immuno-magnetic nanoparticles in growth solution. The QCM immunosensor was fabricated with protein A from *Staphylococcus aureus* and monoclonal anti-*E. coli* O157:H7 antibody. The frequency shift was correlated to the bacterial concentration and LOD was calculated as 23 CFU/mL in phosphate-buffered saline and 53 CFU/mL in milk. They reported that the whole procedure took 4 h.

In another study, a sandwich assay was developed for detection of *E. coli* using Au nanoparticles–modified quartz crystals [38]. A linear correlation was obtained in the 10^2–10^6 CFU/mL concentration range and it was reported that the detectable concentration of PCR product was 1.2×10^2 CFU/mL. In addition, *E. coli* detection was performed in food samples such as apple juice, milk, and ground beef, but this QCM-based sensor response was found less sensitive compared to the one obtained in phosphate buffer.

Aeromonas spp. is a widespread contaminant in environmental sources such as freshwater, sewage, and wastewater. Moreover, the bacterium isn't affected from standard chlorination, thus recolonize in the water [65]. *Aeromonas* spp. have also been found in meat sausages, seafood, cheese, milk, ice cream, and grocery store products. Due to all these reasons, there has been an urgent need to rapidly and sensitively identify *Aeromonas* spp. [47]. Tombelli et al. reported a QCM biosensor for sensitive detection of *A. hydrophila* in extracted cell and PCR product. The biotin-labeled capture DNA probe was immobilized onto streptavidin–gold–modified quartz surface and *A. hydrophila* was then successfully recognized even if in real samples.

B. anthracis is the cause of anthrax, a serious infection among livestock and human beings [66]. Dormant spores of *B. anthracis* can survive in harsh conditions, such as high temperatures, ultraviolet radiation, or high pressures [67]. Hao et al. [27] developed a QCM biosensor for sensitive detection of *B. anthracis*. Similar to other reports for detection of different organisms, DNA probes immobilized on AuNP-modified quartz crystals were used as sensor platform. Two different gene fragments, pag and Ba813, were selected for detection. The linear correlation between the frequency shift of the biosensor and logarithmic number of *B. anthracis* cell concentration was demonstrated from 3.5×10^2 to 3.5×10^7 CFU/mL of cells, for both types of gene fragments of pag and Ba813.

24.3 Conclusion

Biosensing strategies and applications for recognition and monitoring of bacteria and viruses bring several advantages such as practical, time-saving, cost-effective, and real-time analysis in medicine, food, and environment fields.

Moreover, more sensitive and selective detection can be achieved by using portable nanostructured chip and array systems due to the enhanced characteristic surface properties. Continued development in molecular biology, medicine, chemistry, and material science will lead to fabrication of microsensors that are capable of sensing of different bacteria and viruses named as contaminant factors.

Acknowledgment

A.E. would like to express her gratitude to the Turkish Academy of Sciences (TÜBA) as the associate member of TÜBA for its partial support.

References

1. M. Koopmans, E. Duizer, *Int. J. Food Sci. Microbiol.* 90 (2004) 23–41.
2. J.K. Ng, E.S. Selamat, W.T. Liu, *Biosens. Bioelectron.* 23 (2008) 803–810.
3. S.H. Ohk, A.K. Bhunia, *Food Microbiol.* 33(2) (2013) 166–171.
4. P.P. Banada, K. Huff, E. Bae, B. Rajwa, A. Aroonnual, B. Bayraktar, A. Adil, J.P. Robinson, E.D. Hirleman, A.K. Bhunia, *Biosens. Bioelectron.* 24 (2009) 1685–1692.
5. A.D. Taylor, J. Ladd, Q. Yu, S. Chen, J. Homola, S. Jiang, *Biosens. Bioelectron.* 22 (2006) 752–758.
6. M. Grossi, M. Lanzoni, A. Pompei, R. Lazzarini, D. Matteuzzi, B. Riccò, *Biosens. Bioelectron.* 26 (2010) 983–990.
7. E. Palecek, M. Bartosík, *Chem. Rev.* 112 (2012) 3427–3481.
8. J. Wang, *Electroanalysis* 17 (2005) 7–14.
9. A. Erdem, *Talanta* 74 (2007) 318–325.
10. A. Erdem, F. Sayar, H. Karadeniz, G. Guven, M. Ozsoz, E. Piskin, *Electroanalysis* 19 (2007) 798–804.
11. M. Muti, F. Kuralay, A. Erdem, S. Abaci, T. Yumak, A. Sinağ, *Talanta* 82 (2010) 1680–1686.
12. Kh. Ghanbari, S.Z. Bathaie, M.F. Mousavi, *Biosens. Bioelectron.* 23 (2008) 1825–1831.
13. X. Li, M. Cao, H. Zhang, L. Zhou, S. Cheng, J.L. Yao, L.J. Fan, *J. Colloid Interface Sci.* 382 (2012) 28–35.
14. S. Banerjee, D. Konwar, A. Kumar, *Sens. Actuat. B* 171–172 (2012) 924–931.
15. A. Erdem, H. Karadeniz, A. Caliskan, *Electroanalysis* 21 (2009) 464–471.
16. J.N. Coleman, U. Khan, W.J. Blau, Y.K. Gunko, *Carbon* 44 (2006) 1624–1652.
17. A. Erdem, P. Papakonstantinou, H. Murphy, *Anal. Chem.* 78 (2006) 6656–6659.
18. C. Li, E.T. Thostenson, T.W. Chou, *Compos. Sci. Technol.* 68 (2008) 1227–1249.
19. J. Wang, A.N. Kawde, M.R. Jan, *Biosens. Bioelectron.* 20 (2004) 995–1000.
20. F. Valentini, V. Biagiotti, C. Lete, G. Palleschi, J. Wang, *Sens. Actuat. B* 128 (2007) 326–333.

21. T.N. Lien, T.D. Lam, V.T.H. An, T.V. Hoang, D.T. Quang, D.Q. Khieu, T. Tsukahara, Y.H. Lee, J.S. Kim, *Talanta* 80 (2010) 1164–1169.
22. H. Huang, C. Tang, Y. Zeng, X. Yu, B. Liao, X. Xia, P. Yi, P.K. Chu, *Colloids Surf. B* 71 (2009) 96–101.
23. J.W. Liaw, C.H. Huang, *J. Quant. Spectrosc. Radiat. Transfer* 113 (2012) 1446–1453.
24. H. Huang, X. Liu, Y. Zeng, X. Yu, B. Liao, P. Yi, P.K. Chu, *Biomaterials* 30 (2009) 5622–5630.
25. C.G. Aljaro, M.A. Bangar, E. Baldrich, F.J. Munoz, A. Mulchandani, *Biosens. Bioelectron.* 25 (2010) 2309–2312.
26. S. Pal, E.C. Alocilja, F.P. Downes, *Biosens. Bioelectron.* 22 (2007) 2329–2336.
27. R.Z. Hao, H.B. Song, G.M. Zuo, R.F. Yang, H.P. Wei, D.B. Wang, Z.Q. Cui, Z.P. Zhang, Z.X. Cheng, X.E. Zhang, *Biosens. Bioelectron.* 26 (2011) 3398–3404.
28. L. Yang, *Talanta*, 74 (2008) 1621–1629.
29. W. Chunglok, D.K. Wuragil, S. Oaew, M. Somasundrum, W. Surareungchai, *Biosens. Bioelectron.* 26 (2011) 3584–3589.
30. S.D. Mazumdar, M. Hartmann, P. Kampfer, M. Keusgen, *Biosens. Bioelectron.* 22 (2007) 2040–2046.
31. D.A. Boehm, P.A. Gottlie, S.Z. Hua, *Sens. Actuat. B* 126 (2007) 508–514.
32. J. Luo, X. Liu, Q. Tian, W. Yue, J. Zeng, G. Chen, X. Cai, *Anal. Biochem.* 394 (2009) 1–6.
33. M. Varshney, Y. Li, *Biosens. Bioelectron.* 22 (2007) 2408–2414.
34. F.C. Dudak, İ.H. Boyacı, A. Jurkevica, M. Hossain, Z. Aquilar, H.B. Halsall, C.J. Seliskar, W.R. Heineman, *Anal. Bioanal. Chem.* 393 (2009) 949–956.
35. C. RoyChaudhuri, R. Dev Das, *Sens. Actuators A* 157 (2010) 280–289.
36. Z.Q. Shen, J.F. Wang, Z.G. Qiu, M. Jin, X.W. Wang, Z.L. Chen, J.W. Li, F.H. Cao, *Biosens. Bioelectron.* 26 (2011) 3376–3381.
37. C. Poitras, N. Tufenkji, *Biosens. Bioelectron.* 24 (2009) 2137–2142.
38. S.H. Chen, V.C.H. Wu, Y.C. Chuang, C.S. Lin, *J. Microbiol. Methods* 73 (2008) 7–17.
39. C.G. Aljaro, L.N. Cella, D.J. Shirale, M. Park, F.J. Munoz, M.V. Yates, A. Mulchandani, *Biosens. Bioelectron.* 26 (2010) 1437–1441.
40. M.A. Hahn, J.S. Tabb, T.D. Krauss, *Anal. Chem.* 77 (2005) 4861–4869.
41. Y. Huang, X. Dong, Y. Liu, L.J. Li, P. Chen, *J. Mater. Chem.* 21 (2011) 12358–12361.
42. M. Piliarik, L. Parova, J. Homola, *Biosens. Bioelectron.* 24 (2009) 1399–1404.
43. Z.D. Liu, S.F. Chen, C.Z. Huang, S.J. Zhen, Q.G. Liao, *Anal. Chim. Acta* 599 (2007) 279–286.
44. J. Cai, C. Yao, J. Xia, J. Wang, M. Chen, J. Huang, K. Chang, C. Liu, H. Pan, W. Fu, *Sens. Actuat. B* 155 (2011) 500–504.
45. P. Qi, Y. Wan, D. Zhang, *Biosens. Bioelectron.* 39 (2013) 282–288.
46. M. Tichoniuk, D. Gwiazdowska, M. Ligaj, M. Filipiak, *Biosens. Bioelectron.* 26 (2010) 1618–1623.
47. S. Tombelli, M. Mascini, C. Sacco, A.P.F. Turner, *Anal. Chim. Acta* 418 (2000) 1–9.
48. M. Bhattacharya, S. Hong, D. Lee, T. Cui, S.M. Goyal, *Sens. Actuat. B* 155 (2011) 67–74.
49. H. Hsu, W.H. Chen, T.K. Wu, Y.C. Sun, *J. Chromatog. A* 1218 (2011) 1795–1801.
50. X. Liu, Z. Cheng, H. Fan, S. Ai, R. Han, *Electrochim. Acta* 56 (2011) 6266–6270.
51. J.G. Bruno, T. Phillips, M.P. Carrillo, R. Crowell, *J. Fluoresc.* 19 (2009) 427–435.
52. D. Tang, J. Tang, B. Su, G. Chen, *Agrifood. Chem.* 58 (2010) 10824–10830.
53. H.Y. Lin, Y.C. Tsao, W.H. Tsai, Y.W. Yang, T.R. Yan, B.C. Sheu, *Sens. Actuat. A* 138 (2007) 299–305.

54. S.W. Kim, M.G. Kim, J. Kim, H.S. Lee, H.S. Ro, *J. Virol. Methods* 148 (2008) 120–124.
55. A. Erdem, D. Ozkan Ariksoysal, H. Karadeniz, P. Kara, A. Sengonul, A.A. Sayiner, M. Ozsoz, *Electrochem. Commun.* 7 (2005) 815–820.
56. A. Caliskan, A. Erdem, H. Karadeniz, *Electroanalysis* 19 (2009) 2116–2124.
57. M. Muti, F. Kuralay, A. Erdem, *Colloids Surf. B* 91 (2012) 77–83.
58. A. Erdem, M. Muti, H. Karadeniz, G. Congur, E. Canavar, *Colloids Surf. B* 95 (2012) 222–228.
59. M. Muti, S. Sharma, A. Erdem, P. Papakonstantinou, *Electroanalysis* 23 (2011) 272–279.
60. T. Yumak, F. Kuralay, M. Muti, A. Sınag, A. Erdem, S. Abacı, *Colloids Surf. B* 86 (2011) 397–403.
61. A. Erdem, P. Papakonstantinou, H. Murphy, M. McMullan, H. Karadeniz, S. Sharma, *Electroanalysis* 22 (2010) 611–617.
62. A. Erdem, H. Karadeniz, P.E. Canavar, G. Congur, *Electroanalysis* 24 (2012) 502–511.
63. H. Karadeniz, A. Erdem, A. Caliskan, *Electroanalysis* 20 (2008) 1932–1938.
64. H.J. Decker, E. Prohaska, S. Hauck, C. Kößlinger, H. Wolf, *Biosens. Bioelectron.* 14 (1999) 139–144.
65. M. Handfield, P. Simard, R. Letarte, *Appl. Environ. Microbiol.* 62 (1996) 3544–3549.
66. K.A. Edwards, H.A. Clancy, A.J. Baeumner, *Anal. Bioanal. Chem.* 384 (2006) 73–84.
67. L. Mizak, *Przegl. Epidemiol.* 58 (2004) 335–342.

25

Electrochemical DNA Biosensors in Food Safety: Determination of Phenolic Compounds and Antioxidant Capacity in Foods and Beverages

Gennady A. Evtugyn and Anna V. Porfireva

CONTENTS

25.1 Introduction

Antioxidants are chemical compounds able to hinder the oxidation of biological components and hence to prevent many natural and pathological processes related to oxidative stress in human beings. Radical reactions initiated by reactive oxygen species (ROS) are involved in aging processes and degenerative diseases, for example, atherosclerosis and cancer [1]. Supplementations of antioxidants to maintain health and to cure disease are an important strategy in the antioxidant therapy. Natural and synthetic antioxidants are widely used as biologically active additives to food and cosmetics. Some of them show antibacterial properties and prolong the storage period of the foodstuffs. Meanwhile, the excessive amounts of antioxidants provoke the generation of radical species due to pro-oxidative effect and hence could be dangerous [2]. The importance of the antioxidants for food quality assessment and human health calls for the elaboration of fast and reliable techniques for the evaluation of the antioxidant properties of the individual compounds. Total antioxidant activity (AOA), or capacity, of food and beverages is also discussed in terms of their relative benefits and dietary value.

The approaches to assessing antioxidant properties can be subdivided into two main groups. Referring to the foodstuffs, the evaluation of the protecting effect of artificial additives and the contribution of endogenous antioxidants and nutrients are distinguished. A subcategory involves measurement of the AOA in foods, particularly fruits, vegetables, and beverages, but with a view to predict dietary burden and *in vivo* activity [3]. The principles and the opportunities of selected chemical methods for the antioxidant activity (AOA) assessment in foodstuffs have been recently reviewed [4–6].

Both the methodology of the AOA estimation and the quantification of selected antioxidants remain a subject of intensive investigations. Many methods can be used for both tasks, that is, kinetic measurements of the radical quenching [7] and the quantification of the stable radicals by fluorescence [8] and spectrophotometry [9]. In *in vivo* assay, the products of lipid oxygenation and DNA damage were determined by HPLC [10]. Changes in the DNA structure initiated by oxidative treatment were also measured with electrophoresis [11]. Chemical methods for the AOA evaluation are reviewed in [1].

Being sensitive and compatible with other bioassay techniques, common instrumentation methods of the AOA assay are time and labor consuming and cannot be easily adapted for the preliminary testing of the foodstuffs especially outside the walls of laboratory. Therefore, there is an urgent demand in developing easier, faster, and more cost-effective methods compared to the traditional AOA assay. Biosensors are compact analytical devices utilizing biochemical reactions for detecting analytes of biological significance. They exert promising and very attractive opportunities in determination of antioxidant properties of the food additives and natural components of foodstuffs and beverages. In this chapter, DNA sensors for the detection of phenolic antioxidants and general AOA assessment are considered.

25.2 DNA Sensors for the Detection of Damaging Factors

The main concept of the use of the DNA sensors for the detection of antioxidant properties assumes controlled damage of native DNA molecules by various oxidants followed by the estimation of the protecting effect caused by individual antioxidants or extracts from plants or beverages. This protocol mimics specific interactions that take place in a human being under oxidative stress [12]. Certainly, this model does not take into account many other consequences of radical biochemical reactions, for example, lipid and lipoprotein oxidation and enzyme inhibition. Meanwhile, the detection of the DNA damage seems most important for detecting dangerous level of the influence of mutagenic or carcinogenic factors that may cause long-term consequences for the human health. In this chapter, the interactions of DNA with potential sources of radical particles will be first briefly considered and then the assemblies of appropriate DNA sensors for the detection of such factors discussed.

25.2.1 DNA Damage by ROS

There are many model systems applied for the generation of reactive oxidant species. Among others, the Fenton reaction is most frequently used:

$$Fe^{2+} + H_2O_2 \rightarrow Fe^{3+} + OH^- + OH^\bullet \qquad (25.1)$$

The hydroxyl radical $OH^\bullet$ is also involved in various chemical paths producing other reactive species:

$$OH^\bullet + H_2O_2 \rightarrow H_2O + HO_2^\bullet$$
$$HO_2^\bullet + Fe^{3+} \rightarrow Fe^{2+} + O_2 + H^+ \qquad (25.2)$$

Together with superoxide radical ($O_2^{-\bullet}$) and hydrogen peroxide itself, all of these particles that are able to perform reactions such as radical oxidation and hydrogen subtraction are called as ROS.

Besides Fe(II), Cu(II) and to a lesser extent Cr(II) can participate in the generation of the hydroxyl radical. The rate constant for the reaction is higher for copper than iron ion. However, the abundance of iron in biological systems compared to copper makes the first one the main source of $OH^\bullet$ radicals *in vivo* [13].

The $O_2^{-\bullet}$ radicals are easily formed from dissolved oxygen. They are considered as relatively inactive toward DNA [14]. Meanwhile, they promote releasing the Fe^{2+} ions from Fe–S clusters of proteins and from ferritin [15,16]. The activity of superoxide radicals is greatly enhanced by copper ions and especially by the metallocomplexes of transient metals able to partially intercalate the DNA helix.

Classical Fenton reaction (Equation 25.1) can be complicated by various factors both in living beings and model conditions of the laboratory experiment. Thus, some enzymes and cofactors control the reactivity of hydrogen peroxide and its concentration in the cells. Similar factors can enhance the production of reactive species and together with other sources participate in the DNA damage. A scheme of possible processes affecting the Fenton reaction is presented in Figure 25.1 [17]. The concentration of free H_2O_2 in the cells is controlled by the activity of superoxide dismutase (SOD), peroxidase, and catalase. The oxidoreductases produce hydrogen peroxide in the reactions of a substrate oxidation. The endogenous oxidants and reducers participate

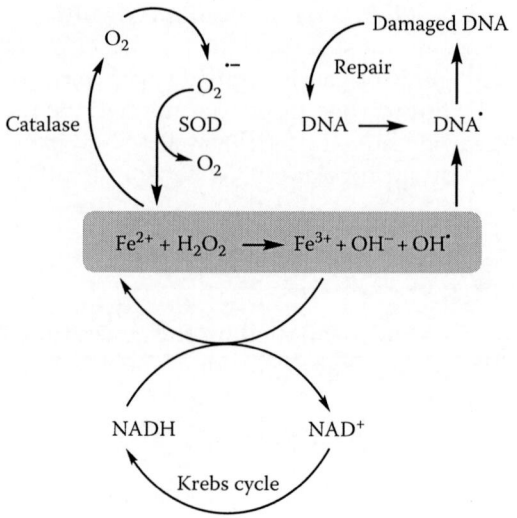

FIGURE 25.1
DNA damage in natural conditions caused by Fenton reaction.

in the following conversion of the DNA•, a product of primary hydrogen subtraction. Some of these reactions can repair the DNA molecules and the others result in severe damage of their structure. The reactivity of Fe^{2+} ions is highly affected by its complexation or involvement in protein molecules.

In laboratory conditions, chelate reagents, for example, EDTA, are used for stabilization of Fe^{2+} ions. Reducing agents, for example, ascorbic acid, can be added to the reaction media to regenerate Fe^{2+} ion [18]. The use of chelating agents differentiates the reactivity of the ROS. This is explained by possible formation of the adducts $[Fe-H_2O_2]^{2+}$ $(Fe-OOH^+)$ [19] or the Fe(IV) compounds [20]. The DNA molecules also accumulate the $Fe^{2+/3+}$ ions near its surface and hence increase possible damage resulted from the Fenton reaction.

Once formed near the DNA molecule, hydroxyl radicals attack both nucleic bases and sugar residues [21]. More than 20 modifications of nucleotides were determined by the GC/MS technique [22,23]. Among others, the guanine is the most easily oxidized 8-oxoguanine that is considered as a clinical biomarker for oxidative damage of DNA [24]. The simplified scheme of 8-oxoguanine formation is presented in the following.

$$\text{(structures: guanine} \xrightarrow{OH^\bullet} \text{intermediate} \rightarrow \text{8-Hydroxyguanine} \rightarrow \text{8-Oxoguanine)}$$

8-Hydroxyguanine   8-Oxoguanine

(25.3)

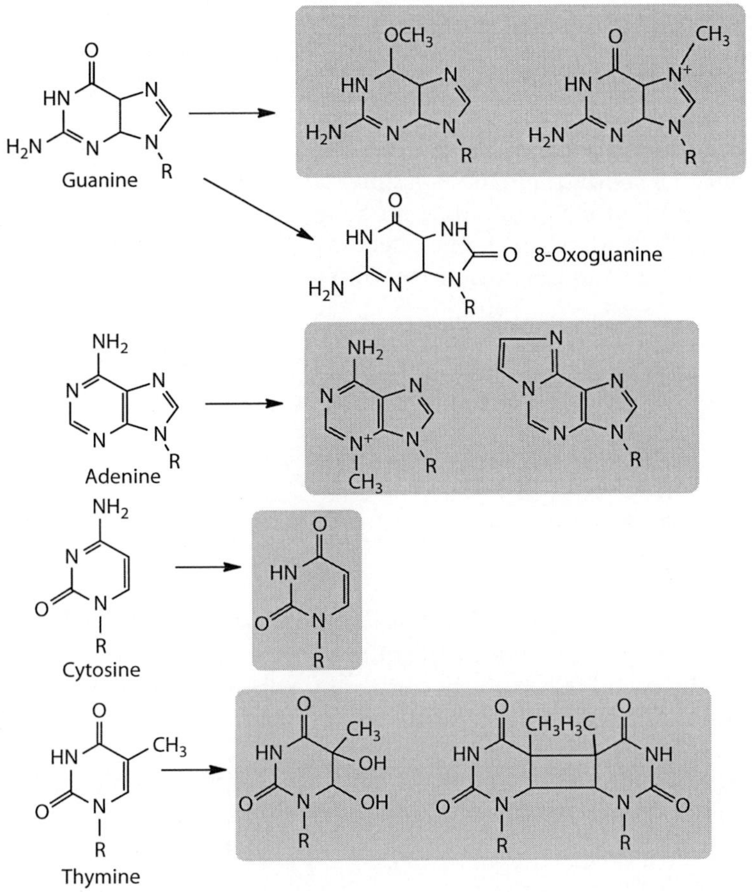

FIGURE 25.2
The products of oxidative degradation of nucleic bases initiated by ROS.

Under reducing conditions, 8-oxoguanine is converted to 2,6-diamino-5-formamido-4-hydroxypyridine by the imidazole ring opening [25]. The products of the oxidation of other nucleic bases are presented in Figure 25.2 [26].

The Fenton reaction also initiates the DNA cleavage [27]. The mechanism proposed involves the attack of the OH• to the 4-C atom of the sugar moiety followed by addition of molecular oxygen and formation of a peroxyl radical. After that, successive rearranges lead to sugar ring cleavage followed by β-elimination. A propenal base is released, and two remaining DNA strands bearing 3'-phosphoglycolate and 5'-phosphate terminal groups, respectively, are formed. The 3'-phosphoglycolate terminus is a marker of the oxidative DNA cleavage [28]. DNA cleavage becomes predominant for the Fenton-like processes involving Cu^{2+}/H_2O_2 system. The reaction is slow enough and the products of the DNA cleavage were monitored by electrophoresis.

Among them, common lesions are apurinic sites resulted from the hydro-lysis of the purine *N*-glycosidic bonds linking the nucleic base to deoxyri-bose, strand breaks (interruptions of the DNA sugar–phosphate backbone), thymine glycol, and nucleic base deamination (e.g., uracil [Equation 25.4] or hypoxanthine). Some of these lesions can be caused by intermediates or by-products of cellular metabolism. Interaction of the DNA with ROS trans-forms supercoiled DNA in single-stranded form. This reaction was moni-tored by electrophoresis [29]. The effect is enhanced by some promoters, for example, Cu^{2+} ions in the presence of thiols or SOD [30,31].

$$\text{(25.4)}$$

25.2.2 Electrochemical DNA Sensors for the Detection of the DNA Damage

There are many strategies for electrochemical detection of the DNA damage. Not all of them are applied for the quantification of the protecting effect of antioxidants. Thus, the analysis of the DNA unwinding and the influence of the DNA damage on the hybridization efficiency are too complicated by various experimental factors, for example, nucleic bases distribution and length of the main chain. The terminal labels introduced in the DNA probe molecules are not so sensitive toward the ROS influence so that the sensitiv-ity of such an assay is insufficient for the aims of the DNA sensors declared. Of other approaches, the following should be considered:

- Monitoring changes in direct or mediated DNA oxidation signals
- Application of electrochemically active intercalators
- Detecting changes in the permeability or charge-transfer properties of the surface DNA layer

The strategies mentioned are based on universal mechanisms of the DNA interaction and charge transduction. Common DNA sensors employ double-stranded (ds-) DNA of different origins. The structure and average molar mass and nucleic base distribution along the primary chain affect the electrochemical behavior of the native and denaturized DNA molecules as well as their response toward genotoxicants [32]. This can be applied for detecting minor conformational changes resulted from the oxidative damage, for example, formation of supercoiled DNA or its unwinding [33]. However, this difference can be omitted in consideration of the DNA damage effects on the biosensor performance assuming rather large number of base pairs in the target DNA molecules.

The electrochemical activity of the DNA on the surface of mercury electrode has been known for about 50 years [34]. The electrode reactions and appropriate mechanisms of the oxidation of nucleic bases have been reviewed in [26,35–37].

Briefly, the DNA damage is mainly detected by the changes in the guanine signals, sometimes confirmed by appearance of the second peak attributed to 8-oxoguanine. Direct observation of guanine signal is possible only on mercury and silver amalgamated and carbonaceous electrodes and is complicated by high overvoltage so that the working potential of about 1 V is required [38]. The elimination of interferences with background current and the resolution of appropriate peaks on voltammograms are achieved by the differential pulse (DPV) or square-wave voltammetry (SWV). Stripping chronopotentiometry based on the recording of the differential curve reflecting the temporal dependence of the potential in galvanostatic regime has been proposed for this purpose [39]. The curves obtained are similar to those of DPV or differential direct current voltammetry with a symmetrical bell shaped peak corresponding to the electrode reaction.

The detection of the DNA damage by guanine signal assumes changes in the availability of the guanine residues in the DNA chain caused by conformational changes, DNA cleavage, and other damages in the reaction with genotoxicant. As in the case of hybridization detection, the guanine signal normally increases with the intensity of the external factor. This distinguishes the response from that observed in the intercalation. As was shown for the DNA physically adsorbed onto graphite electrode, incubation of the biosensor with the anthraquinone and naphthalene derivatives decreased the guanine oxidation current, whereas polychlorobiphenyls increased the peak currents by 30%–40% due to partial unwinding of the DNA helix [39,40]. The relationship between the shift of the guanine signal and the mutagenic effect was confirmed by correlation of the results with those obtained by microbial toxicity tests (*Mutatox*™, Toxalert™, *umu* test) performed for wastewaters and soil extracts [41,42]. The list of other DNA damaging compounds detected by guanine signal includes *s*-triazine herbicides [43], heavy metal ions [44], niclosamides [45], and styrene oxide [46].

The influence of the DNA origin on the detection of damaging factor has been confirmed by the investigation of the interactions of single- and ds-DNA from calf thymus and salmon sperm with benz[a]anthracene and phenanthrene [47]. The guanine oxidation current was most sensitive toward the toxicants with single-stranded DNA from salmon sperm. The detection limits of selected polyaromatic hydrocarbons were in the range from 5 to 50 ng/mL.

The signal of the guanine oxidation can be amplified by the mediators of the electron transfer. Various Co(II), Cu(II), and Ru(II) complexes with flat aromatic heterocyclic ligands (phenanthroline (phen), 2,2′-bipyridine (bpy), etc.) are applied for this purpose [48,49]. The examples of such mediators are presented in Figure 25.3. The behavior of the complexes of transient

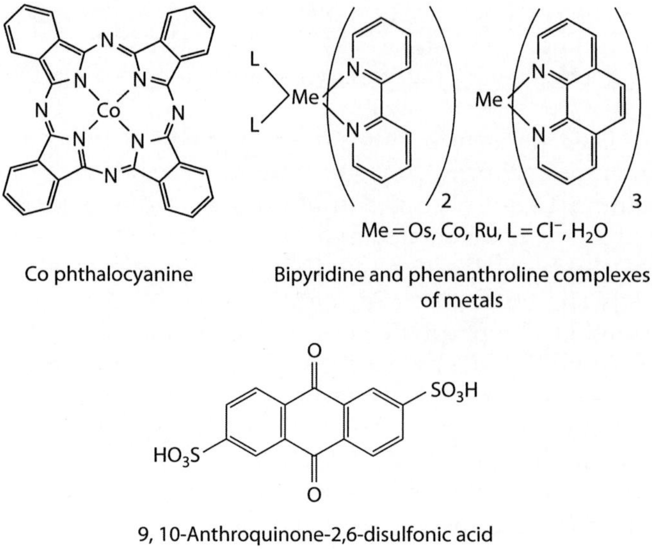
Co
Me
L
L
Me
2
3
Me = Os, Co, Ru, L = Cl⁻, H₂O
Co phthalocyanine
Bipyridine and phenanthroline complexes of metals
SO₃H
HO₃S
9, 10-Anthroquinone-2,6-disulfonic acid

FIGURE 25.3
Electrochemically active intercalators participating in mediated oxidation of guanine residues.

metals is complicated by their participation in the generation of hydroxyl radicals similarly to Fenton reaction as shown in the following for Cu(II) bpy complex [50]:

$$\begin{aligned}
&\mathrm{Cu(bpy)_2^{2+} + e^- \rightarrow Cu(bpy)_2^+}\\
&\mathrm{2Cu(bpy)_2^+ + 2H^+ + O_2 \rightarrow 2Cu(bpy)_2^{2+} + H_2O_2}\\
&\mathrm{H_2O_2 + Cu(bpy)_2^+ \rightarrow OH^- + OH^\bullet + Cu(bpy)_2^{2+}}
\end{aligned} \tag{25.5}$$

Similar processes were observed for bpy complexes of Ru(II) [51–53] and phen complexes of Co(III) [54]. All the complexes mentioned are accumulated on the ds-DNA due to partial intercalation in open circuit. This can be used for amplifying the signal.

The voltammetric signals of mediators make it possible to monitor the efficiency of the interactions between the ROS and the DNA molecules. Thus, the signal of Cu(I/II) ions decreases with the DNA damage because of the lower efficiency of the intercalation of phen and bpy ligands to the DNA helix. Contrary to that, the currents attributed to the reduced forms of Co(III) and Ru(II) complexes increase their signals while accumulated because the ions mentioned retain their redox activity while intercalated in the DNA moiety.

$[Co(bpy)_3]^{3+}$ complex was applied for monitoring the DNA damage caused by Fenton reaction performed in hydrophobic ionic liquid (1-butyl-3-imidazolium hexafluorophosphate) [55]. The comparison of the results

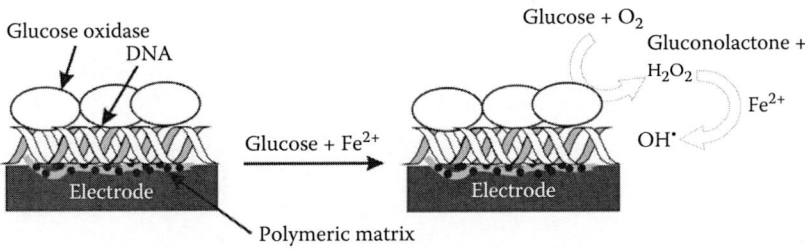

FIGURE 25.4
Measurement of the DNA damage with DNA–glucose oxidase hybrid biosensor.

obtained in the presence of the EDTA as chelating agent and that with no EDTA in the mixture made it possible to conclude that Fe(II) ion was first accumulated on the DNA molecules. After that, the oxidation by hydrogen peroxide via generation of the hydroxyl radical exerting the DNA damage resulted in decrease of the oxidation current attributed to the mediator used.

[Ru(bpy)$_3$]$^{2+}$ was applied as indicator of oxidative DNA damage caused by the ROS generating system consisting of glucose oxidase and Fe^{2+} ions (Figure 25.4) [56]. The addition of glucose and Fe^{2+} ions initiates the enzymatic reaction resulted in formation of H$_2$O$_2$ that generates the hydroxyl radical *in situ* in the DNA microenvironment on the sensor surface. The signal related to the mediated oxidation of guanine residues increases fourfold within the 15 min incubation time.

9,10-anthraquinone-2,6-disulfonic acid retains its redox activity while intercalated in the ds-DNA molecule. The reaction with Fenton reagent decreased the signal recorded by voltammetry due to electrostatic repelling of negatively charged intercalator from the negatively charged DNA immobilized in the chitosan layer onto the transducer surface [57].

The DNA damage can also result in the changes of permeability of the surface layer measured by electrochemical impedance spectroscopy (EIS) [58]. The resistance of the charge transfer increases due to a denser structure of the surface layer and electrostatic repelling of the ferricyanide ions used as redox probe. The changes in the melting curves of the DNA on the solid supports under polarization were also suggested to be used for the damage estimation [59].

25.3 Measurements of Phenolic Antioxidants and AOA with Electrochemical DNA Sensors

In food science, the antioxidant is defined as a substance in foods that when present at low concentrations compared to those of an oxidizable substrate significantly decreases or prevents the adverse effects of reactive

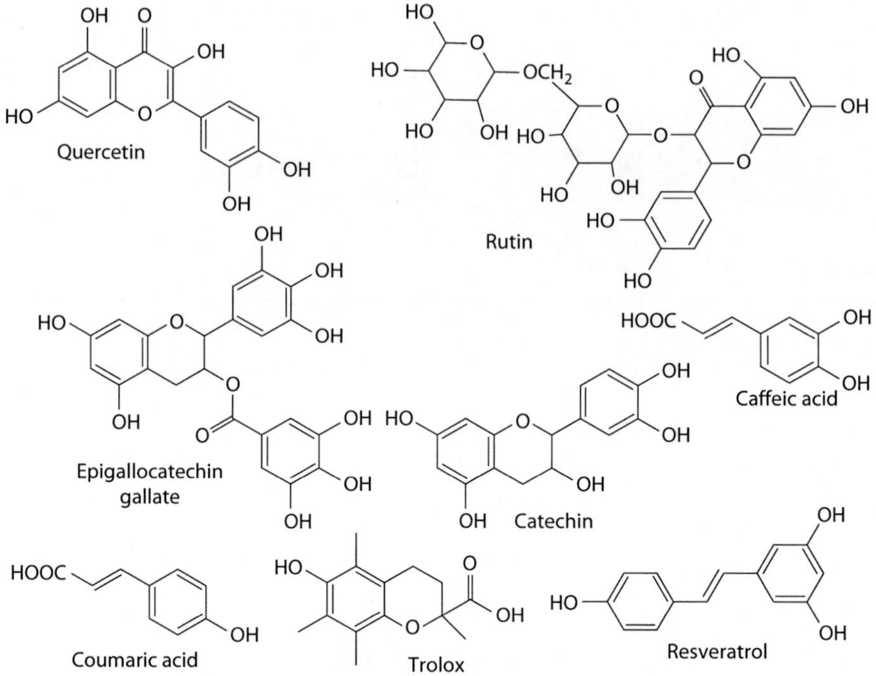

FIGURE 25.5
The examples of individual antioxidants determined with DNA sensors.

species, such as ROS, on normal physiological function in humans [60]. The examples of individual antioxidants tested with the electrochemical DNA sensors are given in Figure 25.5. The application of the DNA sensors assumes the ability of antioxidants to reduce the consequences of the DNA damage within the time comparable with the measurement stage. From general consideration, this does not assume the total estimation of prolonged protecting effect related to the DNA repair functions or the influence of by-products of the cell metabolism affected by the ROS. Meanwhile, the results obtained with the DNA sensors and ROS are usually well correlated with standard techniques elaborated for the AOA estimation (ABTS, DPPH, Trolox, or FRAP assay, see [60] for more details). The possible influence of other sources of radical particles, for example, NO, nitroxyl radicals, and organic hydroperoxides, on biosensor performance is less investigated, and there are no references to the application of the DNA sensors for the antioxidant determination with such genotoxic substances.

In a standard measurement protocol, the antioxidants are added to the working solution together with the Fenton reagent. The signals measured are compared with those obtained in the absence of protecting compounds. From the description of the DNA damage caused by ROS, it can be concluded that the reactions are mostly irreversible and this assumes single use of the biosensors for their detection. The DNA solution can be incubated with the toxic species and antioxidants prior to its immobilization. Although such a procedure complicates the manufacture of the DNA sensor especially in field conditions, it can improve the reproducibility of the response. In other cases, the reaction is performed on the surface of the transducer either in one pot by consecutive addition of an antioxidant, ROS, and a mediator or separately in the stages separated by washing and solution displacement. In the latter case, the antioxidant is first mixed with the Fenton reagent or another source of active species. This decreases the concentration of radical species contacting with the DNA molecules. The signal measurement is performed prior to and after such a mixing, and relative shift of the current is calculated as a measure of protecting effect expressed in the units of an antioxidant concentration. The accuracy of the estimation is commonly worse than 10 rel.% because of the multistage protocol and necessity to compare the signals obtained with several biosensors to average the influence of the damaging factor. This makes stricter the requirements to the robustness of the DNA sensors.

An improvement of metrological characteristics can be achieved by careful control of the reagent mixing and incubation stages in flow-through analysis [61]. In this work, ds-DNA from calf thymus was physically adsorbed onto screen-printed carbon electrode. Fenton reagent was mixed with the individual antioxidant solution (rutin or caffeic acid) and then the DNA damage was estimated in accordance with relative shift of the ferricyanide peak current recorded by direct current voltammetry. The quantification of the protecting effect was performed by relative decay of the signal attributed to the portion of the survived DNA molecules. The DNA sensor described makes it possible to detect 10^{-5}–10^{-9} M of the antioxidants. The relative estimation of the AOA of various types of tea fusions (black, green tea, and gingko) was also performed. Additional modification of the electrode surface with multiwalled carbon nanotubes extended the range of antioxidant concentrations detected and made it possible to detect down to 10^{-9} M rutin by the changes of the charge transfer resistance by EIS [58].

Taking into account single use of the DNA sensors for DNA damage and antioxidant detection, the simplest procedures have been proposed for the DNA immobilization. They include physical adsorption of native or denaturized DNA on the electrode performed in open circuit or under anodic polarization improving the attraction of negatively charged DNA molecules. The efficiency of such protocols is sufficient for several measurements within a day required for the estimation of the relative shifts of the signal prior to and after the contact of the biosensor with the ROS, an antioxidant, and indicator (if required).

Thus, selected flavonoids were tested with such DNA sensors [62]. Carbon paste electrode was first anodized at +1.7 V for 120 s. The signal of the DNA damage was recorded with $[Co(phen)_3]^{3+}$ indicator in DPV mode after 120 s accumulation of indicator in open circuit. The DNA sensor was treated with $[Cu(phen)_2]^{2+}$/ascorbic acid/H_2O_2 mixture able to cleave the DNA strand. The difference in the peak current measured for two identical sensors that were treated with cleavage reagent and that mixed with an antioxidant was calculated as a measure of the protecting effect. Changes in the biosensor signal were attributed to the DNA cleavage and following decrease in the accumulation of an indicator ion. Quercetin, rutin, catechin, and epigallocatechin gallate were tested as model antioxidants. According to the maximal signals reached, the AOA of flavonoids under study decreases in the range: rutin > quercetin > epigallocatechin gallate >> catechin.

The same flavonoids were investigated with screen-printed carbon electrode modified with calf thymus DNA and $[Cu(phen)_2]^{2+}$/ascorbic acid/O_2 scavenging system [63]. Besides the estimation of the protecting effect of flavonoids, their pro-oxidative effect was established. In the presence of dissolved oxygen, the flavonoids provoke the formation of superoxide radical interacting with the DNA molecules. The pro-oxidative effect depends on the reduction strength and changes in order: quercetin > rutin > epigallocatechin gallate > catechin.

Protecting effect of the yeast polysaccharides has been explored using DNA sensor with calf thymus DNA immobilized on screen-printed carbon electrode [64]. Fenton reagent and $[Cu(phen)_2]^{2+}$/H_2O_2 system were used as model toxicants and $[Co(phen)_3]^{3+}$ ion as indicator of the DNA damage. The signal was measured as described earlier [62]. The detection depended on the efficiency of the damaging factors. Thus, no effect was observed in the absence of ascorbic acid added to the Fenton solution to recover the Fe^{2+} ion necessary for the hydroxyl radical generation. For $[Cu(phen)_2]^{2+}$ ion, the polarization of the electrode was suggested for the same purpose. The effect of antioxidants increased with their concentration and reached saturation at about 1–2 mg/mL level. In accordance with maximal signals reached, the following range of activity was established: mannan (C. *krusei*) > extracellular glucomannan (C. *utilis*) > mannan (C. *albicans*) >glucomannan (C. *utilis*). The effect of fungal polysaccharides was expressed in the concentration of the standard antioxidant, that is, Trolox. In spectrophotometric measurements, 10–360 nM concentrations were obtained for different polysaccharides under study.

Herbal extracts from *Peumus boldus*, *Baccharis genistelloides*, *Cymbopogon citratus*, *Foeniculum vulgare*, *Mentha piperita*, and *Camellia sinensis* were tested with the DNA sensor based on screen-printed carbon electrode modified with physically adsorbed ds-DNA from calf thymus [65]. Relative shift of the guanine oxidation current was calculated as a measure of AOA. The radical scavenging of the extracts was independently monitored spectrophotometrically by the DPPH (1,1-diphenyl-2-picrylhydrazyl) assay [66]. This stable radical has maximal absorbance at 517 nm and is converted into a colorless product in the reaction with an antioxidant (Equation 25.6).

$$(25.6)$$

Most of the extracts tested showed AOA comparable to 0.1 mM Trolox as a standard antioxidant. The hierarchy of AOA slightly differed for DNA sensor and DPPH assay. DNA sensor: *B. genistelloides* > *P. boldus* > *F. vulgare* > *C. citratus* > *C. sinensis* > *M. piperita*. DPPH assay: *P. boldus* > *B. genistelloides* > *C. sinensis* > *M. piperita* > *F. vulgare* > *C. citratus*.

The AOA of aqueous extracts of lemon balm (*Melissa officinalis* L.), oregano (*Origanum vulgare* L.), thyme (*Thymus vulgaris* L.), and agrimony (*Agrimonia eupatoria* L.) was determined by DPPH assay and DNA sensor with calf thymus DNA immobilized on the screen-printed carbon electrode [67,68]. In parallel, the total amount of polyphenols was determined in the same extracts by Folin–Ciocalteu agent. The latter one was expressed in the concentration units of tannic acid and was equal to 1532 ± 23, 647 ± 21, 620 ± 20, and 601 ± 27 mg per mL of extract for oregano, lemon balm, thyme, and agrimony, respectively. In blank experiments, the signal of $[Co(phen)_3]^{3+}$ as indicator decreased tenfold, whereas the addition of the extracts protected up to 70% of the DNA cleaved. In all the tests performed, the AOA of the extracts decreased in the range from oregano to lemon balm, thyme, and agrimony. The protecting effect of antioxidants was confirmed by investigation of the rapeseed oil mixed with the spices investigated.

In attempts to increase the specificity of the antioxidant assay, specific systems have been developed for both the generation of reactive species and their detection. Thus, the superoxide anion radical able to specifically oxidize adenine to diimine products has been obtained in the enzymatic reaction of xanthine oxidase (Equation 25.7) [69].

$$(25.7)$$

Voltammetric detection of the DNA sensor response was performed by addition of NADH/Ca^{2+} to the working solution. Electrocatalytic oxidation of NADH enhanced by the Ca presence was detected by appropriate current in DC mode at about 0.14 V. For the determination of AOA, the homo-oligonucleotide dA$_{21}$ was immobilized on the carbon paste electrode by physical adsorption. The electrocatalytic current of NADH oxidation was proportional to the concentration of a standard antioxidant, ascorbic acid, in the range from 10 to 100 μM. A lemon flavor and two different brands of lemon-flavored water samples were tested with the DNA sensor described. The protecting effect of antioxidants ranged from 33% to 67% of the current shift against blank measurement.

Similar results were obtained with glassy carbon electrode modified by adsorbed guanine and adenine nucleotides [70]. Superoxide radical was generated by xanthine–xanthine oxidase system. The peak currents recorded by SWV were attributed to the oxidation of the nucleic bases. In the presence of antioxidants, the peaks increased proportionally to the analyte concentration in the range from 0.1 to 5 mg/L. Ascorbic, gallic, coumaric, and caffeic acids as well as resveratrol were examined as potential standards for the quantification of the AOA value of beverages. A variety of flavors (lemon, tangerine, apple, strawberry, gooseberry, and lime) and flavored waters were tested and their AOA expressed in the concentration units of the individual antioxidants tested.

The UV-initiated DNA damage of ds-DNA adsorbed on the ITO electrode modified with TiO$_2$ nanoparticles has been employed as a model for estimation of the protecting effect of antioxidants [71]. Methylene blue, an electrochemically active intercalator, was used for the signal readout. Its current recorded in SWV mode decreased with the intensity of the DNA damage due to leaching from the DNA layer after irradiation. The generation of the ROS in the system described is presented in the following [72]:

$$TiO_2 + h\nu \rightarrow h_{vb}^{\bullet} + e_{CB}^{-}$$

$$h_{vb}^{\bullet} + H_2O \rightarrow OH^{\bullet} + H^{+}$$

$$h_{vb}^{\bullet} + 2OH_{(ads)}^{-} \rightarrow OH^{\bullet} + OH^{-} \qquad (25.8)$$

$$e_{CB}^{-} + O_2 \rightarrow O_2^{\bullet}$$

$$2O_2^{\bullet} + 2H_2O \rightarrow 2OH^{\bullet} + 2OH^{-} + O_2$$

where v_{vb} are holes created in the valence band of TiO_2 irradiated at 360 nm and e_{CB} are the electrons transferred in the conducting band of TiO_2 films when irradiated at 360 nm. The amounts of hydroxyl radicals generated were monitored independently by fluorescence quenching of terephthalic acid. The DNA sensor proposed was applied for the estimation of antioxidant properties of glutathione and gallic acid in millimolar range of their concentrations. The second-order rate constants were determined after 30 min irradiation and the slopes of calibration curves were compared for both antioxidants.

25.4 Conclusion

The estimation of the antioxidants and AOA of foodstuffs and beverages with DNA sensors offers good opportunities for the preliminary estimation of the aforementioned parameters required for the food safety assessment and dietary value assessment. Protecting effect toward the DNA damage performed in controlled conditions by standard toxicants, for example, Fenton reagent, makes it possible to limit the degree of the DNA damage and hence the necessity in protecting agents suppressing such effect. The concept was confirmed in the texting of selected antioxidants representing the flavonoid family and some simple phenols and ascorbic acid. Herbal extracts rich with the antioxidants were also examined to show the prospects of the biosensors in this area. Nevertheless, some of the methodology problems remain and call to be solved prior to real samples testing.

ROS generation system: Although Fenton reaction is well elaborated, there are some drawbacks, for example, the uncertainty in the rate of the ROS release and the dependence of their reactivity on the microenvironment. Accurate release of the H_2O_2 in glucose oxidase reaction [56] might be an example of successful attempt, but it requires additional experiments to confirm its applicability for antioxidant testing. The problem is that some of the phenolic antioxidants can mediate the enzymatic reaction and hence decrease its activity toward hydroxyl radical formed. The kinetic investigations performed in the case of traditional AOA assays are necessary to control the efficiency of similar interactions with the DNA sensors within the whole incubation period.

Quantification of the results obtained: This can be referred not only to the DNA sensors described but to the AOA concept *per se*. Numerous methods suggested and applied for this parameter utilize the reducers different in their strength and accessibility for oxidant species. Regarding text systems, the estimation of redox species affecting the electron exchange is mostly applied, whereas the hydrogen subtraction is of lesser attention. Even in the framework of traditional quantification systems, only few works provide direct comparison of the results obtained with the DNA sensors and conventional assay techniques. In other approaches, relative shift of the response expressed as percentage of maximal value is mostly used together with a concept of a standard antioxidant (Trolox or ascorbic acid). This is quite convenient if various sources of antioxidants are compared but do not allow comparing the AOA if the measurements were performed in different conditions. Besides the difference in the conditions of the ROS generation and signal readout including the concentrations of ROS generating systems, pH and incubation stages should be taken into account.

Selectivity of the response: There are no attempts to distinguish the signals that can be referred to individual compounds of complex mixtures of antioxidants, for example, plant extracts or juices. In general, the protecting effect can be controlled by various parameters, for example, the nature of indicator system, the structure of the DNA, or the way of the ROS generation. In practice, all of these factors affect the changes in the target interactions that are detected in a similar manner and hence that cannot be used for the signal differentiation. The protocol of the measurement does not allow recording the kinetics of the ROS–antioxidant measurement because the DNA sensor is put into contact with reaction media after the finish of the reactions. As in the case of many other biochemical reactions with low selectivity utilized in biosensor (inhibition by pesticides, enzymatic conversion of readily oxidized phenolics, etc.), multisensory approach can be predicted to be applied in the nearest future. Variations in the DNA sensing layer or indicators used for signal readout can contribute to this problem even though the difficulties related to insufficient robustness of such biosensors and multistage protocol of the response are expected. Together with careful control of the ROS generation, such approaches can offer new opportunities for the application of such DNA sensors in food industry and medicine.

Acknowledgments

GE acknowledges the financial support of the Russian Ministry for Science and Education (Contract No. 6.740.11.0496). AP announces the support of the Russian President stipendium program for young scientists and Russian Foundation for Basic Research (Grant No. 12-03-31737 for young scientists).

References

1. Z.-Q. Liu, *Chem. Rev.* 110 (2010) 5675–5691.
2. B. Halliwell, *Free Radical Res.* 25 (1996) 439–454.
3. S. Beutner, B. Bloedorn, S. Frixel, I. Hernandez Blanco, T. Hoffmann, H.-D. Martin, B. Mayer et al., *J. Sci. Food Agric.* 81 (2001) 559–568.

4. M. Antolovich, P.D. Prenzler, E. Ptsalides, S. McDonald, K. Robards, *Analyst* 127 (2002) 183–198.
5. J.W. Finley, A.-N. Kong, K.J. Hintze, E.H. Jeffery, L.L. Ji, X.G. Lei, *J. Agric. Food Chem.* 59 (2011) 6837–6846.
6. A. Karadag, B. Ozcelik, S. Saner, *Food Anal. Methods* 2 (2009) 41–60.
7. S. Kumari, R.P. Rastogi, K.L. Singh, S.P. Singh, R.P. Sinha, *EXCLI J.* 7 (2008) 44–62.
8. Y. Nomura, H. Fuchigami, H. Kii, Z. Feng, T. Nakamura, M. Kinjo, *Anal. Biochem.* 350 (2006) 196–201.
9. K. Thaipong, U. Boonprakob, K. Crosby, L. Cisneros-Zevallos, D.H. Byrne, *J. Food Comp. Anal.* 19 (2006) 669–675.
10. A. Collins, C. Gedik, N. Vaughan, S. Wood, A. White, J. Dubois, J.F. Rees et al., *Free Radical Biol. Med.* 34 (2003) 1089–1099.
11. C.R. Bertoncini, R. Meneghini, *Nucleic Acids Res.* 23 (1995) 2995–3002.
12. R. Ovádeková, J. Labuda, in: *Utilizing of Bio-Electrochemical and Mathematical Methods in Biological Research* (V. Adam, R. Kizek eds.). Research Signport, Reala, 2007, pp. 173–201.
13. H.B. Dunford, *Coord. Chem. Rev.* 233–234 (2002) 311–318.
14. B.H.J. Bielski, D.E. Cabelli, R.L. Aradi, A.B. Ross, *J. Phys. Chem. Ref. Data* 14 (1985) 1041–1100.
15. D.W. Reif, *Free Radical Biol. Med.* 12 (1992) 417–427.
16. D.H. Flint, J.F. Tuminello, M.H. Emptage, *J. Biol. Chem.* 268 (1993) 22369–22376.
17. E.S. Henle, S. Linn, *J. Biol. Chem.* 272 (1997) 19095–19098.
18. M.F. Barroso, N. de-los-Santos-Álvarez, C. Delerue-Matos, M.B.P.P. Oliveira, *Biosens. Bioelectron.* 30 (2011) 1–12.
19. I. Yamazaki, L.H. Piette, *J. Am. Chem. Soc.* 113 (1991) 7588–7593.
20. S. Goldstein, D. Meyerstein, G. Czapski, *Free Radical Biol. Med.* 15 (1993) 435–445.
21. C.J. Burrows, J.G. Muller, *Chem. Rev.* 98 (1998) 1109–1151.
22. T.G. England, A. Jenner, O.I. Aruoma, B. Halliwell, *Free Radical Res.* 29 (1998) 321–330.
23. M. Dizdaroglu, P. Jaruga, M. Birincioglu, H. Rodriguez, *Free Radical Biol. Med.* 32 (2002) 1102–1115.
24. European Standards Committee on Oxidation and DNA Damage, *Free Radical Res.* 32 (2000) 333–341.
25. H.C. Box, J.B. Dawidzik, E.E. Budzinski, *Free Radical Biol. Med.* 31 (2001) 856–868.
26. M. Fojta, in: *Electrochemistry of Nucleic Acids and Proteins—Towards Electrochemical Sensors for Genomics and Proteomics* (E. Paleček, F. Scheller, J. Wang eds.). Elsevier, Amsterdam, the Netherlands, 2005, Vol. 1, pp. 386–430.
27. A.P. Breen, J.A. Murphy, *Free Radical Biol. Med.* 18 (1995) 1033–1077.
28. R. Meneghini, *Free Radical Biol. Med.* 23 (1997) 783–792.
29. P. Johnson, L.I. Grossman, *Biochemistry* 16 (1977) 4217–4225.
30. C.J. Reed, K.T. Douglas, *Biochem. J.* 275 (1991) 601–608.
31. Y. Ohkuma, S. Kawanishi, *Arch. Biochem. Biophys.* 389 (2001) 49–56.
32. E. Palecek, E. in: *Topics in Bioelectrochemistry and Bioenergetics* (G. Milazzo ed.), Wiley, Chichester, U.K., 1983, pp. 65–155.
33. M. Fojta. V. Stankova, E.P alecek, P. Koscielniak, J. Mitas, *Talanta* 46 (1998) 155–161.
34. E. Palecek, *J. Mol. Biol.* 20 (1966) 263–281.
35. E. Palecek, *Talanta* 56 (2002) 809–819.

36. E. Palecek, M. Fojta, in: *Bioelectronics* (I. Wilner, E. Katz eds.), Wiley-VCH, Weinheim, Germany, 2005, pp. 127–192.

37. E. Paleček, F. Jelen, in: *Electrochemistry of Nucleic Acids and Proteins—Towards Electrochemical Sensors for Genomics and Proteomics* (E. Paleček, F. Scheller, J. Wang eds.). Elsevier, Amsterdam, the Netherlands, 2005, Vol.1, pp. 73–173.

38. R. Fadrna, B. Yosypchuk, M. Fojta, T. Navratil, L. Novotny, *Anal. Lett.* 37 (2004) 399–413.

39. M. Mascini, I. Palchetti, G. Marrazza, *Fresenius J. Anal. Chem.* 369 (2001) 15–22.

40. F. Lucarelli, I. Palchetti, G. Marazza, M. Macini, *Talanta* 56 (2002) 949–957.

41. F. Lucarelli, A. Kicela, I. Palchetti, G. Marrazza, M. Mascini, *Bioelectrochemistry* 58 (2002) 113–118.

42. I. Palchetti, M. Mascini, *Analyst* 133 (2008) 846–854.

43. A.M. Oliveira-Brett, L.A. da Silva, *Anal. Bioanal. Chem.* 373 (2002) 717–723.

44. Q. Zhang, P. Dai, Z. Yang, *Microchim. Acta* 173 (2011) 347–352.

45. F.C. Abreu, M.O.F. Goulart, A.M. Oliveira Brett, *Biosens. Bioelectron.* 17 (2002) 913–919.

46. K. Ramanathan, K. Rogers, *Sens. Actuat. B* 91 (2003) 205–210.

47. M. del Carlo, M. di Marcello, M. Perugini, V. Ponzielli, M. Sergi, M. Mascini, D. Compagnone, *Microchim. Acta* 163 (2008) 163–169.

48. J. Labuda, M. Buckova, M. Vanickova, J. Mattusch, R. Wennrich, *Electroanalysis* 11 (1999) 101–107.

49. J.F. Rusling, *Biosens. Bioelectron.* 20 (2004) 1022–1028.

50. Z.-S. Yang, Y.-L. Wang, Y.-Z. Zhang, *Electrochem. Commun.* 6 (2004) 158–163.

51. L. Zhou, J.F. Rusling, *Anal. Chem.* 73 (2001) 4780–4786.

52. A. Mugweru, J.F. Rusling, *Anal. Chem.* 74 (2002) 4044–4049.

53. B. Wang, J.F. Rusling, *Anal. Chem.* 75 (2003) 4229–4235.

54. J. Yang, Z. Zhang, J.F. Rusling, *Electroanalysis* 14 (2002) 1494–1500.

55. Y. Wang, H. Xiong, X. Zhang, S. Wang, *Sens. Actuat. B* 161 (2012) 274–278.

56. Y. Zu, H. Liu, Y. Zhang, N. Hu, *Electrochim. Acta* 54 (2009) 2706–2712.

57. Y. Liu, N. Hu, *Electroanalysis* 20 (2008) 2671–2676.

58. G. Ziyatdinova, J. Galandova, J. Labuda, *Int. J. Electrochem. Sci.* 3 (2008) 223–235.

59. H. Nasef, V. Beni, C.K. O'Sullivan, *Electrochem. Commun.* 12 (2010) 1030–1033.

60. D. Huang, B. Ou, R.L. Prior, *J. Agric. Food Chem.* 53 (2005) 1841–1856.

61. D. Šimková, E. Beinrohr, J. Labuda, *Acta Chim. Slovac.* 2 (2009) 129–138.

62. O. Korbut, M. Bučková, J. Labuda, P. Gründler, *Sensors* 3 (2003) 1–10.

63. J. Labuda, M. Bučková, L. Heilerová, S. Šilhár, I. Štepánek, *Anal. Bioanal. Chem.* 376 (2003) 168–173.

64. M. Bučková, J. Labuda, J. Sandula, L. Krizková, I. Štepánek, Z. Duracková, *Talanta* 56 (2002) 939–947.

65. L.D. Mello, S. Hernandez, G. Marrazza, M. Mascini, L.T. Kubota, *Biosens. Bioelectron.* 21 (2006) 1374–1382.

66. P. Molyneux, *Songklanakarin J. Sci. Technol.* 26 (2003) 211–219.

67. L'. Heilerová, M. Bučková, P. Tarapčík, S. Šilhár, J. Labuda, *Czech J. Food Sci.* 21 (2003) 78–84.

68. J. Labuda, M. Bučková, L. Heilerová, A. Čaniová-Žiaková, E. Brandšteterová, J. Mattusch, R. Wennrich, *Sensors* 2 (2002) 1–10.

69. M.F. Barroso, N. de-los-Santos-Álvarez, M.J. Lobo-Castañón, A.J. Miranda-Ordieres, C. Delerue-Matos, M.B.P.P. Oliveira, P. Tuñón-Blanco, *J. Electroanal. Chem.* 659 (2011) 43–49.

70. M.F. Barroso, C. Delerue-Matos, M.B.P.P. Oliveira, *Biosens. Bioelectron.* 26 (2011) 3748–3754.
71. J. Liu, C. Roussel, G. Lagger, P. Tacchini, H.H. Girault, *Anal. Chem.* 77 (2005) 7687–7694.
72. K. Nagaveni, M.S. Hegde, N. Ravishankar, G.N. Subbanna, G. Madrao, *Langmuir* 20 (2004) 2900–2907.

26

Biosensors in Quality of Meat Products

Theodoros Varzakas, Georgia-Paraskevi Nikoleli,
Nikolaos Tzamtzis, and Dimitrios P. Nikolelis

CONTENTS

26.1 Introduction

The role of the bioreceptor is typically to convert or accelerate the conversion of the analyte of interest into another chemical species and/or physical property that can be sensed and then transformed into an electrical signal by the transducer. The transducer could be an electrode such as a pH or dissolved oxygen probe. In an ideal situation, where the sample matrix is not too complex, the detection of an analyte with this device would be accomplished without pretreatment or the addition of any reagent. However, the lifetime, stability, reproducibility, and calibration requirements of the biosensor are influenced significantly by the chosen bioreceptors.

26.2 Amperometric Biosensors

An enzyme electrode usually consists of a layer of immobilized enzymes attached to an electrode material, such as gold, platinum, silver, copper, or carbon. The enzyme is chosen to catalyze a reaction that generates a product or consumes a coreactant and can be monitored electrochemically. In amperometric biosensor, a constant potential is applied and the redox current generated is measured. The current response provides a measure of the analyte concentration.

Xanthine, a product of adenine nucleotide degradation in animal tissues, accumulates following death and hence can be used as criteria to check freshness of meat in food industry. Fish meat freshness is very important in food industries to manufacture high-quality products.

A method is described for fabrication of an amperometric xanthine biosensor based on polyvinyl chloride (PVC) membrane-bound xanthine oxidase (XOD). The membrane-bound enzyme oxidizes xanthine into uric acid and H_2O_2, which is split into $2H^+ + O_2 + 2e^-$ at high potential (0.4 V) and measured as current (mA). Sensor showed optimum response within 30 s, at pH 7.0 and 35°C. A linear relationship was observed between current and xanthine concentration ranging from 0.025 to 0.4×10^{-6} M; Km for xanthine and Imax was 0.45×10^{-6} M and 0.002 mA, respectively.

Minimum detection limit of the biosensor was 2.5×10^{-8} M. Biosensor was utilized for determination of xanthine in fish meat and cow and buffalos milk. Biosensor was used 100 times over a period of 45 days with only 30% loss of initial activity, when stored at 4°C [1].

A method is described for construction of an amperometric xanthine biosensor based on graphite rod modified through adsorption of XOD [2]. Enzymatically produced H_2O_2 from xanthine was split into $2H^+ + O_2 + 2e^-$ at 0.6 V and the current was measured, which was directly proportional to xanthine concentration ranging from 1×10^{-7} to 6×10^{-7} M with a detection limit of 1×10^{-7} M. The biosensor exhibited optimum response within 35 s at pH 7.0 and 35°C. It was employed for determination of xanthine in tea leaves (0.9×10^{-5} to 2.5×10^{-5} mmol/g), coffee powder (3.2 μmol/g), and fish meat (90 mmol/g). The content of xanthine in fish meat increased 6.5 times with its storage at room temperature during 15 days. The enzyme electrode could be reused 200 times during the span of 30 days, when stored in reaction buffer at 4°C.

Poehlmann et al. [3] used an electrochemical (amperometric) portable flow-based biosensor for the detection of *Escherichia coli* in meat juice. They immobilized a specific nucleotide sequence to capture the amplified 16S ribosomal RNA that had been isolated from the bacteria using an RNeasy Mini kit. They then used nucleotide conjugated esterase 2 as a reporter enzyme specific for bacterial 16S ribosomal RNA to complete the sandwich. The addition of substrate caused a concentration-dependent signal that was linear between

10^2 and 10^7 CFU/mL, giving a limit of detection (LOD) of 500 CFU/mL in buffer, although the reported sample preparation, for meat juice, and the analysis time were much longer than those described by Chen et al. [4].

26.3 Sample Preparation

Another limitation to the use of portable biosensors, specific for food contaminant analysis, is the necessity of sample preparation. Portable biosensors are ideal for the analysis of liquid samples such as milk, water, and fruit juices for contaminants that have reasonably high permitted levels (e.g., 100 ng/mL for certain drug residues), but when more complicated sample preparation is required, for example, the homogenization of meat samples or the requirement of enrichment steps or other forms of sample concentration, then even the most portable biosensor will be restricted by the preanalysis sample preparation that must be performed in a laboratory [5].

26.3.1 Antibody Sandwich Approach

In sandwich assays, antibodies specific for types of bacteria interact with the bacteria in the sample and a secondary antibody conjugated to a signal-generating moiety, either by itself or with a substrate, can be used to produce a concentration-dependent signal.

This can be achieved using electrochemical biosensors or fiber-optic biosensors that detect fluorescence.

Valadez et al. [6] achieved a LOD of 102 CFU/mL for *Salmonella enterica* in egg and chicken after 6 h enrichment and 1.5 h on their flow-based, optical (fluorescence), fiber-optic biosensor when utilizing a fluorescently labeled secondary antibody. Ohk et al. [7] described an alternative, optical (fluorescence)-based, fiber-optic approach in which they used a capture antibody and a fluorescently labeled aptamer to sandwich *Listeria monocytogenes* with a LOD of 102 CFU/25 g in ready-to-eat meat samples following homogenization in a stomacher. An enrichment period of 18 h and 4 h biosensor detection is required. This is quicker than the microbiological method, but other biosensor methods are faster.

26.3.2 Novel Biological Components and Pathogen Detection

Several articles have emerged showing the use of novel biological components for the detection of bacterial pathogens. Banerjee and Bhunia [8] used mammalian cells (Ped-2E9), with specific sensitivities for pathogens in a nonflow portable optical (absorbance) biosensor that assessed cytotoxicity by measuring the color change caused by alkaline phosphatase activity. They detected

L. monocytogenes in ready-to-eat meats and rice at 10^3–10^4 CFU/mL following enrichment for 4–6 h. Using the mammalian cells, they distinguished pathogenic from nonpathogenic bacteria. The use of mammalian cells has greatly reduced the time of analysis in comparison with the time required for the traditional microbiological methods and allows the distinction between pathogenic and nonpathogenic bacteria. However, it has to overcome problems associated with keeping the cell line alive and free from contamination.

Shabani et al. [9] used bacteriophages for the direct detection of *E. coli* in an array-format electrochemical (conductometric) assay. They achieved a LOD of 10^4 CFU/mL. Using bacteriophages has several advantages in that they are ubiquitous in nature, show highly specific sensitivity for individual bacteria, and are cheaper to produce/use than antibodies [9]. The fact that they are ubiquitous in nature may also be a challenge.

Cheng et al. [10] and Luo et al. [11] both described experiments that measure bioluminescence caused by isolated bacterial ATP in milk or meat juice.

Luo et al. detected *E. coli* and *Staphylococcus aureus* in beef juice at 10^3 CFU/mL levels while employing a mechanical separation of nonbacterial ATP in their disposable optical (luminescence) biosensor. These ATP-based approaches can distinguish between viable and nonviable cells, give results similar to those of traditional plate count tests, do not require culturing steps, and show potential for online monitoring [10,11], although sample preparation is critical in that all living organisms produce ATP and so the specific bacterial ATP needs to be carefully isolated.

Bai et al. [12] reported the detection of 11 foodborne pathogens, in buffer and pork meat, using a microarray approach with biospecific DNA probes immobilized on a sensor surface. The PCR products were denatured and allowed to hybridize with the immobilized probes; thus, the primer region on the biotinylated PCR products was able to bind with their relevant probes. Antibiotin IgG conjugated to HRP was then passed over the surface to complete the sandwich, followed by substrate, which facilitated the detection of the PCR product as a concentration-dependent color change on the surface of the optical (absorbance) biosensor. Bai et al. suggested that real samples will require a long culturing step, 18–24 h, before PCR amplification is performed, but this is surely balanced by the fact that at least 11 pathogens can be detected in one sample run.

Huet et al. [13] took fluoroquinolone detection to another level when they developed [13] and validated [14] a flow-based optical (surface plasmon resonance [SPR]) competitive method that was capable of detecting 13 fluoroquinolones below their maximum residue limits (MRLs) in egg (LOD 1 ng/mL), fish (LOD 1.5 ng/mL), and poultry meat (LOD < 0.5 ng/mL), following buffer extraction. Instead of using serial flow channels, they worked with one channel and developed a very novel "bioactive" antibody. They cleverly engineered their immunogenic conjugate with two different haptens, allowing them to produce a highly cross-reactive antibody.

Multiplexing via selective raising of antibodies can also be observed in the work of Thompson et al. [15] and Connolly et al. [16]. An antibody was originally produced with cross-reactivity to five nitroimidazoles and was utilized in a competitive assay on a flow-based optical (SPR) biosensor to detect the five nitroimidazoles in chicken muscle, at levels below 1 µg/kg, following a solvent extraction, evaporation, and buffer reconstitution [16].

26.4 Biacore Kit

Chloramphenicol (CAP), an effective antibiotic against many microorganisms, is meanwhile banned in the EU for treatment of food-producing animals due to adverse health effects. The Institute for Reference Materials and Measurements (IRMM) developed a certified reference material (CRM) for CAP in pork, intended for validation and method performance verifications of analytical methods. The material will be certified using liquid chromatography–tandem mass spectrometry (LC–MS/MS) and gas chromatography–mass spectrometry (GC–MS) methods and has a target CAP level around the minimum required performance limit (MRPL) of 0.3 µg/kg. To prove that the material can be applied as a quality control tool for screening methods, a commutability study was conducted, involving five commercially available enzyme-linked immunosorbent assay kits and one biosensor assay (Biacore kit). Meat homogenates (cryo-milled wet tissue) with CAP concentrations around the MRPL and the candidate CRM (lyophilized powder) were measured by LC–MS/MS and GC–MS as well as the six screening methods. Pairwise method comparisons of results obtained for the two sample types showed that the CRM can successfully be applied as quality control (QC) sample to all six screening methods [17].

The study suggests that ERM-BB130 is sufficiently commutable with the investigated assays and that laboratories applying one of the investigated kits therefore benefit from using ERM-BB130 to demonstrate the correctness of their results. However, differences among the assays were observed either in the abundance of bias between screening and confirmatory LC and GC methods, the repeatability of test results, or goodness of fit between the methods.

In the case of the modified Biacore assay, the bias was quite consistent among meat samples, and also good consistency between results on meat and the CRM was obtained (relative bias around 10%).

26.5 Potentiometric Biosensors

Potentiometric sensors are based on measuring the potential of an electrochemical cell while drawing negligible or no current. Common examples are the glass Ph electrode and ion selective electrodes for ions such as K^+, Ca^{++}, Na^+, and Cl^-.

A potentiometric biosensor consists of a layer of immobilized enzymes on an electrode and the measurement of potentials at the working electrode is made with respect to a reference electrode. The rate of potential change, rather than steady-state potential values, is often used as the analytical signal for quantification of the substrate.

Potentiometric enzyme electrode for detection of hypoxanthine (Hx) in fish meat is described [18]. The sensor was developed by entrapment of XOD and ferrocene carboxylic acid (Fc) into polypyrrole (PPy) film during galvanostatic polymerization film formation. The responses for Hx were obtained in 0–05 M phosphate buffer (pH 7.1) at 0.0 mV vs Ag/AgCl (3 M KCl). The optimum conditions for the formation of PPy–XOD–Fc film include 0.4 M PPy, 6.2 U/mL XOD, 40 mM Fc, polymerization time of 200 s, and applied current density of 0–5 mA cm^{-2}. The sensor provides a linear response to Hx in concentration range of 5–20 μM, (r = 0.998) and was successfully used for determination of Hx in fish.

26.6 Meat Tenderness

Meat tenderness is one of the most important palatability characteristics for consumers. Besides genetic factors, meat tenderization is influenced by the nature of feeding, age of the animal, degree of stress prior to slaughter, carcass chilling, aging time, and cooking method [19]. Final tenderness is determined by the rate and extent of postmortem proteolysis of key myofibrillar proteins in the muscle. The calpain system (calpain-I, calpain-II, and calpastatin) is the principal contributor to postmortem proteolysis that is closely related to meat tenderness. Among the factors affecting the tenderness, post-rigor calpastatin activity has the largest effect (~40%) on aged beef longissimus muscle [20].

Recently, the development of an optical fiber [21] and a capillary-based biosensor for calpastatin detection in heated meat samples [22] have been reported.

An immunological capacitive biosensor for calpastatin was developed, optimized, and applied for the analysis of meat extract samples [23]. Anticalpastatin antibody was immobilized on a gold electrode modified with a self-assembled monolayer of mercaptoundecanoic acid and protein A from *S. aureus*, and the obtained immunosensor was inserted as the working electrode in an electrochemical cell of a flow injection system. The dynamic range of the sensor was 20–160 ng/mL calpastatin. The electrode could be regenerated and reused for more than 7 days with minimal reduction in sensitivity. For the analysis of real samples, the target analyte was extracted from the longissimus dorsi muscle from beef carcasses directly after slaughtering. The extract was analyzed both with the developed immunosensor and microtiter plate ELISA, and a good correlation was obtained. However, the immunosensor offers advantages of speed, simplicity, sensitivity, and possibility for miniaturization over conventional assays for calpastatin quantification.

26.7 Hypoxanthine Biosensors

The mechanism of amperometric Hx biosensor is based on the direct oxidation of H_2O_2 formed from the enzyme reaction or on O_2 consumption. However, the direct detection of H_2O_2 is usually accomplished by application of anodic potentials (greater than +0.6 V vs. SCE). In contrast, O_2 is usually reduced by application of cathodic potentials (less than 0.5 V vs. SCE) [24].

Several Hx amperometric biosensors have been reported based on the use of various immobilization strategies. In many Hx biosensors, XOD is immobilized in membranes and in Nafion to improve the selectivity of the electrodes. Sol–gel technique was alternatively used for the immobilization of XOD on a graphite–ceramic for Hx biosensor with or without benzyl viologen as a mediator [25].

Hanendez et al. [26] fabricated a Hx biosensor by employing XOD in soluble or immobilized form, in combination with an oxygen electrode and optimized the biosensor to determine the Hx content in pork meat at different postmortem times as a measure of meat freshness. The amperometric signal obtained was related to the oxygen consumed during oxidation of Hx in the soluble or immobilized enzyme. In both cases, a linear relationship between the signal and the Hx was obtained from 8 to 26 µmol L^{-1}. Also, Yano et al. [27] used Hx biosensor for evaluation of meat spoilage and the progress of aging.

Yano et al. [28] set up a chronoamperometric method for Hx and putrescine (Put) determination in meat in which the potential was stepped from 300 to 600 mV. A linear relationship was obtained between 5 and 60 nmol L^{-1} for Put and 0.05 and 1.0 µmol L^{-1} for Hx. The coefficient of variation was 0.75% for 20 nmol L^{-1} Put solution and 2.2% for a meat sample using the Put sensor, 1.09% for 0.25 µmol L^{-1} Hx solution, and 2.6% for a meat sample using the xanthine sensor. Yano et al. [29] also used xanthine sensor to analyze X and Hx in meat, while Cavalheiro and Brajter-Toth [30] used active graphite and carbon fiber surfaces produced by different mechanical/electrochemical methods of surface activation to investigate the amperometric determinations of X and Hx under physiologically relevant conditions.

Most Hx biosensors used different indexes [31] for the determination of fish or meat freshness index, such as Ko value suggested by Fujita et al. [32] to be the ratio of nonphosphorylated ATP metabolites, including xanthine, to the total ATP breakdown products.

26.8 Nanobiosensors

Nanoscaffolding and nanoquenching properties of thiol-capped gold nano-crystals (GNCs), covalently linked to fluorophore-labeled oligonucleotide through metal–sulfur bond, were extensively studied for decades to detect specific sequences and single-nucleotide mismatches.

A novel class of nanobiosensor was developed by integrating a 27-nucleotide *AluI* fragment of swine cytochrome b (cytb) gene to a 3 nm diameter citrate–tannate–coated gold nanoparticle (GNP). The biosensor detected 0.5% and 1% pork in raw and 2.5 h autoclaved pork–beef binary admixtures in a single step without any separation or washing [33]. The hybridization kinetics of the hybrid sensor was studied with synthetic and *AluI* digested real pork targets from moderate to extreme target concentrations and a sigmoidal relationship was found. Using the kinetic curve, a convenient method for quantifying and counting target DNA copy number was developed. The accuracy of the method was over 90% and 80% for raw and autoclaved pork–beef binary admixtures in the range of 5%–100% pork adulteration. The biosensor probe identified a target DNA sequence that was severalfold shorter than a typical PCR template. This offered the detection and quantitation of potential targets in highly processed or degraded samples where PCR amplification was not possible due to template crisis. The assay was a viable alternative approach of qPCR for detecting, quantifying, and counting copy number of shorter-size DNA sequences to address a wide ranging biological problem in food industry, diagnostic laboratories, and forensic medicine.

26.9 Use of Biosensors in Boar Taint Detection

Boar taint is a large problem in the pig husbandry industry. It occurs in meat from entire male pigs and makes it undesirable for sensitive consumers. Surgical castration has been long used to prevent consumers from experiencing taint in meat from male pigs, which is a large problem in the pig husbandry industry. Boar taint is an unpleasant off-odor and off-flavor described with sensory properties characterized as urine-like, animal-like, sweat-like, and fecal-like [34]. The major chemicals responsible for boar taint are skatole (3-methylindole) and androstenone (5α-androst-16-ene-3-one). Other than skatole and androstenone, indole is a third possible chemical involved.

One method for boar taint detection involves harnessing the chemical learning abilities of insect invertebrates for use as biosensors. Previous studies in the honeybee, *Apis mellifera* Linnaeus (Hymenoptera: Apidae), and the parasitoid wasp, *Leptopilina boulardi* Barbotin, Carton & Keiner-Pillault

(Hymenoptera: Figitidae), indicate that learning and reporting of compounds can be concentration dependent [35–37].

It was previously determined that *Microplitis croceipes* can be conditioned to respond to the three main boar taint compounds when presented either as individual odors or in a blend.

The effectiveness and reliability of detection were established when wasps were used as biosensors in a portable device called the "wasp hound" [38] previously developed to record their specific behavioral response [39].

The wasps', *M. croceipes*, ability to learn and respond to particular concentrations of the boar taint compounds, skatole, androstenone, and indole, was tested. Also tested was the wasps' ability to discriminate between known concentrations of indole, skatole, and androstenone in real boar fat samples at room temperature. The wasps were trained using associative learning by providing food-deprived wasps with sucrose–water in the presence of specific odor concentrations. Trained wasps' responses were tested to a range of concentrations of three compounds. The wasps showed unidirectional generalization of learned concentration responses, whereby the direction of concentration generalization was shown to be chemical dependent [40].

Through both positive (sucrose) and negative feeding experiences (water only) with varying compound concentrations, the wasps can also be conditioned to respond to concentrations exceeding a defined threshold, and they were successful in reporting low, medium, and high concentrations of indole, skatole, and androstenone in boar fat at room temperature. The need for threshold detection rather than simple detection of absence/ presence applies to many food quality issues, including the detection of spoilage or pest damage in crops or stored foods. The results of this study show that *M. croceipes'* odor learning and response is concentration dependent and the direction of concentration generalization is chemical dependent.

This is the first demonstration, showing that the direction of concentration generalization is odor dependent.

The ability of the parasitoids to learn and subsequently report specific concentrations of a particular chemical drastically extends their potential use as biosensors. It means that conditioned wasps can now also be used in those applications where threshold detection is required or alternatively in applications that require the detection of elevated concentrations against lower background levels [40].

For biosensors to be used in threshold detection, it is essential that the organism can be conditioned to the threshold concentration and selectively respond to this concentration as well as higher (or in certain instances lower) concentrations. The results of this study show that parasitoid olfactory learning is so sophisticated that both threshold detection and generalization either to higher or lower concentrations are indeed feasible.

26.10 Bioluminescent Bacterial Sensors

Recently, the development of an assay for the detection of tetracyclines in poultry based on a bioluminescent bacterial sensor was reported as an alternative for "classical" microbial screening methods [41]. The sensor bacterium is an *E. coli* strain harboring a plasmid containing a bacterial luciferase operon under the control of a tetracycline-sensitive repressor [42]. The induction of the lux operon, and subsequently the production of a bioluminescent signal, only occurs in the presence of tetracycline. The test is much faster than microbial inhibition tests, as results can be obtained after 3 h, and very inexpensive since it does not require any additional reagents.

Luminescent bacterial biosensors represent an attractive inexpensive, simple, and fast method for screening large numbers of samples. A previously developed cell-biosensor method was subjected to an evaluation study using over 300 routine poultry samples and the results were compared with a microbial inhibition test. The cell-biosensor assay yielded many more suspect samples, 10.2% versus 2% with the inhibition test, which all could be confirmed by LC–MS/MS [43]. Only one sample contained a concentration above the MRL of 100 µg kg^{-1}, while residue levels in most of the suspect samples were very low (<10 µg/kg). The method appeared to be specific and robust. Using an experimental setup comprising the analysis of a series of three sample dilutions allowed an appropriate cutoff for confirmatory analysis, limiting the number of samples and requiring further analysis to a minimum.

26.11 Surface Plasmon Resonance Biosensors

An SPR biosensor inhibition immunoassay for determination of ractopamine (Rac) residue in pork was constructed by immobilizing Rac derivative on the SPR-2004 biosensor chip [44].

Rac belongs to the phenolic group of beta-agonists. Rac was originally used as tocolytics, bronchodilators, and heart tonics in human and veterinary medicine. Subsequently, this compound may be used illegally as growth promoters to acquire economic interest similar to clenbuterol.

After extraction with perchloric acid, pork sample was cleaned by ethyl acetate and analyzed by SPR-2004 biosensor. The LOD was 0.6 µg/kg for pork sample. Recoveries of Rac were higher than 80% with relative standard deviations below 10%. Although the same pretreatment was applied for both the UPLC–MS/MS and the biosensor, the biosensor showed little matrix interference by constructing pure solution and matrix-match calibration curves. Compared with Biacore Q made by Biacore AB, SPR-2004 biosensor made by Chinese Academy of Sciences is of low cost and showed minimal matrix interferences. Accordingly, this biosensor was a promising screening instrument to be used for detection of Rac in supermarkets, food factories, and food regulatory organizations.

SPR biosensor attracted much greater attention lately since it is advantageous over the traditional immunoassays in the virtue of real-time detection, no fluorophore labeling, and minimal matrix effect, compared with conventional immunoassay. It has been used in drug research, food analysis, environmental monitoring, and many other fields [45,46].

Previous studies [47,48] employed Biacore Q SPR biosensor and CM5 sensor chips from Biacore AB (Uppsala, Sweden).

An optical SPR biosensor was sensitive to the presence of *S. typhimurium* in chicken carcass [49]. The Spreeta biosensor kits were used to detect *S. typhimurium* on chicken carcass successfully. A taste sensor like electronic tongue or biosensors was used to basically "taste" the object and differentiated one object from the other with different taste sensor signatures. The SPR biosensor has potential for use in rapid, real-time detection and identification of bacteria and to study the interaction of organisms with different antisera or other molecular species. The selectivity of the SPR biosensor was assayed using a series of antibody concentrations and dilution series of the organism. The SPR biosensor showed promising results to detect the existence of *S. typhimurium* at 1×10^6 CFU/mL. Initial results show that the SPR biosensor has the potential for its application in pathogenic bacteria monitoring. However, more tests need to be done to confirm the detection limitation.

26.12 Dioxin Detection in Meats Using Biosensors

Laschi et al. [50] successfully developed a preliminary disposable electrochemical immunosensor for detection of non-dioxin-like polychlorinated biphenyls (non-DL-PCBs) in ruminant milk, adipose tissue, and meat extracts. These authors used an electrochemical signal as a transducer. An accelerated solvent extractor (ASE) was used for sample extraction. Their results demonstrate that a higher sensitivity of the sensing element to the specific antigen (PCB 28) was observed compared to other congeners.

The following criteria are generally used to assess the potential and limitations of sustainability indicators and applied to assess potential and limitations of methods available to detect contamination of samples with dioxins and DL-PCBs along the food chain of milk, eggs, and meat: (1) validity, (2) simplicity, (3) sensitivity (4) relevance, and (5) economic and technical feasibility.

Dioxins and DL-PCBs are hazardous toxic, ubiquitous, and persistent chemical compounds, which can enter the food chain and accumulate up to higher trophic levels. Their determination requires sophisticated methods, expensive facilities and instruments, well-trained personnel, and expensive chemical reagents. Ideally, real-time monitoring using rapid detection methods should be applied to detect possible contamination along the food chain in order to prevent human exposure. Sensor technology may be promising in this respect. This review gives the state of the art for detecting possible contamination with dioxins and DL-PCBs along the food chain of animal-source foods [51]. The main detection methods applied (i.e., high-resolution gas chromatography combined with high-resolution mass spectrometry (HRGC/HRMS) and the chemical-activated luciferase gene expression method (CALUX bioassay)) each have their limitations. Biosensors for detecting dioxins and related compounds, although still under development, show potential to overcome these limitations. Immunosensors and biomimetic-based biosensors potentially offer increased selectivity and sensitivity for dioxin and DL-PCB detection, while whole cell–based biosensors present interpretable biological results. The main shortcoming of current biosensors, however, is their detection level: this may be insufficient as limits for dioxins and DL-PCBs for food and feedstuffs are in per gram (pg) level. In addition, these contaminants are normally present in fat, a difficult matrix for biosensor detection. Therefore, simple and efficient extraction and cleanup procedures are required that may enable biosensors to detect dioxins and DL-PCBs contamination along the food chain.

References

1. C.S. Pundir, R. Devi, J. Narang, S. Singh, J. Nehra, S. Chaudhry, *J. Food Biochem.* 36 (2012) 21–27.
2. R. Devi, J. Narang, S. Yadav, C.S. Pundir, *J. Anal. Chem.* 67 (2012) 273–277.
3. C. Poehlmann, Y. Wang, M. Humenik, B. Heidenreich, M. Gareis, M. Sprinzl, *Biosens. Bioelectron.* 24 (2009) 2766–2771.
4. S.H. Chen, V.C.H. Wu, Y.C. Chuang, C.S. Lin, *J. Microbiol. Meth.* 73 (2008) 7–17.
5. T.F. McGrath, C.T. Elliott, T.L. Fodey, *Anal. Bioanal. Chem.* 403 (2012) 75–92.
6. A.M. Valadez, C.A. Lana, S.-I. Tu, M.T. Morgan, A.K. Bhunia, *Sensors* 9 (2009) 5810–5824.
7. S.H. Ohk, O.K. Koo, T. Sen, C.M. Yamamoto, A.K. Bhunia, *J. Appl. Microbiol.* 109 (2010) 808–817.
8. P. Banerjee, A.K. Bhunia, *Biosens. Bioelectron.* 26 (2010) 99–106.
9. A. Shabani, M. Zourob, B. Allain, C.A. Marquette, M.F. Lawrence, R. Mandeville, *Anal. Chem.* 80 (2008) 9475–9482.
10. Y. Cheng, Y. Liu, J. Huang, K. Li, W. Zhang, Y. Xian, L. Jin, *Talanta* 77 (2009) 1332–1336.
11. J. Luo, X. Liu, Q. Tian, W. Yue, J. Zeng, G. Chen, X. Cai, *Anal. Biochem.* 394 (2009) 1–6.

12. S. Bai, J. Zhao, Y. Zhang, W. Huang, S. Xu, H. Chen, L.M. Fan, Y. Chen, X.W. Deng, *Appl. Microbiol. Biotechnol.* 86 (2010) 983–990.
13. A.C. Huet, C. Charlier, G. Singh, S.B. Godefroy, J. Leivo, M. Vehniäinen, M.W. Nielen, S. Weigel, P. Delahaut, *Anal. Chim. Acta* 623 (2008) 195–203.
14. A.C. Huet, C. Charlier, S. Weigel, S.B. Godefroy, P. Delahaut, *Food Addit. Contam. A* 26 (2009) 1341–1347.
15. C.S. Thompson, I.M. Traynor, T.L. Fodey, S.R. Crooks, *Anal. Chim. Acta* 637 (2009) 259–264.
16. L. Connolly, C.S. Thompson, S.A. Haughey, I.M. Traynor, S. Tittlemier, C.T. Elliott, *Anal. Chim. Acta* 598 (2007) 155–161.
17. R. Zeleny, H. Emteborg, H. Schimmel, *Anal. Bioanal. Chem.* 398 (2010) 1457–1465.
18. A.T. Lawal, S.B. Adeloju, *Food Chem.* 135 (2012) 2982–2987.
19. D.M. Wulf, J.B. Morgan, J.D. Tatum, G.C. Smith, *J. Anim. Sci.* 74 (1996) 569–576.
20. S.D. Shackelford, M. Koohmaraie, L.V. Cundiff, K.E. Gregory, G.A. Rohrer, J.W. Savell, *J. Anim. Sci.* 72 (1994) 857–863.
21. C.L. Bratcher, S.A. Grant, R.C. Stringer, C.L. Lorenzen, *Biosens. Bioelectron.* 23 (2008) 1429–1434.
22. C.L. Bratcher, S.A. Grant, J.T. Vassalli, C.L. Lorenzen, *Biosens. Bioelectron.* 23 (2008) 1674–1679.
23. K. Zór, R. Ortiz, E. Saatci, R. Bardsley, T. Parr, E. Csöregi, M. Nistor, *Bioelectrochemistry* 76 (2009) 93–99.
24. A.T. Lawal, S.B. Adeloju, *Talanta* 100 (2012) 217–228.
25. J. Niu, J.Y. Lee, *Sensor Actuat. B-Chem.* 62 (2000) 190–198.
26. A.S. Hernández-Cázares, M.C. Aristoy, F. Toldrá, *Food Chem.* 123 (2010) 949–954.
27. Y. Yano, K. Yokoyama, E. Tamiya, I. Karube, *Anal. Chim. Acta* 320 (1996) 269–276.
28. Y. Yano, N. Miyaguchi, M. Watanabe, T. Nakamura, T. Youdou, J. Miyai, M. Numata, Y. Asano, *Food Res. Int.* 28 (1995) 611–618.
29. Y. Yano, N. Kataho, M. Watanabe, T. Nakamura, Y. Asano, *Food Chem.* 52 (1995) 439–445.
30. E.T.G. Cavalheiro, A. Brajter-Toth, *J. Pharm. Biomed. Anal.* 19 (1999) 217–230.
31. J.H.T. Luong, K.B. Male, C. Masson, A.L. Nguyen, *J. Food Sci.* 57 (1992) 77.
32. T. Fujita, Y. Hori, T. Otani, Y. Kunita, S. Sawa, S. Sakai, I. Takagahara, Y. Nakatani, *Agr. Biol. Chem.* 52 (1998) 107–112.
33. M.E. Ali, U. Hashim, S. Mustafa, Y.B. Che Man, M.H.M. Yusop, M. Kashif, T.S. Dhahi, M.F. Bari, M.A. Hakim, M.A. Latif, *J. Nanomater.* 2011 (2011) 1–11.
34. K. Lundström, K.R. Matthews, J.E. Haugen, *Animal* 3 (2009) 1497–1507.
35. L. Kaiser, R. De Jong, *Anim. Learn. Behav.* 23 (1995) 17–21.
36. G.A. Wright, M.G.A. Thomson, B.H. Smith, *P. R. Soc. B* 272 (2005) 2417–2422.
37. G.A. Wright, A. Lutmerding, N. Dudareva, B.H. Smith, *J. Comp. Physiol. A* 191 (2005) 105–114.
38. F. Wackers, D. Olson, G. Rains, F. Lundby, J.-E. Haugen, *J. Food Sci.* 76 (2011) S41–S47.
39. S.L. Utley, G.C. Rains, W.J. Lewis, *ASABE* 50 (2007) 1843–1849.
40. D. Olson, F. Wackers, J.E. Haugen, *J. Food Sci.* 77 (2007) S356–S361.
41. N.E. Virolainen, M.G. Pikkemaat, J.W. Elferink, M.T. Karp, *J. Agric. Food Chem.* 56 (2008) 11065–11070.
42. M.T. Korpela, J.S. Kurittu, J.T. Karvinen, M.T. Karp, *Anal. Chem.* 70 (1998) 4457–4462.

43. M.G. Pikkemaat, M.L.B.A. Rapallini, M.T. Karp, J.W.A. Elferink, *Food Addit. Contam.* 27 (2010) 1112–1117.
44. X. Lu, H. Zheng, X.-Q. Li, X.-X. Yuan, H. Li, L.-G. Deng, H. Zhang et al., *Food Chem.* 130 (2012) 1061–1065.
45. M. Petz, *Monatsh. Chem.* 140 (2009) 953–964.
46. D.R. Shankaran, K.V. Gobi, N. Miura, *Sensor Actuat. B-Chem.* 121 (2007) 158–177.
47. W.L. Shelver, D.J. Smith, *J. Agric. Food Chem.* 51 (2003) 3715–3721.
48. C.S. Thompson, S.A. Haughey, I.M. Traynor, T.L. Fodey, C.T. Elliott, J.P. Antignac, B. Le Bizec, S.R. Crooks, *Anal. Chim. Acta* 608 (2008) 217–225.
49. Y.-B. Lan, S.-Z. Wang, Y.-G. Yin, W.C. Hoffmann, X.-Z. Zheng, *J. Bionic Eng.* 5 (2008) 239–246.
50. S. Laschi, M. Mascini, G. Scortichini, M. Fránek, M. Mascini, *J. Agric. Food Chem.* 51 (2003) 1816–1822.
51. J. Chobtang, I.J.M. de Boer, R.L.A.P. Hoogenboom, W. Haasnoot, A. Kijlstra, B.G. Meerburg, *Sensors* 11 (2011) 11692–11716.

27

Microbial Cells and Enzymes for Assaying the Fermentation Processes of Alcohol Production: Starch, Glucose, Ethanol, BOD

Anatoly N. Reshetilov, V.A. Arlyapov, M.G. Zaitsev, and V.A. Alferov

CONTENTS

27.1 Introduction

Commercial production of ethanol and related control of starch content in initial raw materials, glucose, and ethanol contents in fermentation media, BOD index in distiller's spent grains are topical issues of fermentation biotechnology. Assessment of these components at various stages of production enables optimization of the fermentation process and reduction of material expenses by harmonizing the quality of initial raw materials with the quality and quantity of enzymes and yeast mass used. Constant monitoring of the BOD index in distilleries' wastes makes it possible to significantly reduce the ecological load on the environment, preventing the release of easily oxidized organic substances to water bodies, their pollution, and eutrophication (Henze et al. 2002).

Traditional methods of determining lower alcohols are either insufficiently accurate or are labor-intensive, expensive, and characterized by long assaying times, which prevent their use for constant monitoring of the aforementioned components (Arlyapov et al. 2008). Alcohol concentrations are determined areometrically; however, this method is not highly specific as various impurities in solution contribute to the results of the assay. Gas chromatography, which is a standard alcohol-assay method, and high-performance liquid chromatography used to determine the contents of glucose and other carbohydrates in aqueous solutions require qualified personnel and are rather cost based.

The generally accepted BOD-determination technique requires incubation of oxygen-saturated samples for 5, 10, or 20 days (BOD_5, BOD_{10}, or BOD_{20}, respectively) (Environmental Normative Documents 1997). The lack of promptitude significantly reduces the value of such assays. Ecologically dangerous situations may occur, when releases of polluted waters are left unnoticed or insufficiently clean regenerated wastewaters are released. Modern strict requirements to the state of ecology dictate the necessity of using rapid BOD-control methods.

A characteristic feature in the situation with BOD assays is growing research into "target-oriented" biosensors. Recently, investigators have tended to develop specialized BOD sensors. This tendency is quite logical. It is quite evident that the type of microorganism making the basis of a BOD biosensor for assaying the wastewaters of, for example, a meat-and-milk enterprise will be inefficient in measurements of wastewaters of a fishery plant. We would only like to emphasize the need for developing exactly such target-oriented biosensors. It could be noted in this connection that a highly sensitive biosensor based on the alcohol oxidase (AO) enzyme or, even better, yeast cells can be used at a high profit to assess the BOD of distiller's spent grains. It is known that spent grains may contain up to 0.4%–0.5% of ethanol. This concentration can be reliably detected by alcohol-assay biosensors, because for most biosensors it is a magnitude exceeding the lower limit of detection. And yeast cells used in the basis of such a biosensor would only enable expanding the range of detected compounds, which would bring the assay closer to the determination of the true BOD value.

A topical trend of research is the development of assay techniques, which would simplify and reduce the cost of assaying the aforementioned compounds without loss of accuracy and specificity, as well as enable their constant monitoring in real objects. In this connection, a promising approach is to develop biosensor technologies distinguished by small single-assay times, versatility, and no special sample preparation.

The aim of this work is to present data on recent developments in methods of assaying the contents of starch, glucose, and ethanol and assessing the BOD at various stages of the fermentation process using amperometric biosensors. We present the parameters of biosensors based on the glucose oxidase (GO) enzyme, which in the measuring regime is used together with dissolved α-amylase, the AO enzyme, and whole cells of the yeast *Pichia angusta*. AO was immobilized by a novel method based on the use of benzoquinone and DEAE-dextran. Biosensors in this work are considered in the aggregate and in this combination can be useful in solving practical tasks. These are real-time detection of the level of starch in initial raw materials and determination of the concentration of glucose produced in hydrolysis of starch, the concentration of ethanol, and the BOD of residual distiller's spent grains.

27.2 Development of Biosensors for Assaying the Contents of Starch, Glucose, Ethanol, and the BOD

Biosensors for starch, glucose, and ethanol assays were developed using amylase, GO, and AO enzyme preparations. Wastewaters and process waters of fermentation units contain mainly carbohydrates, alcohols, and amino acids, which is a guiding point in choosing microbial cultures for fabricating the bioreceptor element of the BOD biosensor. To form it, we used the methylotrophic yeast strain *P. angusta* VKM Y-2518 (the aerobic microorganisms *P. angusta* VKM Y-2518 belong to the All-Russian Collection of Microorganisms, FSBIS, G.K. Skryabin Institute of Biochemistry and Physiology of Microorganisms, Russian Academy of Sciences). In the work, these cells were also used to obtain the AO enzyme. Biosensors based on *Pichia* yeasts had been used for rapid assays of alcohols, so it was justified to apply them for the development of specialized receptor elements for assaying alcohol production wastewaters.

Starch, glucose, and ethanol contents were determined and BOD assessed using Clark electrode-based amperometric biosensors. This system enables recording the dependences of biochemical reaction rates on concentrations of substrates, related to the consumption of oxygen during their oxidation. An IPC 2L galvanopotentiostat (Kronas, Russia) integrated with a personal computer and operated by specialized software for recording and processing electrode signals served as biosensors' electronic unit. Sodium–potassium phosphate buffer, pH 7.5, the concentration of the salts 60 mM, which provided for an optimum of bioreceptors' operation, was used for measurements. The measured parameter (biosensor response) was the maximal rate of change of the output signal, which emerged at an addition of substrates (nA/min).

Biosensors based on the AO and GO enzymes were used to assay fermentation intermediates. The results of the biosensor assays were verified by reference methods. Gas chromatography was used as a reference method to assay ethanol in samples of fermentation media. The reference for the glucose assay in solution was iodometric titration. The method is based on the oxidation of aldoses by an alkaline solution of iodine. The unreacted excess of iodine was titered by a solution of sodium thiosulfate.

To determine the content of starch in assayed samples, it was cleaved by a commercial amylase preparation to glucose. The glucose formed in the reaction was assayed by an immobilized GO-based biosensor. The polarimetric method was used as a reference to assay the content of starch in samples. This starch assay method is based on transferring starch to a soluble state and measuring the polarization plane angle of the obtained solution. Polarimetric starch determination methods are easy to handle and require no additional reactions. However, to ensure the accuracy of assay, all optically active compounds capable of distorting the assay results should be removed.

The biorecognition elements of the ethanol biosensor were formed by immobilizing AO on chemically modified nitrocellulose membranes. At the first stage of immobilization, DEAE-dextran was added to the benzoquinone-activated membrane (Figure 27.1).

In immobilization of AO by means of benzoquinone and DEAE-dextran, the binding of protein molecules occurs due to numerous electrostatic interactions with the membrane modified by benzoquinone and DEAE-dextran, as the result of which the enzyme is not inactivated (Zaitsev et al. 2008) (Figure 27.2).

Immobilization of the GO enzyme was done by entrapment in gel based on bovine serum albumin cross-linked by glutaraldehyde.

The characteristics of biorecognition elements based on AO isolated from four methylotrophic yeast strains are given in Table 27.1. Isolated proteins were observed to have a rather large spread of activity. This is a well-known fact—the specific activity of AO varies within a broad range, from 5 to 30 or even 50 units, depending on the yeast species and purification protocol (Pavlishko et al. 2004). A significant spread of specific activity values in

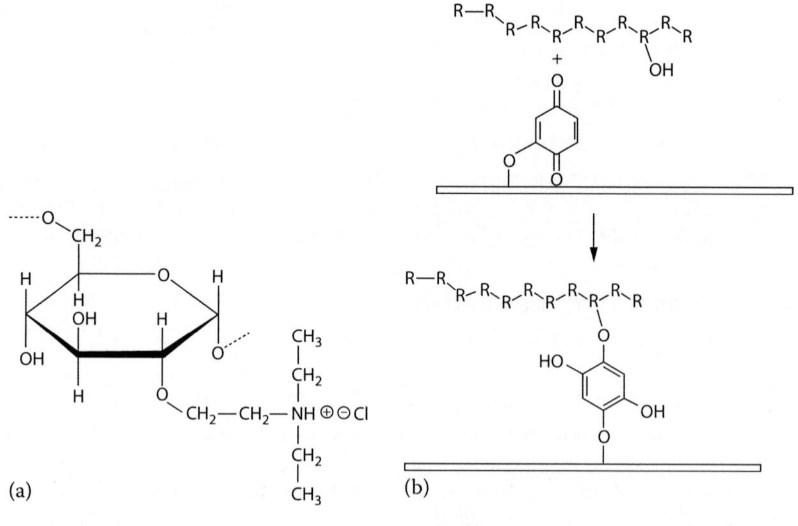

FIGURE 27.1
Structure of DEAE-dextran monomer (R) (a) and addition of DEAE-dextran to benzoquinone-activated membrane (b).

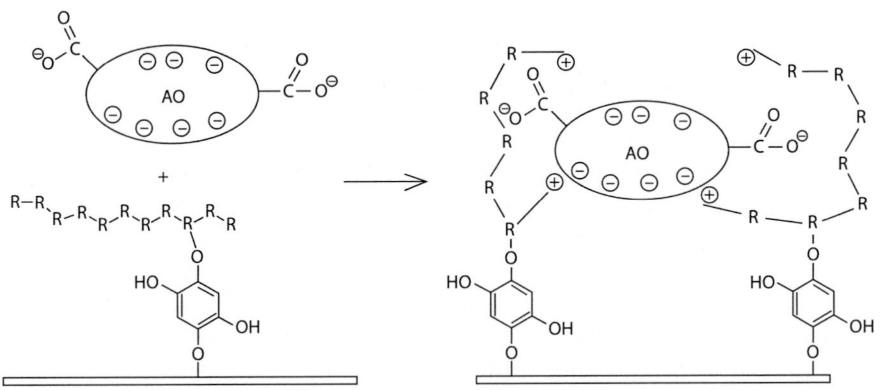

FIGURE 27.2
Immobilization of AO enzyme on the surface of the modified nitrocellulose membrane.

TABLE 27.1

Parameters of the Yeast Alcohol Oxidase-Based Biosensor

Parameter	Biorecognition Elements Based on Yeast Alcohol Oxidases			
	H. polymorpha NCYC 495 ln	*Hansenula* sp.	*P. pastoris*	*C. boidinii*
Specific activity, E	50	22	28	13
Effective Michaelis constant K_M, mM	3.7 ± 0.7	11 ± 2.0	5.6 ± 0.6	4.2 ± 0.4
Stability (operational) time of enzyme-containing membrane, days	8	12	10	2
Reproducibility ($n = 15$; assayed substance, methanol [1 mM])	7%	5.3%	8%	—

chromatographically pure AO preparations can be due to varying contents of the stable semiquinone form in them (up to 60%), which is not reducible by either methanol or else sodium sulfite or dithionite (Pavlishko et al. 2004). It should be noted that AO obtained for formation of biosensors had a sufficiently high value of specific activity, which is indicative of the prospects for using this preparation to produce sensitive biorecognition elements.

The operational time of enzyme-containing membranes (long-time stability of the biosensor) was determined as the time of a 20% biosensor response drop with respect to the maximal value (Table 27.1). A twofold and greater decrease of response occurred on day 20 of storage for all bioreceptors. It is important to note that the maximal responses were obtained using recognition membranes based on AO from *Hansenula polymorpha* NCYC 495 ln. The membrane based on immobilized *H. polymorpha* NCYC 495 ln AO was not inferior by its temporal stability to other immobilized biocatalysts based on commercial AOs (data not shown).

The biocatalytic properties of AOs of various origins proved sufficiently close, though some differences did take place; in particular, there were variations of the Michaelis constant, K_M. For methanol under conditions of homogeneous enzymatic catalysis, the Michaelis constants for AO from *H. polymorpha* were $K_M = 2.2$; for AO from *Candida boidinii*, $K_M = 3.0$; and from *P. pastoris*, $K_M = 3.1$ (Azevedo et al. 2005). The values of the effective Michaelis constants for immobilized enzymes, $K_M{}'$, obtained in the experiment were slightly higher than $K_M{}'$ for AO under conditions of homogeneous enzymatic catalysis. A biosensor with the recognition element based on AO from *H. polymorpha* NCYC 495 ln had a maximal response. This is consistent with the high specific activity of the enzyme based on which the recognition element was fabricated.

Thus, the biocatalyst based on AO isolated from the methylotrophic yeast *H. polymorpha* NCYC 495 ln is not inferior, and by some parameters even superior, to commercially available AO preparations. This is indicative of the prospects of using this immobilized catalyst in the development of ethanol assay biosensors (Dmytruk et al. 2007).

27.3　Characteristics of the Developed Biosensors

Consider the characteristics related to the BOD assessment. The graduation dependence of response on the concentration of assayed substance is the most significant metrological description of the biosensor. As a model, we used a mixture of glucose and glutamic acid (GGA) at a ratio of 1:1 (w/w); it is used as a standard in BOD_5 analyses in the Russian Federation and in international practice (APHA 1992; Environmental Normative Documents 1997).

Bioreceptors based on microbial whole cells are of catalytic type, that is, the biological response in such systems is provided for by the enzymatic microbial reactions. Therefore, the dependence of Figure 27.3a is satisfactorily described by the Michaelis–Menten equation:

$$R = \frac{R_{max}[S]}{K_M + [S]},$$

where R, R_{max}, S, and K_M are, respectively, biosensor signal, maximal biosensor signal, concentration of assayed compound, and the Michaelis constant.

Assay errors are reduced, as a rule, by using a linear segment of the graduation dependence, limited from the aforementioned by the K_M constant ($24\,mg/dm^3$). The analytical and metrological characteristics of a BOD biosensor are often determined using a glucose–glutamate mixture as a model system. In accordance with the norms, a BOD_5 equal to $205\,mg/dm^3$ corresponds to a solution containing $150\,mg/dm^3$ glucose and $150\,mg/dm^3$ glutamic acid

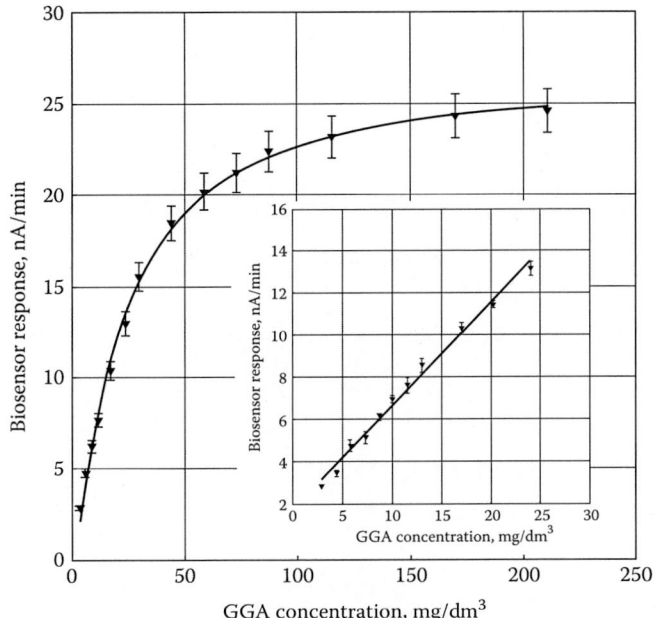

FIGURE 27.3
Dependences of biosensor responses on GGA within a broad range of concentrations. Insert, a linear segment of a response–concentration dependence. The yeast *P. angusta* VKM Y-2518 was used.

$(BOD_5 = 0.68 \times C_{GGA})$. The following characteristics of a BOD biosensor based on immobilized methylotrophic yeasts were obtained for this ratio:

- Relative standard deviation—6.1%
- Long-time stability—26 days
- Sensitivity—0.72 nA · dm³/min · mg
- Single-assay time—10–14 min
- Linear range of response–BOD$_5$ dependence—1.6–16.3 mg/dm³

Graduation dependences of biosensor response on substrate concentration for GO- and AO-based biosensors were plotted. They are given in Figure 27.4.

The characteristics of the biosensors used were determined. As substrate, ethanol was used for the AO-based biosensor and glucose for the GO-based biosensor. The obtained values are given in Table 27.2.

The biosensors based on these enzymes enabled assays of sufficiently low concentrations of glucose and ethanol and could be used to assay fermentation intermediates. The developed breadboards of biosensors were characterized by a high level of operational stability and rapid-test parameters; by their analytical and metrological characteristics, the described models were not inferior to the existing analogues (Ricci et al. 2005; Su et al. 2011; Salimi et al. 2012).

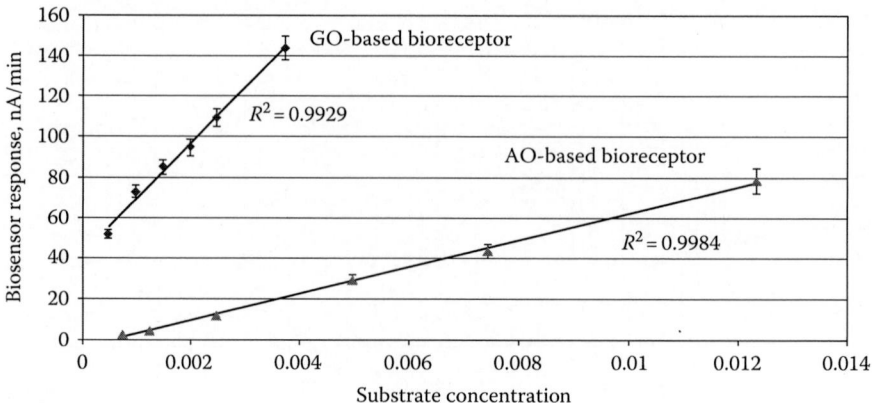

FIGURE 27.4

Responses of the AO- and GO-based biosensors versus substrate concentrations. R^2, the correlation coefficient.

TABLE 27.2

Characteristics of GO- and AO-Based Biosensors for Glucose and Ethanol Assays

Characteristic	AO (Ethanol)	GO (Glucose)
Operational stability (relative standard deviation), %	5	4
Sensitivity, nA·dm³/min·mol	69,000	92,515
Single-assay time, min	5–10	5–10
Linear range of assayed concentrations, mM	0.025–0.5	0.05–0.25

27.4 Applications of Developed Biosensors

27.4.1 Starch Assay

Determination of the content of starch in samples was based on the hydrolysis of starch by amylase to glucose followed by the detection of glucose. Starch was digested by an enzyme preparation of highly active glucoamylase (GA) and Termamyl used in commercial alcohol production (Novozymes A/S, Denmark). An assaying protocol was used, in which a solution of GA was added to the assayed sample and then the content of glucose was assessed by a GO-based biosensor. The range of assayed concentrations of starch was 0.03–0.5 g/L. The assay time did not exceed 8–10 min; the reproducibility (variation coefficient) of signals was ~10%.

To demonstrate the utility of this type of assay, a biosensor based on GO and solution of GA was used to assay starch in samples of potatoes, buckwheat, oat flakes, and wheat flour. Evers' method (Russian State Standard GOST 10845-98) was used as a reference. The content of starch determined by this method in samples of wheat flour was 70%; of rye flour, 56%. The results of the polarimetric assay of starch were classified as true values and were used as initial data for assessing the results of starch assays by the biosensor method.

27.4.2 Glucose and Ethanol Assays

The methods described were used to analyze samples obtained in modeling the ethanol production process. Gas chromatography on a Kristall 5000.2 chromatograph (Special Design Bureau "Khromatek," Russia) was used as the reference method for the ethanol assay. Iodometric titration was the reference method for the glucose assay.

Modelling of the fermentation process and production of fermented mass was performed according to the method described in Yarovenko et al. (2002). For this purpose, use was made of a wheat flour sample diluted in warm distilled water and of Termamyl, a saccharifying enzyme preparation (α-amylase, EC 3.2.1.1, Novozymes A/S, Denmark), which hydrolyzes starch by randomly cleaving α-1,4-glucan bonds. Upon expiration of the required time, the flask was cooled to 60°C, and the enzyme preparation SAN Super 360L (GA, EC 3.2.1.3, Novozymes A/S, Denmark) was added to hydrolyze saccharified mass to glucose; then the flask was again placed into a thermostat for 2 h. Termamyl digested α-1,4-, α-1,6-, and α-1,3-glucan bonds in polysaccharides, successively splitting off glucose residues. After this, the flask was cooled to 30°C, the Superstart yeast preparation was added, and the contents were kept for 120 h in a thermostat for fermentation. The method was aimed to obtain samples with various ethanol contents.

The obtained results (Table 27.3) suggest that the concentration of glucose has a tendency of incessant growth. In all probability, the chosen cultivation time (4 h) is insufficient for the complete hydrolysis of starch

TABLE 27.3

Results of Starch, Glucose, and Ethanol Assays by Biosensors and by Reference Methods

Assay Time (h)	Concentration of Starch in Sample (g/L)		Concentration of Glucose in Sample (mM)		Concentration of Ethanol in Sample (mM)	
	Polarimetric Method	Biosensor Method	Biosensor Method	Titrimetric Method	Biosensor Method	Gas Chromatography
2	47 ± 4	47 ± 3	62 ± 5	66 ± 3	0	0
4	29 ± 3	29 ± 2	230 ± 20	230 ± 10	0	0
18	23 ± 2	22 ± 2	270 ± 20	280 ± 10	1 ± 1	0
23	23 ± 2	21 ± 1	280 ± 20	270 ± 10	15 ± 3	20 ± 10
28	21 ± 2	21 ± 2	280 ± 20	280 ± 10	68 ± 5	70 ± 20
42	19 ± 3	20 ± 1	290 ± 30	290 ± 10	150 ± 10	140 ± 30
67	18 ± 2	18 ± 1	260 ± 20	270 ± 10	250 ± 20	250 ± 40
91	17 ± 2	18 ± 1	220 ± 10	224 ± 9	330 ± 20	350 ± 50
114	17 ± 2	18 ± 1	160 ± 10	163 ± 8	460 ± 30	430 ± 70

contained in the sample; for this reason, residual starch continues to be saccharified simultaneously with fermentation.

The concentration of ethanol begins to grow 14 days after the onset of fermentation. Most probably, yeast cells need this time to adapt to the new conditions of the environment. The concentration of starch continues to go down even after the major part of saccharification is completed, which is consistent with the data on the glucose content in fermentation intermediate samples.

It should be noted that the values of the concentrations of starch, glucose, and ethanol in samples of fermentation intermediates, determined by the biosensor method, coincided with the values of concentrations of these components assayed by the reference methods.

27.4.3 Selectivity in BOD Assays

An important characteristic of an assay is its selectivity, that is, the possibility to determine the content of each component of the assayed item independently of the other components. In biosensor assays, the selectivity is determined by the substrate specificity of biomaterial used to form biosensor's receptor element. In BOD assays, the receptor element should preferably be microbial whole cells possessing a broad substrate specificity (low selectivity). The broad substrate specificity is in this case an advantage, as it leads to increase the accuracy of the assay results (Lei et al. 2006).

The substrate specificity of microorganisms immobilized by adsorption on a glass fiber filter is shown in Figure 27.5. As substrates, easily

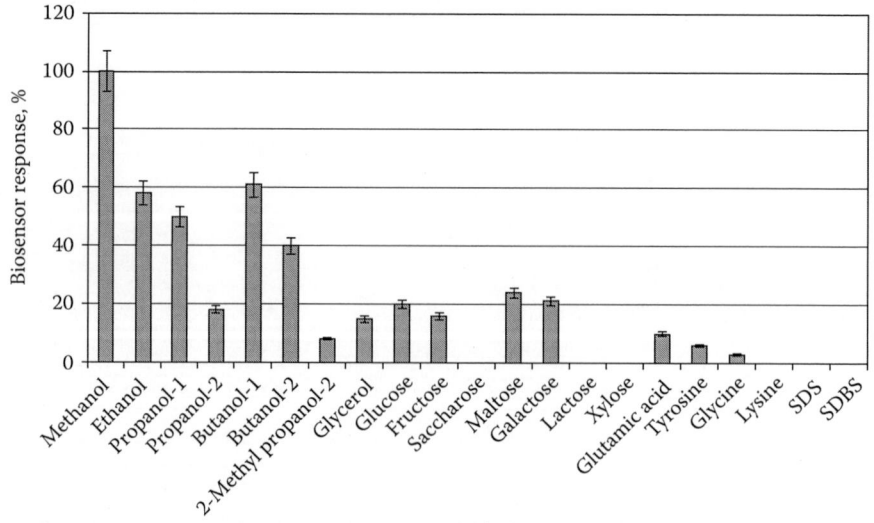

FIGURE 27.5
Substrate specificity of a bioreceptor element based on the immobilized yeast *P. angusta* VKM Y-2518. The sensor responses are expressed as fractions of the response to methanol (which is taken to be 100%). Abbreviations: SDS, sodium dodecyl sulfate; SDBS, sodium dodecyl benzene sulfonate.

oxidized organic substances were chosen; they often occur in wastewaters and process waters of fermentation units.

In operation with a bioreceptor based on *P. angusta* Y-2518 cells, the greatest responses were obtained to methanol and ethanol, linear and branched short-chain alcohols. The ability to oxidize primary, secondary, and branched alcohols makes it possible to use a biosensor based on this strain for BOD assays of wastewaters in distilleries and chemical plants, whose effluents may contain such substances.

27.4.4 BOD$_5$ Assays Using the Developed Biosensors

We analyzed wastewaters of municipal facilities and wheat flour fermentation intermediates, which imitate the composition of wastewaters in fermentation units. The developed breadboard of a BOD biosensor and the standard dilution method were used. Samples of fermentation intermediates were prepared from wheat raw stuff using commercial enzyme preparations Termamyl and SAN Super 360L and the yeast preparation Superstart. The BOD values of wastewaters were determined by the standard dilution method according to the norms valid in the Russian Federation. The content of dissolved oxygen before and after a 5-day incubation was determined by Winkler's iodometric method (Table 27.4).

Thus, the values of BOD$_5$ determined by the biosensor based on the used yeast cells match the BOD$_5$ parameters assayed by the standard technique.

TABLE 27.4

Results of BOD$_5$ Assays of Actual Samples by the Biosensors, mg/dm^3

Samples	*Pichia angusta*-Based Biosensor	Standard Dilution Method
Fermentation intermediates		
91 h	1500 ± 100	1500 ± 100
114 h	1300 ± 100	1300 ± 100
Wastewaters of municipal treatment facilities		
1	100 ± 20	120 ± 20
2	510 ± 40	620 ± 70
3	120 ± 20	150 ± 20
4	90 ± 10	100 ± 10
5	150 ± 20	150 ± 20
6	100 ± 10	96 ± 9
7	130 ± 20	140 ± 20
	100 ± 20	90 ± 10

27.5 Conclusion

Biosensor analyzers for assays of starch, glucose, and ethanol in fermentation products were successfully used under laboratory conditions to determine their characteristics and make measurements. An important element of novelty was the use of a new, previously unknown technique for immobilization of the AO enzyme. The method is considered to be "mild" for enzyme molecules, as it is not based on a covalent bond with matrix elements to which it is cross-linked. The method provided for a broad linear range of detection, from 0.001 to 0.01 mol/L ethanol. Besides, the method enabled bioreceptor stability times up to 10–12 days, which is a sufficiently high value for AO.

The biosensor for assay of starch enabled measurements within the range of 0.03–0.5 g/L; the one based on the immobilized GO enzyme, assays of glucose within 0.05–0.25 mM; and that based on AO, assays of ethanol within the range of 0.025–0.5 mM. The biosensor for express detection of BOD, based on the yeast *P. angusta*, was used to assess this parameter. These yeast cells were shown to have a broad substrate specificity and to be capable of oxidizing substrates in various classes of organic compounds. The range of detectable BOD concentrations was 1.6–16.3 mg/dm^3.

Samples of fermentation intermediates were analyzed. The results of starch, glucose, and ethanol biosensor assays of them proved to be highly consistent with the results obtained by the reference methods (titrimetry, polarimetry, gas chromatography).

The presented biosensor breadboards are based on the use of the simplest type of transducer, the Clark electrode. Comparison of the data of biosensor measurements with those of reference methods indicates the validity of the use of the described devices. In practice, we have not come across the problem of the dependence of biosensor signals on dissolved oxygen concentration. This is primarily due to the fact that in the cuvette technique of measurements, the amount of dissolved oxygen is set up by the buffer medium, while the introduced sample is of a rather small volume.

The biosensors used can be considered to be analogues of analyzers, based on which commercial devices can be made. Such kinds of devices considerably reduce assay times and simplify the technology of assessing the basic components of the fermentation medium. This approach can prove of extreme significance for fermentation manufacture of strong- and low-alcoholic beverages. It can also be vital for protection of the environment in the BOD-based assessment of the quality of alcohol production-waste cleanup.

References

APHA. 1992. *Standard Methods for the Examination of Water and Wastewater, Section 5210*, 18th edn. American Public Health Association, Water Works Association, American Water Environment Federation, Washington, DC, pp. 5.1–5.6.

Arlyapov, V. A., O. N. Ponamoreva, and V. A. Alferov. 2008. A multichannel biosensor for assaying the content of glucose, methanol and ethanol in their joint presence. *Biotekhnologiya* 5:84–91.

Azevedo, A. M., D. M. Prazeres, J. M. Cabral, and L. P. Fonseca. 2005. Ethanol biosensors based on alcohol oxidase. *Biosens. Bioelectron.* 21(2):235.

Dmytruk, K. V., O. V. Smutok, O. B. Ryabova, G. Z. Gayda, V. A. Sibirny, W. Schuhmann, M. V. Gonchar, and A. A. Sibirny. 2007. Isolation and characterization of mutated alcohol oxidases from the yeast *Hansenula polymorpha* with decreased affinity toward substrates and their use as selective elements of an amperometric biosensor. *BMC Biotechnol.* 7:33.

Environmental Normative Documents, Federative 14. 1:2:3:4. 123-97. Quantitative chemical water analysis. The method of measurements of biochemical oxygen demand after n days of incubation (BOD_{compl}) in surface fresh, subsurface (ground), drinking, waste, and purified wastewaters. Moscow, Russia, 1997, 25 pp. (in Russian).

Henze, M., P. Harremoës, J. la Cour Jansen, and E. Arvin. 2002. *Wastewater Treatment: Biological and Chemical Processes*, 3rd edn. Springer-Verlag, Berlin, Germany, 420 pp.

Lei, Yu., W. Chen, and A. Mulchandani. 2006. Microbial biosensors. *Anal. Chim. Acta* 568:200–210.

Pavlishko, H., H. Gayda, and M. Gonchar. 2004. Alcohol oxidase and its bioanalytical application. *Visnyk of L'viv Univ.* (biology series) 35:3.

Ricci, F., D. Moscone, C. S. Tuta et al. 2005. Novel planar glucose biosensors for continuous monitoring use. *Biosens. Bioelectron.* 20:1993–2000.

Salimi, F., M. Negahdary, G. Mazaheri, H. Akbari-dastjerdi, Y. Ghanbari-kakavandi, S. Javadi, S. H. Inanloo, M. Mirhashemi-route, M. H. Shokoohnia, and A. Sayad. 2012. A novel alcohol biosensor based on alcohol dehydrogenase and modified electrode with ZrO_2 nanoparticles. *Int. J. Electrochem. Sci.* 7:7225–7234.

Su, L., W. Z. Jia, C. J. Hou, and Y. Lei. 2011. Microbial biosensors: A review. *Biosens. Bioelectron.* 26:1788–1799.

Yarovenko, V. L., V. A. Marinchenko, V. A. Smirnov et al. 2002. *Alcohol Technology, Chapter 6. Saccharification of cooked mash. Enzymic hydrolysis of starch*, ed. V. L. Yarovenko. Kolos-Press, Moscow, Russia (in Russian), pp. 169–194.

Zaitsev, M. G., V. A. Alferov, T. A. Kuznetsova, T. V. Rogova, A. A. Goryacheva, K. A. Ponamorev, and A. N. Reshetilov. 2008. Characterization of biocatalysts based on immobilized yeast alcohol oxidases as the substrate of biosensors' receptor elements for detection of alcohols. *Izv. Tula State Univ.* (natural sciences). 2:200–207 (in Russian).

28

Biosensors for the Control of Biochemical
Parameters in the Diagnostics of Diseases

Nickolaj F. Starodub

CONTENTS

28.1 Introduction

Clinical biochemistry as part of clinical laboratory diagnostics is based on the qualitative and quantitative determination of a number of biochemical parameters in biological fluids. Its field of activities includes examining the nature of the changes in the number of biochemical quantities at a number of physiological and pathological conditions, as well as the development of methods for their determination. Biosensors are among the recent widespread methods based on modern instrumental electronic devices that are well known. Today, these devices are used in clinical diagnostics to determine such important parameters of the body as glucose, cholesterol, certain immune components, urea, sodium, potassium, calcium, etc. The diagnosis of diabetes to determine the level of glucose in the body with a glucose biosensor is in common practice. It is developed and commercially available in a number of countries as different types of blood glucose meters (Wang, 2001; Eun-Hyung Yoo and Soo-Youn Lee, 2010; Campetelli et al., 2011). However, clinical practice in the diagnosis, assessment of the disease, and its current treatment needs rapid determination of the level of anti-insulin antibodies. Existing analytical devices based on immunochemical analysis provide such an analysis, but it takes a long time to run, it is not cheap, and is quite difficult to be performed directly in the clinic and even more so at the patient's bedside. This can only be achieved in the development of analytical approaches based on the principles of biosensors. In this direction, extensive research was performed, and the results and conclusions about the perspectives of this approach are part of this chapter. On the other hand, there are a number of infectious diseases in humans and animals, the diagnosis of which is based on the definition of specific antigens or antibodies. A striking example of this is viral leukemia in cows, which is prevalent everywhere, in all continents. Moreover, the disease can be transmitted through food from infected animals to humans. Again, there is an arsenal of traditional immunological methods that allow for the biochemical diagnostics of the aforementioned diseases, but they are, as in the first case, deeply routine and cannot provide the requirements of practice for express analysis, low cost, and simplicity. The results of the study demonstrate the advantages of biosensor analysis and they are also presented in this chapter.

28.2 Common Characteristics of Some Biochemical Indexes in the Development of Diabetes

Today, essentially two different types of diabetes are distinguished: insulin-dependent and insulin-independent forms or type 1 and 2. The first type is characterized by absolute or relative insulin deficiency, which is caused by an autoimmune process that is accompanied by a progressive and selective damage to the beta cells of the pancreas in the appropriate group of people. The susceptibility of people to such damage is caused by genetic as well as environmental factors. The final genetic components HLA, DR3, and DR4 loci play a huge role in diabetes, whereas locus HLA-DR2 protects the body from developing the disease. Most of the environmental factors that influence the development of diabetes are not known, but there are highlights such as viral infections (enteroviruses, rubella virus) and certain foods (i.e., it is believed that milk cows at an early age can cause autoimmune process in a certain group of people). It was observed that the autoimmune process, which leads to diabetes, begins much in advance of the manifestation of clinical symptoms. During the process of development in diabetic patients, it can be detected by the increasing titer of various auto antibodies to islet cells, insulin-like protein, and to glutamate decarboxylase. An example of primary prevention of diabetes may serve exceptions drinking milk cows at the early infantile age. Several approaches for secondary prevention of diabetes of this type are under development. It was found that nicotinamide, which is a precursor of nicotinamide adenine nucleotide, may be a factor that contributes to the prevention of disease. Insulin therapy prevents the emergence of diabetes in mice, causing resistance of beta cells to immune attack. Generally, clinical symptoms of diabetes appear after destroying about 80% of the beta cells of the pancreas. For patients with type 1 diabetes, there is no alternative to insulin therapy. Long-term treatment with inadequate doses of insulin results in a lag in growth, pubertal delay, causing microvascular complications and, contributes to mortality (Bougneres et al., 1990; Karvonen et al., 1993; Dahlquist and Mustonen, 1994; Gough, 1996; Lounamaa, 1996; Pozzilli, 1996; Gardner et al., 1997; Greene and Newton, 1997). The most used insulin now is human recombinant forms.

Insulin-independent diabetes or type 2 diabetes is a heterogeneous group of disorders of carbohydrate metabolism. In addition, there is no one universally accepted theory of the pathogenesis of this disease. The genetic basis of this type of diabetes is not in doubt, and genetic determinants in this case are more serious in nature than those in type 1 diabetes. Confirmation of the genetic basis of type 2 diabetes proves the fact that in one-egg twins it develops almost always in both (95%–100%). There is a likelihood of the presence of common defect in the glucose recognition by B cells or peripheral tissues, reducing glucose transport or decreasing response of these cells to glucose stimulation. The risk of developing type 2 diabetes is increasing dramatically in the presence of parents or close relatives with the same disease as well as obesity. In recent years, special attention is paid to the hypothesis of "deficient" phenotype, the essence of which lies in the fact that malnutrition *in utero* or early postnatal period may be the cause of slow development of endocrine pancreatic function and predisposition to diabetes. Under the theory of "exhaustion" of the pancreas, type 2 diabetes occurs as a result of an imbalance between insulin sensitivity and insulin secretion. Until the pancreas can increase insulin secretion to outweigh insulin resistance, glucose tolerance remains at normal (Bougneres et al., 1990; Karvonen et al., 1993; Dahlquist, Mustonen, 1994; Gough, 1996; Lounamaa, 1996; Pozzilli, 1996; Gardner et al., 1997; Greene and Newton, 1997).

The two main mechanisms of glucose apparent in the blood are glucose on an empty stomach directly produced by the liver and glucose absorbed in the gut after eating. However, insulin regulates glucose levels by reducing its production and increasing liver glycogen synthesis, on the one hand, and on the other, increasing its transport and metabolism in peripheral tissues, namely, fats and cells of muscles. In addition, glucose production by liver is controlled with glucagon and catecholamines, which stimulate the release of glucose and function of the liver, thus increasing antagonists of insulin action. Reasons for violation of insulin secretion are (1) reduction of the weight of beta cells of the pancreas, (2) dysfunction of these cells in their unchanged number, and (3) simultaneous display of the factors listed in points 1 and 2.

Considering the phenomenon of insulin resistance in type 2 diabetes, it is necessary to pay attention to a number of groups of substances called as antagonists of insulin: some hormones, insulin antibodies, and antibodies to insulin receptors. Counter-regulatory hormones are growth hormones, cortisol, thyrotropine, placental lactogene, prolactine, adrenocorticotropic hormone, glucagon, catecholamines, and thyroid hormones.

Today we know that the presence of antibodies against insulin shows the development of autoimmune disorders in the body. The identification of this class of compounds and their quantitative analysis has important scientific and practical significance, because it allows, on the one hand, studying the mechanism of various pathological forms of pancreatic cancer, knowledge of which will promote individualization and improve insulin therapy. This result should be provided as soon as possible even with very high sensitivity. On the other hand, the presence of antibodies to insulin may interfere with the determination of its level in the blood and affect the outcome of glycemic control in patients with diabetes. In addition, high levels of antibodies against insulin may contribute to the immune resistance in these patients, requiring the need for dynamic control of the concentration of this type of antibodies in patients with diabetes who receive various insulin preparations. Test for detection of antibodies against insulin can be used to determine the risk in relation to diabetes types 1 and 2. Especially, it should be noted that in most cases there is a need to use express (sometimes the result should be obtained immediately), highly sensitive, and specific methods strictly. Only the last two requirements are met by the classical methods of modern immune-chemical analysis, such as radio-immune (RIA) and immune-enzyme (the ELISA) methods. The latter is used for the detection of antibodies to glutamic acid decarboxylase (Carrol et al., 2003). Its sensitivity reached 20 ng/mL. However, in both cases the expressivity of the immune-chemical methods is insufficient. In full it could be really provided only with the immune biosensorics analysis, which has several advantages over other varieties of immune-chemical methods in respect of the sensitivity, simplicity, expressivity, and economy, and it can be executed directly at the bed of the patient. In the literature, there are data (Bae et al., 2004) about the application of an immune biosensor based on ellipsometry to determine the concentration of insulin in the blood of patients. This biosensor is able to control insulin levels in the range of 10 ng/mL–100 μg/mL.

The main objective for us was the development of a new immune-chemical approach for testing antibodies against insulin. We used a biosensor based on the principles of the effect of surface plasmon resonance (SPR) (Starodub et al., 2007a). Implementation of this approach will enable professionals engaged in the field of diabetology to implement a radically new strategy for diagnosing and studying the pathogenic mechanisms of diabetes, which in its importance in endocrinology is "the number one problem".

28.2.1 Technical Aspects of Biosensor Construction and Analysis Fulfillment

The investigations were fulfilled with the application of the "Plasmon SPR-4M" developed in Ukraine. The principle of its work was described in detail early (Kooyman et al., 1988; Starodub et al., 2001a, 2003, 2007b). The transducer of the immune biosensor was in the form of a glass plate with the gold (20 nm) deposited on the previously formed layer of chromium (3 nm). It was connected with the prism of the measuring device through polyphenyl ether with the refractive index equal to 1.6. Immune globulin and insulin dissolved in 1 mM tris-HCl buffer (pH 8.2) with 0.14 M sodium chloride (PhBS) at the concentration of 1 mg/mL were used as specific antigens. In each experiment, the resonant angle at the successive introduction in the measuring cell was registered: distilled water, specific antigen, solution of bovine serum albumin (BSA in PhBS at the concentration of 1 mg/mL), and antiserum with dilution from 1:10,000 to 1:200. The time of the exposition was 20 min, and after each measurement the cell was washed with distilled water.

28.2.1.1 Choosing Effective Approaches for the Immobilization of Insulin on the Transducer Surface

An important requirement of the biosensor creation was that it should be stable and reproducible. Therefore, one of the main problems when creating any type of biosensor, especially immune, is to provide effective standardized immobilization of biological material and optimization of the functional parameters of the device. Protein molecules can be adsorbed on the metal or silicon surfaces, but the effectiveness of this process is often not high, on the one hand and on the other, the active centers are often blocked at the sorption and it leads to significant loss of the level of the specific interactions on the transducer surface. The level of the nonspecific interactions can be unchanged or even grow and then the ratio of signal to background sharply decreases, and the sensitivity of the immune biosensor falls sharply. To avoid these problems, the transducer surface should be pretreated by thiols or polyelectrolytes, which promotes the state of its standardization (Starodub and Pirogova, 2002; Nabok et al., 2005; Starodub at al., 2005). Also, including the intermediate layers of other molecules causes selective exposure of sites of the sensitive biological structures toward the solution, such as protein A from *Staphylococcus aureus*, or related lectins that interact with Fc-fragments of antibodies, releasing their F(ab)$_2$ fragments for antigen binding (Starodub and Pirogova, 2002; Starodub at al., 2005).

It was shown (Starodub et al., 2007; Palagin et al., 2008; Dz'oma and Starodub, 2009) that modification in the metal surface adsorption of the antigen on the surface of the transducer was stable over time and was not destroyed by washing measuring cell with PhBS. Immobilization antigen was accompanied by changes in the resonant angle within 3200–3500 arc s. In the case of polyelectrolytes, the amount of antigen adsorbed on the surface was slightly larger as well as the biosensor was more stable and reproducible than one with a bulk surface. Application of dodecanethiol also helps to stabilize the immobilized protein layer and increases the density of antigens (such as IgG, and insulin) on the surface (Table 28.1).

TABLE 28.1

Values of Response of SPR Immune Biosensor in Introduction of Insulin in the Measuring Cell Containing Different Types of Transducer Surface

Type of Transducer Surface	Deviation of Resonant Angle, arc. s
Bulk	620 ± 20
Treated by: Polyelectrolyte poly-alylamine hydrochloride (PAA)	890 ± 34
Dodecanethiol	750 ± 30

After a next surface treatment with 1% solution of BSA, significant changes in refractory angle were not observed. This means that the number of free binding sites was minimal, and the concentration of antigen (insulin) was enough to create a dense layer.

The number of physically adsorbed biological molecules is limited by the surface area of the transducer, which is amenable to optical registration. This limitation can be prevented by compounds that can form on the surface of branched structures and thus provide the location of biological material in three-dimensional space. Thus, the amount of immobilized material can be increased, and thus it is possible to further increase the sensitivity of the immune biosensor. Therefore a different scheme was proposed using polyelectrolytes when polycations and polyanions were used not to cover the surface and to form a branched double layer of biological molecules.

Anti-IgG human concentration of 200 µg/mL in PhBS, pH 8.0, was adsorbed spontaneously by the physical sorption on the surface of the transducer, where it was incubated for 30 min at room temperature. After that, the cell was washed with distilled water. Then, the polyelectrolyte solution in the sequence PAA—polystyrene sulfate (PSS)—PAA was made in a cell. The concentration of each over the states was 1 µg/mL in distilled water, and the incubation time was 20 min. Then, the cell was again filled by the rabbit antibody to human IgG, and was kept in this concentration for 20 min and washed with PhBS. After that, the cell was treated with 0.5% glutaraldehyde. After incubation for 10 min, the measuring cell was washed using 0.05 M tris-HCl buffer, changing the pH from 4.0 to 8.0 and destroying electrostatic bonds between IgG and polyelectrolytes. Thus, on the surface of the transducer mono layers are not formed, and on the contrary there is a double layer of antibodies with the covalent bonds between molecules of IgG, formed by glutaraldehyde (Figure 28.1). Then the cell was filled by the solution of the human IgG at a concentration range from 1 ng/mL

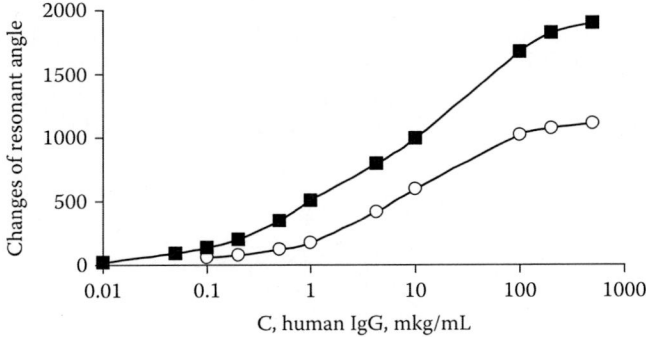

FIGURE 28.1
Response of SPR immune biosensor at the introduction of a solution of the human IgG in a range of concentrations from 10 ng to 500 µg/mL and at the use of different variants of the sensitive surface: 1—double layer of antibodies created by polyelectrolytes; 2—mono layer formed on the bulk gold surface.

to 100 μg/mL in PhBS. The shift of the resonance angle was proportional to the concentration of human IgG in solution. It was shown that the preparation of the sensitive surface in such a way significantly improved the sensitivity, which in this case was 20 ng/mL, with a linear range depending on the quantity of the human IgG in solution: the range of concentrations was from 200 to 100 μg/mL. The sensitivity of the immune biosensor using monolayer antibodies formed on the gold surface was 100 ng/mL; a linear plot graph lies within 1–100 μg/mL (Figure 28.1). Thus, it was found that the formation on the surface of the transducer multilayer (three-dimensional) branched structure of antibodies using polyelectrolyte and glutaraldehyde can significantly increase the sensitivity of SPR immune biosensor.

The same scheme was used for immobilization of insulin. Its solution at a concentration of 0.5 μg/mL in 1 mM Tris HCl buffer, pH 8.2, was introduced into a measuring cell and incubated for 20 min at room temperature. Then the cell was washed with buffer and filled successively with polyelectrolytes PAA—polystyrene sulphate hydrochloride (PSS)—PAA at a concentration of 1 μg/mL in 1 mM Tris HCl buffer with appropriate pH. The cell was washed with buffer solution and treated with insulin, incubated 20 min, and washed with buffer. Then it was filled with 0.5% solution of glutaraldehyde and kept for 15 min. Finally, polyelectrolytes were removed by 50 mM Tris HCl buffer, pH changing from 4.0 to 8.0, to form a two-layer sensitive surface.

It was established that the formation of a double layer of insulin on the transducer surface increased the sensitivity of the immune biosensor to specific antibodies. This is particularly important in the study of the early stages of the disease when the specific antibody titre is low. However, a complication of the process of testing samples of blood serum in the proposed scheme of preparation of the sensitive surface is not appropriate for practice, because the main advantage of biosensor analysis is its expressivity. Thus, the effectiveness of selective immobilization of biological material on the surface previously covered with various chemical agents (thiols, polyelectrolytes) to generate functionally stable uniform optical sensing elements of the immune biosensor based on SPR was analyzed. It was stated that the use of polyelectrolytes for the surface modification transducer of the immune biosensor based on SPR is the most appropriate, affordable, and simple.

28.2.1.2 Basic Algorithm of Insulin Analysis

We have previously stated (Nabok et al., 2005; Starodub et al., 2008) that the detection of low molecular weight compounds strongly depends on the algorithms of the specific analysis, such as "competitive" when immobilized compound competes for binding sites of the selective molecules with one that is in the free state, or vice versa, "compound," which is in the free state and in the conjugated state with other macromolecular substances competes for the binding sites of antibodies immobilized on the surface of the transducer. Another method of analysis is associated with algorithm fulfillment when immobilized antibodies interact with low molecular weight substances and specific free sites to-saturated by the conjugate of the low-molecular substances with corresponding macromolecular compounds, the so-called analysis algorithm—"to saturation." Finally, the simplest analysis algorithm is such that when the transducer surface immobilizes antigen or antibody, they interact directly with the relevant structures that are in the solution, which is uncontrollable.

We conducted a study to determine the optimal algorithm level analysis of specific insulin antibodies in the serum of patients with diabetes. According the results, it was found (Palagin et al., 2008; Dz'oma and Starodub, 2009) that the most acceptable is the last of the above analysis algorithms, namely, the direct determination of anti-insulin antibodies, when insulin immobilized on the surface of the optical transducer.

To find the optimal concentrations of the components' immunochemical reaction for the application in the immune sensor analysis, research was carried out in 0.01 M PhBS (pH 7.4) using three samples of insulin pigs produced by Minsk (Belorussia) and Calbiochem (USA) respectively, recombinant insulin (from Calbiochem), pig antiserum, and antibodies against insulin from various sources. To establish the optimal concentration of insulin required for its immobilization, the preprepared solutions with different contents of protein and the transducer surface processed by polyelectrolyte, as specified in the previous section, were used. The experimental results are presented in Figure 28.2.

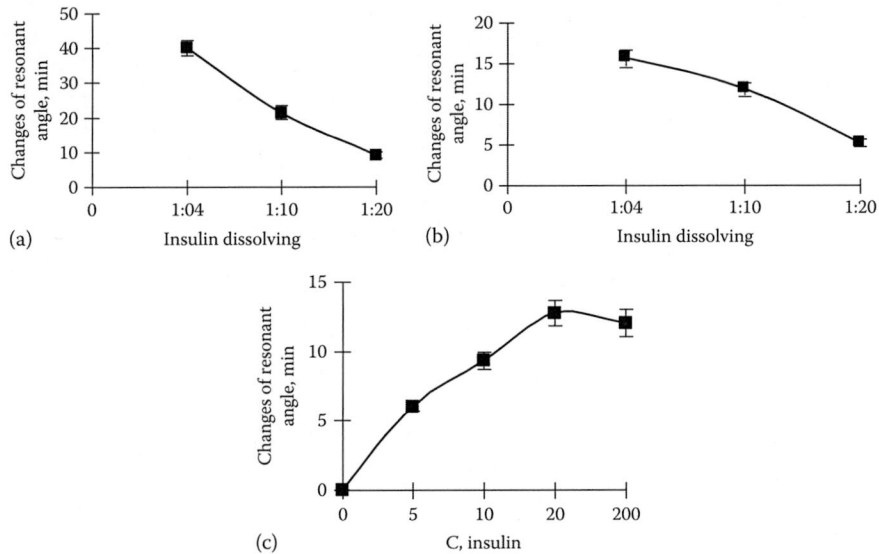

FIGURE 28.2
Dependence of the response of the immune biosensor on concentrations of different types of insulin used for the immobilization on the transducer surface. Above: from Calbiochem (a) and from Minsk (b). (c) Recombinant insulin from Calbiochem, µg/mL.

Therefore, it was stated that the optimal concentration of recombinant insulin should be considered about 20 µg/mL, and for pig insulin produced by Minsk (Belorussia) and Calbiochem (USA) it should be in ratio of dilution of 1:10.

28.2.1.3 Response of the Immune Biosensor with Different Types of Insulin Immobilized on the Transducer Surface before and after Pretreatment by Various Chemical Agents

We have used the following types of transducer surface: bulk and after its treatment by dodecanethiol and polyelectrolytes. It was important to establish with what kind of intensity three selected types of insulin may be immobilized on the previously mentioned transducer surfaces. Experimental results are presented in Figure 28.3.

The concentrations of insulin were taken as mentioned previously.

In all cases of the state of the transducer surface, we registered a sufficiently high level of response, but it was a little intensive if the transducer surface was treated with polyelectrolyte. Next, it was necessary to determine the dependent immune response of the biosensor with different types of transducer surface in the interaction with the specific anti-insulin antibodies. It should simulate the type of immobilized insulin and the state of the transducer surface that are optimal for a biosensor when we evaluate its efficiency on the final result, namely, at the registration of the

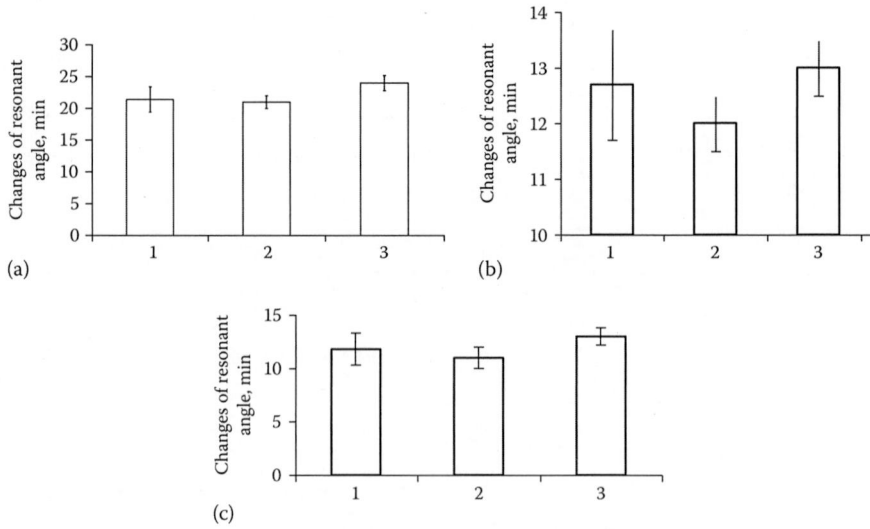

FIGURE 28.3
Response of the immune biosensor at the immobilization of pig and recombinant insulin from Calbiochem ((a) and (b) respectively) and pig insulin from Minsk (c) on the transducer surfaces: bulk (1), treated by dodecanethiol (2) and polyelectrolyte (3).

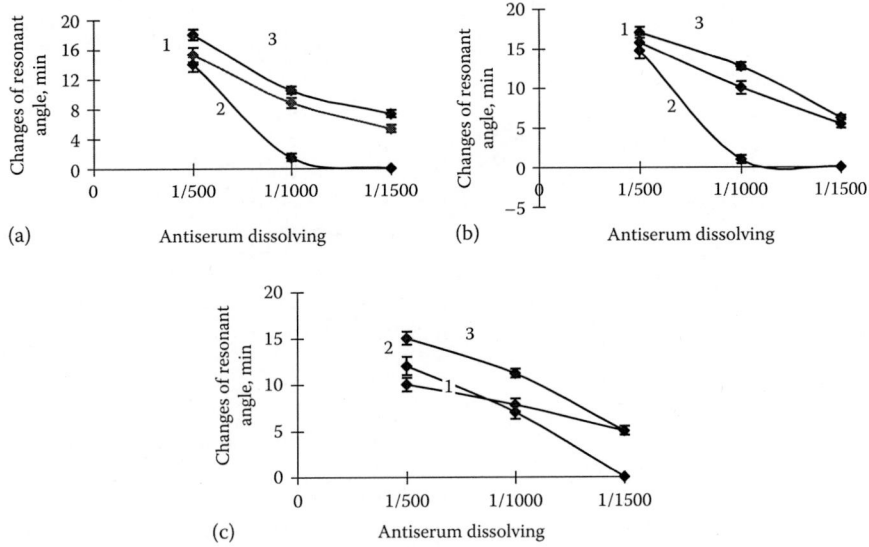

FIGURE 28.4
Response of the immune biosensor at the loading of different concentrations of anti-serum on the transducer surface that contained pig insulin (Calbiochem—(a), Minsk—(c)) and recombinant insulin (b). 1–3—insulin was immobilized on the bulk surface, treated by dodecanethiol and PAA.

formed immune complex. In experiments, the specific anti-insulin serum was used. The results are presented in Figure 28.4.

The analysis of the obtained results testified that in all cases a good response of the immune biosensor can be obtained but the application of PAA for treatment of the transducer surface may give a few high and stable characteristics. The absence of cardinal differences between responses of immune biosensors with bulk and treated transducer surface may be connected with many valence antigenical surfaces of insulin when some antigenic sites may be blocked by surface and others are exposed toward the solution.

28.2.2 Construction of the Calibration Curve for Determination of Antibodies Against Insulin in Blood Serum

In order to analyze blood serum samples in respect of determination of the concentration of antibodies against insulin, it is necessary to have a standard curve. This curve should reflect the dependence of the immune response biosensor on the concentration of antibodies in the model solution. In constructing the calibration curve, we used insulin derived from the company Calbiochem, which, in our opinion, gave the most stable results in previous studies (see previous section). From our colleagues in Minsk (Belorussia), we got 3-type antibodies against insulin (No. 2456, 3973, and 4276). The experimental results are shown in Figure 28.5.

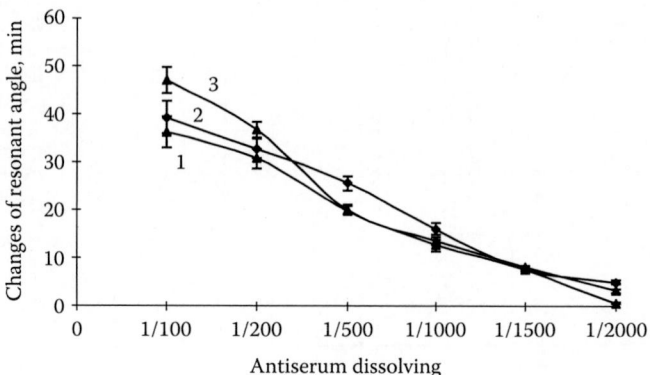

FIGURE 28.5
Response of the immune biosensor in the presence of different concentrations of antibodies to insulin.

As can be seen from Figure 28.5, there is a possibility to analyze the concentration of antibodies against insulin in a fairly wide range. To avoid non-specific reactions in the insulin–insulin interaction, the dilution of serum should be not less than 1:100.

28.2.3 Testing an Immune Biosensor Method in Model Experiments on Clinical Material

Using antigen from the company Calbiochem (as indicated earlier) was defined at the level against insulin antibodies in the sera of patients who died. Preliminary serums were characterized by this index using a standard ELISA method. Moreover, the samples from ill persons were divided by the last method into those that had the maximum and minimum levels of antibodies against insulin. Data obtained by the immune biosensor analysis are presented in Figure 28.6.

As seen from the data presented here, we have a very high correlation immune biosensor response with predefined initial levels of antibodies

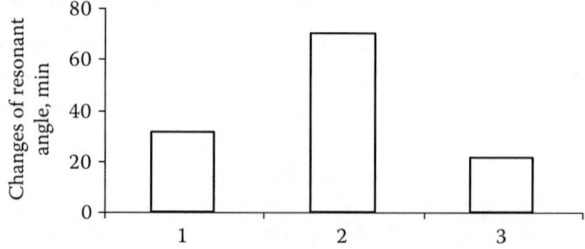

FIGURE 28.6
Response of the immune biosensor at the analysis of antiserums (dissolved in 1:100) from the immunized pig (1), patients with maximal (2) and minimal (3) content of antibodies.

against insulin in samples subjected to analysis. It should be noted that the total analysis time in the case of the immune biosensor method is much less than that required performing a traditional ELISA method. So, if all procedures are needed for immune biosensor analysis, including pre-processing transducer surface and immobilization of insulin, perform simultaneously with the sample analysis. It is necessary to spend more than 30–40 min. However, this time is much less than that needed to perform the traditional ELISA method (4–5 h). Moreover, in the case of the immune biosensor method, there is a possibility to fulfill preliminary procedures as the previous surface transducer treatment, and immobilization of insulin and then pure time of analysis will take only 5–10 min.

28.3 Diagnostics of Bovine Retroviral Leucosis by Using Immune Biosensor Based on SPR

Viral bovine leukemia (VBL) is as a tumor disease of hemolymphopoetical system, which is characterized by proliferation of malignant hematopoietic tissues and disruption of maturation of cells with predominantly intense young forms. All kinds of mammals, birds, and fish suffer from leukemia. It is especially widespread among cattle, which is registered in many countries on all continents. There is evidence that people may be sensitive to this form of leukemia. Available evidence suggests that people, especially men, can get sick with leukemia from consuming milk (Balida and Ferrer, 1977; Donham et al., 1980). The important feature of the disease is a long-term course of the disease without any significant breaches in the health of animals. The leading factor that causes bovine leukemia is virus type C from a family of retroviruses, type oncoviruses. In enzyme reverse transcription of virus containing six proteins, the most important is the surface (shell) glycoprotein gp51 and internal polypeptide p24 (Matthews, 1979).

Prevention and elimination of leukemia is complicated because of the extraordinary spread of the disease. The main reason is the constant control of health of the animal and prevention of introduction in a group of untested animals. Therefore, improvement of diagnostic methods that could allow detecting the disease in its early stages of development is a very important and urgent task.

At present, the main methods of laboratory diagnosis of VBL are reaction immune diffusion (RID), ELISA method, polymerase chain reaction (PCR), and others (Miller et al., 1972; Balida and Ferrer, 1977; Franz et al., 1986; Fechner et al 1996, Blankenstein et al., 1998). For the analysis, generally samples of blood taken from the cervical vein are used. Massive research is conducted every 4 months, and the disease is detected among a group of animals every 10 days during 6 months. At present, reproducing the epizootic picture in full is not possible, because it is not difficult to carry out tests with the same frequency by the existing methods. They have relatively high cost and give rise to the problem of injuring animals during blood sampling. Therefore, one of the important tasks today is to develop a new sensitive and specific, yet simple and rapid, method for the diagnosis of leukemia in the early stages of its development. One such alternative approach is to create instrumental analytical tools based on the principles of biosensors.

Using optical immune biosensors based on surface plasmon resonance (SPR) for express diagnostics of various diseases in humans and animals and environmental monitoring is one of the most promising directions in biosensorics (Starodub et al., 1997, 1999, 2001b, 2002, 2010; Starodub and Starodub, 2001, 2003). The sensitivity of this method is close to the sensitivity of immune-enzyme analysis, but the speed and simplicity of diagnostics of significant components of biological fluids in terms of real time is an indisputable advantage.

These important characteristics of an optical biosensor based on SPR, the sensitivity and specificity of analysis, stability, and reproducibility of the results, depend directly on the state of metalized surface, density structuring, and spatial orientation of biological molecules on it. Thus, the biological material serving as sensing element can be directly immobilized on the metalized surface biosensors. However, the application of different gold film coatings to optimize its performance has been shown (Pirogova and Starodub, 2008).

Our research has focused on developing a method of express diagnostics of leucosis using an immune sensor based on SPR, which is based on determining the level of specific antibodies to this retrovirus in blood serum and milk of animals. Immune SPR biosensor can provide: (1) direct analysis without the use of additional reagents, (2) rapid analysis for 3–5 min, and (3) analysis without drawing blood and using whey milk, and (4) cost-effective analysis (less than the cost of ELISA method).

28.3.1 Conditions of the Determination of the Induced VBL Antibodies in the Serum of Cattle

Immune biosensor analysis uses an SPR device based on the determination of the level of induced VBL antibodies in serum and milk of cattle. This virus should be antigen immobilized on the transducer surface, which provides selective binding of specific antibodies. The use of serially produced antigens was initially tested by industry (R&D, "Leykopol" (Poltava, Ukraine) and Kursk Pharmacological Plant (Russia), and as a rule applied for the traditional immune analysis (RID), it should be additionally purified for increasing the level of specific molecules in the case of application of the immune biosensor. It was achieved by the use of the adsorption chromatography on porous glass.

It was shown that the physical adsorption of antigen on the surface of the transducer is stable over time and is not destroyed at the washing of the measuring cell by buffer. Immobilization of antigen is accompanied by changes in the resonance angle within 1000–1200 angle s. After surface treatment by a solution

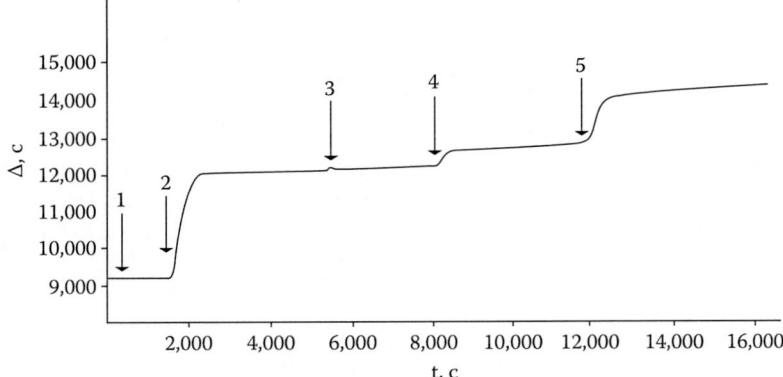

FIGURE 28.7
Response of the immune biosensor after introduction into measuring cell: 1–5—distilled water, antigen, BSA, and antiserum with the dilution of 1:10,000 and 1:1000, respectively. Abscissa— time of analysis, ordinate—changes of the resonant angle.

of BSA, significant changes in angle were not observed. This means that the number of free binding sites was minimal, and the concentration of antigen was enough to create a dense layer. Entering into the measuring cell-specific antiserum from cows clinically sick with leukemia and used as donors' infected cells, the response of biosensor correlates with the degree of dilution (Figure 28.7).

Thus, it was found that the purified antigen can allow determining the levels of specific antibodies in the serum in which it was dissolved in 16,000 times. It should be underlined that the titer of antibodies in the RID was 1:256.

The next step was to investigate a series of samples of blood serum of cows that are dependent on different farms in Poltava region. In the experiment 10 RID-positive (titer of antibodies 1:64–1:128), 18 RID-negative samples, and 15 samples, tentatively designated as RID questionable (such samples formed precipitation lines at a dilution of 1:4 only) were taken. It was found that the immune biosensor analysis could show results at the dilution of sera from the RID-positive samples in 10,000 times. In this case the changes of the resonance angle were in the range of 700–1000 angle. In samples that were determined by the RID as doubtful, the presence of specific antibodies was fixed only at a dilution of sera in 1000 or 500 times. Major differences in the results were recorded in the case of studies of RID-negative sera. Since the level of antibodies in such sera is already beyond the sensitivity of RID, such tests cannot serve as a control. Therefore, sera of cattle from farms where there were no cases of animals with leukemia were used as a control. These sera were obtained from the State Scientific Control Institute of Biotechnology and strains of the Ministry of Agrarian Policy of Ukraine. This serum caused the emergence of an immune biosensor signal, indicating their lack of specific antibodies to antigens VL. It presents considerable interest in the research of the analysis of the blood sera from cows, which were previously vaccinated by preparation of leukemia virus. It turned out that they are capable of inducing specific immune biosensor signal and its presence has a high correlation with vaccination of animals.

Thus, there is a possibility of the express determination of the antibodies induced by VBL in the serum of cattle by an immune biosensor based on SPR. This analysis was more sensitive than the traditionally used method of RID. It was established that RID-positive sera can generate specific signal already at 1:16,000 dilution. However, the optimal serum dilution of blood for the immune biosensor analysis should be considered as 1:500.

28.3.1.1 Optimization of Immobilization Antigen on the Surface of the Transducer

An important requirement of the biosensor is stable and reproducible results. Therefore, one of the main objectives as mentioned in the case of the determination of the anti-insulin antibodies was effective standardized immobilization of biological material and optimization of the functional parameters of the immune biosensor. In this case, we analyzed the efficiency of the immobilization of biological material on the transducer surface through covering it by a variety of biological and chemical agents (lectins, thiols of different origins, polyelectrolytes, and dextrin derivatives). Since the sensitivity of an SPR-based immune biosensor is high enough and sufficient for rapid screening studies on leukemia, for integration antigen on the surface of the sensor, we selected only methods that otherwise were the most simple and showed high efficiency.

28.3.1.2 Modification of the Transducer Surface by Dodecanethiol and Polyalylamine Hydrochloride

Thus, it is shown that if we used polyelectrolytes the amount of antigen adsorbed on the surface was slightly bigger and biosensor response with immobilized antigen was more stable and reproducible than with bulk surface. Application of dodecanethiol also helps stabilize the monolayer of the antigen of VBL. Immobilized antigen thus remained at almost hundred percent of activity within 2 months, whereas the antigen coated on the bulk surface lost activity within 2–3 weeks. Application of antigen on the surface of positively charged polyelectrolyte films increases the sensitivity of biosensors to specific antibodies 15%–20% compared with the case when not using treated surface biosensor transducer. It was also found that the surface modification by thiols had little effect on sensitivity analysis, but at the same time at such modification increased the reproducibility of the obtained results.

In previous studies the scheme of immobilization of antibodies on the surface of the transducer when using polyelectrolyte PAA and PSS (Starodub et al., 2005; Pirogova and Starodub, 2008) was proposed. Three groups of experiments were carried out with cattle blood serum. The first group—a serum positive according to RID and ELISA method with the system IDEX (USA).

Titer of sera in RID was in the range from 1:64 to 1:256. The second group is based on a serum by negative results on RID but positive according to the ELISA method. Finally, the third group—a serum with negative results for RID and ELISA method. Immune biosensor analysis of serums in the first two groups confirmed the presence of these specific antibodies. In the third group of serum samples (negative results for both RID and ELISA method), using immune biosensor analysis the specific antibodies to VBL were found. Therefore, it was established that the formation of a double layer on the transducer surface of the antigen increased by VBL the sensitivity of the biosensor to specific antibodies. This is particularly important in the study of animals at the early stages of the disease when the specific antibody titer is low. Since the surface is prepared in advance, the expressivity analysis is not affected.

28.3.1.3 Modification of the Transducer Surface by Dextran

Such modification can increase the biosensor response by 20% compared with bulk surface. The disadvantage of this method of modification is the high cost of dextrans. To summarize, we compared our results obtained using different types of surfaces (Figure 28.8). We analyzed 10 sera bovine blood, for which the extinction coefficient according to ELISA method was 2.0–2.5 opt. units at a dilution of 1:500. It was confirmed that the most sensitive biosensor had on the surface of which it formed a double layer of antigen using polyelectrolytes. In other cases, such as when using thiol or dextran coatings, significant increase in sensitivity was not observed.

Thus, since the sensitivity and specificity of SPR immune biosensor are high enough for express analysis of samples to detect animals suffering from leukemia, there is no need to use high-cost compounds and complications of algorithm analysis. Using PAA over the states for surface modification transducer immune biosensor based on SPR is the most appropriate, affordable, and simple method.

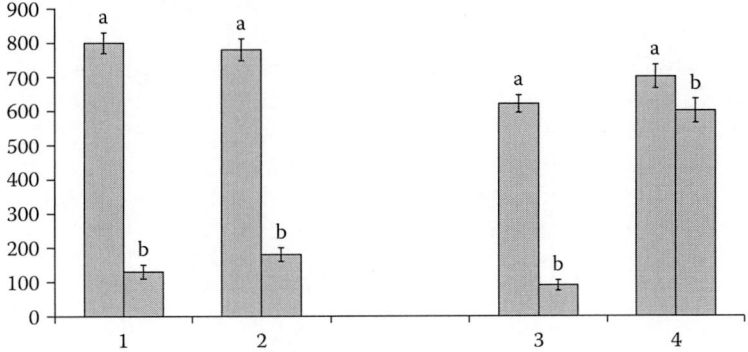

FIGURE 28.8
Value of the response of the immune biosensor at the analysis of serum blood of (a) sick and (b) vaccinated cows in the case of using transducer surfaces modified (1—FGA, 2—lectin of bean (PLA), 3—HPA-lectins) and nonmodified (4) by lectins.

28.3.1.4 Use of Lectins to Determine the Level of Antibodies to Glycoprotein VBL in Serum of Cattle

As noted, the major antigenic protein is VBL surface (shell) glycoproteins gp51, gp70, and p24 internal polypeptide. The vaccine was prepared by R&D Ltd "Leykopol" (Ukraine) and contains mainly internal proteins, including p24 polypeptide. It is widely used in practice. Therefore, the sera of vaccinated animals probably contained mainly antibodies specific to p24. The aim of our research was establishing a further modified immune biosensor based on SPR to provide differentiated determination of antibodies to glycoprotein gp24 in serum. To solve this problem, we used lectins.

We studied the interaction of glycoprotein of VBL with different lectins. Using lectin of lectin wheat germ of *Tuberosum vulgaris* (FGA-R) in comparison with lectin of grape snail of *Helix pomatia* (HPA), the biosensor response increased, indicating increasing density of the formed layer. But in the case of lectins, the shift of the resonance angle was less than at the antigen immobilization directly on the gold surface. This is due to the fact that the lectins bind only glycoproteins, while both the gold surfaces can adsorb all biological molecules that are in solution of antigen.

In special experiments, the sensitivity of the immune biosensors to specific antibodies contained in the blood of sick cows was determined using different lectins immobilized on the transducer surface. It was shown that the response of the immune biosensor using FGA-P-and PLA-lectins likely increases compared with the surface with the bulk gold. We have also investigated the serum of vaccinated calves with high titers of antibodies according to the results of the RID (RID antibody titer was 1:128, 1:256). It was shown that the response of the biosensor in the presence of blood serum and vaccinated animals differed significantly when using a transducer whose surface has been treated previously with lectins (Figure 28.8). The greatest differences were observed when the surface was modified by FGA-R-lectin. In this case, on the unmodified surface the response was almost absent.

Thus, modification of the surface of the transducer using lectins leads to immobilization of glycoproteins on the surface only, thus allowing for the selective determination of antibodies to glycosylated protein content that dominates in the blood serum of sick animals compared to vaccinated animals. In addition, processing the transducer surface by lectins such as FGA-P-and PLA can increase the sensitivity of the immune biosensor to the specific antibodies compared with untreated gold surface.

28.3.2 Comparison of the Immune Sensor Analysis with the Traditional ELISA Method

Results of the immune biosensor analysis were compared with the data obtained in the study of blood samples by the ELISA method. We tested 10 blood sera from clinically sick animals (antibody titer in RID was about 1:256) simultaneously with the SPR-based biosensor and the ELISA method. It was shown that the sensitivity analysis of immune biosensor was approached the one with ELISA method with the specific antibody titer about 1:15,000.

28.3.3 Estimation of Quality of the Immune Biosensor Analysis using SPR as Transducer

Assessment of the quality of immune biosensor analysis was performed using a standard panel of positive and negative leukemic cattle blood serum from the State Scientific Control Institute of Biotechnology and Strains of the Ministry of Agrarian Policy of Ukraine. We tested 20 samples of blood serum. Previously they were analyzed by using commercial test kits (ELISA method and polymerase chain reaction—PCR). The blood serum was diluted 1:500. The obtained results are presented in Table 28.2.

TABLE 28.2

Comparison of Results Obtained by Immune Biosensor, ELISA, and RID Methods in Biochemical Diagnostics of Leucosis in Cows

| Samples | Response of Immune Biosensor | | ELISA Method | | RID Method |
	Δ Resonant Angle, s	Type of Reaction	Coefficient Extinction	Type of Reaction	
1	0	−	0.028	−	−
2	820	+	3.456	+	+
3	0	−	0.00	−	−
4	425	+	2.504	+	±
5	0	−	0.00	−	−
6	530	+	2.512	+	±
7	0	−	0.00	−	−
8	70	±	0.02	−	−
9	360	+	2.444	+	−
10	40	−	0.014	−	−
11	85	±	0.01	−	−
12	370	+	2.444	+	−
13	50	−	0.01	−	−
14	980	+	3.262	+	+
15	540	+	2.860	+	+
16	950	+	3.252	+	+
17	1240	+	3.892	+	+
18	0	−	0.00	−	−
19	820	+	3.173	+	+
20	0	−	0.00	−	−

The high degree of correlation between the data obtained by the immune biosensor, ELISA method, and PCR was stated. Specificity of the analysis by the immune biosensor was 88.9% and sensitivity, 92.3%.

Further research was conducted with blood serum obtained from the OIE and National Reference Laboratory for EBL (Federal Research Institute for Animal Health, Germany). It was a panel of serum samples (21 units), preanalyzed by certified commercial test kits (AGIDT and ELISA method) for antibodies to viral leukemia. We have found a high correlation (92%) between the results of the analysis carried out by immune biosensor, AGIDT, and ELISA method. Only two samples were determined by ELISA as weakly positive, which at the immune biosensor analysis did not give a specific signal.

28.3.4 Definition of Induced VL Antibodies in the Serum of Milk

Questions regarding the detection of specific antibodies in milk are important because research can be fulfilled on the easily accessible body fluids to avoid injury to the animals associated with blood sampling.

We investigated 15 samples of milk RID-positive (titer of specific antibodies in serum at the RID analysis was in the range of 1:64–1:256) and 10 samples from RID-negative cows. It was shown that blood serum RID-positive cows prompted widespread shift of the resonance angle within 1500–1600 angle s and their serum milk 600–650 angle s (Figure 28.9).

The optimal level dilution for the study of specific antibodies in the serum of milk was 1:20. With this dilution, control samples did not cause changes in the resonant angle. However, samples of milk from some RID-negative cows

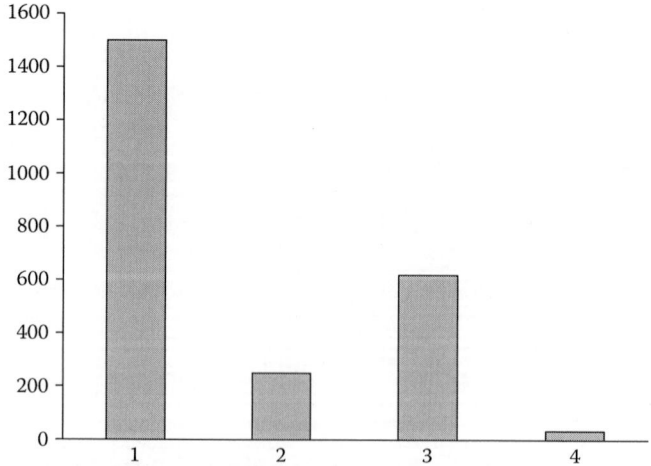

FIGURE 28.9
Values of changes of the resonant angle of the immune biosensor at the analysis of blood (1, 2) and milk (3, 4) serums of RID-positive (1, 3) and -negative (2, 4) cows. Serum blood was diluted in 1:500 and serum milk—in 1:20.

caused a shift of the resonant angle within 60–90 angle s and it is, in our opinion, due to the fact that antibodies present in the sera in such quantities are beyond the sensitivity of RID method.

So, the applicability of immune biosensor analysis of milk for the express screening of leucosis is established. The optimal serum dilution of milk with rapid determination of specific antibodies in milk by SPR-based immune biosensor should be on the level of 1:20.

28.3.5 Immune Biosensors Based on Nanostructures for the Diagnostics of BVL

To continue our investigations in the field of immune biosensor diagnostics of BVL, we proposed a new variant of biosensor based on the structure nano-porous silicon (sNPS) (Starodub and Starodub, 1999; Starodub et al., 2007b, 2011; Kachinskaya et al., 2009; Sitnik et al., 2009).

28.3.5.1 General Abilities of sNPS

The main problem that should be solved at the application of the sNPS as the transducer in the biosensors is providing of the PhL stability during a long time. It depends on the sNPS structure, the content of the inter phase layers, and the porosity level. These abilities of the sNPS may be determined by the method and the regimes of its forming. Our investigations of the sNPS showed that the samples prepared by the method of chemical etching have the stable ChL, conductivity, and photoconductivity characteristics, which were preserved for 5 years.

28.3.5.2 Obtaining sNPS by Chemical Etching

This method is simpler than galvanic anodizing, and it allows forming more thin (<1 μm) homogenous layers of the sNPS. This method is based on the next. At the introduction of some oxidants in the electrolyte based on fluoric acid, the self-solving process for the silicon plate immersed in this solution is started without any external voltage. The electrolyte and HNO_3 (70%) or $NaNO_2$ as oxidants are most often used for such processes at different concentrations of fluoric acid: $HF:HNO_3:H_2O$ = 1:5:10, or 1:3:5, or 4:1:5, or $HF:NaNO_2$ = 100 mL:2 g. Cathode reaction is realized through the reduction of the oxidant by the electrons from the anode:

$$HNO_3 + 3H^+ \rightarrow NO + 2H_2O + 3h^+ \tag{28.1}$$

The complete chemical reaction on the Si surface may be presented by the following equation:

$$3Si + 4HNO_3 + 18HF \rightarrow 3H_2SiF_6 + 4NO + 8H_2O + 3(4-n)h^+ + 3(4-n)e^- \tag{28.2}$$

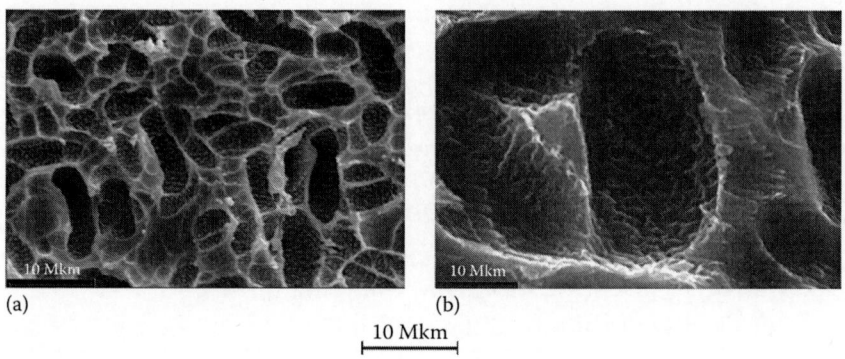

(a)   (b)

|——— 10 Mkm ———|

FIGURE 28.10
General view of the whole surface (a) and its fragment (b) of sNPS obtained by SEM.

In contrast to the electrolytic process, chemical etching is self-regulated and strongly depends on the initial solution content. The oxidant at the etching plays the same role as the anode current density in the electrochemical method. It means that the higher HNO_3 concentration responds to the electrochemical regime with the big current density. At the same time, a huge quantity of HF in the solution leads to a process that is restricted by the accessibility and corresponds to the regime with the low current density as in the case of the electrochemical method. That is why, the content of HNO_3 is the most important parameter in chemical etching.

The position of the PhL spectra maximum does not depend on the oxidant type and is at $\lambda \sim 625$ nm. The method of chemical etching is most adapted to mass manufacturing and is interesting for the preparation of the thin layers of the sNPS for the devices. Nevertheless, at the application of this method it is necessary to solve problems connected with the reproducibility and homogeneity of the layers.

So, due to changes in the surface state and the etching content it is possible to have thin homogeneous luminescent layers of the porous silicon (PS) with a high level of the reproducibility obtained by the chemical method and without the application of the electrical field. The structure of the sNPS was investigated by scanning electronic microscopy (Figure 28.10) and chemical content—Auger spectroscopy (Figure 28.11).

Spectra of the optical reflectance depend on the structure and dimensions of the nanocrystallites, air pores, level of porosity, and the thickness of the layer (Figure 28.12).

The maximal photosensitivity in the visible diapason of the spectra (30–35 mA/mL) is typical for the sNPS layers with the dimensions of the nanocrystallites in 15 nm, and it decreased with the growth of the latter. At the same time, maximal sensitivity to the ultraviolet irradiation was obtained for the sNPS layers with the dimensions of the nanocrystallites in 20–25 nm. sNPS layers obtained by the method of electrochemical etching as well as by chemical etching show the PhL abilities that were typical for this material: the wide region in

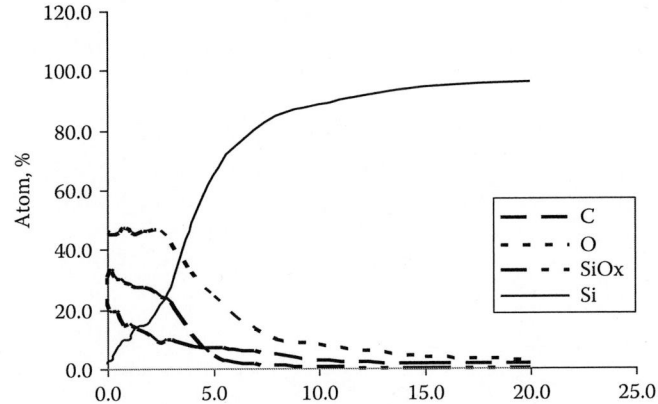

FIGURE 28.11
Dispersion of the elements in the sNPS layer (the speed of the etching—3 nm/min).

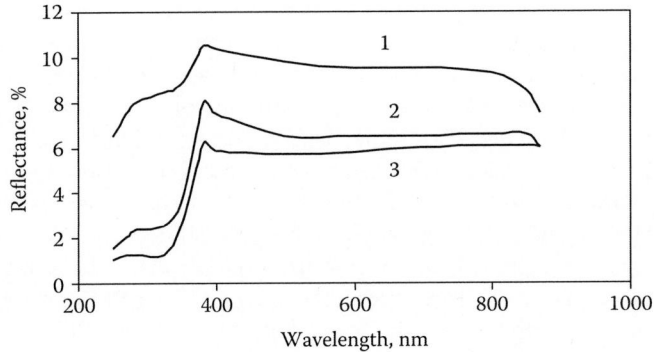

FIGURE 28.12
Spectra of the optical reflectance of the sNPS layer in the dependence on the dimension of the nanocrystallites: 1–5, 2–15, 3–30 nm.

the visible diapason with the intensity that is enough for the observation of the PhL with the naked eye. All samples of sNPS obtained by the chemical method had bright irradiation with the maximum at $\lambda \sim 640$ nm and that prepared by the electrochemical method—700 nm. According to the measurements of the PhL and the photoconductivity of the prepared films, the optimal contents for the chemical treatment and time for the etching were chosen.

28.3.5.3 Photoconductivity of sNPS

Photoelectric processes in the layers of the sNPS that belong to the semiconductor materials accomplished in the result of the photo generation of the electron-hole pairs and following their dividing and recombination. The processes of adsorption on the sNPS surface may arouse new photoelectric effects.

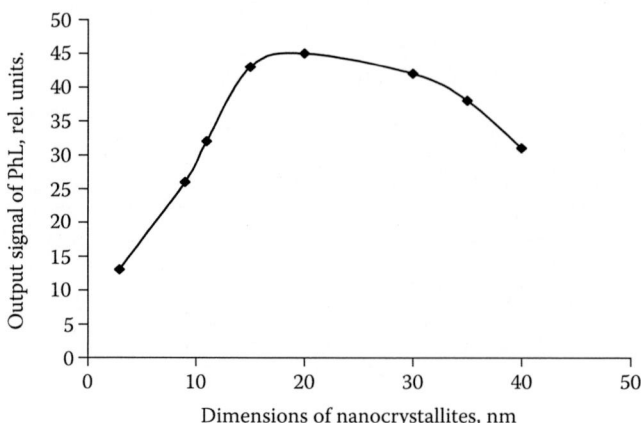

FIGURE 28.13
Dependence of the intensity of the PhL of the sNPS layers at 650 nm on the dimensions of the nanocrystallites.

Nanocrystallites of the silicon with dimensions from 1 to 12 nm are as the silicon regions, which are not dissolved and surrounded by the production of the electrochemical and chemical reactions. At dimensions less than 15–20 nm, the quant-dimensioned effects arise, which lead to the quantization of the energetic spectra of the charge carriers, widening of the prohibited zone up to 1.7–3.4 eV, and to decreasing of the dielectric permeability. The lux-ampere characteristics of the obtained samples have two plots: the linear and the sublinear, which achieves saturation at illumination more than 10,000 lux. The samples with a nanolayer thickness of 15–18 nm have the maximal photosensitivity that has good correlation with the results of the experiments with the PhL (Figure 28.13).

It is necessary to mention that the changing of the etching content and the solution concentrations results in changing of the dynamics of NPS layer growth, the porosity level, the correlation of the dimensions of the crystallites and the holes, the chemical content, and the profile of the dispersion of main admixtures.

28.3.5.4 Some Materials and Chemicals for Immune Chemical Analysis

The layers of the sNPS for the photoresistors were obtained by chemical etching of the monocrystalline silicon in the solution of HF and HNO_3, as described previously. As a preliminary step, the optimal method for the formation of the ohmic contact on the surface of the sNPS from alumina and indium by the magnetron sputtering with the application of the special masks was developed. It was found that the contacts from indium are non-stable due to the mechanical softness of this metal. At the measurements, the contact integrity was destroyed since under it the porous surface was located. In our investigations, the contacts were formed from the alumina with a thickness of about 3 μm.

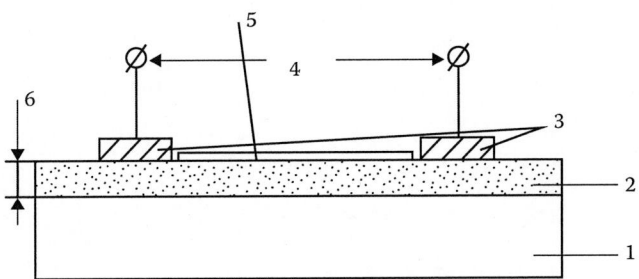

FIGURE 28.14
Scheme of the photoresistor structure based on the sNPS and intended for the analysis of the interactions between biological structures. Where: 1—the crystalline silicon, 2—the sNPS, 3—the electrical contacts (Al with the thickness of ~3 μm), 4—the applied voltage, 5—the biological object, 6—the thickness of the sNPS of 10–40 nm.

As sources of antigens (Ag) we used the mixture of retroviral proteins obtained from the joint venture of the "Leuconad" (Poltava, Ukraine). Blood serums from ill cows were kindly presented by this venture too. Ag was dissolved in 0.05 M tris-HCl buffer (pH 7.3) at different concentrations.

28.3.5.5 Registration of the Specific Immune Complex by the Determination of sPNS Photoconductivity

At the beginning, the specific Ab in the volume of 1 μL was placed on the photoresistor surface between the contacts (Figure 28.14). Then this solution was evaporated at room temperature or at the air stream.

The direct voltage (5 V) from the stabilized power supply was applied to the ohmic contacts, and the current was measured by the digital voltmeter of B7–35 type in the absence of lighting (dark regime) as well as the photocurrent (the difference between the light and dark currents) was registered at the lightening of the sensitive surface by the white spectrum light (source A, illumination of 7000 lux). At the drawing of Ag layer on the sensitive plate and after its drying, the measurements of the dark and light current were repeated. These measurements were made after the immune complex formation (interaction of Ag with specific Ab in the serum blood) too. The control of reaching of the sensor initial state was done according to the reduction of the dark current value after washing the sensitive surface by the buffer solution. The time of the single analysis was 5–10 min only.

28.3.5.6 Registration of the Specific Immune Complex by the PhL of the sPNS

Design of the prototype includes the source of the ultraviolet (UV) radiation (1) with the wavelength of 350 nm, two photodiodes (2 and 3) based on the monocrystalline silicon and placed at the angle of 20°–25° relatively to the plate

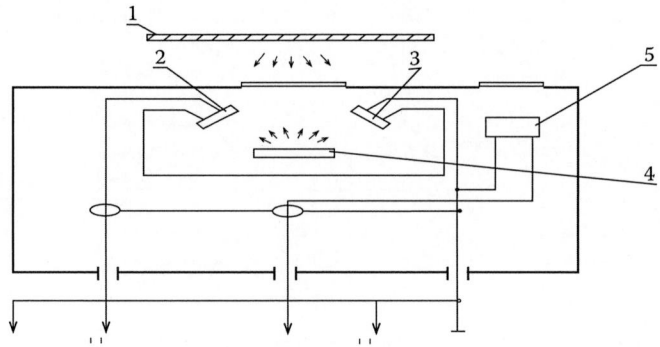

FIGURE 28.15
Design of PhL biosensor (see explanation in the text).

with the layer of the sNPS (4), and the photodiode (5) intended for the determination of the incident UV (Figure 28.15). At the adsorption of the biological molecules, the level of the PhL of the sPNS and the output of the voltage of the consecutive connected photo registers are decreasing. Application of two photo registers of the PhL increases the sensitivity.

To take into account the possible changing of the incident UV, the photodiode (5) is used. The output signal is determined as the relation between the output signal from the photodiodes of 2 and 3 and the output signal from the photodiode of 5. Photodiodes of the n-p-p^+-structures work in the photogenerative regime. Such construction is related to the systems of the differential type.

28.3.5.7 Determination of the Retroviral Specific Ab in the Blood of the ILL Cow by the Measurement of sPNS Photoconductivity

It was stated that the photosensitivity of the sPNS is a little decreased after the immobilization of Ag (crude sample of the retroviral proteins) but at the addition of Ab (serum blood of ill cows) in the dilution of 1:5000 and, in particular, in 1:1000 it sharply decreased. Unfortunately, at the low level of blood dilution (from 1:100 to 1:1) the photosensitivity starts to decrease up to the initial level. Probably, it is connected with the increase in the density of the solution to be analyzed or with other mechanisms of the electronic exchanges between the immune complex and the sNPS surface.

If we take the blood serum from a cow that is not ill, the level of the photocurrent does not change in comparison with the initial one. The same situation was observed if we used the bovine serum albumin instead of the crude samples of retroviral proteins (Figure 28.16).

Therefore, the experimental results give us the possibility of considering that the application of the proposed principle of the immune biosensor may be prospective for the different types of biochemical diagnostics and not for revealing the retroviral infection in cow alone. Of course, it is necessary to

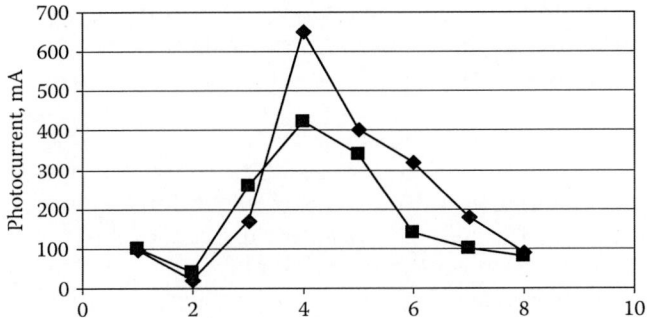

FIGURE 28.16
The level of the photocurrent before (1) and after deposition of Ag (2) and Ab in the different concentrations (3–8—1:5000, 1:1000, 1:100, 1:50, 1:10 and 1:1, respectively). Line A and B are with respect to two different samples of blood serum.

understand what kind of influences the high protein concentration has on the process of recombination in the sPNS. It is necessary to emphasize that the overall time of the analysis is several 12 min only instead of the several hours in the case of the traditionally used ELISA method or several days in the realization of the immune diffusion test.

28.3.5.8 Determination of the Retroviral Specific Ab in the Blood of the Ill Cow by the Method of PhL of sPNS

The deposition of the retroviral proteins on sPNS increases the PhL level, but at the formation of the specific immune complex it decreases. Moreover, the level of the PhL decreasing depends on the concentration of the specific Ab in the blood (Figure 28.17). If we used the nonspecific Ab or serum bovine albumin as Ag, the level of the PhL does not change.

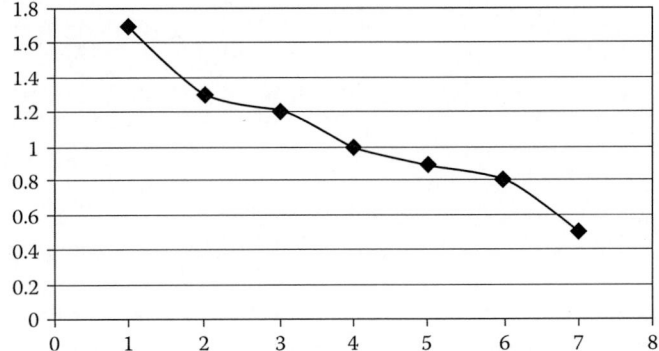

FIGURE 28.17
Dependence of the PhL intensity on the concentration of the specific Ab in the solution to be analyzed (serum blood of ill animals). Abscissa—1—immobilized Ag; 2–7—dilution of blood serum: 1:5000; 1:1000; 1:100; 1:50; 1:10; 1:1, respectively.

In our opinion, red PhL may be connected with the tunnel mechanism of the recombination of the charge bearers at their excitation in the nano-crystallites of oxide or interface. We do not exclude the hydrogen role too for the generation of PhL extinguishing. These conclusions are as result of the coincidence of the possible reasons for the PhL decreasing in the case of the immune complex formation on the sNPS surface. They result in: (1) the changes of the absorbance in the solution at the formation of the specific immune complex on the sNPS surface, (2) the effect of the immune compo-nents or their interaction on the recombinant process of the photocurrent charge in the sNPS. It is well known, the light absorption in the wavelength of the excitation ($\lambda = 350$ нм) and in the wide field of the sNPS PhL is absent in the Ab and Ag solutions as well as in their complexes.

28.4 Conclusion

Serological diagnosis is of great practical importance for primary screening of diabetes and leucosis. The development of serological methods is closely linked to the development of immune biosensors that allow an analysis such as this, and it requires practice and is, sensitive, specific, fast, and cheap. An important advantage of immune biosensors analysis is that it can be done in real time and in the field, at the patient bed, or directly on farms.

Modification of biosensor surface can not only provide effective immo-bilization of biological material and optimize the functional parameters of biosensor but also provide selective analysis, especially in the case of dis-crimination of sick and immunized animals. Use of PAA in the form of a preliminary layer at the immobilization antigen on the surface of the SPR biosensor is most appropriate, affordable, and simple.

Sensitivity analysis by the immune biosensors exceeds that of tradi-tionally used immunological methods including the ELISA method. The optimal serum dilution of blood at the rapid determination of specific anti-bodies in the serum blood patients with diabetes as well as cows in screen-ing for leukemia using optical immune biosensors (based on the SPR and sNPS) should be considered no less than 1:100, or even 1:500. The immune biosensor may be applied for express screening for leucosis through analy-sis of milk. In this case, the optimal serum dilution of milk should be con-sidered 1:20.

The duration of analysis using immune biosensor is only 40 min, includ-ing time for immobilization of antigen on the transducer surface, block-ing free binding sites, and rinsing the measuring cell. The direct testing procedure sera did not exceed 10 min. Also it is necessary to mention that the surface of the biosensor presented interchangeable plates, which can be pre-prepared and used when necessary. The device for immune biosensor analysis can be portable. This creates the conditions for simple, quick, and cheap screening biochemical quantities that have a diagnostic significance.

The application of the sNPS as transducer promotes creating a very simple immune biosensor with stability. The formed specific immune complex on the sNPS surface may be registered by the measurement of its PhL or photoconductivity. According to the results obtained in respect of the application of such immune biosensors for the biochemical diagnostics of bovine leucosis, it is possible to conclude that they will respond to all practical demands, especially, sensitivity, simplicity, rapidity of the analysis, and its fulfillment in field conditions. These biosensors may be applied for registration of any biochemical quantities that may form an immune complex. Further investigations should be directed on studying the mechanisms of biochemical signal registration by the sNPS and on the specification of all concrete moments of the analysis fulfillment.

References

Bae Y.M., Oh B.-K., Lee W.H., and Choi J.-W. Detection of insulin-antibody binding on a solid surface using imaging ellipsometry. *Biosens. Bioelectron.*, 2004, 20(4), 895–902.

Balida V. and Ferrer J.F. Expression of the bovine leukemia virus and its internal antigen in blood lymphocytes. *Proc. Soc. Exp. Biol. Med.*, 1977, 156, 388–391.

Blankenstein P. and Fechner H. Possibilities and limitations for use of the polymerase chain reaction (PCR) in the diagnosis of bovine leukemia virus (BLV) infection in cattle. *Dtsch. Tierarztl. Wochenschr.*, 1998, 105, 408–412.

Bougneres P.F., Landais P., Boisson C., Caret J.C., Frament N., Boitard C., Chaussain J.I., and Bach J.F. Limited duration of remission of insulin dependency in children with recent overt type 1 diabetes treated with low-dose cyclosporin. *Diabetes*, 1990, 39, 1264–1272.

Campetelli G., Zumoffen D., and Basualdo M. Improvements on noninvasive blood glucose biosensors using wavelets for quick fault detection. *J. Sensors.*, 2011, 2011, 1–11, ID 368015, doi:10.1155/2011/368015.

Carrol A.D., Scampavia L., Luo D., Lernmark A., and Ruzicka J. Bead injection ELISA for the determination of antibodies implicated in type 1 diabetes mellitus. *Analyst*, 2003, 128, 1157–1162.

Dahlquist G. and Mustonen L. Childhood onest diabetes-time trends and climatological factors. *Int. J. Epidemiol.*, 1994, 23, 1234–1241.

Donham K.J., Berg J.W., and Sawin R.S. Epidemiologic relationships of the bovine population and human leukemia in Iowa. *Am. J. Epidemiol.*, 1980, 112, 80–92.

Dz'oma Ju.M. and Starodub N.F. Development of the immune biosensor based on the surface plasmon resonance for the express control of auto-immune state of diabetics. *Materials of first International Science Conference of Students, PhD and Jung Scientists Fundamental and Applied Investigations in Biology*, Donezk, Ukraine: Veber, February 23–26, 2009, Vol. 2, pp. 115–116.

Yoo E.-H. and Lee S.-Y. Glucose biosensors: An overview of use in clinical practice. *Sensors*, 2010, 10, 4558–4576; doi:10.3390/s100504558.

Fechner H., Kurg A., Geue L., Blankenstein P., Mewes G., Ebner D., and Beier D. Evaluation of polymerase chain reaction (PCR) application in diagnosis of bovine leukemia virus (BLV) infection in naturally infected cattle. *J. Vet. Med.*, 1996, B43, 621–630.

Franz J., Hampl J., Hofirek B., Skrobak F., Svoboda I., and Granatova M. Radioimmunologic detection of antibodies to bovine leukemia virus. *Vet. Med.* (Praha), 1986, 31, 459–468.

Gardner S.O., Bingley P.J., Sawtell P.A., Weeks S., and Gale E. Rising incidence of insulin dependent diabetes in children aged under 5 years in the Oxford region: Time trend analysis. *BMJ*, 1997, 315, 713–717.

Gough S.C.L. Genetic of insulin-dependent diabetes mellitus. *Baillinre's Clin. Paediat., Childtiood Diabetes*, 1996, 4, 593–608.

Greene S.A. and Newton R.W. Diabetes mellitus in childhood and adolescence. In: Pickup J.C., Williams G. (eds). *Textbook of Diabetes*. 2nd edn. Oxford, Cambridge: Blackwell Science, 1997, Vol. 2(73), pp. 1–73.

Kachinskaya T.S., Shmyryeva O.M., Mel'nichenko M.M., Yurevich E.P., and Starodub M.F. Production of biosensor controls on the basis of nanostructured silicon. *e-J. Surf. Sci. Nanotech.*, 2009, 7, 677–680.

Karvonen M., Puomilehto J., Libman I., and LaPorte R. A review of the recent epidemiological data on the w/oildwide incidence of type 1 diabetes mellitus. *Diabetologia*, 1993, 36, 883–892.

Kooyman R.P.H., Kolkman H., van Gent J., and Greve J. Surface plasmon resonance immunosensors: Sensitivity considerations. *Anal. Chim. Acta*. 1988, 213, 35–45.

Lounamaa R. Epidemiology of childhood-onest IDDM. *Bailliure's Clin. Paediat., Childhood Diabetes*, 1996, 4, 609–626.

Matthews R.E.F. Classification and nomenclature of viruses. *Inter. Virol.*, 1979, 12, 128–296.

Miller J. and Olson C. Precipitation antibody to an internal antigen of the C-type virus associated with bovine lymphosaroma. *J. Nat.Cancer Inst.*, 1972, 49, 1459–1469.

Nabok A.V., Tsargorodskaya A., Hassan A.K., and Starodub N.F. Total internal reflection ellipsometry and SPR detection of low molecular weight environmental toxins. *Appl. Surf Sci*, 2005, 246, 381–386.

Palagin O.V., Romanov V.O., Starodub N.F., Galeljuka I.B., and Skripnik O.V. Virtuaal laboratory for biosensor projecting. *Med. Inform. Eng.*, 2008, 2, 36–40.

Pirogova L.V. and Starodub N.F. Immobilization of the antigen of the retroviruses of cattle on the surface of the immune biosensor. *Biotechnology*, 2008, 2, 52–58.

Pozzilli P. Prevention of insulin-dependent diabetes. *Bailliure's Clin. Paediat., Childhood Diabetes*, 1996, 4, 577–592.

Sitnik Ju.A., Mel'nichenko N.N., Shmyryva A.N., and Starodub N.F. Express methods for diagnostics of retroviral leukosis. In *Proceeding 15-in All Russian Science Conference Students Physician and Junk Scientist*, Kemerovo-Tomsk, Russia, Kemerovo, March 2, April 2, 2009, pp. 449–451.

Starodub N.F., Arthjuch V.P., Pirogova L.V., Nagajeva L.I., Dobrosol G.I., Pavlenko M., and Grotevich V. Express diagnostics of bovine leucosis on the basis of biosensor analysis. *Vet. Med.*, 2001a, 11, 26–27.

Starodub N.F., Demchenko A.V., Piven' N.V., and Goncharik A.V. Development of optical biosensor for express control of anti-insulin antibodies level in blood of diabetics. In *Proceeding of 13th International Conference on Sensor + Test 2007*, Nuremberg, Germany, May 22–24, 2007a, Vol. 1, pp. 439–444.

Starodub N.F., Dibrova T.I., Shirshov Yu.M., and Kostiukevich K.V. Development of sensor based on the surface plasmon resonance for control of biospecific interaction. In: *Eurosensors 11 and 12th Europeans Conference on Solid-State Transducers*, Warsaw, Poland, September 21–24, 1997, Vol. 3, pp. 1429–1432.

Starodub N.F., Dibrova T.L., Shirshov Yu.M., and Kostjukevich K.V. Development of myoglobin sensor based on the surface plasmon resonance. *Ukr. Biochem. J.*, 1999, 71, 33–37.

Starodub N.F., Melnyk V.G., Vasylenko O.D., Shmyriyva O.M., and Bogoljubov M.V. Development and investigation of luminescent impedancemetric biosensor system. In *Proceeding of 13th International Conference on Sensor + Test 2007*, Nuremberg, Germany, May 22–24, 2007b, Vol. 1, pp. 334–338.

Starodub N.F., Nabok A.V., Starodub V.M., Ray A.K., and Hassan A.K. Immobilization of biocomponents for immune optical sensors. *Ukr. Biochem. J.*, 2001b, 73(3), 16–24.

Starodub V.M., Nabok A.V., Starodub N.F., and Torbicz W. Approaches for structured immobilisation of recognising elements on a transducer surface of biosensors. In *Proceedings of NATO Advanced Research Workshop on Nanostructured Materials and Coatings for Biomedical and Sensor Applications*. Gogotsi Y.G., Uvarova I.V. (eds.), Dordrecht, the Netherlands: Kluwer Academic Publication, 2003, pp. 311–325.

Starodub N.F., Pilipeko L.N., Pilipenko I.V., and Egorova A.V. Mycotoxins and other low weight toxins as instrument of bioterrorists: Express instrumental control and some ways to decontaminate polluted environmental objects. *Timisoara Med. J.*, 2008, 58(1–2), 9–18.

Starodub N. and Pirogova L. Improvement of immune component immobilisation on the optical transducers by use of some lectins. In *International Conference Optoelectronics, Optical Sensors and Measuring Techniques*, Germany, May 14–16, 2002, pp. 175–180.

Starodub N.F., Pirogova L.V., Artyukh V.P., and Starodub V.N. Biospecific interactions on the optical transducer surface the base of infection diagnostics. In *NATO ARW Frontiers of Multifunctional Nanosystems*. Buzaneva E. and Scharff P. (eds), 2002, Vol. 57, pp. 369–376.

Starodub N.F., Pirogova L.V., Demchenko A., and Nabok A.V. Antibody immobilisation on the metal and silicon surfaces. The use of self-assembled layers and specific receptors. *Bioelectrochemistry*, 2005, 66(1–2), 111–115.

Starodub N.F., Sitnik J., Mel'nichenko M.M., and Romanov V.O. Nanostructured silicon and SPR based biosensors at the express biochemical diagnostics of the retroviral bovine leucosis. Abstract In *107th International Center of Biocybernetics Seminar Micro and Nanosystems in Biochemical Diagnosis—Principles and Applications*, Warsaw, Poland, May 13–15, 2010.

Starodub N.F., Sitnik J.A., Mel'nichenko M.M., and Shmyryeva O.M. Optical immune biosensors based on the nanostructured silicon and intended the diagnostics of retroviral bovine leucosis. In *The SENSOR + TEST 2011*. Nurenberg, Germany, 2011, pp. 127–132.

Starodub V.M. and Starodub N.F. Optical immune sensors for the monitoring protein substances in the air. In *Eurosensor 12th and 13th European Conference on Solid-State Transducers*, The Hague, the Netherlands, September 12–15, 1999, pp. 181–184.

Starodub V.M. and Starodub N.F. Electrochemical and optical biosensors: Origin of development, achievements and perspectives of practical application. In *Book of NATO Series Novel Processes and Control Technologies in the Food Industry.* Bozoglu F. et al. (eds.), Dordrecht, Kluwer Academic Publication, Amsterdam, the Netherlands, 2001, pp. 63–94.

Starodub N.F. and Starodub V.N. Infectious bovine leucosis and its diagnostics. *Biopolym. Cell*, 2003, 19, 307–316.

Wang J. Glucose biosensors: Forty years of advances and challenges. *Electroanalysis,* 2001, 13(12), 983–988.

Index